T0181036

# Lecture Notes in Computer Science 11537

*Commenced Publication in 1973*
Founding and Former Series Editors:
Gerhard Goos, Juris Hartmanis, and Jan van Leeuwen

More information about this series at http://www.springer.com/series/7407

João M. F. Rodrigues · Pedro J. S. Cardoso ·
Jânio Monteiro · Roberto Lam ·
Valeria V. Krzhizhanovskaya ·
Michael H. Lees · Jack J. Dongarra ·
Peter M. A. Sloot (Eds.)

# Computational Science – ICCS 2019

19th International Conference
Faro, Portugal, June 12–14, 2019
Proceedings, Part II

 Springer

*Editors*
João M. F. Rodrigues
University of Algarve
Faro, Portugal

Pedro J. S. Cardoso
University of Algarve
Faro, Portugal

Jânio Monteiro
University of Algarve
Faro, Portugal

Roberto Lam
University of Algarve
Faro, Portugal

Valeria V. Krzhizhanovskaya
University of Amsterdam
Amsterdam, The Netherlands

Michael H. Lees
University of Amsterdam
Amsterdam, The Netherlands

Jack J. Dongarra
University of Tennessee at Knoxville
Knoxville, TN, USA

Peter M. A. Sloot
University of Amsterdam
Amsterdam, The Netherlands

ISSN 0302-9743          ISSN 1611-3349   (electronic)
Lecture Notes in Computer Science
ISBN 978-3-030-22740-1          ISBN 978-3-030-22741-8   (eBook)
https://doi.org/10.1007/978-3-030-22741-8

LNCS Sublibrary: SL1 – Theoretical Computer Science and General Issues

This Springer imprint is published by the registered company Springer Nature Switzerland AG
The registered company address is: Gewerbestrasse 11, 6330 Cham, Switzerland

# Preface

Welcome to the 19th Annual International Conference on Computational Science (ICCS - https://www.iccs-meeting.org/iccs2019/), held during June 12–14, 2019, in Faro, Algarve, Portugal. Located at the southern end of Portugal, Algarve is a well-known touristic haven. Besides some of the best and most beautiful beaches in the entire world, with fine sand and crystal-clear water, Algarve also offers amazing natural landscapes, a rich folk heritage, and a healthy gastronomy that can be enjoyed throughout the whole year, attracting millions of foreign and national tourists. ICCS 2019 was jointly organized by the University of Algarve, the University of Amsterdam, NTU Singapore, and the University of Tennessee.

The International Conference on Computational Science is an annual conference that brings together researchers and scientists from mathematics and computer science as basic computing disciplines, as well as researchers from various application areas who are pioneering computational methods in sciences such as physics, chemistry, life sciences, engineering, arts, and humanitarian fields, to discuss problems and solutions in the area, to identify new issues, and to shape future directions for research.

Since its inception in 2001, ICCS has attracted an increasingly higher quality and numbers of attendees and papers, and this year was no exception, with over 350 participants. The proceedings series have become a major intellectual resource for computational science researchers, defining and advancing the state of the art in this field.

ICCS 2019 in Faro was the 19th in this series of highly successful conferences. For the previous 18 meetings, see: http://www.iccs-meeting.org/iccs2019/previous-iccs/.

The theme for ICCS 2019 was "Computational Science in the Interconnected World," to highlight the role of computational science in an increasingly interconnected world. This conference was a unique event focusing on recent developments in: scalable scientific algorithms; advanced software tools; computational grids; advanced numerical methods; and novel application areas. These innovative novel models, algorithms, and tools drive new science through efficient application in areas such as physical systems, computational and systems biology, environmental systems, finance, and others.

ICCS is well known for its excellent line-up of keynote speakers. The keynotes for 2019 were:

- Tiziana Di Matteo, King's College London, UK
- Teresa Galvão, University of Porto/INESC TEC, Portugal
- Douglas Kothe, Exascale Computing Project, USA
- James Moore, Imperial College London, UK
- Robert Panoff, The Shodor Education Foundation, USA
- Xiaoxiang Zhu, Technical University of Munich, Germany

This year we had 573 submissions (228 submissions to the main track and 345 to the workshops). In the main track, 65 full papers were accepted (28%); in the workshops, 168 full papers (49%). The high acceptance rate in the workshops is explained by the nature of these thematic sessions, where many experts in a particular field are personally invited by workshop organizers to participate in their sessions.

ICCS relies strongly on the vital contributions of our workshop organizers to attract high-quality papers in many subject areas. We would like to thank all committee members for the main track and workshops for their contribution to ensure a high standard for the accepted papers. We would also like to thank Springer, Elsevier, and Intellegibilis for their support. Finally, we very much appreciate all the local Organizing Committee members for their hard work to prepare this conference.

We are proud to note that ICCS is an A-rank conference in the CORE classification.

June 2019

João M. F. Rodrigues
Pedro J. S. Cardoso
Jânio Monteiro
Roberto Lam
Valeria V. Krzhizhanovskaya
Michael Lees
Jack J. Dongarra
Peter M. A. Sloot

# Organization

## Workshops and Organizers

### Advanced Modelling Techniques for Environmental Sciences – AMES

Jens Weismüller
Dieter Kranzlmüller
Maximilian Hoeb
Jan Schmidt

### Advances in High-Performance Computational Earth Sciences: Applications and Frameworks – IHPCES

Takashi Shimokawabe
Kohei Fujita
Dominik Bartuschat

### Agent-Based Simulations, Adaptive Algorithms, and Solvers – ABS-AAS

Maciej Paszynski
Quanling Deng
David Pardo
Robert Schaefer
Victor Calo

### Applications of Matrix Methods in Artificial Intelligence and Machine Learning – AMAIML

Kourosh Modarresi

### Architecture, Languages, Compilation, and Hardware Support for Emerging and Heterogeneous Systems – ALCHEMY

Stéphane Louise
Löic Cudennec
Camille Coti
Vianney Lapotre
José Flich Cardo
Henri-Pierre Charles

### Biomedical and Bioinformatics Challenges for Computer Science – BBC

Mario Cannataro
Giuseppe Agapito
Mauro Castelli

Riccardo Dondi
Rodrigo Weber dos Santos
Italo Zoppis

## Classifier Learning from Difficult Data – CLD²

Michał Woźniak
Bartosz Krawczyk
Paweł Ksieniewicz

## Computational Finance and Business Intelligence – CFBI

Yong Shi
Yingjie Tian

## Computational Methods in Smart Agriculture – CMSA

Andrew Lewis

## Computational Optimization, Modelling, and Simulation – COMS

Xin-She Yang
Slawomir Koziel
Leifur Leifsson

## Computational Science in IoT and Smart Systems – IoTSS

Vaidy Sunderam

## Data-Driven Computational Sciences – DDCS

Craig Douglas

## Machine Learning and Data Assimilation for Dynamical Systems – MLDADS

Rossella Arcucci
Boumediene Hamzi
Yi-Ke Guo

## Marine Computing in the Interconnected World for the Benefit of Society – MarineComp

Flávio Martins
Ioana Popescu
João Janeiro
Ramiro Neves
Marcos Mateus

**Multiscale Modelling and Simulation – MMS**

Derek Groen
Lin Gan
Stefano Casarin
Alfons Hoekstra
Bartosz Bosak

**Simulations of Flow and Transport: Modeling, Algorithms, and Computation – SOFTMAC**

Shuyu Sun
Jingfa Li
James Liu

**Smart Systems: Bringing Together Computer Vision, Sensor Networks, and Machine Learning – SmartSys**

João M. F. Rodrigues
Pedro J. S. Cardoso
Jânio Monteiro
Roberto Lam

**Solving Problems with Uncertainties – SPU**

Vassil Alexandrov

**Teaching Computational Science – WTCS**

Angela Shiflet
Evguenia Alexandrova
Alfredo Tirado-Ramos

**Tools for Program Development and Analysis in Computational Science – TOOLS**

Andreas Knüpfer
Karl Fürlinger

# Programme Committee and Reviewers

| | | |
|---|---|---|
| Ahmad Abdelfattah | Elisabete Alberdi | Stanislaw |
| Eyad Abed | Marco Aldinucci | Ambroszkiewicz |
| Markus Abel | Luis Alexandre | Ioannis Anagnostou |
| Laith Abualigah | Vassil Alexandrov | Philipp Andelfinger |
| Giuseppe Agapito | Evguenia Alexandrova | Michael Antolovich |
| Giovanni Agosta | Victor Allombert | Hartwig Anzt |
| Ram Akella | Saad Alowayyed | Hideo Aochi |

Rossella Arcucci
Tomasz Arodz
Kamesh Arumugam
Luiz Assad
Victor Azizi Tarksalooyeh
Bartosz Balis
Krzysztof Banas
João Barroso
Dominik Bartuschat
Daniel Becker
Jörn Behrens
Adrian Bekasiewicz
Gebrail Bekdas
Stefano Beretta
Daniel Berrar
John Betts
Sanjukta Bhowmick
Bartosz Bosak
Isabel Sofia Brito
Kris Bubendorfer
Jérémy Buisson
Aleksander Byrski
Cristiano Cabrita
Xing Cai
Barbara Calabrese
Carlos Calafate
Carlos Cambra
Mario Cannataro
Alberto Cano
Paul M. Carpenter
Stefano Casarin
Manuel Castañón-Puga
Mauro Castelli
Jeronimo Castrillon
Eduardo Cesar
Patrikakis Charalampos
Henri-Pierre Charles
Zhensong Chen
Siew Ann Cheong
Andrei Chernykh
Lock-Yue Chew
Su Fong Chien
Sung-Bae Cho
Bastien Chopard
Stephane Chretien
Svetlana Chuprina

Florina M. Ciorba
Noelia Correia
Adriano Cortes
Ana Cortes
Jose Alfredo F. Costa
Enrique
   Costa-Montenegro
David Coster
Camille Coti
Carlos Cotta
Helene Coullon
Daan Crommelin
Attila Csikasz-Nagy
Loïc Cudennec
Javier Cuenca
Yifeng Cui
António Cunha
Ben Czaja
Pawel Czarnul
Bhaskar Dasgupta
Susumu Date
Quanling Deng
Nilanjan Dey
Ergin Dinc
Minh Ngoc Dinh
Sam Dobbs
Riccardo Dondi
Ruggero Donida Labati
Goncalo dos-Reis
Craig Douglas
Aleksandar Dragojevic
Rafal Drezewski
Niels Drost
Hans du Buf
Vitor Duarte
Richard Duro
Pritha Dutta
Sean Elliot
Nahid Emad
Christian Engelmann
Qinwei Fan
Fangxin Fang
Antonino Fiannaca
Christos
   Filelis-Papadopoulos
José Flich Cardo

Yves Fomekong Nanfack
Vincent Fortuin
Ruy Freitas Reis
Karl Frinkle
Karl Fuerlinger
Kohei Fujita
Wlodzimierz Funika
Takashi Furumura
Mohamed Medhat Gaber
Jan Gairing
David Gal
Marco Gallieri
Teresa Galvão
Lin Gan
Luis Garcia-Castillo
Delia Garijo
Frédéric Gava
Don Gaydon
Zong-Woo Geem
Alex Gerbessiotis
Konstantinos
   Giannoutakis
Judit Gimenez
Domingo Gimenez
Guy Gogniat
Ivo Gonçalves
Yuriy Gorbachev
Pawel Gorecki
Michael Gowanlock
Manuel Graña
George Gravvanis
Marilaure Gregoire
Derek Groen
Lutz Gross
Sophia
   Grundner-Culemann
Pedro Guerreiro
Kun Guo
Xiaohu Guo
Piotr Gurgul
Pietro Hiram Guzzi
Panagiotis Hadjidoukas
Mohamed Hamada
Boumediene Hamzi
Masatoshi Hanai
Quillon Harpham

William Haslett
Yiwei He
Alexander Heinecke
Jurjen Rienk Helmus
Alvaro Herrero
Bogumila Hnatkowska
Maximilian Hoeb
Paul Hofmann
Sascha Hunold
Juan Carlos Infante
Hideya Iwasaki
Takeshi Iwashita
Alfredo Izquierdo
Heike Jagode
Vytautas Jancauskas
Joao Janeiro
Jiří Jaroš
Shantenu Jha
Shalu Jhanwar
Chao Jin
Hai Jin
Zhong Jin
David Johnson
Anshul Joshi
Manuela Juliano
George Kallos
George Kampis
Drona Kandhai
Aneta Karaivanova
Takahiro Katagiri
Ergina Kavallieratou
Wayne Kelly
Christoph Kessler
Dhou Khaldoon
Andreas Knuepfer
Harald Koestler
Dimitrios Kogias
Ivana Kolingerova
Vladimir Korkhov
Ilias Kotsireas
Ioannis Koutis
Sergey Kovalchuk
Michał Koziarski
Slawomir Koziel
Jarosław Koźlak
Dieter Kranzlmüller

Bartosz Krawczyk
Valeria Krzhizhanovskaya
Paweł Ksieniewicz
Michael Kuhn
Jaeyoung Kwak
Massimo La Rosa
Roberto Lam
Anna-Lena Lamprecht
Johannes Langguth
Vianney Lapotre
Jysoo Lee
Michael Lees
Leifur Leifsson
Kenneth Leiter
Roy Lettieri
Andrew Lewis
Jingfa Li
Yanfang Li
James Liu
Hong Liu
Hui Liu
Zhao Liu
Weiguo Liu
Weifeng Liu
Marcelo Lobosco
Veronika Locherer
Robert Lodder
Stephane Louise
Frederic Loulergue
Huimin Lu
Paul Lu
Stefan Luding
Scott MacLachlan
Luca Magri
Maciej Malawski
Livia Marcellino
Tomas Margalef
Tiziana Margaria
Svetozar Margenov
Osni Marques
Alberto Marquez
Paula Martins
Flavio Martins
Jaime A. Martins
Marcos Mateus
Marco Mattavelli

Pawel Matuszyk
Valerie Maxville
Roderick Melnik
Valentin Melnikov
Ivan Merelli
Jianyu Miao
Kourosh Modarresi
Miguel Molina-Solana
Fernando Monteiro
Jânio Monteiro
Pedro Montero
James Montgomery
Andrew Moore
Irene Moser
Paulo Moura Oliveira
Ignacio Muga
Philip Nadler
Hiromichi Nagao
Kengo Nakajima
Raymond Namyst
Philippe Navaux
Michael Navon
Philipp Neumann
Ramiro Neves
Mai Nguyen
Hoang Nguyen
Nancy Nichols
Sinan Melih Nigdeli
Anna Nikishova
Kenji Ono
Juan-Pablo Ortega
Raymond Padmos
J. P. Papa
Marcin Paprzycki
David Pardo
Héctor Quintián Pardo
Panos Parpas
Anna Paszynska
Maciej Paszynski
Jaideep Pathak
Abani Patra
Pedro J. S. Cardoso
Dana Petcu
Eric Petit
Serge Petiton
Bernhard Pfahringer

Daniela Piccioni
Juan C. Pichel
Anna
  Pietrenko-Dabrowska
Laércio L. Pilla
Armando Pinto
Tomasz Piontek
Erwan Piriou
Yuri Pirola
Nadia Pisanti
Antoniu Pop
Ioana Popescu
Mario Porrmann
Cristina Portales
Roland Potthast
Ela Pustulka-Hunt
Vladimir Puzyrev
Alexander Pyayt
Zhiquan Qi
Rick Quax
Barbara Quintela
Waldemar Rachowicz
Célia Ramos
Marcus Randall
Lukasz Rauch
Andrea Reimuth
Alistair Rendell
Pedro Ribeiro
Bernardete Ribeiro
Robin Richardson
Jason Riedy
Celine Robardet
Sophie Robert
João M. F. Rodrigues
Daniel Rodriguez
Albert Romkes
Debraj Roy
Philip Rutten
Katarzyna Rycerz
Augusto S. Neves
Apaar Sadhwani
Alberto Sanchez
Gabriele Santin

Robert Schaefer
Olaf Schenk
Ulf Schiller
Bertil Schmidt
Jan Schmidt
Martin Schreiber
Martin Schulz
Marinella Sciortino
Johanna Sepulveda
Ovidiu Serban
Vivek Sheraton
Yong Shi
Angela Shiflet
Takashi Shimokawabe
Tan Singyee
Robert Sinkovits
Vishnu Sivadasan
Peter Sloot
Renata Slota
Grażyna Ślusarczyk
Sucha Smanchat
Maciej Smołka
Bartlomiej Sniezynski
Sumit Sourabh
Hoda Soussa
Steve Stevenson
Achim Streit
Barbara Strug
E. Dante Suarez
Bongwon Suh
Shuyu Sun
Vaidy Sunderam
James Suter
Martin Swain
Grzegorz Swisrcz
Ryszard Tadeusiewicz
Lotfi Tadj
Daniele Tafani
Daisuke Takahashi
Jingjing Tang
Osamu Tatebe
Cedric Tedeschi
Kasim Tersic

Yonatan Afework
  Tesfahunegn
Jannis Teunissen
Andrew Thelen
Yingjie Tian
Nestor Tiglao
Francis Ting
Alfredo Tirado-Ramos
Arkadiusz Tomczyk
Stanimire Tomov
Marko Tosic
Jan Treibig
Leonardo Trujillo
Benjamin Uekermann
Pierangelo Veltri
Raja Velu
Alexander von Ramm
David Walker
Peng Wang
Lizhe Wang
Jianwu Wang
Gregory Watson
Rodrigo Weber dos
  Santos
Kevin Webster
Josef Weidendorfer
Josef Weinbub
Tobias Weinzierl
Jens Weismüller
Lars Wienbrandt
Mark Wijzenbroek
Roland Wismüller
Eric Wolanski
Michał Woźniak
Maciej Woźniak
Qing Wu
Bo Wu
Guoqiang Wu
Dunhui Xiao
Huilin Xing
Miguel Xochicale
Wei Xue
Xin-She Yang

Dongwei Ye
Jon Yosi
Ce Yu
Xiaodan Yu
Reza Zafarani
Gábor Závodszky

H. Zhang
Zepu Zhang
Jingqing Zhang
Yi-Fan Zhang
Yao Zhang
Wenlai Zhao

Jinghui Zhong
Xiaofei Zhou
Sotirios Ziavras
Peter Zinterhof
Italo Zoppis
Chiara Zucco

# Contents – Part II

**Track of Advances in High-Performance Computational
Earth Sciences: Applications and Frameworks**

**Track of Agent-Based Simulations, Adaptive Algorithms and Solvers**

**Track of Applications of Matrix Methods in Artificial Intelligence
and Machine Learning**

**Track of Architecture, Languages, Compilation and Hardware
support for Emerging and Heterogeneous Systems**

# ICCS Main Track

# Synthesizing Quantum Circuits
# via Numerical Optimization

Timothée Goubault de Brugière[1,3,4(✉)], Marc Baboulin[1,3], Benoît Valiron[2,3],
and Cyril Allouche[4]

[1] University of Paris-Sud, Orsay, France
[2] CentraleSupélec, Gif sur Yvette, France
[3] Laboratoire de Recherche en Informatique, Orsay, France
`timothee.goubault@lri.fr`
[4] Atos Quantum Lab, Les Clayes-sous-Bois, France

**Abstract.** We provide a simple framework for the synthesis of quantum circuits based on a numerical optimization algorithm. This algorithm is used in the context of the trapped-ions technology. We derive theoretical lower bounds for the number of quantum gates required to implement any quantum algorithm. Then we present numerical experiments with random quantum operators where we compute the optimal parameters of the circuits and we illustrate the correctness of the theoretical lower bounds. We finally discuss the scalability of the method with the number of qubits.

**Keywords:** Quantum circuit synthesis · Numerical optimization · Quantum simulation · Quantum circuit optimization · Trapped-ions technology

## 1 Introduction

Quantum computing, introduced and theorized about 30 years ago, is still in its infancy: current technological devices can only handle a few qubits. This new paradigm of computation however shows great promises, with potential applications ranging from high-performance computing [8] to machine learning and big data [9]. Quantum algorithms are usually described via quantum circuits, i.e., series of elementary operations in line with the technological specificity of the hardware. The mathematical formalism for quantum computation is the theory of (finite dimensional) Hilbert spaces: a quantum circuit is represented as a unitary operator [16], independently from the machine support on which the algorithm will be executed. Establishing a link between the unitary operators described as matrices and the unitary operators described as circuits is therefore essential, if only to better understand how to design new algorithms. Obtaining the matrix from the circuit can be done either by running the circuit on a quantum hardware (plus some tomography) or via a simulation on a

© Springer Nature Switzerland AG 2019
J. M. F. Rodrigues et al. (Eds.): ICCS 2019, LNCS 11537, pp. 3–16, 2019.
https://doi.org/10.1007/978-3-030-22741-8_1

classical computer [2,24]. Obtaining the circuit from the matrix is more complicated and fits into a more general problematic called *quantum compilation* i.e., how to translate a quantum operator described in an unknown form for the targeted hardware into a sequence of elementary instructions understandable by the machine. Therefore converting a matrix into a circuit is in all circumstances a compilation step because no hardware can directly accept a matrix as input: this particular task of compilation is called *quantum circuit synthesis* and is the central concern of this work. This process must be as automated and optimized as possible in terms of conventional or quantum resources. This task is inevitably hard in the general case: the classical memory required to store the unitary matrix is exponential in the number of qubits and it has been shown that an exponential number of quantum gates is necessary to implement almost every operator [11]. As a result both classical and quantum complexity follow an exponential growth and any methods will be quickly limited by the size of the problem. Yet being able to synthesize quickly and/or optimally any quantum operator or quantum state on a few qubits is a crucial challenge that could be crucial for some applications. In the NISQC (Noisy Intermediate Scale Quantum Computer) era [18], compressing circuits as much as possible will be crucial. And in general, such a procedure can be integrated as a subtask in a peep-hole optimization framework.

*Our Contribution*
The quantum circuits addressed in this paper correspond to the specific technology of trapped-ions quantum computer which requires using specific quantum gates. This technology holds great promise to cross the passage to scale: quantum circuits with trapped-ions technology have great fidelities and the qubits have a long decoherence time. In other words the noise in the results is low and long time computations can be performed before the system interferes with the environment and loses its information. Moreover the particular architecture of trapped-ions quantum circuits makes the problem simpler for numerical optimization because as we will see it only involves the optimization of continuous real parameters. We provide a simple optimization framework and we use it to synthesize generic quantum operators. More specifically, we derive a lower bound on the number of entangling gates necessary to implement any quantum algorithm and propose numerical experiments that confirm the correctness of this bound.

*Plan of the Paper*
In Sect. 2, we give the main definitions related to quantum circuits and quantum gates and we summarize the state of the art in quantum circuit synthesis. Then in Sect. 3 we develop in more details the modeling of a circuit as a parameterized topology and the question of the minimum-sized topology necessary to implement any operator. In Sect. 4 we introduce a generic optimization framework formalizing the circuit synthesis as a classical optimization problem. In Sect. 5 we apply this framework to optimally synthesize generic unitary matrices with the trapped-ions natural set of gates. We also discuss the scalability of this

method and how we can naturally trade the optimality of the final quantum circuit for shorter computational time. We conclude in Sect. 6.

*Notations*

Throughout this paper we will use the following notations. $\mathcal{U}(n)$ denotes the set of unitary matrices of size $n$, i.e. $\mathcal{U}(n) = \{M \in \mathbb{C}^{n \times n} \mid M^\dagger M = I\}$, where $I$ is the identity matrix and $M^\dagger$ is the conjugate transpose of the matrix $M$. The special unitary group $\mathcal{SU}(n)$ is the group of $n \times n$ unitary matrices with determinant 1. $\|x\| = \sqrt{x^\dagger x}$ refers to the Euclidean norm of a vector $x \in \mathbb{C}^n$ and $A \otimes B$ denotes the Kronecker product [5] of two matrices $A$ and $B$.

# 2    Background and State of the Art

## 2.1    Main Notions in Quantum Circuits

The basic unit of information in quantum information is the quantum bit and is formalized as a complex linear superposition of two basis states, usually called the state '0' and the state '1'. More generally when manipulating a set of $n$ qubits we modify a quantum state

$$|\psi\rangle = \sum_{\mathbf{b} \in \{0,1\}^n} \alpha_\mathbf{b} |\mathbf{b}\rangle$$

which is a normalized complex linear superposition of all the possible n-bitstring values. So a quantum state on $n$ qubits can be written as a unit vector in $\mathbb{C}^{2^n}$ and by the laws of quantum physics any operation on this quantum state can be represented as a left-multiplication of its vector representation by a unitary matrix $U \in \mathcal{U}(2^n)$, if we except the measurement operation.

Quantum algorithms are described via quantum circuits—the quantum analog of logical circuits. An example is given in Fig. 1. Each horizontal wire corresponds to one qubit and quantum gates –represented as boxes, or vertical apparatus—are sequentially applied from left to right to different subsets of qubits resulting in a more complex unitary operator. Matrix multiplication $BA$ of two operators $A$ and $B$ enables to apply sequentially the operator $A$ then $B$. To compose two operators $A$ and $B$ acting on different qubits we use the Kronecker product $\otimes$.

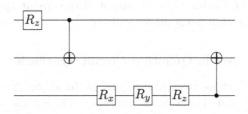

**Fig. 1.** Example of quantum circuit

For a given technology not all quantum gates are realizable and only a few of them are directly implementable: the so-called elementary gates. For example, we have the Hadamard gate $H = \frac{1}{\sqrt{2}} \begin{pmatrix} 1 & 1 \\ 1 & -1 \end{pmatrix}$, or the one-qubit rotations of angle $\theta \in \mathbb{R}$ along the x, y and z axis respectively defined by $R_x(\theta) = \begin{pmatrix} \cos(\theta) & -i\sin(\theta) \\ -i\sin(\theta) & \cos(\theta) \end{pmatrix}, R_y(\theta) = \begin{pmatrix} \cos(\theta) & -\sin(\theta) \\ \sin(\theta) & \cos(\theta) \end{pmatrix}, R_z(\theta) = \begin{pmatrix} e^{-i\theta/2} & 0 \\ 0 & e^{i\theta/2} \end{pmatrix}.$
Fortunately some subsets of elementary gates have been shown to be "universal", which means that any quantum operator can be implemented by a quantum circuit containing only gates from this set. Depending on the technology used, the universal sets available will be different. With superconducting qubits or linear optics, the natural set of gates is the set of one-qubit gates $\mathcal{SU}(2)$, also called local gates because they act on one qubit only, combined with the entangling CNOT (Controlled-NOT) gate which is a NOT gate controlled by one qubit. For instance, in the 2-qubits case, the CNOT gate controlled by the first qubit and applied to the second one can be represented by the matrix $\begin{pmatrix} 1 & 0 & 0 & 0 \\ 0 & 1 & 0 & 0 \\ 0 & 0 & 0 & 1 \\ 0 & 0 & 1 & 0 \end{pmatrix}$.

In this paper we focus on the technology of trapped ions which uses a different universal set of gates, the available gates are:

- local $R_z$ gates,
- global $R_x$ gates i.e. local $R_x$ are applied to every qubit with the same angle,
- the entangling Mølmer–Sørensen gate (MS gate) defined by

$$MS(\theta) = e^{-i\theta(\sum_{i=1}^{n} \sigma_x^i)^2/4}.$$

where $\sigma_x^i$ is the operator $X$ applied to the $i$-th qubit.

Any quantum operator has an abstract representation using a unitary matrix and it is essential to be able to switch from a quantum operator given by a quantum circuit to a quantum operator given by its unitary matrix and vice-versa. From a quantum circuit to a unitary matrix this is the problem of the *simulation of a quantum circuit*. Finding a quantum circuit implementing a unitary matrix is the problem of the *synthesis of a quantum circuit* which is the central concern of this paper. We distinguish two different synthesis problems: the first one consists in implementing a complete unitary matrix and the second one consists in preparing a specific quantum state as output of the circuit applied to the state $|000...00\rangle$: this is the *state preparation problem*.

## 2.2   State of the Art in Quantum Circuit Synthesis

Most quantum algorithms are still designed by hand [6,22] even though some circuits have been found via automatic processes [12]. For the automatic synthesis of quantum circuits there are mainly algebraic methods: we can mention the Quantum Shannon Decomposition (QSD) that gives the lowest number of gates in the general case [20] or the use of Householder matrices to achieve the synthesis in a much lower time [4]. The (H,T) framework is used for one-qubit synthesis [10] although it can be extended to the multi-qubits case [3]. For the particular case of state preparation - the synthesis of one column of the quantum operator - we have devices using multiplexors [15] or Schmidt decomposition [17].

Although the asymptotic complexity of the methods has significantly decreased with the years, some progress can still be made. The motivation behind this article is to use well-known numerical optimization methods in hope of reducing at its maximum the final quantum resources. Using heuristics or classical optimization methods to synthesize circuits is not new. The BFGS (Broyden-Fletcher-Goldfarb-Shanno) algorithm [25] has already been used to synthesize trapped-ions circuits [14] and machine learning techniques have been used in the case of photonic computers [1]. Genetic algorithms have also been used in a general context [13] or for the specific IBM quantum computer [12]. However, these works are purely experimental: the overall optimality of their solution is not discussed.

We tackle the problem of the optimality of the solution by building on the work in [21] that provides a theoretical lower bound of the number of entangling gates necessary in the quantum circuit synthesis problem. The idea is to count the number of degrees of freedom in a quantum circuit and show that this number has to exceed a certain threshold to be sure that an exact synthesis is possible for any operator. To our knowledge numerical methods have not been used in order to address the issue of the lower bound and the more general issue of the minimum quantum resources that are necessary to synthesize a quantum circuit.

## 3   Lower Bounds for the Synthesis of Trapped-Ions Quantum Circuits

The problem of computing the minimal number of quantum gates necessary to implement an operator remains open. Knill [11] showed that for the entire set of quantum operators (up to a zero measure set) we need an exponential number of gates and using a polynomial number of gates is as efficient as using a random circuit in the general case. The special case of circuits built from $\{SU(2), CNOT\}$ has been analyzed in [21], where quantum circuits are modeled as instantiations of circuit topologies consisting of constant gates and parameterized gates. The unspecified parameters are the degrees of freedom (DOF) of the topology. For instance the circuit given in Fig. 1 can be considered as a topology with at most 4 degrees of freedom (one for each rotation) and giving precise angles to the rotations is an instantiation of the topology. As a consequence a topology with $k$ degrees of freedom can be represented by a smooth function

$$f : \mathbb{R}^k \to \mathcal{U}(2^n) \tag{1}$$

that maps the values of angles to the space of unitary matrices of size $2^n$.

We are interested in the image of the function $f$. If a topology $f$ on $n$ qubits can implement any n-qubits operator, i.e., for any operator $U$ on $n$ qubits there exists a vector of angles $x$ such that $f(x) = U$, then we say that the topology is universal. Now what is the minimum number of gates necessary to obtain a universal topology? The best lower bound for this minimum in the case of $\{CNOT, \mathcal{SU}(2)\}$ circuits is given in [21]. In this section, we derive a lower bound in the case of trapped-ions circuits using a similar reasoning.

**Theorem 1.** *A topology composed of one-qubit rotations and parameterized MS gates cannot be universal with fewer than* $\left\lceil \frac{4^n - 3n - 1}{2n + 1} \right\rceil$ *MS gates.*

*Proof.* First we use Sard's theorem [7] to claim that the image of $f$ is of measure 0 if $k = \#DOF < dim(\mathcal{U}(2^n))$. Hence to be sure that we can potentially cover the whole unitary space we need

$$\#DOF \geqslant dim(\mathcal{U}(2^n)) = 4^n \qquad (2)$$

Next we give a normal form to trapped-ion circuits in order to count the number of DOF. MS gates operate on all qubits, they are diagonal in the basis $H^{\otimes n} = \bigotimes_{i=1}^{n} H$ obtained by applying an Hadamard gate to each qubit, the so-called "$|+\rangle / |-\rangle$" basis. We have

$$MS(\theta) = H^{\otimes n} \times D(\theta) \times H^{\otimes n} \qquad (3)$$

with

$$D(\theta) = \text{diag}([e^{(n - Hamm(i))^2 \times \theta}]_{i=0..2^n - 1}) \qquad (4)$$

where $Hamm$ is the Hamming weight in the binary alphabet.

First we can merge the Hadamard gates appearing in Eq. (3) with the local gates so that we can consider that our circuits are only composed of local gates and diagonal gates given by Eq. (4). Then we can write each local gate $U$ as

$$U = R_z(\alpha) \times R_x(-\pi/2) \times R_z(\beta) \times R_x(\pi/2) \times R_z(\gamma) \qquad (5)$$

where $\alpha, \beta, \gamma$ parameterize the unitary matrix $U$. Because the MS gates are now diagonal we can commute the first $R_z$ so that it merges with the next local unitary matrices. By doing this until we reach the end of the circuit we finally get a quantum circuit for trapped-ions hardware with the following basic subcircuit:

- a layer of local $R_z$ gates,
- a layer of global $R_x(\pi/2)$ gates,
- a layer of local $R_z$ gates,
- a layer of global $R_x(-\pi/2)$ gates,
- an MS gate (given in its diagonal form)

This subcircuit is repeated $k$ times for a circuit with $k$ MS gates. Ultimately, the circuit ends with a layer of rotations following the decomposition (5) on each qubit. An example of a quantum circuit on 3 qubits with 2 MS gates is

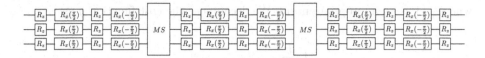

**Fig. 2.** Generic quantum circuit on 3 qubits for the trapped-ions technology

given in Fig. 2. The angles of the $R_z$ rotations are omitted for clarity. The only parameters of such generic circuits are:

- the angles of the $R_z$ rotations,
- the number and the angles of the MS gates.

Each elementary rotation (around the $x, y, z$ axis) is parameterized by one angle so it can only bring one additional degree of freedom, as the MS gates if they are parameterized. Including the global phase, a circuit containing $k$ parameterized MS gates can have at most $(2n + 1) \times k + 3n + 1$ DOF. In the example given Fig. 2 the topology has 24 DOF. To reach universality we must verify Eq. (2), which leads to the lower bound

$$\#MS \geqslant \left\lceil \frac{4^n - 3n - 1}{2n + 1} \right\rceil.$$

This proof easily transposes to the state preparation problem with a few changes:

- a quantum state is completely characterized by $2^{n+1} - 2$ real parameters,
- starting from the state $|0\rangle^{\otimes n}$ we can only add one degree of freedom to each qubit on the first rotations because the first $R_z$ result in an unnecessary global phase.

Consequently the total number of DOF a topology on $n$ qubits with $k$ MS gates can have is at most $(2n + 1)k + 2n$. We get the lower bound

$$\#MS \geqslant \left\lceil \frac{2^{n+1} - 2n - 2}{2n + 1} \right\rceil.$$

To our knowledge this is the first calculus of a lower bound in the context of trapped-ions circuits. In the next section we propose an algorithm based on numerical optimization that achieves accurate circuit synthesis using a number of gates corresponding to the above lower bounds. This gives a good indication of the tightness of these bounds.

## 4　The Optimization Framework

Given a function $f$ representing a topology on $n$ qubits, finding the best possible synthesis of a unitary matrix $U$ with the topology $f$ can be reformulated as solving

$$\arg \min_x \|f(x) - U\| := \arg \min_x g(x),$$

where $\| \cdot \|$ is an appropriate norm. We decompose the cost function $g$ as

$$g(x) = \frac{1}{k} \sum_{i=1}^{k} \left\| f(x) \left| e_{\phi(i)} \right\rangle - u_{\phi(i)} \right\|,$$

where $\phi$ is a permutation of $[\![1, 2^n]\!]$ and $u_j$ denotes the $j$-th column of $U$. In other words we generalize our problem to the synthesis of $k$ given columns of our operator $U$. For simplicity now we can only study the case where $k = 1$: this is the state preparation problem.

Our goal is to rely on various algorithms for non linear optimization. We choose the norm to be the Euclidean norm so that with this choice of norm we have a simple expression of the cost error:

$$g(x) = 2 \times \left( 1 - \mathrm{Re} \left( \langle 0 | f(x)^\dagger | \psi \rangle \right) \right). \tag{6}$$

Hence the cost to compute the error function is equivalent to the simulation cost of the circuit. Starting from the state $|\psi\rangle$ we simulate the circuit $f(x)^\dagger$, then the real part of the first component gives the error. Many methods for simulating a quantum circuit have been designed and we can rely on them to perform efficient calculations.

Since $g$ is $C^\infty$ we can use more performant optimization algorithms if we can compute the gradient and if possible the Hessian. To compute the gradient, we need to give a more explicit formula for $g(x)$. For some $j$ we have

$$\frac{\partial}{\partial x_j} g(x) = -2 \frac{\partial}{\partial x_j} \mathrm{Re} \left( \langle 0 | f(x)^\dagger | \psi \rangle \right),$$

and by linearity we get

$$\frac{\partial}{\partial x_j} g(x) = -2 \, \mathrm{Re} \left( \frac{\partial}{\partial x_j} \langle 0 | f(x)^\dagger | \psi \rangle \right).$$

Let us suppose we have $K$ gates in our circuit. Then we can write

$$f(x)^\dagger = \prod_{i=1}^{K} A_i$$

where $(A_i)_{i=1..K}$ is the set of gates on $n$ qubits that compose the circuit $f$. In all circuits encountered the parameterized gates are of the form $e^{i\theta\Omega}$ where $\Omega$ can be either $X, Y, Z$ (tensored with the identity if necessary) or more complex hermitians (see the MS gate for instance), $\theta$ is the parameter of the gate. We assume that two gates are not parameterized by the same variable. Let $k_1$ be the index of the gate depending on $x_j$, then we have

$$\frac{\partial}{\partial x_j} \langle 0| f(x)^\dagger |\psi\rangle = \frac{\partial}{\partial x_j} \langle 0| \prod_{i=1}^{k_1-1} A_i \times e^{i \times x_j \times \Omega} \times \prod_{i=k_1+1}^{K} A_i |\psi\rangle$$

$$= \langle 0| \prod_{i=1}^{k_1-1} A_i \times \frac{\partial}{\partial x_j} e^{i \times x_j \times \Omega} \times \prod_{i=k_1+1}^{K} A_i |\psi\rangle$$

$$= i \times \langle 0| \prod_{i=1}^{k_1-1} A_i \times \Omega \times A_{k_1} \times \prod_{i=k_1+1}^{K} A_i |\psi\rangle$$

Therefore we have a simple expression for the partial derivatives

$$\frac{\partial}{\partial x_j} g(x) = 2 \operatorname{Im} \left( \langle 0| \prod_{i=1}^{k_1-1} A_i \times \Omega \times A_{k_1} \times \prod_{i=k_1+1}^{K} A_i |\psi\rangle \right).$$

To compute the whole gradient vector we propose the following algorithm:

---

**Algorithm 1.** Computing the gradient of the cost function

---

**Require:** $n > 0, |\psi\rangle \in \mathbb{C}^{2^n}, f : \mathbb{R}^k \to \mathcal{U}(2^n)$,
**Ensure:** Computes $\partial f$
  $\langle \psi_1 | \leftarrow \langle 0|$
  $|\psi_2\rangle \leftarrow f(x)^\dagger |\psi\rangle$
  $m \leftarrow 1$
  // $N$ is the total number of gates in the circuit
  **for** $i = 1, N$ **do**
    // $A_i$ is the $i$-th gate in the circuit
    **if** $A_i$ is parameterized **then**
      // $H_i$ is the Hamiltonian associated to the gate $A_i$
      $df[m] \leftarrow 2 \operatorname{Im} (\langle \psi_1 | H_i |\psi_2\rangle)$
      $m \leftarrow m + 1$
    **end if**
    $\langle \psi_1 | \leftarrow \langle \psi_1 | A_i$
    $|\psi_2\rangle \leftarrow A_i^\dagger |\psi_2\rangle$
  **end for**

---

Therefore computing the gradient is equivalent to simulating two times the circuit. On total we need 3 circuit simulations to compute the error and the gradient.

Note that we need to store two vectors when computing the gradient. If $k > 2^{n-1}$ then we use more memory than if we have only stored the entire matrix. As memory is not a big concern here compared to the computational time, we keep this framework even for $k = 2^n$ for example.

# 5    Numerical Experiments

The experiments have been carried out on one node of the QLM (Quantum Learning Machine) located at ATOS/BULL. This node is a 24-core Intel Xeon(R) E7-8890 v4 processor at 2.4 GHz. Our algorithm is implemented in C with a Python interface and has been executed on 12 cores using OpenMP multi-threading. The uniform random unitary matrices are generated according to the Haar's measure [23]. For the numerical optimization we use the BFGS algorithm provided in the SciPy [19] package.

## 5.1    Synthesizing Generic Circuits

A clear asset of trapped-ion technology is that there is no notion of topology. Contrary to superconducting circuits where we have to deal with the placement of the CNOT gates, we just have to minimize the number of MS gates to optimize the entangling resources. Local rotations are less expensive to realize experimentally [14]. So, as a first approach, we can always apply one-qubit gates on every qubits between two MS gates such that a quantum circuit for trapped ions always follow a precise layer decomposition.

For 50 random unitary matrices on $k \in \{2, 3, 4\}$ qubits, and quantum states on $k \in \{2, 3, 4, 5, 6, 7\}$ qubits, we execute Algorithm 1 with circuits containing various numbers of parameterized MS gates. The stopping criterion for the optimization is the one chosen in SciPy i.e., the norm of the gradient must be below $10^{-5}$. We repeat the optimization process several times per matrix with different starting points to maximize our chance to reach a global minimum. Then we plot, for various numbers of MS gates,

- the error expressed in Formula (6), maximized over the sample of matrices,
- the number of iterations needed for convergence, averaged over the sample of matrices.

We summarize our results in Figs. 3a and 3b corresponding respectively to the circuit synthesis problem on 4 qubits and the state preparation problem on 7 qubits. The results obtained for lower number of qubits are similar. The amount of time required to perform such experiments for larger problems is too important (more than a day).

For both graphs, we observe an exponential decrease of the error with the number of MS gates. This strong decrease shows that there is a range of MS gates count for which we have an acceptable accuracy without being minimal. Although we can save only a small number of MS gates by this trade, we conjecture that for a given precision the number of saved MS gates will increase with the number of qubits. In other words if we want to synthesize an operator up to a given error, the range of "acceptable" number of MS gates will increase with the number of qubits.

What is the experimental lower bound? Interestingly, the number of iterations is a better visual indicator of this lower bound: once the exact number of DOF is

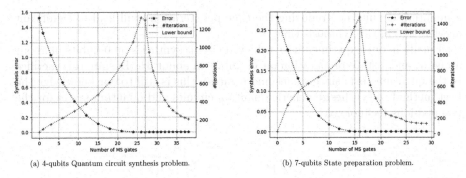

(a) 4-qubits Quantum circuit synthesis problem.      (b) 7-qubits State preparation problem.

**Fig. 3.** Evolution of the synthesis error/number of iterations with the number of MS gates.

reached, we add redundant degrees of freedom into the optimization problem. We make the hypothesis that this leads to a bigger search space but with many more global minima. Hence the search is facilitated and the algorithm can converge in fewer iterations. So the peak of the number of iterations corresponds to the experimental lower bound, which is confirmed by the computed errors.

The experimental lower bounds follow the formulas given in Sect. 3 except for the 2-qubits quantum circuit synthesis case where we need 1 more MS gate to reach universality. This gives strong indications about the correctness of the theoretical lower bounds and the relevance of the approach of counting the degrees of freedom to estimate the resources necessary for implementing a quantum operator.

## 5.2 Tradeoff Quantum/Classical Cost

In the previous experiments, we could synthesize unitary operators on 6 qubits in 7 h and quantum states on 11 qubits in 13 h, both with circuits of optimal size. In the graphs plotted in Figs. 3a and b, we also observe an exponential decrease of the number of iterations after we have reached the optimal number of MS gates. We can exploit this behavior if we want to accelerate the time of the synthesis at the price of a bigger circuit. By adding only a few more MS gates the number of iterations can be significantly reduced, allowing us to address potentially bigger problems in the synthesis. In Fig. 4 we show, for different problem sizes, how the iteration count and the time per iteration evolve when we add 10% more MS gates to the optimal number of gates. We observe that the number of iterations is reduced at most by a factor 5 in the case of 6 qubits for generic operators and 11 qubits for quantum states. More importantly, with such an augmentation in the number of MS gates, the number of iterations only slightly increases, almost linearly, in contrary to circuits of optimal size where the number of iterations increases exponentially. This means that the time per iteration, also increasing exponentially with the number of qubits, is the main limiting factor in the good scalability of the method. Thus any improvement in the classical simulation of a quantum circuit (i.e., reducing the time per iteration) will lead to significant acceleration in the synthesis time.

Finally in our experiments we achieved circuit synthesis on 6 qubits in about an hour, and a state preparation on 11 qubits in about 3 h. Despite this sacrifice in the quantum cost (since we use more gates), this is still to our knowledge the best method to synthesize generic quantum circuits and quantum states for the trapped-ions technology.

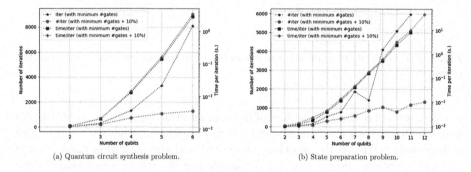

(a) Quantum circuit synthesis problem.        (b) State preparation problem.

**Fig. 4.** Number of iterations and time per iteration for different problem sizes.

## 6   Conclusion

We have explained how a numerical optimization algorithm can be used to synthesize generic quantum circuits adapted to the trapped-ions technology. We have provided a simple algorithm to compute the error and the gradient such that the scalability of our method relies on the simulation cost of the circuits we want to optimize. We have also highlighted a possible tradeoff between the classical time necessary to perform the synthesis and the final size of the circuits resulting in more flexibility in the application of the framework. Finally we have shown that the lower bounds computed experimentally with this framework follow the theoretical results with the hope that it would help for future formal proofs.

As future work we plan to extend our analysis to specific quantum operators for which we expect shorter circuits. We also plan to extend this analysis to $\{CNOT, \mathcal{SU}(2)\}$ based circuits, especially in order to answer the following questions: is the lower bound given in [21] a tight one? Is there a universal topology that reaches that lower bound? We also investigate a way to efficiently compute the Hessian in order to use more complex optimization methods and we plan to address problems with more qubits thanks to GPU and distributed computing.

**Acknowledgement.** This work was supported in part by the French National Research Agency (ANR) under the research project SoftQPRO ANR-17-CE25-0009-02, and by the DGE of the French Ministry of Industry under the research project PIA-GDN/QuantEx P163746-484124.

# References

1. Arrazola, J.M., Bromley, T.R., Izaac, J., Myers, C.R., Bradler, K., Killoran, N.: Machine learning method for state preparation and gate synthesis on photonic quantum computers. Quantum Science and Technology (2018)
2. Bravyi, S., Browne, D., Calpin, P., Campbell, E., Gosset, D., Howard, M.: Simulation of quantum circuits by low-rank stabilizer decompositions. arXiv preprint arXiv:1808.00128 (2018)
3. Giles, B., Selinger, P.: Exact synthesis of multiqubit clifford+$t$ circuits. Phys. Rev. A **87**, 032332 (2013)
4. Goubault de Brugière, T., Baboulin, M., Valiron, B., Allouche, C.: Quantum circuits synthesis using householder transformations (2018, submitted)
5. Graham, A.: Kronecker Products and Matrix Calculus with Application. Wiley, New York (1981)
6. Grover, L.K.: A fast quantum mechanical algorithm for database search. In: Proceedings of the Twenty-eighth Annual ACM Symposium on Theory of Computing, STOC 1996, pp. 212–219. ACM, New York (1996)
7. Guillemin, V., Pollack, A.: Differential Topology, vol. 370. American Mathematical Society (2010)
8. Harrow, A.W., Hassidim, A., Lloyd, S.: Quantum algorithm for linear systems of equations. Phys. Rev. Lett. **103**, 150502 (2009)
9. Kerenidis, I., Prakash, A.: Quantum recommendation systems. In: Papadimitriou, C.H. (ed.) 8th Innovations in Theoretical Computer Science Conference, ITCS 2017. LIPIcs, vol. 67, pp. 49:1–49:21. Schloss Dagstuhl - Leibniz-Zentrum fuer Informatik (2017)
10. Kliuchnikov, V., Maslov, D., Mosca, M.: Fast and efficient exact synthesis of single-qubit unitaries generated by Clifford and T gates. Quantum Inf. Comput. **13**(7–8), 607–630 (2013)
11. Knill, E.: Approximation by quantum circuits. arXiv preprint arXiv:quant-ph /9508006 (1995)
12. Li, R., Alvarez-Rodriguez, U., Lamata, L., Solano, E.: Approximate quantum adders with genetic algorithms: an IBM quantum experience. arXiv preprint arXiv:1611.07851 (2016)
13. Lukac, M., et al.: Evolutionary approach to quantum and reversible circuits synthesis. Artif. Intell. Rev. **20**(3–4), 361–417 (2003)
14. Martinez, E.A., Monz, T., Nigg, D., Schindler, P., Blatt, R.: Compiling quantum algorithms for architectures with multi-qubit gates. New J. Phys. **18**(6), 063029 (2016)
15. Mottonen, M., Vartiainen, J.J.: Decompositions of general quantum gates. arXiv preprint arXiv:quant-ph/0504100 (2005)
16. Nielsen, M.A., Chuang, I.L.: Quantum Computation and Quantum Information. Cambridge University Press, Cambridge (2011)
17. Plesch, M., Brukner, Č.: Quantum-state preparation with universal gate decompositions. Phys. Rev. A **83**, 032302 (2011)
18. Preskill, J.: Quantum Computing in the NISQ era and beyond. Quantum **2**, 79 (2018)
19. Python: Scientific computing tools for Python (SciPy). https://www.scipy.org
20. Shende, V.V., Bullock, S.S., Markov, I.L.: Synthesis of quantum-logic circuits. IEEE Trans. Comput. Aided Des. Integr. Circuits Syst. **25**(6), 1000–1010 (2006)

21. Shende, V.V., Markov, I.L., Bullock, S.S.: Minimal universal two-qubit controlled-not-based circuits. Phys. Rev. A **69**(6), 062321 (2004)
22. Shor, P.W.: Polynomial-time algorithms for prime factorization and discrete logarithms on a quantum computer. SIAM Rev. **41**(2), 303–332 (1999)
23. Stewart, G.: The efficient generation of random orthogonal matrices with an application to condition estimators. SIAM J. Numer. Anal. **17**(3), 403–409 (1980)
24. Vidal, G.: Efficient classical simulation of slightly entangled quantum computations. Phys. Rev. Lett. **91**, 147902 (2003)
25. Wright, S., Nocedal, J.: Numerical optimization. Science **35**(67–68), 7 (1999)

# Application of Continuous Time Quantum Walks to Image Segmentation

Michał Krok[1], Katarzyna Rycerz[2(✉)], and Marian Bubak[2]

[1] IBM Software Laboratory, Kraków, Poland
michal.krok@ibm.com
[2] Department of Computer Science, AGH University of Science and Technology,
Kraków, Poland
{kzajac,bubak}@agh.edu.pl

**Abstract.** This paper provides the algorithm that applies concept of continuous time quantum walks to image segmentation problem. The work, inspired by results from its classical counterpart [9], presents and compares two versions of the solution regarding calculation of pixel-segment association: the version using limiting distribution of the walk and the version using last step distribution. The obtained results vary in terms of accuracy and possibilities to be ported to a real quantum device. The described results were obtained by simulation on classical computer, but the algorithms were designed in a way that will allow to use a real quantum computer, when ready.

**Keywords:** Quantum walk · Image segmentation ·
Quantum algorithms

## 1 Introduction

Image segmentation [26] is one of the most important image processing tools, as it is frequently applied in an early stage of image analysis in order to extract some high-level features of the input image and significantly simplify further computations. Image segmentation is a process of dividing entire image into multiple separate segments – areas inside which pixels expose similar characteristics (e.g. intensity, hue, texture). It might be also defined as an activity of assigning each pixel of an image with a label, in a way that pixels sharing similar traits of interest are labeled alike. Image segmentation has found a broad range of applications: from a very trendy in recent years computer vision (object detection and recognition) [13] to medical diagnostics and treatment planning [14]. Despite many efforts and development of plethora of various methods, image segmentation remains an open research area, as there is still a need for inventing more efficient and more precise solutions.

A very interesting method for segmentation was invented by Grady [9]. The author proposes solution utilizing classical random walks, a very successful computational framework widely used for simulation of some natural and social processes in physics, biology and even economics [22]. The mechanism of random

© Springer Nature Switzerland AG 2019
J. M. F. Rodrigues et al. (Eds.): ICCS 2019, LNCS 11537, pp. 17–30, 2019.
https://doi.org/10.1007/978-3-030-22741-8_2

walk action is quite simple: a particle (walker) explores given space of possible states by moving at each time step to a new state according to current state and given transition probabilities. The walker behavior can be useful for determining some properties of the explored space. Based on the random walk there has been developed a analogous quantum model – quantum random walks. It has been shown that quantum walks can benefit from the extraordinary properties of nanoscale world (quantum superposition, interference or entanglement) to achieve quadratic or even exponential [7] speedup over their classical counterparts for some particular problems. Therefore it seems reasonable to explore this branch of quantum computation in the quest for new, better algorithms.

The aim of this research is to elaborate and examine a new algorithm that applies concept of quantum walks to image segmentation. It is inspired by results from its classical counterpart [9]. In this paper, we present concept, implementation and results of two versions of the solution. The results were obtained by simulation on classical computer, but the algorithms were designed in a way that will allow them to be run on a real quantum computer, when the actual device will be available. This approach seems to be promising, especially in a context of recent work that describes implementation of simple quantum walk on IBM-Q hardware [4].

The rest of the document is organized as follows: Sect. 2 discusses research concerning image segmentation and quantum walks. Section 3 presents elaborated algorithms, while Sect. 5 examines their performance and accuracy. Finally, the effort put in this paper is summarized in Sect. 6.

## 2   Related Works

Image segmentation has been a carefully studied problem over the years and there has been developed a multitude of different solutions [21,26]. One of the simplest are threshold-based algorithms, which assign a label to each pixel according to its intensity value in comparison to one or more thresholds, e.g. frequently used Otsu method [19]. Another approach that has been successfully applied is based on clustering; it determines segments by aggregating pixels in groups (clusters) according to chosen metric [17]. Deep learning and neural networks, especially convolutional neural networks, also proved to be very successful in various image processing tasks, one of which is image segmentation [6].

The interesting class are graph partitioning methods. To this category belong methods like: normalized cuts [23], intelligent scissors [18] and isoperimetric partitioning [10]. The aforementioned Grady algorithm also belongs to the last category of image segmentation solutions. As the input, the user defines (apart from an image) a set of seeds – pixels which already have assigned labels corresponding to the image segment that they belong to. The algorithm then tries to determine for each unmarked pixel, what are the probabilities of reaching each seed as the first by a random walk staring from given pixel. Each pixel is then assigned with a label of the seed that is most probable to be visited first by the walker.

With the development of quantum computing there emerged a new field of research – quantum image processing. In [24] there were presented methods for storing and retrieving as well as segmenting images in a quantum system. However, the proposed algorithm requires specifying objects in image through as set of their vertices, which rarely is a case and obtaining those vertices is a complicated task itself. Two other methods for quantum image segmentations were described in [5,12], but these are the threshold-based solutions which do not take into account the spatial locations of pixels, which results in rather poor outcome.

Since Ashwin and Ashvin [3] showed that quantum walks on line spread quadratically faster than their classical analogues, the quantum model has been studied to find algorithms that would allow for similar speedup. The most spectacular achievement in this field was presented by Childs [7], where the author achieves exponential speedup by applying a continuous time quantum walk in a graph formed of two binary trees. There have been presented several search algorithms that can provide quadratic speedup benefiting from quantum walks e.g. works by Ambainis et al. [2] and Magniez et al. [15].

Over the recent years, there have been developed many various algorithms utilizing quantum walks. Most of the works consider rather theoretical problems, therefore it might be beneficial to elaborate an algorithm that would directly solve a practical, commonly encountered task.

## 3   Walk Model

In this paper we propose two versions of the algorithm for image segmentation utilizing continuous time quantum walks. In continuous time quantum walks (CTQW) transition from the current state to the next one does not take place in discrete steps, but can occur at any moment, according to a $\gamma$ parameter that describes the rate of spread of the walk.

As the input, the elaborated algorithms expect following data:

- Image – two dimensional image of size $M \times N$ with each pixel $p_{ij}$ described by two indices: $i \in \mathbb{Z}_M$ (index of the row) and $j \in \mathbb{Z}_N$ (index of the column). Each pixel can be represented as a single value of luminescence intensity or a vector of intensities in separate channels (e.g. vector of length 3 for the RGB color model).
- Labels $L = \{1, 2, 3, ..., l\}$ – set of labels, one for each segment that the image should be divided into.
- Seeds $S = \{(p_1, l_1), (p_2, l_2), ..., (p_s, l_s)\}$ – set of pixels $p_k$ with already assigned label $l_k$ from the set $L$; there should be multiple seeds for each label.
- Parameter $\beta$ – a parameter responsible for highlighting the stronger differences between pixels; it is used while transforming the image into a graph.
- Parameter $\gamma$ – rate of spread of a continuous time quantum walk, used to construct the evolution operator.
- Parameter $T$ – duration of the quantum walk (number of applications of evolution operator).

Firstly, the input image is converted into a weighted graph with vertices corresponding to pixels of the image. From each vertex there are drawn four edges to the neighboring pixels. Finally, each edge is assigned with a weight describing the level of similarity between adjacent pixels $p_{ij}$ and $p_{kl}$. We use the formula proposed in [9]:

$$w_{ij,kl} = \begin{cases} e^{-\frac{\beta d(p_{ij}, p_{kl})}{\max d(p_{ij}, p_{kl})}}, & \text{if } p_{ij} \text{ and } p_{kl} \text{ are connected,} \\ 0, & \text{otherwise,} \end{cases} \quad (1)$$

where $d(p_{ij}, p_{kl})$ is a metric of pixel similarity defined in our paper as follows:

$$d(p_{ij}, p_{kl}) = \sum_c (p_{ij}[c] - p_{kl}[c])^2, \quad (2)$$

where $c$ are the consecutive channels of pixels.

The main idea to tackle the problem of image segmentation is shown in the Fig. 1 illustrating quantum walks for two example labels: green and red. First, a separate continuous time walk for each label is started and the walker explores the image according to its content. Then, based on which walker was the most willing to explore given region of the image, assign appropriate label to each pixel, which results in dividing the image into segments. The details of constructing the quantum walk described in this paper are presented below.

**Position Space.** The position space for the walk is a Hilbert space of size $M \times N$; each basis state corresponds to a pixel of the image: $\{|0,0\rangle, |0,1\rangle, ..., |0, M-1\rangle, |1,0\rangle, ..., |N-1,0\rangle, |N-1,1\rangle, ..., |N-1, M-1\rangle\}$.

**Initial State.** The initial state of the walker is a superposition of positions of all the seeds for given label. Let the current label be $l_i \in L$ and the $S_{l_i} = \{(p_1, l_i), (p_2, l_i), ..., (p_k, l_i)\} \subset S$ be a subset of seeds corresponding to that label. Then the initial state for walk corresponding to the $i$-th label has the following form:

$$|\psi_i(0)\rangle = \frac{1}{\sqrt{k}} \sum_{j=1}^{k} |p_j\rangle \quad (3)$$

**Evolution Operator.** The most important element of the quantum walk is its evolution operator, that determines the behavior of the walk. The current task is to construct an operator that would encourage walker to explore image according to the weights between neighboring pixel. The desired behavior is to make the walker advance preferably over the higher weights between similar pixels and avoid crossing big gaps in pixel intensities.

The general form of the evolution operator of continuous time quantum walk has the following form (here $i$ means imaginary unit):

$$U(t) = e^{-iHt}, \quad (4)$$

where $t$ denotes time and $H$ is a Hamiltonian matrix. The task is to find an appropriate Hamiltonian matrix H. It has to be a square matrix of size $MN \times$

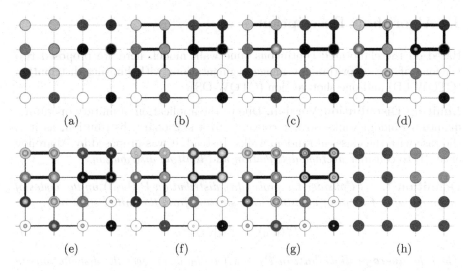

**Fig. 1.** Visualization of the concept outline. The algorithm input is shown in (a) where circles symbolize pixels with their intensity and seeds are marked as two pixels with green and red ring. In (b), image is transformed into weighted graph based on intensities of neighboring pixels (weights are denoted by the edge width). In (c), a walk is started from a superposition of seeds with given (here green) label (colorful dots denote the probability of measuring walker at given pixel) and, in (d), proceeds for several steps. In (e), when the walk ends the limiting distribution (CTQW-LD) or the last state (CTQW-OS) is calculated. In (f), similar procedure is repeated for each (here red) label. In (g), final distributions of each walk are compared and each pixel is assigned with label of the walk that had the highest probability of being measured at given pixel (h). (Color figure online)

$MN$ that would be hermitian (this ensures that the operator $U(t)$ is unitary). It also has to yield satisfying results in terms of position space exploration by the walker according to the content of the image.

The idea is to construct the Hamiltonian based on the weights matrix defined by the Eq. 1 and a free real parameter $\gamma$ that determines the rate of spread of the quantum walk:

$$H_{ij,kl} = \begin{cases} -\gamma w_{ij,kl}, & \text{if } i \neq k \vee j \neq l. \\ \gamma \sum_{k',l'} w_{ij,k'l'}, & \text{if } i = k \wedge j = l. \end{cases} \qquad (5)$$

Please notice, that since weight matrix is symmetrical ($w_{ij,kl} = w_{kl,ij}$) and its values are real numbers, the constructed matrix $H$ is a hermitian matrix and therefore the operator $U(t)$ is unitary.

## 4   Algorithms Description

Based on the elaborated continuous time walk model, here are proposed two versions of the algorithm for image segmentation: with limiting distribution (CTQW-LD) and one shot version (CTQW-OS).

**Limiting Distribution Version.** Due to the application of unitary operators, quantum random walks do not converge to a stationary distribution, as it is the case with the classical random walks. Instead, it was proposed by Aharonov et al. [1] to consider *average distribution* and *limiting distribution*.

**Definition 1.** *[1] Consider the probability distribution $P_t(|\psi_0\rangle)$ on the nodes of the graph after $t$ steps of quantum walk starting from the initial state $|\psi_0\rangle$:*

$$P_t^i(|\psi_0\rangle) = \| \langle i|\psi_t\rangle \|^2. \tag{6}$$

*Then the **average distribution** $\overline{P}_T(|\psi_0\rangle)$ is the mean over the distributions in each time step until $T$:*

$$\overline{P}_T(|\psi_0\rangle) = \frac{1}{T}\sum_{t=0}^{T-1} P_t(|\psi_0\rangle). \tag{7}$$

Notice that the *average distribution* can be understood as a measure of how long the quantum walker spends at each of the position states throughout the first $T$ steps of the walk. It was shown in [1] that, for any initial state $|\psi_0\rangle$, the average distribution converges as the time approaches infinity and the limit is denoted as *limiting distribution*.

**Definition 2.** *[1] **Limiting distribution** of the quantum random walk starting from the initial state $|\psi_0\rangle$:*

$$\pi(|\psi_0\rangle) = \lim_{T\to\infty} \overline{P}_T(|\psi_0\rangle). \tag{8}$$

The *limiting distribution* is not unique, but depends on the initial state.

The version of the algorithm using limiting distribution (CTQW-LD) is shown in the Algorithm 1. For each label the separate quantum walk is performed. After each step of the walk the probability distribution of measuring the walker at given position is recorded. In our simulations we could directly read the walker state after each step, as the calculations were performed using matrix algebra. Also after introducing the optimization (described below) we obtained the state after given number of steps by changing the equation parameters. However, on a quantum device retrieval of the quantum state is not straightforward, This is achieved by performing so called state tomography repeating the walk for given number of steps multiple times and making a measurement which results in finding the walker at some position on the image. The number of samplings grows with the number of basis states (the image size) and the desired accuracy of the reconstructed state. This allows to retrieve the probability distribution

---

**Algorithm 1.** Image segmentation using continuous time quantum walk with limiting distribution solution (CTQW-LD)

---

**Input:** Image $A = (a_{ij}) \in V^{M \times N}$, where $V = \mathbb{R}^c$ and $c$ is number channels for a pixel,

Set of labels $L$,

List of seeds $S = \{(p_1, l_1), ..., (p_s, l_s)\}$, where $p_k \in \mathbb{Z}_M \times \mathbb{Z}_N$ is the seed position and $l_k \in L$ is its label.

**Parameters:** $\beta$ - boundary strengthening parameter,

$\gamma$ - rate of spread of the walker,

$T$ - number of steps of the walk.

**Output:** Segmented image $B = (b_{ij}) \in L^{M \times N}$.

---

1  $W \leftarrow calculate\_weights(A, \beta)$ ;                    // according to the formula 1
2  $H \leftarrow construct\_hamiltonian(W, \gamma)$ ;              // based on the equation 5
3  $U \leftarrow construct\_operator(H)$ ;                          // $U = U(1) = e^{-iH}$
4  **foreach** $(l_i) \in L$ **do**                                // perform walk for each label
5  $\quad$ $S' \leftarrow seeds\_subset(S, l_i)$ ;                  // coordinates of seeds with label $l_i$
6  $\quad$ $|\psi(0)\rangle \leftarrow \frac{1}{\sqrt{|S'|}} \sum_{p \in S'} |p\rangle$ ;   // set initial state to the seed
7  $\quad$ **for** $t \leftarrow 1$ **to** $T$ **do**
8  $\quad\quad$ $|\psi(t)\rangle \leftarrow U |\psi(t-1)\rangle$ ;                         // perform a move
9  $\quad\quad$ $D_t \leftarrow retrieve\_position(|\psi(t)\rangle)$ ; // prob. dist. of measuring walker
$\quad\quad$ at each position
10 $\quad$ **end**
11 $\quad$ $LD_i \leftarrow \frac{1}{T} \sum_{t=1}^{T} D_t$ ;        // limiting distribution of the walk
12 **end**
13 **for** $i \leftarrow 0$ **to** $M - 1$ **do**
14 $\quad$ **for** $j \leftarrow 0$ **to** $N - 1$ **do**
15 $\quad\quad$ $k \leftarrow find\_seed(LD, i, j)$ ;
$\quad\quad$ // find seed that on average has the highest
$\quad\quad$ probability of being measured at pixel $ij$
16 $\quad\quad$ $B_{ij} \leftarrow l_k$ ;                           // assign label to the pixel
17 $\quad$ **end**
18 **end**

---

of the walker after each time step. Upon finishing each walk (the walk for each label) the limiting distribution of the walk is calculated (due to the fact that there is no possibility of running a walk for infinite number of steps, the limiting distribution is approximated with average distribution of T steps). Finally, the result is obtained classically, as each pixel is assigned with label of the seed, for which walker starting from that seed expressed the highest average probability of being measured at that pixel.

**One shot version** of the algorithm (CTQW-OS) is shown in the Algorithm 2. This solution is almost identical to the CTQW-LD, but instead of calculating the limiting distribution of each walk, only the distribution after last state is used. As it will be shown in the next section, this method gives results with a bit

---

**Algorithm 2.** Image segmentation using continuous time quantum walk – one shot solution (CTQW-OS)

---

**Input:** Image $A = (a_{ij}) \in V^{M \times N}$, where $V = \mathbb{R}^c$ and $c$ is number channels for a pixel,

Set of labels $L$,

List of seeds $S = \{(p_1, l_1), ..., (p_s, l_s)\}$, where $p_k \in \mathbb{Z}_M \times \mathbb{Z}_N$ is the seed position and $l_k \in L$ is its label.

**Parameters:** $\beta$ - boundary strengthening parameter,

$\gamma$ - rate of spread of the walker,

$T$ - number of steps of the walk.

**Output:** Segmented image $B = (b_{ij}) \in L^{M \times N}$.

---

1  $W \leftarrow calculate\_weights(A, \beta)$ ;                          // according to the formula 1
2  $H \leftarrow construct\_hamiltonian(W, \gamma)$ ;                 // based on the equation 5
3  $U \leftarrow construct\_operator(H)$ ;                                 // $U = U(1) = e^{-iH}$
4  **foreach** $(l_i) \in L$ **do**                                           // perform walk for each label
5   $S' \leftarrow seeds\_subset(S, l_i)$ ;                     // coordinates of seeds with label $l_i$
6   $|\psi(0)\rangle \leftarrow \frac{1}{\sqrt{|S'|}} \sum_{p \in S'} |p\rangle$ ;             // set initial state to the seed
7   **for** $t \leftarrow 1$ **to** $T$ **do**
8    $|\psi(t)\rangle \leftarrow U |\psi(t-1)\rangle$ ;                             // perform a move
9   **end**
10  $D_i \leftarrow retrieve\_position(|\psi(T)\rangle)$ ;
    // prob. dist. of measuring walker at
     each position after last step
11 **end**
12 **for** $i \leftarrow 0$ **to** $M - 1$ **do**
13  **for** $j \leftarrow 0$ **to** $N - 1$ **do**
14   $k \leftarrow find\_seed(D, i, j)$ ;
     // find seed that has the highest probability of being
      measured at the end of the walk at pixel $ij$
15   $B_{ij} \leftarrow l_k$ ;                                      // assign label to the pixel
16  **end**
17 **end**

---

worse segmentation accuracy, but for careful choice of parameters the difference is ommitable. However, it does not require obtaining the probability distribution after each step, which is a tremendous profit, especially if the algorithm would be executed on a real quantum device. Quantum state tomography, which is used to retrieve the quantum state of the walker is an expensive operation, but in the CTQW-OS version of the algorithm it needs to be performed only once for each label, what is a great advantage over the CTQW-LD algorithm.

**Optimization.** In order to improve the performance of simulations of the algorithms on classical computers there have been introduced some optimizations. First, each walk was performed in parallel in separate processes. Additionally, it was tested that sampling the walker distribution not after each step, but for about every $100^{th}$ step in the range $[1, T]$ gives almost undisturbed results. Therefore, instead repeating evolution operator $U$, we used the numerical solver of Lindblad master equation [25].

## 5   Evaluation and Results

The evaluation of the proposed solutions could not, obviously, be performed on the real quantum device, as the requirements e.g. the number of qubits needed to express the walker state, the level of customization the quantum operators etc. are beyond the capabilities of currently accessible quantum computers e.g. IBM-Q [11] or Rigetti [20]. Therefore, there had to be performed a simulation of quantum computations on a classical machine. The project code has been prepared using Python language (version 3.4.3) with *QuTiP* open-source Python library for simulation of quantum systems (version 4.2.0). The program was executed on a computer with processor Intel Core i5-4210M and 16 GB of RAM. We used the Berkeley Segmentation Dataset and Benchmark (BSDS500) [16] dataset which contains a multitude of photographs, with corresponding contour images performed manually by people.

The most important parameter for the proposed solution is the set of seeds that needs to be provided separately. We considered one set created manually, by careful choice of evenly spread pixels with a bit higher concentration around borders, especially the weak and unclear boundaries (these sets contained about 0.2% pixels of the whole image); and three sets that were prepared automatically by drawing uniformly a fraction of pixels from the image and assigning them with the labels based on the ground truth segmentation from BSDS500 dataset (with the concentrations at the level of 0.2%, 0.5% and 1%, respectively). There are also other three free parameters:

**Parameter** $\beta$ – it is used during weights calculation; the greater its value, the more significant the stronger differences between pixels are. So the high $\beta$ parameter makes the walker stay closer to its seed rather than disperse and explore further regions. However, the change of this parameter does not significantly influence the outcome segmentation.

**Parameter** $\gamma$ – the rate of spread of the walker. The greater the value of this parameter, the greater is the area explored by the walker upon a single application of evolution operator. This means that one could set $\gamma$ to a high value to limit the number of steps required. Indeed, increase of this parameter results in obtaining better outcome for smaller number of steps. But for relatively high values of $\gamma$, as the number of steps grows, the segmentation accuracy drops almost as rapidly as it reaches its peak value.

**Parameter** $T$ – determines the number of steps (number of applications of the evolution operator). Too small values result in poor image exploration by walkers and appearance of some blank areas. Too high values result in walkers diffusion to neighboring segments that reduces the segmentation quality over time. It is

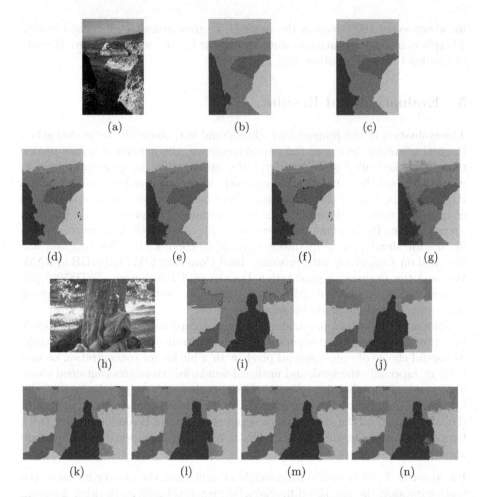

**Fig. 2.** Comparison of results on an example of color and grayscale test images. The original images are shown in (a, h) and the manual segmentations in (b, i). Results of Grady's algorithm is shown in (c, j). Results of CTQW-LD with configuration: $\beta = 150.0, \gamma = 0.001, T = 5000$ is shown in (d, k), CTQW-LD with the most optimal configuration: $\beta = 150.0, \gamma = 0.001, T = 20000$ in (e, l), CTQW-OS with the most optimal configuration: $\beta = 150.0, \gamma = 0.001, T = 5000$ in (f, m) and finally, CTQW-OS with configuration: $\beta = 150.0, \gamma = 0.001, T = 20000$ in (g, n).

especially important for the CTQW-OS algorithm that strongly suffers from too high $T$ value.

The optimal configurations for both of algorithms have been chosen as follows: $\beta = 150.0, \gamma = 0.001, T = 20000$ for CTQW-LD solution and $\beta = 150.0, \gamma = 0.001, T = 5000$ for CTQW-OS method.

In addition to the ground truth segmentation obtained from the manually created image outlines from BSDS500 dataset, we have also prepared a reference implementation of the classical random walks algorithm [9]. Each segmentation method was then compared against the ground truth images and finally there was calculated the percentage of pixels with consentaneous labels.

The representative results of proposed algorithms are presented in the Fig. 2, which shows segmentations of two sample images: a seashore landscape in color in the Fig. 2a and a man under a tree in grayscale in the Fig. 2h. Then there are reference segmentations: the Fig. 2b and the Fig. 2i are ground truth images included in BSDS500 dataset, while the Fig. 2c and the Fig. 2j were prepared using Grady's algorithm. The last four images for each photo were obtained with algorithms proposed in this paper: the Fig. 2d and k present results of CTQW-LD solution with configuration: $\beta = 150.0, \gamma = 0.001, T = 5000$, Fig. 2e and l come from CTQW-LD algorithm with optimal configuration: $\beta = 150.0, \gamma = 0.001, T = 20000$, while Fig. 2f. Figure 2m were obtained with CTQW-OS method with optimal configuration: $\beta = 150.0, \gamma = 0.001, T = 5000$. Figure 2g and n show CTQW-OS with configuration: $\beta = 150.0, \gamma = 0.001, T = 20000$. As it can be seen, 5000 steps is too little for the CTQW-LD algorithm (visible non-covered black areas in Fig. 2e and l). For CTQW-OS, if the walk duration is too long, the walks from adjacent segments tend to diffuse causing deterioration of the segmentation quality (Fig. 2g, n). As could be expected CTQW-OS is a bit inferior to CTQW-LD. Nevertheless, the difference is not obvious, due to careful choice of parameters (especially number of steps), while the gain in terms of performance is tremendous. CTQW-LD looks comparably, if not better than Grady's method (please notice the difference around the head of the man under tree). Obviously, all of these solutions are imperfect which can be clearly seen in the areas of weak and blurred margins (like, the line of horizon or leaves of the tree).

**Table 1.** Average accuracy (over test images) of developed algorithms as well as Grady's method for different sets of seeds.

| | Manual seeds 0.2% | Automatic seeds 0.2% | Automatic seeds 0.5% | Automatic seeds 1.0% |
|---|---|---|---|---|
| Grady ($\beta = 90.0$) | 93.50 | 90.25 | 93.50 | 94.88 |
| Grady ($\beta = 150.0$) | 93.63 | 90.38 | 93.50 | 94.88 |
| CTQW-LD ($\beta = 150.0, \gamma = 0.001, T = 5000$) | 87.30 | 88.23 | 93.34 | 94.89 |
| CTQW-LD ($\beta = 150.0, \gamma = 0.001, T = 20000$) | 93.40 | 90.89 | 93.48 | 94.72 |
| CTQW-OS ($\beta = 150.0, \gamma = 0.001, T = 5000$) | 87.18 | 88.22 | 93.19 | 94.65 |
| CTQW-OS ($\beta = 150.0, \gamma = 0.001, T = 20000$) | 91.84 | 90.24 | 91.76 | 92.76 |

Table 1 presents average accuracy (over test images) of developed algorithms in comparison to Grady's method for different sets of seeds. It clearly shows that the segmentation accuracy increases with the size of seeds set. Manually chosen seeds give better results than the same amount of automatically drawn ones, due to their careful selection. More accurate of the proposed solutions is CTQW-LD algorithm, which gives almost as good results as reference algorithm or even outperforms it in some cases. The other method provides slightly worse accuracy. The last two rows show how sensitive to the walk duration is the CTQW-OS solution; the accuracy drops significantly if the number of steps is too high. Nonetheless, for optimal configuration of parameters the results of both developed methods might be considered satisfactory. The only unfavorable aspect is the quite considerable amount of seeds needed to obtain a decent segmentation accuracy. As the 0.2% seems to be a bit too little, there has to be specified about 0.5% seeds, which for image of size $320 \times 480$ means that over seven hundred of pixels need to be labeled.

The average execution times for test images of size $320 \times 480$ and optimal configurations were as follows: about 5–6 min for CTQW-LD and around a minute in case of CTQW-OS method. Along with the growth of $\gamma$ or $T$ the duration of the calculations increases linearly. For comparison Grady algorithm for the same samples requires about 30 s. This means that developed solutions are comparable both in terms of accuracy as well as performance to Grady algorithm. Of course, these are the execution times of simulation on classical computer which would not practically matter during transition to a quantum device, nevertheless the proposed results, especially CTQW-OS, are expected to do well also in quantum conditions.

## 6    Summary, Conclusion and Future Work

In this paper, there have been presented two novel quantum algorithms for performing image segmentation based on continuous time quantum walks: one calculating the limiting distribution of the walk and the other utilizing only its last state. Both methods have been closely examined and tested using simulation of quantum computations on a classical computer. The obtained results are satisfactory, of the same accuracy and similar performance as the reference classical Grady's [9] algorithm. A crucial difference between simulation and the actual execution of the algorithm on a quantum computer is the fact, that during simulation, there is always access to the whole quantum state. In real conditions the actual state of the particle is not known until measurement, which reduces it to one of the basis states according to probability distribution determined by the current state. So, to gain the knowledge on the walker state after a given number of steps, one needs to perform the same walk multiple times and measure it in order to reconstruct the probability distribution. In these conditions the CTQW-LD solution is not very effective in quantum case – in order to obtain the limiting distribution the experiment has to be repeated multiple times for different numbers of steps. For this reason, it would be vastly beneficial to apply the CTQW-OS algorithm, which requires the construction of the probability distribution only once per walk.

The segmentation presented in this work is a seeded-based method. User has to specify a set of pixels with assigned segment to which they belong which is not an automatic approach. Also to achieve good results a considerable amount of seeds needs to be provided. Therefore, future work in this field might concern elaboration of either a way to limit the number of required seeds or a method for automatic specifying the seeds, similar to those currently applied in medical diagnostic images segmentation [8]. This would allow to significantly increase the user experience while using the proposed solutions for image segmentation.

The future work on this matter could consider translating the algorithm (for some small size images) on a quantum device - firstly on a quantum simulator like IBM Q simulator or Atos QLM and further to make an attempt to run it on a quantum computer (e.g. IBM Q or Rigetti). This would require disassembling the evolution operator into basic one and two qubit quantum gates, as only those are handled by current quantum devices, as well as taking into account error correction strategies to mitigate the risk of quantum state decoherence.

**Acknowledgements.** This research was partly supported by the National Science Centre, Poland – project number 2014/15/B/ST6/05204.

# References

1. Aharonov, D., Ambainis, A., Kempe, J., Vazirani, U.: Quantum walks on graphs. In: Proceedings of the Thirty-Third Annual ACM Symposium on Theory of Computing - STOC 2001, pp. 50–59 (2000). https://doi.org/10.1145/380752.380758. http://arxiv.org/abs/quant-ph/0012090
2. Ambainis, A., Kempe, J., Rivosh, A.: Coins make quantum walks faster (2004). http://arxiv.org/abs/quant-ph/0402107
3. Ashwin, N., Ashvin, V.: Quantum walk on the line. Technical report (2000)
4. Balu, R., Castillo, D., Siopsis, G.: Physical realization of topological quantum walks on IBM-Q and beyond (2017). https://doi.org/10.1088/2058-9565/aab823. http://arxiv.org/abs/1710.03615
5. Caraiman, S., Manta, V.I.: Image segmentation on a quantum computer. Quantum Inf. Process. **14**(5), 1693–1715 (2015). https://doi.org/10.1007/s11128-015-0932-1
6. Chen, L.C., Papandreou, G., Kokkinos, I., Murphy, K., Yuille, A.L.: DeepLab: semantic image segmentation with deep convolutional nets, atrous convolution, and fully connected CRFs. IEEE Trans. Pattern Anal. Mach. Intell. **40**(4), 834–848 (2017). https://doi.org/10.1109/TPAMI.2017.2699184
7. Childs, A.M., Cleve, R., Deotto, E., Farhi, E., Gutmann, S., Spielman, D.A.: Exponential algorithmic speedup by quantum walk, pp. 59–68 (2002). https://doi.org/10.1145/780542.780552. http://arxiv.org/abs/quant-ph/0209131
8. Fasihi, M.S., Mikhael, W.B.: Overview of current biomedical image segmentation methods. In: Proceedings of the 2016 International Conference on Computational Science and Computational Intelligence, CSCI 2016, pp. 803–808 (2017). https://doi.org/10.1109/CSCI.2016.0156
9. Grady, L.: Random walks for image segmentation. IEEE Trans. Pattern Anal. Mach. Intell. **28**(11), 1768–1783 (2006). https://doi.org/10.1109/TPAMI.2006.233
10. Grady, L., Schwartz, E.L.: Isoperimetric graph partitioning for image segmentation. IEEE Trans. Pattern Anal. Mach. Intell. **28**(3), 469–475 (2006). https://doi.org/10.1109/TPAMI.2006.57
11. Hebenstreit, M., Alsina, D., Latorre, J.I., Kraus, B.: Compressed quantum computation using a remote five-qubit quantum computer. Phys. Rev. A **95**(5) (2017). https://doi.org/10.1103/PhysRevA.95.052339

12. Li, H.S., Qingxin, Z., Lan, S., Shen, C.Y., Zhou, R., Mo, J.: Image storage, retrieval, compression and segmentation in a quantum system. Quantum Inf. Process. **12**(6), 2269–2290 (2013). https://doi.org/10.1007/s11128-012-0521-5
13. Li, J., Li, X., Yang, B., Sun, X.: Segmentation-based image copy-move forgery detection scheme. IEEE Trans. Inf. Forensics Secur. **10**(3), 507–518 (2015). https://doi.org/10.1109/TIFS.2014.2381872
14. Ma, Z., Manuel, J., Tavares, R.S., Natal, R., Mascarenhas, T.: A review of algorithms for medical image segmentation and their applications to the female pelvic cavity. Comput. Methods Biomech. Biomed. Eng. **13**(2), 235–246 (2010). https://doi.org/10.1080/10255840903131878
15. Magniez, F., Nayak, A., Roland, J., Santha, M.: Search via quantum walk. SIAM J. Comput. **40**(1), 142 (2011). https://doi.org/10.1137/090745854. http://arxiv.org/abs/quant-ph/0608026
16. Martin, D., Fowlkes, C., Tal, D., Malik, J.: A database of human segmented natural images and its application to evaluating segmentation algorithms and measuring ecological statistics. In: Proceedings of the IEEE International Conference on Computer Vision, vol. 2, pp. 416–423 (2001). https://doi.org/10.1109/ICCV.2001.937655
17. Mat Isa, N., Salamah, S., Ngah, U.: Adaptive fuzzy moving K-means clustering algorithm for image segmentation. IEEE Trans. Consum. Electron. **55**(4), 2145–2153 (2009). https://doi.org/10.1109/TCE.2009.5373781. http://ieeexplore.ieee.org/document/5373781/
18. Mortensen, E.N., Barrett, W.A.: Interactive segmentation with intelligent scissors. Graph. Models Image Process. **60**(5), 349–384 (1998). https://doi.org/10.1006/gmip.1998.0480. http://linkinghub.elsevier.com/retrieve/pii/S1077316998904804
19. Otsu, N.: A threshold selection method from gray-level histograms. IEEE Trans. Syst. Man Cybern. B Cybern. **9**(1), 62–66 (1979). https://doi.org/10.1109/TSMC.1979.4310076. http://ieeexplore.ieee.org/document/4310076/
20. Otterbach, J.S., Manenti, R., Alidoust, N., Bestwick, et al.: Unsupervised machine learning on a hybrid quantum computer (2017). http://arxiv.org/abs/1712.05771
21. Pal, N.R., Pal, S.K.: A review on image segmentation techniques. Pattern Recogn. **26**(9), 1277–1294 (1993). https://doi.org/10.1016/0031-3203(93)90135-J
22. Sharma, S.: Vishwamittar: Brownian motion problem: Random walk and beyond. Resonance **10**(8), 49–66 (2005). https://doi.org/10.1007/BF02866746
23. Shi, J., Malik, J.: Normalized cuts and image segmentation. IEEE Trans. Pattern Anal. Mach. Intell. **22**(8), 888–905 (2000). https://doi.org/10.1109/34.868688
24. Venegas-Andraca, S.E., Ball, J.L.: Processing images in entangled quantum systems. Quantum Inf. Process. **9**(1), 1–11 (2010). https://doi.org/10.1007/s11128-009-0123-z
25. Weinberg, S.: Quantum mechanics without state vectors. Phys. Rev. A **90**, 042102 (2014). https://doi.org/10.1103/PhysRevA.90.042102
26. Yuheng, S., Hao, Y.: Image segmentation algorithms overview. CoRR abs/1707.0 (2017). http://arxiv.org/abs/1707.02051

# Synchronized Detection and Recovery
# of Steganographic Messages
# with Adversarial Learning

Haichao Shi[1,2], Xiao-Yu Zhang[1], Shupeng Wang[1(✉)], Ge Fu[3(✉)],
and Jianqi Tang[3]

[1] Institute of Information Engineering, Chinese Academy of Sciences, Beijing, China
{shihaichao,zhangxiaoyu,wangshupeng}@iie.ac.cn
[2] School of Cyber Security,
University of Chinese Academy of Sciences, Beijing, China
[3] National Computer Network Emergency Response Technical Team/Coordination
Center of China, Beijing, China
fg@cert.org.cn, tangjianqi1974@163.com

**Abstract.** In this work, we mainly study the mechanism of learning
the steganographic algorithm as well as combining the learning process with adversarial learning to learn a good steganographic algorithm.
To handle the problem of embedding secret messages into the specific
medium, we design a novel adversarial module to learn the steganographic algorithm, and simultaneously train three modules called generator, discriminator and steganalyzer. Different from existing methods,
the three modules are formalized as a game to communicate with each
other. In the game, the generator and discriminator attempt to communicate with each other using secret messages hidden in an image. While
the steganalyzer attempts to analyze whether there is a transmission of
confidential information. We show that through unsupervised adversarial training, the adversarial model can produce robust steganographic
solutions, which acts like an encryption. Furthermore, we propose to
utilize supervised adversarial training method to train a robust steganalyzer, which is utilized to discriminate whether an image contains secret
information. Extensive experiments demonstrate the effectiveness of the
proposed method on publicly available datasets.

**Keywords:** Steganography · Steganalysis · Adversarial learning

## 1 Introduction

Steganography aims to conceal a payload into a cover object without affecting
the sharpness of the cover object. The image steganography is the art and science of concealing covert information within images, and is usually achieved by

---

H. Shi and X.-Y. Zhang contributed equally to this study and share the first authorship.

J. M. F. Rodrigues et al. (Eds.): ICCS 2019, LNCS 11537, pp. 31–43, 2019.
https://doi.org/10.1007/978-3-030-22741-8_3

modifying image elements, such as pixels or DCT coefficients. Steganographic algorithms are designed to hide the secret information within a cover message such that the cover message appears unaltered to an external adversary. On the other side, steganalysis aims to reveal the presence of secret information by detecting the abnormal artefacts left by data embedding and recover the secret information of the carrier object. For a long period of time, many researchers have been involved in developing new steganographic systems. Meanwhile, the development of steganalytic tools are also gradually growing. Inspired by the ideas from adversarial learning, we model the relationship between these two aspects and optimize them simultaneously.

Learning methods have been widely studied in the computer vision community [18–28], among which adversarial learning is based on the game theory and is combined with unsupervised way to jointly train the model. Shi et al. [29] propose a novel strategy of secure steganography based on generative adversarial networks to generate suitable and secure covers for steganography, in which the decoding process was not considered. In this paper, we not only encode the secret messages into the images, but also decode them utilizing a discriminator network. Then, we utilize the steganalysis network to detect the presence of hidden messages. Through unsupervised training, the generator plays the role of a sender, which is utilized to generate steganographic images as real as possible. As the discriminator plays the role of a receiver, it not only differentiates between real images and generated images, but also extracts the secret messages. The steganalysis network, as a listener of the whole process, incorporates supervised learning with adversarial training to compete against state-of-the-art steganalysis methods.

In summary, this paper makes the following contributions:

- We incorporate the generative adversarial network with steganography, which proves to be a good way of encryption.
- We integrate unsupervised learning method and supervised learning method to train the generator and steganalyzer respectively, and receive robust steganographic techniques in an unsupervised manner.
- We also utilize the discriminative network to extract the secret information. Experiments are conducted on widely used datasets, to demonstrate the advantages of the proposed method.

The rest of the paper is structured as follows. In Sect. 2, we discuss the related work of steganography and adversarial networks. In Sect. 3, we elaborate the proposed method. Experiments are conducted in Sect. 4 to demonstrate the effectiveness of the proposed method. In Sect. 5, we draw conclusions.

## 2   Related Work

### 2.1   Steganography

The image-based steganography algorithm can be split into two categories. The one is based on the spatial domain, the other is based on the DCT domain. In our work, we mainly focus on the spatial domain steganography. The state-of-the-art steganographic schemes mainly concentrate on embedding secret information within a medium while minimizing the perturbations within that medium. On the contrary, steganalysis is to figure out whether there is secret information or not in the medium [5,6].

Least Significant Bit (LSB) [11,12] is one of the most popular embedding methods in spatial domain steganography. If LSB is adopted as the steganography method, the statistical features of the image are destroyed. And it is easy to detect by the steganalyzer. For convenience and simple implementation, the LSB algorithm hides the secret to the least significant bits in the given images channel of each pixel. Mostly, the modification of the LSB algorithm is called ±1-embedding. It randomly adds or subtracts 1 from the channel pixel, so the last bits would match the ones needed.

Besides the LSB algorithm, some sophisticated steganographic schemes choose to use a distortion function which is used for selecting the embedding localization of the image. This type of steganography is called the content-adaptive steganography. The minimization of the distortion function between the cover image $\mathcal{C}$ and the steganographic image $\mathcal{S}$ is usually required. These algorithms are the most popular and the most secure image steganography in spatial domain, such as HUGO (Highly Undetectable steGO) [4], WOW (Wavelet Obtained Weights) [2], S-UNIWARD (Spatial UNIversal WAvelet Relative Distortion) [3], etc.

$$d(\mathcal{C}, \mathcal{S}) = f(\mathcal{C}, \mathcal{S}) * |\mathcal{C} - \mathcal{S}| \tag{1}$$

where $f(\mathcal{C}, \mathcal{S})$ is the cost of modifying a pixel, which is variable in different steganographic algorithms.

HUGO is a steganographic scheme that defines a distortion function domain by assigning costs to pixels based on the effect of embedding some information within a pixel. It uses a weighted norm function to represent the feature space. WOW is another content-adaptive steganographic method that embeds information into a cover image according to textural complexity of regions. It is shown in WOW that the more complex the image region is, the more pixel values will be modified in this region. S-UNIWARD introduces a universal distortion function that is independent of the embedded domain. Despite the diverse implementation details, the ultimate goals are identical, i.e. they are all devoted to minimizing this distortion function to embed the information into the noise area or complex texture rather than the smooth image coverage area.

## 2.2  Adversarial Learning

In recent years, Generative Adversarial Networks (GANs) have been successfully applied to image generation tasks. The method that generative adversarial networks generate images can be classified into two categories in general. The first is mainly exploring image synthesis tasks in an unconditioned manner that generates synthetic images without any supervised learning schemes. Goodfellow et al. [7] propose a theoretical framework of GANs and utilize GANs to generate images without any supervised information. However, the early GANs has somewhat noisy and blurry results and the gradient will be vanished when training the networks. Later, Radford et al. [14] propose a deep convolutional generative adversarial networks (DCGANs) for unsupervised representation. To solve the situation of gradient vanishing, WGAN [13] is proposed using the Wasserstein distance instead of the Jensen-Shannon divergence, to make the dataset distribution compared with the learning distribution from G.

Another direction of image synthesis with GANs is to synthesize images by conditioning on supervised information, such as text or class labels. The Conditional GAN [15] is one of the works that develop a conditional version of GANs by additionally feeding class labels into both generator and discriminator of GANs. Info-GAN [16] introduces a new concept, which divides the input noise $z$ into two parts, one is the continuous noise signal that cannot be explained, and the other is called $C$. Where $C$ represents a potential attribute that can be interpreted as a facial expression, such as the color of eyes, whether with glasses or not, etc. in the facial tasks. Recently, Reed et al. [17] utilize GANs for image synthesis using given text descriptions, which enables translation from character level to pixel level.

Adversarial learning has been applied to steganographic and cryptographic problems. In Abadis [1] work, they integrate two neural networks to an adversarial game to encrypt a secret message. In this paper, we are devoted to training a model that can learn a steganographic technique by itself making full use of the discriminator and recovering secret messages synchronously.

# 3　Adversarial Steganography

This section mainly discusses our proposed adversarial steganographic scheme. In Sect. 3.1, we elaborate the network architecture of our model. Then, we formulate the problem of steganography and steganalysis to point out the foundation of the proposed steganographic scheme.

## 3.1　Network Architecture

Our model consists of three components, generator, discriminator and steganalyzer. The generator is used to generate steganographic images. Besides discriminating between real images and generated images, the discriminator is also utilized to extract the secret messages. And the steganalyzer is utilized to distinguish covers from steganographic images.

**Generator.** The generator receives the cover images and secret messages. Before learning the distribution of the images, the secret messages are first embedded into the cover images. The steganographic images embedded with secret messages are then put into the generator. We use a fully connected layer, which can cover anywhere of the image not merely a fixed region. Then four fractionally-strided convolution layers, and finally a hyperbolic tangent function layer. The architecture is shown in Fig. 1.

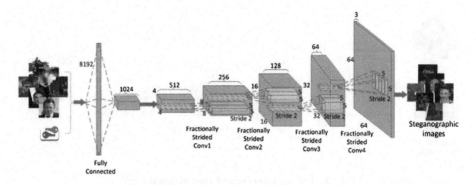

**Fig. 1.** The architecture of the generator.

**Discriminator.** The discriminator is mainly responsible for extracting the secret messages. Besides, it can also help to optimize the visual quality of the generated images. We use four convolutional layers and a fully connected layer. In detail, we additionally add a decoding function at the end of the network to extract the secret messages. The decoding function is acted as an interface between the steganographic images and the secret messages. It analyzes the modification of pixels in images, and recovers the secret messages. As shown in Fig. 2 is the architecture of discriminator.

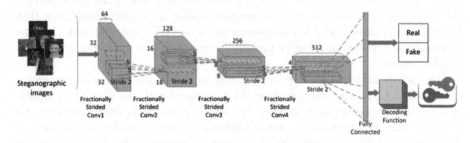

**Fig. 2.** The architecture of the discriminator.

**Steganalyzer.** The inputs of steganalyzer are both covers and steganographic images. Its architecture is similar to the discriminator, as is illustrated in Fig. 3. With distinct, we first use a predefined high-pass filter to make a filtering operation, which is mainly for steganalysis. And then we use four convolutional layers. Finally, we use a classification layer, including a fully connected layer and a soft-max classifier.

### 3.2 Optimization Objective

Our training scheme includes three parties: generator, discriminator and steganalyzer. In this game, the generator conveys a secret medium to the discriminator,

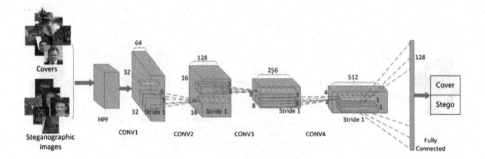

**Fig. 3.** The architecture of the steganalyzer.

which the steganalyzer attempts to eavesdrop in the conveying process and find out whether there are some special information containing in the process. Generally speaking, the generator generates a steganographic image within the cover image embedded with a secret message. Then the generator passes it to the next network namely discriminator, concentrating on decoding and recovering the message. When an image is input to the steganalyzer, a confidence probability will be output by steganalyzer on how much likely the image is hidden with a secret message.

For the whole process of information conveying is transparent to the discriminator, so the decoding can be easily achieved. In the game of the three parties, the generator is trained to learn to produce a steganographic image such that the discriminator can decode and recover the secret message, and such that the steganalyzer can output a better probability whether it is a cover or a steganographic image.

The whole architecture can be depicted in Fig. 4, the generator receives a cover image, $C$, and a secret message, M. Combined with the above two inputs, it outputs a steganographic image, $S_C$, which is simultaneously given to the discriminator and steganalyzer. Once receiving a steganographic image, the discriminator decodes it and attempts to recover the message. For the game is under generative adversarial networks, the discriminator can also improve the visual quality of the generated images. In addition to the steganographic images, the cover images are also input into the steganalyzer, whose output is a probability of classifying the steganographic images and covers.

Let $\omega_G, \omega_S, \omega_S$ denote the parameters of generator, discriminator and steganalyzer, respectively. We utilize $G(\omega_G, M, C)$ represents generators output given an image $C$ and secret message M, $D(\omega_D, S_C)$ represents discriminators output on steganographic images $S_C$, and $S(\omega_S, C, S_C)$ represents steganalyzers output on covers and steganographic images. Then we use $L_G$, $L_D$, $L_S$ represent the loss of three networks respectively. The above formulations can be expressed as follows:

$$D(\omega_D, S_C) = D(\omega_D, G(\omega_G, M, C)) \tag{2}$$

$$S(\omega_S, C, S_C) = S(\omega_S, C, G(\omega_G, M, C)) \tag{3}$$

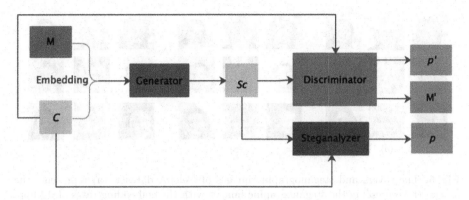

**Fig. 4.** The whole architecture of the training game. M and M′ represents the secret message and extracted message from the steganographic image respectively, $C$ represents the cover image, and $S_C$ represents the steganographic image. $p$ is the probability of classifying the covers and steganographic images. $p′$ is the probability of classifying the real images and generated images.

where $D(\omega_D, S_C)$ is output of the discriminator, $S_{(\omega_S}, C, S_C)$ is the output of the steganalyzer.

For the generator's loss, we correspond it with the other two networks. And we add constraints to normalize the loss.

$$L_G(\omega_G, \mathrm{M}, C) = \lambda_G \cdot d(C, S_C) + \lambda_D \cdot L_D(\omega_D, S_C) + \lambda_S \cdot L_S(\omega_S, C, S_C) \quad (4)$$

where $d(C, S_C)$ is the Euclidean distance between the covers and the steganographic images. And $\lambda_G$, $\lambda_D$, $\lambda_S$ are the calculation loss weights of the generator, discriminator and steganalyzer respectively. These variables are varying from $[0, 1]$.

Also, we set discriminator's loss to be the norm form. Here, we use the $\ell_2$ norm.

$$L_D(\omega_G, \omega_D, \mathrm{M}, C) = ||\mathrm{M}, D(\omega_D, S_C)||_2 = ||\mathrm{M}, D(\omega_D, G(\omega_G, \mathrm{M}, C))||_2 = ||\mathrm{M}, \mathrm{M}'||_2 \quad (5)$$

where the form $||\mathrm{M}, \mathrm{M}'||_2$ is the distance between M and M′.

And we set steganalyzer's loss to be binary cross-entropy of logistic regression.

$$L_S(\omega_G, \omega_S, C, S_C) = -\frac{1}{n} \sum_x \ln S(\omega_S, C, S_C) + (1 - y)\ln(1 - S(\omega_S), C, S_C) \quad (6)$$

where $y = 1$ if $x = C$, which means that the images are classified as covers, $y = 0$ if $x = S_C$, which means the steganographic images are identified. We apply adversarial learning techniques to discriminative tasks to learn the steganographic algorithm. Under unsupervised adversarial training, the model can produce steganographic techniques, which acts like an encryption.

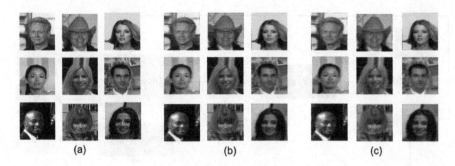

(a)                    (b)                    (c)

**Fig. 5.** The covers and steganographic images of CelebA dataset. (a) represents the covers, (b) represents the steganographic images with the embedding rates of 0.1 bpp, (c) represents the steganographic images with the embedding rates of 0.4 bpp.

## 4   Experiments

Experiments are conducted on two image datasets: celebrity faces in the wild (CelebA) [8] and BossBase dataset. We perform the center crop of size 64 × 64 for both two datasets. For the choice of M, we concatenate a random paragraph message to generalize our model to random message, with each sample of each dataset. And we consider different kinds of embedding rates, which vary from 0.1 bpp (bits per pixel) to 0.4 bpp. Generally, most steganographic algorithms can successfully hide secret messages approximately at 0.4 bpp. All experiments are performed in TensorFlow [9], on a workstation with a Titan X GPU.

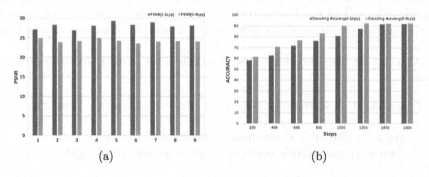

(a)                            (b)

**Fig. 6.** (a) represents the contrast of PSNR of embedding rates at 0.1 bpp and 0.4 bpp, where the number in horizontal axis represents the numerical order of images in Fig. 5 of each part from left to right, up to bottom. (b) represents the contrast of decoding accuracy between 0.1 bpp and 0.4 bpp.

## 4.1   CelebA Dataset

CelebA dataset contains 202,599 images of celebrity faces from 10,177 unique identities. We train the model using RMSProp optimization algorithm with the learning rate of $2 \times 10^{-4}$. At each batch we jointly train the generator, discriminator and steganalyzer.

From the results, we can see that there is no noticeable image quality decrease between 0.1 bpp and 0.4 bpp. As is shown in Fig. 5. And we also calculate the PSNR (Peak Signal to Noise Ratio) to evaluate the steganographic effect. As shown in Fig. 6.

From the charts, we can see that the discriminator can hardly decode the secret messages in the first few rounds of training. For the visual quality of the generated images is low and the discriminator is randomly guessing the messages essentially. After 800 training steps, the discriminator can decode the messages correctly with an average success of 90% at 0.4 bpp. After 1,000 training steps, the discriminator approaches to convergence gradually and decodes the messages with success rate above 95%. We can also conclude that the discriminator performs better under embedding rate at 0.4 bpp than 0.1 bpp when decoding the messages.

## 4.2   BossBase Dataset

In addition to the experiments on the CelebA dataset, we also train our model on the BossBase dataset, which is a standard steganography dataset consisting of 10,000 grayscale images of various scenes. For the images from this dataset do not come from a single distribution, so the dataset can perform worse than the experiments on the CelebA dataset.

(a)                              (b)                              (c)

**Fig. 7.** The covers and steganographic images of BossBase dataset. (a) represents the covers, (b) represents the steganographic images with the embedding rates of 0.1 bpp, (c) represents the steganographic images with the embedding rates of 0.4 bpp.

Qualitative results on BossBase are shown in Fig. 7. Contrast of PSNR and decoding accuracy at 0.1 bpp and 0.4 bpp is shown in Fig. 8. While the decoding

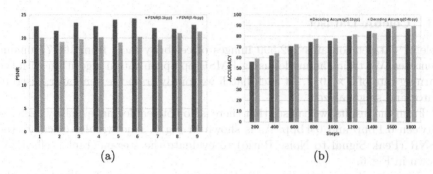

**Fig. 8.** (a) represents the contrast of PSNR of embedding rates at 0.1 bpp and 0.4 bpp, where the number in horizontal axis represents the numerical order of images in Fig. 7 of each part from left to right, up to bottom. (b) represents the contrast of decoding accuracy between 0.1 bpp and 0.4 bpp.

rates suffers slightly from the image quality, the discriminator is still able to learn to decode the messages successfully. This also implies the effectiveness of the generator, which is able to learn a good steganographic algorithm.

We can conclude from the two groups of experiments that for the datasets with distinct distribution, different embedding rates can influence the similarity between the original images and the embedded images.

### 4.3  Comparison with the State-of-the-Arts

We train in such a way that the steganalyzer can detect the covers and steganographic images with low rates. Thus, on the basis of the experiment, we additionally conduct another experiments compared with the state-of-the-art steganographic algorithms, which shows competitive results. Note that Steganalyzer* denotes the proposed steganalysis method, and Generator* denotes the proposed steganography method.

**Table 1.** Accuracy of distinguishing between cover and steganographic images for the GNCNN and Steganalyzer* at an embedding rate of 0.4 bpp.

| Steganographic algorithm | CelebA | | BossBase | |
|---|---|---|---|---|
| | Steganalyzer | | Steganalyzer | |
| | GNCNN | Steganalyzer* | GNCNN | Steganalyzer* |
| HUGO | 0.93 | 0.91 | 0.72 | 0.68 |
| S-UNIWARD | 0.91 | 0.89 | 0.75 | 0.71 |
| Generator* | 0.94 | 0.92 | 0.83 | 0.82 |

For both datasets, we compare our scheme against steganographic algorithms including HUGO [4] and S-UNIWARD [3]. As shown in Table 1, Steganalyzer* performs competitively with other methods. By training Steganalyzer* in a supervised fashion, the experiments show that it has the capacity to become a strong steganalyzer, competing against established techniques like GNCNN [10], and so is a good choice for the steganalyzer. Furthermore, for both the CelebA and BossBase datasets, we use 10,000 samples and split them in half, to create a training set and a test set. We then train the Generator* on the training set and test the model on the testing set. Following this, additional 5,000 steganographic images are created utilizing each steganography algorithm. Thus, the cover set and steganographic images set are all consisting of 5,000 images respectively.

As demonstrated in the experiments, Steganalyzer* performs competitively against the GNCNN, and the Generator* also performs well against other steganographic techniques. The experimental results show that on the one hand, the generated images are good at visual quality. On the other hand, the generated images are harder to detect, which shows the security of the generated images as covers.

## 5   Conclusion

In this paper, an adversarial steganography architecture with generative adversarial networks is proposed. On the basis of generative adversarial networks, we leverage the adversarial structure to form an effective steganographic method. Encouraging results are received from experiments conducted on the widely used datasets in comparison with several state-of-the-art methods. It is also worth studying the adaption of the adversarial steganography architecture in adaptive steganographic algorithm. Furthermore, it is interesting to resort the networks to a game-theoretic formulation when we characterize the interplay between the steganographer and the steganalyst.

**Acknowledgements.** This work was supported by the National Natural Science Foundation of China (Grant 61871378), and the Open Project Program of National Laboratory of Pattern Recognition (Grant 201800018).

## References

1. Abadi, M., Andersen, D.G.: Learning to protect communications with adversarial neural cryptography. CoRR, abs/1610.06918 (2016)
2. Holub, V., Fridrich, J.J.: Designing steganographic distortion using directional filters. In: IEEE International Workshop on Information Forensics and Security, WIFS 2012, Costa Adeje, Tenerife, Spain, pp. 234–239 (2012)
3. Holub, V., Fridrich, J.J., Denemark, T.: Universal distortion function for steganography in an arbitrary domain. EURASIP J. Inf. Secur. **2014**, 1 (2014)
4. Pevný, T., Filler, T., Bas, P.: Using high-dimensional image models to perform highly undetectable steganography. In: Böhme, R., Fong, P.W.L., Safavi-Naini, R. (eds.) IH 2010. LNCS, vol. 6387, pp. 161–177. Springer, Heidelberg (2010). https://doi.org/10.1007/978-3-642-16435-4_13

5. Xu, G., Wu, H., Shi, Y.: Structural design of convolutional neural networks for steganalysis. IEEE Signal Process. Lett. **23**(5), 708–712 (2016)
6. Xu, G., Wu, H., Shi, Y.Q.: Ensemble of CNNs for steganalysis: an empirical study. In: Proceedings of the 4th ACM Workshop on Information Hiding and Multimedia Security, IH&MMSec 2016, Vigo, Galicia, Spain, pp. 103–107 (2016)
7. Goodfellow, I., et al.: Generative adversarial nets. In: Ghahramani, Z., Welling, M., Cortes, C., Lawrence, N.D., Weinberger, K.Q. (eds.) Advances in Neural Information Processing Systems, vol. 27, pp. 2672–2680. Curran Associates Inc. (2014)
8. Liu, Z., Luo, P., Wang, X., Tang, X.: Deep learning face attributes in the wild. In: IEEE International Conference on Computer Vision, ICCV 2015, Santiago, Chile, pp. 3730–3738 (2015)
9. Abadi, M., Agarwal, A., Barham, P., et al.: Tensorflow: large-scale machine learning on heterogeneous distributed systems. CoRR, abs/1603.04467 (2016)
10. Qian, Y., Dong, J., Wang, W., Tan, T.: Deep learning for steganalysis via convolutional neural networks. In: Media Watermarking, Security, and Forensics, San Francisco, CA, USA, Proceedings, p. 94090J (2015)
11. Mielikainen, J.: LSB matching revisited. IEEE Signal Process. Lett. **13**(5), 285–287 (2006)
12. Ker, A.D.: Resampling and the detection of LSB matching in color bitmaps. In: Security, Steganography, and Watermarking of Multimedia Contents VII, San Jose, California, USA, Proceedings, pp. 1–15 (2005)
13. Arjovsky, M., Chintala, S., Bottou, L.: Wasserstein generative adversarial networks. In: Precup, D., Teh, Y.W. (eds.) Proceedings of the 34th International Conference on Machine Learning, volume 70 of Proceedings of Machine Learning Research, pp. 214–223. International Convention Centre, Sydney, Australia (2017). PMLR
14. Radford, A., Metz, L., Chintala, S.: Unsupervised representation learning with deep convolutional generative adversarial networks. CoRR, abs/1511.06434 (2015)
15. Odena, A., Olah, C., Shlens, J.: Conditional image synthesis with auxiliary classifier GANs. In: Proceedings of the 34th International Conference on Machine Learning, ICML 2017, Sydney, NSW, Australia, pp. 2642–2651 (2017)
16. Chen, X., Duan, Y., Houthooft, R., Schulman, J., Sutskever, I., Abbeel, P.: Infogan: interpretable representation learning by information maximizing generative adversarial nets. In: Advances in Neural Information Processing Systems 29: Annual Conference on Neural Information Processing Systems, Barcelona, Spain, pp. 2172–2180 (2016)
17. Reed, S.E., Akata, Z., Yan, X., Logeswaran, L., Schiele, B., Lee, H.: Generative adversarial text to image synthesis. In: Proceedings of the 33rd International Conference on Machine Learning, ICML 2016, New York City, NY, USA, pp. 1060–1069 (2016)
18. Zhang, X., Shi, H., Zhu, X., Li, P.: Active semi-supervised learning based on self-expressive correlation with generative adversarial networks. Neurocomputing **2019**(01), 083 (2019). https://doi.org/10.1016/j.neucom
19. Zhang, X.: Simultaneous optimization for robust correlation estimation in partially observed social network. Neurocomputing **205**, 455–462 (2016)
20. Zhang, X., Wang, S., Yun, X.: Bidirectional active learning: a two-way exploration into unlabeled and labeled dataset. IEEE Trans. Neural Netw. Learn. Syst. **26**(12), 3034–3044 (2015)
21. Zhang, X., Wang, S., Zhu, X., Yun, X., Wu, G., Wang, Y.: Update vs. upgrade: modeling with indeterminate multi-class active learning. Neurocomputing **162**, 163–170 (2015)

22. Zhang, X.: Interactive patent classification based on multi-classifier fusion and active learning. Neurocomputing **127**, 200–205 (2014)
23. Zhang, X., Shi, H., Li, C., Zheng, K., Zhu, X., Duan, L.: Learning transferable self-attentive representations for action recognition in untrimmed videos with weak supervision. In: Proceedings of the 33rd AAAI Conference on Artificial Intelligence, pp. 1–8 (2019)
24. Zhang, X., Xu, C., Cheng, J., Lu, H., Ma, S.: Effective annotation and search for video blogs with integration of context and content analysis. IEEE Trans. Multimedia **11**(2), 272–285 (2009)
25. Liu, Y., Zhang, X., Zhu, X., Guan, Q., Zhao, X.: ListNet-based object proposals ranking. Neurocomputing **267**, 182–194 (2017)
26. Li, C., Liu, Q., Liu, J., Lu, H.: Ordinal distance metric learning for image ranking. IEEE Trans. Neural Networks **26**(7), 1551–1559 (2015)
27. Li, C., Wang, X., Dong, W., Yan, J., Liu, Q., Zha, H.: Joint active learning with feature selection via cur matrix decomposition. IEEE Trans. Pattern Anal. Mach. Intell. **41**(6), 1382–1396 (2019)
28. Li, C., Wei, F., Dong, W., Wang, X., Liu, Q., Zhang, X.: Dynamic structure embedded online multiple-output regression for streaming data. IEEE Trans. Pattern Anal. Mach. Intell. **41**(2), 323–336 (2019)
29. Shi, H., Dong, J., Wang, W., Qian, Y., Zhang, X.: SSGAN: secure steganography based on generative adversarial networks. In: Zeng, B., Huang, Q., El Saddik, A., Li, H., Jiang, S., Fan, X. (eds.) PCM 2017. LNCS, vol. 10735, pp. 534–544. Springer, Cham (2018). https://doi.org/10.1007/978-3-319-77380-3_51

# Multi-source Manifold Outlier Detection

Lei Zhang[1]([✉]), Shupeng Wang[1]([✉]), Ge Fu[2]([✉]), Zhenyu Wang[1], Lei Cui[1], and Junteng Hou[1]

[1] Institute of Information Engineering, CAS, Beijing 100093, China
{zhanglei1,wangshupeng,wangzhenyu,cuilei,houjunteng}@iie.ac.cn
[2] CNCERT/CC, Beijing 100029, China
fg@cert.org.cn

**Abstract.** Outlier detection is an important task in data mining, with many practical applications ranging from fraud detection to public health. However, with the emergence of more and more multi-source data in many real-world scenarios, the task of outlier detection becomes even more challenging as traditional mono-source outlier detection techniques can no longer be suitable for multi-source heterogeneous data. In this paper, a general framework based the consistent representations is proposed to identify multi-source heterogeneous outlier. According to the information compatibility among different sources, Manifold learning are combined in the proposed method to obtain a shared representation space, in which the information-correlated representations are close along manifold while the semantic-complementary instances are close in Euclidean distance. Furthermore, the multi-source outliers can be effectively identified in the affine subspace which is learned through affine combination of shared representations from different sources in the feature-homogeneous space. Comprehensive empirical investigations are presented that confirm the promise of our proposed framework.

**Keywords:** Multi-source · Manifold learning · Heterogeneous · Outlier detection

## 1 Introduction

Recent years have witnessed significant advances in multi-source learning. Many multi-source techniques have been developed to assist people in extracting useful information from rapidly growing volumes of multi-source data [25,26]. However, unlike the previous work that focused on the study of mining general patterns from different sources, it is worth to note that no existing efforts have focused on the detection of multi-source heterogeneous outliers. In particular, this detection is generally performed to identify abnormal heterogeneous observation from different sources.

Furthermore, detecting outliers is more interesting and useful than identifying normal instances [4,11,17]. For example, to protect the properties of customers, an electronic commerce detection system can monitor the customers' financial

© Springer Nature Switzerland AG 2019
J. M. F. Rodrigues et al. (Eds.): ICCS 2019, LNCS 11537, pp. 44–58, 2019.
https://doi.org/10.1007/978-3-030-22741-8_4

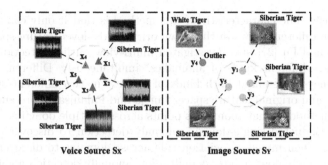

**Fig. 1.** Multi-source heterogeneous outlier

activities in order to identify abnormal consuming behavior of credit card as criminal activities (outlier). Consequently, multi-source outlier detection is an important task in data mining, with many practical applications ranging from fraud detection to public health. Many outlier detection methods have been proposed over the past decades, including mono-source [6,12,13,18] and multi-view [10,14,21,27] outlier detection.

**Mono-source Outlier Detection.** Recently, some researchers have investigated many machine learning methods [6,12,13,18] to deal with outlier detection problems in mono-source data. Knorr et al. pointed out in [12] that the identification of outliers can lead to the discovery of truly unexpected knowledge in various actual application fields. Meanwhile, they proposed and analyzed several algorithms for finding DB-outliers. In [13], Li et al. presented a representation-based method to calculate the reverse unreachability of a point to evaluate to what degree this observation is boundary point or outlier. Rahmani and Atia [18] proposed two randomized algorithms for two distinct outlier models namely, sparse and independent outlier models based robust principal component analysis. k-Nearest Neighbors (kNN) [6] defines the distance of a given data point to its kth nearest neighbors as the outlier. The greater the value of the score, the most likely the outlier.

However, with the emergence of more and more multi-source data in many real-world scenarios, the task of outlier detection becomes even more challenging as traditional mono-source outlier detection techniques can no longer be suitable for multi-source heterogeneous data. For example, as shown in Fig. 1, it is greatly difficult in the single voice source to discovery the outlier $x_4$, i.e., White tiger, hidden in the cluster of Siberian tiger. However, according to the distribution of the corresponding Image source, the heterogeneous outlier $y_4$ can be easily identified. Thus, it is necessary to develop an effective outlier detection method for multi-source heterogeneous data.

**Multi-view Outlier Detection.** To overcome the drawback of traditional mono-source methods, several efforts [10,14,21,27] have been devoted to identifying multi-source outliers. In [21], Li et al. proposed a multi-view outlier detection framework to detect two different types of outliers from multiple views simultaneously. The characteristic of this method is that it only can detect the outliers from different views in their own original low-level feature spaces. Similarly, Zhao and Fu [27] also investigated multi-view outlier detection problem to detect two different kinds of anomalies simultaneously. Different from Li's approach, they represented both kinds of outliers in two different spaces, i.e. latent space and original feature space. Janeja and Palanisamy presented in [10] a two-step method to find anomalous points across multiple domains. This technique first conducted single-domain anomaly detection to discover outliers in each domain, then mined association rule across domains to discover relationship between anomalous points. A multi-view anomaly detection algorithm was developed by Liu and Lam in [14] to find potentially malicious insider activity across multiple data sources.

**Difficulties and Challenges.** Generally, these existing methods tend to identify multi-source outliers from different feature spaces by using association analysis across sources. Most of these methods are, however, designed for detecting the Class-outliers [21,27] in spaces of the original attributes, and discovering Attribute-outliers [21,27] in combinations of underlying spaces. Consequently, these methods will face an enormous challenge in the real-world applications for the following reason. Due to the attempt of identifying multi-source outliers in different original feature spaces, it is extremely difficulty for the above-mentioned approaches to capture much more complementary information from different sources. It will lead to a low recognition rate for multi-source outliers. Furthermore, it has been proved in [26] that the consistent representations for multi-source heterogeneous data will be more favorable for fully exploiting the complementarity among different sources. Thus, it is inevitably an urgent problem to detect all kinds of multi-source outliers from different sources in a consistent feature-homogeneous space.

## 1.1 Main Contributions

The key contributions of this paper are highlighted as follows:

- To detect multi-source heterogeneous outliers, a general Multi-Source Manifold Outlier Detection (MMOD) framework based the consistent representations for multi-source heterogeneous data is proposed.
- Manifold learning is integrated in the framework to obtain a shared-representation space, in which the information-correlated representations are close along manifold while the semantic-complementary instances are close in Euclidean distance.
- According to the information compatibility among different sources, an affine subspace is learned through affine combination of shared representations from different sources in the feature-homogeneous space.
- Multi-source heterogeneous outliers can be effectively identified in the affine subspace under the constraints of information compatibility among different sources.

## 1.2 Organization

The remainder of this paper is organized as follows: In Sect. 1.3, the notations are formally defined. We present a general framework for detecting multi-source heterogeneous outliers in Sect. 2.1. Furthermore, Sect. 2.2 provides an efficient algorithm to solve the proposed framework. Experimental results and analyses are reported in Sect. 3. Section 4 concludes this paper.

## 1.3 Notations

In this section, some important notations are summarized into Table 1 for convenience.

**Table 1.** Notations

| Notation | Description |
|---|---|
| $S_x$ | Source $X$ |
| $S_y$ | Source $Y$ |
| $X_N \in \mathbb{R}^{n_1 \times d_x}$ | Normal samples in $S_x$ |
| $Y_N \in \mathbb{R}^{n_1 \times d_y}$ | Normal samples in $S_y$ |
| $X_S \in \mathbb{R}^{n_2 \times d_x}$ | Suspected outliers in $S_x$ |
| $x_i \in \mathbb{R}^{d_x}$ | The $i$-th sample from $S_x$ |
| $y_i \in \mathbb{R}^{d_y}$ | The $i$-th sample from $S_y$ |
| $n_1$ | Number of normal data |
| $n_2$ | Number of suspected outliers |
| $d_x$ | Dimensionality of $S_x$ |
| $d_y$ | Dimensionality of $S_y$ |
| $(x_i, y_i)$ | The $i$-th multi-source datum |
| $\|\cdot\|_F$ | Frobenius norm |
| $\nabla f(\cdot)$ | Gradient of smooth function $f(\cdot)$ |

## 2 Detecting Multi-source Heterogeneous Outliers

Here we propose a general framework to detect heterogeneous outliers in multi-source datasets.

## 2.1 The Proposed MMOD Model

In the light of the existing multi-source methods' shortcomings, we focus on a particularly important problem of identifying all kinds of multi-source heterogeneous outliers in a consistent feature-homogeneous space.

In general, outliers are located around the margin of the data set with high density, such as a cluster. Furthermore, Elhamifar [7] has pointed out that each

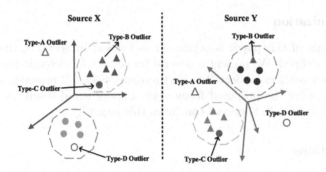

**Fig. 2.** 4 types of heterogeneous outliers.

data point in a union of subspaces can be efficiently represented as a linear or affine combination of other points in the dataset. Meanwhile, Bertsekas has also proved that all convex combinations are geometrically within the convex hull of the given points [2]. Consequently, the negative components in the representation correspond to the outliers outside the convex combination of its neighbors [5,20].

Following the above-mentioned theoretical results, we propose a novel Multi-Source Manifold Outlier Detection (MMOD) framework based the consistent representations for multi-source heterogeneous data. The main goal of the proposed framework is unified detection of outliers from heterogeneous datasets in a feature-homogeneous space, in order to avoid wasting the complementary information among different sources and improve the recognition rate for multi-source outliers.

In this paper, we focus on detecting all kinds of heterogeneous outliers (See Fig. 2) from multi-source heterogeneous data. In particular, we aim to identify four types of heterogeneous outliers that are defined below.

**Definition 1. Type-A outliers** have consistent abnormal behaviors in each source.

**Definition 2. Type-B outliers** are deviant instances that show normal clustering results in one source but abnormal cluster memberships in another source.

**Definition 3. Type-C outliers** own abnormal clustering results in each source.

**Definition 4. Type-D outliers** refer to exceptional samples that exhibit normal clustering results in one source but abnormal behavior in another source.

Given a normal dataset $X_N = \{x_1, x_2, \cdots, x_{n_1}\} \in \mathbb{R}^{d_x \times n_1}$ and a sample $c \in \mathbb{R}^{d_x}$, an affine space $H = \{w \in \mathbb{R}^{n_1} | X_N w = c\}$ can be spanned by its neighbors from Source $X$. Note that $w$ is the representation of $c$ in the affine space, and $w_i$ is the component of the representation $w$ of $c$. Generally, outliers are located around the margin of the dataset with high density, such as a cluster.

It is known that if $0 \le w_i \le 1$, then the point $c$ will be within (or on the boundary) of the convex hull. If any $w_i$ is less than zero or greater than 1, then the point will lie outside the convex hull. Thus, data representation can uncover the intrinsic data structure. Obviously, heterogeneous outliers can be identified according to the following principle.

1. $c$ is normal point if $0 \le w_i \le 1$.
2. $c$ is abnormal point (outlier) if any $w_i < 0$ or $w_i > 1$.

Specifically, the new distance metrics are defined as follows to learn a Mahalanobis distance [23]:

$$\mathcal{D}_{M_X}(x_i, x_j) = (x_i - x_j)^T M_X (x_i - x_j), \tag{1}$$

$$\mathcal{D}_{M_Y}(y_i, y_j) = (y_i - y_j)^T M_Y (y_i - y_j), \tag{2}$$

where $M_X = A^T A$ and $M_Y = B^T B$ are two positive semi-definite matrices. Thus, the linear transformations $A$ and $B$ can be applied to each pair of co-occurring heterogeneous representations $(x_i, y_i)$.

Then the proposed approach can be formulated as follows:

$$\Psi_1: \quad \begin{aligned} & \min_{A,B,M,W} \quad \| X_N A M B^T Y_N^T \|_F^2 + \alpha \| X_N A - Y_N B \|_F^2 \\ & \text{s.t.} \quad A^T A = I \quad \text{and} \quad B^T B = I \quad \text{and} \quad M \succeq 0 \\ & \qquad X_S A = W^T Y_N B \end{aligned} \tag{3}$$

where $A \in \mathbb{R}^{d_x \times k}$, $B \in \mathbb{R}^{d_y \times k}$, $k$ is the dimensionality of the feature-homogeneous subspace, and $\alpha$ is a trade-off parameter. The first item of the objective function in the model $\Psi_1$ is to measure the smoothness between different linear transformations $A$ and $B$ to extract the information correlation among heterogeneous representations. Moreover, the motivation of introducing the second item in the objective function is to capture the semantic complementarity among different sources. The orthogonal constraints $A^T A = I$ and $B^T B = I$ are added into the optimization to effectively remove the correlations among different features in the same source, the positive semidefinite restraint $M \in \mathbb{S}_+^{k \times k} \succeq 0$ can ensure a well-defined pseudo-metric. To identify multi-source heterogeneous outliers, the affine hull constraint $X_S A = W^T Y_N B$ based on data representation is added into the model $\Psi_1$ to learn an affine subspace. The matrices $W \in \mathbb{R}^{n_1 \times n_2}$ encodes the neighborhood relationships between points in the affine subspace, $w_i \in \mathbb{R}^{n_1}$ is the representation of $x_S^i \in \mathbb{R}^{d_x}$ in the affine subspace, respectively.

Note that solving the problem $\Psi_1$ in Eq. (3) directly is a challenging task for two main reasons. First, it is difficult to seek the solution that satisfies the convex hull constraint. Second, the orthogonal constraints are not smooth, which makes it even more difficult to compute the optimum. Thus, we propose to use Lagrangian duality to augment the objective function with a weighted sum of the convex hull constraint to obtain a solvable problem $\Psi_2$ as follows:

$$\Psi_2: \quad \begin{aligned} & \min_{A,B,M,W} \quad \| X_N A M B^T Y_N^T \|_F^2 + \alpha \| X_N A - Y_N B \|_F^2 + \beta \| X_S A - W^T Y_N B \|_F^2 \\ & \text{s.t.} \quad A^T A = I \quad \text{and} \quad B^T B = I \quad \text{and} \quad M \succeq 0 \end{aligned} \tag{4}$$

In Sect. 2.2, an efficient algorithm is proposed to solve the problem $\Psi_2$.

## 2.2   An Efficient Solver for $\Psi_2$

Here we provide an efficient algorithm to solve $\Psi_2$.

The optimization problem $\Psi_2$ in Eq. (4) can be simplified as follows:

$$\min_{Z \in \mathcal{C}} \quad F(Z) = \| \cdot \|_F + \alpha \| \cdot \|_F + \beta \| \cdot \|_F, \tag{5}$$

where $F(\cdot)$ is a smooth objective function, $Z = [A_Z \ \ B_Z \ \ M_Z \ \ W_Z]$ symbolically represents the optimization variables, and $\mathcal{C}$ is the closed domain with respect to each variable:

$$\mathcal{C} = \{Z | A_Z^T A_Z = I, B_Z^T B_Z = I, M_Z \succeq 0\}. \tag{6}$$

Obviously, the optimization problem in Eq. (5) is non-convex. However, Ando and Zhang have testified in [1] that the alternating optimization method can effectively solve non-convex problem. They have also pointed out that this method usually did not lead to serious problems since given the local optimal solution of one variable, the solution of other variables would still be globally optimal.

Additionally, the problem in Eq. (5) is separately convex with respect to each optimization variable. Furthermore, as $F(\cdot)$ is continuously differentiable with Lipschitz continuous gradient [15] with respect to each variable, respectively. Thus, through combining Accelerated Projected Gradient (APG) [15] method and alternating optimization approach [1], the problem in Eq. (5) can be effectively solved.

However, the non-convex optimization problem in Eq. (5) is generally difficult to optimize due to the orthogonal constraints. Guo and Xiao have pointed out in [8] that Gradient Descent Method with Curvilinear Search (GDMCS) in [24] can effectively solve non-convex optimization problem for a local optimal solution as long as the Armijo-Wolfe conditions are satisfied.

Furthermore, since the objective function in Eq. (5) is smooth, the gradient of the objective function with respect to $A, B$ can be easily computed, respectively. In each iteration of the gradient descent procedure, given the current feasible point $(A, B)$, the gradients can be computed as follows:

$$G_1 = \nabla_A F(A, B), \tag{7}$$
$$G_2 = \nabla_B F(A, B). \tag{8}$$

We then compute two skew-symmetric matrices:

$$F_1 = G_1 A^T - A G_1^T, \tag{9}$$
$$F_2 = G_2 B^T - B G_2^T. \tag{10}$$

It is easy to see $F_1^T = -F_1$ and $F_2^T = -F_2$. The next new point can be searched as a curvilinear function of a step size variable $\tau$, such that

$$Q_1(\tau) = (I + \tau F_1/2)^{-1}(1 - \tau F_1/2)A, \tag{11}$$
$$Q_2(\tau) = (I + \tau F_2/2)^{-1}(1 - \tau F_2/2)B. \tag{12}$$

It is easy to verify that $Q_1(\tau)^T Q_1(\tau) = I$ and $Q_2(\tau)^T Q_2(\tau) = I$ for all $\tau \in \mathbb{R}$. Thus we can stay in the feasible region along the curve defined by $\tau$. Moreover, $dQ_1(0)/d\tau$ and $dQ_2(0)/d\tau$ are equal to the projections of $(-G_1)$ and $(-G_2)$ onto the tangent space $\mathcal{C}$ at the current point $(A, B)$. Hence $\{Q_1(\tau), Q_2(\tau)\}_{(\tau \geq 0)}$ is a descent path in the close neighborhood of the current point. We thus apply a similar strategy as the standard backtracking line search to find a proper step size $\tau$ using curvilinear search, while guaranteeing the iterations to converge to a stationary point. We determine a proper step size $\tau$ as one satisfying the following Armijo-Wolfe conditions [24]:

$$F(Q_1(\tau), Q_2(\tau)) \leq F(Q_1(0), Q_2(0)) + \rho_1 \tau F_\tau'(Q_1(0), Q_2(0)), \tag{13}$$

$$F_\tau'(Q_1(\tau), Q_2(\tau)) \geq \rho_2 F_\tau'(Q_1(0), Q_2(0)). \tag{14}$$

Here $F_\tau'(Q_1(\tau), Q_2(\tau))$ is the derivative of $F$ with respect to $\tau$,

$$\begin{aligned}
F_\tau'(Q_1(\tau), Q_2(\tau)) = \\
-tr((\nabla_A F(Q_1(\tau), Q_2(\tau)))^T (I + \frac{\tau}{2} F_1)^{-1} F_1 (\frac{A + Q_1(\tau)}{2})) \\
-tr((\nabla_B F(Q_1(\tau), Q_2(\tau)))^T (I + \frac{\tau}{2} F_2)^{-1} F_2 (\frac{B + Q_2(\tau)}{2})).
\end{aligned} \tag{15}$$

Therefore,

$$\begin{aligned}
F_\tau'(Q_1(0), Q_2(0)) &= -tr(G_1^T (G_1 A^T - A G_1^T) A) \\
&\quad - tr(G_2^T (G_2 B^T - B G_2^T) B) \\
&= -\frac{\|F_1\|_F^2}{2} - \frac{\|F_2\|_F^2}{2}.
\end{aligned} \tag{16}$$

Accordingly, it is appropriate to use the gradient descent method to solve the problem $\Psi_2$ in Eq. (5).

The APG algorithm is a first-order gradient method, which can accelerate each gradient step on the feasible solution to obtain an optimal solution when minimizing a smooth function [16]. This method will construct a solution point sequence $\{Z_i\}$ and a searching point sequence $\{S_i\}$, where each $Z_i$ is updated from $S_i$.

Furthermore, a given point $s$ in the APG algorithm needs to be projected into the set $\mathcal{C}$:

$$proj_\mathcal{C}(s) = arg \min_{z \in \mathcal{C}} \|z - s\|_F^2 / 2. \tag{17}$$

Weinberger et al. proposed a Positive Semi-definite Projection (PSP) [23] to minimize a smooth function while remaining positive semi-definite constraints. It will project optimal variables into a cone of all positive semi-definite matrices after each gradient step. The projection is computed from the diagonalization of optimal variables, which effectively truncates any negative eigenvalues from the gradient step, setting them to zero. Then we can use the PSP to solve the problem in Eq. (17).

Finally, to solve the problem in Eq. (5), the projection $Z = [A_Z \; B_Z \; M_Z \; W_Z]$ of a given point $S = [A_S \; B_S \; M_S \; W_S]$ onto the set $\mathcal{C}$ is defined by:

$$proj_{\mathcal{C}}(S) = arg \min_{Z \in \mathcal{C}} \|Z - S\|_F^2/2. \tag{18}$$

By combining APG, GDMCS, and PSP, we can solve the problem in Eq. (18). The overall algorithm is given in Algorithm 1, where the function **Schmidt**$(\cdot)$ denotes the GramSchmidt process.

---

**Algorithm 1.** Multi-Source Manifold Outlier Detection (**MMOD**)

---

**Input:** $F(\cdot)$, $Z_0 = [A_{Z_0} \; B_{Z_0} \; M_{Z_0} \; W_{Z_0}]$, $\tau_1$, $0 < \rho_1 < \rho_2 < 1$, $s=1$, $t=1$, $p=1$,
 $\quad q=1$, $\eta_1 > 0$, $\theta_1 > 0$, $\lambda_1 > 0$, $\mu_1 > 0$.

**Output:** $Z^*$.

1: Set $A_{Z_1} = A_{Z_0}$, $B_{Z_1} = B_{Z_0}$, $M_{Z_1} = M_{Z_0}$, and $W_{Z_1} = W_{Z_0}$.

2: **for** $i = 1,2,\cdots,max\text{-}iter$ **do**

3:     Fix $B$, $M$, $W$ and approximately solve for $A$.

4:     Define $F_{\eta,A_S}(A_Z) = F(A_S) + \langle \nabla F(A_S), A_Z - A_S \rangle + \eta \|A_Z - A_S\|_F^2/2$.

5:     **for** $j = 1,2,\cdots,h_1$ **do**

6:        Set $a_j = (s-1)/s$.

7:        Compute $A_{S_i} = (1 + \alpha_j)A_{Z_i} - \alpha_j A_{Z_{i-1}}$.

8:        Compute $\nabla_{A_S} F(A_{S_i})$.

9:        **while** (true)

10:           Compute $\widehat{A_S} = A_{S_i} - \nabla_{A_S} F(A_{S_i})/\eta_i$.

11:           Compute $[\widehat{A_S}] = \boldsymbol{Schmidt}(\widehat{A_S})$.

12:           Set $[A_{Z_{i+1}}] = \boldsymbol{GDMCS}(F(\cdot), \widehat{A_S}, \tau_1, \rho_1, \rho_2)$.

13:           **if** $F(A_{Z_{i+1}}) \leq F_{\eta_i, A_{S_i}}(A_{Z_{i+1}})$, then break;

14:           **else** Update $\eta_i = \eta_i \times 2$.

15:           **end-if**

16:        **end-while**

17:        Update $s = \left(1 + \sqrt{1+4s^2}\right)/2$, $\eta_{i+1} = \eta_i$.

18:     **end-for**

19:     Fix $A$, $M$, $W$ and approximately solve for $B$.

20:     Define $F_{\theta,B_S}(B_Z) = F(B_S) + \langle \nabla F(B_S), B_Z - B_S \rangle + \theta \|B_Z - B_S\|_F^2/2$.

21:     **for** $j = 1,2,\cdots,h_2$ **do**

22:        Set $a_j = (t-1)/t$.

23:        Compute $B_{S_i} = (1 + \alpha_j)B_{Z_i} - \alpha_j B_{Z_{i-1}}$.

24:        Compute $\nabla_{B_S} F(B_{S_i})$.

25:        **while** (true)

26:           Compute $\widehat{B_S} = B_{S_i} - \nabla_{B_S} F(B_{S_i})/\theta_i$.

27:           Compute $[\widehat{B_S}] = \boldsymbol{Schmidt}(\widehat{B_S})$.

28:           Set $[B_{Z_{i+1}}] = \boldsymbol{GDMCS}(F(\cdot), \widehat{B_S}, \tau_1, \rho_1, \rho_2)$.

29:           **if** $F(B_{Z_{i+1}}) \leq F_{\theta_i, B_{S_i}}(B_{Z_{i+1}})$, then break;

30:           **else** Update $\theta_i = \theta_i \times 2$.

31:           **end-if**

32:     end-while
33:        Update $t = \left(1+\sqrt{1+4t^2}\right)/2$, $\theta_{i+1} = \theta_i$.
34:  end-for
35:  Fix $A$, $B$, $W$ and approximately solve for $M$.
36:  Define $F_{\lambda, M_S}(M_Z) = F(M_S) + \langle \nabla F(M_S), M_Z - M_S \rangle + \lambda \|M_Z - M_S\|_F^2/2$.
37:  for $j = 1,2,\cdots, h_3$ do
38:     Set $a_j = (p-1)/p$.
39:     Compute $M_{S_i} = (1 + \alpha_j)M_{Z_i} - \alpha_j M_{Z_{i-1}}$.
40:     Compute $\nabla_{M_S} F(M_{S_i})$.
41:     while (true)
42:        Compute $\widehat{M_S} = M_{S_i} - \nabla_{M_S} F(M_{S_i})/\lambda_i$.
43:        Compute $[M_{Z_{i+1}}] = \boldsymbol{PSP}(\widehat{M_S})$.
44:        if $F(M_{Z_{i+1}}) \leq F_{\lambda_i, M_{S_i}}(M_{Z_{i+1}})$, then break;
45:        else Update $\lambda_i = \lambda_i \times 2$.
46:        end-if
47:     end-while
48:     Update $p = \left(1+\sqrt{1+4p^2}\right)/2$, $\lambda_{i+1} = \lambda_i$.
49:  end-for
50:  Fix $A$, $B$, $M$ and approximately solve for $W$.
51:  Define $F_{\mu, W_S}(W_Z) = F(W_S) + \langle \nabla F(W_S), W_Z - W_S \rangle + \mu \|W_Z - W_S\|_F^2/2$.
52:  for $j = 1,2,\cdots, h_4$ do
53:     Set $a_j = (q-1)/q$.
54:     Compute $W_{S_i} = (1 + \alpha_j)W_{Z_i} - \alpha_j W_{Z_{i-1}}$.
55:     Compute $\nabla_{W_S} F(W_{S_i})$.
56:     while (true)
57:        Compute $W_{Z_{i+1}} = W_{S_i} - \nabla_{W_S} F(W_{S_i})/\lambda_i$.
58:        if $F(W_{Z_{i+1}}) \leq F_{\mu_i, W_{S_i}}(W_{Z_{i+1}})$, then break;
59:        else Update $\mu_i = \mu_i \times 2$.
60:        end-if
61:     end-while
62:     Update $q = \left(1+\sqrt{1+4q^2}\right)/2$, $\mu_{i+1} = \mu_i$.
63:  end-for
64: end-for
65: Set $Z^* = [A_{Z_{i+1}} \ B_{Z_{i+1}} \ M_{Z_{i+1}} \ W_{Z_{i+1}}]$.

## 3   Experimental Evaluation

Our experiments are conducted on three publicly available multi-source datasets, namely, UCI Multiple Features (UCI MFeat) [3], Wikipedia [19], and MIR Flickr [9]. The statistics of the datasets are given in Table 2, and brief descriptions of the chosen feature sets in the above-mentioned datasets are listed in Table 3.

Note that all the data are normalized to unit length. Each dataset is randomly separated into a training set and a test set. The training samples account for 80% of each original dataset, and the remaining ones act as the test data.

**Table 2.** Statistics of the multi-source datasets

| Dataset | Total attributes | Total classes | Total samples |
|---------|------------------|---------------|---------------|
| UCI MFeat | 123 | 10 | 2000 |
| Wikipedia | 258 | 10 | 2866 |
| MIR Flickr | 5857 | 38 | 25000 |

**Table 3.** Brief descriptions of the feature sets

| Dataset | Feature set | Total attributes | Total labels | Total instances |
|---------|-------------|------------------|--------------|-----------------|
| UCI MFeat | fou $(S_x)$ | 76 | 10 | 2000 |
| | zer $(S_y)$ | 47 | 10 | 2000 |
| Wikipedia | Image $(S_x)$ | 128 | 10 | 2866 |
| | Text $(S_y)$ | 130 | 10 | 2866 |
| MIR Flickr | Image $(S_x)$ | 3857 | 38 | 25000 |
| | Text $(S_y)$ | 2000 | 38 | 25000 |

Such a partition of each dataset is repeated five times and the average performance is reported. The 100 outliers are generated from Gaussian noise with the same dimension as normal samples in each source. We mix these outlier data into each dataset. Some key parameters of all the methods in our experiments are tuned using the 5-fold cross-validation based on the AUC (area under the receiver operating characteristic curve) on the training set. Particularly, the LIBSVM classifier serves as the benchmark for the tasks of classification in the experiments.

### 3.1 Comparison of Multi-view Outlier Detection Methods

The purpose of comparing the proposed MMOD model and multi-view outlier detection methods, such as Li's method [21], Zhao's method [27], Janeja's method [10], and Liu's method [14] is to show the importance of identifying multi-source outliers in a consistent feature-homogeneous space. Due to the attempt of identifying multi-source outliers in different original feature spaces, it is extremely difficulty for the other compared approaches to capture much more complementary information from different sources. It will lead to a low recognition rate for multi-source outliers. To validate this point, we further compare the recognition rate of MMOD with the above-mentioned multi-view outlier detection methods. The parameter settings in the compared methods are the same as in their original literatures.

The proposed MMOD model identifies multi-source outliers in a consistent feature-homogeneous space. As shown in Table 4, MMOD can improve more effectively the recognition rate for multi-source outliers than Li's method, Zhao's method, Janeja's method, and Liu's method. It means that MMOD can capture the information compatibility among different sources more effectively.

**Table 4.** Comparison of multi-view outlier detection methods

| Method | Dataset | | |
|--------|----------|-----------|-----------|
|        | UCI MFeat | Wikipedia | MIR Flickr |
| Li's   | 0.7492 | 0.7822 | 0.8053 |
| Zhao's | 0.7192 | 0.7618 | 0.7891 |
| Janeja's | 0.7197 | 0.7359 | 0.8096 |
| Liu's  | 0.8013 | 0.8296 | 0.8394 |
| **MMOD** | **0.8977** | **0.8671** | **0.8824** |

## 3.2  Comparison of Mono-source Outlier Detection Approaches

To evaluate the performance of outlier detection, we compare our method with some representative state-of-the-art mono-source methods such as Li's method [13], Rahmani's method [18], and kNN [6] in three multi-source datasets. Basic metric, Precision (P), is used to evaluate the ability of each algorithm. For Li's method, Rahmani's method, and kNN, we first use CCA [22] to project the multi-source data into a feature-homogeneous space and then apply these methods to retrieve the most likely outlier. For MMOD, we tune the regularization parameters on the set $\{10^i | i = -2, -1, 0, 1, 2\}$. For Li's method and Rahmani's method, the experiment settings follow the original works [13,18], respectively. The parameter $k$ in kNN is selected from the set $\{2*i + 1 | i = 5, 10, 15, 20, 25\}$.

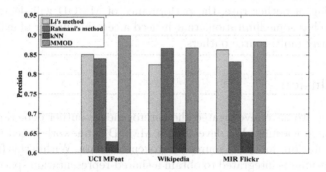

**Fig. 3.** Comparisons of outlier detection approaches

From Fig. 3, we can see that MMOD achieves significant gains, and can almost detect all the outliers. This observation indicates that MMOD will be more favorable to detect multi-source heterogeneous outliers because of fully taking into account the information compatibility and semantic complementarity among different sources.

## 3.3 Comparison in Different Outlier Rates

To test the performance of the proposed MMOD in different outlier rates, we further compare the recognition rate of MMOD with other multi-view outlier detection methods such as Li's method [21], Zhao's method [27], Janeja's method [10], and Liu's method [14] in the larger MIR Flickr dataset. We tune the outlier rates on the set {10%, 15%, 20%, 25%}.

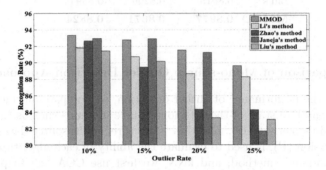

**Fig. 4.** Comparison in different outlier rates

We can see from Fig. 4 that MMOD is superior to other multi-view out-lier detection methods in recognition rate. This observation further confirms that MMOD can effectively identify multi-source outliers. Nevertheless, with the increasing of outlier rate, the performance of MMOD will degrade. Thus, MMOD also has some limitations that it need a certain number of existing sam-ples to identify multi-source outliers.

# 4    Conclusion

In this paper, we have investigated the heterogeneous outlier detection problem in multi-source learning. We developed a MMOD framework based the consis-tent representations for multi-source heterogeneous data. Within this framework, Manifold learning is integrated to obtain a shared-representation space, in which the information-correlated representations are close along manifold while the semantic-complementary instances are close in Euclidean distance. Meanwhile, an affine subspace is learned through affine combination of shared representa-tions from different sources in the feature-homogeneous space according to the information compatibility among different sources. Finally, multi-source hetero-geneous outliers can be effectively identified in the affine subspace.

**Acknowledgments.** This work was supported by National Natural Science Founda-tion of China (No. 61601458, 61602465).

# References

1. Ando, R.K., Zhang, T.: A framework for learning predictive structures from multiple tasks and unlabeled data. J. Mach. Learn. Res. **6**(3), 1817–1853 (2005)
2. Bertsekas, D.P.: Convex Optimization Theory. Athena Scientific (2009)
3. Breukelen, M.V., Duin, R.P.W., Tax, D.M.J., Hartog, J.E.D.: Handwritten digit recognition by combined classifiers. Kybernetika -Praha- **34**(4), 381–386 (1998)
4. Chandola, V., Banerjee, A., Kumar, V.: Anomaly detection: a survey. ACM Comput. Surv. **41**(3), 1–58 (2009)
5. Chen, D., Lv, J., Yi, Z.: A local non-negative pursuit method for intrinsic manifold structure preservation. In: Proceedings of AAAI Conference on Artificial Intelligence, pp. 1745–1751 (2014)
6. Cover, T.M., Hart, P.E.: Nearest neighbor pattern classification. IEEE Trans. Inf. Theory **13**(1), 21–27 (2002)
7. Elhamifar, E., Vidal, R.: Sparse subspace clustering: algorithm, theory, and applications. IEEE Trans. Pattern Anal. Mach. Intell. **35**(11), 2765–2781 (2013)
8. Guo, Y., Xiao, M.: Cross language text classification via subspace co-regularized multi-view learning. In: Proceedings of ACM International Conference on Machine Learning, pp. 915–922 (2012)
9. Huiskes, M.J., Lew, M.S.: The MIR Flickr retrieval evaluation. In: Proceedings of ACM International Conference on Multimedia Information Retrieval, pp. 39–43 (2008)
10. Janeja, V., Palanisamy, R.: Multi-domain anomaly detection in spatial datasets. Knowl. Inf. Syst. **36**(3), 749–788 (2013)
11. Knorr, E.M., Ng, R.T.: Algorithms for mining distance-based outliers in large datasets. In: Proceedings of International Conference on Very Large Data Bases, pp. 392–403 (1998)
12. Knorr, E.M., Ng, R.T., Tucakov, V.: Distance-based outliers: algorithms and applications. VLDB J. **8**(3), 237–253 (2000)
13. Li, X., Lv, J., Yi, Z.: An efficient representation-based method for boundary point and outlier detection. IEEE Trans. Neural Net. Learn. Syst. **29**(1), 51–62 (2016)
14. Liu, A., Lam, D.: Using consensus clustering for multi-view anomaly detection. In: IEEE Symposium on Security and Privacy Workshops, pp. 117–124 (2012)
15. Nesterov, Y.: Introductory lectures on convex optimization. Appl. Optim. **87**(5), 236 (2004)
16. Nesterov, Y.: Smooth minimization of non-smooth functions. Math. Program. **103**(1), 127–152 (2005)
17. Pimentel, M.A.F., Clifton, D.A., Clifton, L., Tarassenko, L.: A review of novelty detection. Signal Process. **99**(6), 215–249 (2014)
18. Rahmani, M., Atia, G.: Randomized robust subspace recovery and outlier detection for high dimensional data matrices. IEEE Tran. Signal Process. **65**(6), 1580–1594 (2017)
19. Rasiwasia, N., et al.: A new approach to cross-modal multimedia retrieval. In: Proceedings of ACM International Conference on Multimedia, pp. 251–260 (2010)
20. Roweis, S.T., Saul, L.K.: Nonlinear dimensionality reduction by locally linear embedding. Science **290**(5500), 2323–2326 (2000)
21. Sheng, L., Ming, S., Yun, F.: Multi-view low-rank analysis for outlier detection. In: Proceedings of SIAM International Conference on Data Mining, pp. 748–756 (2015)

22. Sun, L., Ji, S., Ye, J.: Canonical correlation analysis for multilabel classification: a least-squares formulation, extensions, and analysis. IEEE Trans. Pattern Anal. Mach. Intell. **33**(1), 194–200 (2011)
23. Weinberger, K.Q., Saul, L.K.: Distance metric learning for large margin nearest neighbor classification. J. Mach. Learn. Res. **10**(1), 207–244 (2009)
24. Wen, Z., Yin, W.: A feasible method for optimization with orthogonality constraints. Math. Program. **142**(1–2), 397–434 (2013)
25. Zhang, L., et al.: Collaborative multi-view denoising. In: Proceedings of ACM SIGKDD International Conference on Knowledge Discovery Data Mining, pp. 2045–2054 (2016)
26. Zhang, L., Zhao, Y., Zhu, Z., Wei, S., Wu, X.: Mining semantically consistent patterns for cross-view data. IEEE Trans. Knowl. Data Eng. **26**(11), 2745–2758 (2014)
27. Zhao, H., Fu, Y.: Dual-regularized multi-view outlier detection. In: Proceedings of the International Joint Conference on Artificial Intelligence, pp. 4077–4083 (2015)

# A Fast *k*NN-Based Approach for Time Sensitive Anomaly Detection over Data Streams

Guangjun Wu[1], Zhihui Zhao[1,2], Ge Fu[3(✉)], Haiping Wang[1(✉)], Yong Wang[1], Zhenyu Wang[1], Junteng Hou[1,2], and Liang Huang[3]

[1] Institute of Information Engineering, Chinese Academy of Sciences, Beijing, China
{wuguangjun,zhaozhihui,wanghaiping,wangyong,wangzhenyu, houjunteng}@iie.ac.cn
[2] School of Cyber Security, University of Chinese Academy of Sciences, Beijing, China
[3] National Computer Network Emergency Response Technical Team/Coordination, Center of China (CNCERT/CC), Beijing, China
{fg,huangliang}@cert.org.cn

**Abstract.** Anomaly detection is an important data mining method aiming to discover outliers that show significant diversion from their expected behavior. A widely used criteria for determining outliers is based on the number of their neighboring elements, which are referred to as Nearest Neighbors (NN). Existing *k*NN-based Anomaly Detection (*k*NN-AD) algorithms cannot detect streaming outliers, which present time sensitive abnormal behavior characteristics in different time intervals. In this paper, we propose a fast *k*NN-based approach for Time Sensitive Anomaly Detection (*k*NN-TSAD), which can find outliers that present different behavior characteristics, including normal and abnormal characteristics, within different time intervals. The core idea of our proposal is that we combine the model of sliding window with Locality Sensitive Hashing (LSH) to monitor streaming elements distribution as well as the number of their Nearest Neighbors as time progresses. We use an $\epsilon$-approximation scheme to implement the model of sliding window to compute Nearest Neighbors on the fly. We conduct widely experiments to examine our approach for time sensitive anomaly detection using three real-world data sets. The results show that our approach can achieve significant improvement on recall and precision for anomaly detection within different time intervals. Especially, our approach achieves two orders of magnitude improvement on time consumption for streaming anomaly detection, when compared with traditional *k*NN-based anomaly detection algorithms, such as exact-Storm, approx-Storm, MCOD etc, while it only uses 10% of memory consumption.

**Keywords:** Anomaly detection · Data streams · LSH · Time sensitive

This work was supported by the National Key Research and Development Program of China (2016YFB0801305).

J. M. F. Rodrigues et al. (Eds.): ICCS 2019, LNCS 11537, pp. 59–74, 2019.
https://doi.org/10.1007/978-3-030-22741-8_5

# 1    Introduction

Anomaly (a.k.a. outlier) detection [4] is an important data mining method in applications, such as DDoS detection [12] and medical health monitoring [23]. The task is to identify instances or patterns that do not conform to their expected behaviors [9]. Anomaly detection over data streams is a popular topic in computational science because of its inherent vagueness in definition of outliers, like how to define regular behavior, to what degree an outlier needs to be inconsistent with the regular behaviors when confronting the one-pass and infinite high-speed data streams [21].

Anomaly detection can be classified into supervised [20] and unsupervised [16] in principle. Supervised scenarios are not appropriate for data steams processing due to lack of label information about outliers for one-pass streaming elements. Unsupervised anomaly detection does not require any label information and it is appropriate for data steams processing. Currently, unsupervised anomaly detection methods over data streams can be classified into three categories: (1) Statistical based, (2) Clustering based, and (3) $k$ Nearest Neighbors ($k$NN) based [22].

$k$NN-based approaches have drawn the most attention of researchers to find anomalies over data streams. $k$NN-based approaches assume that the proximity of outliers to their nearest neighbors deviates significantly from the proximity of normal data to their neighbors. Given a dataset $D$ and a threshold $k$, if data point $x$ has less than $k$ neighbors in $D$, $x$ is marked as outlier. This kind of approaches have gained popularity as they are easy to implement on the fly. However, current $k$NN-based anomaly detection method cannot detect outliers with time sensitive abnormal distribution. For example, in DDoS applications, like port scanning, SATAN scanning or IP half-way scanning, attackers send scanning packets in different time intervals and this kind of pre-attack cannot be detected by current $k$NN-based methods [24,26].

We briefly describe challenges of anomaly detection over data streams, which are addressed in our work:

(1) *Abnormal distribution detection.* When detecting outliers, $k$NN-based methods only focus on whether the NN number reaches the threshold but do not care the changes of NN number with time. However, there may be some cases where the NN number is normal but the distribution is significantly uneven during a time period. This kind of anomaly cannot be detected by only counting NN, so that we need time sensitive anomaly detection algorithm.

(2) *Concept drift.* Most existing data stream anomaly detection techniques ignore one important aspect of stream data: arrival of novel anomalies. In dynamically changing and non-stationary environments, the data distribution can change over time yielding the phenomenon of concept drift [13]. Concept drift is aware of different time intervals, but traditional anomaly detection techniques don't take time into consideration, and as a result many novel anomalies cannot be detected.

(3) *Real-time requirements.* Streaming applications require fast-response and real-time processing from volumes of high-speed and multidimensional

dynamic streaming data over time. However, storing all the streaming data requires unbearably large memory so that it is necessary to reduce memory requirements. Additionally, algorithms communicating with DRAM (main memory) are just too slow for the speed at which the data is generated. Ideally, we need ultra-low memory anomaly detection algorithms that can entirely sit in the cache which can be around 2–10 times faster than accessing main memory [11].

The contributions of this paper are as follows:

(1) We propose a *k*NN-based approach *k*NN-TSAD for Time Sensitive Anomaly Detection, which combines an $\epsilon$-approximation scheme Deterministic Waves (DW) with LSH to monitor streaming elements distribution as well as the number of their Nearest Neighbors as time progresses. We use LSH to count NN number without massive calculation of distances between data points. We use DW to monitor distribution of streaming elements during different time intervals. Thus, our anomaly detection method is time sensitive.
(2) The time and space efficiency of our approach is very high and it well meets the real-time requirements. We combine DW with LSH by the way that every item of the hash array is implemented as a DW window and data update can be very efficient and cost very low. Furthermore, we use LSH to search NN which does not need massive range queries and distance calculation, and it does not need to store neighbor information for every element.
(3) We provide a comparison of our algorithm with other five traditional *k*NN-based algorithms based on three real-world datasets. Our experiment shows that *k*NN-TSAD is two orders of magnitude faster than other algorithms but only uses 10% of memory consumption. Moreover the recall and precision of *k*NN-TSAD are higher than others over datasets with abnormal distribution.

## 2 Preliminaries

In the section, we first give the problem definition, and then we will introduce the LSH structure, SRP and the DW model used in this paper.

### 2.1 Problem Definition

In this paper, we take data distribution into account and we detect *k*NN-based outliers as well as outliers with abnormal distribution. We give the definition of *k*NN-based time sensitive outlier in the data streams in Definition 1. The used notation is summarized in Table 1.

**Definition 1.** *(kNN-based Time Sensitive Outlier in Data Streams). Given a data stream $DS$, the current window $W$, a point $x_t$ in $DS$ and two threshold parameters $\alpha$ and $\beta$, we aim to decide whether $x_t$ is an outlier or not. We denote the answer by $NN(x_t, W)$ and $V(x_t, W)$. Formally, if $NN(x_t, W) < \alpha\mu$ or $V(x_t, W) > \beta\delta$, $x_t$ is an outlier, otherwise not.*

**Table 1.** Frequently used notations.

| Notation | Description |
|----------|-------------|
| $DS$ | Data stream |
| $x_t$ | The data point observed at time $t$ |
| $W$ | Current sliding window |
| $w$ | Number of time intervals of every sliding |
| $\alpha, \beta$ | Thresholds parameters |
| $NN(x_t, W)$ | Number of neighbors of $x_t$ in the current window |
| $\mu$ | Mean of $NN(x_t, W)$ till current window |
| $V(x_t, W)$ | Variance of the current window |
| $\delta$ | Mean of $V(x_t, W)$ till current window |

## 2.2 Locality Sensitive Hashing

Locality Sensitive Hashing (LSH) is a popular sublinear time algorithm for efficient approximate nearest neighbors search. LSH leverages a family of functions where each function hashes the points in such a way that under the hash mapping, similar points have a high probability of getting the same hash value so that to be mapped into the same bucket. The probability of collision for an LSH hash function is proportional to the similarity distance between the two data vectors, that is $Pr[h(x) = h(y)] \propto sim(x, y)$.

**Definition 2.** *(LSH Family). A family $\mathcal{H}$ is called $(S_0, cS_0, u_1, u_2) - sensitive$ if for any two data points $x, y \in \mathbb{R}^d$ and $h$ chosen uniformly from $\mathcal{H}$ satisfies the following:*

*(1) If $sim(x, y) \geq S_0$ then $Pr[h(x) = h(y)] \geq u_1$.*
*(2) If $sim(x, y) \leq cS_0$ then $Pr[h(x) = h(y)] \leq u_2$.*

## 2.3 Signed Random Projections

Usually, the family of hash functions of LSH are generated from random projections, the intuition is that points that are nearby in the space will also be nearby in all projections [10, 15]. Given a vector $x$, it utilizes a random vector $w$ with each component generated from i.i.d. normal (i.e., $w_i \sim N(0, 1)$), and only stores the sign of the projection. We define a hash function $h_w(x)$ as follows:

$$h_w(x) = sign(w^T x) \tag{1}$$

And corresponding to this vector $w$ we define the projection function as follows:

$$sign(w^T x) = \begin{cases} 0 & w^T x \geq 0 \\ 1 & w^T x < 0 \end{cases} \tag{2}$$

For vectors $x$ and $w$, the probability $Pr[h(x) = h(w)] = 1 - \frac{\theta(x, w)}{\pi}$.

## 2.4 Deterministic Waves

Deterministic Waves [14] is a $\epsilon$-approximation scheme for the Basic Counting problem for any sliding window up to a prespecified maximum window size $W$. The most important thing that distinguishes DW from other sliding-window models is that, for the sum of integers in $[0, 1, ..., R]$ it improves the per-item processing time from $O(1)$ amortized ($O(\log N)$ worst case) to $O(1)$ worst case. The space complexity of DW is $O(\frac{1}{\epsilon} \log^2(\epsilon W))$.

Consider a data stream of integers and a desired positive $\epsilon < 1$. The algorithm maintains two counters: $pos$, which is the current length of the stream, and $rank$, which is the current integers sum in the stream. The procedure of estimating the sum of integers in a window of size $n \leq W$ is as follows. Let $s = max(0, pos - n + 1)$, we estimate the sum of integers in stream positions $[s, pos]$. The steps are:

(1) Let $p_1$ be the maximum position stored in the wave that is less than $s$, and $p_2$ be the minimum position stored in the wave that is greater than or equal to $s$. (If no such $p_2$ exists, return $\hat{x} := 0$ as the exact answer.) $v$ is the integer value at $p_2$. Let $r_1$ and $r_2$ be the $rank$ of $p_1$ and $p_2$ respectively.
(2) Return $\hat{x} := rank - \frac{r_1 + r_2 - v}{2}$.

# 3   kNN-TSAD Algorithm

In this section we first walk through an overview of the framework, and then describe the procedure in detail.

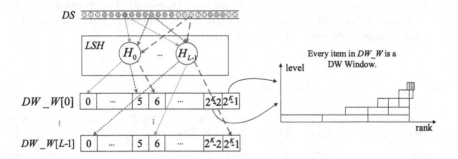

**Fig. 1.** Framework overview of kNN-TSAD.

## 3.1 Overview of the Framework

Our kNN-TSAD algorithm combines the functionality of LSH and sliding windows which supports time-based sliding windows queries, so that it can be used for compactly summarizing high-speed and multidimensional streams over sliding windows. It replaces every item of the LSH array with sliding window structures (i.e., DW windows) covering the last $W$ time units of data streams.

Figure 1 illustrates the overview of our proposed framework. $DW\_W$ is a two-dimensional LSH array of $L$ rows and $2^K$ columns. $K$ and $L$ are pre-specified hyper-parameters. $H_i, i \in 0, 1, \ldots, L-1$ are a group of hash functions, and results of $H_i$ are corresponding to the $i$th row of $DW\_W$. Every item of the LSH array is a DW window. There are two main phases of our $k$NN-TSAD algorithm.

**Insert Phase.** The framework performs inserting a data point as follows:

(1) The nearest neighbors are searched by applying LSH on the data point.
(2) The data point is inserted into corresponding neighbourhoods which are implemented as DW windows, and expired data are deleted from DW windows.
(3) The global mean count and the mean variance of active sliding window are updated.

**Query Phase.** The framework performs an anomaly query as follows:

(1) NN number of current data point and variance of current window are calculated.
(2) NN number and variance of the current data point are compared with mean values, and then an anomaly decision result is given.

### 3.2  Nearest Neighbors Searching

In $k$NN-based methods for anomaly detection, NN searching is the most important part. Existing methods search neighbors by computing distances of every two data points, and then compare to a distance threshold. This process is quite time consuming. Instead, utilization of LSH can significantly reduce time overhead and improve efficiency of NN searching.

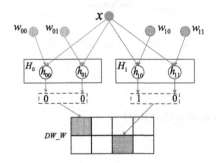

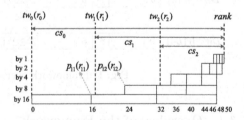

**Fig. 2.** An example of NN searching using LSH.

**Fig. 3.** An example of time sensitive monitoring.

We randomly generate $K \times L$ vectors $w_{ij}, i \in 0, 1, \ldots, L-1, j \in 0, 1, , K-1$, with each component generated from i.i.d. normal (i.e., $w_{ij} \sim N(0,1)$).

Then using these random vectors we define $K \times L$ independent signed random projections $h_{ij}, i \in 0, 1, \ldots, L - 1, j \in 0, 1, \ldots, K - 1$, given in Eqs. 1 and 2 and each of them gives one-bit output. So that we can finally generate $L$ $K$-bit hash functions $H_i(x) = [h_{i1}(x); h_{i2}(x); \ldots; h_{iK}(x)], i \in 0, 1, \ldots, L - 1$, which respectively correspond to the indexes in $L$ rows of hash array $DW\_W$. Final NN number of a data point $x$ is the average of all $H_i(x)$. By this way the collision probability is significantly reduced from $p = 1 - \frac{\theta(x,y)}{\pi}$ to $p = Pr[H(x) = H(y)] = p^K$, if and only if $h_j(x) = h_j(y)$ for all $j \in 0, 1, \ldots, K - 1$, $H(x) = H(y)$. Figure 2 shows an example of NN searching process of a data point.

### 3.3   Time Sensitive Monitoring

To monitor the sliding windows in more detail, we preset a time interval whose length is $\frac{1}{w}$ of the window size. Utilizing the recorded position queue in the DW model, we calculate the average counts of each time interval based on different time range within the sliding window and then work out their variance. The variance will significantly deviate from average level if distribution of data in current sliding window is abnormal.

We have illustrated the procedure of estimating the count of DW in Sect. 2.4, we use the same method to calculate the count of every time interval.

---

**Algorithm 1.** $k$NN-based Anomaly Detection.

---

**Input**: Data stream $DS$, Array parameters $L$ and $K$, Threshold parameter $\alpha$
**Hash Initialize**: Generate $L \times K$ independent SRP hash functions
$DW\_W$ = new DW$[L][2^K]$, $\mu = 0$, n=0
As a data point $x_t \in DS$ comes
/*Insert: add the coming data $x_t \in DS$ to the structure. */
c = 0
**for** $i = 0$ to $L - 1$ **do**
      Add $x_t$ to $DW\_W[i][H_i(x_t)]$ and delete expired data points
      $c = c + \frac{1}{L} DW\_W[i][H_i(x_t)].count$
**end for**
$\mu = \frac{1}{n+1}(n\mu + c)$
$n$++
/*Query: determine whether $x_t \in DS$ is an outlier or not. */
**if** there are expired data **then**
      **if** $c < \alpha\mu$ **then**
            Report anomaly
      **end if**
**else**
      **for** $i = 0$ to $L - 1$ and $j = 0$ to $2^K - 1$ **do**
            Get the count $c$ of $DW\_W[i][j]$
            **if** $c < \alpha\mu$ **then**
                  Report anomaly
            **end if**
      **end for**
**end if**

---

Let $tw_i, i \in 0, 1, \ldots, w-1$ be the left bound of every time interval, and $tw_0$ be the left bound time of the sliding window. Let $p_{i1}$ be the maximum position stored in the wave that is before $tw_i$, and $p_{i2}$ be the minimum position stored in the wave that is after or at $tw_i$. Let $r_{i1}$ and $r_{i2}$ be the counts of $p_{i1}$ and $p_{i2}$ respectively. So that $r_i := \frac{r_{i1}+r_{i2}-1}{2}$, $i \in 0, 1, \ldots, w-1$, and $rank$ is the current exact count. Then we use different time range to compute average count of each time interval $cs_i$ and $cs_i = \frac{rank-r_i}{w-i}$. For example, as shown in Fig. 3, $p_{11}$ is the maximum position before $tw_1$ and $p_{12}$ is the minimum position after $tw_1$. Counts of the two positions are 16 and 24 respectively, so that we estimate the count at $tw_1$ which is $r_1 = \frac{16+24-1}{2} = 20$. The current count $rank$ is 50, thus $cs_1 = \frac{rank-r_1}{2} = 15$, $cs_0 = \frac{rank-r_0}{3} = 16$ and $cs_2 = \frac{rank-r_2}{1} = 16$. Finally we get $\bar{cs} = \frac{15+16+16}{3} = 15$, and then the current variance $\sigma^2 = 2$.

### 3.4   Efficient Update

Updating operation of our algorithm is to straightforward increment (or decrement) the counters when a data point is added (or removed). We only need to add or delete data points in the corresponding DW window and update the global mean values $\mu$ and the mean variance $DW\_W[i][H_i(x)].\delta$ of current sliding window. This operation does not need lots of distance computation to update neighbor information like traditional $k$NN-based algorithms.

As a new data point $x_t \in DS$ arrives, we add it to the DW window at corresponding index $H_i(x)$ of each row so that every DW window keeps the number of hits to this particular window. We get the neighbors count $c$ of $x_t$, $c = \frac{1}{L} \sum_{i=0}^{L-1} DW\_W[i][H_i(x)]$. Then we calculate average count of every time interval $cs_j, j \in [0, w-1]$ so as to calculate the variance $\sigma^2$. We compute the mean count $\mu$,

$$\mu = \frac{1}{n+1}\left(n\mu + \frac{1}{L}\sum_{i=0}^{L-1} DW\_W[H_i(x)]\right) \tag{3}$$

and the mean variance $DW\_W[i][H_i(x)].\delta$ of past part of $DW\_W[i][H_i(x)]$,

$$DW\_W[i][H_i(x)].\delta = \frac{DW\_W[i][H_i(x)].count \times DW\_W[i][H_i(x)].\delta + \sigma^2}{DW\_W[i][H_i(x)].count + 1}. \tag{4}$$

Deviation from the mean values indicate anomaly. It turns out that we can dynamically update the mean values $\mu$ and $DW\_W[i][H_i(x)].\delta$ on the fly, as we observe new coming data.

---

**Algorithm 2.** Abnormal Distribution Detection.

---

**Input**: Data stream $DS$, Array parameters $L$ and $K$, Number of time intervals $w$, $\beta$

**Hash Initialize**: Generate $L \times K$ independent SRP hash functions

$DW\_W = $ new $DW[L][2^K]$

As a data point $x_t \in DS$ comes

$\sigma^2 = 0$

**for** $i = 0$ to $L - 1$ **do**

  /*Insert: add the coming data $x_t \in DS$ to the structure.*/

  Add $x_t$ to $DW\_W[i][H_i(x_t)]$ and delete expired data points

  Get the average counts $cs_0, cs_2, ..., cs_{w-1}$ based on different time range

  $\bar{cs} = \frac{1}{w} \sum_{q=0}^{w-1} cs_q$

  $\sigma^2 = \frac{1}{w} \sum_{q=0}^{w-1} (cs_q - \bar{cs})^2$

  $DW\_W[i][H_i(x)].\delta = \frac{DW\_W[i][H_i(x)].count \times DW\_W[i][H_i(x)].\delta + \sigma^2}{DW\_W[i][H_i(x)].count}$

  /*Query: determine whether there is a distribution anomaly or not.*/

  **if** $\sigma^2 > \beta DW\_W[i][H_i(x)].\delta$ **then**

    Report anomaly

  **end if**

**end for**

---

### 3.5  Anomaly Detection

Our algorithm serves for data streams which are high-speed and one-pass. We use two types of queries to accurately conduct anomaly detection.

(1) *kNN-based anomaly detection queries.* $k$NN-based anomaly detection is a global procedure. A global mean count $\mu$ is calculated to get the threshold $\alpha\mu$. In inserting phase we compute current count and compare with the threshold, and if a DW window satisfies $c < \alpha\mu$, all data in that window are marked as outliers. The process is shown in Algorithm 1.

(2) *Distribution anomaly detecion queries.* Distribution anomaly detection is a local procedure. Every DW window keeps a mean variance $DW\_W[i][H_i(x)].\delta$, and with it we get a threshold $\beta DW\_W[i][H_i(x)].\delta$ of every DW window. In inserting phase we compute current variance $\sigma^2$ and compare with the threshold, if a DW window satisfies $\sigma^2 > \beta DW\_W[i][H_i(x)].\delta$, there exists a distribution anomaly. The process is shown in Algorithm 2.

### 3.6  Complexity Analysis

**Time Complexity:** For a new coming data point, we need to compute $KL$ hashes, $Lw$ DW counts and a variance. For random projections based LSH, signed random projection is a special case. We can calculate $KL$ LSH hashes of the data vector, with dimensions $d$, in time $O(d \log d + KL)$. Using DW model, each item is processed in $O(1)$ worst case. At each time instant, it can provide an estimate in $O(1)$ time. Finding the time division nodes needs to traverse the

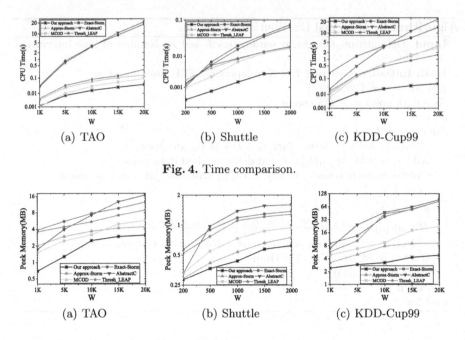

**Fig. 4.** Time comparison.

**Fig. 5.** Memory comparison.

position queue of DW, the time is $O(\frac{1}{\epsilon}(\log(\epsilon W)))$, thus the total time we need is $O(d \log d + \frac{1}{\epsilon}(\log(\epsilon W)))$.

**Space Complexity:** It is presented in [14] that an $\epsilon$-approximation DW window for basic counting uses only $O(\frac{1}{\epsilon} \log^2(\epsilon W))$ memory bits of workspace, and here we need $2^K \times L$ DW windows in the two-dimensional array. $K$ and $L$ are preset constant parameters, and in all our experiments we use $K = 15$ and $L = 50$ for all the three datasets. However, it is unlikely that the DW windows will get too many hits, the number of hits of all our experiments is no more than 3000. Therefore, with these small constant values, the memory cost is extremely low, and the required memory is $O(\frac{1}{\epsilon} \log^2(\epsilon W))$.

## 4  Experimental Evaluation

### 4.1  Datasets

We choose the following three real-world and synthetic datasets for our evaluation: (1) TAO, (2) Statlog Shuttle and (3) KDD-Cup99 HTTP. The first dataset contains 575,648 records with 3 attributes. It is available at Tropical Atmosphere Ocean project [3]. The second dataset shuttle dataset [2] describes radiator positions in a NASA space shuttle with 9 attributes. It was designed for supervised anomaly detection. In the original datasets, about 20% of the data are regarded as anomaly. The third dataset is KDD-Cup99 HTTP [1]. KDD-Cup99 HTTP dataset is the largest benchmark for unsupervised anomaly detection evaluation. It contains simulated normal and attack traffic on an IP level in a computer network environment in order to test intrusion detection systems. We use HTTP traffic only and also limit DoS traffic from the dataset. We sample part of it and the features of "protocol" and "port" information is removed. There are totally 395,423 instances with 6,315 labeled anomalies.

## 4.2  Experimental Methodology

Our experiments are conducted in a 3.30 GHz core Windows machine with 8 GB of RAM. Experiments focus on four aspects: (1) time efficiency; (2) memory consumption; (3) recall and precision; (4) time sensitivity property. We use five $k$NN-based anomaly detection algorithms to compare with our proposal. We summarize the time and space complexities of each algorithm and provide comparison in Table 2. $S$ denotes the slide size which characterizes the speed of the data stream, $k$ is the neighbor count threshold, and $\eta$ denotes the fraction of the window stored in micro-clusters in the MCOD algorithm. For fair evaluation and reproducibility, we implement all the algorithms in Java.

**Table 2.** Comparison of $k$NN-based anomaly detection algorithms.

| Algorithm | Time complexity | Space complexity |
|---|---|---|
| Our proposal | $O(d \log d + \frac{1}{\epsilon}(\log(\epsilon W))$ | $O(\frac{1}{\epsilon} \log^2(\epsilon W))$ |
| Exact-Storm | $O(W)$ | $O(W)$ |
| Approx-Strom | $O(W^2/S)$ | $O(W^2/S + W)$ |
| AbstractC | $O(W \log k)$ | $O(kW)$ |
| MCOD | $O(kW \log k + (1 - \eta)W \log((1 - \eta)W))$ | $O(\eta W + (1 - \eta)kW)$ |
| Thresh_LEAP | $O(W^2 \log S/S)$ | $O(W^2/S)$ |

As shown in Fig. 4, when $W$ increases, the CPU time for each algorithm increases as well, due to a larger number of data points to be processed in every window. We can see among the six algorithms $k$NN-TSAD is consistently efficient. The reason is that adding and removing data points in $k$NN-TSAD are very efficient as well as neighbor searching. It does not need to carry out range queries and calculate distances to update neighbor information like other $k$NN-based algorithms, thus $k$NN-TSAD saves much time.

Figure 5 reports the memory consumption of the evaluated algorithms when varying the window size $W$. The memory requirement increases with $W$ consistently across different datasets. The five traditional methods store neighborhood information for every data point in current window, which requires large amount of memory. However, every item of the $k$NN-TSAD array (i.e. every DW window) is a neighborhood with no need to extra store neighborhood information. Thus $k$NN-TSAD consumes much less memory compared with other methods.

Exact-Storm, AbstractC, MCOD and Thresh_LEAP calculate all the neighbours to decide whether a data point is an outlier. They only differ in neighborhood searching and storing, such that they get almost the same anomaly detection result. So in terms of recall and precision we solely need to test on one of them, we choose MCOD. Approx-Storm adapts exact-Storm with two approximations to reduce the number of data points and the space for neighbor stored in each window. We plot the recall and precision of $k$NN-TSAD, MCOD and approx-Storm to make a comparison (see Figs. 6 and 7). We can see approx-Storm gets the lowest recall and precision of the three. MCOD calculate exact number of neighbors of every data points so that it gives the highest recall and precision. $k$NN-TSAD is between the other two algorithms on dataset Shuttle and KDD-Cup99. But on dataset TAO, as seen in Figs. 6(a) and 7(a), $k$NN-TSAD gets the best anomaly detection result. This is because that part of abnormal data in the TAO dataset appears in a concentrated way during a period of time. In this case the number of neighbors exceeds the threshold due to high frequency. $k$NN-TSAD divides the sliding window into several time intervals and calculates the variance of their counts to monitor count distribution. Figure 8 shows the variation of variance over the time period when outliers appear. Obviously between 3900 s and 6000 s, the variance experienced a process of sharp increase and decrease indicating that there appears an anomaly. However this cannot be detected by other $k$NN-based algorithms, and this is why $k$NN-TSAD get the highest recall and precision over TAO.

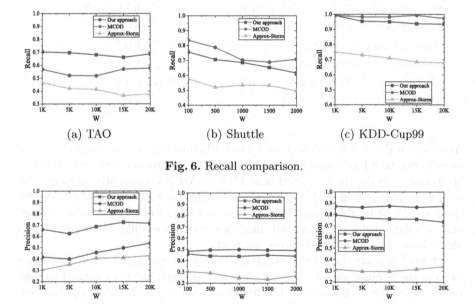

(a) TAO            (b) Shuttle            (c) KDD-Cup99

**Fig. 6.** Recall comparison.

(a) TAO            (b) Shuttle            (c) KDD-Cup99

**Fig. 7.** Precision comparison.

To summarize, $k$NN-TSAD provides the highest performance over dataset TAO whose anomaly detection depends on the data distribution, and this indicates the time sensitive property of our approach. Moreover, $k$NN-TSAD cost the least time and space over all datasets compared with other algorithms and shows its excellent time and space efficiency. LSH and DW introduce calculation error to the algorithm, so that the recall and precision are slightly lower than MCOD which count exact NN number. But when compared with the approximate method Approx-Storm, the recall and precision are significantly higher.

## 5   Related Work

Anomaly detection has been studied for years in the communities of multidimensional data sets [5] and the metric spaces [25]. Usually, similarity is used to decide whether an object is an outlier or not. Extensively, the problem of anomaly detection has been studied by the statistics community [17], where the objects are modeled as a distribution, and outliers are those whose deviation from this distribution is more than others. However, for high-dimensional data, statistical techniques fail to model the distribution accurately. Additionally, for large databases these techniques fail to scale well either.

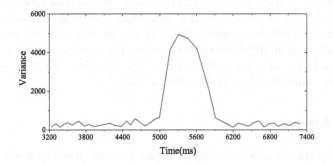

**Fig. 8.** Variation of variance over time.

The problem of anomaly detection has been also addressed by data mining communities. Techniques that focus on identifying Nearest Neighbors-based patterns are popular in data mining and other related communities. $k$ Nearest Neighbors-based outliers were first introduced by Knorr and Ng [18]. Given parameters $k$ and $R$, an object is marked an outlier if there are less than $k$ objects in the input data set lying within distance $R$ from it. However, this algorithm operates in a static fashion. This means that if there happen some changes in the data, the algorithm must be executed from scratch, leading to performance degradation while updating frequently. For example in the streaming case, where objects arrive in a streaming fashion [7].

In [6], an exact algorithm exact-Storm was proposed to efficiently detect outliers over data streams. It stores up to $k$ preceding neighbors and the number of succeeding neighbors of $x$. $x$ is an outlier if $x$ has less than $k$ neighbors. An approximate algorithm approx-Storm is derived from the former one. Up to $\rho W$ $(0 < \rho < 1)$ safe inliers are preserved for each window, and it only store the ratio between the number of $x$'s preceding neighbors which are safe inliers and the number of safe inliers in the window. Such it can estimate the number of $x$'s neighbours. Kontaki et al. [19] proposed the MCOD algorithm which aims to reduce the number of distance computations. MCOD stores the neighboring data points in micro-clusters, and a micro-cluster is composed of no less than $k+1$ data points. Some data points may not fall into any micro-clusters, if they have less than $k$ neighbors, they are outliers. Cao et al. [8] proposed a framework named LEAP which encompasses two general optimization principles. First, the "minimal probing" principle uses a lightweight probing operation to gather minimal yet sufficient evidence for outlier detection. Second, the "lifespan-aware prioritization" principle leverages the temporal relationships among stream data points to prioritize the processing order among them during the probing process. Yang et al. [27] proposed an algorithm Abstract-C which maintains a compact summary of the neighbourships, namely the count of the neighbors for each data point. Instead of storing a list of preceding neighbors or number of succeeding neighbors for each data point $x$, a sequence is used to store the number of neighbors in every window that $x$ participates in.

$k$NN-based anomaly detection methods rely on neighbors search to decide whether a data point is an outlier. And neighbors searching relies on calculating the similarity between data points which is time consuming. In this paper, we use LSH to measure similarity between data points. Data points with high similarity are more likely to enter the same buckets after hashing. There is no need to calculate the similarity of every two data points.

## 6    Conclusion

Anomaly detection applications need to provide real-time monitoring and the capability of detecting anomalies over continuous data streams using limited resources. In this paper, we leverage the features of LSH and DW and propose a $k$NN-based time sensitive anomaly detection approach $k$NN-TSAD over data streams. Our approach can detect abnormal data distribution over different time intervals which cannot be found by traditional $k$NN-based methods. Moreover, our approach demonstrates high time and space efficiency for solving the problem of data stream processing. However, we can see from the experiments that the recall and precision of our approach are slightly lower than exact algorithms in some cases. In the future, we plan to design a more efficient monitoring strategy to improve and optimize our approach to get higher recall and precision.

# References

1. KDD-Cup99 HTTP. http://kdd.ics.uci.edu/databases/kddcup99/kddcup99.html
2. Statlog Shuttle. https://archive.ics.uci.edu/ml/datasets/Statlog+(Shuttle)
3. TAO. http://www.pmel.noaa.gov
4. Aggarwal, C.C.: Data Streams: Models and Algorithms, vol. 31. Springer, London (2007). https://doi.org/10.1007/978-0-387-47534-9
5. Aggarwal, C.C., Yu, P.S.: Outlier detection for high dimensional data. In: ACM Sigmod Record, vol. 30, pp. 37–46. ACM (2001)
6. Angiulli, F., Fassetti, F.: Detecting distance-based outliers in streams of data. In: Proceedings of the Sixteenth ACM Conference on Conference on Information and Knowledge Management, pp. 811–820. ACM (2007)
7. Babcock, B., Babu, S., Datar, M., Motwani, R., Widom, J.: Models and issues in data stream systems. In: Proceedings of the Twenty-First ACM SIGMOD-SIGACT-SIGART Symposium on Principles of Database Systems, pp. 1–16. ACM (2002)
8. Cao, L., Yang, D., Wang, Q., Yu, Y., Wang, J., Rundensteiner, E.A.: Scalable distance-based outlier detection over high-volume data streams. In: IEEE 30th International Conference on Data Engineering (ICDE), pp. 76–87. IEEE (2014)
9. Chandola, V., Banerjee, A., Kumar, V.: Anomaly detection: a survey. ACM Comput. Surv. (CSUR) **41**(3), 15 (2009)
10. Charikar, M.S.: Similarity estimation techniques from rounding algorithms. In: Proceedings of the Thiry-Fourth Annual ACM Symposium on Theory of Computing, pp. 380–388. ACM (2002)
11. Conway, P., Kalyanasundharam, N., Donley, G., Lepak, K., Hughes, B.: Cache hierarchy and memory subsystem of the AMD Opteron processor. IEEE Micro **30**(2), 16–29 (2010)
12. Dang, Q.V.: Outlier detection on network flow analysis. arXiv preprint arXiv:1808.02024 (2018)
13. Gama, J., Žliobaitė, I., Bifet, A., Pechenizkiy, M., Bouchachia, A.: A survey on concept drift adaptation. ACM Comput. Surv. (CSUR) **46**(4), 44 (2014)
14. Gibbons, P.B., Tirthapura, S.: Distributed streams algorithms for sliding windows. In: Proceedings of the Fourteenth Annual ACM Symposium on Parallel Algorithms and Architectures, pp. 63–72. ACM (2002)
15. Goemans, M.X., Williamson, D.P.: 879-approximation algorithms for MAX CUT and MAX 2SAT. In: Proceedings of the Twenty-Sixth Annual ACM Symposium on Theory of Computing, pp. 422–431. ACM (1994)
16. Görnitz, N., Kloft, M., Rieck, K., Brefeld, U.: Toward supervised anomaly detection. J. Artif. Intell. Res. **46**, 235–262 (2013)
17. Härdle, W., Simar, L.: Applied Multivariate Statistical Analysis. Springer, Heidelberg (2007). https://doi.org/10.1007/978-3-662-45171-7
18. Knox, E.M., Ng, R.T.: Algorithms for mining distance based outliers in large datasets. In: Proceedings of the International Conference on Very Large Data Bases, pp. 392–403. Citeseer (1998)
19. Kontaki, M., Gounaris, A., Papadopoulos, A.N., Tsichlas, K., Manolopoulos, Y.: Continuous monitoring of distance-based outliers over data streams. In: IEEE 27th International Conference on Data Engineering (ICDE), pp. 135–146. IEEE (2011)
20. Leung, K., Leckie, C.: Unsupervised anomaly detection in network intrusion detection using clusters. In: Proceedings of the Twenty-Eighth Australasian Conference on Computer Science, vol. 38, pp. 333–342. Australian Computer Society, Inc. (2005)

21. Sadik, M.S., Gruenwald, L.: DBOD-DS: distance based outlier detection for data streams. In: Bringas, P.G., Hameurlain, A., Quirchmayr, G. (eds.) DEXA 2010. LNCS, vol. 6261, pp. 122–136. Springer, Heidelberg (2010). https://doi.org/10.1007/978-3-642-15364-8_9

22. Salehi, M., Rashidi, L.: A survey on anomaly detection in evolving data: [with application to forest fire risk prediction]. ACM SIGKDD Explor. Newslett. 20(1), 13–23 (2018)

23. Saneja, B., Rani, R.: An efficient approach for outlier detection in big sensor data of health care. Int. J. Commun. Syst. 30(17), e3352 (2017)

24. Sezari, B., Möller, D.P., Deutschmann, A.: Anomaly-based network intrusion detection model using deep learning in airports. In: 17th IEEE International Conference on Trust, Security and Privacy in Computing and Communications/12th IEEE International Conference on Big Data Science and Engineering (TrustCom/BigDataSE), pp. 1725–1729. IEEE (2018)

25. Tao, Y., Xiao, X., Zhou, S.: Mining distance-based outliers from large databases in any metric space. In: Proceedings of the 12th ACM SIGKDD International Conference on Knowledge Discovery and Data Mining, pp. 394–403. ACM (2006)

26. Viet, H.N., Van, Q.N., Trang, L.L.T., Nathan, S.: Using deep learning model for network scanning detection. In: Proceedings of the 4th International Conference on Frontiers of Educational Technologies, pp. 117–121. ACM (2018)

27. Yang, D., Rundensteiner, E.A., Ward, M.O.: Neighbor-based pattern detection for windows over streaming data. In: Proceedings of the 12th International Conference on Extending Database Technology: Advances in Database Technology, pp. 529–540. ACM (2009)

# $n$-gram Cache Performance in Statistical Extraction of Relevant Terms in Large *Corpora*

Carlos Goncalves[1,2](✉) (iD), Joaquim F. Silva[2](iD), and Jose C. Cunha[2](iD)

[1] Instituto Superior de Engenharia de Lisboa, Lisbon, Portugal
cgoncalves@deetc.isel.pt
[2] NOVA Laboratory for Computer Science and Informatics, Caparica, Portugal
{jfs,jcc}@fct.unl.pt

**Abstract.** Statistical extraction of relevant $n$-grams in natural language *corpora* is important for text indexing and classification since it can be language independent. We show how a theoretical model identifies the distribution properties of the distinct $n$-grams and singletons appearing in large *corpora* and how this knowledge contributes to understanding the performance of an $n$-gram cache system used for extraction of relevant terms. We show how this approach allowed us to evaluate the benefits from using Bloom filters for excluding singletons and from using static prefetching of nonsingletons in an $n$-gram cache. In the context of the distributed and parallel implementation of the LocalMaxs extraction method, we analyze the performance of the cache miss ratio and size, and the efficiency of $n$-gram cohesion calculation with LocalMaxs.

**Keywords:** Large *corpora* · Statistical extraction · Multiword terms · Parallel processing · $n$-gram cache performance · Cloud computing

## 1  Introduction

Multiword expressions in natural language texts are $n$-grams (sequences of $n \geq 1$ consecutive words). Statistical extraction of relevant expressions, useful for text indexing and classification, can be language-independent. Thus it can be included in initial stages of extraction pipelines, followed by language-specific syntactic/semantic filtering. The increased availability of large *corpora* [1,2] due to the Web growth challenges statistical extraction methods. We focus on $n$-gram distribution models and parallel and distributed tools for extracting relevant expressions from large *corpora*. LocalMaxs [3,4], a multiphase statistical extraction method, has a $1^{st}$ phase for collecting $n$-gram frequency statistics, a $2^{nd}$ phase for calculating an $n$-gram cohesion metric, and a $3^{rd}$ phase for applying an

Acknowledgements to FCT MCTES and NOVA LINCS UID/CEC/04516/2019.

J. M. F. Rodrigues et al. (Eds.): ICCS 2019, LNCS 11537, pp. 75–88, 2019.
https://doi.org/10.1007/978-3-030-22741-8_6

$n$-gram relevance filtering criterion. The computational complexity of methods as LocalMaxs depends on $n$-gram distribution properties. Thus we proposed [5] a theoretical model predicting the $n$-gram distribution as a function of *corpus* size and $n$-gram size ($n \geq 1$), validated empirically for estimating the numbers of distinct $n$-grams, $1 \leq n \leq 6$, with English and French *corpora* from 2 Mw ($10^6$ words) to 1 Gw ($10^9$ words) [6]. It allows to identify how the numbers of distinct $n$-grams tend asymptotically to *plateaux* as the *corpora* grow toward infinity. Due to the large numbers of distinct $n$-grams in large *corpora*, the memory limitations become critical, motivating optimizations for space efficient $n$-gram data structures [7]. We pursue an orthogonal approach using parallel computing for acceptable execution times, overcoming the memory limitations by data partitioning with more machines, and using data distribution for scalable storage of the $n$-grams statistical data [8,9]. Guided by the theoretical model estimates, we developed a parallel architecture for LocalMaxs: with an on-demand dynamic $n$-gram cache to keep the $n$-gram frequency data, used for supporting the cohesion calculations in the $2^{nd}$ phase; a distributed in-memory store as a repository of the $n$-gram global frequency values in the *corpus* and the cohesion and relevance values; and a workflow tool for specifying multiphase methods; supported by a distributed implementation with a configurable number of virtual machines. LocalMaxs execution performance for extracting relevant 2-grams and 3-grams from English *corpora* up to 1 Gw was shown scalable, with almost linear relative speed-up and size-up, with up to 48 virtual machines on a public cloud [8,9]. However, that implementation achieves low efficiency relative to a single ideal sequential machine because the on-demand dynamic $n$-gram cache is unable to overcome the communication overheads due to the $n$-gram references missing in the cache, requiring the remote fetching of the $n$-gram global frequency counts. To improve the $n$-gram cache efficiency, we discuss two new aspects, as extensions to the LocalMaxs parallel architecture. The first one consists in filtering the singleton $n$-grams. To evaluate this, we extend the theoretical model to predict the distribution of singleton $n$-grams, $1 \leq n \leq 6$, applying this to English *corpora* from a few Mw to infinity. Then we show that this singletons filtering with Bloom filters [10] leads to a reduction of the $n$-gram cache miss ratio, but it depends on the evolution of the numbers of singletons as the *corpus* size grows. The second improvement relies on the static prefetching of the $n$-grams statistical data into the cache. This, for a multiphase method, can be performed completely in the $1^{st}$ phase (collecting $n$-gram statistics), so that during a subsequent phase where the $n$-gram cache is used for cohesion metric and relevance calculation, there is no cache miss overhead. For LocalMaxs, this leads to a virtually 0% cache miss ratio for any *corpus* sizes. In the paper we discuss background (Sect. 2), the theoretical model and the distribution of singleton $n$-grams (Sect. 3), the two $n$-gram cache improvements (Sect. 4 ) and the obtained results (Sect. 5).

## 2   Background

Relevant expressions, e.g. "United Nations", can be used to summarize, index or cluster documents. Due to their semantic richness, their automatic extraction from raw text is of great interest. Extraction approaches can be linguistic, statistical or hybrid [11,12]. Most of the statistical ones are language-neutral [13], using metrics as Mutual Information [14], Likelihood Ratio [15], $\Phi^2$ [16]. Among the latter, LocalMaxs [3,4] extracts multiword relevant expressions [17].

**LocalMaxs.** It relies on a generic cohesion metric, called "glue", (as $SCP_f$ Eq. (1) below; Dice [4]; or Mutual Information), and on a generic relevance criterion (as Eq. (2) below), that, for a given input set of *n*-grams ($n \geq 2$), identifies the ones considered relevant, according to the strength of their internal co-occurrence:

$$SCP_f(w_1 \cdots w_n) = \frac{f(w_1 \cdots w_n)^2}{\frac{1}{n-1}\sum_{i=1}^{n-1} f(w_1 \cdots w_i) \times f(w_{i+1} \cdots w_n)} \tag{1}$$

where $f(w_1 \cdots w_i)$ is the frequency of the *n*-gram $(w_1 \cdots w_i)$, $i \geq 1$, in the *corpus*. The denominator has the frequencies of all "leftmost and rightmost sub *n*-grams" (that, for simplicity, are abbreviated as "sub *n*-grams") of sizes from 1 to $n-1$ contained in $(w_1 \cdots w_n)$. E.g., considering the 5-gram "European Court of Human Rights", the sub *n*-grams whose frequencies are needed for the glue calculation are: the 1-grams, "European" and "Rights"; the 2-grams, "European Court" and "Human Rights"; the 3-grams, "European Court of" and "of Human Rights"; and the 4-grams, "European Court of Human" and "Court of Human Rights".

**LocalMaxs Relevance Criterion.** Let $W = (w_1 \ldots w_n)$ be an *n*-gram and $g(.)$ a generic cohesion metric. Let $\Omega_{n-1}(W)$ be the set of $g(.)$ values for all contiguous $(n-1)$-grams within the *n*-gram $W$; Let $\Omega_{n+1}(W)$ be the set of $g(.)$ values for all contiguous $(n+1)$-grams containing *n*-gram $W$. $W$ is relevant expression iff:

$$\forall_x \in \Omega_{n-1}(W), \forall_y \in \Omega_{n+1}(W)$$
$$length(W) = 2 \wedge g(W) > y \quad \vee \quad length(W) > 2 \wedge g(W) > \tfrac{x+y}{2} \tag{2}$$

For the example $W = (European\,Court\,of\,Human\,Rights)$, the sets are: $\Omega_{n-1}(W) = \{g(European\,Court\,of\,Human), g(Court\,of\,Human\,Rights)\}$; and $\Omega_{n+1}(W) = \{g(Y)\}$, such that $Y = (w_L\,W)$ or $Y = (W\,w_R)$ where symbols $w_L$ and $w_R$ stand for unigrams appearing in the *corpus*, and $Y$ is the $(n+1)$-gram obtained from the concatenation of $w_L$ or $w_R$ with $W$.

**Parallel LocalMaxs Architecture.** Figure 1a shows the logical dependencies of LocalMaxs for extracting relevant *n*-grams, $2 \leq n \leq 5$. For a given maximum *n*-gram size $n_{MAX}$ the relevant *n*-grams, $2 \leq n \leq n_{MAX}$, are identified in the *corpus* in three phases: (1) counting all *n*-gram occurrences, $1 \leq n \leq (n_{MAX}+1)$; (2) calculating the glue ($g_{2 \cdots (n_{MAX}+1)}$) for all distinct *n*-grams, $2 \leq n \leq (n_{MAX}+1)$; and (3) applying a relevance criterion to all distinct nonsingleton *n*-grams, $2 \leq n \leq n_{MAX}$. The workflow is executed [8,9] by a collection of virtual machines, each with one controller (for LocalMaxs functions: count, glue, relevance), one server (for storing the *n*-gram data), and local *n*-gram caches (Fig. 1b).

In phase one, the *n*-gram counting is performed in parallel by different controllers acting on equal-size input *corpus* partitions. It generates the distinct

$n$-gram tables, one for each $n$-gram size, containing the total counts of all the $n$-gram occurrences in the *corpus*. These tables, partitioned by $n$-gram hashing, are stored in a distributed collection of servers, thus supporting a repository of the global $n$-gram frequency counts in the *corpus* (in the end of phase one). For $K$ machines, each server $S(j)$ in each machine $j$: $1 \leq j \leq K$, keeps a local $n$-gram table $(D_i(j))$, for $n$-grams of size $i$: $1 \leq i \leq (n_{MAX} + 1)$. The set of local $n$-gram tables $(1 \leq i \leq (n_{MAX} + 1))$ within each server $S(j)$ is: $\{D_1(j), D_2(j), \cdots, D_i(j), \cdots, D_{n_{MAX}+1}(j)\}$. The set of distinct $n$-grams of size $i$ in the *corpus* $(D_i)$ is the union of the disjoint local $n$-gram tables $D_i(j)$ in all servers $1 \leq j \leq K$. In each machine $(j)$, there is one controller $(Ctrl(j))$ co-located with one local server $S(j)$. Phase two input consists of a set of distinct $n$-grams whose glues must be calculated. These $n$-grams and their frequency counts are found, by each machine controller, in the local server $n$-gram tables. However, the frequencies of the sub $n$-grams required for glue calculation of each distinct $n$-gram must be fetched from the global distributed repository. So, in this phase the repeated sub $n$-gram references used by the glue calculations justify a per machine $n$-gram cache for each $n$-gram size $(C_1, ..., C_{n_{MAX}})$ (Fig. 1b). Each local cache entry has the frequency of a distinct $n$-gram. In the end of phase two all the distinct $n$-gram entries in the global repository become updated with their glue values. The input to phase three, for each machine controller, consists of the local $n$-gram tables updated by phase two, used to evaluate the $n$-gram relevance, finally stored in the local $n$-gram table. At the end, for all tables $(D_i(j))$ of the global repository, each entry has: an unique $n$-gram identification, its global frequency, its glue value, and its relevance flag (yes/no). As the *corpus* data is unchanged during LocalMaxs execution, a static work distribution leads to a balanced load in all phases since the local table sizes are approximately equal, $|D_i(j)| \approx (|D_i|/K)$ (for each $n$-gram size $i$, $1 \leq i \leq n$; machine $j$, $1 \leq j \leq K$), and the controller input partitions are of equal sizes.

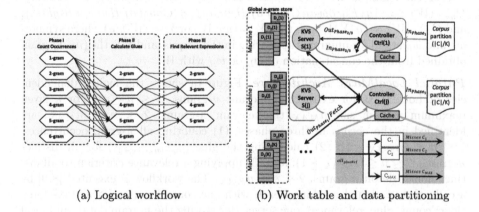

(a) Logical workflow          (b) Work table and data partitioning

**Fig. 1.** Parallel LocalMaxs architecture

## 3   A Theoretical Model for *n*-gram Distribution

We review a theoretical model [5] for the efficient estimation of the number of distinct *n*-grams ($n \geq 1$), for any *corpus* size, for each given language. Here the model is extended for predicting the number of singleton *n*-grams.

***Distinct n-grams.*** By Zipf-Mandelbrot Law [18,19] and Poisson distribution:

$$f(r,c,n) = \left((1 + \beta(n))^{\alpha(n)} \times f(1,c,n)\right) \times \frac{1}{(r + \beta(n))^{\alpha(n)}} \tag{3}$$

where $f(r,c,n)$ is the absolute frequency of the $r^{th}$ most frequent *n*-gram of size $n$ in a *corpus* $C$ of $c = |C|$ words. The most frequent *n*-gram of size $n$ is ranked $r = 1$ with frequency $f(1,c,n)$, and the least frequent *n*-gram of size $n$ has rank $r = D(c,n)$, i.e., the number of distinct *n*-grams of size $n$ for a *corpus* $C$. For each language, $\alpha(n)$ and $\beta(n)$ are approximately constant (in Eq. (3)). As confirmed empirically, the relative frequency, $p_1(n)$, of the first ranked *n*-gram tends to be constant wrt the *corpus* size: $f(1,c,n) = p_1(n) \times c$. Thus, $\left((1 + \beta(n))^{\alpha(n)} \times f(1,c,n)\right)$ in Eq. (3) is constant for each *corpus* size, hence the frequency of each rank follows a power law with $\alpha(n) > 0$. Let random variable $X$ be the number of occurrences of *n*-gram $w$ in rank $r$ in a *corpus*, in language $l$, by Poisson distribution, the probability of $w$ occurring at least once is:

$$Pr(X \geq 1) = 1 - e^{-\lambda} \tag{4}$$

where $\lambda$ is the Poisson parameter, the expected frequency of *n*-gram $w$ in that *corpus*. For each rank $r$, we have $\lambda = f(r,c,n)$. Thus, $Dist(l,c,n)$, the expected number of distinct *n*-grams of size $n$ in a *corpus* of size $c$ in language $l$, is:

$$Dist(l,c,n) = \sum_{r=1}^{v(n)} \left(1 - e^{-f(r,c,n)}\right) = v(n) - \sum_{r=1}^{v(n)} e^{-\left(\left(\frac{1+\beta(n)}{r+\beta(n)}\right)^{\alpha(n)} \times (p_1(n) \times c)\right)} \tag{5}$$

For each *n*-gram size $n$ there is a corresponding language *n*-gram vocabulary of specific size $v(n)$, which in our interpretation includes all different word flexions as distinct. The parameters $\alpha$, $\beta$, $p_1$, $v$ were estimated empirically for the English language, for 1-grams to 6-grams [5], using a set of Wikipedia *corpora* from 2 Mw to 982 Mw (Table 1). In Fig. 2a, the curves for the estimates ($Es$) of the numbers of distinct *n*-grams ($1 \leq n \leq 6$) are shown dotted and for the observed data ($Obs$) are filled, corresponding to a relative error ($Es/Obs-1$) generally below 1% [5]. Above well identified *corpus* size thresholds, for each *n*-gram size, the number of distinct *n*-grams reaches an asymptotic *plateau* determined by the finite vocabulary size, at a given time epoch. Any further *corpus* increase just increases the existing *n*-grams frequencies.

***Number of Singleton n-grams.*** From the Poisson distribution, the number of distinct *n*-grams with frequency $k \geq 0$ is estimated as:

$$W(k,c,n) = \sum_{r=1}^{r=v(n)} \frac{\lambda_r^k \times e^{-\lambda_r}}{k!} = \sum_{r=1}^{r=v(n)} \frac{f(r,c,n)^k \times e^{-f(r,c,n)}}{k!} \tag{6}$$

**Table 1.** Best $\alpha$, $\beta$, $v$ (number of $n$-grams) and $p_1$ for the English *corpora*

|       | unigrams            | bigrams             | trigrams            | trigrams            | pentagrams            | hexagrams             |
|-------|---------------------|---------------------|---------------------|---------------------|-----------------------|-----------------------|
| $\alpha$ | 1.3466           | 1.1873              | 0.9800              | 0.8252              | 0.8000                | 0.8000                |
| $\beta$  | 7.7950           | 48.1500             | 21.8550             | 0.4200              | $-0.4400$             | 0.6150                |
| $v$      | $1.95 \times 10^8$ | $7.08 \times 10^8$ | $3.54 \times 10^9$ | $9.80 \times 10^9$ | $5.06 \times 10^{10}$ | $3.92 \times 10^{11}$ |
| $p_1$    | 0.05037          | 0.00827             | 0.00239             | 0.00238             | 0.00238               | 0.00067               |

where $\lambda_r = f(r, c, n)$. For $k = 1$ it estimates the number of singletons (Fig. 2b), $1 \leq n \leq 6$. The number of singletons increases with the *corpus* size, as new ones keep appearing until a maximum, and vanishes gradually due to the vocabulary finiteness. Singletons keep a significant proportion of the distinct $n$-grams for a wide range: e.g., proportions fall below 80% only for *corpora* around 8 Mw, 1 Gw, 4 Gw, 16 Gw, 131 Gw, respectively, for 2-grams, 3-grams, 4-grams, 5-grams, 6-grams. Singleton 1-gram proportion is above 55% for *corpora* up to 16 Gw.

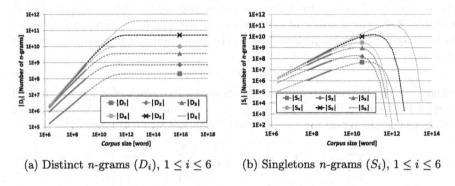

(a) Distinct $n$-grams $(D_i)$, $1 \leq i \leq 6$    (b) Singletons $n$-grams $(S_i)$, $1 \leq i \leq 6$

**Fig. 2.** Estimates (dotted) and empirical data (filled)

## 4    $n$-gram Cache System

Caching has been widely studied: in linguistics [20], Zipf distributions [21], Web search [22], or mining [23]. An $n$-gram cache is useful to statistical methods. In [9] a dynamic on-demand $n$-gram cache exploits repetitions in texts, reducing access overheads to a remote store in LocalMaxs $2^{nd}$ phase. Overheads were further reduced [9]: by an $n$-gram cache warm-up using combined metric calculations; and by using more machines, thus reducing the per machine number of $n$-gram misses (albeit non linearly) and the miss time penalty. That reduction is still not enough. Thus, we discuss two new improvements, validated experimentally for English *corpora* up to 1 Gw. Firstly (Sect. 4.2), we filter the large proportions of singletons in the *corpus*, out of the $n$-gram cache, using Bloom filters [10]. In [24], alternatives for Bloom filters, caching and disk/in-memory storage were evaluated but focused on performance and scalability in text mining. Distinctively we developed an *n*-gram cache for $n \geq 1$ and analyzed Bloom filters efficiency depending on the numbers of singletons, from small *corpus* sizes up to infinity. Secondly (Sect. 4.3), using static prefetching we achieved a 0% *n*-gram cache miss ratio.

**An *n*-gram Cache in LocalMaxs $2^{nd}$ Phase.** For each glue calculation, references to sub *n*-grams are generated, which are submitted as cache input references to check if they are already in the local *n*-gram cache, otherwise they must first be fetched from the global *n*-gram repository. The set of references, $all_{glue_{g_n} Ref}(j)$, contains all sub *n*-gram occurrences for glue calculation $(g_n)$ of the distinct *n*-grams of size $n$ in table $D_n(j)$ in machine $j$. The set of distinct sub *n*-grams (sizes 1 to $(n-1)$), found within $all_{glue_{g_n} Ref}(j)$, is $D_{all_{1 \cdots (n-1)}}(j) = D_{1_{in} D_2 \cdots D_n}(j) \cup D_{2_{in} D_3 \cdots D_n}(j) \cup \cdots \cup D_{(n-1)_{in} D_n}(j)$. Each set $D_{i_{in} D_n}, 1 \leq i \leq (n-1)$, contains the distinct sub *n*-grams of size $i$, occurring within the *n*-grams in $D_n$ table (Eq. (1)). For a single machine, $D_{i_{in} D_n}$ is $D_i, 1 \leq i \leq (n-1)$, the set of distinct *n*-grams of size $i$ in the *corpus*. For multiple machines, each one handles the distinct sub *n*-gram references in its local tables ($D_2(j), D_3(j)$, etc.).

## 4.1 Dynamic On-Demand n-gram Cache

A dynamic on-demand *n*-gram cache, assumed unbound, is able to contain all the distinct sub *n*-grams of each local *n*-gram table. We analyzed the cold-start (first occurrence) misses behavior, for an initially empty cache. If there is not enough memory, the cache capacity misses are also handled. To reduce the cold misses overhead we built an *n*-gram cache warm-up [9] using combined glues: whereas the single glue calculation for 2-grams $(g_2)$ only requires access to the 1-gram cache, for the combined glues of 2-grams up to 6-grams $(g_{2 \cdots 6})$ the 1-gram cache $(C_1)$ is reused five times, the 2-gram cache $(C_2)$ is reused four times, and so on for caches $C_3, C_4, C_5$. In a single machine, the global miss ratio $(mr)$ of an unbound cache system with subcaches $C_1$ to $C_5$ used for glue $g_{2 \cdots 6}$, is:

$$mr = \frac{D_{all_{1 \cdots 5}}}{all_{glue_{g_2 \cdots g_6} Ref}} = \frac{\sum_{i=1}^{5} |D_i|}{\sum_{i=2}^{6} 2 \times (i-1) \times |D_i|} \tag{7}$$

The miss ratio decreases with the *corpus* size and increases with the glue calculation complexity (*n*-gram size). Using the theoretical model for a single machine, we predicted the evolution of the miss ratio of the dynamic on-demand *n*-gram cache (Fig. 3a) for glue $g_{2 \cdots 6}$: it varies from around 11%, for *corpus* size close to 10 Mw, to an asymptotic limit of around 1.5% in the *plateaux* (beyond 1 Tw). Results were experimentally validated for English *corpora* up to 1 Gw.

**Effect of Multiple Machines ($K > 1$).** Due to a multiplication effect of the nonsingletons (mostly the common ones e.g. "the", "and") cited by multiple distinct *n*-grams spread across multiple machines [9], the number of distinct sub *n*-grams for glue calculation in each machine is not reduced by a factor of $K$ wrt the number of distinct *n*-grams in the *corpus*, unlike the number of cache references per machine that is reduced as $1/K$ compared to the case of a single machine. Thus, the per machine miss ratio of a dynamic on-demand $n$-gram cache increases with $K$ for each *corpus* size. Indeed we have shown [9] that, for each $n$-gram size $n$, the miss ratio follows a power trend: $mr(K) \propto K^{b(n)}$, $0 < b(n) < 1$.

### 4.2  Bloom Filters for Singletons in an On-Demand n-gram Cache

Singletons can be filtered by Bloom filters [10], trained with the nonsingletons occurring in the *corpus*. For the majority of singletons the Bloom filter says: "definitely not in set". For all the nonsingletons and a minority of singletons it says: "possibly in set". The percentage of false positives is kept low enough by adequate Bloom filter implementation. During phase one of LocalMaxs each server (Sect. 2) generates a Bloom filter for each local $n$-gram table. At the end of phase one, after the servers have updated all the $n$-gram frequencies, the filters were trained with all the nonsingletons. In the beginning of phase two, each machine controller gets a copy of the trained Bloom filters.

*Single Machine Case ($K = 1$).* The proportion of the total number of singletons ($S_{all}$) wrt the total number of distinct $n$-grams in the *corpus* ($D_{all}$) is:

$$SF_{all} = \frac{S_{all}}{D_{all}} = \frac{\sum_{i=1}^{5} S_i}{\sum_{i=1}^{5} D_i} \tag{8}$$

also illustrating the $SF_{all}$ ratio for the case of glue $g_{2\ldots6}$. Thus:

$$\frac{mr}{mr_{BF}} = \frac{\frac{D_{all}}{all_{glueRef}}}{\frac{(D_{all}-S_{all})}{all_{glueRef}}} = \frac{1}{1 - SF_{all}} \tag{9}$$

where $mr$ and $mr_{BF}$ are, respectively, the miss ratio without and with Bloom filters. The case of glue $g_{2\ldots6}$ is illustrated in Fig. 3 where the miss ratios of the individual caches $C_1$ to $C_5$ are shown (dotted), as well as the global miss ratio of the cache system $C_{1+\ldots+5}$ (filled). The curves result from the model predictions, and were experimentally validated for *corpus* sizes up to 1 Gw.

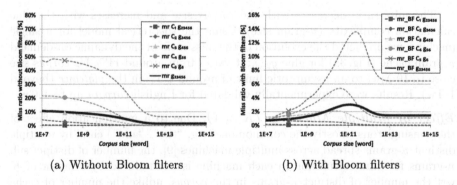

(a) Without Bloom filters          (b) With Bloom filters

**Fig. 3.** Dynamic on-demand cache $mr$ for $K = 1$, glue $g_{2\ldots6}$. Different $Y$-scales

Due to the composition of the individual miss ratios of the caches $C_1$ to $C_5$, the miss ratio with Bloom filters (Fig. 3b) for glue $g_{2...6}$ is mostly dominated by the larger populations of the larger *n*-gram sizes. It varies from about 1% for the smallest *corpus* (about 10 Mw) to 1.5% in the *plateau*. The reduction, wrt not using Bloom filters, is due to the increased filtering effectiveness in handling the large global proportion of singleton *n*-grams, $1 \leq n \leq 5$. The miss ratio has a non monotonic behavior, with a single peak of about 3% at around 100 Gw, however for *corpora* until 1 Gw it remains always below about 1.3%.

**Effect of Multiple Machines** *(K > 1)*. The per machine miss ratio is:

$$mr_{BF} = \frac{\hat{D}_{all} - \hat{S}_{all}}{\hat{all}_{glueRef}} = \frac{\hat{NS}_{all}}{\hat{all}_{glueRef}} = \frac{NS_{K=1} \times K^{-b_{NS}}}{all_{glueRef}/K} = \frac{NS_{K=1} \times K^{1-b_{NS}}}{all_{glueRef}}$$

$$(10)$$

where $\hat{D}_{all} = \left( \sum_{j=1}^{K} D_{all_{1...(n-1)}}(j) \right)/K$ is the per machine number of distinct sub *n*-grams of sizes 1 to $n-1$; $\hat{S}_{all}$ is the per machine number of singleton sub *n*-grams of sizes 1 to $n-1$. Due to their multiplicative effect with $K$ (Sect. 4.1), the number of per machine nonsingletons follows $\hat{NS}_{all} \propto K^{-b_{NS}}, 0 < b_{NS} < 1$ ($b_{NS}$ empirically determined). Thus $mr_{BF}$ increases with $K$. Table 2 shows experimental values of the miss ratios without and with Bloom filters, for a LocalMaxs implementation using $K = 16$ machines in a public cloud [25], compared to a single machine case, for *corpora* of sizes 205 Mw and 409 Mw, when calculating glue $g_{234}$.

**Table 2.** Distinct *n*-grams, singletons, cache references (numbers of *n*-grams); Cache miss ratio (%) without/with Bloom filters

| | $K = 1$ | | $K = 16$ | |
|---|---|---|---|---|
| | $|C| = 205\,\text{Mw}$ | $|C| = 409\,\text{Mw}$ | $|C| = 205\,\text{Mw}$ | $|C| = 409\,\text{Mw}$ |
| $\hat{D}_{all_{1...3}}$ | $115,803,664$ | $201,335,533$ | $22,075,227$ | $38,276,819$ |
| $\hat{S}_{all_{1...3}}$ | $92,600,190$ | $158,713,260$ | $13,874,669$ | $23,569,998$ |
| $\hat{all}_{glueRef_{g_2 \cdots g_4}}$ | $1,222,207,756$ | $2,236,879,374$ | $76,386,941$ | $139,802,828$ |
| $mr$ | $9.47\%$ | $9.00\%$ | $28.90\%$ | $27.38\%$ |
| $mr_{BF}$ | $1.90\%$ | $1.91\%$ | $10.74\%$ | $10.52\%$ |

Overall, Bloom filters lead to a reduction in the cache size and in the miss ratio, both determined by the singleton ratio ($SF_{All}$). As this ratio tends to zero the Bloom filter effect diminishes progressively.

## 4.3   n-gram Cache with Static Prefetching

Whenever one can identify the set of distinct *n*-grams in a *corpus* and their frequencies in a $1^{st}$ phase, one can anticipate their fetching into the *n*-gram cache before the execution starts in a $2^{nd}$ phase. The Fixed Frequency Accumulation

Set (FAset) $FA$ for ensuring a static hit ratio $h_S(n, FA)$ is the minimal subset of distinct $n$-grams of a given size $n$, whose cumulative sum of frequencies is a percentage of the number of occurrences of the $n$-grams of size $n$:

$$h_S(n, FA) = \frac{\sum_{ng \in FA} freq_{inCorpus}(ng)}{|Set_{All_n}|} \tag{11}$$

where $ng$ is a distinct $n$-gram within the FAset and $freq_{inCorpus}(ng)$ is its frequency in the *corpus* $C$; and $|Set_{All_n}| = |C| - (n-1)$ for $n \geq 1$. When applying this concept to an $n$-gram cache one must consider, as the denominator of Eq. (11), the set of cache input references $all_{glue}Ref_{n-gram}$ instead of the set $Set_{All_n}$. For glue $g_2$ of the 2-grams in the $D_2$ table, LocalMaxs requires access to all the subunigram occurrences (in a total of $all_{g_2}Ref_{1-gram} = 2 \times |D_2|$). The FAset to be loaded in cache $C_1$ is the subset of the elements in the set $D_{1_{in}D_2}$ (Sect. 2) whose accumulated sum of frequencies of occurrences within the 2-grams of the $D_2$ table ensures a desired static hit ratio ($h_{S_{C_1}}$) for the 1-grams cache. For a combined glue, e.g. $g_{234}$, using caches $C_1$, $C_2$ and $C_3$ (1-grams, ..., 3-grams), let $freq_{in\ all_{g_{234}}Ref_{i-gram}}(ng)$ ($1 \leq i \leq 3$) be the frequency of a distinct $n$-gram $ng$ occurring in the set of cache input references $all_{g_{234}Ref_{i-gram}}$. To ensure a target static hit ratio $h_S$ (or miss ratio $mr_S$) the FAset must enforce the following proportion of hits ($nbrHits$) wrt the total number of cache references ($|all_{g_{234}Ref}| = \sum_{i=1}^{3} all_{g_{234}Ref_{i-gram}}$, for glue $g_{234}$):

$$h_S = \frac{nbrHits}{|all_{glueRef}|} = \frac{\sum_{i=1}^{3} \sum_{ng \in FAset} freq_{in\ all_{g_{234}}Ref_{i-gram}}(ng)}{|all_{g_{234}Ref}|} = 1 - mr_S \tag{12}$$

Options for selecting the distinct $n$-grams for the FAset are: (i) All the distinct $n$-grams; (ii) Only the nonsingletons; (iii) A subset of the distinct $n$-grams. Option (i) seems the best but there is no need to include the singletons, which suggests option (ii). If there is not enough memory for all the nonsingletons in the cache, option (iii) must be taken ensuring the maximum number of hits per $n$-gram, under the existing memory constraints [17]. The LocalMaxs workflow (Fig. 1a) allows to completely calculate the FAsets for each $n$-gram size in phase one, overlapped with the $n$-gram counting, using dedicated threads. As the machine allocation to LocalMaxs tasks in all phases is made before execution starts, one can also prefetch the FAsets into the corresponding machines in phase one. Thus the FAset calculation and prefetching times are hidden from the total execution time, as far as the additional thread overheads are kept small. In option (ii), by prefetching all distinct nonsingletons completely in phase one, the nonsingleton miss overheads in phase two are eliminated, leading to a 0% overall miss ratio.

*Multiple Machines Case ($K > 1$).* The FAset size per machine decreases with $K$ as the number of distinct sub $n$-grams per machine [17]. But, unlike the dynamic on-demand cache, the miss ratio with static prefetching can be kept constant wrt $K$ by adjusting the per machine FAset according to the number of machines, e.g., for a 0% miss ratio, all the nonsingletons in the per machine distinct *n*-gram tables must be always included in the local FAset.

***Experimental Results.*** We compared the communication and glue calculation times of static prefetching of all nonsingletons *versus* on-demand caching. In each machine $(j)$, phase two takes a total time $T_2(j)$ consisting of time components for: input $T_{input}(j)$; local glue calculation $T_{Glue}(j)$; sub *n*-gram fetch $T_{comm}(j)$; glue output $T_{output}(j)$. The input/output consists of local machine interactions between the co-located server and controller (Fig. 1b), being the same in both cache cases. Table 3 shows the communication and glue times of $g_{234}$ (machine average), for two *corpus* sizes, in LocalMaxs phase two [9,17] in a public cloud [25] with 16 machines (each 64 GB RAM, 4 vCPU@1.5 GHz).

**Table 3.** Glue and communication times $(min{:}sec)K = 16$: Dynamic *vs.* Static cache

| | $\hat{T}_{comm} + \hat{T}_{Glue}$ (Dynamic \| Static) | $\hat{T}_{RemoteFetch}$ (Dynamic \| Static) |
|---|---|---|
| $\|C\| = 205\,Mw$ | 07:20 \| 01:41 | 05:30 \| — |
| $\|C\| = 409\,Mw$ | 14:21 \| 03:27 | 11:00 \| — |

$\hat{T}_{comm}$ includes the per machine times for *n*-gram cache fetch: local access and remote $(\hat{T}_{RemoteFecth})$. $\hat{T}_{Glue}$ is the local per machine glue calculation time. For the static prefetching cache $\hat{T}_{RemoteFecth}$ is zero. The cache static prefetching time of the nonsingletons is accounted for in phase one, overlapped with counting.

## 4.4   Cache Alternatives

For glue $g_{2...6}$ and three *corpus* sizes, Table 4 shows the cache miss ratio and size, and the efficiency of the glue calculation (values shown as triples $\{(mr); (Size); (E)\}$) for a single machine. This efficiency (with $K = 1$) reflects the ratio of the communication overheads suffered by a single real machine *versus* an ideal machine, i.e., $E = T_0/T_1$. Miss ratio and size are analyzed first, followed by the efficiency.

***Cache Miss Ratio and Size.*** These values result from the model predictions of the numbers of distinct *n*-grams and singletons (Sect. 3). The values for the 8 Mw and 1 Gw *corpora* agree with the empirical data from real English *corpora* [6]. The first line shows miss ratio and size expressions. Remaining lines show: (i) For the on-demand cache, its miss ratio (cache system $C_1, ..., C_5$), from 11.06% in the 8 Mw *corpus* to 2.11% in the 1 Tw *corpus*, and its size (the number of distinct *n*-grams) – Sect. 4.1; (ii) For the dynamic cache with Bloom filter, its miss ratio, from 0.76% in the 8 Mw *corpus* to 2.08% in the 1 Tw *corpus* (where the singletons have practically disappeared, Fig. 2b), and its size (the number of nonsingletons) – Sect. 4.2; (iii) For the static prefetching case of the FAset filled with all the nonsingletons – Sect. 4.3, the miss ratios of 43.3%, 27.7% and 0.04%, respectively, for the 8 Mw, 1 Gw and 1 Tw *corpora*, are due to the singleton misses, not involving any fetching overhead, leading to a miss ratio of 0%.

**Table 4.** Cache alternatives ($K = 1$, $g_{2...6}$) — Miss ratio, Cache size (number of $n$-grams in units of $M = 10^6$ or $G = 10^9$), Efficiency

| Corpus size | Dynamic $(mr\ \%)$; (Size); (E %) | Dynamic with Bloom filter $(mr\ \%)$; (Size); (E %) | Static $(mr\ \%)$; (Size); (E %) |
|---|---|---|---|
| Generic | $\left(\frac{D_{All}}{all_g}\right)$; $(D_{All})$; $(E_D)$ | $\left(\frac{NS_{All}}{all_g}\right)$; $(NS_{All})$; $(E_{BF})$ | $(mr_S)$; $(|FA_{set}|)$; $(E_S)$ |
| Small (8 Mw) | (11.06); (23$M$); (6) | (0.76); (1.6$M$); (49) | (0*); (1.6$M$); (100) |
| Large (1 Gw) | (9.02); (1.8$G$); (7) | (1.32); (0.27$G$); (34) | (0**); (0.27$G$); (100) |
| Very large (1 Tw) | (2.11); (65$G$); (20) | (2.08); (64$G$); (20) | (0***); (64$G$); (100) |

$D_{All} = |D_{All_{in}\,C}|$; $S_{All} \equiv |S_{All_{in}\,C}|$; $all_g \equiv |all_{g_{2...6}\,Ref}|$

$|FA_{set}| = |D_{All_{in}\,C}| - |S_{All_{in}\,C}| = |NS_{All_{in}\,C}| \implies$ $(43.3\% \to 0^*)$, $(27.7\% \to 0^{**})$, $(0.04\% \to 0^{***})$

***Efficiency.*** The glue efficiency $E$, for $K \geq 1$ wrt the glue computation in an ideal (no overheads) sequential machine ($T_0 = \left(\sum_{i=2}^{6} |D_i|\right) \times t_{glue}$, for $g_{2...6}$), is:

$$E = \frac{T_0}{K \times \hat{T}} = \frac{T_0}{K \times \left(\frac{T_0}{K} + \hat{T}_{comm}\right)} = \frac{1}{1 + \frac{1}{G}} = \frac{1}{1 + \frac{all_{g_{2...6}\,Ref}}{\sum_{i=2}^{6} |D_i|} \times \frac{t_{fetch}}{t_{glue}} \times mr}$$

$$\tag{13}$$

where $t_{glue}$ is the per $n$-gram local glue time; $\hat{T}$ is the per machine execution time; $\hat{T}_{comm} = (all_{g_{2...6}\,Ref}/K) \times t_{fetch} \times mr$ is the per machine $n$-gram misses communication time; $t_{fetch}$ is the per $n$-gram remote fetch time; and $G = (T_0/K)/\hat{T}_{comm}$ is the computation-to-communication granularity ratio, which includes: (i) the algorithm-dependent term $f_a = \left(\sum_{i=2}^{6} |D_i|\right)/all_{g_{2...6}\,Ref}$, i.e. the number of glue operations per memory reference, being approximately constant with the *corpus* size, for each glue, e.g., around 0.10 for $g_{2...6}$; (ii) the measured implementation-dependent ratio $f_i = \frac{t_{fetch}}{t_{glue}} \approx 20$, staying almost constant wrt the *corpus* size and number of machines used ($1 \leftrightarrow 48$); (iii) and $1/mr$. Thus, in LocalMaxs $G = \frac{f_a \times f_i}{mr} \approx \frac{0.10/20}{mr} = \frac{0.005}{mr}$. For example, $E \geq 90\% \implies G \geq 10 \implies mr \leq 0.05\%$, which can only be achieved by a static prefetching cache (Sect. 4.3). Indeed Table 4 shows that for the on-demand cache the efficiency values are very low, even with Bloom filters where $E \leq 50\%$ always. In general, other methods, exhibiting higher values of the algorithm-dependent term $f_a$, will require less demanding (i.e., higher) miss ratio values: e.g., if the $f_a$ term is around 100, then $mr = 50\%$ would be sufficient to ensure $E = 90\%$.

## 5 Conclusions and Future Work

We found out that for the statistical extraction method LocalMaxs the miss ratio of a dynamic on-demand $n$-gram cache is lower bounded by the proportion of distinct $n$-grams in a *corpus*. The proportion of distinct $n$-grams wrt the total number of cache references, i.e. the miss ratio, decreases monotonically with the *corpus* size tending to an asymptotic *plateau*, e.g., ranging from 11%

(for the smaller *corpora*) to 1.5% (in the *plateaux* region) for English *corpora* when considering a single machine. However, these miss ratio values imply very low efficiency of the glue calculation wrt an ideal sequential machine, from 6% to 26%. This is due to the significant amount of cold-start *n*-gram misses needed to fill up the *n*-gram cache with the frequencies of all distinct *n*-grams. To overcome these limitations we have shown that Bloom filters or static prefetching can significantly improve on the cache miss ratio and size. Bloom filters benefits were found related to the distribution of the singletons along the entire *corpus* size spectrum. By extending a previously proposed theoretical model [5], we found out that the number of singletons first increases with the *corpus* size until a maximum and then it decreases gradually, tending to zero as the singletons disappear for very large *corpora*, e.g., in the Tw region for the English *corpora*. This behavior of the singletons determines the effectiveness of the Bloom filters which achieve a reduction of the miss ratio, namely, to the range from 1% (for the smaller *corpora*) to 1.5% (in the *plateaux* region) for English *corpora* for a single machine. However, the corresponding efficiency is still low, always below 49%. Hence, using an *n*-gram cache with static prefetching of the nonsingletons is of utmost importance for higher efficiency. We have shown that, in a multiphase method like LocalMaxs where it is possible during a $1^{st}$ phase (in overlap with the *n*-gram frequency counting), to anticipate and prefetch the set of *n*-grams needed, then one can ensure a 0% miss ratio in a $2^{nd}$ phase for glue calculation, leading to 100% efficiency. For a static prefetching cache (sec. 4.3), it is possible, by design, to keep a constant miss ratio, leading to a constant efficiency wrt the number of machines. The above improvements were implemented within the LocalMaxs parallel and distributed architecture (Sect. 2), experimentally validated for *corpora* up to 1 Gw. Although this study was conducted in the context of LocalMaxs, the main achievements apply to other statistical multiphase methods accessing large scale *n*-gram statistical data, thus potentially benefiting from an *n*-gram cache. For *corpora* beyond 1 Gw we conjecture that the global behavior of the *n*-gram distribution, as predicted, remains essentially valid, as the model relies on the plausible hypothesis of a finite *n*-gram vocabulary for each language and *n*-gram size, at each temporal epoch. We will proceed with this experimentation for *corpora* beyond 1 Gw, although fully uncut huge Tw ($10^{12}$ words) *corpora* are not easily available yet [26].

# References

1. Google Ngram Viewer. https://books.google.com/ngrams
2. Lin, D., et al.: New tools for web-scale n-grams. In: LREC (2010)
3. da Silva, J.F., Dias, G., Guilloré, S., Pereira Lopes, J.G.: Using *LocalMaxs* algorithm for the extraction of contiguous and non-contiguous multiword lexical units. In: Barahona, P., Alferes, J.J. (eds.) EPIA 1999. LNCS (LNAI), vol. 1695, pp. 113–132. Springer, Heidelberg (1999). https://doi.org/10.1007/3-540-48159-1_9
4. da Silva, J.F., et al.: A local maxima method and a fair dispersion normalization for extracting multiword units. In: Proceedings of the 6th Meeting on the Mathematics of Language, pp. 369–381 (1999)

5. da Silva, J.F., et al.: A theoretical model for n-gram distribution in big data corpora. In: IEEE International Conference on Big Data, pp. 134–141 (2016)
6. Parallel LocalMaxs. http://cjsg.ddns.net/~cajo/phd/
7. Arroyuelo, D., et al.: Distributed text search using suffix arrays. Parallel Comput. **40**(9), 471–495 (2014)
8. Goncalves, C., et al.: A parallel algorithm for statistical multiword term extraction from very large corpora. In: IEEE 17th International Conference on High Performance Computing and Communications, pp. 219–224 (2015)
9. Goncalves, C., et al.: An n-gram cache for large-scale parallel extraction of multiword relevant expressions with LocalMaxs. In: IEEE 12th International Conference on e-Science, pp. 120–129. IEEE Computer Society (2016)
10. Bloom, B.H.: Space/Time trade-offs in hash coding with allowable errors. Commun. ACM **13**(7), 422–426 (1970)
11. Daille, B.: Study and implementation of combined techniques for automatic extraction of terminology. In: The Balancing Act: Combining Symbolic and Statistical Approaches to Language. MIT Press (1996)
12. Velardi, P., et al.: Mining the web to create specialized glossaries. IEEE Intell. Syst. **23**(5), 18–25 (2008)
13. Pearce, D.: A comparative evaluation of collocation extraction techniques. In: 3rd International Conference on Language Resources and Evaluation (2002)
14. Church, K.W., et al.: Word association norms, mutual information, and lexicography. Comput. Linguist. **16**, 22–29 (1990)
15. Dunning, T.: Accurate methods for the statistics of surprise and coincidence. Comput. Linguist. **19**, 61–74 (1993)
16. Church, K.W., et al.: Concordance for parallel texts. In: 7th Annual Conference for the new OED and Text Research, pp. 40–62 (1991)
17. Goncalves, C.: Parallel and distributed statistical-based extraction of relevant multiwords from large corpora. Ph.D. dissertation, FCT/UNL (2017)
18. Zipf, G.K.: The Psychobiology of Language: An Introduction to Dynamic Philology. MIT Press, Cambridge (1935)
19. Mandelbrot, B.B.: On the theory of word frequencies and on related Markovian models of discourse. In: Structures of Language and its Mathematical Aspects, vol. 12, pp. 134–141. American Mathematical Society (1961)
20. Kuhn, R.: Speech recognition and the frequency of recently used words: a modified Markov model for natural language. In: Proceedings of the 12th Conference on Computational Linguistics, COLING 1988, vol. 1, pp. 348–350. ACM (1988)
21. Breslau, L., et al.: Web caching and Zipf-like distributions: evidence and implications. In: Eighteenth Annual Joint Conference of the IEEE Computer and Communications Societies, INFOCOM 1999, vol. 1, pp. 126–134, March 1999
22. Baeza-Yates, R., et al.: The impact of caching on search engines. In: Proceedings of the 30th Annual International ACM SIGIR Conference on Research and Development in Information Retrieval, SIGIR 2007, pp. 183–190. ACM (2007)
23. Yang, Q., et al.: Web-log mining for predictive web caching. IEEE Trans. Knowl. Data Eng. **15**(4), 1050–1053 (2003)
24. Balkir, A.S., et al.: A distributed look-up architecture for text mining applications using MapReduce. In: International Conference for High Performance Computing, Networking, Storage and Analysis (SC), pp. 1–11 (2011)
25. Luna Cloud. http://www.lunacloud.com
26. Brants, T., et al.: Large language models in machine translation. In: Proceedings of the Joint Conference on Empirical Methods in Natural Language Processing and Computational Natural Language Learning, pp. 858–867 (2007)

# Lung Nodule Diagnosis via Deep Learning and Swarm Intelligence

Cesar Affonso de Pinho Pinheiro[1], Nadia Nedjah[1(✉)],
and Luiza de Macedo Mourelle[2]

[1] Department of Electronics Engineering and Telecommunications,
Faculty of Engineering, State University of Rio de Janeiro, Rio de Janeiro, Brazil
cesaraffonso05@hotmail.com, nadia@eng.uerj.br
[2] Department of Systems Engineering and Computation, Faculty of Engineering,
State University of Rio de Janeiro, Rio de Janeiro, Brazil
ldmm@eng.uerj.br

**Abstract.** In general, it is difficult to perform cancer diagnosis. In particular, pulmonary cancer is one of the most aggressive type of cancer and hard to be detected. When properly identified in its early stages, the chances of survival of the patient increase significantly. However, this detection is a hard problem, since it depends only on visual inspection of tomography images. Computer aided diagnosis methods can improve a great deal the detection and, thus, increasing the surviving rates. In this paper, we exploit computational intelligence techniques, such as deep learning, convolutional neural networks and swarm intelligence, in order to propose an efficient approach that allows identifying carcinogenic nodules in digital tomography scans. We use 7 different swarm intelligence techniques to approach the learning stage of a convolutional deep learning network. We are able to identify and classify cancerous pulmonary nodules successfully in the tomography scans of the Lung Image Database Consortium and Image Database Resource Initiative (LIDC-IDRI). The proposed approach, which consists of training Convolutional Neural Networks using swarm intelligence techniques, proved to be more efficient than the classic training with Back-propagation and Gradient Descent. It improves the average accuracy from 93% to 94%, precision from 92% to 94%, sensitivity from 91% to 93% and specificity from 97% to 98%.

**Keywords:** Deep learning · Swarm intelligence ·
Lung cancer detection · Convolutional neural networks

## 1 Introduction

This work is proposed to contribute to the community of researchers who attempt to improve the identification techniques of this evil that carries such expressive numbers and ends a lot of victim's life every day. Methods like the one designed in this work are very useful to medical doctors, because they are able, in a short

© Springer Nature Switzerland AG 2019
J. M. F. Rodrigues et al. (Eds.): ICCS 2019, LNCS 11537, pp. 89–101, 2019.
https://doi.org/10.1007/978-3-030-22741-8_7

period of time, to save human lives with greater efficiency. Therefore, computer aided diagnosis tools have turned out to be increasingly essential for medical's routines.

This work aims to explain its contribution by first presenting in Sect. 2 all of the main works that were used as knowledge basis and inspiration. Then, in Sects. 3 and 4 it gives through the conceptual background required for its development. It presents an introduction to the main concepts, such as Deep Learning and Swarm Intelligence. Later, in Sect. 5 we describe the proposed model of the nodules detection problem. Subsequently, in Sect. 6 we explain all the methodologies we used in this work. This includes the image processing, image classification and network training. Finally, Sect. 7 presents the results obtained in this work and discusses the relations between different swarm-training algorithms so that the final conclusion can be presented in Sect. 8.

## 2    Related Work

In [1], a Computer-Aided Diagnosis tool (CAD), which is based on convolutional neural network (CNN) and deep learning, is described to identify and classify lung nodules. Transfer Learning is used in this work. It consists of using a previously trained CNN model and it only adjusts some of the network last layer's parameters in order to make them fit the scope of the application. The authors used the ResNet network, designed by Microsoft. Among 1971 competitors, this model was ranked 41st in the competition organized by Microsoft. It thus confirmed that the method is really effective for this kind of application.

In [2] the authors used the same method as [1]. The difference between those works is the chosen networks for the use of Transfer Learning. In this work, the U-Net convolutional network was used, which significantly increased the performance of the nodule detection problem. The U-Net CNN specializes in pattern recognition tomography scans. The results reported in this work show a better performance than those shown in [1], allowing a higher accuracy, sensitivity and specificity.

In [3], the Particle Swarm Optimization algorithm was used by the authors to develop and train a CAD system based on an artificial neural network, which is able to identify and classify carcinogenic nodules in mammography digital images. The authors had to design a model dedicated to the extraction of features using image processing techniques, because of the lack of data. Once the attributes were extracted, the classification was performed according to two different approaches: the first one uses a neural network trained using backpropagation and the second one exploits a neural network trained using Particle Swarm Optimization (PSO). The former approach reached 88.6% of accuracy, 72.7% of sensitivity and 93.6% of specificity. The latter achieved 95.6% of accuracy, 87.2% of sensitivity and 97.3% of specificity. Thus, the results reported in this work justify the exploitation of swarm intelligence based techniques to discover the weights of the neural network, are able to enhance the classification performance in diagnosis tool.

# 3   Deep Learning with Convolutional Neural Networks

Convolutional neural networks are deep learning models that are structured as input layer, which receives the images; several layers, that execute image processing operation, aiming at feature mapping; and a classification neural network, which uses the obtained feature mapping and returns the classification result. The mapping layers implement three main operations, which are convolution, that is basically a matrix scan to detect feature similarities; pooling, that is an operation applied to extract the most relevant information of each feature map; and the activation layers, that are mostly used to reduce the linearity of these networks and increase its performance.

Besides the feature layers, CNNs include a fully connected layer. Based on the received feature maps, this classical multilayer perceptron analyzes the input and outputs the final classification result.

When Transfer Learning is used, a pre-trained CNN is selected and used to identify the application patterns after a new training. At this stage, the multilayer perceptron network must be retrained to allow the correct classification of the application data. The chosen network for this work was the U-Net [4], as it is specialized in medical imaging pattern recognition.

# 4   Swarm Intelligence Algorithms

Swarm intelligence denotes a set of computational intelligence algorithms that simulate the behavior of colonies of animals to solve optimization problems. In order to perform neural networks training via swarm intelligence based algorithms, the following steps are implemented: (i) define the number of coefficients and that of the network layers; (ii) characterize the search space of the weights values; (iii) define a cost function to be used to qualify the solutions; (iv) select a swarm intelligence algorithm to be applied. Subsequently, a swarm of particles is initialized, wherein each particle is represented by the set of the weights required by the classification network. Afterwards, the quality of each particle of the swarm is computed and moved in the search space, depending on the rules of the selected optimization algorithm. This process is iterated until a good enough solution for the application is found. Algorithm 1 describes the thus explained steps.

In this application, 7 swarm intelligence algorithms were used. They are the Particle Swarm Optimization (PSO) [5], Artificial Bee Algorithm (ABC) [6], Bacterial Foraging Optimization (BFO) [7], Firefly Algorithm (FA) [8], Firework Optimization Algorithm (FOA) [9], Harmony Search Optimization (HSO) [10] and Gravitational Search Algorithm (GSO) [11].

# 5   Proposed Methodology

Given the swarm algorithms, the convolutional neural network is trained using each of the 7 methods and also using the traditional method with Back-Propagation and the Gradient Descent. Once the experiments have been executed, the performance of every selected algorithm is investigated regarding

**Algorithm 1.** Neural network training algorithm based on swarm intelligence approach

---

$e = \#$epochs;
$n = \#$particles;
**Initialize** the particle positions of the first swarm;
**for** $i := 1$ to $e$ **do**
    **for** $j = 1$ to $n$ **do**
        **Select** the $j$-th particle to be used as the set of weights for the classification network;
        **Evaluate** the idealized cost function of that particle;
    **end for**
    **Apply** the swarm rules to obtain a new swarm;
**end for**
**Select** the best particle as the final set of weights for the network;

---

accuracy, sensitivity, precision and specificity. During the experimentation phase, some of the network parameters were also tested to evaluate the impact and possibly result in performance improvements. Among the parameters chosen for these tests are the size of the Pooling matrix and the network's activation function. After obtaining the results, a ordered Table was generated to choose the best algorithms for this application, according to each performance indicator.

# 6   Implementation Issues

In this work, the medical-image database LIDC-IDRI [12] was used to train the nodule classification model. It contains over 250,000 images of CT scans of lungs. All the images in this database were analyzed by up to four medics, which gave their diagnosis on each case.

The LUNA16 database [13], is a lung CT-image database which was originated from the LIDC-IRDI. This set was created by selecting only the images of the LIDC-IRDI that contains complete information. This database is also used in this work to provide a richer information.

## 6.1   Data Preprocessing

Training the network with all points in each image would highly increase the computational effort. To cope with this problem, square cuts of $50 \times 50$ pixels were selected around each nodule pointed by the specialists. Figure 1 shows an example of a square cut.

With this configuration, these images are ready to feed the model. However, the distribution of classes is still heavily unbalanced for the training. The current dataset has about 550,000 annotations, and about 1300 were classified as nodules. To prevent training problems, negative classifications are randomly reduced and data augmentation is applied on positive classifications. Data augmentation routines include rotating the images in $90°$, horizontal and vertical inverting.

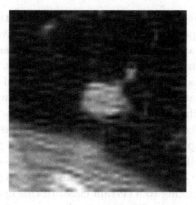

**Fig. 1.** Nodule cropped image

Once these routines are properly applied, the dataset is left as 80% negative and 20% positive regarding image labels, which still may not be the perfect balance, but over-increasing the positive classifications could lead into variance problems. After this class balancing, data preprocessing is thus completed.

### 6.2 The Model

The U-Net [4] convolutional neural network is the base of the model developed in this work. The architecture of the network is shown in Fig. 2.

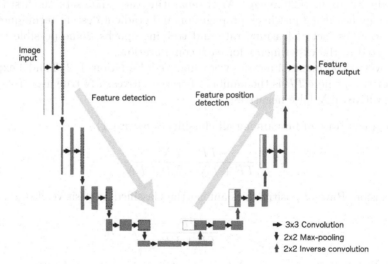

**Fig. 2.** U-Net architecture

The U-Net is composed of two stages. The first stage is focused on finding-out what features are present in the image. The second stage works on stating where these features are located on the image. This network uses $3 \times 3$ filters to apply convolution operations. It also uses ReLU activation functions, $2 \times 2$ Max-pooling matrices, and $2 \times 2$ inverse convolution filters.

To fit the U-Net [4] network to the application images, its layers were remodeled to work with the $50 \times 50 \times 1$ size images generated during preprocessing. After the mapping operations, the network outputs a feature map, which is ready to be analyzed by classification neural network. The maps are used as inputs for a fully connected neural network built with 100 units. This classification network receives the feature map and outputs a two-dimensional Softmax vector, which is a vector with probabilistic values, ranging from 0 to 1 and that, all together, sum up to 1. The value of each position in this vector represents the chance of the following input belonging to a certain class. In this approaching, the highest value will be chosen as the classification result.

To this network a Dropout layer was also added, which adds a technique used to control the over-fitting problem, an increasing error due to over-training. Dropout is used to turn some activations layers to zero regarding a probabilistic decision. This operation is used make the network redundant, as it must classify correctly even without some of its activation layers. This technique is an ally against over-fitting.

## 6.3   Training and Testing

With the model and the dataset ready to be used, training could be started. The training dataset had about 6900 classified images. The validation set was then selected with about 1620 images. After separating these data sets, the first training was applied using just back-propagation and gradient descent techniques. To choose the best set of learning rate and training epochs, some possible values were tested in 100 experiments for each configuration.

In every experiment, 4 metrics were analyzed. Equations 1, 2, 3 and 4 explain these metrics, where $TP$ is the number of true positives, $TN$ true negatives, $FP$ false positives $FN$ false negatives:

- Accuracy: Rate of hits among all classifications (Eq. 1)

$$\frac{TP + TN}{TP + TN + FP + FN} \tag{1}$$

- Precision: Rate of positive hits among the classified as positive (Eq. 2)

$$\frac{TP}{TP + FP} \tag{2}$$

- Specificity: Rate of negative hits among the real negatives (Eq. 3)

$$\frac{TN}{TN + FP} \tag{3}$$

– Sensitivity: Rate of positive hits among the real positives (Eq. 4)

$$\frac{TP}{TP + FN} \tag{4}$$

Figure 3 shows the mean results obtained on each tested configuration of hyper-parameters. The number of epochs and the learning rate were tested in pairs, ranging from 4 different values for each hyper-parameters. Table 1 shows the results obtained on these experiments.

**Table 1.** Pairs of hyper-parameters experimented

| Case | Epochs | Learning rate | Case | Epochs | Learning rate |
|------|--------|---------------|------|--------|---------------|
| $S_1$ | 30 | 0.0001 | $S_9$ | 300 | 0.0001 |
| $S_2$ | 30 | 0.001 | $S_{10}$ | 300 | 0.001 |
| $S_3$ | 30 | 0.01 | $S_{11}$ | 300 | 0.01 |
| $S_4$ | 30 | 0.1 | $S_{12}$ | 300 | 0.1 |
| $S_5$ | 100 | 0.0001 | $S_{13}$ | 1000 | 0.0001 |
| $S_6$ | 100 | 0.001 | $S_{14}$ | 1000 | 0.001 |
| $S_7$ | 100 | 0.01 | $S_{15}$ | 1000 | 0.01 |
| $S_8$ | 100 | 0.1 | $S_{16}$ | 1000 | 0.1 |

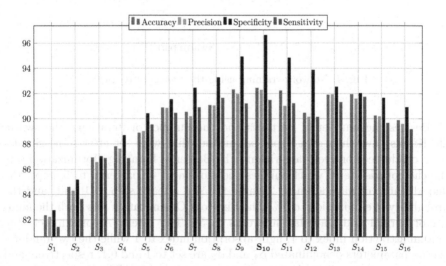

**Fig. 3.** Impact of the number of epochs and the learning rate

The model that used 300 epochs and a 0.001 learning rate presented a slightly better performance regarding accuracy and precision. This configuration also performed better in specificity. With these results, this configuration was chosen for this application. After testing the model with Back-propagation and gradient descent, it was ready to be tested with swarm intelligence.

## 6.4   Training with Swarming

The swarm algorithms used in this work are composed of 100-coordinate vector particles. Each swarm had 300 particles and was trained over 300 iterations per experiment. Figure 4 explains the training of a neural network via swarm intelligence, where each particle is a full configuration of weights for the classifier network.

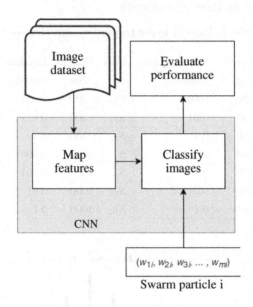

**Fig. 4.** Network training using the swarm of particles

Besides the number of particles and the number of iterations, the swarm algorithms have their own hyper-parameters that were set after some simulations conducted for every algorithm. The Bacterial Foraging Optimization sets the chemotaxis steps, which mimics the movement of bacteria, to 5 steps, the max distance of navigation to 2 units, the size of each step to 1 unit, and the probability of elimination to 5%. The Firework Optimization sets both the number of normal and Gaussian fireworks to 50. The Firefly Algorithm sets the mutual attraction index to 1, light absorption index to 1 while the two randomization parameters denominated $\alpha_1$ and $\alpha_2$ are set to 1 and 0.1, respectively and two parameters to adjust the Gaussian curve to 0 and 0.1. The Harmony Search Algorithm uses the pitch adjustment rate of 0.5, the harmony consideration rate to 0.5, and the bandwidth to 0.5. The Particle Swarm Optimization sets the inertia coefficient to 0.49, the cognitive coefficient to 0.72 and the social coefficient to 0.72. The Gravitational Search Algorithm uses an initial gravitational constant of 30 and an initial acceleration $\alpha$ of 10. The Artificial Bee Algorithm uses only two parameters: the number of particles and iterations parameters.

The swarm optimization algorithms are used to train and test the classification model over 100 experiments. In these experiments, accuracy, precision, sensitivity and specificity are used to evaluate the classification performance. To look for possible improvements, some of the U-Net parameters were also modified. These parameters are the size of the max-pooling matrices ($3 \times 3$, $4 \times 4$ and $5 \times 5$) and the activation functions. Regarding the activation functions, the hyperbolic tangent was also tested to substitute the ReLU function.

## 7   Performance Evaluation

In order to find out the best swarm strategy for training the CNNs for applications such as the one under consideration here, the performance evaluation of each algorithm was computed. The results obtained proved that swarm algorithms provide high efficiency.

Another included test is the use of the hyperbolic tangent instead of the ReLU as activation function. This testing is conducted to compare the efficiency of these functions in this application. Although the ReLU function is the most used function in convolutional neural networks, the hyperbolic tangent function was also tested with a swarm-trained model.

Figures 6, 7, 8 and 9 present the performance averages (in percentage) of each algorithm in the four metrics at the end of 100 experiments. Each figure presents all the results obtained in one metric with models tested at the following conditions: ReLU with pooling matrix sizes of $2 \times 2$ ($C_1$), $3 \times 3$ ($C_2$), $4 \times 4$ ($C_3$) and $5 \times 5$ ($C_4$), and Hyperbolic Tangent functions with pooling matrix sizes of $2 \times 2$ ($C_5$), $3 \times 3$ ($C_6$), $4 \times 4$ ($C_7$) and $5 \times 5$ ($C_8$).

After these experiments, a decrease in the performance metrics was observed as the max-pooling matrices size increased. Thus, it is possible to conclude that, for this application, the bigger the max-pooling matrix, the worse the model performs. This behavior may be caused by the loss of information in the feature map as these matrices grow bigger.

Based on these results, we can say that the TanH function performed a little worse than the ReLU. It is known that the latter decreases the linearity of data processing in CNNs, providing a better performance. Considering different configurations of the activation function and max-pooling matrix size, Fig. 5 shows the average results for accuracy, precision, sensitivity and specificity. Besides showing the better performance of the ReLU models, we can see that the smaller size of the max-pooling matrix also contributes to the enhancement of the performance.

Table 2 shows the best results obtained in each performance metric, stating which algorithm achieved that result and comparing it to the Back-propagation model. Taking these results into account, one can state the real effectiveness of using swarm intelligence algorithms to train convolutional neural networks for detecting pulmonary nodules. The experiments showed that at least 5 out of 7 swarm-trained models were superior when compared to the back-propagation

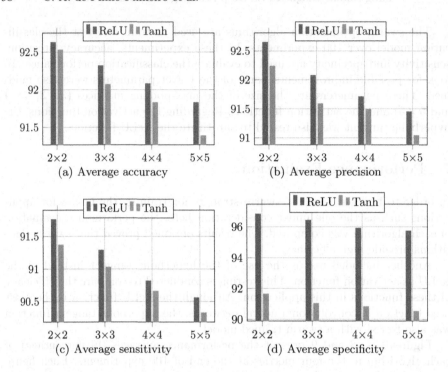

**Fig. 5.** Average performance comparison when using ReLU *vs.* TanH regarding all considering techniques

**Table 2.** Performance comparison regarding swarm *vs.* back-propagation experiments

| Metric | Back-propagation | Swarm-based training | Algorithm |
| --- | --- | --- | --- |
| Accuracy | 92.80 | 93.71 | PSO |
| Precision | 92.29 | 93.53 | GSA |
| Sensitivity | 91.48 | 92.96 | PSO |
| Specificity | 96.62 | 98.52 | HA |

models. The PSO algorithm reached the best performance in accuracy and sensitivity, the Harmony Search in specificity and the Gravitational Search in precision. When comparing to [3], the same behavior was observed, where the swarm-trained methods were vastly superior against back-propagation ones. From these results, problem's nature must be analyzed to choose the best algorithm.

Figures 6, 7, 8 and 9 demonstrate the behaviors of the top three algorithms used in this work under all tested conditions regarding accuracy, precision, sensitivity and specificity. From these behaviors, its possible to observe the tendency of decreasing performance as the pooling matrix size increases as well as a slight superiority between ReLU models over Hyperbolic Tangent cases.

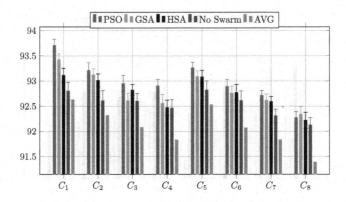

**Fig. 6.** Accuracy performance of top 3 algorithms, no swarm model and average of the results obtained by all algorithms

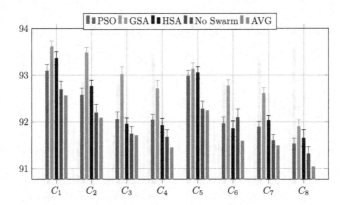

**Fig. 7.** Precision performance of the top 3 algorithms, no swarm model and average of the results obtained by all algorithms

Figures 6, 7, 8 and 9 show that the best performance for specificity is obtained by Harmony Search algorithm. Therefore, we can conclude that this algorithm is well suited for assuring that lung nodules classified as non-cancerous are in fact non-cancerous. However, when precision is concerned, the best algorithm is Gravitational Search, which means that it is suited for assuring that lung nodules classified as cancerous are in fact cancerous. Moreover, concerning accuracy and sensitivity, the best algorithm is PSO, which means that it is well suited for assuring that cancerous lung nodules are positively classified.

Classifying a cancerous patient as healthy is worse than classifying a non-cancerous patient as healthy. So, the false negative rate is the most important factor. Accuracy is the second most important, which outputs the overall model performance. Based on this observation, PSO can be considered the best algorithm for training the lung nodule classifier.

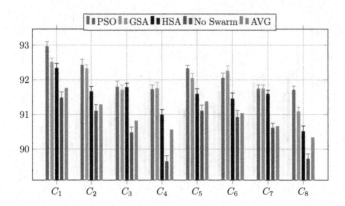

**Fig. 8.** Sensitivity performance of the top 3 algorithms, no swarm model and average of the results obtained by all algorithms

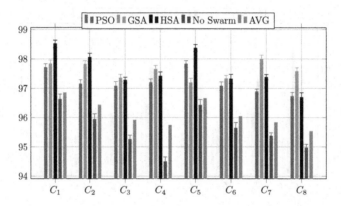

**Fig. 9.** Specificity performance of the top 3 algorithms, no swarm model and average of the results obtained by all algorithms

## 8    Conclusion

With the results obtained, it is possible to confirm the efficiency of adopted training strategy, based on the usage of swarm intelligence techniques. It achieved an improvement of the average accuracy from 92.80% to 93.71%, precision from 92.29% to 93.53%, sensitivity from 91.48% to 92.96% and specificity from 96.62% to 98.52%. During the performed simulation, we investigated the impact of the activation function. We verified the performance of the Rectified Linear Unit (ReLU) function as well as the Hyperbolic Tangent function. We also investigated the impact of different max-pooling functions on the performance of the network. We concluded that the ReLU models achieved a better performance than hyperbolic tangent based models. With respect to the max-pooling matrix size, we proved that the larger the matrix is, the worse the performance obtained.

Regarding the performance of the investigated swarm intelligence algorithms, three out of the seven exploited methods provided the best performances. PSO, HSA and GSA allowed the achievement of the best performances, regarding the four considered metrics. It is noteworthy to point out that labeling a non-cancerous nodule as cancerous is a bad decision. However, identifying a cancerous nodule as non-cancerous is worse as it affects the following treatment of the corresponding patient. Based this observation, one would elect the PSO technique as the best one simply because it achieved higher accuracy yet it provided the lowest rate of false negatives.

As a future work, we intend to apply our approach to images of other kind of tumors so as to generalize the obtained results.

# References

1. Kuan, K., et al.: Deep learning for lung cancer detection: tackling the kaggle data science bowl 2017 challenge. arXiv preprint arXiv:1705.09435 (2017)
2. Chon, A., Balachandra, N., Lu, P.: Deep convolutional neural networks for lung cancer detection. Standford University (2017)
3. Jayanthi, S., Preetha, K.: Breast cancer detection and classification using artificial neural network with particle swarm optimization. Int. J. Adv. Res. Basic Eng. Sci. Technol. 2 (2016)
4. Ronneberger, O., Fischer, P., Brox, T.: U-net: convolutional networks for biomedical image segmentation. In: Navab, N., Hornegger, J., Wells, W.M., Frangi, A.F. (eds.) MICCAI 2015. LNCS, vol. 9351, pp. 234–241. Springer, Cham (2015). https://doi.org/10.1007/978-3-319-24574-4_28
5. Kennedy, J., Eberhart, R.: Particle swarm optimization (1995)
6. Karaboga, D., Basturk, B.: A powerful and efficient algorithm for numerical function optimization: artificial bee colony (ABC) algorithm. J. Global Optim. 39(3), 459–471 (2007)
7. Passino, K.M.: Bacterial foraging optimization. Int. J. Swarm. Intell. Res. 1(1), 1–16 (2010)
8. Ritthipakdee, T., Premasathian, J.: Firefly mating algorithm for continuous optimization problems, July 2017
9. Tan, Y., Zhu, Y.: Fireworks algorithm for optimization. In: Tan, Y., Shi, Y., Tan, K.C. (eds.) ICSI 2010. LNCS, vol. 6145, pp. 355–364. Springer, Heidelberg (2010). https://doi.org/10.1007/978-3-642-13495-1_44
10. Geem, Z.W., Kim, J.H., Loganathan, G.: A new heuristic optimization algorithm: harmony search. SIMULATION 76(2), 60–68 (2001)
11. Khajehzadeh, M., Eslami, M.: Gravitational search algorithm for optimization of retaining structures. Indian J. Sci. Technol. 5(1) (2012)
12. Cancer Imaging Archive: The lung image database consortium image collection (LIDC-IDRI). https://wiki.cancerimagingarchive.net/display/Public/LIDC-IDRI. Accessed 2 Feb 2018
13. Jacobs, C., Setio, A.A.A., Traverso, A., van Ginneken, B.: Lung nodule analysis 2016 (LUNA16). https://luna16.grand-challenge.org/. Accessed 2 Feb 2018

# Marrying Graph Kernel with Deep Neural Network: A Case Study for Network Anomaly Detection

Yepeng Yao[1,2], Liya Su[1,2], Chen Zhang[1], Zhigang Lu[1,2(✉)], and Baoxu Liu[1,2]

[1] Institute of Information Engineering, Chinese Academy of Sciences, Beijing, China
{yaoyepeng,suliya,zchen,luzhigang,liubaoxu}@iie.ac.cn
[2] School of Cyber Security,
University of Chinese Academy of Sciences, Beijing, China

**Abstract.** Network anomaly detection has caused widespread concern among researchers and the industry. Existing work mainly focuses on applying machine learning techniques to detect network anomalies. The ability to exploit the potential relationships of communication patterns in network traffic has been the focus of many existing studies. Graph kernels provide a powerful means for representing complex interactions between entities, while deep neural networks break through new foundations for the reason that data representation in the hidden layer is formed by specific tasks and is thus customized for network anomaly detection. However, deep neural networks cannot learn communication patterns among network traffic directly. At the same time, deep neural networks require a large amount of training data and are computationally expensive, especially when considering the entire network flows. For these reasons, we employ a novel method *AnoNG* to marry graph kernels to deep neural networks, which exploits the relationship expressiveness among network flows and combines ability of neural networks to mine hidden layers and enhances the learning effectiveness when a limited number of training examples are available. We evaluate the proposed method on two real-world datasets which contains low-intensity network attacks and experimental results reveal that our model achieves significant improvements in accuracies over existing network anomaly detection tasks.

**Keywords:** Network anomaly detection · Deep neural network · Graph kernel · Communication graph embedding

## 1 Introduction

The Internet has become an important part of our society in many areas, such as science, economics, government, business, and personal daily lives. Further,

---

Y. Yao and L. Su—The two lead authors have contributed equally to this study.

© Springer Nature Switzerland AG 2019
J. M. F. Rodrigues et al. (Eds.): ICCS 2019, LNCS 11537, pp. 102–115, 2019.
https://doi.org/10.1007/978-3-030-22741-8_8

an increasing amount of critical infrastructure, such as the industrial control systems and power grids, is managed and controlled via the Internet. However, it is important to note that today's cyberspace is full of threats, such as distributed denial of service attack, information phishing, vulnerability scanning, and so on. Recent news also points out that attackers can build large organization consisting of different types of devices and bypass existing defenses by changing communication patterns [6].

Despite these difficulties, network anomaly detection is still making progress. Network anomaly detection systems identify network anomalies by monitoring and analyzing network traffic and searching for anomalous flows with malicious behaviors. Despite having high accuracy rates and low false alarm rates, these systems are easily evaded by recent collaborative-based anomalies, even anomalies with the similar packets as legitimate traffic. Moreover, existing detection modules usually consider less about the communication relationship, such as the communication pattern between inner and outer networks, so that it is difficult to describe anomalies from single flow itself in network.

Collective network anomalies are major kinds of network anomaly nowadays [12]. According to the communication volume of the traffic flows, individual steps can be roughly classified into two types, namely high-intensity attacks and low-intensity attacks. High-intensity attacks such as Smurf [9] transmit packets at a high rate causing a sudden surge in traffic volume whereas low-intensity attacks such as Slowloris [19] are high-level attacks that do not rely on intensity and volume to breakdown network service. Other low-intensity attacks do not attempt to completely disrupt the service but rather degrade the quality of service over a long period of time to achieve economic damage or reduce the performance of servers, indicates that all the packets involved in the attack are not detected. In complex attacks, an attacker floods the target system with low-intensity, highly targeted, and application-specific traffic.

Recent research has shown that the following challenges that are not tackled in previous researches. We briefly describe two of the modern challenges for network anomaly detection that we will address in this work.

**Challenge 1: Low-Intensity Volume of Network Anomaly Traffic Might Seem Innocuous.** Most traditional methods analyze statistical changes of traffic volume. The statistical features of the network flows contain information such as the number of flows, the number of packets of the flow, the average length of packets of the flow, the network protocol flags, and the duration of the flow. If a statistical mode mismatch in those values is detected, the event is reported as an anomaly. For example, if the statistical features of the destination port numbers suddenly increase, it is assumed that some hosts conduct port scanning. However, the anomalous activities that may not increase the number of packets are not visible in a large network. Those volume-based anomaly detection techniques, however, cannot detect low-intensity attacks.

**Challenge 2: More Advanced Evasion Techniques Can Conceal Real Attacks.** Even though some flow features have the ability to capture anomalous traffic that has statistical variations of traffic volume, they cannot capture correlations in communications between hosts because they focus on the statistics of traffic sent by individual hosts. For example, command and control communications, the major source of advanced cyber anomalies, consist of attackers that originate attack commands. The attackers make up a cluster where they communicate with each other. With traffic statistics-based detection methods, we can detect partial anomalies but cannot detect or capture the overall anomalies, especially for the concealed network attacks.

By exploiting the above two challenges, we adopt communication graph embedding method to represent the structural similarities of communication patterns by calculating the similarities between graphs generated by network flows. For the reason that network anomalies may not occur as one single action on a single host, there may be many small seemingly innocuous multi-hosts behaviors. Attackers might do several small probe actions that are temporally spaced or they might initiate a massive denial of service attack. Each change of anomalies has a specific pattern of a particular feature set.

In this paper, we investigate the benefit and efficiency of applying the graph kernel embedding approach to improve the deep neural networks. Our proposed approach aims to improve the accuracy of network anomaly detection process and analyze through an attributed communication graph by using communication graph features, which stores the representation of the status of the participating networks in a specific time interval. Our contribution in this paper is threefold:

- To understand the new network threat, we propose a novel network flow analytics architecture for low-intensity network anomaly that can evade other detection approach easily;
- To detect anomaly flows and improve the accuracy of network anomaly detection process, we develop the first work to systematically combine graph kernels with deep neural networks for network anomaly detection;
- To evaluate the proposed method, we build a working system to hold a sliding time window view of the entire network based on the structural similarity analysis of network flows. In this way, it is optimal for the application of machine learning techniques. We thoroughly evaluate our system with two real-world datasets and demonstrate that the proposed method effectively outperforms the traditional machine learning solutions and the baseline deep neural network detection solutions in terms of accuracy.

The rest of the paper is organized as follows. Backgrounds and previous work related to this paper are discussed in Sect. 2. Section 3 described the threat model and major approaches. The proposed anomaly detection framework is discussed in Sect. 4 followed by the empirical evaluation in Sect. 5. Last, Sect. 6 concluded this paper.

## 2    Background and Related Work

We first lay out the background information of network evasion attacks against anomaly detection models. Then we review related prior research.

## 2.1    Network Evasion Attack Against Anomaly Detection Models

Since attackers have become more organized, skilled and professional, research in network anomaly detection must always take into account the presence of an evasion attack. While some detection papers are providing discussion on evasion resistance, there is rarely any practical method to increase evasion resilient capabilities of the anomaly detection models. In fact, the purpose of evasion attacks is to move an anomalous flow to a benign one, or to attempt to mimic a specific benign point. Therefore, the attacker attempts to modify the features of the anomalous flow so that the anomaly detection model marks the flow as benign.

The limitation of evasion attacks is performed in two ways. First, there could be features that are not modifiable to the extent required to perform the attack. There is also a limitation on how difference the anomalous activities can be changed, because the activities still need to follow their original, anomalous purposes. For example, the target IP addresses of network flows cannot be modified arbitrarily, so the communication patterns of anomalous activities are relatively stable. The evasion attack also needs understandings of what is considered benign in the production detection model, which is usually not available.

## 2.2    Deep Neural Network Based Network Anomaly Detection

Deep neural networks have been shown tremendous performance on a variety of application areas, such as image, speech and video analysis tasks, while they are high-dimensional inputs and have high computational requirements. For network anomaly detection, deep neural networks increase the detection rate of known anomalies and reduce the false positive rate of unknown anomalies.

Alom et al. [1] train deep belief network models for identifying any kind of unknown anomalies in dataset and evaluated the performance on intrusion detection datasets. Their proposed system not only detects anomalies but also classifies them and achieves 97.5% accuracy for only fifty iterations. Alrawashdeh et al. [2] explore the anomaly detection capabilities of restricted Boltzmann machine and deep belief network. They outperform the former works in both detection speed and accuracy and achieve a detection rate of 97.9% presenting machine learning approaches for predicting anomalies with reasonable understood. As for deep belief neural network [11], it achieves accuracy as high as 99% combining neural networks with semi-supervised learning in a smaller percentage of labeled samples.

In the case of Convolutional Neural Network (CNN)-based and Long Short-Term Memory (LSTM)-based anomaly detection, Zhang et al. [20] propose a specially designed CNN to detect network anomalies. They find only some preprocessing is needed whereas the tedious feature extraction is done by the CNN itself and the experimental results show that the designed CNN model has a good performance. Chawla et al. [4] employ recurrent neural network (RNN) as their model for host-based intrusion detection systems which determine legitimate behavior based on sequences of system calls. Kim et al. [7] propose a system-call language-modeling method on host-based intrusion detection systems. Their method learns the semantic meaning and interactions of each system call which has never been considered before and the validity and effectiveness are shown on public benchmark datasets.

## 2.3  Graph Based Anomaly Detection

Several graph-based and machine learning techniques have been investigated over the last two decades for detecting network anomaly detection. Our previous work [18] proposed an effective method for deep graph feature extraction to distinguish network anomaly in network flows.

The graph kernel approach is a rapid feature extraction scheme and its run-time scales only linearly in the number of edges of the graphs and the length of the subgraph sequence, by converting a high-dimensional graph into a feature vector space. The scalar product of the two graphs in space measures their similarity. Generally, graph kernels are utilized to define the similarity function of two graphs. One effective way is the Weisfeiler-Lehman (WL) kernel [16]. However, WL kernel only support discrete features and costs linear memory of training examples. Other methods include depth graph kernel [17] and graph invariant kernel [13] comparing based on the existence or counting of the small substructures, such as the shortest path [3], graphlets, subtrees, and other graph invariants [13]. We adopt general shortest path kernel as one possible labeling procedure to compute receptive fields.

## 3  Threat Model and Our Approach

To better understand the efficiency of the method we proposed in this paper, we describe the threat model and the defense model. The threat that we study as a use case is a kind of the denial of service attack that exhausts the resources using low-intensity traffic flows described in [8], seen in Fig. 1. We consider an outside attacker that can send arbitrary network flows to a server or link that is vulnerable to network anomalies. The attacker can exploit the vulnerability by sending carefully crafted flows that will consume a significant amount of the resources. The attackers' goal is to occupy available resources by sending multiple flows in parallel at a low intensity.

In low-intensity attacks, the attacker can use controller servers to attack the targets. To achieve this goal, the attack assigns botnets controlled by his controller servers to send legitimate, low-intensity network flows towards certain target servers. Since anomalous flows are well-formed, they cannot be easily distinguished from legitimate flows through statistical features, such as the average packet size. The attacker can also send legitimate flows to hide attack trace among legitimate network flows. The attacker does not send numerous attack flows within a very short time interval, because volumetric attacks with a high intensity can be easily detected by network-based anomaly detection, and attacks with low-intensity are already sufficient to affect the network.

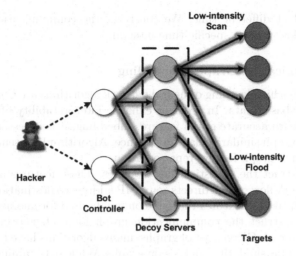

**Fig. 1.** A use case for threat model: a low-intensity version of the network anomalies.

### 3.1 Problem Definition

Given a set of network flows, we can detect several kinds of anomalous network flows, but some kinds of anomalies can be evaded by modifying several statistical flow features. In fact, similar communication patterns represents on communication graphs, for example, when the network flows come from the similar sources and go towards the similar destinations.

Communication graphs are used as abstraction of network communication patterns. Hosts are represented by vertices, while communication between them is indicated by edges, weighted by the statistical features of exchanged data. Given a directed multigraph $G < V, E >$, the purpose of the anomaly detection is to relate a set of vertices to a set of edges. An anomaly at time interval $[t, t+1]$ attacks a set of host $E_{A(t)} \in E$, affecting the connectivity of a set of vertices $V_{A(t)} \in V$. As described in [8], the goal of the anomaly detection is to find the malicious host set and detect the malicious communication flows until $t + 1$.

Assume that $p(e)$ expresses the probability of $e$ acting as a malicious host at time interval $t$, $p(v)$ is the probability of $v$ being an target at time interval $[t, t+1]$, and $p(e, v)$ is the probability of $e$ attacking $v$ at this time interval. The goal is to seek to minimize the total possibility of $p(e)$ over all time intervals. In the neural networks model, system needs to extract the information about $p(v)$ and seek to minimize $p(v)$. However, attackers are able to do little changes to evade the single time interval detection without much effort. While it is difficult to evade from the whole communication graph pattern which graph kernels have the ability to describe, that is, graph kernels can capture the relationship between $p(v)$ and $p(e)$, that is $p(e, v)$.

For estimating the similarity between communication patterns on a pair of communication graphs, we adopt graph kernel functions. It relies on graph comparison. It is clear that there is no link between these network flows because they happened at different time. We construct the communication graph for each network flow using a specific time interval.

## 3.2 Communication Graph Embedding

Most existing machine learning or neural network algorithms are generally applicable to vector-based data. In order to enhance interpretability of deep neural networks and learn accurate graph feature embedding, it is necessary to extract the features of graph similarity data in advance. Algorithm 1 presents the details of communication graph feature representation.

First, we extract the statistical flow features. These features are extracted from each network flow, which directs source IP address to destination IP address and contains statistical features that are commonly used for anomaly detection.

Then we construct the communication graph for each network flow by a specific time interval. Given a set of graphs, many algorithms have been designed and widely used for similarity analysis on graphs. While determining all paths is NP-hard, finding special subsets of paths is not necessarily. Determining longest paths in a graph is also NP-hard. To overcome this problem, the shortest path similarity measures [3] are selected in this paper, which capture diverse aspects of the network communication pattern and can be efficiently computed.

According to the sorted hosts and the length of shortest paths, the shortest path compares the similarity of two communication graphs. Using the network flows of each pair of hosts noted as $i$ and $j$ constructs the similarity metrics. The shortest path $d(i, j)$ represents the shortest distance from host $i$ to $j$. With low resolution the vertices are more likely to interact which is what our model wants to detect recorded as $p(h, v)$ (seen in Sect. 3.1).

At the same time, the shortest path also describes the communication situation of pair of hosts as toward or away from each other in the network flow graph. As discussed in [5], given the small world property of networks, we can limit the distance searched before stopping and supposing no other paths.

## 4 Graph Kernels Combine Deep Neural Networks Framework

We propose AnoNG framework by combining deep neural networks and graph kernels from two different angles. We adopt the LSTM and CNN as our baseline deep neural network models because they are effective for network anomaly detection. There are two ways to extend the kernel vectors for DNNs. We can append the kernel vectors either to input vectors, or to the softmax classifier. To determine which strategy works best, we perform a pilot study. We found that the first method causes the dimension of features turn to be too large, so we use dimensional reduction method before inputting the DNN. While for the second method, the softmax classifier tends to heavily rely on the kernel vectors, and the result is similar to the one only given kernel features, so we design a weight variable to calculate the final results. The architecture of AnoNG framework can be seen in Fig. 2.

---

**Algorithm 1.** Communication Graph Embedding

---

**Input:**

    A network communication graph set $\{\mathcal{G}_1, \mathcal{G}_2, \ldots, \mathcal{G}_N\}$ with $N$ communication graphs;

**Output:**

    A $|N| \times |N|$ kernel matrix $K_{SP}$ as similarity features with $K_{SP_N}$ assigning to $\mathcal{G}_N$ as its embedding feature for similarity measurement;

1: Computing the shortest path graphs $S < V, E >$ of each network communication graph: Where the weight of each edge in $E$ is equal to the length of the shortest path between the corresponding vertices in $\mathcal{G} < V, E >$

2: Computing the graph kernel matrices $K_{SP}$ for each pair of graphs: For each pair of graphs $\mathcal{G}_1$ and $\mathcal{G}_2$ with $S_1 < V, E >$ and $S_2 < V', E' >$

$$K_{SP}(\mathcal{G}_1, \mathcal{G}_2) = \sum_{e \in E} \sum_{e' \in E'} \kappa(e, e')$$

    where $\kappa$ is shortest path kernel defined on the edges of $S_1$ and $S_2$, which measures the similarity between local sub-structures centered at $e$ and $e'$

3: Computing the kernel matrix for the $N$ graphs: Compute the kernel matrices $K_{SP}$.

---

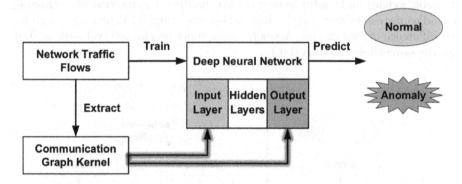

**Fig. 2.** Overview of the network anomaly detection framework using deep neural networks and graph kernels (AnoNG).

## 4.1 Integrating Graph Kernels at Input Level

At the input level, we use the evaluation outcomes of graph kernels as features fed to DNN models as Fig. 3(a). The number of graph features for network anomaly detection can be different in detailed situation. We extract network flow features according to effectiveness, and graph kernels' number can change depending on computational efficiency.

Because of the conflicting situations illustrated above, we cannot directly add or multiply the kernel vectors to flow features. To solve this issue, we extend the flow features with kernel features to form aggregated features as the input of DNNs.

## 4.2 Integrating Graph Kernels at Output Level

At the output level, graph kernels are used to improve the output of DNNs in Fig. 3(b). Since the graph kernels used in the input level only extend the statistical network flow features, we can also dig further expression by DNNs. So we combine the softmax classifier results of DNNs as the final results.

For instance, the softmax classifier of graph kernels can output the probability distribution $K_p = (K_{p1}, K_{p2})$ of two classes. DNN outputs the probability distribution $N_p = (N_{p1}, N_{p2})$ for two classes. We define the final probability of the label value for a network flow is:

$$p = a_1 K_p + a_2 N_p \tag{1}$$

where $a_1 + a_2 = 1$ and $K_p$ is the probability distribution produced by the graph kernels and $N_p$ is the probability distribution produced by the baseline deep neural network model. And $a_1$ and $a_2$ are trainable weights indicating the overall confidence for anomalous or benign labels. Here we assign 0.1 and 0.9 for original trainable weights and use 0.1 as interval. We modify 0.1 as interval when training $a_1$ and $a_2$ instead of more precise interval because using 0.1 is precise enough for the difference of two classes. Actually, using more precise interval such as 0.01 has the same effect as interval 0.1.

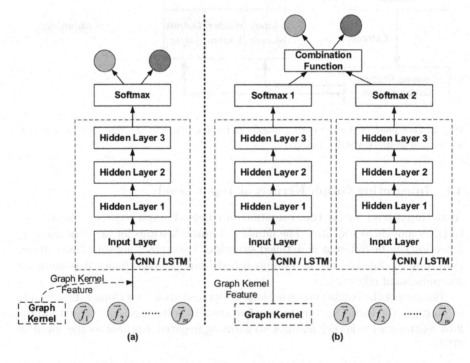

**Fig. 3.** Combination of graph kernels and deep neural networks.

# 5    Evaluations

In this section, we evaluate the proposed AnoNG framework. To evaluate the anomaly detection performance of our proposed AnoNG detection module, we adopt the UNSW-NB15 [10] dataset and the CIC-IDS-2017 [15] dataset, respectively. Both the datasets labeled the network anomaly flows, which means we have ground truth of the traffic. Then, we divide each dataset into training and test data sets using a ratio of 70% and 30%.

## 5.1    Dataset Description

*UNSW-NB15 Dataset:* This dataset is created by establishing a synthetic environment at the UNSW cybersecurity lab. The data presents a mixture of the real modern normal and the contemporary synthesized attack activities of the network traffic. It had 47 features including new low-intensity attack scenarios and 9 significant series of the cyberattack. We select only 8 statistical network flow features from this dataset, namely *dur, sbytes, dbytes, Sload, Dload, Spkts, Dpkts, Sintpkt, Dintpkt.*

*CIC-IDS-2017 Dataset:* This dataset is a newer dataset that contains a more recent form of anomalies such as high-intensity high level layer attacks generated using botnets and low-intensity attacks generated using Slowloris tool. We also select 8 statistical network flow features from this dataset, namely *Flow Duration, Total Fwd Packets, Total Backward Packets, Total Length of Fwd Packets, Total Length of Bwd Packets, Flow Bytes/s, Flow Packets/s, Average Packet Size.*

## 5.2    Experimental Environment

We implemented a proof-of-concept version of our AnoNG framework (see Sects. 3 and 4). Python3, Scikit-learn and Keras with TensorFlow backend are used as the software frameworks, which are run on the Ubuntu 16.04 64-bit OS. The server is a DELL R720 with 16 CPU cores, 128 GB of memory and NVIDIA Quadro GPU. To training the deep learning methods, we set the batch size to 128, the learning rate is set to 0.001, the epochs of training as 10, and the dropout rate is set to 0.5.

For evaluation, we report precision, recall, and F1-score, which depend on the four terms true positive (TP), true negative (TN), false positive (FP) and false negative (FN) for the anomaly detection results achieved by all methods.

We conduct three experiments: baseline DNNs, integrating graph kernels at input level of DNNs, and integrating graph kernels at output level of DNNs on two real-world datasets. We select $t$ as 60 s as the time window to construct communication graphs as it was shown as a suitable classification point. All experiments were repeated 10 times and we report the means of evaluation matrices for all algorithms. The best results are bolded. For the compared methods, both LSTM and CNN are considered as baselines. To provide a better overview of the performance of the baseline approach on the statistical network flow features, the overall precision, recall and F1-score are presented in Table 1.

We set dimensions of the CNN layers to input-128-256-128-output for both datasets. Dimensions of the LSTM layers are set to input-128-256-128-output, where the output is the statistical feature dimension, which varies between datasets. All layers are densely (fully) connected. To training the deep learning methods, we set the batch size to 128, and set the learning rate to 0.001. The epochs of training is set to 10, and the dropout rate is set to 0.5. The ADAM optimizer is adopted for variant of gradient descent.

Both the datasets are divided into two sets: training set and test set, which are the mixture of both anomalies and benign ones. Then, a deep neural network model is trained to predict on the training set and the prediction errors on the training set are fit to a multivariate Gaussian using maximum likelihood estimation. The loss function that we adopt is the sigmoid cross entropy. The threshold for discriminating between anomalous and benign values is then determined via by maximizing the accuracy value with respect to the threshold. At last, the trained model is then used to predict on the test set and the results are recorded and compared.

### 5.3    Performance on DNN with Graph Kernels at Input Level (GKIL)

Table 1 details the results of the experiments. According to our experimental results, we believe that the AnoNG framework is successful because it consistently generates highly accurate results in both real-world datasets used to identify anomaly and benign flows. From the results, we can see that AnoNG significantly outperforms baseline methods. We also compare AnoNG's accuracy on the network anomaly detection task against the random forests classifier proposed in [14,15].

It is noted that the communication graph features improve the recall increasing from 0.90105 to 0.99757 and the F1-score increasing from 0.93292 to 0.96990 of CNN model in UNSW-NB15 dataset, as well as the precision increasing from 0.95420 to 0.98847 and the F1-score increasing from 0.95719 to 0.98776 of LSTM model in CIC-IDS-2017 dataset.

### 5.4    Performance on DNN with Graph Kernels at Output Level (GKOL)

To understand the necessity of graph kernels, we further measured the performance of the GKOL algorithm. The results for the GKOL algorithm of our validation are equally promising.

After training and obtaining representations of statistical network flow features and graph kernels, for each DNN model, we select trainable weights of $a_1$ from 0.1 to 0.9. The results of precision, recall and F1-score are shown in Fig. 4, respectively, from which we can observe that GKOL is consistently above baseline methods, and achieves statistically significant improvements on all metrics.

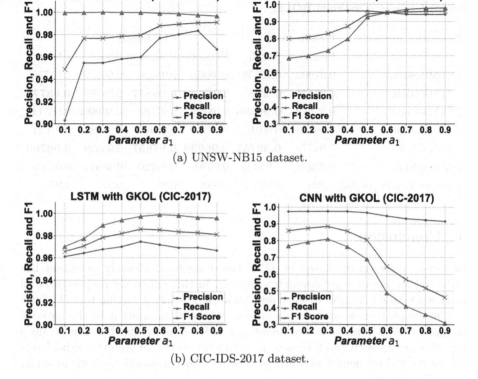

(a) UNSW-NB15 dataset.

(b) CIC-IDS-2017 dataset.

**Fig. 4.** Precision, recall and F1-score with respect to parameter $a_1$

Take $a_1 = 0.6$ as an example, LSTM with GKOL outperforms best performance in CIC-IDS-2017 dataset. Therefore, we can draw the conclusion that GKOL maintains a more decent performance in anomaly detection tasks compared with other methods.

## 5.5   Discussions

AnoNG is designed based on learning the features of network communication graphs. For one thing, the evaluation for this approach verifies its capability to effectively detect existing low-intensity attacks in large-scale networks as its profile involves communication patterns which can be extracted via constructing the communication graphs. For another, graph kernel embedding aims at find small variations on communication patterns between benign and anomalous flows. If the variances between these features are not high, the performance of this technique will decrease, the feature evaluation methods can be further adopted to select the most highly varied features by testing these features and their principal components.

**Table 1.** Precision, recall and F1-score over the datasets of different methods. The best performance is shown in **BOLD**.

| Evaluation methods | UNSW-NB15 | | | CIC-IDS-2017 | | |
|---|---|---|---|---|---|---|
| | Precision | Recall | F1-score | Precision | Recall | F1-score |
| LSTM baseline | 0.97999 | 0.99635 | 0.98810 | 0.95420 | 0.96020 | 0.95719 |
| LSTM+GKIL | 0.97576 | 0.99085 | 0.98325 | **0.98847** | 0.98705 | **0.98776** |
| LSTM+GKOL | 0.97682 | **0.99896** | 0.98776 | 0.97182 | **0.99882** | 0.98514 |
| CNN baseline | 0.96713 | 0.90105 | 0.93292 | 0.86467 | 0.96081 | 0.91021 |
| CNN+GKIL | 0.94372 | **0.99757** | **0.96990** | **0.97657** | 0.95929 | **0.96786** |
| CNN+GKOL | 0.95294 | 0.95305 | 0.95300 | 0.94720 | 0.48977 | 0.64568 |
| Random forests [14, 15] | 0.999 | 0.910 | 0.953 | 0.98 | 0.97 | 0.97 |

# 6    Conclusion

In recent years, graph-based or DNN-based network anomaly detection has attracted more and more research attention. In this paper, we propose a novel framework called AnoNG to solve new network anomaly detection tasks by integrating graph kernels with deep neural networks, then apply it to the network anomaly detection problem and investigate different methods of combining deep neural networks with graph kernels. Experiments show that AnoNG outperforms the existing deep neural network models on two real-world network anomaly detection datasets.

**Acknowledgments.** This research is supported by the National Natural Science Foundation of China (No. 61702508, No. 61802394, No. 61802404) and strategic priority research program of CAS (No. XDC02040100, No. XDC02030200, No. XDC02020200). This research is also supported by Key Laboratory of Network Assessment Technology, Chinese Academy of Sciences and Beijing Key Laboratory of Network Security and Protection Technology. We thank the anonymous reviewers for their insightful comments on the paper.

# References

1. Alom, M.Z., Bontupalli, V., Taha, T.M.: Intrusion detection using deep belief networks. In: 2015 National Aerospace and Electronics Conference (NAECON), pp. 339–344. IEEE (2015)
2. Alrawashdeh, K., Purdy, C.: Toward an online anomaly intrusion detection system based on deep learning. In: 2016 15th IEEE International Conference on Machine Learning and Applications (ICMLA), pp. 195–200. IEEE (2016)
3. Borgwardt, K.M., Kriegel, H.P.: Shortest-path kernels on graphs. In: Fifth IEEE International Conference on Data Mining (ICDM 2005), p. 8 IEEE (2005)
4. Chawla, A., Lee, B., Fallon, S., Jacob, P.: Host based intrusion detection system with combined CNN/RNN model. In: Proceedings of Second International Workshop on AI in Security (2018)

5. Chiu, C., Zhan, J.: Deep learning for link prediction in dynamic networks using weak estimators. IEEE Access **6**, 35937–35945 (2018)
6. Ionut, A.: New DDoS attack method obfuscates source port data. https://www.securityweek.com/new-ddos-attack-method-obfuscates-source-port-data. Accessed 10 March 2019
7. Kim, G., Yi, H., Lee, J., Paek, Y., Yoon, S.: LSTM-based system-call language modeling and robust ensemble method for designing host-based intrusion detection systems. arXiv preprint arXiv:1611.01726 (2016)
8. Liaskos, C., Kotronis, V., Dimitropoulos, X.: A novel framework for modeling and mitigating distributed link flooding attacks. In: IEEE INFOCOM 2016-The 35th Annual IEEE International Conference on Computer Communications, pp. 1–9. IEEE (2016)
9. Mirkovic, J., Reiher, P.: A taxonomy of DDoS attack and DDoS defense mechanisms. ACM SIGCOMM Comput. Commun. Rev. **34**(2), 39–53 (2004)
10. Moustafa, N., Slay, J.: UNSW-NB15: a comprehensive data set for network intrusion detection systems (UNSW-NB15 network data set). In: 2015 Military Communications and Information Systems Conference (MilCIS), pp. 1–6. IEEE (2015)
11. Nadeem, M., Marshall, O., Singh, S., Fang, X., Yuan, X.: Semi-supervised deep neural network for network intrusion detection (2016)
12. Navarro, J., Deruyver, A., Parrend, P.: A systematic survey on multi-step attack detection. Comput. Secur. **76**, 214–249 (2018)
13. Orsini, F., Frasconi, P., De Raedt, L.: Graph invariant kernels. In: Twenty-Fourth International Joint Conference on Artificial Intelligence (2015)
14. Rahul, V.K., Vinayakumar, R., Soman, K., Poornachandran, P.: Evaluating shallow and deep neural networks for network intrusion detection systems in cyber security. In: 2018 9th International Conference on Computing, Communication and Networking Technologies (ICCCNT), pp. 1–6. IEEE (2018)
15. Sharafaldin, I., Lashkari, A.H., Ghorbani, A.A.: Toward generating a new intrusion detection dataset and intrusion traffic characterization. In: ICISSP, pp. 108–116 (2018)
16. Shervashidze, N., Schweitzer, P., Leeuwen, E.J., Mehlhorn, K., Borgwardt, K.M.: Weisfeiler-lehman graph kernels. J. Mach. Learn. Res. **12**(Sep), 2539–2561 (2011)
17. Yanardag, P., Vishwanathan, S.: Deep graph kernels. In: Proceedings of the 21th ACM SIGKDD International Conference on Knowledge Discovery and Data Mining, pp. 1365–1374. ACM (2015)
18. Yao, Y., Su, L., Lu, Z.: DeepGFL: deep feature learning via graph for attack detection on flow-based network traffic. In: MILCOM 2018–2018 IEEE Military Communications Conference (MILCOM), pp. 579–584. IEEE (2018)
19. Zargar, S.T., Joshi, J., Tipper, D.: A survey of defense mechanisms against distributed denial of service (DDoS) flooding attacks. IEEE Commun. Surv. Tutorials **15**(4), 2046–2069 (2013)
20. Zhang, M., Xu, B., Bai, S., Lu, S., Lin, Z.: A deep learning method to detect web attacks using a specially designed CNN. In: Liu, D., Xie, S., Li, Y., Zhao, D., El-Alfy, E.-S.M. (eds.) ICONIP 2017. LNCS, vol. 10638, pp. 828–836. Springer, Cham (2017). https://doi.org/10.1007/978-3-319-70139-4_84

# Machine Learning for Performance Enhancement of Molecular Dynamics Simulations

JCS Kadupitiya, Geoffrey C. Fox, and Vikram Jadhao[(✉)]

Intelligent Systems Engineering, Indiana University, Bloomington, IN 47408, USA
{kadu,gcf,vjadhao}@iu.edu

**Abstract.** We explore the idea of integrating machine learning with simulations to enhance the performance of the simulation and improve its usability for research and education. The idea is illustrated using hybrid OpenMP/MPI parallelized molecular dynamics simulations designed to extract the distribution of ions in nanoconfinement. We find that an artificial neural network based regression model successfully learns the desired features associated with the output ionic density profiles and rapidly generates predictions that are in excellent agreement with the results from explicit molecular dynamics simulations. The results demonstrate that the performance gains of parallel computing can be further enhanced by using machine learning.

**Keywords:** Machine learning · Molecular dynamics simulations · Parallel computing · Scientific computing · Clouds

## 1 Introduction

In the fields of physics, chemistry, bioengineering, and materials science, it is hard to overstate the importance of parallel computing techniques in providing the needed performance enhancement to carry out long-time simulations of systems of many particles with complex interaction energies. These enhanced simulations have enabled the understanding of microscopic mechanisms underlying the macroscopic material and biological phenomena. For example, in a typical molecular dynamics (MD) simulation of ions in nanoconfinement [2,13] (Fig. 1), $\approx 1$ ns of dynamics of $\approx 500$ ions on one processor takes $\approx 12$ h of runtime, which is prohibitively large to extract converged results for ion distributions within reasonable time frame. Performing the same simulation on a single node with multiple cores using OpenMP shared memory reduces the runtime by a factor of 10 ($\approx 1$ h), enabling simulations of the same system for longer physical times. Using MPI dramatically enhances the simulation performance: for systems with thousands of ions, speedup of over 100 can be achieved, enabling the generation of the needed data for evaluating converged ionic distributions. Further, a

Supported by National Science Foundation through Awards 1720625 and 1443054.

© Springer Nature Switzerland AG 2019
J. M. F. Rodrigues et al. (Eds.): ICCS 2019, LNCS 11537, pp. 116–130, 2019.
https://doi.org/10.1007/978-3-030-22741-8_9

**Fig. 1.** Sketch of ions represented by blue and red spheres confined by material surfaces. (Color figure online)

hybrid OpenMP/MPI approach can provide even higher speedup of over 400 for similar-size systems enabling state-of-the-art research with simulations that can explore the dynamics of ions with less controlled approximations, accurate potentials, and for tens or hundreds of nanoseconds over a wider range of physical parameters.

Despite the employment of the optimal parallelization model suited for the size and complexity of the system, scientific simulations remain time consuming. This is particularly evident in the area of using simulations in education where real-time simulation-driven responses to students in classroom settings are desirable. While in research settings, instantaneous results are generally not expected and simulations can take up to several days, it is often desirable to rapidly access expected trends associated with relevant physical quantities that could have been learned from past simulations or could be predicted with reasonable accuracy based on the history of data generated from earlier simulation runs. As an example, consider a simulation framework, Ions in Nanoconfinement, that we deployed as a web application on nanoHUB to execute the aforementioned simulations of ionic systems [14]. In classroom usage, we have observed that the fastest simulations can take about 10 min to provide the converged ionic densities while the slowest ones (typically associated with larger system sizes and more complex ion-ion interaction energies) can take over 3 h. Similarly, for research applications, not having rapid access to expected overall trends in the key simulation output quantities (e.g., the variation of contact density as a function of ion concentration) can make the process of starting new investigations unwieldy and time-consuming. Primary factors that contribute to this scenario are the time delays resulting from the combination of waiting time in a queue on a computing cluster and the actual runtime for the simulation.

In this paper, we explore the idea of integrating machine learning (ML) layer with simulations to enhance the performance and improve the usability of simulations for both research and education. This idea is inspired by the recent development and use of ML in material simulations and scientific software applications

[4,6,9,17,24]. We employ a particular example, hybrid OpenMP/MPI parallelized MD simulations of ions in nanoconfinement [13,14] to illustrate this idea. We demonstrate that an artificial neural network (ANN) based regression model, trained on data generated via these simulations, successfully learns pre-identified key features associated with the output ionic density profile (the contact, mid-point, and peak densities). The ML approach entirely bypasses simulations and generates predictions that are in excellent agreement with results obtained from explicit MD simulations. The results demonstrate that the performance gains of parallel computing can be enhanced using data-driven approaches such as ML which improves the usability of the simulation framework by enabling real-time engagement and anytime access.

## 2  Background and Related Work

### 2.1  Coarse-Grained Simulations of Ions in Nanoconfinement

The distribution of ions often determines the assembly behavior of charged or neutral nanomaterials such as nanoparticles, colloids, or biological macro-molecules. Accurate knowledge of this ionic structure is exploited in many applications [8] including the design of double-layer supercapacitor and the extraction of metal ions from wastewater. As a result, extracting the distribution of ions in nanoconfinement created by material surfaces has been the focus of recent experiments and computational studies [13,18,21,22,25]. From a modeling standpoint, the surfaces are often treated as planar interfaces considering the size difference between the ions and the confining material particles, and the solvent is coarse-grained to speed-up the simulations. Such coarse-grained simulations have been employed to extract the ionic distributions over a wide range of electrolyte concentrations, ion valencies, and interfacial separations using codes developed in individual research groups [2,3,13] or using general purpose software packages such as ESPRESSO [16] and LAMMPS [19].

Generally, the average equilibrium ionic distribution, resulting from the competing inter-ionic steric and electrostatic interactions as well as the interactions between the ions and surfaces, is a quantity of interest. However, in many cases, the density of ions at the point of closest approach of the ion to the interface (contact density), the peak density, or the density at the center of the slit (mid-point density) directly relate to important experimentally-measurable quantities such as the effective force between the confining surfaces or the osmotic pressure [21,25]. It is thus useful to compute the variation of the contact, peak, and mid-point densities as a function of the solution conditions or ionic attributes.

### 2.2  Machine Learning for Enhancing Simulation Performance

Recent years have seen a surge in the use of ML to accelerate computational techniques aimed at understanding material phenomena. ML has been used to predict parameters, generate configurations in material simulations, and classify material properties [4,9,17,24]. For example, Fu et al. [17] employed ANN to select efficient updates to accelerate Monte Carlo simulations of classical Ising spin models near critical parameters associated with the phase transition. Similarly, Botu et al. [4] employed kernel ridge regression to accelerate MD method for nuclei-electron systems by learning the selection of probable configurations in MD simulations, which enabled bypassing explicit simulations for several steps.

The integration of ML layer for performance enhancement of scientific simulation frameworks deployed as web applications is relatively far less explored. nanoHUB is the largest online resource for education and research in nanotechnology [15]. This cyberinfrastructure hosts over 500 simulation tools and serves 1.4 million users worldwide. Our survey indicated that only one simulation tool on nanoHUB [10] employs ML-based methods to enhance the performance and usability of the simulation software. This simulation tool employs a deep neural network to bypass computational limitations in extracting transfer times associated with the excitation energy transport in light-harvesting systems [10].

## 3    Framework for Simulating Ions in Nanoconfinement

We developed a framework for simulating ions in nanoscale confinement (referred here as the nanoconfinement framework) that enabled a systematic investigation of the self-assembly of ions characterized by different ionic attributes (e.g., ion valency) and solution conditions (e.g., ion concentration and inter-surface separation) [11–14,23]. The framework has been employed to extract the ionic structure in electrolyte solutions confined by planar and spherical surfaces. Results have elucidated the microscopic mechanisms involving the ionic correlations and steric effects that determine the distribution of ions [11–13].

In the work presented here, we focus on ions confined by unpolarizable surfaces where the simulations are relatively faster [13], thus easing the training and testing of the ML model. We identify the following system attributes as key parameters that determine the self-assembly of ions: inter-surface separation or confinement length $h$, ion valencies ($z_p$, $z_n$ associated with the positive and negative ions), electrolyte concentration $c$, and ion diameter $d$. We ignore the effects arising due to the solvent-induced asymmetry in ionic sizes, and assign the positive and negative ions with the same diameter. Additional details regarding the model and simulation method can be found in the original paper [13].

The nanoconfinement framework employs a code written in C++ for simulating ions near unpolarizable interfaces. The code, available as an open-source repository on GitHub [14], is accelerated using a hybrid parallel programming technique (see Sect. 3.1). This framework is also deployed as a web application (Ions in Nanoconfinement) on nanoHUB [14] (see Sect. 3.2). We have verified that the ionic density profiles obtained using our code agree with the densities extracted via simulations performed in LAMMPS. We note that the ML model proposed in this work to predict the desired key ionic densities is agnostic to the code engine that enables the MD simulations for generating the training dataset.

## 3.1    Hybrid OpenMP/MPI Parallelization

The nanoconfinement framework is based on a code that employs the velocity-Verlet algorithm to update the positions and velocities at each simulation step. First, the velocities of all ions are updated for half timestep $\Delta/2$, following which the positions of all ions are updated for $\Delta$. At this point, the forces on all ions are computed, and finally the velocities for all ions are updated for the next $\Delta/2$. Once the system reaches equilibrium, the positions of the ions are stored periodically to extract density profiles.

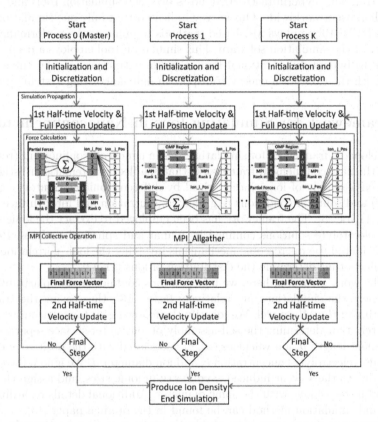

**Fig. 2.** Hybrid model implemented in the Force Calculation block using MPI and OpenMP to accelerate the nanoconfinement framework.

To improve the runtime of the simulation, the framework employs the hybrid master-only model that uses one MPI process per node and OpenMP on the cores of the node, with no MPI calls inside the parallel regions. This hybrid model is applied for the force and energy calculation subroutines, and it enables the domain decomposition under a two-level mechanism. On the MPI level, coarse-grained domain decomposition is performed using boundary conditions

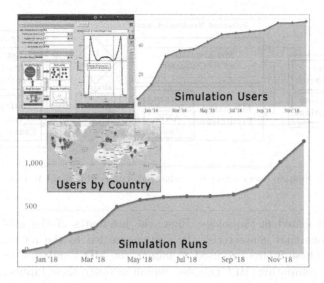

**Fig. 3.** A screenshot of the GUI (top left) and usage statistics of the nanoconfinement framework deployed on nanoHUB.

as explained in Fig. 2. The second level of domain decomposition is achieved through OpenMP loop level parallelization inside each MPI process. This multilevel domain decomposition has advantages over pure MPI or pure OpenMP, when cache performance is taken into consideration. This strategy also provides the maximum access locality, a minimum number of cache misses, non-uniform memory access (NUMA) traffic and inter-node communication [20].

### 3.2 Deployment on nanoHUB

The nanoconfinement framework is deployed as a web application (Ions in Nanoconfinement) [14] on the nanoHUB cyberinfrastructure. nanoHUB provides user-friendly, web-based access for executing simulation codes to researchers, students, and educators broadly interested in nanoscale engineering [15]. In less than 1 year of its launch, the Ions in Nanoconfinement application has nucleated 58 users worldwide and has been run over 1200 times [14] (Fig. 3). This application is designed to launch simulations that use virtual machines or supercomputing clusters depending on user-selected inputs, and it has been employed to teach graduate courses at Indiana University. Advances in the framework that enable real-time results can significantly enhance the experience of users that employ this application in both education and research. The current version of "Ions in Nanoconfinement", like almost all other applications on nanoHUB, does not support this feature.

## 4 ML-Enabled Performance Enhancement

We now describe an approach based on ML to enhance the overall performance and usability of the nanoconfinement framework. Figure 4 shows the overview

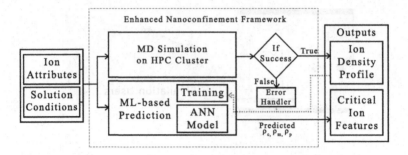

**Fig. 4.** System overview of the enhanced nanoconfinement framework.

of the implemented methodology. First, the attributes of the ions and solution conditions that characterize the system are fed to the nanoconfinement framework. These inputs are used to launch the MD simulation on the high-performance computing (HPC) cluster. Simultaneously, these inputs are also fed to the ML-based density prediction module to predict the contact, mid-point, and peak densities. The outputs of the MD simulation are the ion density profiles that characterize the ionic structure near unpolarizable interfaces. Error handler aborts the program and displays appropriate error messages when a simulation fails due to any pre-defined criteria. In addition, at the end of the simulation run, ionic density values are saved for future retraining of the ML model. ML model is retrained for every 2500 new simulation runs.

After reviewing and experimenting with many ML techniques for parameter tuning and prediction including polynomial regression, support vector regression, decision tree regression, and random forest regression, the artificial neural network (ANN) was adopted for predicting critical features associated with the output ionic density. Figure 5 shows the details of this ANN-based ML model. The data preparation and preprocessing techniques, feature extraction and regression techniques as well as their validation are discussed below.

### 4.1 Data Preparation and Preprocessing

Prior domain experience and backward elimination using the adjusted R squared is used for creating the training data set. Five input parameters that significantly affect the dynamics of the system are identified: confinement length $h$, positive valency $z_p$, negative valency $z_n$, salt concentration $c$, and the diameter of the ions $d$. Future work will explore the training with additional input parameters such as temperature and solvent permittivity that are fixed in this initial study to room temperature (298 K) and water permittivity ($\approx 80$) respectively.

Contact density $\rho_c$, mid-point (center of the slit) density $\rho_m$, and peak density $\rho_p$ associated with the final (converged) distribution for positive ions were selected as the output parameters. Few discrete values for each of the input/output parameters were experimented with and swept over to create and run 6,864 simulations for training the ML model. The range for ionic system

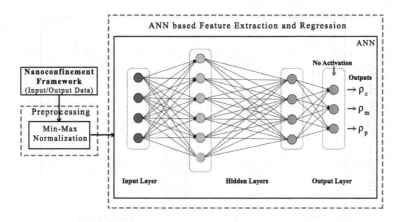

**Fig. 5.** ANN-based regression model to enhance the nanoconfinement framework.

parameters was selected based on physically meaningful and experimentally-relevant values: $h \in (3.0, 4.0)$ nm, $z_p \in 1, 2, 3$ (in units of electronic charge $|e|$); $z_n \in -1, -2$ (in units of $|e|$); $c \in (0.3, 0.9)$ M, and $d \in (0.5, 0.75)$ nm. All simulations were performed for over $\approx 5$ ns. The entire data set was separated into training and testing sets using a ratio of 0.7:0.3. Min$-$max normalization filter was applied to normalize the input data at the preprocessing stage.

## 4.2 Feature Extraction and Regression

The ANN algorithm with two hidden layers (Fig. 5) was implemented in Python for regression of three continuous variables in the ML model. Outputs of the hidden layers were wrapped with the *relu* function; the latter was found to converge faster compared to the sigmoid function. No wrapping functions were used in the output layers of the algorithm as ANN was trained for regression.

By performing a grid search, hyper-parameters such as the number of first hidden layer units, second hidden layer units, batch size, and the number of epochs were optimized to 17, 9, 25, and 100 respectively. Adam optimizer was used as the backpropagation algorithm. The weights in the hidden layers and in the output layer were initialized to random values using a normal distribution at the beginning. The mean square loss function was used for error calculation. To stop overtraining the network, a drop out mechanism for hidden layer neurons was employed during the training time. ANN implementation, training, and testing were programmed using scikit-learn, Keras, and TensorFlow ML libraries [1,5,7]. Rest of the regression methods were implemented using scikit-learn ML libraries. For the kernel ridge regression, radial basis function kernel was used.

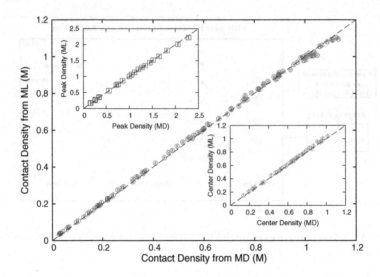

**Fig. 6.** Accuracy comparison between ML predictions and MD simulation results for the contact densities (red circles) of ions in systems characterized by inputs selected from the following ranges of parameters: $h \in (3.0, 4.0)$ nm, $z_p \in 1, 2, 3$, $z_n \in -1, -2$, $c \in (0.3, 0.9)$ M, and $d \in (0.5, 0.75)$ nm. Top-left and bottom-right insets show the comparison for the peak (blue squares) and mid-point (green diamonds) densities respectively for a subset of the selected systems. Black dashed lines with a slope of 1 represent perfect correlation. All densities are shown in units of molars. (Color figure online)

## 5   Results

### 5.1   Bypassing Simulations with ML-Enabled Predictions

We experimented with 6 regression models to predict the key output density features identified above: contact density ($\rho_c$), mid-point density ($\rho_m$), and peak density ($\rho_p$). These models were tested on 2060 sets of input parameters ($h, z_p, z_n, c, d$). These sets were comprised of parameter values within the range for which the models were trained; see Sect. 4.1. Table 1 shows the success rate and the mean square error (MSE) for testing data sets. The success rate was calculated based on the error bars associated with the density values obtained via MD simulations: ML prediction was considered successful when the predicted density value was within the error bar of the simulation estimate. Simulations were run for sufficiently long times (over $\approx 5$ ns) to obtain converged density estimates and error bars. MSE values are calculated using k-fold cross-validation techniques with k = 20. ANN based regression model predicted $\rho_c$, $\rho_m$ and $\rho_p$ accurately with a success rate of 95.52% (MSE $\approx 0.0000718$), 92.07% (MSE $\approx 0.0002293$), and 94.78% (MSE $\approx 0.0002306$) respectively. ANN outperformed all other non-linear regression models (Table 1).

Figure 6 shows the comparison between the predictions made by the ML model and the results obtained from MD simulations for the contact, mid-point,

**Table 1.** Comparison of regression models for the prediction of output density values.

| Model | Contact density | | Midpoint density | | Peak density | |
|---|---|---|---|---|---|---|
| | Success % | MSE | Success % | MSE | Success % | MSE |
| Polynomial | 61.04 | 0.0129300 | 60.84 | 0.0187700 | 61.87 | 0.0100400 |
| Kernel-Ridge | 78.86 | 0.0030900 | 76.57 | 0.0041200 | 75.93 | 0.0049800 |
| Support vector | 80.11 | 0.0012700 | 79.55 | 0.0024900 | 81.98 | 0.0010600 |
| Decision tree | 68.44 | 0.0084600 | 64.54 | 0.0094900 | 62.47 | 0.0110700 |
| Random forest | 74.15 | 0.0045700 | 70.85 | 0.0078900 | 75.09 | 0.0040800 |
| ANN based | 95.52 | 0.0000718 | 92.07 | 0.0002293 | 94.78 | 0.0002306 |

and peak densities associated with positive ions. For clarity, results are shown for a randomly selected subset of the entire testing dataset described in Sect. 4.1. $\rho_c$, $\rho_m$ and $\rho_p$ predicted by the ML model were found to be in excellent agreement with those calculated using the MD method; data from either approach fall on the dashed lines which indicate perfect correlation.

## 5.2 Rapid Access to Trendlines Using ML

Trendlines exhibiting the variation of $\rho_c$, $\rho_m$, and $\rho_p$ for a wide range and combinations of the five input parameters ($h$, $z_p$, $z_n$, $c$, $d$) were extracted using the ML model. In Figs. 7 and 8, we show a small selected subset of these trends.

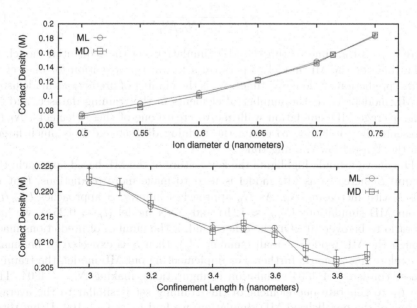

**Fig. 7.** (Top) Trendlines for contact density vs. ion diameter for systems with $h = 4$ nm, $z_p = 3$, $z_n = -1$, and $c = 0.85$ M. (Bottom) Trendlines for contact density vs. confinement length for systems with $z_p = 2$, $z_n = -1$, $c = 0.9$ M, and $d = 0.553$ nm.

Figure 7 shows the variation of the contact density $\rho_c$ with the ion diameter $d$ and confinement length $h$. Figure 7 (top) illustrates how $\rho_c$ varies with $d \in (0.5, 0.75)$ nm at constant $h = 4$ nm, $z_p = 3$, $z_n = -1$, and $c = 0.85$ M. Figure 7 (bottom) illustrates the variation in $\rho_c$ when the confinement length $h$ is tuned between 3.0 and 4.0 nm, with other parameters held constant ($z_p = 2$, $z_n = -1$, $c = 0.9$ M, and $d = 0.553$ nm). Circles represent ML predictions and squares show MD results. ML predictions are within the errorbars generated via MD simulations and follow the simulation-predicted trends for both cases. Results demonstrate that contact density varies rapidly when the diameter is changed but exhibits a slower variation when the confinement length is varied. We find that the same ML model is able to track the distinct variations while exhibiting different resolution (sensitivity) criteria.

Figure 8 shows similar comparison between ML and MD results for the variation of $\rho_c$, $\rho_m$ and $\rho_p$ vs. salt concentration $c$. Excellent agreement is seen between the two approaches. We note that the prediction time for the ML model to obtain these densities is in the order of a few seconds while the simulations can take up to hours to compute one contact density with similar accuracy level.

## 5.3  Speedup

Traditional speedup formulae associated with parallel computing methods need to be adapted for evaluating the speedup associated with the ML-enhanced simulations. We propose the following simple formula which is illustrative at this preliminary stage. We define the ML speedup as:

$$S = \frac{t_{sim}}{t_p + t_{tr} \cdot N_{tr}/N_p}, \tag{1}$$

where $t_{sim}$ is the time to run the MD simulation via the sequential model, $t_p$ is the time for the ML model to perform a forward propagation for one set of inputs (prediction or "lookup" time), $N_p$ is the number of predictions made using the ML model, $N_{tr}$ is the number of elements in the training dataset, and $t_{tr}$ is the average MD simulation walltime to create one of these elements. $N_{tr}t_{tr}$ represents the total time to create the training dataset and it is much larger than the TensorFlow training time.

The above formula highlights the key feature of the ML-based approach: the speedup $S$ increases as ML model is used to make more predictions, that is, $S$ rises with increasing $N_p$. As $N_p$ approaches infinity, $S$ approaches $t_{sim}/t_p$; for our MD simulations ($t_{sim} \approx 12$ h) and ANN model ($t_p \approx 0.25$ s), we find this ratio to be over $10^5$. On the other hand, if the number of predictions made through the ML model is small (smaller $N_p$), then $S$ is expected to be small. We explore this scenario further. For implementing our ML model, the training dataset consisted of 4804 simulation configurations, making $N_{tr} = 4804$. The time $t_{tr}$ to generate one element of this training set is similar to the average runtime of the parallelized MD simulation; we find $t_{tr} \approx t_{sim}/100$. Using these relations and noting $t_p \ll t_{sim}$, a lower bound on speedup can be derived as the

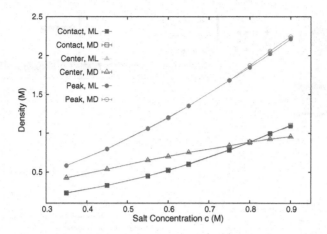

**Fig. 8.** Contact, peak, and center-of-the-slit (mid-point) density vs. salt concentration associated with the distribution of ions for systems characterized with $h = 3.2$ nm, $z_p = 1$, $z_n = -1$, and $d = 0.714$ nm. Closed symbols represent ML predictions and open symbols with error bars are MD simulation results.

result for $N_p = 1$. For this case, we find the "speedup" $S \approx t_{sim}/(t_{tr} N_{tr}) \approx 10^{-2}$. Finally, when the number of predictions $N_p$ are similar to the number of elements in the training dataset $N_{tr}$, then Eq. 1 yields $S \approx t_{sim}/t_{tr}$, which is equivalent to the speedup associated with the traditional parallel computing approach.

## 6   Outlook and Future Work

Based on the aforementioned investigations, we propose to design and integrate an ML layer with the nanoconfinement framework. This enhanced application will be deployed on nanoHUB using the Jupyter python notebook interface. Figure 9 shows a sketch of the proposed GUI. Users will be able to click both "Run with MD" button and "Predict with ML" button simultaneously or separately depending on the desired information. "Predict with ML" will activate the ML layer and predict $\rho_c$, $\rho_m$, and $\rho_p$ almost instantaneously. These ML-predicted values will be shown in three text boxes and will appear as markers on the density profile plot. If users also select "Run with MD", the entire density profile will be added at the end of the simulation. For illustration purposes, Fig. 9 shows the final density plot using this integrated MD + ML approach for the input parameters $h = 3.0$ nm, $z_p = 1$, $z_n = -1$, $c = 0.9$ M, and $d = 0.714$ nm.

In this initial study, we focused on a particular example framework to illustrate the idea of using ML-based methods to enhance the performance and usability of scientific simulations. The results from this investigation are encouraging and we intend to explore these ideas in the future to provide a richer set of predictions (finer-grained density profile) for the nanoconfinement framework, and extend the method to other simulation techniques. Here, the training data

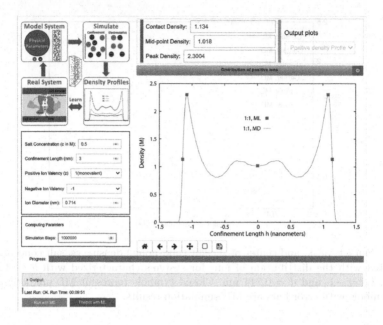

**Fig. 9.** Proposed GUI for integrating an ML layer with the nanoconfinement framework deployed on nanoHUB. The GUI includes text boxes (top) showing the ML-predicted contact, mid-point, and peak densities for an example ionic system. These results also appear as markers on the plot showing the density profile generated by MD simulations.

was created as a distinct step during the process of generating the ML model. Future approaches will involve implicit training of the ANN while the simulations are used in research and education. We expect that the usefulness of the ML-enabled enhancements demonstrated in this work strengthen the case for scientific simulation applications to be designed and developed with an ML wrapper that both optimizes the application execution and learns from the simulations.

# References

1. Abadi, M., et al.: TensorFlow: a system for large-scale machine learning. OSDI **16**, 265–283 (2016)
2. Allen, R., Hansen, J.P., Melchionna, S.: Electrostatic potential inside ionic solutions confined by dielectrics: a variational approach. Phys. Chem. Chem. Phys. **3**, 4177–4186 (2001)
3. Boda, D., Gillespie, D., Nonner, W., Henderson, D., Eisenberg, B.: Computing induced charges in inhomogeneous dielectric media: application in a monte carlo simulation of complex ionic systems. Phys. Rev. E **69**(4), 046702 (2004)
4. Botu, V., Ramprasad, R.: Adaptive machine learning framework to accelerate ab initio molecular dynamics. Int. J. Quantum Chem. **115**(16), 1074–1083 (2015)
5. Buitinck, L., et al.: API design for machine learning software: experiences from the scikit-learn project. arXiv:1309.0238 (2013)

6. Butler, K.T., Davies, D.W., Cartwright, H., Isayev, O., Walsh, A.: Machine learning for molecular and materials science. Nature **559**(7715), 547 (2018)
7. Chollet, F., et al.: Keras (2015)
8. Feng, G., Qiao, R., Huang, J., Sumpter, B.G., Meunier, V.: Ion distribution in electrified micropores and its role in the anomalous enhancement of capacitance. ACS Nano **4**(4), 2382–2390 (2010)
9. Ferguson, A.L.: Machine learning and data science in soft materials engineering. J. Phys. Condens. Matter **30**(4), 043002 (2017)
10. Häse, F., Kreisbeck, C., Aspuru-Guzik, A.: Machine learning for quantum dynamics: deep learning of excitation energy transfer properties. Chem. Sci. **8**(12), 8419–8426 (2017)
11. Jadhao, V., Solis, F.J., Olvera de la Cruz, M.: Simulation of charged systems in heterogeneous dielectric media via a true energy functional. Phys. Rev. Lett. **109**, 223905 (2012)
12. Jadhao, V., Solis, F.J., Olvera de la Cruz, M.: A variational formulation of electrostatics in a medium with spatially varying dielectric permittivity. J. Chem. Phys. **138**(5), 054119 (2013)
13. Jing, Y., Jadhao, V., Zwanikken, J.W., Olvera de la Cruz, M.: Ionic structure in liquids confined by dielectric interfaces. J. Chem. Phys. **143**(19), 194508 (2015)
14. Kadupitiya, K., Marru, S., Fox, G.C., Jadhao, V.: Ions in nanoconfinement, December 2017. https://nanohub.org/resources/nanoconfinement, online on nanoHUB; source code on GitHub at https://github.com/softmaterialslab/nanoconfinement-md
15. Klimeck, G., McLennan, M., Brophy, S.P., Adams III, G.B., Lundstrom, M.S.: nanohub.org: advancing education and research in nanotechnology. Comput. Sci. Eng. **10**(5), 17–23 (2008)
16. Limbach, H.J., Arnold, A., Mann, B.A., Holm, C.: ESPResSo - an extensible simulation package for research on soft matter systems. Comp. Phys. Comm. **174**(9), 704–727 (2006)
17. Liu, J., Qi, Y., Meng, Z.Y., Fu, L.: Self-learning monte carlo method. Phys. Rev. B **95**, 041101 (2017)
18. Luo, G., et al.: IoN distributions near a liquid-liquid interface. Science **311**(5758), 216–218 (2006)
19. Plimpton, S.: Fast parallel algorithms for short-range molecular dynamics. J. Comput. Phys. **117**(1), 1–19 (1995)
20. Rabenseifner, R., Hager, G., Jost, G.: Hybrid MPI/OpenMP parallel programming on clusters of multi-core SMP nodes. In: 2009 17th Euromicro International Conference on Parallel, Distributed and Network-Based Processing, pp. 427–436. IEEE (2009)
21. dos Santos, A.P., Netz, R.R.: Dielectric boundary effects on the interaction between planar charged surfaces with counterions only. J. Chem. Phys. **148**(16), 164103 (2018)
22. Smith, A.M., Lee, A.A., Perkin, S.: The electrostatic screening length in concentrated electrolytes increases with concentration. J. Phys. Chem. Lett. **7**(12), 2157–2163 (2016)
23. Solis, F.J., Jadhao, V., Olvera de la Cruz, M.: Generating true minima in constrained variational formulations via modified lagrange multipliers. Phys. Rev. E **88**(5), 053306 (2013)

24. Spellings, M., Glotzer, S.C.: Machine learning for crystal identification and discovery. AIChE J. **64**(6), 2198–2206 (2018)
25. Zwanikken, J.W., Olvera de la Cruz, M.: Tunable soft structure in charged fluids confined by dielectric interfaces. Proc. Nat. Acad. Sci. **110**(14), 5301–5308 (2013)

# 2D-Convolution Based Feature Fusion for Cross-Modal Correlation Learning

Jingjing Guo[1,2], Jing Yu[1,2(✉)], Yuhang Lu[1,2], Yue Hu[1], and Yanbing Liu[1]

[1] Institute of Information Engineering, Chinese Academy of Sciences, Beijing, China
[2] School of Cyber Security, University of Chinese Academy of Sciences,
Beijing, China
{guojingjing,yujing02,luyuhang,huyue,liuyanbing}@iie.ac.cn

**Abstract.** Cross-modal information retrieval (CMIR) enables users to search for semantically relevant data of various modalities from a given query of one modality. The predominant challenge is to alleviate the "heterogeneous gap" between different modalities. For text-image retrieval, the typical solution is to project text features and image features into a common semantic space and measure the cross-modal similarity. However, semantically relevant data from different modalities usually contains imbalanced information. Aligning all the modalities in the same space will weaken modal-specific semantics and introduce unexpected noise. In this paper, we propose a novel CMIR framework based on multi-modal feature fusion. In this framework, the cross-modal similarity is measured by directly analyzing the fine-grained correlations between the text features and image features without common semantic space learning. Specifically, we preliminarily construct a cross-modal feature matrix to fuse the original visual and textural features. Then the 2D-convolutional networks are proposed to reason about inner-group relationships among features across modalities, resulting in fine-grained text-image representations. The cross-modal similarity is measured by a multi-layer perception based on the fused feature representations. We conduct extensive experiments on two representative CMIR datasets, i.e. English Wikipedia and TVGraz. Experimental results indicate that our model outperforms state-of-the-art methods significantly. Meanwhile, the proposed cross-modal feature fusion approach is more effective in the CMIR tasks compared with other feature fusion approaches.

**Keywords:** 2D-convolutional network · Inner-group relationship · Feature fusion · Cross-modal correlation · Cross-modal information retrieval

---

Y. Hu—Co-corresponding author.
This work is supported by the National Key Research and Development Program (Grant No. 2017YFB0803301).

J. M. F. Rodrigues et al. (Eds.): ICCS 2019, LNCS 11537, pp. 131–144, 2019.
https://doi.org/10.1007/978-3-030-22741-8_10

# 1  Introduction

With the explosive increase of online multimedia data, such as image, text, video, and audio, there is a great demand for intelligent search across different modalities. Cross-modal information retrieval (CMIR), which aims to enable queries of one modality to retrieve semantically relevant information of other modalities, plays an important role for realizing intelligent search. Since data from different modalities are heterogenous in feature representations, it's challenging to directly measure the cross-modal similarity. The typical solution for the problem of "heterogeneous gap" is to learn a common semantic space and align the features of different modalities in such space [13,15,17]. The cross-modal similarity can be directly measured by computing the feature distance in the common semantic space.

However, the common semantic space is equal for all the modalities, which ignores the fact that semantically relevant data from different modalities contains imbalanced information. For example, an image and its textual description convey the same semantics, but containing unequal amount of information. The image usually contains complex visual scenes which cannot be completely described by a few words. On the other hand, the text describes more background content beyond the visual information. Not all the fine-grained information between text and image can be aligned exactly. Therefore, projecting different modalities into the same space will loss some important modality-specific information and introduce some extra noise.

Recently, a few novel works explore to capture the cross-modal correlations based on feature fusion and compute the cross-modal similarity directly without common space learning [10,22,24]. Yu *et al.* [24] have demonstrated the advantages of element-wise product feature fusion for cross-modal correlation modeling. They propose a dual-path neural networks to learn visual features and textual features respectively. Then element-wise product is utilized to fuse two modal features for the successive distance metric learning. However, such simple feature fusion approach is not capable to generate expressive text-image features that can fully capture the complex correlations across modalities. More effective feature fusion approaches [3,7] have been proposed in the visual question answering (VQA) tasks and proved to be effective to improve the answer accuracy. Cross-modal information retrieval also need specific feature fusion approach to capture complex relationship between multi-modal features. How to design effective feature fusion approach for cross-modal information retrieval tasks has not been well studied.

In this paper, we propose a **2D-Convo**lution based **F**eature **F**usion (**2D-ConvFF** for short) approach to explore more complex interactions between multi-modal features for enhancing their semantic correlations and improving the performance of cross-modal information retrieval. A dual-path neural network is proposed to respectively extract the image features and text features. Then 2D-ConvFF is utilized for multi-modal feature fusion. Specifically, the learnt multi-modal features are preliminarily fused by a cross-modal feature matrix. Then we propose the 2D-convolutional networks to reason about inner-group relationships across different modal features for fine-grained text-image feature fusion. Cross-modal similarity is measured by a multi-layer perception given the fused features. The model is trained by a ranking-based pairwise loss function. Compared with previous works, our proposed 2D-ConvFF can leverage the advantages of convolutional networks to capture inner-group relations across multi-modal features with small amount of parameters.

The main contribution of this work is the proposed 2D-convolutional networks for enhancing multi-modal feature fusion and eventually improving the cross-modal information retrieval performance. This also leads to the views of how to nicely capture more complex correlations across multi-modal data to learn more informative representations for cross-modal information retrieval, which are problems not fully explored yet.

## 2   Related Work

The mainstream solution for cross-modal information retrieval is to learn a common semantic space for multi-modal data, and directly measure their similarity. Traditional statistical correlation analysis methods, such as CCA [17] and its variants [15,16], are representative solutions for such case, which learn a subspace that maximizes the pairwise correlations between two modalities. Considering the importance of semantic information, some methods explore the semantics of category to constrain the common semantic space learning, such as semi-supervised methods [20,25] and supervised methods [18,21]. With the advance of deep learning in multimedia applications, deep neural networks (DNN) are used in cross-modal information retrieval [1,23]. This kind of methods generally construct a dual-path neural network to represent different modalities and jointly learn their correlations in an end-to-end mode. He et al. [4] apply two convolution networks to map images and texts into a common space and measure the cross-modal similarity by cosine distance.

Apart from common semantic space learning, another kind of solutions are based on correlation measurement. These methods are designed to train an end-to-end network to directly predict cross-modal correlations between different modal data [10,22,24] by multi-modal feature fusion. Wang et al. [22] propose to use two branch networks to project the images and texts to the same dimension. Then they fuse the two branches via element-wise product and the model is trained with an regression loss to predict the similarity score. Concatenation, element-wise addition and element-wise product are the basic approaches in cross-modal feature fusion [10,22,24]. It is very difficult for such a simple model to mine complex fine-grained relationships between multi-modal features.

Recently, some works exploit to fuse multi-modal features via more complex interactions. Huang et al. [5] measure multiple local similarities within a few timesteps, and aggregate them with hidden states to obtain a final matching score as the desired global similarity. Bilinear models, such as Multi-modal Compact Bilinear (MCB) [3], Multi-modal Low-rank Bilinear (MLB) [7] and Multi-modal Factorized Bilinear (MFB) [7], have been proposed for multi-modal feature fusion, especially for the visual question answering (VQA) tasks. Despite the remarkable progress of previous studies, these methods mainly capture the

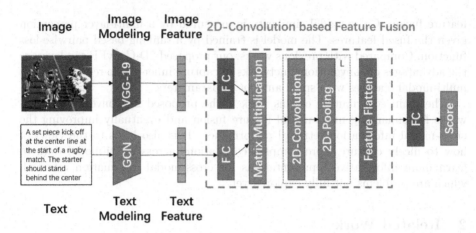

**Fig. 1.** This overall framework of the 2D-Convolution based Feature Fusion (2D-ConvFF). Given a image and a text, we use VGG-19 to get the image feature vector and use GCN to get the text feature vector. FC represents the fully connected layer. After that, we get two feature vectors which have the same dimensions. Matrix Multiplication denotes that we use two feature vectors get a feature matrix by matrix multiplication. Next, we use $L$ successive convolution layers and max-pooling layers mine internal relations of feature groups between multi-modal features. Feature Flatten flattens the above fused feature matrix. Finally, we use a fully connected layer to attain the similarity score.

pairwise correlations between multi-modal features with high computational complexity. In our work, we propose a 2D-Convolution based Feature Fusion approach to explore inner-group interactions between multi-modal features with relatively small amount of parameters for improving cross-modal information retrieval.

## 3    Methodology

In this section, we introduce the novel cross-modal information retrieval framework based on cross-modal feature fusion. The overall framework is illustrated in Fig. 1. It is a dual-path neural network to simultaneously learn the feature representations of the texts and images. Then the proposed 2D-convolution based feature fusion method is applied to fuse the text features and image features. The cross-modal similarity score is measured by a multi-layer perception given the fused features. Our framework mainly contains four components: text modeling, image modeling, 2D-convolution based feature fusion and the objective function. We will describe each component in detail in the following sections.

### 3.1    Text Modeling

Most of existing deep neural network methods for cross-modal information retrieval use word-level features, e.g. bag-of-words and word embeddings [2,11], or

sequential features, e.g. RNN [12], to represent texts. Besides word-level semantics, the semantic relations between words are also informative. Kipf *et al.* [8] propose Graph Convolutional Network (GCN), which has good performance in modeling the relational semantics of texts and improve the accuracy of text classification. We follow the extended work [24], which combines the structural information and semantic information, to represent a text by a word-level featured graph
(bottom in Fig. 1). We firstly extract the most common words from the text corpus and use a pre-trained word2vec [11] embedding to represent each word. Then
each extracted word is corresponding to a vertex in the graph. We construct the
edge set by computing $k$-nearest neighbors of each vertex based on the cosine similarity between word word2vec embeddings. Each text is represented by a bag-
of-words vector and the word frequency serves as the 1-dimensional feature of
the corresponding vertex. Then, we adopt Graph Convolutional Network (GCN)
to enhance the text features based on the word-level relational connections and
obtain the text representations. Given a text $T$, we learn the text representation
$f_T$ by the GCN model $V_T(\cdot)$, formally denoted as: $f_T = V_T(T)$.

## 3.2    Image Modeling

Convolutional neural network has achieved a great success in image recognition
and other computer vision tasks. Simonyan *et al.* [8] show the increase of convolutional layer depth is beneficial to the improvement of accuracy. Based on
this study, in the image modeling path (top in Fig. 1), we utilize the pre-trained
VGG-19 [19] to extract the features from output of the last fully connected layer
as the image features. Given the fixed image features, a fully connected layer is
used to fine-tune the feature representations according to the downstream task.
Given an image $I$, we learn the image representation $f_I$ based on the pre-training
VGG-19 model with fine-tuning stage $V_I(\cdot)$ denoted as: $f_I = V_I(I)$.

## 3.3    2D-Convolution Based Feature Fusion

The proposed 2D-Convolution based feature fusion method (middle in Fig. 1)
mainly consists of two parts: feature matrix construction and feature fusion.
The detailed procedure is illustrated in Fig. 2.

**Feature Matrix Construction.** According to Sects. 3.1 and 3.2, we obtain
the text feature $f_T \in \mathbb{R}^m$ and the image feature $f_I \in \mathbb{R}^n$. Since $f_T$ and $f_I$ have
different dimensions, we apply two fully connected layer $W_t$ and $W_i$ to project
$f_T$ and $f_I$ to the same dimension. The feature mapping of the text is formally
defined as:

$$\tilde{f}_T = (f_T)^T W_t \tag{1}$$

where $W_t \in \mathbb{R}^{m \times d}$ is the trained parameter of textual fully connected layer and
$\tilde{f}_T \in \mathbb{R}^d$ is the refined text feature. Similarly, we define the feature mapping of
the image as follows:

$$\tilde{f}_I = (f_I)^T W_i \tag{2}$$

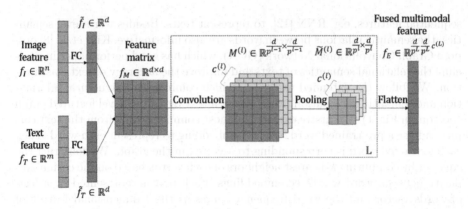

**Fig. 2.** Illustration of 2D-Convolution based Feature Fusion. We firstly use two fully connected layers to map the image features and text features to the same dimension. The two feature vectors are multiplied to form a feature matrix. Then the 2D-convolutional network with multiple convolutional and pooling operations are applied to enhance the cross-modal correlations. Finally, we flatten the enhanced multi-modal feature matrix and obtain the fused text-image features.

where $W_i \in \mathbb{R}^{n \times d}$ is the trained parameter of visual fully connected layer, and $\tilde{f}_I \in \mathbb{R}^d$ is the refined image feature. We utilize vector multiplication to form the feature matrix, the operation is defined as follows:

$$f_M = (\tilde{f}_T)^T \tilde{f}_I \tag{3}$$

where $f_M \in \mathbb{R}^{d \times d}$ is the feature matrix, which captures pairwise relationships between each dimension of the text feature and the image feature.

**2D-Convolutional Network.** We introduce a multi-layer convolutional network for multi-modal feature fusion. The constructed feature matrix can only capture pairwise relationships. Except pairwise relationships, there exists inner-group relationships between the image feature and the text feature which conveys more complex cross-modal correlations. A straightforward solution is to fuse inner-group features in the feature matrix by convolutional operations.

Given the feature matrix $f_M$, we construct $L$ convolutional layers each followed by a ReLU activation and a max-pooling layer. The $l_{th}$ convolution layer has $c^{(l)}$ convolution kernels with the kernel size $n^{(l)} \times n^{(l)}$. The strides of all convolutions are set to 1 and zero-padding is adopted to guarantee that the output after each convolution has the same size as the input. The output feature matrix of the last convolution layer is flattened and forms the fused text-image feature. The feature fusion procedure is formally defined as follows.

For the first convolution layer, the $k_{th}$ kernel, denoted as $w^{(1,k)}$ scans over the feature matrix $M^{(0)} = f_M$ to generate a feature map $M^{(1,k)}$ defined by:

$$M_{i,j}^{(1,k)} = ReLU\left(\sum_{x=0}^{n^{(1)}-1}\sum_{y=0}^{n^{(1)}-1} w_{x,y}^{(1,k)} \cdot M_{i+x,j+y}^{(0)} + b^{(1,k)}\right) \tag{4}$$

where $n^{(1)}$ denotes the size of the kernel in the first convolution layer. The size of all convolution kernels in the same convolutional layer are the same. $b^{(1,k)}$ denotes the bias of the $k_{th}$ kernel.

After the ReLU activation function, we apply $p \times p$ max-pooling to form the feature map $\tilde{M}^{(1,k)}$:

$$\tilde{M}_{i,j}^{(1,k)} = \max_{0 \le x < p} \max_{0 \le y < p} M_{i\cdot p+x, j\cdot p+y}^{(1,k)} \tag{5}$$

where $p$ denotes the size of the pooling kernel. After the first convolution layer, the output feature maps are fed to the next convolution layer as the input, denoted as $M^{(l)}$, where $l$ is the number of the convolution layer. The $(l+1)_{th}$ layer of convolution is formally defined as:

$$M_{i,j}^{(l+1,k')} = ReLU\left(\sum_{k'=0}^{c^{(l+1)}-1}\sum_{x=0}^{n^{(l+1)}-1}\sum_{y=0}^{n^{(l+1)}-1} w_{x,y}^{(l+1,k')} \cdot \tilde{M}_{i+x,j+y}^{(l,k)} + b^{(l+1,k')}\right) \tag{6}$$

$$\tilde{M}_{i,j}^{(l+1,k')} = \max_{0 \le x < p} \max_{0 \le y < p} M_{i\cdot p+x, j\cdot p+y}^{(l+1,k')} \tag{7}$$

where $c^{(l)}$ denotes the number of feature maps in the $l_{th}$ convolution layer. $M^{(l)} \in \mathbb{R}^{\frac{d}{p^{l-1}} \times \frac{d}{p^{l-1}}}$ and $\tilde{M}^{(l)} \in \mathbb{R}^{\frac{d}{p^l} \times \frac{d}{p^l}}$. The number of the convolution layers is set to $L$. Finally, We flatten $\tilde{M}^{(L)}$ to get the fused text-image feature $f_E \in \mathbb{R}^{\frac{d}{p^L}\frac{d}{p^L}c^{(L)}}$ as the output of feature fusion process.

## 3.4   Objective Function

After 2D-convolutional networks, we obtain a fused text-image feature vector. The fused feature vector is fed into a fully connected layer to get the similarity score. The objective function is a rank-based pairwise loss function [9], which maximizes the mean similarity score $u^+$ between text-image pairs from the same semantic concept and minimize the mean similarity score $u^-$ between pairs from different semantic concepts. At the same time, our model also minimises the variance of pairwise similarity score for both matching $\sigma^{2+}$ and non-matching $\sigma^{2-}$ pairs. The loss function is as follows:

$$Loss = (\sigma^{2+} + \sigma^{2-}) + \lambda \max(0, m - (u^+ - u^-)) \tag{8}$$

where $m$ is the threshold between the mean distributions of matching similarity and non-matching similarity, and $\lambda$ is used to adjust the influences of mean and variance.

# 4 Experiment

## 4.1 Datasets

To evaluate the performance of our proposed model, we accomplish the general cross-modal retrieval tasks: text-query-images and image-query-texts. Experiments are conducted on two benchmark datasets: English Wikipedia (Eng-Wiki for short) [17] and TVGraz [13]. Eng-Wiki, collected from Wikipedia's "featured articles", contains 2,866 text-image pairs, where 2,173 pairs for training and 693 pairs for test. TVGraz is a collection of webpages, which contains 2,360 text-image pairs, where 1,885 pairs for training and 475 pairs for test. Both of the two datasets contains 10 categories and each text-image pair belongs to one category. We utilize VGG-19 [19] to model the images and represent each image as a 4,096-dimensional vector. Meanwhile, we adopt GCN model with the same structure as [24] to model the texts and represent each text as a 10,055-dimensional vector in Eng-Wiki and a 8,172-dimensional vector in TVGraz, respectively.

## 4.2 Evaluation and Implementation

In all the experiments, mean average precision (MAP) [17] is used to evaluate the retrieval results. Higher MAP indicates better retrieval performance. We implement our framework in Tensorflow. Following the same strategy as [24] for constructing training set, we selected 40,960 text-image pairs as positive samples and 40,960 text-image pairs as negative samples from the training set of the Eng-Wiki dataset. For the TVGraz dataset, 40,000 positive samples and 40,000 negative samples are randomly selected from the corresponding training set. We set the regularization weight 0.005, learning rate 0.001 with an Adam optimization, and 50 epochs for training. The dropout ratio at the input of the last fully connected layer is 0.4 for the Eng-Wiki dataset and 0.2 for the TVGraz dataset. We set the same parameters for loss function as in [24]. In the text modeling and image modeling paths, the feature dimension after fully connected layer is set to 128 for both modalities on the two datasets.

## 4.3 The Influence of 2D-Convolution Settings

To evaluate the influence of the 2D-convolutional networks in our proposed framework, we vary the number of convolution layers as well as the size of kernels to form several variant models based on our proposed framework. Except for the convolution part, other structures in the framework (as shown in Fig. 1) are fixed for all the variant models. By default, each convolutional layer is followed by a ReLU activation layer and a $2 \times 2$ max-pooling layer. The variant models include:

- **baseline**: we fuse the image feature and text feature by directly multiplying their feature vectors to get the feature matrix, without convolution layers. Then a non-linear mapping is followed to obtain the similarity score.

**Table 1.** Retrieval results of different 2D-convolution settings on the Eng-Wiki dataset.

| Method | Text query | Image query | Average |
|---|---|---|---|
| Baseline | 0.780 | 0.406 | 0.593 |
| 2D-ConvFF (16-1 × 1) | 0.827 | 0.448 | 0.637 |
| 2D-ConvFF (16-3 × 3) | **0.874** | 0.394 | 0.634 |
| 2D-ConvFF (16-1 × 1, 32-1 × 1) | 0.863 | 0.419 | 0.641 |
| 2D-ConvFF (16-1 × 1, 32-3 × 3) | 0.836 | 0.449 | 0.643 |
| 2D-ConvFF (16-3 × 3, 32-3 × 3) | 0.846 | **0.457** | **0.651** |

**Table 2.** Retrieval results of different 2D-convolution settings on the TVGraz dataset.

| Method | Text query | Image query | Average |
|---|---|---|---|
| Baseline | 0.869 | 0.913 | 0.891 |
| 2D-ConvFF (16-1 × 1) | 0.861 | 0.921 | 0.892 |
| 2D-ConvFF (16-3 × 3) | 0.875 | 0.906 | 0.891 |
| 2D-ConvFF (16-1 × 1, 32-1 × 1) | 0.884 | 0.921 | 0.902 |
| 2D-ConvFF (16-1 × 1, 32-3 × 3) | 0.899 | 0.926 | 0.912 |
| 2D-ConvFF (16-3 × 3, 32-3 × 3) | **0.938** | **0.933** | **0.935** |

- **2D-ConvFF** (16-1 × 1): we add one convolution layer in the 2D-ConvFF module. The convolution layer has 16 kernels and that the kernel size is $1 \times 1$, which can capture the pairwise relationships between each dimension of the image feature and the text feature.
- **2D-ConvFF** (16-3 × 3): this model has the same structure as the model 2D-ConvFF (16-1 × 1), only differing in the kernel size. The kernel size in this model is $3 \times 3$.
- **2D-ConvFF** (16-1 × 1, 32-1 × 1): we add two convolution layers in the 2D-ConvFF module. The first convolution layer has 16 kernels while the second convolution layer has 32 convolution kernels. The size of all the kernels is $1 \times 1$.
- **2D-ConvFF** (16-1 × 1, 32-3 × 3): this model has the same structure as the model 2D-ConvFF (16-1 × 1, 32-1 × 1), only differing in the kernel size. The kernel size of the second convolution layer in this model is $3 \times 3$.
- **2D-ConvFF** (16-3 × 3, 32-3 × 3): this model has the same structure as the model 2D-ConvFF (16-1 × 1, 32-1 × 1), only differing in the kernel size. The kernel size in this model is $3 \times 3$.

Table 1 shows the retrieval results on the Eng-Wiki dataset. Compared with the baseline model, we obtain about 4% improvement when adding only one convolution layer. The performance of $1 \times 1$ kernels is slightly superior than that of $3 \times 3$ kernels. Furthermore, we achieve another 1% improvement when adding two convolution layers. The aforementioned results indicate that the convolution layers can effectively capture inner-group relationships of multi-modal features and enhance the expressiveness of the text-image features, thus promoting the retrieval performance. Two convolution layers with $3 \times 3$ kernels achieve the best performance on the Eng-Wiki dataset. Table 2 presents the retrieval results on the TVGraz dataset and we come out with similar conclusion as the Eng-Wiki dataset. We observe that adding convolution layers can achieve higher MAP scores compared with feature fusion by only multiplication. Meanwhile, using two convolution layers are more effective than only one convolution layer. This is because that two convolution layers can capture more complex cross-modal correlations within lager receptive field. In summary, 2D-ConvFF(16-3×3, 32-3× 3) achieves the best performance on both datasets. In the following experiments, we use this model to compare with state-of-the-art methods.

## 4.4    Comparison with State-of-the-Art Methods

We compare our model with several state-of-the-art models. These models are widely cited works in this field, which include unsupervised models, supervised models and semi-supervised models. CCA [17] and CM [13] are unsupervised methods which adopts pairwise constrains to maximize the correlation between multi-modal features. AUSL [25] is a semi-supervised model that leverages both labelled and unlabelled data. SM [13], SCM [13], GMLDA [18], GMMFA [18], ml-CCA [15], TCM [14], LCFS [21], LGCFL [6], JFSSL [20] and GIN [24] are supervised models that use the semantic category information to capture the correlation between paired text-image pairs. Compared with GIN, when we process the training data to generate positive and negative samples, we remove some of the repeated samples to guarantee the quality of the training data. For fair comparison, we retrain the GIN model with new data and use the results to compare with our model.

**Table 3.** Comparison with state-of-the-art methods on the Eng-Wiki dataset.

| Method | Text query | Image query | Average |
|---|---|---|---|
| CCA [17] | 0.187 | 0.216 | 0.202 |
| SCM [17] | 0.234 | 0.276 | 0.255 |
| TCM [14] | 0.293 | 0.232 | 0.266 |
| LCFS [21] | 0.204 | 0.271 | 0.238 |
| LGCFL [6] | 0.316 | 0.378 | 0.347 |
| ml-CCA [15] | 0.287 | 0.353 | 0.312 |
| CMLDA [18] | 0.289 | 0.316 | 0.302 |
| CMMFA [18] | 0.296 | 0.316 | 0.306 |
| AUSL [25] | 0.332 | 0.397 | 0.364 |
| JFSSL [20] | 0.410 | **0.467** | 0.439 |
| GIN [24] | 0.789 | 0.432 | 0.610 |
| 2D-ConvFF (ours) | **0.846** | 0.457 | **0.651** |

Table 3 shows the MAP scores on the Eng-Wiki dataset. Our proposed model remarkably outperforms state-of-the-art methods in all measures except for the

**Table 4.** Comparison with state-of-the-art methods on the TVGraz dataset.

| Method | Text query | Image query | Average |
|---|---|---|---|
| CM [13] | 0.450 | 0.460 | 0.455 |
| SM [13] | 0.585 | 0.619 | 0.602 |
| SCM [17] | 0.696 | 0.693 | 0.695 |
| TCM [14] | 0.706 | 0.694 | 0.695 |
| GIN [24] | 0.885 | 0.903 | 0.894 |
| 2D-ConvFF (ours) | **0.938** | **0.933** | **0.935** |

image query, which is slightly inferior to JFSSL. Compared with the second best model GIN, 2D-ConvFF is about 5.7%, 2.5%, and 4.1% improvement on text query, image query, and the average performance, respectively. It proves the effectiveness of our proposed model on the most widely compared cross-modal retrieval dataset. Table 4 shows the results on the TVGraz dataset. We observe that our model achieve the best results on all the measures, which respectively gains about 5.3%, 3%, and 4.1% improvement on text query, image query, and average performance, compared with state-of-the-art model GIN. GIN use element-wise product to fuse features among the text and image, which only can attain simple correlations across modalities. While 2D-ConvFF capture the inner-group fine-grained correlations across different modalities by the convolution networks. We can conclude that our proposed feature fusion method is a great benefit to the cross-modal retrieval tasks. The complex cross-modal interactions can be well modeled by the convolution networks.

## 4.5   Comparison with Baseline Models

Besides our proposed model, we implement another four baseline models to evaluate the influence of different feature fusion methods on the final cross-modal retrieval performance. All the experiments are conducted on the Eng-Wiki dataset. The other four baseline models are respectively combined with four feature fusion methods, including concatenation, element-wise addition, element-wise product, and vector multiplication.

**Table 5.** Comparison with baseline models on the Eng-Wiki dataset.

| Method | Text query | Image query | Average |
|---|---|---|---|
| Concatenation | 0.103 | 0.088 | 0.096 |
| Element-wise addition | 0.111 | 0.083 | 0.097 |
| Element-wise product | 0.789 | 0.432 | 0.610 |
| Vector multiplication | 0.780 | 0.406 | 0.593 |
| 2D-ConvFF (ours) | **0.846** | **0.457** | **0.651** |

The MAP scores are listed in Table 5. It's obvious that simply using linear models, i.e. concatenation and element-wise addition, to fuse the image feature and the text feature leads to extremely low MAP scores. This is due to the fact that the feature distributions of different modalities vary dramatically. Linear models may not be able to capture such complex correlations across modalities and construct informative text-image features. Compared with linear models, non-linear models (i.e. element-wise product), vector multiplication, and 2D-ConvFF obtain significant improvement on the CMIR performance. The performance of element-wise product is comparable with vector multiplication. Our proposed 2D-ConvFF further outperforms element-wise product by about 4% and achieves the best performance on all the measures. It indicates that 2D-ConvFF achieves more effective fusion of multi-modal features by capturing their complex relationships sufficiently. As a result, our 2D-ConvFF can generate more informative image-text feature representations and improve the retrieval results remarkably.

## 5    Conclusion

In this paper, we propose a 2D-Convolution based Feature Fusion (2D-ConvFF) framework for cross-modal information retrieval. By introducing convolution operations on the cross-modal feature matrix, we capture more complex correlations between the features of multi-modal data, resulting in more informative fused text-image representations. Experimental results on both Eng-Wiki and TVGraz datasets indicate that our proposed 2D-ConvFF can achieve significant improvements on the general cross-modal information retrieval tasks comparing with state-of-the-art methods. Besides, baseline study further proves that our 2D-ConvFF is superior to the existing feature fusion methods. In the future work, we will further explore multi-modal correlations by fusing more explicit features, such as image regions and keywords, to make the model more explainable.

## References

1. Castrejon, L., Aytar, Y., Vondrick, C., Pirsiavash, H., Torralba, A.: Learning aligned cross-modal representations from weakly aligned data. In: IEEE Conference on Computer Vision and Pattern Recognition (CVPR), pp. 2940–2949 (2016)
2. Devlin, J., Chang, M.W., Lee, K., Toutanova, K.: BERT: pre-training of deep bidirectional transformers for language understanding. arXiv: 1810.04805 (2018)
3. Fukui, A., Park, D.H., Yang, D., Rohrbach, A., Darrell, T., Rohrbach, M.: Multimodal compact bilinear pooling for visual question answering and visual grounding. In: Conference on Empirical Methods in Natural Language Processing (EMNLP), pp. 457–468 (2016)
4. He, Y., Xiang, S., Kang, C., Wang, J., Pan, C.: Cross-modal retrieval via deep and bidirectional representation learning. IEEE Trans. Multimedia (TMM) 18(7), 1363–1377 (2016)
5. Huang, Y., Wang, W., Wang, L.: Instance-aware image and sentence matching with selective multimodal LSTM. In: IEEE Conference on Computer Vision and Pattern Recognition (CVPR), pp. 2310–2318 (2017)

6. Kang, C., Xiang, S., Liao, S., Xu, C., Pan, C.: Learning consistent feature representation for cross-modal multimedia retrieval. IEEE Trans. Multimedia (TMM) **17**(3), 276–288 (2017)
7. Kim, J.H., On, K.W., Lim, W., Kim, J., Ha, J.W., Zhang, B.T.: Hadamard product for low-rank bilinear pooling. In: International Conference on Learning Representations (ICLR) (2017)
8. Kipf, T.N., Welling, M.: Semi-supervised classification with graph convolutional networks. In: International Conference on Learning Representations (ICLR) (2017)
9. Kumar, B.G.V., Carneiro, G., Reid, I.: Learning local image descriptors with deep siamese and triplet convolutional networks by minimizing global loss functions. In: IEEE Conference on Computer Vision and Pattern Recognition (CVPR), pp. 5385–5394 (2016)
10. Lu, Y., Yu, J., Liu, Y., Tan, J., Guo, L., Zhang, W.: Fine-grained correlation learning with stacked co-attention networks for cross-modal information retrieval. In: Liu, W., Giunchiglia, F., Yang, B. (eds.) KSEM 2018. LNCS (LNAI), vol. 11061, pp. 213–225. Springer, Cham (2018). https://doi.org/10.1007/978-3-319-99365-2_19
11. Mikolov, T., Chen, K., Corrado, G., Dean, J.: Efficient estimation of word representations in vector space. In: International Conference on Learning Representations (ICLR), pp. 1–12 (2013)
12. Mikolov, T., Karafiát, M., Burget, L., Cernocký, J., Khudanpur, S.: Recurrent neural network based language model. In: Annual Conference of the International Speech Communication Association (INTERSPEECH), pp. 1045–1048 (2010)
13. Pereira, J.C., et al.: On the role of correlation and abstraction in cross-modal multimedia retrieval. IEEE Trans. Pattern Anal. Mach. Intell. (TPAMI) **36**(3), 521–535 (2014)
14. Qin, Z., Yu, J., Cong, Y., Wan, T.: Topic correlation model for cross-modal multimedia information retrieval. Pattern Anal. Appl. (PAA) **19**(4), 1007–1022 (2016)
15. Ranjan, V., Rasiwasia, N., Jawahar, C.V.: Multi-label cross-modal retrieval. In: IEEE International Conference on Computer Vision (ICCV), pp. 4094–4102 (2015)
16. Rasiwasia, N., Mahajan, D., Mahadevan, V., Aggarwal, G.: Cluster canonical correlation analysis. In: International Conference on Artificial Intelligence and Statistics (AISTATS), pp. 823–831 (2014)
17. Rasiwasia, N., et al.: A new approach to cross-modal multimedia retrieval. In: ACM International Conference on Multimedia (ACM MM), pp. 251–260 (2010)
18. Sharma, A., Kumar, A., Daume, H., Jacobs, D.W.: Generalized multiview analysis: a discriminative latent space. In: IEEE Conference on Computer Vision and Pattern Recognition (CVPR), pp. 2160–2167 (2012)
19. Simonyan, K., Zisserman, A.: Very deep convolutional networks for large-scale image recognition. In: International Conference on Learning Representations (ICLR) (2015)
20. Wang, K., He, R., Wang, L., Wang, W., Tan, T.: Joint feature selection and subspace learning for cross-modal retrieval. IEEE Trans. Pattern Anal. Mach. Intell. (TPAMI) **38**(10), 2010–2023 (2016)
21. Wang, K., He, R., Wang, W., Wang, L., Tan, T.: Learning coupled feature spaces for cross-modal matching. In: IEEE International Conference on Computer Vision (ICCV), pp. 2088–2095 (2013)
22. Wang, L., Li, Y., Huang, J., Lazebnik, S.: Learning two-branch neural networks for image-text matching tasks. IEEE Trans. Pattern Anal. Mach. Intell. (TPAMI) **41**(2), 394–407 (2018)

23. Yan, F., Mikolajczyk, K.: Deep correlation for matching images and text. In: IEEE Conference on Computer Vision and Pattern Recognition (CVPR), pp. 3441–3450 (2015)
24. Yu, J., et al.: Modeling text with graph convolutional network for cross-modal information retrieval. In: Pacific-Rim Conference on Multimedia (PCM), pp. 862–871 (2005)
25. Zhang, L., Ma, B., He, J., Li, G., Huang, Q., Tian, Q.: Adaptively unified semi-supervised learning for cross-modal retrieval. In: International Joint Conference on Artificial Intelligence (IJCAI), pp. 3406–3412 (2017)

# DunDi: Improving Robustness of Neural Networks Using Distance Metric Learning

Lei Cui[1], Rongrong Xi[1(✉)], Zhiyu Hao[1], Xuehao Yu[2], and Lei Zhang[1]

[1] Institute of Information Engineering, Chinese Academy of Sciences, Beijing, China
{cuilei,xirongrong,haozhiyu,zhanglei1}@iie.ac.cn
[2] State Grid Information and Telecommunication Branch, Beijing, China

**Abstract.** The deep neural networks (DNNs), although highly accurate, are vulnerable to adversarial attacks. A slight perturbation applied to a sample may lead to misprediction of the DNN, even it is imperceptible to humans. This defect makes the DNN lack of robustness to malicious perturbations, and thus limits their usage in many safety-critical systems. To this end, we present DunDi, a metric learning based classification model, to provide the ability to defend adversarial attacks. The key idea behind DunDi is a metric learning model which is able to pull samples of the same label together meanwhile pushing samples of different labels away. Consequently, the distance between samples and model's boundary can be enlarged accordingly, so that significant perturbations are required to fool the model. Then, based on the distance comparison, we propose a two-step classification algorithm that performs efficiently for multi-class classification. DunDi can not only build and train a new customized model but also support the incorporation of the available pre-trained neural network models to take full advantage of their capabilities. The results show that DunDi is able to defend 94.39% and 88.91% of adversarial samples generated by four state-of-the-art adversarial attacks on the MNIST dataset and CIFAR-10 dataset, without hurting classification accuracy.

**Keywords:** Robustness · Deep neural network · Metric learning

## 1 Introduction

Over the past few years, the machine learning systems, especially the one cooperated with the deep neural network (DNN), have reached promising achievement in many fields including image recognition, speech translation, and policy decision [11]. Due to the advance, they are being widely deployed in many real-world applications such as self-driving, face authentication and medical diagnosis. Despite the success, however, recent studies have shown that the deep neural network models, although highly accurate, are vulnerable to adversarial attacks [4,13,16]. That is, the model can be fooled to mispredict a modified sample which is intentionally perturbed by the attacker, so-called adversarial sample.

© Springer Nature Switzerland AG 2019
J. M. F. Rodrigues et al. (Eds.): ICCS 2019, LNCS 11537, pp. 145–159, 2019.
https://doi.org/10.1007/978-3-030-22741-8_11

These adversarial samples can cause unexpected behaviour of DNNs, and thus are problematic when DNNs are used in safety-critical systems. For example, an adversarial traffic sign may fool the self-driving system to determine a wrong policy and even worse cause car crash [14], and an adversarial human face will allow an attacker to pass the face authorization [21]. Therefore, to provide the ability to defend adversarial samples (or robustness in other literature), is one key factor in deploying DNNs in many safety-critical systems.

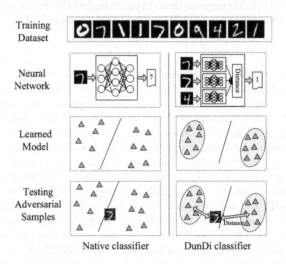

**Fig. 1.** Comparison of traditional model and DunDi model. DunDi employs the Triplet Network to train a model that puts similar samples together and enlarges the distance between different samples, which in turn enlarges the distance between samples and model's boundary. The testing samples will be classified exploiting the distance metric. As we can see, an adversarial sample can move across the boundary to lead to misprediction of the Native classifier, but it will be correctly classified by DunDi since it is closer to the samples of the proper label.

Several approaches have been proposed for defending adversarial attacks in recent years, such as adversarial training [3,22,23], defensive distillation [17] and deep contractive networks [7]. In addition to defense, some approaches tend to detect the adversarial samples and exclude them upon testing [12]. Unfortunately, some works show that these models are still vulnerable to adversarial samples [5]. Recent studies have shown that the DNNs are not robust mainly because the natural samples lie very close to the model's decision boundary [4,13]. Or in other words, the distance between the sample and boundary is short, so that the sample can easily move across the boundary by patching with slight perturbations. Inspired by this, we propose DunDi[1] to defend against adversarial samples

---

[1] DunDi is a kind of magic in ancient Eastern legends, and it can transform far to near or near to far, thus avoiding enemy attacks.

by enlarging the distance between samples and boundary, so that a much larger magnitude of perturbations are required to move samples across the boundary, as shown in Fig. 1. To achieve this, the key idea behind DunDi is a metric learning model which has been used in many fields [20]. Specifically, we employ a recently proposed metric learning model named as Triplet Network [8]. The Triplet Network takes as input three samples $\{x_a, x_p, x_n\}$ each time in which $x_a$ and $x_p$ are similar while $x_a$ and $x_n$ are different. Then, it learns to pull together the samples of the same label (i.e., reducing the distance between $x_a$ and $x_p$) meanwhile pushing away the samples of different labels (i.e., enlarging the distance between $x_a$ and $x_n$). Consequently, it enlarges the distance between samples and model's boundary, which in turn amplifies the required perturbations for crafting adversarial samples. Exploiting the trained metric model that provides distance computation between samples, we then design a two-step algorithm for efficient multi-class classification. DunDi can build a new model with customized DNN topology. In addition, it supports the incorporation of the pre-trained DNNs, to take full use of their advantages such as high accuracy.

We implement DunDi on Keras/Tensorflow platform. The experimental results show that DunDi is able to defend a significant fraction of adversarial samples. Specifically, it correctly predicts 94.39% and 88.91% of adversarial samples [5] respectively on the MNIST and CIFAR-10 dataset. To conclude, this paper makes the following contributions:

- First, we propose a metric learning based method to enlarge the distance between samples and model's boundary, which in turn amplifies the magnitude of malicious perturbations required to craft adversarial samples, for improving the robustness of DNN to adversarial attacks.
- Second, we design a two-step classification algorithm for classifying samples in a multi-class scenario efficiently based on the distance metric.
- Third, we implement DunDi and evaluate it on two datasets to show its effectiveness when suffering different types of adversarial attacks.

## 2 Background and Related Work

### 2.1 Adversarial Attacks

Adversarial attacks intentionally design a new sample by adding slight perturbations to a sample that is correctly classified by the model, so that the model misclassifies the new sample, so-called adversarial sample. Many recent works tend to carefully perturb samples for improving the attack success rate, including fast gradient sign method (FGSM), iterative FGSM (iFGSM), Jacobian-based saliency map algorithm (JSMA), projection method (DeepFool), etc. We will give a brief introduction to these methods below.

**Notations:** Let $x$ be a sample, $f$ be the trained classifier, $y$ be the predicted label of input $x$ (i.e., $y = f(x)$), and $y*$ be the true label of $x$. During the training stage, the model $f$ will be continuously updated for minimizing the value of loss function $L(y, y*)$ given a set of $x$ and $y*$.

**FGSM:** FGSM [6] computes the gradient of the loss function $L$ with respect to a sample $x$, generates perturbations formed by the sign of the gradient multiplied by a parameter $\epsilon$, and finally crafts an adversarial sample $x'$ by adding perturbations to $x$, as shown in Eq. 1. $\epsilon$ refers to the specified magnitude of perturbations. If it is small enough, the crafted adversarial sample and the original sample will be indistinguishable from the human view.

$$x' = x + \epsilon sign(\nabla_x L(y, y*)) \tag{1}$$

A later work, named as iterative FGSM or iFGSM, extends FGSM by applying FGSM multiple times with a finer perturbation. In each step, iFGSM clips pixel values of intermediate results of the previous step. The advantage over FGSM is that it achieves smaller distortion and higher attack success rate [9].

**JSMA:** JSMA is a targeted attack approach, i.e., it specifies a target label $y_t$ and tends to perturb $x$ so that the model misclassifies the crafted sample $x'$ to $y_t$ [15]. To achieve this, JSMA iteratively searches for pixels or pairs of pixels in $x$ to change such that the probability of the target label is increased and the probability of all other labels are decreased.

**DeepFool:** DeepFool is an untargeted attack algorithm based on the theory of projection to the closest separating hyperplane in classification [14], so that it can compute minimal perturbations to sufficiently fool the classifier.

### 2.2   Defenses

**Adversarial Training.** This method augments the training dataset with a number of correctly labelled adversarial samples and then retrains the neural network model. With weights updating for minimizing the value of loss function, especially that of adversarial samples, the retrained neural network model will be more robust to adversarial samples [22,23].

**Defensive Distillation.** This strategy firstly trains a model using hard class labels. Then, exploiting the initial model's softmax probability outputs, it trains a second model. This method can smooth the model's surface in the directions would be exploited by attackers, making it difficult to discover gradients for crafting adversarial samples [17].

**Gradient Masking.** Observed that most adversarial attack techniques generate adversarial samples using the gradients of the neural network, gradient masking is proposed to eliminate the gradients, so that the attackers fail to find appropriate directions that can be used for perturbing the samples [19,24]. However, a recent study has shown that this method fails to work in many scenarios because the attacker can still discover gradients in a high dimension space.

## 3   Design and Implementation of DunDi

In this section, we first formulate the robustness problem by the distance between samples and model's boundary and then present an overview of the proposed

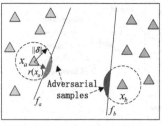 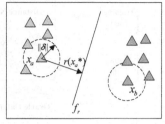

(a) Models are not robust.          (b) Model is more robust.

**Fig. 2.** Samples, boundaries, and adversarial samples. $f_a$ and $f_b$ are two accurate but not robust classifiers, while $f_r$ is more robust because the distance between samples and boundary is larger. The green and red regions denote the distribution of adversarial samples with respect to $x_a$ and $x_b$ respectively. (Color figure online)

solution. Finally, we introduce two essential parts, i.e., how to enlarge the distance for improving robustness and how to classify samples with the distance metric.

### 3.1 Enlarging Distance for Robustness

We first introduce some notations related to the robustness of neural networks [4], and then explain why increasing the distance would improve robustness.

**Perturbation.** Let $x$ be a sample, $f$ be the trained classification model. $\delta$ refers to the perturbations applied to $x$ for generating a new sample $x'$, i.e., $x' = x + \delta$. If the predicted label of $x'$ is different from that of $x$, i.e., $f(x) \neq f(x')$, then an adversarial sample $x'$ with respect to $\delta$ is successfully crafted.

**Magnitude of Perturbation.** The magnitude of perturbations $\delta$ is denoted by $\parallel \delta \parallel_S$. It can be computed in many norms such as $\ell_1$, $\ell_2$ and $\ell_\infty$ in $S$ which denotes the space of admissible perturbations.

**Robustness of One Sample.** The robustness of one sample $x$, denoted by $r(x)$, can be measured by the minimal perturbations changing the label of $x$, i.e., $r(x) = min \parallel \delta \parallel_S$ subject to $f(x + \delta) \neq f(x)$. The smaller value of $r(x)$ implies that it requires slight modifications at $x$ for changing its label, thereby enabling the perturbed samples difficult to be detected.

**Robustness Over the Classifier.** To measure the robustness over the classifier $f$, denoted by $R(f)$, we can simply compute the expectation of $r(x)$ over all the samples, i.e., $R(f) = \mathbb{E}_X(r(x))$.

Figure 2(a) illustrates the notations associated with two classes of samples and several classifiers. For a sample $x_a$, the dashed cycle shows the space of admissive perturbations, i.e., $S$. A large number of new samples can be generated by adding perturbations within the cycle to $x_a$. Given a classifier $f_a$, the robustness of $x_a$, denoted by $r(x_a)$ in Fig. 2(a), refers to the shortest distance for moving $x_a$ across the boundary of $f_a$.

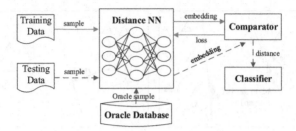

**Fig. 3.** Overview. The green arrow shows the training stage, while the red dashed arrow shows the testing stage. (Color figure online)

Recent DNNs have shown the ability to train accurate classifiers. However, high accuracy does not mean robustness. As shown in Fig. 2(a), two classifiers $f_a$ and $f_b$ are accurate because they can correctly classify the samples. Unfortunately, they are not robust. Any sample $x'_a$ in the green region, which perturbs $x_a$ by adding perturbations with a magnitude greater than $r(x_a)$, can lead to misclassification of $f_a$. The same is true for sample $x_b$ and classifier $f_b$. The reason is that the sample lies too close to the classifier's boundary, so that it can be easily perturbed for moving across the boundary. Inspired by this, an accurate and robust classifier should keep the samples far away from the boundary for defending adversarial attacks. Figure 2(b) presents a more robust classifier $f_r$, i.e., a larger magnitude of perturbations (i.e., $r(x_a*)$) is required to fool the classifier due to a larger distance between $x_a$ and $f_r$. Such considerable perturbations would surpass the allowed perturbing space $S$, making the crafted sample be easily detected. This gives us a hint that enlarging the distance between samples and boundary would help to improve the robustness of the model.

### 3.2  Overview of DunDi

As stated above, the distance is one of the keys to improve the robustness of neural network models [4,13]. To this end, we propose to enlarge the distance between samples and model's boundary with metric learning. Figure 3 presents a brief overview of DunDi on describing the process of training stage and testing stage. The critical component is a Distance Neural Network (Distance NN). It is responsible for producing an embedding, which is a numeric vector representing learned features for each input sample. The embeddings should satisfy two requirements. First, for samples of the same label, the distance of their embeddings should be small. Second, for samples of different labels, the distance of associated embeddings should be large. To achieve this, the Distance NN during the training stage takes as input samples and produces embeddings. Then, the Comparator computes the distance between embeddings, computes the loss between the expected distance (i.e., specified by the user) and the predicted distance (i.e., computed between generated embeddings), and then leverages the

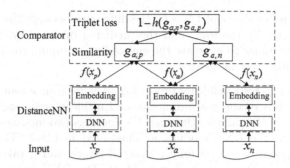

**Fig. 4.** Design of Distance NN and Comparator based on Triplet Network. $f$ denotes the embedding for each sample, $g$ denotes the similarity between two embeddings, and $1 - h$ denotes the triplet loss.

loss to update the weights of Distance NN with feedback propagation. Finally, with the aim to minimize the loss, the Distance NN will learn to generate embeddings following the two requirements.

For testing, DunDi maintains an Oracle Database storing a certain number of samples for each class in the dataset, which are considered as oracle samples for classification. Given a testing sample, the Classifier leverages Distance NN to produce the embedding, employs Comparator to compute the distance between the testing sample and oracle samples, and finally employs the proposed multi-class classification algorithm to predict the label.

### 3.3 Enlarging Distance Using Metric Learning

In DunDi, we enlarge the distance with metric learning (also known as similarity learning), which tends to learn from samples a similarity function that measures how similar two objects are. Specifically, we employ the idea of recently proposed Triplet Network to train the model [8]. The Triplet Network takes as input a tuple, which contains three samples along with a ranking of similarities between them. It can learn to compute similarities that are distributed in a wide range, i.e., [0,1], enabling the model to outperform other metric learning methods such as Siamese Network that only operates either 0 (same samples) or 1 (different samples) [2]. As shown in Fig. 4, the input of Distance NN is a tuple denoted by $\{x_a, x_p, x_n\}$, in which $x_a$ (anchor sample) is more similar to $x_p$ (positive sample) than it is to $x_n$ (negative sample). DunDi aims to put the positive pair closer (i.e., $x_a$ and $x_p$) meanwhile pushing the negative one further (i.e., $x_a$ and $x_n$), for following the ranking of similarities provided in the input tuple. To achieve this, it computes the triplet loss of three input samples, which acts as the loss to be minimized during the training stage.

As shown in Fig. 4, given the tuple $\{x_a, x_p, x_n\}$, the Distance NN produces an embedding for each sample, denoted by $f(x)$. With these embeddings, the Comparator calculates the similarity of each pair, denoted by $g(f(x_a), f(x_p))$ (or $g_{a,p}$ for short) for the positive pair and $g_{a,n}$ for the negative pair. Then, it computes the predicted discrepancy between $g_{a,p}$ and $g_{a,n}$, denoted by $h(g_{a,p}, g_{a,n})$, and finally computes the triplet loss by the difference between the expected discrepancy (specified by the user) and the predicted discrepancy. The following parts will describe how to generate the embedding, compute the similarity and compute the triplet loss in detail.

**Distance NN.** It is responsible for producing an embedding for an input sample. The embedding is a vector in a $n$-dimensional Euclidean space, representing the learned features of the sample. Distance NN contains two modules, the Embedding module and the DNN module, as illustrated in Fig. 4. The Embedding module is composed of several fully connected layers, and it takes the weights of the last layer as the generated embedding. The DNN module can be the convolutional neural network or other DNNs. It is worth noting that the DNN module can incorporate the pre-trained DNN models to take full advantage of their capabilities. For a pre-trained DNN model, we simply remove its softmax layer and then connect it to the Embedding module to produce the embedding.

**Similarity Computation.** There exist many methods to compute the similarity (or, in this case, the distance) between two embeddings, e.g., Euclidean, Manhattan and Cosine distance. In DunDi, we leverage the Euclidean distance to measure the similarity, i.e., $g_{a,p} = \sqrt[2]{(a_1 - p_1)^2 + (a_2 - p_2)^2 + ... + (a_n - p_n)^2}$.

**Loss Function.** As stated before, the tuple $\{x_a, x_p, x_n\}$ implies that $x_a$ and $x_p$ are more similar than $x_a$ and $x_n$. For simplicity, we assign $x_a$ and $x_p$ with samples of the same label while $x_n$ a different label, so that the expected discrepancy, denoted by $h_e(g_{a,n}, g_{a,p})$, is 1. Let $h_p(g_{a,n}, g_{a,p})$ denote the predicted discrepancy, then the triplet loss value is computed by $1 - h_p(g_{a,n}, g_{a,p})$. The Comparator will return the loss back to the Distance NN for optimizing the weights of the neural network model. By iteratively optimization, the triplet loss will continue to decrease until convergence. We leverage the *sigmoid* function to compute the predicted discrepancy, i.e., $h_p = sigmoid(g_{a,n} - g_{a,p})$, so that its value as well the triplet loss can be mapped to a range of 0 to 1.

### 3.4  Multi-class Classification Algorithm

As for testing, it only requires a pair of samples each time, one of which is the testing sample and the other from the labelled dataset. We employ the Distance NN to produce embeddings and then computes the similarity between the two. Exploiting the similarities between the testing sample and other samples, we can classify the testing sample using the idea of nearest neighbour clustering, i.e., it will be assigned with the label of the closest class of samples.

Unfortunately, this approach will introduce significant time overhead, mainly because there exist multiple classes each of which has thousands of samples in the dataset for similarity computation. To mitigate this problem, instead of using the whole dataset, we use a small number ($n$) of labelled samples of each class as oracle samples. These samples are randomly selected to cover the diversity of samples. We then propose a two-step multi-class classification algorithm for reducing the computation cost, as shown in Algorithm 1. The first step attempts

---

**Input:** $text\_x$
**Output:** $predicted\_label$
$selected\_labels \leftarrow \{\}$, $similarities \leftarrow \{\}$ ;
**for** $label \in all\_labels$ **do**
  $label\_sim \leftarrow 0$;
  **for** $sample \in oracle\_samples[label][0:m]$ **do**
    | $sim \leftarrow sim(test\_x, sample)$, $label\_sim \leftarrow label\_sim + sim$;
  **end**
  $similarities \leftarrow similarities \cup label\_sim$;
**end**
$selected\_labels \leftarrow argtop\_k(similarities)$;
$predicted\_label \leftarrow 0$, $similarities \leftarrow \{\}$ ;
**for** $label \in selected\_labels$ **do**
  $label\_sim \leftarrow 0$ ;
  **for** $sample \in oracle\_samples[label][0:n]$ **do**
    | $sim \leftarrow sim(test\_x, sample)$, $label\_sim \leftarrow label\_sim + sim$;
  **end**
  $similarities \leftarrow similarities \cup label\_sim$;
**end**
$predicted\_label = arg\max(similarities)$

---

**Algorithm 1.** Two-step classification algorithm

to exclude the most unlikely labels. It compares the testing sample with $m$ ($m < n$) oracle samples of each class (# 2–6) and selects the top $k$ labels showing the highest similarities (#7). Then, the second step compares the testing sample with $n$ samples of each of the selected $k$ labels (#9–13) and finally assigns it with the label with the highest similarity (#14). We empirically set the value of $n$, $m$, and $k$ to 20, 3, and 5 respectively, which achieves well trade-off between the accuracy and testing time.

## 3.5 Implementation

We implement DunDi using Keras on top of TensorFlow [1]. Within the paper, we focus on image classification, and thus we use the convolutional neural network as the DNN model in the Distance NN. Unlike traditional CNN topologies, which are configured with a softmax layer to help output a single label or a probability distribution over several labels, the CNN of DunDi has no softmax layer but is connected to the Embedding module. The Embedding module is a simple neural network composed of two fully-connected layers, and it produces a 128 length vector which is the final embedded representation of the input sample. As for the pre-trained CNN model, we firstly employ it to produce intermediate vectors and feed them into the Embedding module. Then we only update the weights of the Embedding module for minimizing the triplet loss value during training.

In Comparator, we compute the Euclidean similarity between two embeddings. Then, we compute the predicted discrepancy between the similarities of

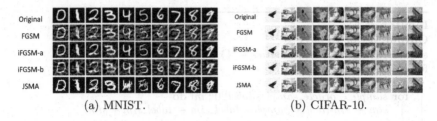

(a) MNIST.                                    (b) CIFAR-10.

**Fig. 5.** Adversarial samples of different attacks.

two pairs, i.e., positive pair and negative pair. Finally, we use the Sigmoid function to normalize the predicted discrepancy and compute the triplet loss by the difference between the expected discrepancy and predicted discrepancy. The triplet loss will be minimized during training, as mentioned before. In Classifier, we compare the testing sample and samples in the Oracle Database, and then employ the two-step classification algorithm for image classification.

## 4   Evaluation

### 4.1   Experimental Setup

**Dataset.** We evaluate DunDi on MNIST and CIFAR-10 dataset. It is worth noting that in DunDi, the input upon training is a tuple of three images, in which the first two are of the same class (positive pair) and the third a different class (negative pair). Consider that there exist thousands of images in the dataset, we randomly select images of each class and pack them into tuples. Finally, we generate 360K tuples for training and use 10K original testing images for testing.

**Models.** As for MNIST, we use three CNN models of different architectures [10,18] as Native models for comparison. DunDi uses the same topology of these models as the DNN module and then retrains the entire Distance NN, named as DunDi_R (Retrained). In addition, it also incorporates the pre-trained models available in DeepXplore [18] and only trains the Embedding module, named as DunDi_P (Pre-trained). Similarly, for CIFAR-10, we employ the VGG-16 model to retrain a DunDi_R model, and we also incorporate a pre-trained VGG-16 model to build DunDi_P. We by default set the number of oracle images to 10.

**Adversarial Samples.** We generate adversarial samples with the method proposed in [5], which supports four types of attacks, i.e., FGSM, JSMA, and two variants of iFGSM (iFGSM-a which stops iterating when misclassification is achieved, and iFGSM-b which runs for a certain number of iterations that is well beyond the average misclassification point). For each attack, we generate one adversarial sample with respect to each testing image and eventually get 40,000 adversarial images for each dataset. Figure 5 shows several adversarial samples generated by these attack approaches.

We run the experiments on one Tesla K40 GPU configured with 12 GB of RAM. We train the model of DunDi_R with VGG-16 for 200 epochs and train others for 10 epochs. The following sections will report the experimental results.

## 4.2 Accuracy of Models

We first present the accuracy of different models. As can be seen in Table 1, for the MNIST dataset, the accuracy of the three native models is 98.78%, 98.99% and 99.05% respectively, which match the reported accuracy of previous works, e.g., 98.3%, 98.9% and 99.05% respectively in [18]. The results of both retrained and pre-trained DunDi models are comparable with the accuracy of the native models. For example, the accuracy is 99.05%, 99.05% and 98.94% for Model3, DunDi_R3 and DunDi_P3 respectively when the number of oracle images is 20. These results imply that the proposed classification algorithm based on the distance metric is feasible. In addition, with the increasing number of oracle images, the accuracy increases as well. For example, the accuracy is 98.86%, 99.0% and 99.05% respectively when the number is 1, 10 and 20 for DunDi_R3. This is mainly because more oracle images are capable of representing the diversity of features for each class of images, so that the variance is reduced and the accuracy can be improved accordingly. The results on CIFAR-10 show similar results. It should be pointed out that the accuracy of DunDi_P is higher than that of Native Model, e.g., 93.1% against 89.7%. We suspect this is because the distance between images across different classes is much larger than that within the same class, so that the distance comparison can help to hide the variances between

**Table 1.** Accuracy of different models. The Native Models are irrelevant to oracle images, so their results are only reported on column '1'.

| Dataset | Model name | Number of oracle images | | |
|---|---|---|---|---|
| | | 1 | 10 | 20 |
| MNIST | Native Model1 | 98.78% | N/A | N/A |
| | Native Model2 | 98.99% | N/A | N/A |
| | Native Model3 | 99.05% | N/A | N/A |
| | DunDi_R1 | 97.88% | 98.53% | 98.65% |
| | DunDi_R2 | 98.59% | 98.96% | 98.98% |
| | DunDi_R3 | 98.86% | 99.0% | 99.05% |
| | DunDi_P1 | 98.79% | 98.96% | 99.04% |
| | DunDi_P2 | 98.07% | 98.59% | 98.71% |
| | DunDi_P3 | 98.5% | 98.82% | 98.94% |
| CIFAR-10 | Native Model | 89.7% | N/A | N/A |
| | DunDi_R | 91.16% | 91.37% | 91.53% |
| | DunDi_P | 93.1% | 93.28% | 93.32% |

images of the same class, particularly for CIFAR-10 where the images exhibit large variances. We leave the further exploration as our future work.

## 4.3  Robustness of Models

We measure the robustness by the defense success rate, which refers to the number of correctly predicted adversarial samples to the total number of adversarial samples. Figure 6 compares the defense success rate of various DunDi models and a recently proposed method which employs Bayesian uncertainty estimates and density estimates to detect adversarial samples (named as Bayesian) [5]. As we can see, all the DunDi models achieve high defense success rate for the four types of adversarial attacks. For MNIST, the average defense success rate of the six DunDi models is 98.07%, 91.45%, 90.91% and 97.13% respectively for iFGSM-a, iFGSM-b, FGSM, and JSMA attack. As for CIFAR-10, the average value of DunDi models is 90.46%, 94.0%, 80.7% and 90.49% respectively. The accuracies are comparable with the results reported in the Bayesian approach. These results prove the effectiveness of the proposed DunDi approach which improves robustness by enlarging the distance between samples and model's boundary.

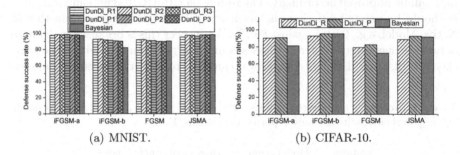

(a) MNIST.                    (b) CIFAR-10.

**Fig. 6.** Defense success rate of different models.

## 4.4  Varying Parameters

In this section, we show the effects on accuracy and robustness of varying configurations used in DunDi, including the size of training dataset, the number of oracle images[2]. For brevity, here we only report the results on MNIST.

**Varying Dataset Size.** Table 2 illustrates the accuracy and robustness on the training dataset of different sizes (or, in this case, the number of tuples). It can be seen that the accuracy increases with the increasing size of the dataset, e.g., it increases from 96.1% to 98.8% when the size increases from 18,000 to 360,000.

---

[2] Here we only report the results of varying number of oracle images (i.e., $n$) because it plays a more significant role than $m$ and $k$ in the multi-class classification algorithm.

**Table 2.** Accuracy and defense success rate of DunDi_R1 with varying dataset size.

| Dataset size | 18,000 | 36,000 | 180,000 | 360,000 |
|---|---|---|---|---|
| Accuracy | 96.1% | 96.6% | 98.0% | 98.8% |
| Defense success rate | 91.90% | 92.02% | 92.35% | 93.06% |

The defense success rate, which denotes the average value of DunDi_R1 under four adversarial attacks, also shows an overall increasing trend. Thus, the model can achieve higher accuracy and robustness when training on more tuples.

**Varying Number of Oracle Images.** Figure 7 compares the robustness with varying number of oracle images. As expected, the defense success rate increases with the increasing number of oracle images, mainly because more oracle image can capture the diversity of features. It is worth noting that the defense success rate is high even if there is only 1 oracle image for each label, e.g., the value of DunDi_P1 is 97.55%, 88.42%, 87.94% and 97.1% respectively for the four adversarial attacks, as shown in Fig. 7(b). This implies that the model can learn representative features to characterize the distance between images.

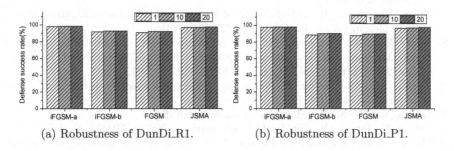

(a) Robustness of DunDi_R1.            (b) Robustness of DunDi_P1.

**Fig. 7.** Robustness with varying number of oracle images.

## 5   Conclusions

The deep neural networks are vulnerable to adversarial attacks, which limits the usage of DNN in safety-critical systems. This paper presents DunDi to improve the robustness of neural networks. DunDi employs the metric learning approach to put the samples of the same class together and push the samples of different classes away, so that the distance between samples and model's boundary can be enlarged as well. The increase of distance will require a larger magnitude of perturbations for generating adversarial samples to move across the boundary, which yet will be easily detected. The experimental results show that DunDi can defend a significant fraction of adversarial samples without losing classification accuracy. In the future, we plan to extend DunDi with deeper CNN models such as ResNet-50 and evaluate it with real-world images of ImageNet.

**Acknowledgement.** Thanks for the valuable comments from anonymous reviewers of ICCS2019 and researchers of the George Washington University. This work has been supported by the National Natural Science Foundation of China (grant no. 61602465 and 61601458), National Key Research and Development Program of China (grant no. 2016QY04W0804), and Beijing Natural Science Foundation (grant no. 4172069).

# References

1. Abadi, M., Barham, P., Chen, J., Chen, Z., et al.: TensorFlow: a system for large-scale machine learning. OSDI **16**, 265–283 (2016)
2. Chopra, S., Hadsell, R., LeCun, Y.: Learning a similarity metric discriminatively, with application to face verification. In: CVPR, pp. 539–546 (2005)
3. Cui, L., Hao, Z., Wang, D., Li, Y.: Detecting adversarial samples using heatmap of neural networks. In: AAAI-19 Workshop on Engineering Dependable and Secure Machine Learning (2019)
4. Fawzi, A., Dezfooli, S.M.M., Frossard, P.: A geometric perspective on the robustness of deep networks. IEEE Sign. Process. Mag. (2017)
5. Feinman, R., Curtin, R.R., Shintre, S., Gardner, A.B.: Detecting adversarial samples from artifacts. arXiv preprint arXiv:1703.00410 (2017)
6. Goodfellow, I.J., Shlens, J., Szegedy, C.: Explaining and harnessing adversarial examples. In: ICLR (2015)
7. Gu, S., Rigazio, L.: Towards deep neural network architectures robust to adversarial examples. arXiv preprint arXiv:1412.5068 (2014)
8. Hoffer, E., Ailon, N.: Deep metric learning using triplet network. In: Feragen, A., Pelillo, M., Loog, M. (eds.) SIMBAD 2015. LNCS, vol. 9370, pp. 84–92. Springer, Cham (2015). https://doi.org/10.1007/978-3-319-24261-3_7
9. Kurakin, A., Goodfellow, I., Bengio, S.: Adversarial machine learning at scale. arXiv preprint arXiv:1611.01236 (2016)
10. LeCun, Y.: The MNIST database of handwritten digits
11. LeCun, Y., Bengio, Y., Hinton, G.: Deep learning. Nature **521**(7553), 436 (2015)
12. Metzen, J.H., Genewein, T., Fischer, V., Bischoff, B.: On detecting adversarial perturbations. In: ICLR (2017)
13. Moosavi-Dezfooli, S.-M., Fawzi, A., Fawzi, O., Frossard, P.: Universal adversarial perturbations. In: CVPR, pp. 86–94 (2017)
14. Moosavi Dezfooli, S.M., Fawzi, A., Frossard, P.: DeepFool: a simple and accurate method to fool deep neural networks. In: CVPR (2016)
15. Papernot, N., McDaniel, P., et al.: The limitations of deep learning in adversarial settings. In: Security and Privacy (EuroS&P), pp. 372–387 (2016)
16. Papernot, N., McDaniel, P., Goodfellow, I., et al.: Practical black-box attacks against deep learning systems using adversarial examples. arXiv preprint (2016)
17. Papernot, N., McDaniel, P., Wu, X., et al.: Distillation as a defense to adversarial perturbations against deep neural networks. In: Security and Privacy (SP), pp. 582–597 (2016)
18. Pei, K., Cao, Y., Yang, J., Jana, S.: DeepXplore: automated whitebox testing of deep learning systems. In: SOSP, pp. 1–18 (2017)
19. Ross, A.S., Doshi-Velez, F.: Improving the adversarial robustness and interpretability of deep neural networks by regularizing their input gradients. arXiv preprint arXiv:1711.09404 (2017)
20. Schroff, F., Kalenichenko, D., Philbin, J.: FaceNet: a unified embedding for face recognition and clustering. In: CVPR, pp. 815–823 (2015)

21. Sharif, M., Bhagavatula, S., Bauer, L., Reiter, M.K.: Adversarial generative nets: neural network attacks on state-of-the-art face recognition. arXiv preprint arXiv:1801.00349 (2017)
22. Szegedy, C., Zaremba, W., Sutskever, I., et al.: Intriguing properties of neural networks. In: ICLR (2014)
23. Tramèr, F., Kurakin, A., Papernot, N., et al.: Ensemble adversarial training: attacks and defenses. arXiv preprint arXiv:1705.07204 (2017)
24. Tramèr, F., Kurakin, A., Papernot, N., Goodfellow, I., et al.: Ensemble adversarial training: attacks and defenses. In: ICLR (2018)

# Autism Screening Using Deep Embedding Representation

Haishuai Wang[1](✉), Li Li[2], Lianhua Chi[3], and Ziping Zhao[4]

[1] Fairfield University, Fairfield, CT 06825, USA
hwang@fairfield.edu
[2] Harvard Medical School, Boston, MA 02115, USA
Li_Li@hms.harvard.edu
[3] La Trobe University, Melbourne, Australia
L.Chi@latrobe.edu.au
[4] Tianjin Normal University, Tianjin, China
zhaoziping@tjnu.edu.cn

**Abstract.** Autism spectrum disorder (ASD) is a developmental disorder that affects communication and behavior. An early diagnosis of neurodevelopmental disorders can improve treatment and significantly decrease associated healthcare cost, which reveals an urgent need for the development of ASD screening. However, the data used for ASD screening is heterogenous and multi-source, resulting in existing screening tools for ASD screening are expensive, time-intensive and sometimes fall short in predictive accuracy. In this paper, we apply novel feature engineering and feature encoding techniques, along with a deep learning classifier for ASD screening. Algorithms were created via a robust deep learning classifier and deep embedding representation for categorical variables to diagnose ASD based on behavioral features and individual characteristics. The proposed algorithm is effective compared with baselines, achieving 99% sensitivity and 99% specificity. The results suggest that deep embedding representation learning is a reliable method for ASD screening.

**Keywords:** Autism spectrum disorder · Deep learning · ASD screening · Categorical embedding

## 1 Introduction

Autistic Spectrum Disorder (ASD) is a mental disorder characterized by difficulties with social interaction and communication, and by restricted and repetitive behavior [7]. In the United States, there are 3 million people who have ASD, and around 1 out of 68 children are diagnosed with ASD [24]. ASD is a neurodevelopment condition associated with significant healthcare costs, and early diagnosis can significantly reduce the cost and improve the quality of life of children with ASD. The increase in the number of ASD cases and the cost impact of ASD push forward the research of effective screening methods [14,15,20]. Most tools

© Springer Nature Switzerland AG 2019
J. M. F. Rodrigues et al. (Eds.): ICCS 2019, LNCS 11537, pp. 160–173, 2019.
https://doi.org/10.1007/978-3-030-22741-8_12

for autism screening are based on score-sheets with questions for the parent or the medical practitioner [2], and the summation score is compared with predetermined thresholds to produce results. For example, the Modified Checklist for Autism in Toddlers (M-CHAT) [5] is a checklist-based screening tool for autism with children between the ages of 16 and 30 months. The Child Behavior Checklist (CBCL) [4] is a parent-completed screening tool. However, the diagnostic process for ASD is costly and time-consuming [12]. Recently, machine learning based approaches have been showing a great direction on objective evaluation of neuropsychiatric disorders [3,13,26]. Machine learning or deep learning based approaches might allow for detecting ASD diagnosis automatically and are able to provide a map of high-risk populations [19].

Due to the significance of ASD screening, we propose to develop, validate and assess efficient deep learning algorithms that identify ASD versus non-ASD. The ASD screening algorithm will be based on data consolidated from heterogeneous sources, including questionnaire and demographics. There are some existing forecasting algorithms/tools being used for ASD screening based on these kinds of data [21–23]. However, these approaches more focused on improving or applying machine learning algorithms whereas ignored to propose effective feature engineering methods for the heterogenous ASD data. For instance, the ASD data are usually a mix of numerical and categorical features, which pose a challenge for directly applying classifiers as they can only deal with numerical inputs by design. Nevertheless, most of the existing methods applied one-hot encoding to deal with categorical variables for the ASD data, and then input one-hot vectors (or dummy variables) to machine learning models or feed them into neural networks. Since one-hot encoding usually yields sparse vectors, it potentially limits model performance and screening accuracy. Although categorical data is very common in medical datasets, there are no existing works to effectively handle or represent categorical medical data. Thus, an effective feature encoding method is required to improve screening accuracy.

Nowadays, deep learning has been one of the most prominent machine learning techniques [15,25], and it has the capability to model data with non-linear structures and learn a high-level representation of features. In this paper, we present deep embedding representation for categorical variables, along with neural network as a classifier for ASD screening. Specifically, we first learn the categorical feature representation from an embedding layer, then we combine the learned embeddings and continuous variables as input to a dense layer, followed by a non-linear activation layer. We use several fully connected layers on top of the embedding layer. We use DENN for short to denote the model name. The DENN model outperforms the existing methods on the ASD screening data.

## 2  Data and Analysis

### 2.1  Data Description and Data Exploration

The data we used is the Autism Screening Adult Data Set provided by the UCI Machine Learning Repository [8]. This is a new dataset related to autism screening of adults that contains 704 observations with 21 variables to be

**Table 1.** Basic descriptive statistics for the variables (we truncated the variable names of the A1_Score to A10_Score variables to A1 to A10).

|      | A1 | A2 | A3 | A4 | A5 | A6 | A7 | A8 | A9 | A10 | Age | Result |
|------|-----|-----|-----|-----|-----|-----|-----|-----|-----|-----|-----|-----|
| Mean | 0.723 | 0.453 | 0.459 | 0.497 | 0.498 | 0.285 | 0.417 | 0.651 | 0.325 | 0.574 | 29.698 | 4.883 |
| Std | 0.447 | 0.498 | 0.498 | 0.500 | 0.500 | 0.451 | 0.493 | 0.476 | 0.468 | 0.494 | 16.507 | 2.498 |
| Min | 0.000 | 0.000 | 0.000 | 0.000 | 0.000 | 0.000 | 0.000 | 0.000 | 0.000 | 0.000 | 17.000 | 0.000 |
| Max | 1.000 | 1.000 | 1.000 | 1.00 | 1.00 | 1.000 | 1.000 | 1.000 | 1.000 | 1.000 | 383.00$^*$ | 10.000 |

* Indicate a possible outlier value

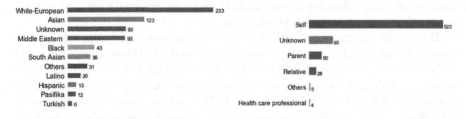

**Fig. 1.** Statistical distribution of ethnicity. **Fig. 2.** Statistical distribution of relation.

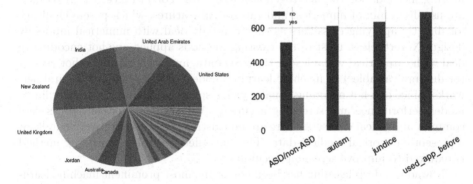

**Fig. 3.** Statistical distribution of country (only show names of major countries).

**Fig. 4.** Statistical distributions on three categorical variables (*autism, used_app_before, jaundice*), and the class label (*ASD/non-ASD*).

utilised for further analysis especially in determining influential autistic traits and improving the classification of ASD cases. The raw data contains ten binary variables (AQ-10-Child) representing the screening questions, and the categorical variables of *gender, ethnicity, jaundice, autism, country_of_res, used_app_before, age_desc, relation* and *Class/ASD*. There are also two numeric variables named *age* and *result*. The 21 variables, representing behavioural features and individuals characteristics, that have proved to be effective in detecting the ASD cases from controls in behaviour science [6,22,23]. Some basic descriptive statistics for the variables are shown in Table 1 and Figs. 1, 2, 3 and 4.

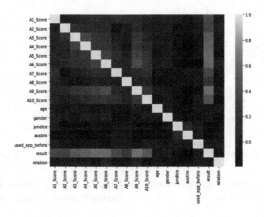

**Fig. 5.** Correlation heatmap for the dataset.

## 2.2   Data Cleaning and Preprocessing

After obtaining the statistic information of the variables from data exploration, we need to clean the data of unwanted information to prepare it as appropriate input features for our machine learning algorithms.

*Cleaning Missing Values and Outliers:* In this data, there are missing values in the age variable. Since there are only two missing age values, we simply remove all of the observations containing the missing age values. As we can see from Table 1, there is an impossibly large maximum value of 383 in the age variable. Given that there are already many typographical errors present in the data, it is reasonable to assume that this is because of a typing error and the intended value should be 38. After preprocessing the missing values and outliers, we visualize the data to explore potential challenges and solutions for a learning model.

## 2.3   Data Visualization and Analysis

Before proceeding to apply any machine learning algorithm, we visualize and analyze the data to provide a better solution for ASD screening. In Fig. 5, we compare the correlation among the variables to demonstrate how close two variables are to having a linear relationship with each other. From Fig. 5, we can observe that there is a high correlation between AQ-10-Child score and result, which indicates they have a linear relationship. However, other variables have a non-linear relationship (as shown in Fig. 5). Therefore, a good machine learning model using these variables should consider both linear and non-linear relationships.

To obtain the first impression of the internal connections of some of the variables that are presented in the data set, we visualize the connection between several attributes and display how they are related with the target class in Figs. 6, 7 and 8. Figures 6 and 7 show a similar relationship between *gender* and *result* vs. *jaundice* and *result*. From both cases, the individual with a higher *result* score is

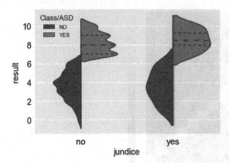

**Fig. 6.** The connection visualization between *jaundice* and *result*.

**Fig. 7.** The connection visualization between *gender* and *result*.

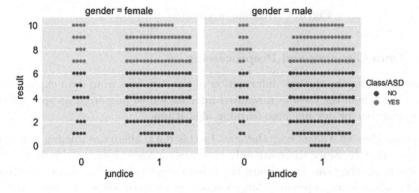

**Fig. 8.** The connection visualization among *jaundice*, *gender* and *result*.

more likely to have ASD, independent of the other variables. The variables *gender* and *jaundice* do not seem a major influence in deciding ASD. Nevertheless, from Fig. 8, we can observe that when *jaundice* presents at birth, the individual with a higher *result* score will have ASD regardless of their gender. In Fig. 9, we plot the data in terms of *age*, *gender*, *jaundice* and *relation* to visualize their distribution on ASD class and non-ASD class. They have similar distribution on the data belongs to different class.

The purpose is to develop effective algorithms for ASD screening and diagnosis based on the data described above. However, from the above analysis, the variables in the dataset consist of two data types: continuous (e.g., age) and categorical (e.g., gender). Dealing with continuous numeric data is often easier than categorical data given that it can be fed into most of machine learning models after normalization. However, naively applying machine learning algorithms with integer representation for categorical variables does not work well. Since categorical variables are known to hide and mask lots of interesting information in a dataset and they might even be the most important variables in a model, we will present an advanced technique called deep embedding representation to deal with categorical variables in neural networks for ASD screening.

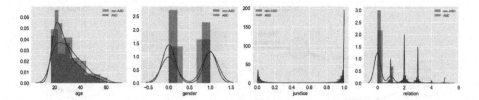

**Fig. 9.** Distribution plot for some of the input variables to machine learning models.

## 3   Prediction Methodology

We aim to use the dataset described in Sect. 2 to predict whether these patients actually have autism accurately. A main challenge to develop an algorithm is that the data contains both continuous numerical variables and categorical variables. Therefore, we first propose a deep categorical embedding method for feature engineering, and then present a neural network classifier using the embedded features for ASD screening. The model is denoted as DENN for short. The overall architecture of DENN is illustrated in Fig. 10.

### 3.1   Deep Embedding Representation for Categorical Variables

For the continuous numerical data, we use identity mapping to map numerical values to feature vectors. Since neural network can only deal with numerical inputs by design, the categorical features can not be input to neural network directly. One-hot encoding is a commonly used method for converting a categorical variable into a continuous variable. By one-hot encoding, the new representation is a vector with one element being one and all others being zero, so it is also called dummy coding or the 1-of-$c$ encoding where $c$ is the number of total possible categories of a categorical feature. If we have $P$ categorical features and the $i$-th feature can take $c_i$ values, this encoding will result in dimensionality of $d$, such that,

$$d = \sum_{i=1}^{P} c_i \tag{1}$$

Although one-hot encoding is simple and common, it often yields a very sparse and high dimensional representation of the data, and it ignores the informative relations between them because it treats different values of categorical variables completely independent of each other. In this section, we present an advanced method that aims to capture the different categories with a much smaller dimension and represents the categorical features more efficiently.

We map categorical variables in a function approximation problem into Euclidean spaces, which is to build a vector embedding to every category type, such that $e_i : x_i \mapsto \mathbf{x}_i$. For each categorical variable, we initialise a $m \times D$ embedding matrix $\boldsymbol{E}$ as

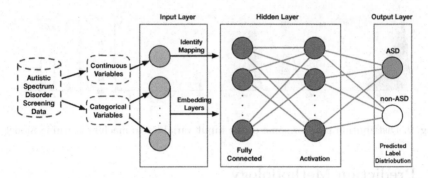

**Fig. 10.** Overall architecture of DENN.

where $m$ is the number of total possible categories of a categorical variable, hyperparameter $D$ is the desired dimension for embedding representation, which is usually less than $m$.

$$E = \begin{pmatrix} x_{11} & x_{12} & \cdots & x_{1D} \\ x_{21} & x_{22} & \cdots & x_{2D} \\ \vdots & & \ddots & \vdots \\ x_{m1} & x_{m2} & \cdots & x_{mD} \end{pmatrix} = (x_{ij}) \in \mathbb{R}^{m \times D} \tag{2}$$

Then we add an embedding layer in a deep neural network to do a lookup for a given value from the embedding matrix $E$, which returns a vector $\mathbf{x}_k = (x_{kj}) \in \mathbb{R}^{1 \times D}$. The new representation $\mathbf{x}_k$ along with the numerical variables would then be fed into the next layer in the neural networks, and all the embeddings are updated and learned through backpropagation.

## 3.2   Neural Network Architecture for ASD Screening

As shown in Fig. 10, the embedded categorical variables are concatenated with numerical features as new feature vectors that can be fed into a dense layer. There are several dense layers in our neural network architecture, and each dense layer followed by an activation layer. We used ReLu as the activation function to introduce non-linearity. After the ReLu activation, we have $f(z) = max(0, z)$ that gives an output $z$ if $z$ is positive and 0 otherwise.

Then, an output layer is placed after the last hidden layer. From hidden layer to output layer is a sigmoid function to output the predicted probability of a feature vector $\mathbf{x}$ belonging to class 1:

$$Pr(y = 1 | \mathbf{x}, \mathbf{w}_o) = \frac{1}{1 + exp(-\mathbf{w}_o^T \mathbf{x})} \tag{3}$$

where $\mathbf{w}_o$ is the weights of the output layer (the subscript $o$ represents the parameters in the output layer).

Since this is a binary classification problem, we encoded the class label as one-hot vector and set categorical cross entropy as the loss function, as follows:

$$\mathcal{L} = -\frac{1}{N} \sum_{i=1}^{N} \sum_{t=1}^{K} (y_{i,t} * log(\hat{y}_{i,t})) \tag{4}$$

where $N$ is the total number of training samples, $\hat{y}$ is the predicted label, and $K$ is the total number of classes ($K$ equals 2 for ASD screening). To learn and optimize the parameters of the model, we minimize the loss function $\mathcal{L}$. The loss minimization and parameter optimization can be performed through the backpropagation using mini-batch stochastic gradient descent.

# 4  Experiments and Evaluation

## 4.1  Datasets and Setup

We trained the proposed approach on the Autistic Spectrum Disorder Screening Data from the University of Irvine data collection[1]. After preprocessing, this dataset contains 702 samples, including 513 patients or children that have been screened for autism (i.e., 189 cases and 513 controls). We randomly selected 80% as training data and the remaining 20% as testing data.

Since our model is based on supervised learning, the input of the models utilizes a training dataset of cases (the number is 155) and controls (the number is 406) that have already been diagnosed. Usually, the cases and controls have been generated using a screening tool such as ADOS-R, ADI-R, etc., in a clinic by a behaviorist, clinical psychologist, or a licensed clinician specialized in that tool. In our experiment settings, we use two fully connected layers (1000 and 500 neurons respectively) on top of the embedding layer. The neural network is trained for 50 epochs.

**Table 2.** Confusion Matrix for ASD screening.

| | | Predicted class | |
|---|---|---|---|
| | | ASD | non-ASD |
| Actual Class | ASD | True Positive (TP) | False Negative (FN) |
| | non-ASD | False Positive (FP) | True Negative (TN) |

## 4.2  Evaluation Criteria

The experiments are designed to validate whether the ASD screening model, which combines deep embedding representation and neural networks, can achieve better performance than using one-hot encoding and shallow machine learning

---

[1] https://archive.ics.uci.edu/ml/datasets/Autistic+Spectrum+Disorder+Screening+Data+for+Children++.

models. All the experiments were implemented in Python and Keras framework. The source codes used for this experiments and results can be found in this github repository[2].

Following the most common procedure for evaluating models for ASD screening, we use Sensitivity (Recall), Specificity (Precision), F-measure (F1-Score) Receives Operating Characteristic (ROC) Curve, and Area Under ROC Curve (AUC) to evaluate the ASD screening approaches. To compute the measurements, we can use a confusion matrix to summarize the performance *w.r.t.* various models, as shown in Table 2. For classification problem, if the sample is positive and it is classified as positive, it is counted as a true positive (Eq. 5); If the sample is negative while it is classified as positive, it is counted as a false positive (Eq. 6).

$$TP = \frac{Correctly\ classified\ positives\ samples}{Total\ no.\ of\ positives} \tag{5}$$

$$FP = \frac{Incorrectly\ classified\ negative\ samples}{Total\ no.\ of\ negative\ samples} \tag{6}$$

Following formulas are used to measure above mentioned performance measures are shown below.

$$Specificity = \frac{\sum_{n=1}^{N} TN_n}{\sum_{n=1}^{N} TN_n + FP_n}, \ Sensitivity = \frac{\sum_{n=1}^{N} TP_n}{\sum_{n=1}^{N} TP_n + FN_n} \tag{7}$$

$$F1 = 2 \times \frac{Specificity \times Sensitivity}{Specificity + Sensitivity} \tag{8}$$

*Baselines.* We also conducted comparison experiments with different methods, including Random Forest, Support Vector Machine (SVM), Gradient Boosting, and Neural Networks without embedding layer. These machine learning-based baselines have been applied and reported [1,2,9,11,19,23] for ASD screening based on the same dataset. The inputs for all of the baselines are one-hot-encoded features. The baselines are implemented using the scikit-learn library of python and default parameters.

### 4.3   Experimental Results

Table 3 and Figs. 11, 12 and 13 present the performance of ASD screening based on different machine learning models on the test dataset. In comparison to the baselines on the test dataset, we observed that the deep embedding representation model performs better than baselines in terms of specificity, sensitivity and f1 score. The baseline using neural network without embedding layer (denoted as NN for short) performs better than other baselines but worse than the DENN model. This is because the deep embedding representation is more efficient than one-hot encoding for categorical feature representation, and neural network fits

---

[2] https://github.com/BlindReview/ASD_Prediction.

better than shallow machine learning models on the autism dataset. As we can observe from the ROC curves in Fig. 11, under the same false positive rate, we are able to identify ASD with high true positive rate, which is better than that of in the baselines.

**Table 3.** Comparison of different methods *w.r.t.* specificity, sensitivity and F1-score.

| Method | Specificity | Sensitivity | F1-Score |
|---|---|---|---|
| DENN | **0.99** | **0.99** | **0.99** |
| NN | 0.94 | 0.94 | 0.94 |
| Random Forest | 0.92 | 0.91 | 0.91 |
| SVM | 0.73 | 0.73 | 0.73 |
| Gradient Boosting | 0.85 | 0.85 | 0.85 |

We visualize the confusion matrix of the DENN model on the test data in Fig. 12. From Fig. 12, we can observe that there is only one sample is misclassified on the test data. The deep embedding representation with neural networks is able to achieve almost 100% accuracy for ASD screening on the dataset. However, other baselines have higher misclassification rate for ASD group because there are fewer ASD samples in the dataset. The baselines are not able to learn enough information due to their limited learning capability. From the experimental results, we can conclude that effective categorical feature embedding method can help to improve the learning ability and model performance. The data has both linearity and nonlinearity nature, resulting in DENN achieves the best performance, followed by neural network based approach.

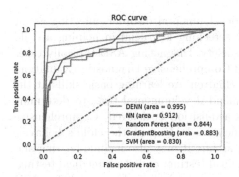

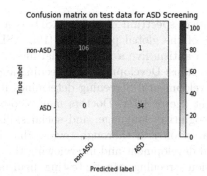

**Fig. 11.** ROC curve *w.r.t.* various ASD screening approaches.

**Fig. 12.** Confusion matrix of DENN model on the test data for ASD screening.

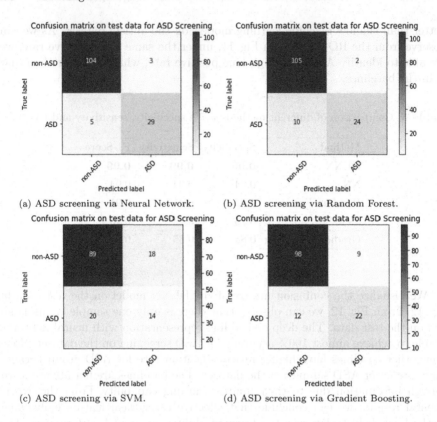

(a) ASD screening via Neural Network.    (b) ASD screening via Random Forest.

(c) ASD screening via SVM.    (d) ASD screening via Gradient Boosting.

**Fig. 13.** Confusion matrix on the test data *w.r.t.* various baseline methods.

## 5    Related Work

Autism spectrum disorder (ASD) is a developmental disorder, affecting about 1% of the global population [19]. ASD screening is crucial in helping a child with autism live a more normal life in society. Diagnosing ASD typically takes two steps: Developmental Screening and Comprehensive Diagnostic Evaluation. Developmental Screening determines if children are learning basic skills at the right age or later. Doctors may recommend developmental tests to determine if cognitive, language, and social skills acquisition is delayed. Comprehensive evaluation is a thorough review that includes looking at the child's behavior and development and interviewing the parents. It may also include hearing and vision screening, genetic testing, neurological testing, and other medical testing.

To date, behavior-based tests are the standard clinical approach for ASD diagnosis [7,19], hence, traditional tools for autism screening are based on score-sheets with questions for the parent or the medical practitioner [2], and the summation score is compared with predetermined thresholds to produce results. For example, the Modified Checklist for Autism in Toddlers (M-CHAT) [5] and the Child Behavior Checklist (CBCL) [4]. A number of other existing clinical diagnosis methods have also been used for ASD identification, such as Autism Diagnostic Interview-Revised (ADI-R) [18] and Autism Diagnostic Observation Schedule-Revised (ADOS-R) [17], have shown superior performance. Yet those methods rely on handcrafted rules that employ mathematical summation formulas of scores to come up with the appropriate diagnosis [23]. Moreover, the majority of existing ASD screening tools require substantial time to produce a complete diagnosis. Therefore, they are time-consuming and costly.

Several studies have recently employed machine learning and deep learning to improve the diagnosis process for ASD. Researchers have adopted several supervised machine learning techniques (such as Neural Network, Decision Trees, Logistic Regression, Random Forest, Support Vector Machines (SVM), k-Nearest Neighbors (kNN), Naive Bayes) to solve the classification problem of predicting whether an individual with certain characteristics has ASD [1,2,9,19,23]. The study [11] applied a machine learning algorithm to identify ASD from attention deficit hyperactivity disorder based on a 65-item Social Responsiveness Scale. Another study [10] combines the Social Responsiveness Scale with the ADI-R score to train their models to distinguish ASD from controls. More recently, studies [3,15,16] have developed machine learning models using the Autism Brain Imaging Data Exchange (ABIDE) towards the automated diagnosis of ASD based on brain neuroimaging data.

The main purposes of the existing machine learning-based ASD diagnostic tools were to improve diagnosis accuracy, and speed up diagnosis time to provide timely access to healthcare services. However, the existing machine learning or deep learning based approaches for ASD either rely on neuroimaging data or simply applying traditional learning algorithms without considering the latent feature characteristics. Therefore, new effective feature representation methods for machine learning based ASD diagnosis is significant to improve the diagnosis performance.

## 6    Conclusion

Autism spectrum disorder (ASD) is a developmental disability that can cause significant social, communication and behavioral challenges. ASD screening is significant because it enables early intervention. Early treatment is more effective than later treatment for ASD. In this paper, we are able to predict autism in patients with about 99% accuracy based on the UCI ASD screening data. Since there are plenty of health applications are going to have categorical data, we also learned how to deal with categorical data using deep embedding representation along with a neural network as a classifier for ASD screening. Compared with other models for ASD screening, the DENN model is more efficient and accurate for ASD screening.

# References

1. Abbas, H., Garberson, F., Glover, E., Wall, D.P.: Machine learning for early detection of autism (and other conditions) using a parental questionnaire and home video screening. In: 2017 IEEE International Conference on Big Data (Big Data), pp. 3558–3561. IEEE (2017)
2. Abbas, H., Garberson, F., Glover, E., Wall, D.P.: Machine learning approach for early detection of autism by combining questionnaire and home video screening. J. Am. Med. Inf. Assoc. **25**(8), 1000–1007 (2018)
3. Abraham, A., et al.: Deriving reproducible biomarkers from multi-site resting-state data: an autism-based example. NeuroImage **147**, 736–745 (2017)
4. Achenbach, T.M., Rescorla, L.A.: Manual for the ASEBA Preschool Forms and Profiles, vol. 30. University of Vermont, Research Center for Children, Youth And Families, Burlington, VT (2000)
5. Allely, C.S., Wilson, P.: Diagnosing autism spectrum disorders in primary care. Practitioner **255**(1745), 27–31 (2011)
6. Altay, O., Ulas, M.: Prediction of the autism spectrum disorder diagnosis with linear discriminant analysis classifier and k-nearest neighbor in children. In: 2018 6th International Symposium on Digital Forensic and Security (ISDFS), pp. 1–4. IEEE (2018)
7. Association, A.P., et al.: Diagnostic and Statistical Manual of Mental Disorders (DSM-5®). American Psychiatric Publication (2013)
8. Asuncion, A., Newman, D.: UCI machine learning repository (2007)
9. van den Bekerom, B.: Using machine learning for detection of autism spectrum disorder (2017)
10. Bone, D., Goodwin, M.S., Black, M.P., Lee, C.C., Audhkhasi, K., Narayanan, S.: Applying machine learning to facilitate autism diagnostics: pitfalls and promises. J. Autism Dev. Disord. **45**(5), 1121–1136 (2015)
11. Duda, M., Ma, R., Haber, N., Wall, D.: Use of machine learning for behavioral distinction of autism and ADHD. Transl. Psychiatry **6**(2), e732 (2016)
12. Galliver, M., Gowling, E., Farr, W., Gain, A., Male, I.: Cost of assessing a child for possible autism spectrum disorder? An observational study of current practice in child development centres in the UK. BMJ Paediatr. Open **1**(1) (2017)
13. Ghiassian, S., Greiner, R., Jin, P., Brown, M.R.: Using functional or structural magnetic resonance images and personal characteristic data to identify adhd and autism. PloS one **11**(12), e0166934 (2016)
14. Guo, X., Dominick, K.C., Minai, A.A., Li, H., Erickson, C.A., Lu, L.J.: Diagnosing autism spectrum disorder from brain resting-state functional connectivity patterns using a deep neural network with a novel feature selection method. Front. Neurosci. **11**, 460 (2017)
15. Heinsfeld, A.S., Franco, A.R., Craddock, R.C., Buchweitz, A., Meneguzzi, F.: Identification of autism spectrum disorder using deep learning and the abide dataset. NeuroImage Clin. **17**, 16–23 (2018)
16. Li, H., Parikh, N.A., He, L.: A novel transfer learning approach to enhance deep neural network classification of brain functional connectomes. Front. Neurosci. **12**, 491 (2018)
17. Lord, C., et al.: The autism diagnostic observation schedule–generic: a standard measure of social and communication deficits associated with the spectrum of autism. J. Autism Dev. Disord. **30**(3), 205–223 (2000)

18. Lord, C., Rutter, M., Le Couteur, A.: Autism diagnostic interview-revised: a revised version of a diagnostic interview for caregivers of individuals with possible pervasive developmental disorders. J. Autism Dev. Disord. **24**(5), 659–685 (1994)
19. Parikh, M.N., Li, H., He, L.: Enhancing diagnosis of autism with optimized machine learning models and personal characteristic data. Front. Comput. Neurosci. **13** (2019)
20. Rad, N.M., et al.: Deep learning for automatic stereotypical motor movement detection using wearable sensors in autism spectrum disorders. Signal Process. **144**, 180–191 (2018)
21. Thabtah, F.: ASDTests. A mobile app for ASD screening (2017)
22. Thabtah, F.: Autism spectrum disorder screening: machine learning adaptation and DSM-5 fulfillment. In: Proceedings of the 1st International Conference on Medical and Health Informatics, pp. 1–6. ACM (2017)
23. Thabtah, F.: Machine learning in autistic spectrum disorder behavioral research: a review and ways forward. Inform. Health Soc. Care, 1–20 (2018)
24. de la Torre-Ubieta, L., Won, H., Stein, J.L., Geschwind, D.H.: Advancing the understanding of autism disease mechanisms through genetics. Nature Med. **22**(4), 345 (2016)
25. Wang, H., Cui, Z., Chen, Y., Avidan, M., Abdallah, A.B., Kronzer, A.: Predicting hospital readmission via cost-sensitive deep learning. IEEE/ACM Trans. Comput. Biol. Bioinform. (2018)
26. Yahata, N., et al.: A small number of abnormal brain connections predicts adult autism spectrum disorder. Nature Commun. **7**, 11254 (2016)

# Function and Pattern Extrapolation with Product-Unit Networks

Babette Dellen[1(✉)], Uwe Jaekel[1], and Marcell Wolnitza[1,2]

[1] Department of Mathematics and Technology, RheinAhrCampus Remagen,
University of Applied Sciences Koblenz, Joseph-Rovan-Allee 2,
53424 Remagen, Germany
[2] Third Institute of Physics - Biophysics, Georg-August-University Göttingen,
Friedrich-Hund-Platz 1, 37077 Göttingen, Germany
{dellen,jaekel,wolnitza}@hs-koblenz.de

**Abstract.** Neural networks are a popular method for function approximation and data classification and have recently drawn much attention because of the success of deep-learning strategies. Artificial neural networks are built from elementary units that generate a piecewise, often almost linear approximation of the function or pattern. To improve the extrapolation of nonlinear functions and patterns beyond the training domain, we propose to augment the fundamental algebraic structure of neural networks by a product unit that computes the product of its inputs raised to the power of their weights. Linearly combining their outputs in a weighted sum allows representing most nonlinear functions known in calculus, including roots, fractions and approximations of power series. We train the network using stochastic gradient descent. The enhanced extrapolation capabilities of the network are demonstrated by comparing the results for a function and pattern extrapolation task with those obtained using the nonlinear support vector machine (SVM) and a standard neural network (standard NN). Convergence behavior of stochastic gradient descent is discussed and the feasibility of the approach is demonstrated in a real-world application in image segmentation.

**Keywords:** Product units · Neural network · Function extrapolation

## 1 Introduction

Complex machine learning problems that have been intractable in the past are now being solved using deep neural networks performing fast computations on parallel hardware [9–11,15,20]. Standard neural networks eventually perform a piecewise, almost linear approximation of the function or pattern that is to be learned [8,13,14,16,18,19,21]. This limits extrapolation of nonlinear relationships into areas of the feature space that are not covered by the training data. However, in many applications, extrapolating and making predictions from data is exactly what is required. In the past, solutions of this problem have been sought by nonlinearly transforming the data implicitly or explicitly into a higher

© Springer Nature Switzerland AG 2019
J. M. F. Rodrigues et al. (Eds.): ICCS 2019, LNCS 11537, pp. 174–188, 2019.
https://doi.org/10.1007/978-3-030-22741-8_13

dimensional space [5, 17] or by adding nonlinear units to the network representing terms of higher-order polynomials [4]. The disadvantage is that either the basis of this nonlinear space or a nonlinear kernel has to be provided beforehand. Alternatively, it would have to be adapted to the problem at hand by considering an exponentially increasing repertoire of combinatorial combinations of them, leading quickly to a computationally intractable situation.

In this paper, we follow a different approach: We expand the algebraic capabilities of the neural network by integrating input multiplication at the operational level through product units [6, 12], representing a computational multiplicatory analogon to the classical summation unit, the McCulloch Pitts neuron (McP) [13] (see Fig. 1a–b). We combine the output of several product units by means of a classical summation unit into a generalized algebraic operations network (GAON) (see Fig. 1c), which then can be used to build larger networks (see Fig. 1d).

Product units have been first introduced in the past to learn the higher-order input combinations of nonlinear problems such as the parity problem [6, 12]. However, for the rather simple networks investigated in these studies, difficulties were arising when using backpropagation to train those networks. For example, it was reported that the parity-8 problem could not be trained using backpropagation [12]. In that particular case, the neural network consisted of a single product unit. It was assumed that the solution space for product units is too convoluted, giving rise to many local minima. Weight initialization turned out to be especially difficult for product-unit networks. This discouraging result probably explains why the idea of product-unit neural networks was not taken much further in the following years.

Our work differs from those approaches in several ways: (i) We only use positive inputs to avoid difficulties otherwise arising for exponents representing roots. This is achieved by taking the absolute value of the input before feeding it into the product unit. For problems containing negative values, we first feed the input to a standard summation layer (which is trained together with the nonlinear network) to transform the input data into the positive domain. (ii) When solving classification problems, the exponents (or weights) of the product unit can only take positive values. This avoids exploding terms for small input data, but does not limit the applicability of the approach (see Discussion) (iii) We consider only networks consisting of substructures of several product units, i.e., the GAON (see Fig. 1c). This property allows a larger range of functions to be described by the network. (iv) For the parity-8 problem and a real-world labeling problem with mutliple classes we choose a network in which the output of several GAONs is fed into a standard summation layer with softmax activation. Decision hypersurfaces emerge as intersections of GAON functions in the data space. Different from standard neural networks, those hypersurfaces can be highly nonlinear. (v) We explore especifically the extrapolation capabilities of product-unit networks and compare them to the ones of standard neural networks [8, 13, 14, 16, 21] and the nonlinear support-vector machine [5, 17].

For the problems and the network architectures that we studied for this work, we did not encounter problems with convergence. This is not in contradiction with the studies described before [6, 12]. Our network is larger and employs the product units in a different way.

## 2  Methods

In this work, we propose to augment the algebraic structure of neural networks by allowing generalized multiplicatory operations to take place at the level of their basic computational units. Artificial neural networks consist of simple computational units that receive inputs from many other units of the same kind (see Fig. 1a) [8, 13, 18] and form a weighted sum of these inputs that is passed through a threshold or other rectifying function to still other units of the same kind, which process their inputs in like manner. This elementary unit is known as the McCulloch Pitts neuron (McP) [13]. In the standard neural network, these units are arranged in layers, and information is flowing only in a single direction through the layers [19]. Learning is usually performed by adjusting the weights through error backpropagation after definition of a suitable loss (or cost) function [21]. Let $\mathbf{x} = (x_1, \ldots, x_n)$ be an n-dimensional input vector that is supplied to an McP unit; then the net output of this unit (prior to any rectification step) is defined by $o_{\text{sum}}(\mathbf{x}) = \sum_{i=1}^{n} w_i x_i$, where the $w_i$, $i = 1, \ldots, n$ are the weights of the connections (see Fig. 1a). Solutions can be thought of as a division of the data space by a hyperplane [18], with class regions corresponding to the resulting half spaces. If one seeks to separate regions that are defined by highly nonlinear boundaries, many hyperplanes are needed to carve the hypervolume corresponding to the class-defining region. This can be achieved by arranging many McP neurons in a layer and combining their outputs again using an McP unit, representing the output unit of the resulting *multilayer perceptron* [19].

In contrast to the classical approach, we aim at separating the class-defining regions by nonlinear hypersurfaces. To achieve this, we use product units [6, 12], similar to the McP unit, that perform algebraic multiplication instead of summation (see Fig. 1b, left panel), yielding the output $o_{\text{mult}}(\mathbf{x}) = \prod_{i=1}^{n} x_i^{w_i}$, where the $w_i$, $i = 1, \ldots, n$ are real numbers, representing the weights of the connections.

From a computational point of view, the operation performed by the multiplication unit can be implemented through an equivalent network consisting of a balanced arrangement of McP units with logarithmic and exponential activation functions (see Fig. 1b, right panel). Switching from summation to multiplication can be achieved by taking the logarithmic values of the inputs, i.e., $\log x_1, \ldots, \log x_n$, and applying the exponential function to the net output, yielding

$$o_{\text{mult}}(\mathbf{x}) = \exp\left(\sum_{i=1}^{n} w_i \log x_i\right)$$

$$= \prod_{i=1}^{n} x_i^{w_i}.$$

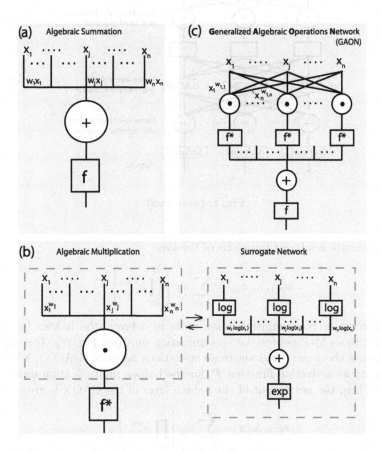

**Fig. 1.** (a) Algebraic summation by the classical McCulloch-Pitts (McP) unit. (b) Left panel: Algebraic multiplication unit: Inputs are raised to the power of their weights and then multiplied, followed by an activation $f^*$ [6,12]. Right panel: Algebraic multiplication (gray broken line) can be implemented by a surrogate network with alternating log and exp activation functions. Here, the absolute values of the inputs are taken and a small shift is added to avoid zero values in the input. (c) Generalized algebraic operations network (GAON) built from elementary summation and multiplication units. (d) Larger network including a hidden layer of GAONs. The first dense layer allows transforming the input data into a suitable domain. Several parallel GAONs implement nonlinear input-output relations and their output is passed on a dense summation layer performing softmax activation for classification purposes.

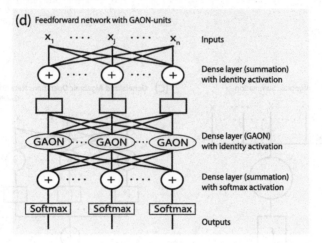

**Fig. 1.** (*continued*)

To approximate nonlinear functions of the type

$$s(x_1, ..., x_n) = \sum_{k=1}^{m} w_k \left[ \prod_{i=1}^{n} x_i^{w_{k,i}} \right],$$

we arrange $m$ or more multiplication units in a layer (the hidden layer) and connect it to an McP neuron, i.e., a summation unit (see Fig. 1c). This arrangement defines the generalized algebraic operations network (GAON). We choose the identity as activation function $f^*$ for the hidden multiplication units of the GAON. Then, the net output of the hidden layer of the GAON is given by

$$o_{\text{GAON}}(\mathbf{x}) = \sum_{k=1}^{m} w_k \left[ \prod_{i=1}^{n} x_i^{w_{k,i}} \right],$$

where the $w_{k,i}$ are weights from the input units to the hidden units of the GAON, and the $w_k$ are the weights from the hidden units to the output node of the GAON. The final output of the GAON is then $\hat{y} := f(o_{\text{GAON}}(\mathbf{x}))$, where $f$ is the activation function of the output unit. For function approximation tasks, we choose $f$ to be the identity. For classification tasks, we choose the activation function to be $f(b) = 1/(1 + e^{-b})$. This form of network not only allows representation of any polynomial of $n$-input variables, but also fractions and roots. Functions that can be described by power series (such as cosine and sine) could potentially be approximated by the network if a large enough number of hidden multiplication units is used. Training is performed using error backpropagation [19,21]. The loss functions and further details are provided in the appendix. Furthermore, we integrate GAON-units into larger networks (see Fig. 1d). First, the input is fed into a dense layer of summation units with identity activation. A bias input is included. Then, the output of the dense layer is processed by a dense layer of parallel GAON-units. The GAON-layer output provides input to another dense layer of summation units with softmax activation. We use this network to solve multi-label classification tasks.

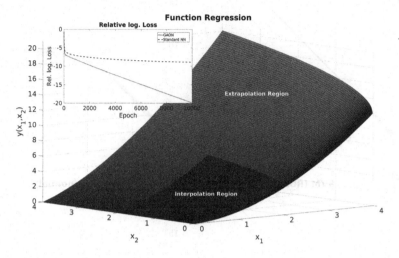

**Fig. 2.** Surface plot of the function $y = x_1\sqrt{x_2}+x_1^{1.8}$. Blue colors indicate the data used for training of the methods, the red colors represent the test data used for evaluation. Inset: The relative logarithmic values of the loss function for the regression task during training are plotted for the standard NN and the GAON. (Color figure online)

## 3    Results

### 3.1    Function Regression

We first test the network of Fig. 1c having $m = 2$ hidden units on a function-regression task. For this purpose, we generate a data set consisting of feature vectors $(x_1, x_2)$ and output values $y = x_1\sqrt{x_2} + x_1^{1.8}$. This function is plotted in Fig. 2. The input values of the training data are in the range of $x_1, x_2 \in [0, 2]$, while those of the test data cover the larger ranges $x_1, x_2 \in [0, 4]$ . The respective areas of the function are plotted in blue (training) and red (testing) color. In this way we can evaluate the extrapolation capacity of the method. Using stochastic gradient descent, the algorithm converges to a stable solution after less than 100 epochs, as seen in the inset of Fig. 2, where the relative logarithmic loss is plotted as a function of the training epochs. We observed similar convergence behavior for other choices of the input-output relationship. For comparison, we use a standard neural network with one hidden layer of 10 hidden McP units (standard NN). The standard NN is trained in the same way as the GAON on the same training set (see appendix) [19]. As seen in Fig. 2, the value of the loss function of the standard NN remains larger than that of the GAON even after many epoch. We should note however that the learning rate of the GAON is

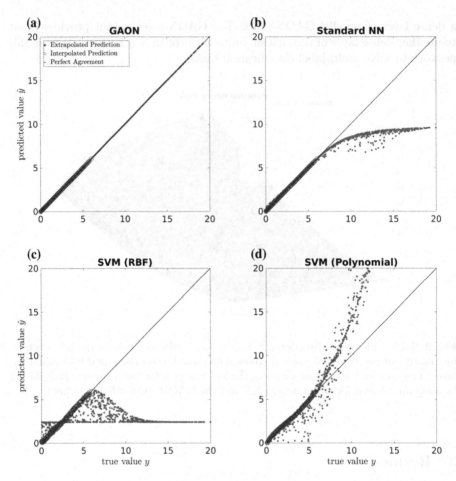

**Fig. 3.** Function-regression task. The function $y(x_1, x_2) = x_1\sqrt{x_2} + x_1^{1.8}$ is learned and the predicted value is plotted against the true value for (a) the GAON (b), the standard NN, (c) the SVM with RBF kernel, and (d) polynomial kernel. (Color figure online)

slower than that of the standard NN (by a factor of $\approx 0.1$), because the GAON is more sensitive to weight changes than the standard NN (see appendix). This results in a larger number of iterations per epoch for the GAON.

We further compare the extrapolation capabilities of our method with those of the standard NN and the nonlinear support-vector machine (SVM), the latter for both polynomial and radial-basis-function kernels (RBF) [3,19]. The results are shown in Fig. 3. To evaluate the results, we plot the predicted output $\hat{y}$ as a function of the true value $y$ for the different methods. In case of perfect agreement, the points should lie on a line, i.e., $\hat{y} = y$ (shown in black). The results are plotted as blue circles for interpolated values (i.e., values that lie within the range of the training data) and as red dots for extrapolated values (situated outside that range). It can be seen that the results from the GAON lie close to the line of perfect agreement for both interpolated and extrapolated predictions (Fig. 3a). For the standard NN (Fig. 3b) and the SVMs (Fig. 3c-d), the extrapolated predictions mostly disagree with the observed values.

## 3.2   Classification

We further test the network on a data-classification task. Here, we use $m = 10$ hidden units for the GAON, but similar results are obtained for other choices of $m$ (not shown). In this task, vectors belonging to class $+1$ (TRUE) have end points lying within an area of symmetrical shape, namely a circle, a rectangle, and a diamond, defined by the implicit function $|x_1|^p + |x_2|^p \leq r^p$ with $p = 0.5, 2, 20$, respectively. The vectors lying outside of the shape belong to class 0 (FALSE). Since the shapes are symmetric, it suffices to train and test the methods on the absolute values of the input data. To specifically test their extrapolation capabilities, we train the methods only on partial shapes. The respective region of feature space used for training is underlaid in gray color in the plots (see Fig. 4). We repeat the training and testing process 100 times, varying randomly drawn training sets and computing the frequency with which the data point is assigned to the class TRUE. The frequency of occurrence is color-coded in shades of blue and presented together with the correct class boundaries of the classification problem (black line). In Fig. 4a, the results from the GAON are shown for the different shapes. In areas where the patterns have to be extrapolated, our method outperforms the standard NN (Fig. 4b) and the two SVMs (Fig. 4c–d) for all cases.

**Table 1.** Classification results for the different methods and data sets. The number of support vectors is averaged over 100 trials.

| Method | Accuracy [%] | | | |
|---|---|---|---|---|
| | Total | Interpolation | Extrapolation | # Hidden Units/Support Vectors |
| **p = 0.5** | | | | |
| GAON | 97.68 ± 0.12 | 99.64 ± 0.01 | 91.82 ± 0.48 | 10 |
| Standard NN | 97.66 ± 0.04 | 99.65 ± 0.01 | 91.69 ± 0.16 | 10 |
| SVM (RBF) | 94.65 ± 0.04 | 99.04 ± 0.02 | 81.49 ± 0.14 | 352.97 ± 7.23 |
| SVM (Polynomial) | 94.40 ± 0.37 | 99.43 ± 0.02 | 79.30 ± 1.46 | 8.49 ± 0.08 |
| **p = 2** | | | | |
| GAON | 98.80 ± 0.09 | 99.72 ± 0.01 | 96.04 ± 0.35 | 10 |
| Standard NN | 97.80 ± 0.11 | 99.72 ± 0.01 | 92.03 ± 0.44 | 10 |
| SVM (RBF) | 97.30 ± 0.03 | 99.50 ± 0.01 | 90.69 ± 0.13 | 239.13 ± 4.90 |
| SVM (Polynomial) | 98.22 ± 0.24 | 99.61 ± 0.02 | 94.03 ± 0.93 | 6.63 ± 0.09 |
| **p = 20** | | | | |
| GAON | 99.22 ± 0.02 | 99.29 ± 0.01 | 99.00 ± 0.06 | 10 |
| Standard NN | 98.46 ± 0.11 | 99.74 ± 0.01 | 94.63 ± 0.45 | 10 |
| SVM (RBF) | 97.76 ± 0.03 | 99.40 ± 0.02 | 92.83 ± 0.12 | 258.01 ± 5.64 |
| SVM (Polynomial) | 98.85 ± 0.05 | 99.44 ± 0.02 | 97.07 ± 0.16 | 9.42 ± 0.16 |

The accuracy of the classifications results obtained with the different methods for the three shapes is presented in Table 1. The GAON achieves higher accuracies than the standard NN and the two SVMs in the extrapolation task for $p = 0.5$, $p = 2$ and $p = 20$. The GAON outperforms all three methods in the extrapolation task. In the interpolation task, the standard NN and the GAON achieve comparable accuracies.

### 3.3   Real-World Application

As an illustrative example for a real-world application of product-unit networks, we considered the problem of completing incomplete labelings of images as they occur frequently in image-segmentation tasks. In Fig. 5a the image of a tobacco plant during plant growth is shown [2]. Using graph-based image segmentation [7] and removing segments below a critical size threshold, an imcomplete labeling of the image is obtained (see Fig. 5b). Areas without label are depicted in black color. The number of labels corresponds to the number of prominent leaves in the images plus the background. The leaf pointing north is particularly poorly segmented by the method. Completing the labeling of the image can be posed as a classification task, where the pixel coordinates and their respective class labels ranging from 0 to 5 provide the input training data for the classifier. Pixels, for which no label could be assigned, are not a part of the training data.

We use these data to train the product-unit network shown in Fig. 1d. The two-dimensional input data, i.e., the pixel coordinates, provide the input to a dense layer consisting of two summing units without activation. A bias is provided to each of the summing units to allow shifts of the data. The output of the two summing units is fed to a layer of six GAON-units. The output of the GAON-layer then is passed on to another layer of six summing units with softmax activation. For the loss function, Tensorflow's sparse categorical crossentropy is used [1]. Using gradient descent with batches of 40 data points, we obtained an accuracy of 0.9979 after 2000 epochs. We use the neural network to extrapolate class labels into the previously unlabeled regions and obtain the completed segmentation shown in Fig. 5c. The segments now assume the nonlinear, approximately elliptic shape of the leaves.

## 4   Discussion

In this work, we augmented the algebraic structure of neural networks with a multiplicatory elementary unit [6,12] that computes the products of its inputs raised to the power of their weights, representing the multiplicatory analogon to the classical McP-neuron. To evaluate the extrapolation capabilities of the network, we used both a function-extrapolation task and a pattern-classification-extrapolation task. We could demonstrate that the GAON outperforms the standard NN and the nonlinear support vector machine for the given tasks. We presume that the gain in performance originates from the property that the

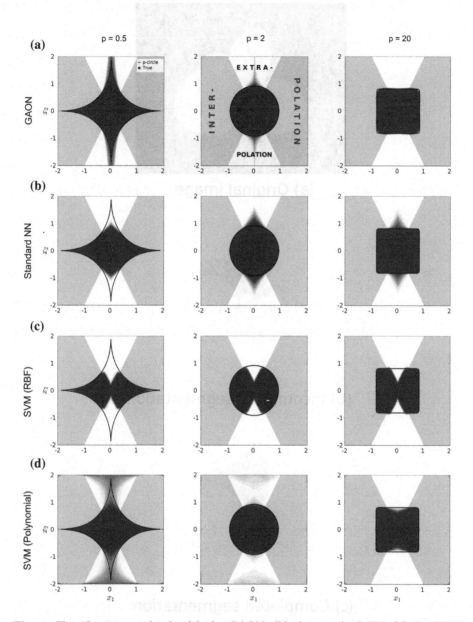

**Fig. 4.** Classification results for (a) the GAON, (b) the standard NN, (c) the SVM with RBF kernel, and (d) polynomial kernel for the different data sets. The frequency of occurrence of the class label TRUE is color coded in shades of blue for 100 trials and shown together with the correct class boundaries (black line).

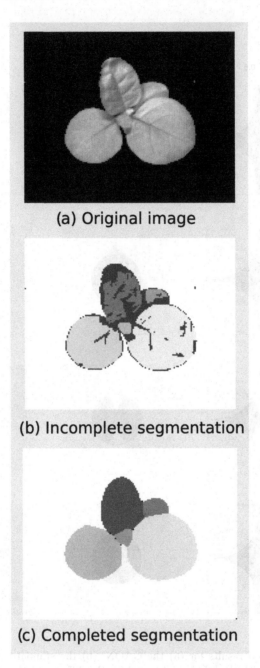

(a) Original image

(b) Incomplete segmentation

(c) Completed segmentation

**Fig. 5.** (a) Grayscale image of a tobacco plant [2]. (b) Incomplete segmentation obtained with a graph-based image-segmentation method [7]. (c) Completed segmentation obtained with a product-unit network trained with the incomplete-segmentation result (for better visualization of the results, the color code of the segment labels has been slightly modified).

multiplier network can learn arbitrary functions of the form

$$\sum_{k=1}^{m} w_k \prod_{i=1}^{n} x_i^{w_{k,i}},$$

being generated by the outputs of the hidden units $\prod_{i=1}^{n} x_i^{w_{k,i}}$, allowing extrapolation into unknown domains. The net outputs of the hidden units of the standard NN are linear functions and as such do not represent a generating set, hindering extrapolation of the pattern. The nonlinear support vectors machine is constrained by the choice of the kernel function; the polynomial kernel SVM performs excellently for the circular shape in our example, but in this case the kernel represents an almost perfect fit for the characteristics of the pattern.

Our proposed network is structurally simple and the training straightforward. The use of adjustable weights that are appearing as exponents of the inputs in the network allows the GAON to learn arbitrary functions from the data. This property makes it fundamentally different from polynomial classifiers, where a fixed set of predefined polynomial basis functions is chosen for the discrimination function and combined in a weighted sum [19]. Related to this, Bayesian neural networks employing higher-order polynomials [4] have been proposed in the past, for which however similar restrictions apply.

We restricted ourselves to positive weights in the data classification task. This prevents failure when input values are near zero. However, this does not limit the applicability of the method by any means. We explain this giving an example: The inequality $\frac{1}{x_1} + x_2^2 \leq 1$ contains a negative exponent, but it can be recast by multiplying both sides with $x_1$ (given that we work with positive input data only, which can always be achieved by shifting). This yields the new inequality $1 + x_1 x_2^2 \leq x_1$, or $x_1 x_2^2 - x_1 \leq -1$, which still represents the same classification problem as the original inequality. Hence, there exists an infinite set of discrimination functions with positive exponents that describe the classification problem. More than that, we suspect that this property allows the network to escape local minima by moving toward a more advantageous discriminant function, securing convergence of the method. This conjecture might further explain why the multiplier network is robust to changes in the number of hidden units when solving classification tasks. However, when solving regression tasks, extrapolation performance depends on the number of hidden units, and we observed that extrapolation improves when the number of hidden units matches the number of basis functions of the function space required for properly describing the input-output relationship. In all cases our method leads to a very sparse representation of functions and patterns that allows, unlike nonlinear SVMs, a direct interpretation of its components in terms of mulitplicative relationships.

We further demonstrated that product units can be intergrated in larger networks (see Fig. 1d) using the framework Tensorflow [1] by extrapolating labels in image segmentation into unlabeled regions (see Fig. 5). This is particular useful when the shapes of the objects that are to be segmented are unknown prior to the task. Our approach can be applied to other labeling problems as well, e.g., the assignment of disparity labels to pixels in stereo vision.

To address concerns regarding convergence of gradient-descent training, we revisited the parity-8 problems [12] and trained the same network that was used to solve the image-segmenation task (see Fig. 1d) for the parity-8 problem. The method converged to solutions with accuracy 1. The authors of [12] searched for a single set of weights that would classify the parity-8 data correctly by considering a network that basically consisted of a single product unit only. In a larger product-unit network, there presumably exists more than one weight combination that classifies the data correcty. We reckon that this has an impact on convergence.

**Acknowledgements.** We thank John W. Clark for his help in improving the manuscript through his comments. B.D. and U.J. contributed in equal parts to the ideation, design and implementation of the method. M.W. contributed to the implementation, integration and the training/testing of the method.

# A    Appendix

All methods are implemented in MATLAB. We use a training set size of 1000, drawn uniformly from randomly distributed values in the range $0, \ldots, +2$ for the regression task and $-2, \ldots, +2$ for the classification task. Every input is mapped onto the first quadrant by using only absolute values. To test the inter- and extrapolation capabilities of our network, we require the examples of the training set to satisfy the following condition: $|x_1| > 0.5|x_2|$. This defines our interpolation region (represented as the white space in Fig. 4). The examples of the test set are allowed to take on any value in the range indicated above. We use self-coded functions for the GAON and standard NN and built-in functions from Matlab for the SVMs.

We also implemented the GAON in Python within the Keras-API of the Tensorflow library [1] to enable fast computation with graphics processing units (GPUs) and make use of its modularity. We used this framework to build the larger product-unit network used for multi-label classification.

## A.1    GAON

For training in function-regression tasks, we use $\sum_{j=1}^{N}(\hat{y}_j - y_j)^2$ as loss function, where $y_j$ is the true value corresponding to the input vector $\mathbf{x}_j$ and $\hat{y}_j$ the predicted value. For network training in classifications tasks, we use the cross entropy $\epsilon = -\sum_{j=1}^{N}[y_j \log \hat{y}_j + (1 - y_j) \log (1 - \hat{y}_j)]$ as loss function. $N$ is the number of training vectors.

For our runs with the GAON we use a constant learning rate of $l_r = 10^{-3}$ and train the network for 10000 epochs with stochastic gradient descent, using one training example per iteration. One epoch is defined as a complete process over the whole training set (corresponding to 1000 iterations in our setup). The initial values for the weights in the first layer are drawn randomly from a uniform distribution, whereas those of the second layer are normally distributed. Additionally we restrict the weights of the first layer to be only positive in the training process (see discussion).

For the results of the classifications (Fig. 4 and Table 1) we repeat the whole training and testing process 100 times with varying training sets and present the mean along with the standard error of the mean.

## A.2  Standard Neural Network

For our runs with the standard neural network we use a constant learning rate of $l_r = 10^{-2}$. We train the network with stochastic gradient descent for 10000 epochs with one example per iteration. The initial values for the weights are drawn in the same as in the GAON method, but here we do not restrict the weights in any layer. We used the same loss functions as for the GAON.

## A.3  Support Vector Machine

For our runs with the SVMs we use the Matlab-function *fitrsvm* for the regression and *fitcsvm* for the classification task. Our models include standardization (centering and devision by standard deviation) of the input and an automatic selection of the scaling factor by a heuristic procedure. The polynomial SVM has an order of four.

# References

1. Abadi, M., et al.: TensorFlow: Large-scale machine learning on heterogeneous systems (2015). https://www.tensorflow.org/
2. Agostini, A., Alenyà, G., Fischbach, A., Scharr, H., Wörgötter, F., Torras, C.: A cognitive architecture for automatic gardening. Comput. Electron. Agric. **138**, 69–79 (2017)
3. Bose, B., Guyon, I., Vapnik, V.: A training algorithm for optimal margin classifiers. In: Proceedings of the 5th Annual Workshop on Computational Learning Theory, pp. 144–152 (1992)
4. Clark, J.W., Gernoth, K.A., Dittmar, S., Ristig, M.L.: Higher-order probabilistic perceptrons as bayesian inference engines. Phys. Rev. E **59**, 6161–6174 (1999)
5. Cortes, C., Vapnik, V.: Support vector network. Mach. Learn. **20**, 273–297 (1995)
6. Durbin, R., Rumelhart, D.E.: Product units: a computationally powerful and biologically plausible extension to backpropagation networks. Neural Comput. **1**(1), 133–142 (1989)
7. Felzenszwalb, P.F., Huttenlocher, D.P.: Efficient graph-based image segmentation. Int. J. Comput. Vis. **59**(2), 167–181 (2004)
8. Hubel, D., Wiesel, T.: Receptive fields, binocular interaction and functional architecture in the cat's visual cortex. J. Physiol. Lond. **160**, 54–106 (1962)
9. Husain, F., Dellen, B., Torras, C.: Scene Understanding Using Deep Learning. Academic Press, San Francisco (2017)
10. Husain, F., Schulz, H., Dellen, B., Torras, C., Behnke, S.: Combining semantic and geometric features for object class segmentation of indoor scenes. IEEE Robot. Autom. Lett. **2**, 49–55 (2017)

11. LeCun, Y., Bengio, Y., Hinton, G.: Deep learning. Nature **521**, 436–444 (2015)
12. Leerink, L.R., Giles, C.L., Horne, B.G., Jabri, M.A.: Learning with product units. Adv. Neural Inf. Process. Syst. **7**, 537 (1995)
13. McCulloch, W.S., Pitts, W.: A logical calculus of the ideas immanent in nervous activity. Bull. Math. Biophys. **5**, 115–133 (1943)
14. Minsky, M.L., Papert, S.: Perceptrons : An Introduction to Computational Geometry. MIT press, Cambridge (1988)
15. Rajaraman, S.K., Antani, S., Candemir, S., Xue, Z., Kohli, M., Thoma, G.: Comparing deep learning models for population screening using chest radiography. SPIE Med. Imaging **10575E**, 1–11 (2018)
16. Rosenblatt, F.: The perceptron - a probabilistic model for information storage and organization in the brain. Psychol. Rev. **65**, 386 (1958)
17. Scholkopf, B., Smola, A.J.: Learning with Kernels: Support Vector Machines, Regularization, Optimization, and Beyond. MIT Press, Cambridge (2001)
18. Theodoridis, S., Koutroumbas, K.: Linear Classifiers, 4th edn. Academic Press, Boston (2009)
19. Theodoridis, S., Koutroumbas, K.: Nonlinear Classifiers, 4th edn. Academic Press, Boston (2009)
20. Silver, D., et al.: Mastering the game of go with deep neural networks and tree search. Nature **529**, 484–489 (2016)
21. Werbos, P.J.: Beyond regression : new tools for prediction and analysis in the behavioral sciences. Harward University (1974)

# Fast and Scalable Outlier Detection with Metric Access Methods

Altamir Gomes Bispo Junior and Robson Leonardo Ferreira Cordeiro[✉]

University of São Paulo, São Carlos, São Paulo, Brazil
`robson@icmc.usp.br`

**Abstract.** It is well-known that the existing theoretical models for outlier detection make assumptions that may not reflect the true nature of
outliers in every real application. With that in mind, this paper describes
an empirical study performed on **unsupervised outlier detection**
using 8 algorithms from the state-of-the-art and 8 datasets that refer
to a variety of real-world tasks of high impact, like spotting cyberattacks, clinical pathologies and abnormalities in nature. We present the
lowdown on the results obtained, pointing out to the strengths and weaknesses of each technique from the **application specialist's point of
view**, which is a shift from the designer-based point of view that is commonly considered. Interestingly, many of the techniques had unfeasibly
high runtime requirements or failed to spot what the specialists consider
as outliers in their own data. To tackle this issue, we propose *MetricABOD*: a novel angle-based outlier detection algorithm that makes the
analysis up to **thousands of times faster**, still being in average **26%
more accurate** than the most accurate related work. This improvement
is essential to enable outlier detection in many real-world applications
for which the existing methods lead to unexpected results or unfeasible runtime requirements. Finally, we studied two real collections of text
data to show that our *MetricABOD* works also for adimensional, purely
metric data.

**Keywords:** Applied computational sciences · Complex data ·
Data Mining · Unsupervised outlier detection · Metric Access Methods

## 1 Introduction

An important task with high-impact real-world consequences is the forecasting and detection of extreme events and exception cases, like frauds in the
financial sector, cyberattacks, clinical pathologies and abnormalities in nature.
Such phenomena are known as outliers. The volume of data being collected by

This work was partially supported by the São Paulo Research Foundation (FAPESP) –
Grants No 2018/05714-5 and 2016/17078-0, by the Coordination for Improvement of
Higher Education Personnel (CAPES) – Finance Code 001, and by the National Council
for Scientific and Technological Development (CNPq).

ⓒ Springer Nature Switzerland AG 2019
J. M. F. Rodrigues et al. (Eds.): ICCS 2019, LNCS 11537, pp. 189–203, 2019.
https://doi.org/10.1007/978-3-030-22741-8_14

enterprises from a variety of scientific areas is increasing exponentially over time, thus forcing analysts to use automatic procedures to detect outliers [6,8].

Unfortunately, the present literature lacks ample and thoughtful comparison between the many existing outlier detection methods. One recent work [5] succeeded to reduce this gap, but, according to its authors, it is a "meta-analysis". That is, their work aims to be as generalist as possible and, as such, it does not focus on any application nor domain and does not discuss accuracy in real applications from the point of view of the specialist users, i.e., "what the application specialists expect from the data". The major merit of their work lies in that it presents useful and very needed pointers to guide further investigation over a diversity of application domains. However, it is still unclear how accurate most of the existing methods are when spotting what the specialists consider as outliers in datasets of distinct natures, such as those collected in the many real-world applications that can benefit from the detection of outliers.

It is well-known in the literature that the existing theoretical models for unsupervised outlier detection make certain assumptions that may not reflect the true nature of specific outliers in every single application [1,5]. With that in mind, we performed an empirical study to evaluate 8 state-of-the-art outlier detection algorithms using 8 datasets from a variety of real-world applications with ground truth data manually created by specialist users; we were focused on verifying whether or not the algorithms are able to spot in an automatic and timely manner what the specialists selected as outliers. Interestingly, many of the algorithms had unfeasibly high runtime requirements or failed to return what the specialists expect from their data. This paper's main contributions are:

1. **Empirical evaluation:** we evaluated 8 state-of-the-art outlier detection methods from the **application specialist's point of view**, and report results focused on quantifying how useful they can be for a variety of real world tasks of high impact, such as spotting breast cancer, heart and thyroid anomalies, detecting cyberattacks, and detecting musk species;
2. *MetricABOD*: we carefully designed a new angle-based algorithm that makes the analysis up to **thousands of times faster**, still being in average **26% more accurate** than the most accurate related work. The main innovation is a new usage for tree-based data structures known as Metric Access Methods (MAM) [17], which were originally designed to index complex data, such as images, audio, large graphs and fingerprints. This improvement was essential to enable outlier detection in many of the datasets that we studied, for which the existing methods lead to unexpected results or unfeasible runtime requirements. Finally, we studied real collections of text data to show that our *MetricABOD* works also for adimensional, purely metric data.

The rest of this paper is organized as follows: background concepts (Sect. 2), related works (Sect. 3), empirical evaluation (Sect. 4), proposed algorithm (Sect. 5) and conclusions (Sect. 6).

# 2     Background

There are many definitions for outliers in the literature. The distance-based one [10] is very commonly used: *"An object P in a dataset T is a DB(f, ξ)-outlier if at least one fraction f of the objects in T have a distance to P that is greater than ξ."* Here, term $DB(f, \xi)$-outlier is a shorthand notation for a Distance-Based outlier with supporting parameters $f$ and $\xi$. This definition is independent of the data distribution and it is also intuitive, simple and practical. Due to such qualities, it serves as a basis for several outlier detection algorithms [3,4].

In spite of that, there is not a single, universally accepted definition for what an outlier is, neither an unanimous way to compute outlierness scores. Outliers matter to statisticians since long ago. Research works in the domain of Statistics routinely assume a parametric distribution for the data, following a fixed set of parameters. However, it is not safe to assume, a priori, any parametric distribution for a dataset in most real-world scenarios. And even if the distribution is assumed as non-parametric, the existing unsupervised outlier detection algorithms must also assume a specific definition of outliers within the distribution to create a model, and then use the model to search for data objects that fit this definition. Obviously, different algorithms employ distinct models, thus they may obtain different results. Since there is no consensus, it is not unusual that two given algorithms' results clash [12].

Let us highlight that the application specialist user expects results of his/her true interest, despite the model and algorithm at hand. At this point, we make a shift from the designer-based point of view that is routinely seen in the literature, to the user-based one. Even elaborate and theoretically sound techniques may not translate to acceptable results for the end user. Also, many real-world datasets from which spotting outliers is desirable have millions of objects and hundreds of attributes. Outlier detection in these data is a demanding task [6]. In fact, it is truly an unfeasible task in most cases. The main challenges are: (a) very large runtime requirements, even for approximate methods; (b) low availability of ground truth to assess a method's accuracy; (c) data of high dimensionality and its unwanted effects, and; (d) sensitivity to input parameter values.

The next subsections discuss the main undesirable effects of dealing with data of high dimensionality; we also present the basics of Metric Access Methods.

## 2.1     Curse of High Dimensionality

As the number of dimensions of a dataset increases, the variance of distances between any two data objects tends to zero [1]. Figure 1 illustrates this fact by presenting dissimilarity matrices for two toy datasets with 10 objects each. The objects were randomly picked from a multivariate normal distribution. Figures 1a and b respectively refer to data of low and high dimensionality, i.e., 5 and 100 dimensions. Darker colors represent smaller distances; lighter colors refer to larger distances. As it can be seen, it is much harder to discriminate the distinct dissimilarities in the high-dimensional scenario than it is in the low-dimensional one. This simple example illustrates the evanescence of the variance of pairwise dissimilarities between objects, as the dimensionality increases.

In this fashion, according to the distance-based outlier definition, and also to most of the other existing definitions, all objects in a dataset of high dimensionality are potential outliers. To tackle this problem with subspace selection or dimensionality reduction is not obvious. According to [1], the main difficulties are: (a) circular references between neighborhoods and subspaces; (b) noise with its masking and dilution effects, and; (c) rarity-based subspace outliers. State-of-the-art solutions to the problem perform angle-based analysis, rather than the traditional distance-based one, as we discuss later in the paper.

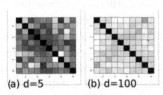

**Fig. 1.** Dissimilarity matrix for two toy datasets with 10 random objects each. (a) 5 attributes; (b) 100 attributes.

**Fig. 2.** 2-dimensional objects at: (a) level $h$ of a tree-based MAM structure; (b) level $h - 1$ of the same structure.

## 2.2 Metric Access Methods

Metric Access Methods (MAM) [17] allow efficient queries upon datasets embedded in a *metric space*. A metric space is formally defined by a data domain $X$ and a distance function $d : X \times X \to \mathbb{R}$, such that the rules as follows apply:

(i) Positivity: for all $x, y \in X$, $0 \le d(x, y) < \infty$
(ii) Non-degenerated: for all $x, y \in X$, $d(x, y) = 0 \leftrightarrow x = y$
(iii) Symmetry: for all $x, y \in X$, $d(x, y) = d(y, x)$
(iv) Triangle inequality: for all $x, y, z \in X$, $d(x, y) \le d(x, z) + d(z, y)$

A metric space may not allow sorting objects. Thus, MAM structures must use distances only; nothing else. State-of-the-art MAMs are tree-based. Every tree node stores a subset of objects. One of them is selected to be the node's representative, which is also known as the pivot. The representative is usually the object with the shortest average distance to all others. The node's region of representation is a hyper-sphere centered at the representative object with a radius that is greater than or equal to the distance from it to any other object in the same node. Figure 2 illustrates a tree-based MAM upon objects in a two-dimensional space. We assume that each node stores up to 4 objects. In Fig. 2a, four nodes with representatives $A$, $B$, $C$ and $D$ are defined out of all objects at the lowest level $h$. At level $h - 1$, only these representatives are considered; as it can be seen in Fig. 2b, object $D$ is selected to represent the previous level's representatives into a single root node. When using a MAM, the triangle inequality property and the representatives serve as a basis to prune unnecessary distance calculations, thus leading to efficient data manipulation.

# 3 Related Works

To evaluate the distances from objects to their $k$-th nearest neighbors is one of the simplest approach to spot outliers. The *KNN-Outlier* method ranks the objects' respective distances to their $k$-th nearest neighbors, or KNN distances, from longest to shortest, and gives higher outlierness scores to objects with higher KNN distances. It is the basis for many other works, like *Orca* [3] and *RBRP* (Recursive Bin Partitioning and Re-projection) [7].

*LOF* (Local Outlier Factor) [4] is another well-known outlier detection method. It introduced the concepts of core distance and reachability distance, inspiring many posterior works, such as *LOCI* (Local Correlation Integral) [15], *aLOCI* (Approximated Local Correlation Integral) [15] and *ABOD* (Angle-Based Outlier Detection) [11]. According to *LOF*, a *core object* is an object that has at least a certain number of neighbours in its $\xi$-neighborhood. An object $x$ is *density-reachable* from object $y$ if there is a chain of objects $p_1, p_2, ...p_q$, where $p_1 = y$ and $p_q = x$, such that $p_{i+1}$ is in the $\xi$-neighborhood of $p_i$ for $i \in \{1, 2, ... q - 1\}$. The *core-distance* of an object $x$ is the smallest distance $\xi$ that makes it a core object. The *reachability-distance* of $x$ is the smallest distance $\xi$ that puts it in the $\xi$-neighbourhood of a core object $y$. LOF employs these concepts to calculate outlierness scores for all data objects according to their $\xi$-neighbourhoods' density of objects in the feature space.

From here on, we briefly describe methods that were developed having in mind the problems inherent to high dimensional data. Two basic approaches have been explored in the literature: subspace analysis and angle-based analysis. A subspace is a representation of the original space that excludes some of the dimensions/attributes present in the latter. Since a proper subspace has lower dimensionality than that of the original space, the data projection can alleviate the dilution and noise effects present in higher dimensional spaces and may also improve the contrast for outlier detection methods that use distances as a basis.

Unfortunately, the number of distinct subspaces to analyze grows exponentially with regard to the number of attributes of a dataset. Therefore, extensive subspace search is not feasible. Aggarwal and Yu [2] faced this issue using an evolutionary algorithm. Their method works by iteratively improving candidate sets of subspaces until a final set of feasible subspace candidates is generated. OUTRES [14] works by performing statistical tests upon subspace projections of dataset objects, starting from 1-dimensional projections and adding new dimensions one-by-one. Only subspace projections that pass their statistical test are considered. OUTRANK (Outlier Ranking) [13] enumerates as outliers the objects that overlap less, considering their presence or absence inside clusters found in distinct subspaces. HiCS (High Contrast Subspaces) [9] selects feasible subspaces by performing a statistical test upon each candidate subspace. HiCS itself does not measure the outlierness of objects; it merely retrieves and aggregates the results obtained by a coupled outlier detection method, such as LOF. Thus, HiCS is described by its authors as a "meta" outlier detection method. Finally, note that all of the aforementioned methods allow to set thresholds in an attempt to prune uninteresting subspaces.

ABOD (Angle-Based Outlier Detection) [11] uses a different approach. It was inspired by previous research works in the domain of text comparison that had verified that angles are more resilient than distances to the "curse of high dimensionality". The authors of ABOD noted that this fact remains true also when detecting outliers from high-dimensional data in a general context, and not only for text. For each object $P$, ABOD calculates the angles between every possible pair of objects $(x, y)$ using $P$ as the pivot, such that $x \neq y \neq P$. The variance of angles is then attributed to $P$ as its outlierness score. Small variances of angles indicate outliers; larger variances refer to inliers. The intuition here is that outliers tend to be in the borders of the feature space, so they have neighbors concentrated in one specific direction, while the inliers' neighbours are spread all over the space. Figure 3 illustrates this idea considering an inlier (Fig. 3a) and an outlier (Fig. 3b). In both cases, colored semi-circles represent some of the angles associated to pairs of objects when taking the object of interest $P$ as the pivot. Figure 4 complements this example by plotting the corresponding angle values and variances.

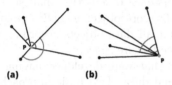

**Fig. 3.** ABOD's angle-based analysis for an inlier (left) and an outlier (right).     **Fig. 4.** Angle values and their variances (outlierness scores) for data in Fig. 3.

ABOD may not be limited in practice by data dimensionality. However, due to scalability issues, its use is clearly impractical for data of high cardinality. It computes angles for every possible triple of objects, so ABOD is $O(n^3)$ where $n$ is the number of objects. LB-ABOD(Lower Bound-based ABOD) [11], FastA-BOD [11] and FastVOA (Fast Variance Of Angles) [16] improve upon ABOD focused on tackling its scalability issues. Unfortunately, they either obtain only minor speedup or degrade accuracy considerably.

## 4   Unsupervised Outlier Detection at Work

Many of the state-of-the-art methods in outlier detection validate their results only on synthetic data of low cardinality. Some methods are also tested for one or two real applications using data with up to a few thousand objects, but not for a broad range of applications in which diversity takes place, nor using datasets that are as large as the ones required for commercial use. Therefore, it is still unclear how efficient and effective these methods are to be used in the real world.

In this section, we shrink this limitation considering a variety of real-world tasks of high impact, like spotting cyberattacks, clinical pathologies and abnormalities in nature. We studied the behaviour of 8 state-of-the-art algorithms face to 8 datasets of distinct natures, that is, data from diverse real applications with varying cardinality and dimensionality. To make it possible, we assume that the ground truth created by specialist users is correct, and verify whether or not the algorithms are able to obtain similar results in an automatic and timely manner. The main motivation here is that outlier detection must be useful in practice, not only in academia. Specifically, we focus on answering the questions as follows:

1. How effective and efficient are the state-of-the-art algorithms to be used in diverse real world applications?
2. What are the best applications for each algorithm?
3. Is there any algorithm with remarkably high accuracy in a general context?

### 4.1  System Configuration

The algorithms studied are: KNN-Outlier, LOF, ABOD, FastABOD, LB-ABOD, FastVOA, aLOCI and HiCS. We used the Elki (https://elki-project.github.io) implementation for all algorithms, except for FastVOA that was evaluated with an implementation of our own. The algorithms were tested in a Xeon E5-2640v3 machine with 16 GB of RAM, running Debian 9 64-bit. The number of samples $k$ was set to 100 in both FastABOD and LB-ABOD. For LB-ABOD, we also set parameter $l$ as two times the number of outliers known to exist in the ground truth. The same selection was used for parameter $k$ in KNN-Outlier, LOF and HiCS. Finally, FastVOA was tuned with $t = 100$, $s1 = 1,600$ and $s2 = 10$, following its authors recommendation.

The evaluation was performed by asking each algorithm for the $o$ objects with the highest outlierness scores, where $o$ is the number of outliers present in the ground truth. Considering correctly detected outliers as true positives $(tp)$, the accuracy of each algorithm was calculated as $\frac{tp}{o}$. Every experiment was ran 5 times. We report average accuracy and average runtime.

### 4.2  Datasets

Table 1 summarizes the datasets studied. They are available at the ODDS repository (http://odds.cs.stonybrook.edu/). All datasets come from real-world applications and include ground truth created by specialist users. For brevity, they are described in the following with focus on the practical benefits of spotting outliers in each case. Detailed descriptions are found in the data source website.

- **BreastW** and **Mammography:** two datasets with features from medical exams of breast cancer suspicious cases. They were collected in separate from two distinct locations. Spotting outliers here means detecting severe health problems, that is, distinguishing between the benign and the malignant cases.
- **Cardio:** features extracted from Fetal Heart Rate (FHR) and Uterine Contraction (UC) of cardiotocograms classified by expert obstetricians. To spot outliers in this scenario is important to prevent and treat heart malfunctions.
- **Annthyroid:** features from medical thyroid exams. Here, to detect outliers refers to the identification of the hypothyroid disease in human patients.

– **Satimage-2:** features from satellite imagery. Here, to spot outliers means uncovering abnormal and unexpected patterns of topography, land use, etc.
– **Shuttle:** features collected from the use of a space shuttle. Identifying outliers here means spotting potentially dangerous in-flight abnormalities.
– **Http (KDDCUP99):** log files of network communication. To spot outliers here means distinguishing between cyber attacks and regular transactions.
– **Musk:** features extracted from molecules classified as musk or non-musk by human experts in chemistry. Spotting outliers in such kind of data helps discovering new material that may be valuable to pharmaceutical corporations.

**Table 1.** Summary of datasets.

| Dataset | # Axes | # Objects | # Outliers | Dataset | # Axes | Objects | # Outliers |
| --- | --- | --- | --- | --- | --- | --- | --- |
| BreastW | 9 | 683 | 239 | Mammography | 6 | 11,183 | 260 |
| Cardio | 21 | 1,831 | 176 | Annthyroid | 6 | 7,200 | 534 |
| Satimage-2 | 36 | 5,803 | 71 | Shuttle | 9 | 49,097 | 3,511 |
| Http (KDDCUP99) | 3 | 567,479 | 2,211 | Musk | 166 | 3,062 | 97 |

## 4.3 Results

Table 2 reports accuracy and runtime results for the algorithms and datasets that we studied[1]. Runtime results are in minutes. As it was expected, some of the algorithms had unfeasibly high runtime requirements for our largest datasets Shuttle and Http (KDDCUP99), with $\sim 50k$ and $\sim 500k$ objects respectively. Therefore, we represent with "N/A" the cases that exceeded a timeout limit of 24 h. With regard to accuracy, note that many of the state-of-the-art techniques failed to return what the application specialists expect from their own data. LOF, HiCS, FastABOD and FastVOA had the lowest average accuracies, overall. On the other hand, aLOCI, ABOD, LB-ABOD and KNN-Outlier performed considerably better, being the most accurate methods in average. Let us highlight that the original ABOD method obtained the second best accuracy, but its use is unfortunately prohibitive due to scalability issues. In fact, ABOD is the worst methods in terms of runtime. The variation LB-ABOD is slightly faster without losing much in accuracy, although its runtime requirements are still far from being acceptable for most real-world uses. Finally, FastABOD is tens of times faster than ABOD and LB-ABOD in average, but its accuracy degrades considerably, that is, it spots correct outliers in only 34% of the times.

It is **important** to note that bad results over specifc datasets occurred for all methods, even for KNN-Outlier and ABOD that are overall the most accurate ones. Remarkable examples are: KNN-Outlier with the Http network logs and ABOD with the Musk chemical data. This scenario was already expected; it simply corroborates the fact that the theoretical outlier detection models make assumptions that may not reflect the true nature of outliers in every application.

---

[1] For brevity, the last column of Table 2 reports the results obtained with our proposed *MetricABOD* algorithm. They will be discussed latter in the paper; see Sect. 5.1.

**Table 2.** Accuracy/runtime (in minutes) results.

| Dataset | ABOD | LABOD | FABOD | LOF | KNN | aLOCI | HiCS | FVOA | MABOD |
|---|---|---|---|---|---|---|---|---|---|
| BreastW | **0.95/0.04** | **0.95/0.04** | 0.94/0.00 | 0.32/0.00 | 0.94/0.00 | 0.62/0.00 | 0.22/1.16 | 0.47/0.16 | **0.95/0.00** |
| Cardio | 0.52/1.23 | 0.52/1.25 | 0.36/0.01 | 0.23/0.00 | 0.51/0.00 | 0.04/0.00 | 0.32/10.0 | 0.19/0.46 | **0.61/0.00** |
| Sat-2 | 0.88/58.9 | 0.88/61.9 | 0.29/0.06 | 0.07/0.02 | **0.91/0.02** | 0.00/0.00 | 0.11/36.5 | 0.11/1.59 | 0.90/0.01 |
| Shuttle | *N/A* | *N/A* | 0.31/9.24* | 0.19/0.70 | 0.34/0.68 | **0.88/9.28** | *N/A* | 0.44/20.8 | 0.58/0.09 |
| Annth | 0.25/123 | 0.24/81.8 | 0.26/0.09 | **0.32**/0.02 | 0.24/0.01 | 0.10/0.00 | 0.15/28.7 | 0.15/2.06 | 0.24/0.01 |
| Mammo | **0.28**/253* | 0.11/223* | 0.26/0.28 | 0.23/0.02 | 0.27/0.04 | 0.08/0.03 | 0.21/60.9 | 0.17/3.42 | 0.26/0.03 |
| Http | *N/A* | *N/A* | *N/A* | 0.04/24.0 | 0.03/58.7 | **0.98/0.24** | 0.05/194 | 0.00/272* | **0.98**/2.98 |
| Musk | 0.07/7.50 | 0.07/7.33 | 0.02/0.03 | 0.00/0.03 | **1.00**/0.03 | 0.98/0.00 | 0.95/8.89 | 0.05/0.80 | 0.77/0.01 |
| **Average** | 0.49/73.9 | 0.46/62.5 | 0.34/1.38 | 0.17/3.09 | 0.53/7.43 | 0.46/1.19 | 0.28/48.5 | 0.19/37.6 | **0.66/0.39** |

\* result obtained on one AMD EPYC 7571 machine with 128GB RAM, due to main memory exceed.

# 5  MetricABOD

The results reported so far indicate that ABOD is one of the most accurate outlier detectors in a general context. Unfortunately, ABOD's cubic time complexity on the number of objects makes its use impractical for most real-world scenarios. Data dimensionality does not really play a factor in ABOD's runtime, even less with precomputation of inner products to make angle calculations faster. Thus, one must reduce the number of calculations **per object** to speed-up ABOD.

As it was discussed before, ABOD uses every data object $P$ as the pivot for $O(n^2)$ angle calculations, where $n$ is the total number of objects. Specifically, the outlierness score of an object $P$ is the variance of angles between every possible pair of objects $(x, y)$ where $P$ is the pivot. Obviously, $x$, $y$ and $P$ must be distinct objects. Figures 3 and 4 from the previous Sect. 3 illustrate this process.

To calculate angles only for pairs $(x, y)$ taken from a **sample dataset** is one clever way to achieve better scalability, still obtaining an outlierness score for **every** data object. It turns the original $O(n^3)$ complexity into $O(n.m^2)$ with $n \gg m$, where $n$ and $m$ are the full dataset cardinality and the sample size respectively. Obviously, the use of samples leads to approximate outlierness scores that may negatively impact the accuracy of results. So, the question is:

– *How to select appropriate objects to be in the sample?*

Random sampling is a quick and unburdensome way to answer this question. In fact, it has already been evaluated in one of the existing ABOD sequels, the FastABOD algorithm. However, in high-dimensional spaces, random sampling sports unbearably magnified sampling errors, thus leading to skewed variances of angles. Due to this fact, FastABOD's authors suggest the use of one distinct sample for each data object: its own $k$ nearest neighbours. In other words, FastABOD computes approximate outlierness scores for each object $P$ by using $P$ as the pivot for angle calculations among its own $k$-nearest neighbours. Unfortunately, $k$-nn sampling considerably degrades accuracy, as it could be observed in the results of the previous section.

This section presents a novel sampling strategy to improve ABOD by taking advantage of tree-based Metric Access Methods (MAM). As it was described in Sect. 2.2, tree-based MAMs have nodes that follow an hierarchical organization. Each node stores a set of objects; one of them is selected to be the node's representative, which is usually the object with the shortest average distance to the others. The leaf nodes store all dataset objects; their representatives are stored redundantly in nodes of the immediate higher tree level, from which representatives are also selected. This organization continues recursively up to the root of the tree; see Fig. 2 for an illustrative example.

The nodes of a tree-based MAM are built in a bottom-up fashion, like in a B-tree, so to minimize the number of levels, the number of nodes per level and the overlap among the nodes' hyper-spheres of representation in the feature space. This fact is **essential** to our proposal: it means that the representatives of nodes in any tree level preserve key characteristics of the full dataset, with more or less details according to the level. The highest degree of detail is in the leaf nodes' representatives. Since the representative is the object with the shortest average distance to the others, it is akin to the first moment or center of mass of the node's hyper-sphere. So, it is similar to the median for the node's objects. In fact, as it happens with random sampling, there is a strong correspondence between the representatives' statistical moments and those of the full dataset. Note, however, that low density regions in the feature space end up underrepresented with random sampling, while it never happens with MAM representatives. The representatives also compare favorably with FastABOD's $k$-nn sampling. Although the $k$ nearest neighbours of an object account for most of its variance of angles, to use them as references does not necessarily discriminate outliers correctly, as it was shown in the previous section. For example, FastABOD fails for the case of an outlier object whose $k$ nearest neighbours spread in many directions, while the rest of the dataset, i.e., the vast majority of it, concentrates in a single direction. Note that this case would not be a problem with MAM representatives; most of them would also concentrate in one direction, just like in the full data.

With that in mind, we **propose** to use the leaf nodes' representatives from any tree-based MAM as samples to speed-up ABOD. Algorithm 1 gives the full pseudo-code; let us call it the *MetricABOD* algorithm. We argue that the leaf node's representatives lead to a reasonable estimate for the exact variances of angles, and report latter in the paper experimental results that support our claim. In fact, the representatives even improved accuracy for the datasets that we studied; see details at Sect. 5.1. Figure 5 illustrates how our *MetricABOD* works. To easy comprehension, the same toy data of our running example from Fig. 2 is reused here. There are $n = 15$ objects with two attributes each. At first, a tree-based MAM is created. Four circles in the 2-dimensional space illustrate its leaf nodes. The representative objects are $A$, $B$, $C$ and $D$. In this setting, the approximate outlierness score for an object of interest $P$ is computed per $P$ as the pivot for angle calculations among pairs of objects $(x, y)$, such that $x, y \in \{A, B, C, D\}$ and $x \neq y \neq P$. Colored semi-circles illustrate some of the angles to be computed. Specifically, only $C_2^4 = \frac{4!}{2!.(4-2)!} = 6$ angle calculations are performed to process $P$. Note that the original ABOD algorithm would require $C_2^{n-1} = \frac{14!}{2!.(14-2)!} = 91$ calculations for the same case.

In a general scenario, our *MetricABOD* avoids computing $C_2^{n-1} - C_2^{m-1}$ angles **per object**, with $n \gg m$, where $n$ and $m$ are respectively the full dataset cardinality and the number of leaf nodes, i.e., the sample cardinality. So, it turns ABOD's $O(n^3)$ overall complexity into $O(n.m^2)$. Note that the time complexities to build and traverse a tree-based MAM are respectively $O(n \log n)$ and $O(n)$ for the most typical scenarios, so the additional tree-related cost does not modify the aforementioned overall complexity of *MetricABOD*. Finally, it is worth noting that the sample size $m$ is linearly correlated with the degree of the tree, i.e., the maximum capacity of a node, so $m$ can be easily tuned for any case of use.

## 5.1  Results: Comparison with Related Works

This section describes the experiments performed to evaluate our proposed *MetricABOD* algorithm. The state-of-the-art Slim-tree [17] MAM was used as the supporting tree structure in all experiments. *MetricABOD* was validated on the same 8 real world datasets of the previous Sect. 4 using the same methodology and system configuration. For a fair comparison with the related works, the tree degree was tuned to generate nearly $m = 100$ leaf nodes' representatives in every experiment; note that it is the same value used for the related works'

---

**Algorithm 1.** The *MetricABOD* algorithm

---

**Require: Data** – input dataset;
**Ensure: Result** – pairs (**P, Score**) with the objects of **Data** sorted as per their corresponding outlierness scores;
  1: Build a tree-based MAM for dataset **Data**. Let it be **Tree**;
  2: **Sample** = $\emptyset$;
  3: **for** each leaf node **n** of **Tree do**
  4:   Let **r** be the representative object of **n**;
  5:   **Sample = Sample** $\cup$ {**r**};
  6: **end for**
  7: **Result** = $\emptyset$;
  8: **for** each object **P** in set **Data do**
  9:   **Angles** = $\emptyset$;
 10:   **for** each object **x** $\neq$ **P** in set **Sample do**
 11:     **for** each object **y** $\neq$ **x** $\neq$ **P** in set **Sample do**
 12:       **Angle**$_{(x,y)}$ = the angle $\theta_{\overrightarrow{Px}\overrightarrow{Py}}$ between the difference vectors $\overrightarrow{Px}$ and $\overrightarrow{Py}$;
 13:       **Angles = Angles** $\cup$ {**Angle**$_{(x,y)}$};
 14:     **end for**
 15:   **end for**
 16:   **Score** = the variance of angles in **Angles**;
 17:   **Result = Result** $\cup$ {(**P, Score**)};
 18: **end for**
 19: Sort **Result** in descending order with regard to the **Score** values.

---

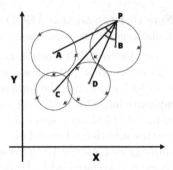

**Fig. 5.** Our proposed angle-based analysis for an object $P$. Circles in the 2-dimensional space illustrate the leaf nodes of any tree-based MAM. The new *MetricABOD* algorithm uses only the representatives $A$, $B$, $C$ and $D$ to compute the outlierness score of $P$.

similar parameter $k$. As it was done for the other algorithms, we were particularly interested in evaluating *MetricABOD*'s ability to spot what the application specialists expect from their data in an automatic and timely manner.

The last column of Table 2 reports the results obtained with *MetricABOD*. Note that the tree-related costs are **included** in the runtime results. As it was expected, *MetricABOD* coped with low-to-high dimensional data being considerably faster than the original ABOD method; surprisingly, it even improved ABOD's accuracy for the datasets that we studied. In fact, our proposed algorithm made the analysis up to **thousands of times faster** when compared with the 8 related works evaluated, still being in average **26% more accurate** than the most accurate related work, i.e., KNN-outlier. These results indicate that our *MetricABOD* is the best option, among those, which were considered, for use with a broad range of applications in which diversity takes place, especially when considering datasets that are as large as those required for commercial use.

To better understand why *MetricABOD* outperformed ABOD in accuracy, we investigated the case with the highest discrepancy; that is, the analysis of the Musk dataset. Ground truth shows that there are 97 outliers in this data. ABOD and *MetricABOD* respectively identified 7% and 77% of them. There are 29 common objects among the 97 most outlying objects indicated by each technique, and only 7 of them are true outliers. Interestingly, the common objects are among ABOD's most outlying objects, while they figure among the least outlying ones for *MetricABOD*, thus further corroborating the superiority of our proposal's results for this dataset. Let us highlight that Musk has the largest dimensionality among all datasets studied, with 166 attributes. Due to this fact, we conjecture that the dimensionality was high enough to affect ABOD's results, even with its angle-based analysis that is less susceptible than distances to the curse of the high dimensionality. On the other hand, there are evidences in the literature [18] indicating that to spot appropriate representatives for a dataset is somehow similar to reducing its dimensionality. Thus, we believe that the tree nodes' representatives obtained from Musk reduced the effects of the high dimensionality, thus enabling the angle-based analysis to obtain better results.

## 5.2    Results: Spotting Outliers in Adimensional Data

This section describes additional experiments performed with two real collections of text data. We aim at showing that the new *MetricABOD* algorithm works also for adimensional, purely metric data. As we discussed before, *MetricABOD* must compute angles among pairs of objects $(x, y)$ using other object $P$ as the pivot. But, *how to compute angles in an adimensional data space?* To tackle the problem, we used a simple angle estimation strategy based on distances only: given $x$, $y$ and $P$, the corresponding angle was computed from a triangle in the 2-dimensional space, whose side sizes are distances $d(x, y)$, $d(x, P)$ and $d(y, P)$.

The datasets studied are:

- **NSF abstracts:** the NSF Research Award Abstracts 1990-2003 from the UCI repository (archive.ics.uci.edu/ml/datasets/). It has 129, 000 abstracts describing awards granted by the National Science Foundation for basic research from 1990 to 2003. The abstracts were represented as sets of steamed words, and compared using the Jaccard distance function;
- **Brazilian names:** a set of 2, 755 first names used in Brazil. It is available at: https://gabrielrb.net/2011/10/18/dados-prontos-em-formato-sql-e-csv/. The L-Edit distance function was applied to compare the names.

We ran *MetricABOD* in both datasets, thus obtaining the corresponding outlierness scores. Since there is no ground truth for these data, accuracy was empirically evaluated by verifying whether or not the highest scores are in objects far away from the others. Figure 6 ranks the Brazilian names according to their average distances, on metric space, to all other names. Figure 7 ranks the NSF abstracts in a similar way. In each illustration, orange vertical lines indicate the top-20 outliers. Five of these lines report the actual names or abstracts' NSF identifiers; they are the top-5 outliers. Let us emphasize that in most cases the outliers are far away from any other object; note the log scale in the horizontal axes. Finally, as Brazilian citizens, we also argue that the most outlying names are indeed very distinct from names of people that are regularly used in Brazil.

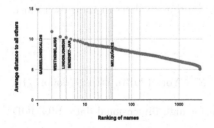
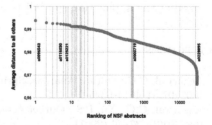

**Fig. 6.** Rank of Brazilian names as per their average L-Edit distances to all others. Orange lines are the top-20 outliers; lines with text are the top-5 ones. (Color figure online)

**Fig. 7.** Rank of NSF abstracts as per their average Jaccard distances to all others. Orange lines are the top-20 outliers; lines with text are the top-5 ones. (Color figure online)

# 6   Conclusion

This paper described an empirical study performed on unsupervised outlier detection using 8 datasets that refer to a variety of real-world tasks of high impact, like spotting cyberattacks, clinical pathologies and abnormalities in nature, and 8 algorithms from the state-of-the-art. We presented the lowdown on the results obtained, pointing out to the strengths and weaknesses of each technique from the **application specialist's point of view**, which is a shift from the designer-based point of view that is commonly considered. Interestingly, many of the techniques had unfeasibly high runtime requirements or failed to spot what the specialists consider as outliers in their own data. To tackle this issue, we carefully designed *MetricABOD*: a novel ABOD-based algorithm that made the analysis up to **thousands of times faster**, still being in average **26% more accurate** than the most accurate related work. The main innovation is a new usage for tree-based Metric Access Methods that were originally designed to index complex data. This improvement is essential to enable outlier detection in many real-world applications for which the existing methods lead to unexpected results or unfeasible runtime requirements. Additionally, we studied two real collections of text data to show that our *MetricABOD* works also for adimensional, purely metric data.

Finally, we must highlight that all theoretical models existing for unsupervised outlier detection make assumptions that may not reflect the true nature of outliers in every single application [1,5]. The results reported in our work corroborate this fact; they demonstrate how a given method can be very accurate in specific cases and still fail in others, in the sense of returning or not what the application specialist expects from his/her data. Here, one must remember that specific data mining tasks all have biases that are not well-understood; they may be hindering some of the techniques that we studied. Due to this fact, supervised outlier detection can be handy in tasks where unsupervised methods do not perform acceptably. Further discussion can also be conducted as to the validity of the ground truth used and their representativeness as outliers. Nevertheless, a **few** exceptions apart, it is always desirable that outlier detection techniques return what the application specialists understand as outliers in their own data.

# References

1. Aggarwal, C.C.: Outlier Analysis. Springer, New York (2013). https://doi.org/10.1007/978-1-4614-6396-2
2. Aggarwal, C.C., Yu, P.S.: Outlier detection for high dimensional data. SIGMOD Rec. **30**(2), 37–46 (2001)
3. Bay, S.D., Schwabacher, M.: Mining distance-based outliers in near linear time with randomization and a simple pruning rule. In: ACM SIGKDD, pp. 29–38 (2003)
4. Breunig, M.M., Kriegel, H.P., Ng, R.T., Sander, J.: LOF: identifying density-based local outliers. SIGMOD Rec. **29**(2), 93–104 (2000)
5. Campos, G.O., et al.: On the evaluation of unsupervised outlier detection: measures, datasets, and an empirical study. DMKD **30**(4), 891–927 (2016)

6. Fan, J., Li, R.: Statistical challenges with high dimensionality: feature selection in knowledge discovery. In: Congress of Mathematicians, pp. 595–622 (2006)
7. Ghoting, A., Parthasarathy, S., Otey, M.E.: Fast mining of distance-based outliers in high-dimensional datasets. DMKD 16(3), 349–364 (2008)
8. Johnstone, I.M., Titterington, D.M.: Statistical challenges of high-dimensional data. Philos. Trans. A 367(1906), 4237–4253 (2009)
9. Keller, F., Muller, E., Bohm, K.: HiCS: high contrast subspaces for density-based outlier ranking. In: IEEE ICDE, pp. 1037–1048 (2012)
10. Knorr, E.M., Ng, R.T.: Algorithms for mining distance-based outliers in large datasets. In: VLDB, pp. 392–403 (1998)
11. Kriegel, H.P., Schubert, M., Zimek, A.: Angle-based outlier detection in high-dimensional data. In: ACM SIGKDD, pp. 444–452 (2008)
12. Marques, H.O., Campello, R.J.G.B., Zimek, A., Sander, J.: On the internal evaluation of unsupervised outlier detection. In: SSDBM, pp. 7:1–7:12 (2015)
13. Muller, E., Assent, I., Steinhausen, U., Seidl, T.: OutRank: ranking outliers in high dimensional data. In: IEEE ICDE Workshop, pp. 600–603 (2008)
14. Muller, E., Schiffer, M., Seidl, T.: Statistical selection of relevant subspace projections for outlier ranking. In: IEEE ICDE, pp. 434–445 (2011)
15. Papadimitriou, S., Kitagawa, H., Gibbons, P.B., Faloutsos, C.: LOCI: fast outlier detection using the local correlation integral. In: IEEE ICDE, pp. 315–326 (2003)
16. Pham, N., Pagh, R.: A near-linear time approximation algorithm for angle-based outlier detection in high-dimensional data. In: ACM SIGKDD, pp. 877–885 (2012)
17. Traina Jr., C., Traina, A., Faloutsos, C., Seeger, B.: Fast indexing and visualization of metric data sets using slim-trees. IEEE TKDE 14(2), 244–260 (2002)
18. Zimek, A., Schubert, E., Kriegel, H.P.: A survey on unsupervised outlier detection in high-dimensional numerical data. ASA Data Sci. 5(5), 363–387 (2012)

# Deep Learning Based LSTM and SeqToSeq Models to Detect Monsoon Spells of India

Saicharan Viswanath[1], Moumita Saha[2(✉)], Pabitra Mitra[3],
and Ravi S. Nanjundiah[2,4,5]

[1] National Institute of Technology Karnataka, Surathkal, India
vsaicharan1998@gmail.com
[2] Centre for Atmospheric and Oceanic Sciences, Indian Institute of Science,
Bangalore, India
moumita.saha2012@gmail.com, ravisn@gmail.com
[3] Indian Institute of Technology Kharagpur, Kharagpur, India
pabitra@gmail.com
[4] Divecha Centre for Climate Change, Indian Institute of Science, Bangalore, India
[5] Indian Institute of Tropical Meteorology, Pune, India

**Abstract.** Monsoon spells are important climatic phenomenon modulating the quality and quantity of monsoon over a year. India being an agricultural country, identification of monsoon spells is extremely important to plan agricultural policies following the phases of monsoon to attain maximum productivity. Monsoon spells' detection involve analyzing and predicting monsoon at daily levels which make it more challenging as daily-variability is higher as compared to monsoon over a month or an year. In this article, deep-learning based long short-term memory and sequence-to-sequence models are utilized to classify monsoon days, which are finally assembled to detect the spells. Dry and wet days are classified with precision of 0.95 and 0.87, respectively. Break spells are observed to be forecast with higher accuracy than the active spells. Additionally, sequence-to-sequence model is noted to perform superior to that of long-short term memory model. The proposed models also outperform traditional classification models for monsoon spell detection.

**Keywords:** Active spell · Break spell · Long short term memory ·
Sequence-to-sequence model · Attention mechanism · Classification ·
Spells' detection

## 1 Introduction

Indian monsoon is mainly governed by south-west monsoon, occurring between June and September. The quality of monsoon during this period is extremely

Supported by Centre for Atmospheric and Oceanic Sciences at Indian Institute of Science, Indian Institute of Technology Kharagpur and National Institute of Technology Karnataka.

important for growth and prosperity of the country. The summer rainfall not only regulates the agricultural productivity but also assists in fresh water renewal, and nourishes the flora and fauna of the subcontinent. The characteristics and variability of Indian monsoon are widely studied in literature owing to its importance to the country [4, 6, 10–12].

Monsoon variability is high and its prediction resembles a provocative area in the field of climate research. The distribution of rainfall over the country on spatial basis as well as the temporal basis during the monsoon period are non-uniform. There can be continues period of days when it is completely dry or it can be heavily raining for a stretch of days. The monsoon process bears high variability and uncertainty. Dry days can lead to low monsoon year or even a drought year. Similarly, continues wet days with high rainfall values can lead to a excess rainfall year or even a disastrous flood. Thus, classifying monsoon on daily scale at a lead is extremely important to frame the policies and prevention measures for the country.

The article aims to identify the spells during monsoon months. Monsoon spells are defined by meteorologist considering various climatic properties like cloud nature and properties, different synoptic conditions, rainfall anomalies, position of monsoon trough, and wind strength and directions [5, 9, 19]. Monsoon spells are categorized into two phases – break and active spells. Break spells are defined as minimum three days or more at continuum having no rainfall or standard deviation below the mean rainfall. On a similar note, active spells refer to three or more days at continuum when the rainfall amount is at least standard deviation above the mean rainfall. Rajeevan et al. [18] proposed the standards for monsoon spells, characterized them to establish that the active spells are of less duration than the break spells. Active, weak, and break spells are also studied utilizing the horizontal wind shear [16]. Kumar and Dessai [13] have worked towards identifying breaks of Indian summer monsoon and showed their results consistent with conventional methods. Mandke et al. [14] have highlighted the variation of break and active spells with enhanced carbon-dioxide concentration, using global climate models. Ensemble models of National Centre for Environmental Prediction (NCEP) are also explored to forecast the monsoon spells of India [1]. Linear discriminant analysis seems to assists in the prediction of spells for Indian sub-continent [25]. Rao et al. [20] have studied the climatological characteristics of rainfall over India during break and active spells. The future projection of Indian monsoon's break and active spells are simulated by Sudeepkumar et al. [28] and characterized for different regions considering the climate change scenario.

Deep-learning based approach is proposed to detect the monsoon spells of India. Data-driven machine-learning or deep-learning methods have shown their efficiency in different climate problems owing to availability of huge climatic data [3, 7, 15, 21–23]. Deep-learning based autoencoder and stacked autoencoder have been used to predict aggregate Indian monsoon [24], regional Indian monsoon [27] at annual scale with high accuracy. Deep learning method is also utilized to predict early and late temporal phases of monsoon [26]. In this article, we propose identification of break and active monsoon spells of India using long short-term memory (LSTM) and sequence-to-sequence (Seq2Seq) models. Both LSTM and Seq2Seq models are different flavors of recurrent neural network (RNN). The prime reason for choosing recurrent neural networks is that they are capable of capturing the temporal variation in data, which is essential for time series prediction. LSTM networks uses special units in addition to standard RNN units, which help them to remember or forget information over long sequences (as required for modeling). Identification of spell is modeled as a classification problem, where we first attempt to classify each day as dry, wet, or normal day. Finally, the classification results at daily scale are summed up to discover the break or active monsoon spells. Spells are detected at a lead of one day to five days.

The continuing article is ordered as follows. The input data-set, output rainfall classes, and the detailed preprocessing steps are discussed in Sect. 2. Section 3 elaborates the proposed approach with brief description of LSTM and Seq2Seq models, and various experimental set-ups. Section 4 highlights the experimental outcomes and analysis, along with the performance metrics used. Lastly, Sect. 5 presents the conclusion and future scopes of the work.

## 2    Data and Preprocessing

### 2.1    Data Sources

Monsoon spell detection involves considering daily rainfall during June-September for the central region of Indian subcontinent, which extends the geographical lattitide of 21 °N - 27 °N, and longitude of 72 °E - 85 °E. India Meteorological Department provides the daily rainfall at $1° \times 1°$ spatial resolutions [17]. Four climatic variables, namely, sea level pressure (SLP), air temperature (AT), v-wind (VWND), and u-wind (UWND) are considered as input variable to classify the monsoon days. All the input variables are considered at surface and the daily values are collected from reanalysis data prepared by NCAR [8], provided at $2.5° \times 2.5°$ spatial resolutions for central India. All the data are collected for 1948–2014.

### 2.2    Preprocessing: Normalizing Features

All the input variables and rainfall data are normalized. Normalization is performed with removing the mean, followed by dividing the residual with the standard deviation of data (Eq. 1). Both the mean and standard deviation are computed considering the training data. This is done to prevent leaking of information from validation and test sets. Normalizing data helps in avoiding training problems like local optima and weight decay. It also leads to faster convergence to global optima.

$$X_{\text{norm}} \leftarrow \frac{X - Mean(X_{\text{train}})}{Std(X_{\text{train}})} \tag{1}$$

## 2.3  Prepropessing: Daily Rainfall Class Assignment

After normalizing the rainfall to mean zero and unit standard-deviation, daily rainfall are labeled among three rainfall classes – dry, wet or normal. A sample or day is labeled as dry class if the normalized rainfall is below −1, and it is labeled as wet class if the normalized rainfall is above +1. Finally, days or samples which receive rainfall having rainfall in middle of these two extremes are marked as normal class. These class labels are used further for classification task.

## 3  Methodology

The proposed method for predicting the break and active monsoon spells of India consists of the following steps – (a) flattening the spatial-temporal input data, (b) classifying the days into dry, wet, or normal class using LSTM or Seq2Seq models, (c) Summing up the classified days to detect break or active monsoon spells. The summary of the proposed approach to detect the monsoon spells (break or active) is shown in Fig. 1.

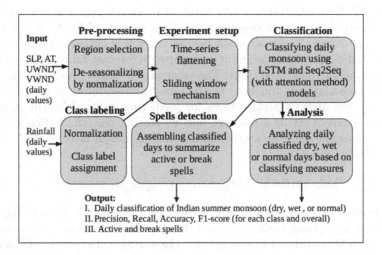

**Fig. 1.** Block-flow of the proposed method for identifying break and active monsoon spells of India with LSTM and Seq2Seq models

## 3.1  Input Time Series Flattening

The input consists of dimension equivalent to $TimeStepHist \times NumFeatures$, where each row or tuple represents the value of input variables on a particular day and each column represents the distinct input variable. As an example, if we are using rainfall, AT and SLP of four previous days history (day 1 to day 4) as our input to classify the rainfall of succeeding day (day 5), then our input will have dimension $4 \times 3$, where four indicates the number of past history of days used ($TimeStepHist$) and three indicates the number of input features ($NumFeatures$). Considering all the examples of rainfall classification tuples lead to adding a third dimension to the input ($NumExamples \times TimeStepHist \times NumFeatures$). However, for feeding the input to the proposed model, it is required to flatten the input tensor into two-dimension matrix, which is arranged in form of $NumExamples \times (TimeStepHist \times NumFeatures)$ (i.e. the last two dimension are flattened to single dimension as feature1 at timehist1, feature2 at timehist1, ......, feature1 at timehist2, feature2 at timehist2, and so on).

### 3.2    Classification Using Long Short Term Memory Network Model and Sequence-to-Sequence Model

Classification of the rainfall days during June-September at a lead is performed using two mutants of recurrent neural network, namely, sequence-to-sequence and long short term memory models.

**Long Short Term Memory Model:** Long short term memory networks (LSTM) is an extension and flavor of recurrent neural network (RNN), used for learning long-term dependencies. RNN consists of feedback connection between the layers. The motivation behind using this network lies in presence of sequential observations, where it is required to forecast future time-series by learning from the past time-steps. The value of the climate variables for a certain time-step may get influenced by the same variable values at the past. This signify the use of RNN architectures for predicting spells of Indian monsoon. The hidden layers act as a memory unit by storing the information captured from previous states of the sequential input. The hidden layer of RNN can be described as in Eq. 2.

$$\text{hid}_i = f(\text{RecWeight}_{\text{hid}} h_{i-1} + \text{Weight}_{\text{hid}} x_i + \text{bias}_{\text{hid}}) \tag{2}$$

The function $f$ corresponds to the activation function for introducing nonlinearities (eg. sigmoid, tanh, etc.), and the mapping from hidden to the output layer is mainly considered as softmax for classification. Despite of all its capabilities, RNN are prone to the *vanishing gradient problem*, and hence unable to learn long term dependencies. On the other hand, LSTM can learn and adapt the existing long distance dependencies. They are capable of remembering the useful information over a longer sequences and forgets the other through special structures called gates. A LSTM cell is shown in Fig. 2. It consists of an input gate (ig), a forget gate (fg), an output gate (og) and a memory cell (mc)

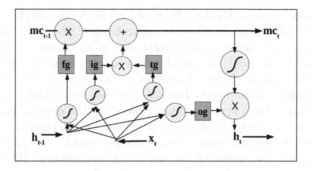

**Fig. 2.** A long short term memory cell

(shown in Eq. 3). These gates can learn to modulate the flow of data, depending on the input and the hidden states, which helps in imitating long-range dependencies.

$$
\begin{aligned}
ig_t &= sigmoid(\text{RecWeight}_{ig}h_{t-1} + \text{Weight}_{ig}x_t + \text{bias}_{ig}) \\
fg_t &= sigmoid(\text{RecWeight}_{fg}h_{t-1} + \text{Weight}_{fg}x_t + \text{bias}_{fg}) \\
og_t &= sigmoid(\text{RecWeight}_{og}h_{t-1} + \text{Weight}_{og}x_t + \text{bias}_{og}) \\
tg_t &= hypTan(\text{RecWeight}_{mc}h_{t-1} + \text{Weight}_{mc}x_t + \text{bias}_{mc}) \\
mc_t &= fg_t \circ mc_{t-1} + ig_t \circ tg_t \\
hidden_t &= og_t \circ hypTan(mc_t),
\end{aligned}
\tag{3}
$$

where $fg_t$, $ig_t$, $og_t$, $mc_t$ are the activation at the forget gate, input gate, output gate, and memory cell at time instant $t$. $\text{RecWeight}_r$ and $\text{Weight}_r$ are the learned weight of recurrent and input connections and $\text{bias}_r$ are the respective biases. The subscript $r$ may be any of the forget, input, output, or memory cell. The activation function used are sigmoid and hyperbolic tangent (hypTan). Finally $\circ$ symbolizes for element-wise multiplication of the matrices, and $hidden_t$ represents the hidden state vector or the output of the LSTM unit at time instant $t$.

The proposed LSTM classification model consists of two LSTM layers, followed by a fully connected layer with soft-max activation for classifying the daily rainfall.

(a) Input layer- The input to the first LSTM layer is of the form (*BatchSize* × *TimeStepHist* × *NumFeatures*).
(b) Hidden layer- There are two hidden layers with 80 neurons in each LSTM cell. The activation function used is hyperbolic tangent.
(c) Dense softmax layer- A fully connected layer connecting the last output of LSTM with softmax activation to classify monsoon days into classes (wet, normal, or dry).

The hyper-parameters *BatchSize* and *TimeStepHist* are fixed empirically, by varying different length of history at lead over the validation period.

**Sequence-to-Sequence Model:** Sequence to Sequence (Seq2Seq) is a composite model used to learn time-series dependencies. The model is built with two LSTM units, one imitating the functioning of an encoder, while the other imitate a decoder. As usual, the encoder tries to learn from the input sequence, and it produces a complex hidden learned state or context, which aims to apprehend the variability of the input. The decoder tries to produce the sequence of output, considering the learned complex context by the encoder and the previous output state. Attention mechanism as proposed by Bahdanau et al. [2] adds more power to the present sequence-to-sequence model. During decoding, the attention mechanism permits the decoder to concentrate over different encoder outputs with addition to the last encoder state. A set of attention weights is calculated, which is then multiplied with the encoder outputs to create context vectors. The context vectors contain information pertaining to a certain part of the sequence in input, and thus helping the decoder in producing a more accurate output.

The proposed Seq2Seq monsoon classification model, consists of input similar to LSTM (discussed previously), two LSTM units with 100 neurons each, in encoder and decoder parts of the model and finally, a dense soft-max layer. The attention mechanism is also applied to improve the model's performance. The proposed architecture and working of Seq2Seq model is shown in Fig. 3.

The training of Seq2Seq classification model is performed as following. The input sequence is fed to the first encoder layer. The last cell of the encoder stores the contextual information related to the sequence, which is fed to the decoder. During decoding steps, the decoder output of the previous time-step $t-1$ is provided as an input at the succeeding time-step $t$. This method is known as *greedy sampling* and makes the model more robust to learn from previous mistakes. An alternative to this procedure is to feed the correct input to the

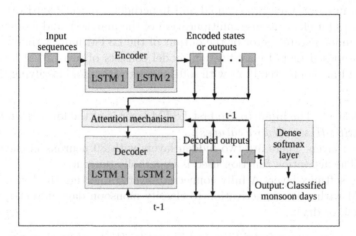

**Fig. 3.** Sequence-to-sequence model with attention mechanism for daily rainfall classification

decoder at every time-step, even if the decoder have made a mistake at previous step. This method is known as *guided training*. We have implemented both the training methods for Seq2Seq model, however, *greedy sampling* performed better for the rainfall classification task.

**Experimental Setup-Sliding Window:** Real time classification of dry, wet, or normal days requires continuous learning of sequential input. The number of time-steps to be learned increases with time and the computational complexity becomes exponential. Additionally, the near time-steps influence the current output value to a greater extent as compared to the time-steps at far past. For these reasons, a sliding window method is used (Fig. 4) in training process of the proposed approach. A sliding window of size *TimeStepsHistory* is chosen, which indicates a fixed number of previous time-steps at lead is used to predict the following time-step.

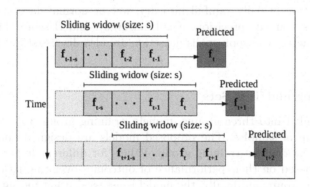

**Fig. 4.** Sliding window training approach (eg. prediction of $f_t$ instant is provided using past time-steps from $f_{t-1}$ to $f_{t-1-s}$, i.e. TimeStepsHistory $= s$ time-steps (sliding window size))

**Experimental Setup-Hyper Parameters:** For LSTM and Seq2Seq models, the optimal values for various hyper-parameters are ascertained empirically. The values are selected considering their performance on the validation data set. Hyper-parameters for the model include the count of layers, the count of neurons in LSTM cell, the length of input sequence history (*TimeStepsHist*) or sliding window size to be considered, and weighting component. The learning rate value is set to $1 * e^{-3}$ for training. Mini-batch is considered with size of 32 for gradient descent. Adam optimizer is used, and finally, L2 regularization is implemented with the regularization parameter equal to $1 * e^{-3}$ to avoid over-fitting of the model.

**Experimental Setup-Loss Function:** A soft-max cross entropy loss is used for modeling the multi-class classification problem. However, there is a problem of class imbalance for our study. Out of the total days considered, 15% are wet days, 18% are dry days and rest 67% resembles the normal days. To solve the issue of class imbalance, weighted soft-max cross entropy loss is used, which allows to use different scores for different classes, and hence a mis-classification of the less common classes can be penalized more to improve prediction accuracy. For Seq2Seq model, the cost function is considered as a amalgamation of two losses – (a) loss 1: average loss of each time step prediction, (b) loss 2: loss of the prediction calculated at the last time step. The losses are combined as shown in Eq. 4, where $\beta$ represents the weighting factor.

$$\text{overall loss} = \beta * \text{loss } 1 + (1 - \beta) * \text{loss } 2 \tag{4}$$

## 4 Experimental Results and Analysis

The classification of daily rainfall into dry, wet or normal days are presented with measures, namely, precision, recall, accuracy, and f1-score. The outcome of break and active monsoon spells detection are also discussed in the following section.

### 4.1 Training and Test Sets

Data are divided into three groups, namely, training group, validation group, and test group of samples. The proposed models are designed and optimized utilizing the data in the training group. Values for different hyper-parameters are selected based on their performance of outcomes over the data samples of the validation group, and finally, the model is assessed using the samples in the test group. The input consists of 8174 time-steps, among which 80% is used for training the models, 10% for setting the hyper-parameters (validation) and the remaining 10% for test and analysis. In terms of years, approximately, fifty-three years of sixty-seven are consumed for training, seven for validation and seven in testing. As required by the problem, only days of June-September of all the years represent the tuples. Among 8174 days under consideration, the number of wet and dry days are observed as 1211 and 1470, which comprised 14.8% and 18% of total number of days under study.

### 4.2 Performance Metrics

The multi-class monsoon classification problem is evaluated with precision and recall of each class (dry, wet or normal). Precision is computed as the count of accurately classified positive class among all the results that it has classified as positive. Similarly, recall is computed as the count of rightly classified positive class out of observed or actual positive class examples. The overall classification result is presented in terms of accuracy and f1-score, where each of these measures are calculated as average of same measures for individual classes. F1-score refers to the harmonic mean of recall and precision. F1-score is better as its value approaches 1 and worst at value 0.

## 4.3 Classification Results of Daily Monsoon

The performances of the LSTM and the Seq2Seq model in classifying the daily rainfall classes is presented in terms of performance metrics (discussed in previous section) for the test period 2008–2014. The classification is performed considering a lead of one to five days. Conventional classifiers, namely, the support vector machine (SVM) and K-nearest neighbor (KNN) are also designed for evaluating the proposed models against them. For the SVM classifier model, various kernels are implemented but the linear kernel is observed to provide best results. For the KNN model, the number of neighbors is considered as a hyper-parameter and varied with lead times. The classification results by the proposed models along with the two traditional models are presented for a lead of one to five days in Tables 1(a–e), respectively. It is observed that classification performance is best for least lead of one day and it degrades gradually as the lead is increased from one to five days. Dry days are noted to be classified with better accuracy as compared to the wet days. The proposed LSTM classifier provides an accuracy of 92.73% and f1-score of 0.953 at a lead of one day. Similarly, the Seq2Seq classifier classify monsoon days with 92.50% accuracy and 0.953 f1-score. Moreover, it classify the dry and wet days with precision of 0.952 and 0.856, respectively. The proposed Seq2Seq model classify the daily rainfall with best accuracy among all the four models. SVM classifier with linear kernel performs comparable, but the KNN classifier can not classify the monsoon days with good accuracy. The proposed models are observed to perform superior to the SVM and KNN classifiers for all the leads. The variation of accuracy and f1-score of the classification by proposed LSTM and Seq2Seq models, and traditional SVM and KNN classification models are shown in Fig. 5a and b, respectively.

## 4.4 Break and Active Spells Detection

The classification models predict the occurrence of dry or wet days, which is extended to the prediction of break or active spell. The classified days are summed up to present the break or active spells. We summarize a break or an active spell only when we obtain continuous three or more days (as per their definition) of dry or wet classified days, respectively. There are sixteen break and twenty-one active spells in the test period (2008–2014).

Table 2 highlights the total count of observed spells and count of correctly identified spells by LSTM and Seq2Seq models at lead of three days. It is noted that LSTM and Seq2Seq models have correctly identified thirteen and sixteen out of twenty-one active spells, respectively. Similarly, the corresponding models have correctly identified thirteen and fourteen break spells, respectively. Identification of break spells appear to be predominant to the identification of active spells (similar to the trend of dry and wet days classification). A detailed evaluation

**Table 1.** Precision and recall for classifying dry, wet, or normal days and overall accuracy and f1-score of classification at different leads (a–e: prediction at a lead of one to five days) for 2008–2014

(a) Classification at lead 1

| Models | SVM | KNN | LSTM | Seq2Seq |
|---|---|---|---|---|
| Dry day classification | | | | |
| Precision | 0.931 | 0.810 | 0.951 | 0.952 |
| Recall | 0.955 | 0.942 | 0.950 | 0.982 |
| Wet day classification | | | | |
| Precision | 0.841 | 0.742 | 0.870 | 0.856 |
| Recall | 0.835 | 0.751 | 0.884 | 0.884 |
| Normal day classification | | | | |
| Precision | 0.944 | 0.923 | 0.961 | 0.967 |
| Recall | 0.947 | 0.877 | 0.962 | 0.953 |
| Overall classification | | | | |
| Accuracy | 90.53 | 82.50 | 92.73 | 92.50 |
| F1-score | 0.935 | 0.875 | 0.954 | 0.953 |

(b) Classification at lead 2

| Models | SVM | KNN | LSTM | Seq2Seq |
|---|---|---|---|---|
| Dry day classification | | | | |
| Precision | 0.869 | 0.756 | 0.885 | 0.921 |
| Recall | 0.936 | 0.770 | 0.936 | 0.961 |
| Wet day classification | | | | |
| Precision | 0.807 | 0.651 | 0.803 | 0.851 |
| Recall | 0.752 | 0.462 | 0.842 | 0.801 |
| Normal day classification | | | | |
| Precision | 0.925 | 0.820 | 0.944 | 0.943 |
| Recall | 0.921 | 0.87 | 0.922 | 0.936 |
| Overall classification | | | | |
| Accuracy | 86.70 | 74.23 | 87.73 | 90.50 |
| F1-score | 0.903 | 0.795 | 0.910 | 0.926 |

(c) Classification at lead 3

| Models | SVM | KNN | LSTM | Seq2Seq |
|---|---|---|---|---|
| Dry day classification | | | | |
| Precision | 0.793 | 0.633 | 0.809 | 0.909 |
| Recall | 0.808 | 0.596 | 0.840 | 0.834 |
| Wet day classification | | | | |
| Precision | 0.784 | 0.319 | 0.782 | 0.796 |
| Recall | 0.672 | 0.252 | 0.652 | 0.744 |
| Normal day classification | | | | |
| Precision | 0.871 | 0.741 | 0.882 | 0.911 |
| Recall | 0.897 | 0.781 | 0.901 | 0.932 |
| Overall classification | | | | |
| Accuracy | 81.60 | 56.43 | 82.43 | 87.20 |
| F1-score | 0.849 | 0.667 | 0.859 | 0.890 |

(d) Classification at lead 4

| Models | SVM | KNN | LSTM | Seq2Seq |
|---|---|---|---|---|
| Dry day classification | | | | |
| Precision | 0.696 | 0.541 | 0.670 | 0.751 |
| Recall | 0.701 | 0.509 | 0.670 | 0.808 |
| Wet day classification | | | | |
| Precision | 0.536 | 0.219 | 0.614 | 0.686 |
| Recall | 0.363 | 0.192 | 0.421 | 0.578 |
| Normal day classification | | | | |
| Precision | 0.784 | 0.721 | 0.823 | 0.836 |
| Recall | 0.832 | 0.742 | 0.855 | 0.862 |
| Overall classification | | | | |
| Accuracy | 67.20 | 49.36 | 70.23 | 75.76 |
| F1-score | 0.745 | 0.615 | 0.769 | 0.814 |

(e) Classification at lead 5

| Models | SVM | KNN | LSTM | Seq2Seq |
|---|---|---|---|---|
| Dry day classification | | | | |
| Precision | 0.641 | 0.434 | 0.681 | 0.744 |
| Recall | 0.605 | 0.318 | 0.601 | 0.707 |
| Wet day classification | | | | |
| Precision | 0.333 | 0.301 | 0.525 | 0.531 |
| Recall | 0.272 | 0.231 | 0.34 | 0.413 |
| Normal day classification | | | | |
| Precision | 0.741 | 0.684 | 0.761 | 0.823 |
| Recall | 0.797 | 0.765 | 0.845 | 0.842 |
| Overall classification | | | | |
| Accuracy | 57.16 | 47.30 | 65.56 | 69.93 |
| F1-score | 0.684 | 0.603 | 0.717 | 0.759 |

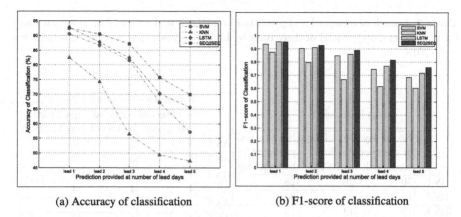

(a) Accuracy of classification          (b) F1-score of classification

**Fig. 5.** Comparison of (a) accuracy and (b) f1-score in monsoon classification by proposed LSTM and Seq2Seq, with conventional SVM and KNN classifiers at lead of one to five days for 2008–2014

is presented elaborating the actual length of each spells (as number of days) along with the continuous days predicted as respective classes (wet or dry) by the proposed models. The observed length of spells and corresponding predicted period of days for break and active spells during 2008–2014 are shown in Table 3a and b, respectively. It is noted that the Seq2Seq model predict the dry and wet spells more accurately in terms of length of spells. Although the LSTM model detects dry spells quite well, but it fails to detect many of the active spells.

**Table 2.** Observed and correctly predicted count of break and active monsoon spells at a lead of three days by LSTM and Seq2Seq models for test-period 2008–2014

| Models | Observed count of break spells | Predicted count of break spells | Observed count of active spells | Predicted count of active spells |
|--------|-------------------------------|--------------------------------|--------------------------------|---------------------------------|
| LSTM   | 16 | 13 | 21 | 13 |
| Seq2Seq | 16 | 14 | 21 | 16 |

The observed length and the predicted length of break and active monsoon spells are also presented in Fig. 6a and b, respectively. Actual length of observed break or active spells in terms of number of continuous days is shown by the bars in the figure, and the symbols represent the predicted spell length by the proposed LSTM and Seq2Seq models.

**Table 3.** Length of observed and predicted break (BS) and active (AS) monsoon spells (as number of days) at lead of three days by LSTM and Seq2Seq models for test-period 2008–2014

(a) Break spells

|  | No. of days in observed break spell | No. of days predicted by LSTM | No. of days predicted by Seq2Seq |
|---|---|---|---|
| BS 1 | 30 | 28 | 28 |
| BS 2 | 10 | 6 | 8 |
| BS 3 | 13 | 13 | 13 |
| BS 4 | 7 | 7 | 6 |
| BS 5 | 5 | 3 | 4 |
| BS 6 | 14 | 13 | 13 |
| BS 7 | 16 | 15 | 15 |
| BS 8 | 6 | 5 | 5 |
| BS 9 | 9 | 7 | 7 |
| BS 10 | 9 | 6 | 4 |
| BS 11 | 15 | 11 | 12 |
| BS 12 | 4 | 2 | 3 |
| BS 13 | 3 | 0 | 0 |
| BS 14 | 5 | 3 | 4 |
| BS 15 | 3 | 0 | 2 |
| BS 16 | 6 | 4 | 5 |

(b) Active spells

|  | No. of days in observed active spell | No. of days predicted by LSTM | No. of days predicted by Seq2Seq |
|---|---|---|---|
| AS 1 | 3 | 2 | 3 |
| AS 2 | 3 | 2 | 2 |
| AS 3 | 3 | 3 | 3 |
| AS 4 | 4 | 4 | 4 |
| AS 5 | 3 | 3 | 3 |
| AS 6 | 4 | 1 | 2 |
| AS 7 | 4 | 1 | 3 |
| AS 8 | 8 | 5 | 6 |
| AS 9 | 6 | 3 | 3 |
| AS 10 | 3 | 0 | 2 |
| AS 11 | 6 | 3 | 5 |
| AS 12 | 4 | 3 | 3 |
| AS 13 | 4 | 3 | 4 |
| AS 14 | 4 | 4 | 4 |
| AS 15 | 3 | 2 | 2 |
| AS 16 | 11 | 7 | 9 |
| AS 17 | 5 | 3 | 4 |
| AS 18 | 6 | 5 | 5 |
| AS 19 | 3 | 1 | 3 |
| AS 20 | 4 | 3 | 4 |
| AS 21 | 4 | 2 | 2 |

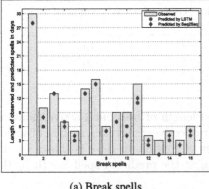

(a) Break spells

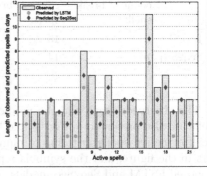

(b) Active spells

**Fig. 6.** Observed and predicted length of monsoon spells by LSTM and Seq2Seq models during test period 2008–2014 for (a) break spells, and (b) active spells

# 5  Conclusions

The prediction of break and active monsoon spells of India is performed during June-September. Two different models, namely the LSTM classification model and the Seq2Seq model with attention mechanism are proposed. Both the models performed better than traditional SVM and KNN classifiers. One of the important observations is that detecting active spells is harder than detecting dry spells. The possible explanation can be that an average dry spell lasted longer compared to an average active spell, thus making it less random and easier to detect. Another major problem faced during the classification process is that the data set is imbalanced. Weighted soft-max loss is implemented but the weights add to number of hyper-parameters thus making the model harder to tune. Even though the hyper-parameters are selected considering performance over the validation set, grid search or other similar methods can be implemented to detect the optimal set of hyper-parameters. LSTM classification model and the Seq2Seq model can also be greatly improved by acquiring more data. Another possible area of extension can be use of convolution neural network (CNN) for multivariate time series. Although CNN's have extensively used as feature extractors in images, they can also be used to extract features from climatic time-series. Hybrid of CNN and LSTM can also be used for further prospect. Finally, exploration of spatial-temporal nature of climate data may lead to better understanding, and classifying or predicting the climatic phenomenon.

# References

1. Abhilash, S., Sahai, A., Pattnaik, S., Goswami, B., Kumar, A.: Extended range prediction of active-break spells of Indian summer monsoon rainfall using an ensemble prediction system in ncep climate forecast system. Int. J. Climatol. **34**(1), 98–113 (2014)
2. Bahdanau, D., Cho, k., Bengio, Y.: Neural machine translation by jointly learning to align and translate. In: International Conference on Learning Representations. (2015). arXiv:1409.0473
3. Chattopadhyay, S., Chattopadhyay, G.: Comparative study among different neural net learning algorithms applied to rainfall time series. Meteorol. Appl. **15**(2), 273–280 (2008)
4. Fan, K., Liu, Y., Chen, H.: Improving the prediction of the East Asian summer monsoon: new approaches. Weather Forecast. **27**(4), 1017–1030 (2012)
5. Gadgil, S., Joseph, P.V.: On breaks of the Indian monsoon. J. Earth Syst. Sci. **112**(4), 529–558 (2003)
6. Gadgil, S., Rajeevan, M., Nanjundiah, R.S.: Monsoon prediction - why yet another failure? Curr. Sci. **88**(9), 1389–1400 (2005)
7. Hong, W.C.: Rainfall forecasting by technological machine learning models. Appl. Math. Comput. **200**(1), 41–57 (2008)
8. Kalnay, E., et al.: The NCEP/NCAR 40-year reanalysis project. Bull. Am. Meteorol. Soc. **77**(3), 437–471 (1996)
9. Kiran, V.R., Rajeevan, M., Rao, S.V.B., Rao, N.P.: Analysis of variations of cloud and aerosol properties associated with active and break spells of Indian summer monsoon using MODIS data. Geophys. Res. Lett. **36**(9), L09706 (2009). https://doi.org/10.1029/2008GL037135

10. Kripalani, R.H., Kulkarni, A.: Rainfall variability over South-East Asia-connections with Indian monsoon and ENSO extremes: new perspectives. Int. J. Climatol. **17**(11), 1155–1168 (1997)
11. Kucharski, F., Bracco, A., Yoo, J.H., Molteni, F.: Atlantic forced component of the Indian monsoon interannual variability. Geophys. Res. Lett. **35**(4), L04706 (2008). https://doi.org/10.1029/2007GL033037
12. Kumar, K.K., Rajagopalan, B., Hoerling, M., Bates, G., Cane, M.: Unraveling the mystery of Indian monsoon failure during El Niño. Science **314**(5796), 115–119 (2006)
13. Kumar, M.R.R., Dessai, U.R.P.: A new criterion for identifying breaks in monsoon conditions over the Indian subcontinent. Geophys. Res. Lett. **31**(18), L18201 (2004)
14. Mandke, S.K., Sahai, A.K., Shinde, M.A., Joseph, S., Chattopadhyay, R.: Simulated changes in active/break spells during the Indian summer monsoon due to enhanced CO2 concentrations: assessment from selected coupled atmosphere-ocean global climate models. Int. J. Climatol. **27**(7), 837–859 (2007)
15. Mekanik, F., Imteaz, M.A., Gato-Trinidad, S., Elmahdi, A.: Multiple regression and artificial neural network for long-term rainfall forecasting using large scale climate modes. J. Hydrol. **503**, 11–21 (2013)
16. Prasad, V.S., Hayashi, T.: Active, weak and break spells in the Indian summer monsoon. Meteorol. Atmos. Phys. **95**(1–2), 53–61 (2007)
17. Rajeevan, M., Bhate, J., Kale, J.D., Lal, B.: High resolution daily gridded rainfall data for the Indian region: analysis of break and active monsoon spells. Curr. Sci. **91**(3), 296–306 (2006)
18. Rajeevan, M., Gadgil, S., Bhate, J.: Active and break spells of the Indian summer monsoon. J. Earth Syst. Sci. **119**(3), 229–247 (2010)
19. Ramamurthy, K.: Monsoon of India: some aspects of the break in the Indian southwest monsoon during July and August. Forecast. Manual 18.3, pp. 1–57 (1969). IMD Poona
20. Rao, T.N., Saikranthi, K., Radhakrishna, B., Rao, S.V.B.: Differences in the climatological characteristics of precipitation between active and break spells of the Indian summer monsoon. J. Clim. **29**(21), 7797–7814 (2016)
21. Saha, M., Chakraborty, A., Mitra, P.: Predictor-year subspace clustering based ensemble prediction of Indian summer monsoon. Adv. Meteorol. **2016**(9031625), 1–12 (2016)
22. Saha, M., Mitra, P.: Identification of Indian monsoon predictors using climate network and density-based spatial clustering. Meteorol. Atmos. Phys. **128**(5), 1–14 (2018)
23. Saha, M., Mitra, P., Chakraborty, A.: Fuzzy clustering-based ensemble approach to predicting Indian monsoon. Adv. Meteorol. **2015**(32835), 1–12 (2015)
24. Saha, M., Mitra, P., Nanjundiah, R.S.: Autoencoder-basedidentification of predictors of Indian monsoon. Meteorol. Atmos. Phys. **128**, 613–628 (2016). https://doi.org/10.1007/s00703-018-0637-y
25. Saha, M., Mitra, P., Nanjundiah, R.S.: Prediction of active and break spells of Indian summer monsoon using linear discriminant analysis. In: International Conference on Advances in Pattern Recognition, pp. 1–6 (2017)
26. Saha, M., Mitra, P., Nanjundiah, R.: Predictor discovery for early-late Indian summer monsoon using stacked autoencoder. Proc. Comput. Sci. **80**, 565–576 (2016)
27. Saha, M., Mitra, P., Nanjundiah, R.: Deep learning for predicting the monsoon over the homogeneous regions of India. J. Earth Syst. Sci. **126**(4), 1–18 (2017)
28. Sudeepkumar, B.L., Babu, C.A., Varikoden, H.: Future projections of active-break spells of Indian summer monsoon in a climate change perspective. Glob. Planet. Change **161**, 222–230 (2018)

# Data Analysis for Atomic Shapes
# in Nuclear Science

Mehmet Cagri Kaymak[1]([✉]), Hasan Metin Aktulga[1], Ron Fox[2],
and Sean N. Liddick[2,3]

[1] Department of Computer Science and Engineering, Michigan State University,
East Lansing, MI 48824, USA
kaymakme@msu.edu
[2] National Superconducting Cyclotron Laboratory, Michigan State University,
East Lansing, MI 48824, USA
[3] Department of Chemistry, Michigan State University,
East Lansing, MI 48824, USA

**Abstract.** We consider the problem of detecting a unique experimental signature in time-series data recorded in nuclear physics experiments aimed at understanding the shape of atomic nuclei. The current method involves fitting each sample in the dataset to a given parameterized model function. However, this procedure is computationally expensive due to the nature of the nonlinear curve fitting problem. Since data is skewed towards non-unique signatures, we offer a way to filter out the majority of the uninteresting samples from the dataset by using machine learning methods. By doing so, we decrease the computational costs for detection of the unique experimental signatures in the time-series data. Also, we present a way to generate synthetic training data by estimating the distribution of the underlying parameters of the model function with Kernel Density Estimation. The new workflow that leverages machine learned classifiers trained on the synthetic data are shown to significantly outperform the current procedures used in actual datasets.

**Keywords:** Machine learning · Density estimation · Nuclear physics

## 1 Introduction

Although many people assume that the atomic nucleus adopts a spherical shape, the actual nuclear shapes could vary [10]. Deformed nuclei are quite common at certain regions of proton and neutron numbers. Further, some regions of proton and neutron number are categorized as exhibiting signs of shape-coexistence phenomena in which different nuclear states can be interpreted as having different mean square charge radii. A key experimental indicator of the phenomena is the presence of strong transitions between states with identical angular momentum and parity. Such transitions between spin 0 states proceed through the emission of an atomic electron and when the excited spin-0 state is at a low enough energy,

© Springer Nature Switzerland AG 2019
J. M. F. Rodrigues et al. (Eds.): ICCS 2019, LNCS 11537, pp. 219–233, 2019.
https://doi.org/10.1007/978-3-030-22741-8_16

the state will live for a longer period of time before decaying. We describe the workflow and associated challenges through a recent experiment to explore the spin-0 to spin-0 transitions observed in the mass 32 region of the nuclear chart. The experiment starts with a beam of rare isotopes produced at the National Superconducting Cyclotron Laboratory (NSCL), which in this case is a beam containing $_{32,33}$Na isotopes and numerous neighboring mass nuclei. The $_{32,33}$Na nuclei moving at relativistic speeds are brought to rest inside a large volume active detector. Since the $_{32,33}$Na isotopes are radioactive, they will eventually "beta decay" (a process which changes a neutron into a proton) into an iso-tope of $_{32,33}$Mg. In rare cases, the decay will leave the resulting isotope with an excess of energy which it will shed within a few hundred nanoseconds through the emission of an energetic atomic electron from the spin-0 to spin-0 transition.

The rare isotopes are brought to rest in a large volume, unsegmented CeBr$_3$ detector. The unsegmented CeBr$_3$ scintillator is readout using a Hamamatsu 13700 position sensitive photomultiplier tube (PSPMT) which is segmented into an array of 16 × 16 pixels for a total of 256 pixels. Each pixel and a signal corre-sponding to the sum over the entire detector is connected to individual digitizing electronics channels running between 250 and 500 MSPS. An onboard firmware controls when data will be recorded to the onboard memory. For these experi-ments, information related to the energy of the signal includes time, the channel that recorded it, and a waveform which records the history of the detector volt-age as a function of time. An example of a recorded waveform (*i.e.*, time-series data of detector voltage) from the detector is shown in Fig. 1. What is sought is essentially a unique signature, *i.e.*, double pulses (to the right of the figure cor-responding to the science case in which the first pulse represents the beta-decay electron and the second pulse represents the transition of interest), which are extremely rare in the presence of other "uninteresting" signal shapes, *i.e.*, single pulses (to the left).

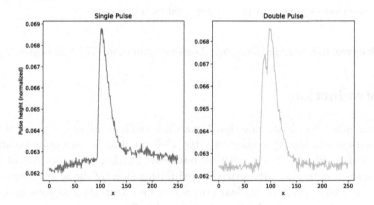

**Fig. 1.** Different pulses from the dataset.

Whenever a double pulse is detected, the amplitude of both pulses and the time difference between them must be calculated for further physical analysis, which is accomplished by fitting a known functional form to the observed signal, called lmfit, as explained in further detail below. The lmfit procedure is computationally demanding, and in the absence of a better alternative, it is used for detecting the double pulses among the plethora of uninteresting signals. Consequently, the existing practice cannot enable real-time analysis of the vast number of traces obtained during experiments. This is highly problematic because only a single experimental session can typically be secured at NSCL per year, and during an experiment scientists need to be able to determine whether they have been able to capture a sufficient number of the rare double pulse signals. It should be noted that since each signal is independent, their processing can easily be parallelized across a number of processors which would potentially ensure real-time processing guarantees with moderate resource investments, but NSCL is soon to be upgraded to the Facility for Rare Isotope Beams (FRIB), which will increase the data collection rates by two to three orders of magnitude. As such, it is critical to have a highly accurate yet inexpensive computational tool to detect double pulses, and this constitutes our main goal in this paper.

In what follows, we first briefly describe the time-series dataset from nuclear physics experiments in Sect. 2. The proposed machine learning based framework is presented in Sect. 3 which explains how realistic-looking data is generated synthetically, preprocessing of the data and classification methods used to detect unique signatures. Finally, the framework is evaluated according to its recall, precision and speedup rates for different scenarios in Sect. 4, and we review the related work in Sect. 5.

## 2   Dataset

The dataset used in this study includes 14,985,016 samples. The snapshots recorded by the data acquisition system consist of 250 2-byte unsigned integers, denoting the energy intensities captured by the detector. With the exception of any noise (which can cause significant distortions), pulses are expected to exhibit patterns given by the following equations:

$$G(x) = \frac{A_1 e^{-k_1(x-T_1)}}{1 + e^{-k_2(x-T_1)}} + O, \tag{1}$$

$$F(x) = \frac{A_1 e^{-k_1(x-T_1)}}{1 + e^{-k_2(x-T_1)}} + \frac{A_2 e^{-k_3(x-T_2)}}{1 + e^{-k_4(x-T_2)}} + O, \tag{2}$$

where $G$ and $F$ serve as models for the single pulse and double pulse data, respectively. Effectively, $F$ replicates the pulse term in $G$, albeit with a different set of parameters. Parameters involved in these models are explained in Table 1. In case of a double pulse, rise rates $(K_1, K_3)$ and amplitudes $(A_1, A_2)$ of both signals can be significantly different from each other, whereas decay rates $(K_2, K_4)$ are almost identical across all samples because decay rate is a property of the detector itself. The time difference between the two signals $(T_2 - T_1)$ has a distribution which favors short time differences, meaning the two signals are more likely to overlap with each other rather than being two separate pulses. As the time difference decreases, or the relative amplitudes between the two signal heights become significantly different, it becomes increasingly difficult to decide if a trace is a single pulse or a double pulse trace, underlining the main challenge faced by nuclear physicists.

Currently used `lmfit` software uses non-linear least-squares curve fitting method with the Levenberg-Maequardt algorithm [16] implemented in the GSL (GNU Scientific Library). The target is to minimize the summation of the residuals by adjusting the parameters of the model:

$$\hat{\beta} \equiv argmin_\beta \sum_{i=1}^{N} (y_i - f(x_i, \beta))^2 \tag{3}$$

The Levenberg-Maequardt algorithm is an iterative procedure and works by calculating the Jacobian matrix for each step which increases the computational complexity of the algorithm.

**Table 1.** Parameters

| Parameter | Definition |
| --- | --- |
| $O$ | Offset |
| $T_1$ | Starting position of the first pulse |
| $K_1$ | Rise rate of the first pulse |
| $A_1$ | Amplitude of the first pulse |
| $K_2$ | Decay rate of the first pulse |
| $T_2$ | Starting position of the second pulse |
| $K_3$ | Rise rate of the second pulse |
| $A_2$ | Amplitude of the second pulse |
| $K_4$ | Decay rate of the second pulse |

As noted earlier, single pulses always constitute the majority of the captured snapshots in all experiments. Detailed information about the specific data set that we are working with is given in Table 2.

**Table 2.** Dataset

| Type | Count | Percentage |
| --- | --- | --- |
| Single pulse | 14968801 | 99.9% |
| Double pulse | 16215 | 0.1% |

## 3    Realtime Beta Decay Detection Framework

To enable real time processing of traces obtained in beta decay experiments, we propose a novel framework based on machine learning techniques. Since there is no need for single pulse data to be analyzed using lmfit, our proposed framework acts as an inexpensive filter whose main duty is to reduce the number of traces to be analyzed by lmfit as much as possible without missing any of the rare beta decay events.

Arguably, an important precedent to a good machine learning algorithm is a high quality training dataset. The fact that our dataset is highly skewed towards single pulse data presents a challenge at this point. There are different ways to handle skewed distribution of the classes. Weights corresponding to classes could be adjusted [19] or data could be resampled to have a fair balance of the classes [4]. For this problem, since we have a good analytical model for the different trace types, we choose to estimate the underlying parameter distributions of the double/single pulses and generate synthetic pulses. Models in Eqs. 1 and 2 are used to capture the underlying parameters with high fidelity and generate synthetic data by feeding these parameters back into the equations. To obtain higher quality synthetic data, we use Kernel Density Estimation (KDE) techniques.

A shortcoming in using the analytical models for pulse generation is that we only can produce clean traces without any noise. Since the synthetic dataset should be representative of the real dataset to have better generalization performance, noise must be modeled too. By adding the modeled noise into the generated traces, one can obtain more realistic synthetic data for training.

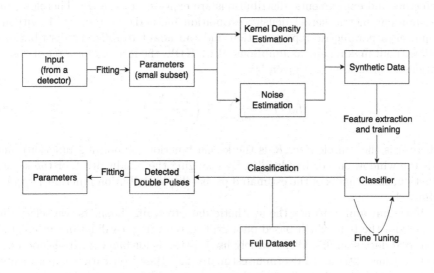

**Fig. 2.** Proposed method.

Figure 2 gives a visual overview of this framework. The first step is determining distribution of parameters in Eq. 2 using `lmfit` on a small sample dataset which indeed can be obtained by scientists during their preparatory phase before the actual experiments begin. These distributions are then fed into our analytical models in conjunction with the noise model and synthetic traces are generated. Synthetic data are used to enhance the small real dataset obtained during preparations, and machine learned classifiers are training using the enhanced dataset. These classifiers are then used the filter out the uninteresting samples from the dataset, significantly accelerating the overall procedure. We note that classifiers may emit false positives, which can then be eliminated by `lmfit` in the final analysis stage.

## 3.1 Synthetic Data Generation

Generating synthetic data to train a classifier is used in many domains including computer vision [18] and speech recognition [12]. We have chosen this approach due to the skewed distribution of classes in actual datasets. Note that there is no need for generating single pulse signals as the majority of traces collected are already traces with a single pulse. Consequently, we draw the single pulse data from the actual dataset itself, and synthetically generate the double pulse traces by approximating the underlying parameter distribution of this rare event.

**Kernel Density Estimation (KDE).** Histograms of individual parameters describing traces with double pulses are given in Fig. 3. As can be seen, it is hard to describe these distributions analytically. Also, due to the nature of the detectors and experiments, distributions are expected to change. Therefore, we use non-parametric kernel density estimation methods, as they work without requiring a parametric representation and can adapt to different distributions well without making any assumptions. With KDE, the formula for a set of observations $X$ at a point $x$ is given by:

$$f_h(x) = \frac{1}{nh} \sum_{i=1}^{n} K\left(\frac{x - X_i}{h}\right),$$

(4)

where $n$ is the sample size, $K$ is the kernel function (smoothing function) and $h$ is the window size or bandwidth. By changing the bandwidth $h$ or the kernel function, smoothness of the estimated probability distribution function $f$ can be adjusted.

Since our aim is to use the synthetic data to train classifiers, we select the kernel and the bandwidth based on a grid search using a validation set created from real data samples. Other widely used methods for bandwidth selection are Scott estimate [21] and Silverman estimate [22]. Based on earlier experiments, we have concluded that selecting the parameters based on the performance on the validation set yields better performance.

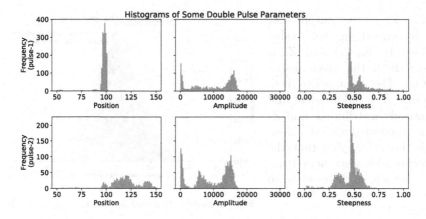

**Fig. 3.** Histograms of the selected parameters. The top row for the parameters corresponding to the first pulse of double pulse samples and the bottom row for the second pulse of double pulse samples

Using the KDE technique described above, new samples are generated to create a more balanced training dataset for which we generate 50,000 samples for the double pulse class. However, parameters require strict bounding. Position variable of the pulses should be in the $[0, 250]$ range. Also, decay time and steepness cannot be less than 0. To impose these boundary conditions, samples drawn from the KDE are checked for these conditions and the ones which violate any of the conditions are redrawn till all the conditions are met.

**Noise Estimation.** Noise is a significant component of the traces obtained from the detectors. Depending on the detector, noise levels can actually be high compared to the pulse(s). To create a high quality synthetic dataset, we modeled the noise as well. While it is hard to separate the noise from the signals, we observe that the part of the trace before the rise of the first peak gives a good representation of the noise in the rest of the signal. Since this initial part of the trace before the pulse is expected to be flat, it is straightforward to detect the noise there. For the intensity values in this initial segment, we calculate the differences from the mean value in that range; create a histogram by aggregating the differences across all traces in the training data. As seen in Fig. 4, for the particular dataset that we work with, the noise could be modeled as an additive Gaussian distribution:

$$Z_i \sim \mathcal{N}(0, \sigma^2)$$
$$Y_i = X_i + Z_i$$

$$(5)$$

where $Y_i$ is the output, $X_i$ is the underlying true value defined by Eq. (2) and $Z_i$ is drawn from the given distribution. The assumptions here are that $Z_i$ is independent and identically distributed, and $Z_i$ is not correlated with $X_i$. Basically, Eq. (5) implies that the detector adds some noise coming from a fixed distribution to the true values coming from the event.

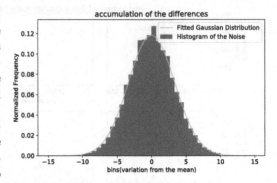

**Fig. 4.** Noise estimated for the experimental data

Based on the analysis via the given method, we found out that the noise for the particular experiment could be modeled as $\mathcal{N}(0, 11.34)$. We note that in cases where the noise does not fit into a simple analytical form such as the Gaussian distribution, one can use the KDE technique described above for accurately modeling the noise as well.

### 3.2 Preprocessing of the Data

**Normalization.** Since the offset $O$ and amplitude variables $A_1$ and $A_2$ cause variations between traces that are otherwise structurally similar, data needs to be normalized to minimize the effects of these variations. By doing so, classifiers are expected to become less susceptible to non-structural variations. We have chosen to use the L2-norm normalization for this purpose.

**Dimensionality Reduction.** In the dataset, each sample consists of 250 points. As can be seen in Fig. 1, majority of these points do not present any information about the class which the sample belongs to. Including unnecessary points increase the complexity of the classification process. We investigated the use of Principal Component Analysis (PCA) which is useful to represent the $d$-dimensional data in a lower dimensional space while keeping the maximum amount of variance. PCA is defined as an orthogonal linear transformation. By lowering the dimensionality, our expectation is that the classification process would get faster and it would require less space. To not have data leakage, we calculated the covariance matrix required for PCA by only using the training data. As will be discussed below, for the classifiers that use synthetic data, the covariance matrix is calculated on synthetic data only.

### 3.3 Classification Methods

Recently, neural network methods such as Convolutional Neural Networks (CNN) or different variations of Recurrent Neural Networks (RNN) have gained immense popularity for classification tasks. They are both proven to be successful on tasks related to time series analysis such as speech recognition [8,17], and human activity recognition [25]. However, they are computationally expensive since they operate on raw data and require high number of layers for accuracy.

As discussed before, speed is crucial for our purposes due to the high data acquisition rate and high number of signals that need to be classified and analyzed. Also, the system could tolerate a reasonable number of false positives because the lmfit method will be used to further clean unwanted traces and analyze all traces that are detected to have double pulses. Under these circumstances, we decided to test Random Forest (RF) [14], Support Vector Machines (SVM) [23] and Gradient Boosted Classifiers (GBC) [5]. For the RF, Classification and Regression Tree (CART) algorithm is used. It basically splits the nodes based on Gini index which is a way to calculate impurity [3]. Since a single decision tree may not be robust to outliers and noise, we trained a Random Forest classifier which is an ensemble learning method. Multiple decision trees are grown together and after the training, the classification is done by voting between the trained trees. GBC is another ensamble learning method. For GBC, Friedman Mean Squared Error is used as the splitting criteria [6]. The training procedure for GBC is sequential unlike RF where the trees could be grown in parallel. Each newly added tree is generated to correct the errors caused by the previous one. SVM tries to separate the dataset in order to classify the data into two groups by passing a linearly separable hyper-plane through the data space. However, most of the datasets are complex and not easily separable by a hyperplane. To solve that kernel trick could be used. The idea is that data could become linearly separable after projecting the input vectors to a higher dimensional feature space. Finally, all classifiers (RF, GBC and SVM) are tuned with grid search and the tuning is done based on precision and recall scores for double pulses.

## 4    Evaluation of the Proposed Framework

We present results from different classifiers tested in our proposed framework. We basically test three different scenarios to show how the framework performs under different conditions. In all scenarios, it is assumed that experimental non-unique signatures (single pulses) are available. For the first two cases, we assume that we have double pulse samples available as well, but in different quantities. For the last case, only single pulses and their parameter distributions are used to explore the scenario where there is no double pulse data available. We used two different splits for our evaluations for the first two scenarios. The first one inspects cases where there is a limited amount of real data for training purposes, such as during the brief preparation period leading up to the actual experiments, and the second one inspects cases with abundant real data, for instance, while repeating an experiment or on-the-fly tuning of ML models during the experiment which typically takes a week or so. For the first split, 10% of the available dataset is used for training (or generating synthetic data), another 10% is used for validation (tuning the parameters) and the remaining is used for testing. For the second split, these percentages are 60-20-20, respectively. In the third case where there is no experimental double pulse data available beforehand, synthetic double pulses are generated for training purposes by drawing the parameters from various uniform distributions. Since $T_1$, $K_1$, $K_2$, $K_3$ and $K_4$ depend on the properties of the detector but not the actual experiment, the parameter distributions of the single pulses for $T_1$, $K_1$ and $K_2$ (1) are used to model the listed parameters. $T_2$, $A_1$ and $A_2$ are drawn from the uniform distributions of unif($T_1$, 250), unif(50, 16383) and unif(50, 16383), respectively, as determined by the experimental setup (16383 is the largest possible amplitude value that the detector could record). Using uniform distributions allows us to avoid any bias in the models when we assume there is no actual double pulse data available.

For each split, two different evaluations are performed. The first evaluation finds parameters of the double pulses in the training data, then estimates the distribution of the parameters and the noise. Then new parameters are drawn from the fitted distribution and synthetic data is generated by feeding the parameter to the functional for the double pulse traces. For the second evaluation, the training data is used directly without adding any synthetic data. Note that for the synthetic data based evaluation, only double pulse samples are generated with the described method since it is easy to acquire single pulse samples and using actual single pulses yield better results than synthetically generated ones. For the third case, as we assume there is no experimental data available for the double pulses, only single pulse parameters from Eq. (1) are used to model the detector dependent parameters and the rest are drawn from uniform distributions.

All computations have been performed on Laconia, a cluster with over 400 compute nodes at Michigan State University's High Performance Computing Center. Each of the base compute nodes on Laconia has 28 cores, located on two fourteen-core Intel Xeon E5-2680v4 Broadwell 2.4 GHz processors, and has 128 GB DDR3 2133 MHz ECC memory. Each core possesses a 64 KB L1 cache (32 KB instruction, 32 KB data), a 256 KB L2 cache. Between fourteen cores of a single "Broadwell" processor, a 35 MB L3 cache is shared. At the time of the experiments, the Laconia nodes ran CentOS version 6.8 distribution of GNU/linux for x86_64 architectures, kernel version 2.6.32-696.20.1, and glibc version 2.12-1.192. The software was built using the GNU Compiler Collection version 6.2 with the -O3 flag. For all experiments, we restricted our computations to a single CPU core on a Laconia node. For the lmfit, GSL library is used and for the classification procedures sklearn package is used. Since both inference and lmfit steps are highly parallelizable due to the independence of the samples, single core speed is a good indicator for performance.

## 4.1   Classification Results

As mentioned above, two important metrics for our purposes are recall rate and speedup on the real data compared to the baseline lmfit method. Since our main goal is to create a filtering mechanism for the real data, we present results for the testing dataset coming from the real data. As the data is extremely skewed, we cannot use accuracy as a comparison metric. If we use a classifier which classifies everything as a single pulse, based on Table 2, the accuracy would be around 99.999% without doing any meaningful work. Therefore, we choose to compare the classifiers based on recall rates which shows how accurately a classifier detects double pulse traces. We present results from SVM, Random Forest and Gradient Boosting Classifier based classifiers trained on real data and synthetic data created with Kernel Density Estimation. The parameters of the KDE was selected based on the performance of the classifiers trained with it. When only 10% of the data (limited data) is used to generate synthetic data, 0.1 and Gaussian are chosen as KDE bandwidth and kernel respectively. When 40% of the data (abundant data) is used to generate synthetic data, the chosen parameters are 0.001 and Gaussian kernel. Also, the peremeters of the classifiers and number of components for PCA are chosen based on the performance in the validation set. For cases where there is experimental data available for double pulses, projecting the data onto the space created with 15 principal components yields better performance for the majority classifiers. As such, we fixed the corresponding parameter to 15 to limit the search space for PCA. On the other hand, for the last case, since the overall variance is high in the training data due to the uniform distributions, the first 50 components are used. Other hyperparameters required for the classifier and KDE have chosen based on their performance on the validation set.

Results for different datasets are reported in Table 3. Speedups with respect to the baseline `lmfit` method are calculated according to the following formula:

$$speedup = \frac{1}{T_c} + \frac{1}{T_{LMFit}} * (P_{double}/PR_c) \qquad (6)$$

where $T_c$ and is the throughput of the classifier, $T_{LMFit}$ is the throughput for the `lmfit` method, $P_{double}$ is the percantage of the double pulses in the dataset and $PR_c$ is the precision rate of the classifier. Note that this speedup formula accounts for the cost of eliminating the false positives produced by our proposed classifiers using the `lmfit` method at the end.

As seen in these results, Random Forest method performs better in terms of recall rate and gives a reasonable speedup. Even though Gradient Boosting Classifier is a lot faster than Random Forest for inference, their speedups are relatively close. The reason is that GBC has lower precision than Random Forest and triggers the `lmfit` method more often to eliminate false positives. SVM has the lowest precision and the lowest throughput (with the exception of synthetic training dataset generated from a fully uniform distribution) compared to the other two methods. Based on these results, Random Forest method is a reasonable candidate to use for classification part of the proposed framework. The random forest models trained on abundant data synthetic data generated from abundant data have the highest recall rate which is 0.9995%.

When we further examine the results, it can be seen that in the absence of sufficient data, generating synthetic data for training purposes helps classifiers to learn more about the unique signature. When 60% of the data is used for the training, it results in better or comparable recall rate for the classifiers.

**Table 3.** Classification results

| Training data | Model | Recall | Precision | Speedup |
|---|---|---|---|---|
| Synthetic data with uniformly distributed $T_2$, $A_1$ & $A_2$ | RF | 0.9406 | 0.4315 | 37.8 |
| | SVM | 0.8448 | 0.6970 | 52.2 |
| | GBC | 0.9174 | 0.4944 | 44.0 |
| Synthetic data generated from limited data | RF | 0.9993 | 0.4566 | 40.0 |
| | SVM | 0.9914 | 0.1156 | 10.2 |
| | GBC | 0.9945 | 0.4667 | 41.5 |
| Synthetic data generated from abundant data | RF | 0.9995 | 0.4514 | 39.5 |
| | SVM | 0.9935 | 0.1012 | 8.9 |
| | GBC | 0.9951 | 0.5019 | 44.6 |
| Limited data | RF | 0.9843 | 0.1266 | 11.3 |
| | SVM | 0.9476 | 0.0323 | 2.9 |
| | GBC | 0.9741 | 0.0992 | 8.9 |
| Abundant data | RF | 0.9995 | 0.3266 | 28.8 |
| | SVM | 0.9943 | 0.0917 | 8.1 |
| | GBC | 0.9989 | 0.3708 | 33.0 |

For precision, classifiers trained on real data still lack behind the ones trained on synthetic data. One reason might be that in the 60% of the data, there are less than 10,000 double pulse samples. However, for the synthetic data, we are generating 50,000 double pulse samples, which is more than the number of double pulses in the whole dataset. When there is no double pulse available for the training and the given uniform distributions are used to generate double pulse samples, the recall rate of the classifiers are relatively lower unsurprisingly but the precision rates are high potentially related to the low recall rates. This causes the speedups to be high since precison and speedup are correlated based on Eq. 6. Also, since the actual data is skewed towards single pulses and the hyperparameters are chosen to maximize the recall rate as the classifiers primary purpose is to filter out non-unique signatures while keeping almost all of the double pulses, the precision rates vary a lot. This situation renders majority of the SVM models useless as low precision rates decrease their speedups.

## 5    Related Work

Different pulse shape classification (PSD) and discrimination methods using machine learning based approaches have been proposed in the literature [1,11,20]. More broadly, in nuclear physics use of machine learning techniques have been explored for other problems such as track classification [13] and jet classification [9]. In [20], it is shown that the support vector machine (SVM) method yields better results than the charge-integration PSD method that originated in analog systems. However, in our dataset, as shown above SVM performs worse compared to the ensemble tree based methods in terms of recall rate. In [11], a neural network architecture called auto-encoders is used to learn how to represent a given trace in a smaller space. Finally, another network is trained to classify transformed traces. We note that we also experimented with neural networks in this study and although neural networks were highly accurate in the classification task, they were significantly slower than the methods we presented above.

To be able to deploy a supervised machine learning method, sufficient amount of training data should be available beforehand. To overcome that issue, simulated data is used for track classification in [13] where the authors show that even though simulated data help training supervised classifiers, results are more promising when experimental data is used. Our results support the same conclusion and we offer a novel way to increase performance when simulated data is used for training.

Finally, we note that unlike the typical classification tasks, it is not practical to compare the performance of the classifiers reported in this study to others, because the detector, its setup and the way experiments are conducted render a classifier developed for a given context ineffective for another context.

# 6    Conclusion

We proposed a novel framework to enable realtime analysis of data from beta decay experiments performed at MSU's National Superconducting Cyclotron Laboratory. The framework uses computationally inexpensive machine learning techniques to accelerate the classification bottlenecks with use of the existing lmfit method. We also developed a synthetic training set generation tool to address the issue of limited (or non-existent) double pulse samples observed in actual training datasets beforehand. In conclusion, we observe that generating synthetic data significantly improves the recall rates for the classifiers used, if the real dataset is limited in size and/or the unique signature we are looking for is extremely rare in the dataset. By modeling the distribution of the underlying parameters via KDE and generating new samples by feeding parameters drawn from KDE to the functional form of the wanted pattern, we have been able to develop a high-performance filtering mechanism to speedup the overall data analysis process. For the classification schemes with highest recall rates, i.e., RF and GBC, we observe speedups up to $44x$ compared to the current method. Coupled with use of parallelism on a server with moderate computing power, the proposed framework can easily provide real-time analysis guarantees.

As future work, in terms of data generation, a Generative Adversarial Network [7] could be used. It could generate more realistic samples without even providing a functional form for the signature pattern and estimating the distribution of its parameters. In terms of further performance improvements, to ensure real-time analysis with the rate of data collection that will be possible at the upcoming Facility for Rare Isotope Beams, an $850M facility funded by the US Department of Energy, our proposed framework can be ported to FPGAs and GPUs. There are several examples in the literature which demonstrate significant performance boosts by the use of these accelerators, see for instance [15,24] and [2].

# References

1. Balmer, M.J., Gamage, K.A., Taylor, G.C.: Comparative analysis of pulse shape discrimination methods in a 6li loaded plastic scintillator. Nucl. Instrum. Methods Phys. Res. Sect. A **788**, 146–153 (2015)
2. Bauer, S., Köhler, S., Doll, K., Brunsmann, U.: FPGA-GPU architecture for kernel SVM pedestrian detection. In: 2010 IEEE Computer Society Conference on Computer Vision and Pattern Recognition-Workshops, pp. 61–68. IEEE (2010)
3. Breiman, L., Friedman, J., Olshen, R., Stone, C.J.: Classification and regression trees. Wadsworth statistics/probability series (1984)
4. Chawla, N.V.: C4.5 and imbalanced data sets: investigating the effect of sampling method, probabilistic estimate, and decision tree structure. In: Proceedings of the ICML, vol. 3, pp. 66 (2003)
5. Friedman, J.H.: Greedy function approximation: a gradient boosting machine. Ann. Stat. **29**, 1189–1232 (2001)
6. Friedman, J.H.: Stochastic gradient boosting. Comput. Stat. Data Anal. **38**(4), 367–378 (2002)
7. Goodfellow, I., et al.: Generative adversarial nets. In: Advances in Neural Information Processing Systems, pp. 2672–2680 (2014)
8. Graves, A., Mohamed, A.R., Hinton, G.: Speech recognition with deep recurrent neural networks. In: 2013 IEEE International Conference on Acoustics, speech and signal processing (ICASSP), pp. 6645–6649. IEEE (2013)
9. Guest, D., Cranmer, K., Whiteson, D.: Deep learning and its application to LHC physics. Ann. Rev. Nucl. Part. Sci. **68**, 161–181 (2018)
10. Heyde, K., Wood, J.L.: Shape coexistence in atomic nuclei. Rev. Mod. Phys. **83**(4), 1467 (2011)
11. Holl, P., Hauertmann, L., Majorovits, B., Schulz, O., Schuster, M., Zsigmond, A.: Deep learning based pulse shape discrimination for germanium detectors. arXiv preprint arXiv:1903.01462 (2019)
12. Ko, T., Peddinti, V., Povey, D., Seltzer, M.L., Khudanpur, S.: A study on data augmentation of reverberant speech for robust speech recognition. In: 2017 IEEE International Conference on Acoustics, Speech and Signal Processing (ICASSP), pp. 5220–5224. IEEE (2017)
13. Kuchera, M.P., et al.: Machine learning methods for track classification in the AT-TPC. arXiv preprint arXiv:1810.10350 (2018)
14. Liaw, A., Wiener, M., et al.: Classification and regression by randomforest. R news **2**(3), 18–22 (2002)
15. Mishina, Y., Murata, R., Yamauchi, Y., Yamashita, T., Fujiyoshi, H.: Boosted random forest. IEICE Trans. Inf. Syst. **98**(9), 1630–1636 (2015)
16. Moré, J.J.: The levenberg-marquardt algorithm: implementation and theory. In: Watson, G.A. (ed.) Numerical Analysis. LNM, vol. 630, pp. 105–116. Springer, Heidelberg (1978). https://doi.org/10.1007/BFb0067700

17. Palaz, D., Magimai-Doss, M., Collobert, R.: Analysis of CNN-based speech recognition system using raw speech as input. In: Sixteenth Annual Conference of the International Speech Communication Association (2015)

18. Richardson, E., Sela, M., Kimmel, R.: 3D face reconstruction by learning from synthetic data. In: 2016 Fourth International Conference on 3D Vision (3DV), pp. 460–469. IEEE (2016)

19. Rosenberg, A.: Classifying skewed data: importance weighting to optimize average recall. In: Thirteenth Annual Conference of the International Speech Communication Association (2012)

20. Sanderson, T., Scott, C., Flaska, M., Polack, J., Pozzi, S.: Machine learning for digital pulse shape discrimination. In: 2012 IEEE Nuclear Science Symposium and Medical Imaging Conference Record (NSS/MIC), pp. 199–202. IEEE (2012)

21. Scott, D.W.: On optimal and data-based histograms. Biometrika **66**(3), 605–610 (1979)

22. Silverman, B.W.: Density Estimation for Statistics and Data Analysis. Routledge, Boca Raton (2018)

23. Suykens, J.A., Vandewalle, J.: Least squares support vector machine classifiers. Neural Process. Lett. **9**(3), 293–300 (1999)

24. Van Essen, B., Macaraeg, C., Gokhale, M., Prenger, R.: Accelerating a random forest classifier: Multi-core, GP-GPU, or FPGA? In: 2012 IEEE 20th International Symposium on Field-Programmable Custom Computing Machines, pp. 232–239. IEEE (2012)

25. Yang, J., Nguyen, M.N., San, P.P., Li, X., Krishnaswamy, S.: Deep convolutional neural networks on multichannel time series for human activity recognition. In: IJCAI, vol. 15, pp. 3995–4001 (2015)

# A Novel Partition Method for Busy Urban Area Based on Spatial-Temporal Information

Zhengyang Ai[1,3], Kai Zhang[2], Shupeng Wang[1(✉)], Chao Li[2(✉)], Xiao-yu Zhang[1], and Shicong Li[2]

[1] Institute of Information Engineering, Chinese Academy of Sciences, Beijing, China
{aizhengyang,wangshupeng,zhangxiaoyu}@iie.ac.cn
[2] National Computer Network Emergency Response Technical Team/Coordination Center of China, Beijing, China
{zhangkai,lichao,lishicong}@cert.org.cn
[3] School of Cyber Security, University of Chinese Academy of Sciences, Beijing, China

**Abstract.** Finding the regions where people appear plays a key role in many fields like user behavior analysis, urban planning, etc. Therefore, how to partition the world, especially the urban areas where people are crowd and active, into regions is very crucial. In this paper, we propose a novel method called Restricted Spatial-Temporal DBSCAN (RST-DBSCAN). The key idea is to partition busy urban areas based on spatial-temporal information. Arbitrary and separated shapes of regions in urban areas would be then obtained. Besides, we would further get busier region earlier by RST-DBSCAN. Experimental results show that our approach yields significant improvements over existing methods on a real-world dataset extracted from Gowalla, a location-based social network.

**Keywords:** Area partition · Density-based clustering · Social link mining · Location-based social network

## 1 Introduction

With the rapid development of internet, cyberspace has become an important field for entertainment, consumption, etc. If we associate the cyberspace with geospatial, i.e. linking the online with offline, a mass of user behaviors could be mined, which would be of great help for recommendation system, friendship prediction, urban planning, etc. [1]. The first step for mapping cyberspace to geospatial is to find the regions where users locate. Therefore, how to divide the world, especially urban areas, becomes an important and necessary intermediate link. Traditionally, three methods are widely used for area partition, i.e. **address method**, **rigid method** and **cluster method**. However, these methods all have limitations, which would be introduced in detail as follows.

© Springer Nature Switzerland AG 2019
J. M. F. Rodrigues et al. (Eds.): ICCS 2019, LNCS 11537, pp. 234–246, 2019.
https://doi.org/10.1007/978-3-030-22741-8_17

Address method refers to partitioning areas with physical address, like Starbucks, KFC, etc. In fact, as addresses often reveal user behaviors, like shopping, having dinner, entertainment, etc., this method is commonly used to study either user behavior prediction or product recommendation [2–4]. Although achieving high precision, address method has limitation in the scale of regions. And it's also too difficult for the method to find interactions between two users. For instance, the sparsity of user interactions in Foursquare and Yelp are greater than 99.99% [5]. As a result, it is inappropriate to do researches related to user interactions using this method.

Grid method refers to partitioning areas with gird. The metric for partitioning areas is usually kilometer or degree of latitude and longitude. For instance, Backstrom et al. divided the US into $0.01 \times 0.01$ degree regions (about 0.4 square miles) [6], while Cho et al. discretized the world into $25 \times 25$ km cells [7]. This partitioning method is simple and practicable, and obtained regions could cover an ideal range to find interactions between users. Therefore, it is often used to study user social relationships [8–10]. However, its disadvantages are also obvious. The method could not partition areas based on practical significance or independent meaning, and always divides a function unit, e.g. a shop or a store, into two or more separate girds, missing the interaction information of users. Besides, it is hard to set a proper size for girds, since little size may miss interactions, while big size may cover too many user traces and make the extraction of interactions too hard to fulfill.

Cluster method refers to partitioning areas by clustering locations in the form of GPS, and density-based clustering (DBSCAN) is a representative method. Zhou et al. designed a new approach based on DBSCAN, with which arbitrary shapes could be obtained, and areas could be partitioned according to practical significance or independent meaning as a result [11]. Besides, reasonable parameter value is easy to set for DBSCAN when partitioning areas. By this way, DBSCAN outperforms other cluster methods evidently, and is widely used later when dealing with personal locations [12,13]. However, as DBSCAN works based on density, a cluster would "spread" infinitely when density meets the requirement. As a result, intensive locations in busy urban areas would be clustered into a same group with DBSCAN. Figure 1 shows the hot map of user check-ins in the busy area of Austin, US with the dataset extracted from Gowalla, a location-based social network. Intensive check-ins and locations as shown in the figure, make it difficult to partition the area with DBSCAN. In fact, this area is taken as a region when clustered with 0.001 degree. Therefore, DBSCAN doesn't work well when dealing with a mass of locations, especially with those in busy urban areas.

In this paper, we propose a method for partitioning busy urban areas with spatial-temporal information in LBSN (Location-Based Social Network). On one hand, arbitrary shapes would be obtained by designing method with density-based clustering, overcoming the stiff of grid method. On the other hand, a boundary is designed ingeniously for limiting the scale of clusters, overcoming the infinite "spread" of DBSCAN. Our approach is validated using real user data and show a good performance.

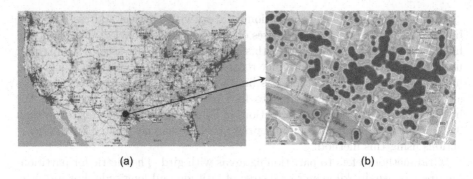

(a)                                        (b)

**Fig. 1.** The hot map of users' check-ins in the busy urban area of Austin, US, around the Texas government.

Our contributions are concluded as follows:

- We propose a new method, **RST-DBSCAN**, for partitioning busy urban areas based on spatial-temporal information, with which arbitrary and separated shapes of regions would be obtained reasonably;
- We propose a new time mapping method to connect temporal information with spatial information. The method makes it possible to partition areas with 3-dimensional spatial-temporal information;
- We further propose to cluster from the locations which own greater density. In this way, we can get busier regions earlier, making the approach more reasonably;
- We visually and quantitatively evaluate our approach with a location-based social network, Gowalla. Results show that the performance of RST-DBSCAN outperforms that of competitors.

The rest of this paper is organized as follows. Section 2 introduces the design of RST-DBSCAN. Section 3 describes experiments. Finally, Sect. 4 gives the conclusion.

## 2    Design of the Model

In this section, we introduce our method in the following steps. Definitions for designing RST-DBSCAN would be introduced at first. Then we introduce the ideas for designing RST-DBSCAN in the view of space and time respectively. Finally, we describe the method with schematic and pseudocode.

### 2.1    Definitions

DBSCAN is a well-known cluster method and works by clustering points which satisfy the density conditions into a group. Therefore, two important concepts, $\varepsilon$, the radius of a circle, and *MinPts*, the minimum number of neighbor points within that circle, are needed [14]. In this section, points are employed to denote locations in 2-dimensional spatial space. Then the circle with radius $\varepsilon$ (named **neighbor region** here) means the scale covered by a location, and *MinPts* means the minimum number of location's neighbors. Besides, we make definitions for designing RST-DBSCAN as follows.

### Definition 1: Candidate core location

If the neighbor region of a location, $p$, covers at least *MinPts* locations, $p$ is a candidate core location.

### Definition 2: Neighbor location

The neighbors of $p$, denoted by $N_\varepsilon(p)$, is defined by

$$N_\varepsilon(p) = \{q \in S \mid dist(p, q) \leq \varepsilon\} \tag{1}$$

Here, $S$ is the set of all locations, and $q$ is any location in the set.

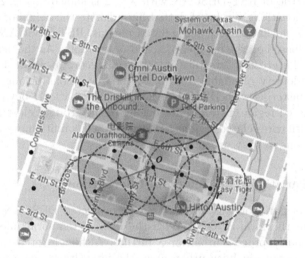

**Fig. 2.** The schematic of RST-DBSCAN

### Definition 3: Core location

The location which starts a clustering process is taken as core location.

### Definition 4: Location directly density reachable

For a core location $p$ and a location $q$, we say that $q$ is directly density reachable from $p$ if $q$ is the neighbor location of $p$.

### Definition 5: Location density reachable

For a location $p$ and a location $q$, we say $q$ is density reachable from $p$ if there is a chain of locations $q_1, q_2, ..., q_n$, such that $q_1 = q, q_n = p$, and $q_{i+1}$ is directly density reachable from $q_i$. Here $1 \leq i \leq n$.

**Definition 6: Core-related location**

Assume location n is density reachable from core location $p$ and satisfies the following condition

$$\{n \in S \mid dist(p, n) \leq \lambda \cdot \varepsilon\} \tag{2}$$

Then location n is a core-related location for $p$. Besides, the circle region with radius $\lambda \cdot \varepsilon$ is called as **core-related region** of $p$.

What needs to note is that once a candidate core location is covered by a core location, it becomes a core-related location.

## 2.2    Area Partition with Spatial Information

For the sake of understanding, we introduce how RST-DBSCAN works with spatial information at first in this section, and then the application of time information would be introduced in the next section.

The schematic of RST-DBSCAN with $\lambda = 2$ is shown in Fig. 2. Here black points denote locations, circles of dotted lines denote the neighbor regions, and the circles of solid line denote core-related regions. RST-DBSCAN decides the core locations based on the number of neighbor locations. As the density around location $o$ is the greatest, we take it as the first core location. Location $s$, $r$ and $t$ are all density reachable from $o$, so these locations belong to the same group when clustering with DBSCAN. However, as location $t$ beyond the range of $o$'s core-related region, it would not belong to the group if clustered by RST-DBSCAN. With core-related regions as the restricted condition, we would obtain separated regions in busy urban areas.

## 2.3    Area Partition with Temporal Information

In Sect. 2.2, we partition busy urban area by restricted DBSCAN with spatial information in the form of GPS. With the help of temporal information, we could continuously process urban area further. Although intensive check-in is an important feature in busy urban area, the intensity varies a lot over time. Figure 3 shows the hot map of users' check-ins around the Texas government at different time periods. As can be seen, most of check-ins appear at commercial districts in the daytime, while most of them appear at resident area at night. In more detail, Fig. 4 shows the situation of check-ins in a few blocks away. As can be seen, check-ins are active in shopping mall in the afternoon and at dusk, while more users appear in bars at early morning. Therefore, conclusions can be drawn that users check-ins also reflect their living habits, and area partitioned with temporal information would contain more information.

However, how to use temporal information for partitioning area is still a challenging issue. Previous studies often partition time with hour periods, which is not reasonable enough. For instance, a restaurant is active from 5:10 PM to 20:45 PM, and it is hard to partition this time period with traditional method. Therefore, we bring out a new time-mapping method to connect temporal information with spatial information. The formula of time-mapping is

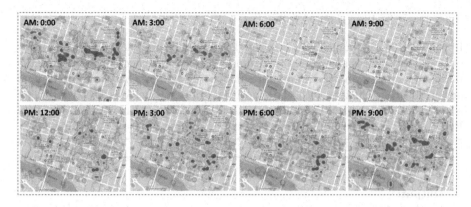

**Fig. 3.** The hot map of users' check-ins around the Texas government at different time periods

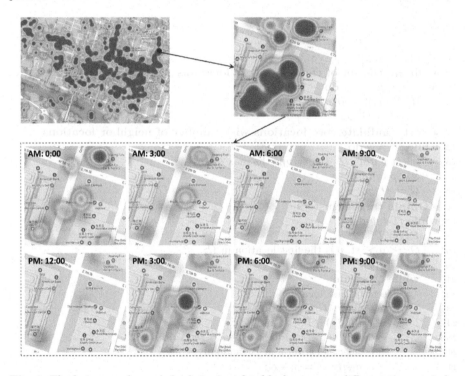

**Fig. 4.** The hot map of users' check-ins in a few blocks away at different time periods.

$$l_{Time} = \vartheta \cdot \varepsilon \cdot \left( T_{hour} + \frac{T_{min}}{60} + \frac{T_{min}}{3600} \right) \tag{3}$$

Here $\vartheta$ is time-mapping parameter. When $\vartheta = 1$, 1 hour corresponds to $\varepsilon$ degree in the spatial scale. Then we could deal with temporal information with RST-DBSCAN, with which arbitrary and reasonable time period would be obtained.

## 2.4  Description of the Model

In this section, we would introduce the design of our approach. The pseudocode of RST-DBSCAN is shown in Algorithm 1. We take the check-in dataset $D$, the radius of a circle $\varepsilon$ and the minimum number of neighbors *MinPts* as input, and obtain location clusters with the algorithm. Here $D$ includes user coordinate of latitude and longitude and check-in time. Specific implementation details of RST-DBSCAN algorithm are as follows.

---

**Algorithm 1.** RST-DBSCAN

---

**Input:**

$\varepsilon$; *MinPts*;$\vartheta$; Dataset $D$

**Output:**

Location clusters;

1  • **Initialization**
2    $\Omega' = \Phi$;
3    $k = 0$;
4    $\Omega =<>$;
5    $l_{Time}= \vartheta \cdot \varepsilon \cdot (T_{hour} + \frac{T_{min}}{60} + \frac{T_{min}}{3600})$;
6    $\Gamma = \{D, l_{Time}\}$;
7  • **Obtain the set of candidate core locations**
8  **for** $j = 1, 2, \ldots, m$ **do**
9      **if** $|N_\varepsilon(x_j)| \geq MinPts$ **then**
10          $\Omega' = \Omega' \cup x_j$ ;
11  • **Sort candidate core locations with number of neighbor locations**
12  **while** $\Omega' \neq \Phi$ **do**
13      select $p$ from $\Omega'$, st. $N_\varepsilon(p)$is the max in $\Omega'$;
14      add $p$ to $\Omega$;
15      $\Omega' = \Omega'\backslash p$;
16  • **Obtain target clusters**
17  **while** $\Omega \neq \Phi$ **do**
18      $\Gamma_{old} = \Gamma$;
19      $Q =< o >$, st. $o$ is the first location in $\Omega$;
20      $\Gamma = \Gamma\backslash\{o\}$;
21      **while** $Q \neq \Phi$ **do**
22          select the first location $q$ in $Q$;
23          **if** $|N_\varepsilon(q)| \geq MinPts$ **then**
24              $\Delta = N_\varepsilon(q) \cap \Gamma$;
25              **while** *unprocessed location exists in* $\Delta$ **do**
26                  select location $r$ in $\Delta$ randomly;
27                  $dist(r, o) = |r - o|$;
28                  **if** $dist(r, q) > \lambda * \varepsilon$ **then**
29                      $\Delta = \Delta\backslash r$;
30              add locations in $\Delta$ to $Q$;
31              $\Gamma = \Gamma\backslash\Delta$;
32      $C_k = \Gamma_{old}\backslash\Gamma$;
33      $k = k + 1$;
34      $\Omega = \Omega\backslash C_k$;

---

(1) Initialization

Set the number of clusters as 0. Compute the mapping result of time and obtain 3-dimensional dataset with unprocessed locations, $\Gamma$. The dataset of candidate core locations $\Omega'$ and a queue $\Omega$ are initialized as empty, respectively. Here we say a location has been processed if it has been clustered into to a group, unprocessed otherwise.

(2) Obtain the set of candidate core locations

Traverse all the locations. Take location $x$ for example. Calculate the number of $x$'s neighbors, $|N_\varepsilon(x)|$, and then confirm whether $x$ belongs to $\Omega'$ or not according to the numerical value of $|N_\varepsilon(x)|$ and $MinPts$.

(3) Sort locations with their number of neighbors

Sort locations in $\Omega'$ in descending order, according to their number of neighbors. Then a queue of candidate core locations $\Omega$ is formed. Location with more neighbors in $\Omega$ would be in more forward position.

(4) Obtain target clusters

Extract the first candidate core location $q$ in $\Omega$. Drop $q$ if it has been processed, otherwise set it as the core location. Find all locations which are unprocessed and density reachable from $q$, and remove those that beyond the scope of $q'$ core-related region. Then we would obtain a cluster with $q$ as the center. Mark these locations as processed. Repeat the process above until $\Omega$ is empty, and all locations will be clustered.

With above steps, a region where locations belong to the same cluster is formed at last. Based on practical significance and independent meaning, RST-DBSCAN could partition the area into arbitrary and separated shapes of regions.

## 3  Experiments

In this section, we carry on two experiments to verify the effectiveness of our approach. One is to partition a busy urban area with **RST-DBSCAN**. The other is to mine social links with user interactions in the regions obtained by RST-DBSCAN, as area partition plays an important role on social link mining.

### 3.1  Evaluation Index

We adopt three performance metrices, precision $(P)$, recall $(R)$, F1-measure $(F_1)$ to estimate the model, which can be calculated as follows. Let $TP$, $TN$, $FP$ and $FN$ denote the numbers of true positives, true negatives, false positives and false negatives respectively.

$$P = \frac{TP}{TP + FP} \tag{4}$$

$$R = \frac{TP}{TP + FN} \tag{5}$$

$$F_1 = 2\frac{P * R}{P + R} \tag{6}$$

As can be seen, F1-measure is the comprehensive value of precision and recall. Besides, links in social network are typically sparse, so F1-measure plays a more important role than accuracy when evaluating model. As a result, F1-measure is taken as a main performance index here.

**Table 1.** Comparison of DBSCAN and RST-DBSCAN when the value of $\varepsilon$ varies.

| $\varepsilon$ | Region number/DBSCAN | Region number/RST-DBSCAN |
|---|---|---|
| 0.005 | 1 | 4 |
| 0.001 | 1 | 14 |
| 0.0005 | 12 | 45 |

## 3.2 Experiment 1

**Dataset**

Experiments are performed on a publicly available dataset, which is extracted from a location-based social network namely Gowalla. It collects user check-ins and social links from 2009.2 to 2010.10 [7]. As most check-ins appear in Austin, US, we choose the urban area of Austin, around the government of Texas, as the target. The detailed scope is 30.262° N -30.270° N, 97.730° W-97.747° W, and 5,173 users check in 72,131 times at 1,450 different locations.

**Comparative Approach**

**DBSCAN** is taken as the comparative approach in this part.

**Experimental Setup**

We set $MinPts = 1$ for both DBSCAN and RST-DBSCAN. $\lambda$ is set to 3 for RST-DBSCAN. $\varepsilon$ is set in the range of $0.005, 0.001, 0.0005$.

**Evaluation Results**

When the value of $\varepsilon$ varies, the results are shown in Table 1. As can be seen, for $\varepsilon = 0.005$ and 0.001, DBSCAN couldn't partition this area, while RST-DBSCAN divides the area into 4 and 14 regions respectively. For $\varepsilon = 0.005$, the area could be partitioned by DBSCAN, while the number is still far less than that by RST-DBSCAN. In fact, as 0.001 degree is roughly equal to 0.1 km, smaller size even couldn't cover a unit like a store or a restaurant. Therefore, 0.001 is nearly the minimum size for rationality, and our approach outperforms DBSCAN substantially.

More visually, Fig. 5 shows 1,450 different locations in this area, represented by red points. When $\varepsilon = 0.001$, these locations belong to a same group clustered by DBSCAN. When clustered with RST-DBSCAN, the area is divided into 14 regions and results are shown in Fig. 6. Here we mark the locations with 14 different colors, and locations with the same color belong to the same group. We draw borders of these groups manually for viewing convenience.

## 3.3  Experiment 2

### Dataset

The dataset extracted from Gowalla is also employed in this experiment. As inadequate information would make adverse effects on the experiment, we select users whose check-ins and friends are all at the top 2% ( i.e. the time of check-ins is above 186, and the number of friends is above 52). There are 761 users which are matched with the conditions and there exists 8,828 links among them.

### Comparative Approaches

**Address** method, **Grid** method and **DBSCAN** are taken as comparative approaches in this part.

**Fig. 5.** Locations in the urban area of Austin.

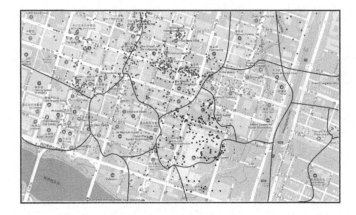

**Fig. 6.** The visual result of area partition with RST-DBSCAN.

We obtain regions with these methods at the first step, and compute the interactions of users in these regions to mine unknown social links. Scellatopropose et al. have proved that social links could be mined by user interactions in spatial space and designed several features for mapping relations [15]. We choose **CR** as the feature in this experiment, which is represented as the number of common regions two persons both check in. We calculate CR and assume there exists social links among users whose CR are at the $k$ top. Recall, precision and F1-measure are taken as the metrics to evaluate our approach.

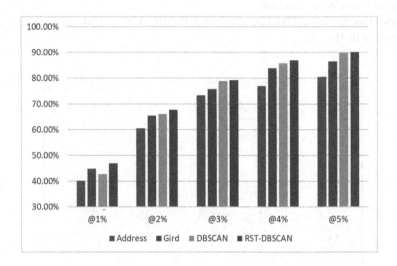

**Fig. 7.** The recall of social link mining by top-k.

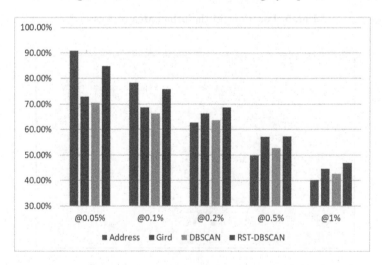

**Fig. 8.** The precision of social link mining by top-k.

**Experimental Setup**

We set $MinPts = 1$, $\varepsilon = 0.001$ and $\lambda = 3$ for RST-DBSCAN. The $MinPts$ and $\varepsilon$ of DBSCAN is the same for a fair comparison. We set gird method with $0.01 \times 0.01$ degree, as this is the most effective value and is often used when studying relationships between user interactions in spatial and social link mining.

**Evaluation Results**

Results are shown in Figs. 7 and 8. As can be seen, recall@$k$ of RST-DBSCAN outperforms the other methods, and the advantage of RST-DBSCAN is more obvious when $k$ is smaller. At the same time, precision@$k$ of RST-DBSCAN also outperform that of **Grid** and DBSCAN. An interesting phenomenon is that precision@$k$ of **Address** is the greatest when $k$ is small enough. In fact, it is normal as address is much more precise than other methods, and two persons meet at smaller regions frequently means it is more possible for them to be friends. However, as **Address** method commonly covers small regions, a mass of interactions between two persons are easily missed, leading a rapid drop of Precision@$k$ when $k$ increases. Therefore, compared to others, **Address** is not a stable method when mining social links.

Specially, we compute the F1-measure when $k = 1$, and the values of **Address**, **Grid**, DBSCAN, RST-DBSCAN are 40.14%, 44.72%, 42.71%, 46.95% respectively, and the F1-measure of RST-DBSCAN outperforms others' by 6.81%, 2.24%, 4.24%. As RST-DBSCAN could partition busy urban areas more reasonably, especially for commercial areas, better results would be obtained when mining social links.

## 4 Conclusion

In this paper, we present a new area partition method, called RST-DBSCAN, for partitioning busy urban areas with spatial-temporal information in LBSN. Our approach is able to divide the areas into arbitrary and separated regions based on practical significance or independent meaning reasonably. Comprehensive experiments are conducted on a real-world dataset, and results show that our approach performs better than competitors.

## References

1. Zheng, Y.: Computing with Spatial Trajectories. Springer, Heidelberg (2011). https://doi.org/10.1007/978-1-4614-1629-6
2. Liu, Y., Wei, W., Sun, A., Miao, C.: Exploiting geographical neighborhood characteristics for location recommendation, pp. 739–748 (2014)
3. Lian, D., Ge, Y., Zhang, F., Yuan, N.J.: Content-aware collaborative filtering for location recommendation based on human mobility data, pp. 261–270, (2015)
4. Liu, Q., Wu, S., Wang, L., Tan, T.: Predicting the next location: a recurrent model with spatial and temporal contexts. In: Thirtieth AAAI Conference on Artificial Intelligence, pp. 194–200 (2016)

5. Hu, B., Ester, M.: Social topic modeling for point-of-interest recommendation in location-based social networks. In: IEEE International Conference on Data Mining, pp. 845–850 (2014)
6. Backstrom, L., Sun, E., Marlow, C.: Find me if you can: improving geographical prediction with social and spatial proximity. In: WWW, pp. 61–70 (2010)
7. Cho, E., Myers, S.A., Leskovec, J.: Friendship and mobility: user movement in location-based social networks. In: ACM SIGKDD International Conference on Knowledge Discovery and Data Mining, San Diego, CA, USA, pp. 1082–1090, August 2011
8. Pham, H., Shahabi, C., Liu, Y.: EBM: an entropy-based model to infer social strength from spatiotemporal data. In: ACM Sigmod International Conference on Management of Data (2013)
9. Zhang, K., Yun, X., Zhang, X.Y., Zhu, X., Chao, L., Wang, S.: Weighted hierarchical geographic information description model for social relation estimation. Neurocomputing **216**, 554–560 (2016)
10. Zhang, X.Y., Zhang, K., Yun, X., Wang, S., Yuan, Q.: Location-based correlation estimation in social network via collaborative learning. In: IEEE INFOCOM 2016 - IEEE Conference on Computer Communications Workshops (INFOCOM WKSHPS) (2016)
11. Zhou, C., Frankowski, D., Ludford, P., Shekhar, S., Terveen, L.: Discovering personal gazetteers: an interactive clustering approach. In: ACM International Workshop on Geographic Information Systems, pp. 266–273 (2004)
12. Zhou, C., Bhatnagar, N., Shekhar, S., Terveen, L.: Mining personally important places from GPS tracks. In: IEEE International Conference on Data Engineering Workshop, pp. 517–526 (2007)
13. Killijian, M.O.: Next place prediction using mobility Markov chains. In: The Workshop on Measurement, Privacy, and Mobility, p. 3 (2012)
14. Han, J., Kamber, M.: Data Mining: Concepts and Techniques, 2nd edn. China Machine Press, Beijing (2006)
15. Scellato, S., Noulas, A., Mascolo, C.: Exploiting place features in link prediction on location-based social networks. In: ACM SIGKDD International Conference on Knowledge Discovery and Data Mining (2011)

# Mention Recommendation with Context-Aware Probabilistic Matrix Factorization

Bo Jiang[1], Zhigang Lu[1,2(✉)], Ning Li[1], and Zelin Cui[1]

[1] Institute of Information Engineering, Chinese Academy of Sciences, Beijing, China
{jiangbo,luzhigang,lining6,cuizelin}@iie.ac.cn
[2] School of Cyber Security, University of Chinese Academy of Sciences,
Beijing, China

**Abstract.** Mention as a key feature on social networks can break through the effect of structural trapping and expand the visibility of a message. Although existing works usually use rank learning as implementation strategy before performing mention recommendation, these approaches may interfere with the influening factor exploration and cause some biases. In this paper, we propose a novel Context-aware Mention recommendation model based on Probabilistic Matrix Factorization (CMPMF). This model considers four important mention contextual factors including topic relevance, mention affinity, user profile similarity and message semantic similarity to measure the relevance score from users and messages dimensions. We fuse these mention contextual factors in latent spaces into the framework of probabilistic matrix factorization to improve the performance of mention recommendation. Through evaluation on a real-world dataset from Weibo, the empirically study demonstrates the effectiveness of discovered mention contextual factors. We also observe that topic relevance and mention affinity play a much significant role in the mention recommendation task. The results demonstrate our proposed method outperforms the state-of-the-art algorithms.

**Keywords:** Mention recommendation · Social network ·
Probabilistic Matrix Factorization · Contextual information

## 1 Introduction

With the advent and development of web applications, social networks have become popular, especially for information sharing. Through these services, a lot of user behaviors like following, posting and retweeting are fully recorded. Particularly, retweeting is the most important mechanism that forms a diffusion network in the way of virus on social networks. Hence, understanding the mechanism of information diffusion is especially critical problem for many social applications including user interest modeling [16,25], influential spreaders identifying [2,28] and social recommendations [20,23]. Recently, various studies on

© Springer Nature Switzerland AG 2019
J. M. F. Rodrigues et al. (Eds.): ICCS 2019, LNCS 11537, pp. 247–261, 2019.
https://doi.org/10.1007/978-3-030-22741-8_18

(a) examples of mention on social network

| User | Content |
|------|---------|
| russiaFIFA | welcome to world cup in russia @messi @ronaldo @neymar |
| Bob | @alydesigns this is a new magic function for iphone and mac |
| Candy | @julieebaby awe i love you too!!!! 1 am here i miss you |

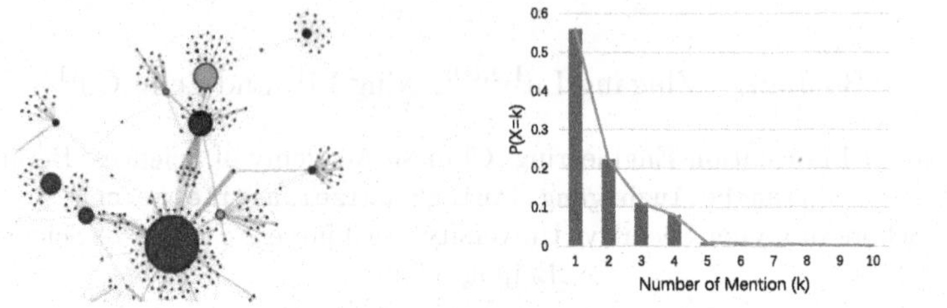

(b) diffusion network of a mentioned message     (c) distribution of number of mentions in a message

**Fig. 1.** The observation and analysis of mention on social network.

information diffusion have been proposed including the influence factors investigation [1,15] and user's spreading behaviors prediction [29]. However, these studies assume that information diffusion based on retweeting behaviors build on the underlying following network among users, which may cause only be connected users can receive the message. Thus, the scale of information diffusion would be limited due to the effect of structural trapping.

To break through the structural trapping, social networks offer mention function which can improve the visibility scope of message. Figure 1 gives an illustration of mention on social network. We observe from Fig. 1(a) that mention allows a user to introduce other users in a message by the form of @username. Further, Fig. 1(b) illustrates the diffusion network of a mentioned message, forming a large-scale cascade. Meanwhile, only one or two users are mentioned in a message at most cases in Fig. 1(c). Hence, mention plays important roles in both expanding information diffusion and improving social relationship. The problem of whom-to-mention have attracted more and more attention in recent years, including ranking-based recommendation [10,21,22,27,30], link prediction [8] and unbalance assignment [4]. The above methods consider different factors in the mention task from the aspects of content [5,10,21], social influence [10,17], spatiotemporal information [4,21], and user's interests [6,17]. Despite the vast and growing studies on mention behaviors, none of models jointly consider both topic relevance and interaction histories as well as homophily influence.

In this paper, we propose a novel Context-aware Mention recommendation model based on Probabilistic Matrix Factorization to recommend the right users. To provide more accurate mention recommendation, we first quantify topic relevance on the basis of interest match between messages and users as well as mention affinity based on interaction histories, and then force the two entries onto the product of user and message latent feature matrices. Meanwhile, we introduce two similarity regularization constraint terms from user and content dimensions on the basis of homophily assumption. Finally, we collaboratively factorize these social contextual factors under probabilistic matrix factorization. A real-world mention dataset is conducted from Weibo. The experiment results show that our proposed model outperforms the baseline models. Furthermore, mention contextual factors can boost the performance of recommendation, which demonstrates the effectiveness of discovered contextual factors.

This work makes the following three contributions:

- We propose a novel context-aware mention recommendation model based on probabilistic matrix factorization. This model considers topic relevance, mention affinity, user profile and message semantic similarities as important contextual factors to mention the candidates.
- We also observe topic relevance and mention affinity play a much significant role in the task, and incorporating user profile and message semantic similarities indeed can improve the performance of mention recommendation.
- Comprehensive experiments on the real-world dataset clearly validate that our proposed model outperforms state-of-the-art comparison methods, which proves the effectiveness of discovered mention contextual factors.

The remainder of this paper is organized as follows: Sect. 2 reviews related work. Section 3 detailed describes the proposed models. We empirically evaluate our proposed method on a real-world dataset in Sect. 4, including a comparison to baseline methods. We conclude the paper in Sect. 5.

## 2 Related Work

### 2.1 Social Recommendation Methods

A great deal of work with social recommendation has been proposed by considering contextual information. For example, SoRec [11] uses network structure information and rating records to solve the data sparsity and poor prediction accuracy problems based on probabilistic matrix factorization. Similarly, Context MF [9] considers user preference and social influence to improve the accuracy of social recommendation. STE [12] utilizes social trust restrictions by fusing users' tastes and their trusted friends' favors together on the recommender systems. mTrust [19] studies multi-faceted trust relationships between users for rating prediction. SocialMF [7] incorporates the mechanism of trust propagation into the matrix factorization approach for recommendation in social networks. SocialReg [13] imposes social regularization terms to constrain the objective functions based on users' social friend information. TBPR [24] studies the effects of distinguishing strong and weak ties by using neighbourhood overlap to approximate tie strength in social recommendation. In a word, incorporating social contextual information indeed can improve the recommendation performance successfully.

## 2.2   Mention Behavior Modeling

Who-to-mention can be viewed as a recommendation task. For instance, Wang et al. [22] use ranking support vector regression to recommend the candidates with interest match, user relationship and social influence features. Tang et al. [21] employ ranking support vector machine as the solution by utilizing content, social, location and time features. To solve the mention overload problem, Zhou et al. [30] propose a personalized ranking model by considering multi-dimensional relations among users and tweets to generate the personalized mention list. Li et al. [10] utilize probabilistic factor graph model with mention relationship as edges and candidates as nodes to deal with overwhelmed information. Gong et al. [5] propose a topical translation-based method to predict the mentioned users by considering both content of microblog and histories of candidate users. Huang et al. [6] design an end-to-end memory network by incorporating users' interests with external memory. Ma et al. [14] propose a cross-attention memory network by using user's interests with external memory and the cross-attention mechanism to extract both textual and visual information.

Other works also tackle the problem as a classification prediction task. For example, Jiang et al. [8] use link prediction to predict mention behaviors by using user, textual, social tie and temporal information features. Similarly, Bao et al. [3] propose the response prediction and formulate it as a binary classification task by using three factors from structure, influence, and content. Besides, this problem can be modeled as an unbalance assignment problem using Hungarian method to find the optimal users in the appropriate time by Ding et al. [4].

Different from the above studies, this work proposes a novel context-aware mention recommendation model based probabilistic matrix factorization, which can incorporate topic relevance and mention affinity as well as homophily influence for mentioning the appropriate users.

## 3   Mention Recommendation Model

In this section, we first formulate the problem of mention recommendation based on Probabilistic Matrix Factorization. Then, we describe how to incorporate mention contextual factors, and discuss how we learn the hidden variables.

### 3.1   Mention Formulation by Probabilistic Matrix Factorization

Suppose that we have $M$ users and $N$ messages. Let $R \in \mathbb{R}^{M \times N}$ be the mentioning matrix, where the observed mentions are indicated by 1 values, and missing entries are assumed to be 0 values. $U \in \mathbb{R}^{K \times M}$ and $V \in \mathbb{R}^{K \times N}$ be user and message latent feature matrices respectively, where $K$ is the dimension of latent factors. The preference of user $u_i$ is represented by vector $U_i \in \mathbb{R}^{K \times 1}$ and the characteristic of message $m_j$ is represented by vector $V_j \in \mathbb{R}^{K \times 1}$. The dot product of $U_i$ and $V_j$ can approximate the mention behavior between $u_i$

and $m_j$: $\hat{R}_{ij} \approx U_i^T V_j$. Mention recommendation based on Probabilistic Matrix Factorization (PMF) [18] solve the following problem

$$\mathcal{L} = \min_{U,V} \sum_{i=1}^{M} \sum_{j=1}^{N} I_{ij}(R_{ij} - g(U_i^T V_j))^2 + \lambda(\|U\|_F^2 + \|V\|_F^2) \tag{1}$$

where $I_{ij}$ is an indicator function, $I_{ij}$ is 1 if $u_i$ is mentioned in $m_j$ and 0 otherwise. $g(x) = 1/(1+exp(-x))$ is the logistic function that maps $U_i^T V_j$ to (0,1), ($\|U\|_F^2 + \|V\|_F^2$) can avoid overfitting, $\|\cdot\|_F$ denotes the Frobenius norm of the matrix.

Since $R$ is highly sparse, it is impossible to accurately recommend the right users only rely on the observed mention behaviors. However, we argue that incorporating mention contextual factors (e.g., topic relevance, mention affinity, user profile similarity and message semantic similarity) can alleviate the data sparsity and improve the performance of mention recommendation. Based on these ideas, we propose a novel mention recommendation model.

### 3.2 Mention Contextual Factors

In this section, we introduce the social contextual factors of influenced mention behaviors. Here, we use the terms "mentioner" refer to the publisher of message, and "mentionee" refer to the possible mentioned user in a message.

**Modeling Message-Mentionee Topic Relevance Feature.** From the perspective of target user, one is likely to be accepted and retweeted the notification message if he/she is interested in the content of message, otherwise the message will be viewed as spam and be ignored. Hence, we argue that the topic relevance of message and user is an important factor in the mention decision-making process. Here, we denote the topic distribution of message $m_j$ as

$$T_t = (p(z_1|m_j), p(z_2|m_j), \cdots, p(z_k|m_j)) \tag{2}$$

where $p(z_i|m_j)$ can be learnt in training BTM model [26], which can solve the problem of sparse word co-occurrence patterns at document-level.

User's topic interests can be reflected in user-generated content. Due to data sparseness, we aggregate all short texts from the same user to form a long pseudo-document before performing BTM. The user $u_i$'s topic interests is defined as

$$T_{u_i} = (p(z_1|u_i), p(z_2|u_i), \cdots, p(z_k|u_i)) \tag{3}$$

where $z_i$ is the $i$-th topic interest of user $u_i$. Then, we use Jensen-Shannon divergence to measure topic distribution distinguishable between $u_i$ and $m_j$ as

$$JSD(T_{u_i}\|T_{m_j}) = \frac{1}{2}D(T_{u_i}\|\bar{T}) + \frac{1}{2}D(T_{m_j}\|\bar{T}) \tag{4}$$

where $T_{u_i}$ and $T_{m_j}$ are the topic distribution of user $u_i$ and message $m_j$ respectively, and $\bar{T}$ is the average result of $T_{u_i}$ and $T_{m_j}$. $D(\cdot\|\cdot)$ is the KL divergence and is calculated by $D(T_{u_i}||\bar{T}) = \sum_{k=1}^{K} log\frac{T_{u_i}(k)}{\bar{T}(k)}T_{u_i}(k)$. A smaller JSD value means a greater topic relevance between user and message, indicating the user is more likely to be interested in and retweet the message.

We introduce a user-message topic relevance matrix $W \in \mathbb{R}^{M \times N}$, which consists of $JSD(T_{u_i}||T_{m_j})$. and then force the matrix to the approximated predictions of the observed entries, controlled by their association strengths as

$$\mathcal{L}_1 = \min \sum_{i=1}^{M}\sum_{j=1}^{N} W_{ij}U_i^T V_j \tag{5}$$

In Eq. (5), a large value of $W_{ij}$ indicates message $m_j$ is strongly matching relation with the topic interests of user $u_i$, thus $u_i$ is more likely to be mentioned in $m_j$.

**Modeling Mentioner-Mentionee Mention Affinity Feature.** Mention can form a strong affinity relationship among users. For example, a user who accepted mention notifications from the same user is more likely to be mentioned again in the future. In this paper, we define the mention affinity from $u$ to $v$ as

$$A_{u \to v} = \frac{|\mathcal{N}(u \to v)|}{|\mathcal{N}(u)|} \tag{6}$$

where $\mathcal{N}(u)$ is the set of messages that user $u$ uses mention when posting and $|\mathcal{N}(u)|$ is the number of posting mention messages of user $u$ in the set $\mathcal{N}(u)$. $\mathcal{N}(u \to v)$ is the set of messages that user $u$ mentions user $v$ and $|\mathcal{N}(u \to v)|$ is the number of messages of user $u$ mention $v$ in the set $\mathcal{N}(u \to v)$.

Similarly, we also construct a user-user mention affinity matrix $S \in \mathbb{R}^{M \times N}$, which consists of the user-user pairs mention affinity score. To model the strength of mention preference, we force the mention affinity matrix to the approximated predictions of the observed entries, controlled by their mention strengths as

$$\mathcal{L}_2 = \min \sum_{i=1}^{M}\sum_{j=1}^{N} S_{ij}U_i^T V_j \tag{7}$$

Similarly, in Eq. (7), a large value of $S_{ij}$ indicates that user $u_i$ is strongly interacted with user $u_j$, thus $u_i$ is more likely to mention $u_j$ in the near future.

**Modeling Mentionee-Mentionee Profile Similarity Feature.** Our observation find that users with similar social status and similar topic interests are likely to be mentioned in the same message. We also assume that users are similar in hidden user space have similar preferences. The user's profile information consists of topic interest and social status. The user's topic interests can be profiled by the set of messages posted by users. The user's social status can be described by two aspects: one is the social features including the number of messages, friends, followers and mutual fans, and the other one is the behavior features like the average number of retweetings and comments per message.

We first use the same strategy and topic model to the vectorized user as the above descried. Next, we employ the cosine similarity to calculate the profile similarity score between the published user $u_i$ and the mentioned user $u_j$ as

$$S_{topic}(i,j) = \frac{\mathcal{U}(i)\mathcal{U}(j)}{\|\mathcal{U}(i)\| \, \|\mathcal{U}(j)\|} \tag{8}$$

where $\mathcal{U}(i) = <\mathcal{U}_i^{topic}, \mathcal{U}_i^{social}>$ is the combination vector of topic interests and social status. $\mathcal{U}_i^{topic}$ is the learned topic vector representations and $\mathcal{U}_i^{social}$ is the learned social status representations for user $u_i$, respectively. Additionally, we argue that incorporating user clustering module by their profile information can reduce the noisy data and improve the performance of mention recommendation. Here, to cluster users, we use the K-means clustering algorithm, and obtain the user cluster set $H(u)$ in observed spaces.

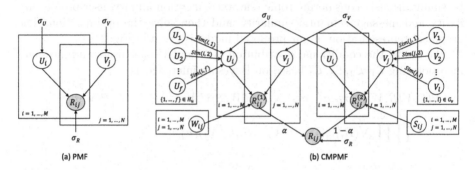

(a) PMF                                          (b) CMPMF

**Fig. 2.** The graphical models representation of probabilistic matrix factorization (PMF) and context-aware mention probabilistic matrix factorization (CMPMF).

We believe that two users with a similar profile are likely to be mentioned in the same message. To model a similar profile between $u_i$ and $u_f$, we force user's personal preferences $U_i$ and $U_f$ close to each other as

$$\mathcal{L}_3 = \min \sum_{i=1}^{M} \sum_{f \in H(u)} sim(i,f)\|U_i - U_f\|_F^2 \tag{9}$$

where a large value of $sim(i,f)$ indicates that user's personal preferences $U_i$ and $U_f$ should be very close, while a small value of $sim(i,f)$ indicates that the distance of user's personal preferences $U_i$ and $U_f$ could be large.

**Modeling Message-Message Semantic Similarity Feature.** Similarly, messages with similar topics are likely to mention the same users, and the messages similarities in observed spaces are consistent with the latent space. We use the same method to the vectorized message, and then using message clustering to reduce the noisy data and improve the performance of mention recommendation. We also use the K-means clustering algorithm to cluster messages, and generate the message cluster set $G(v)$ in observed spaces.

Two messages with a similar topic distributions are likely to mention the similar users. To model a similar topic distributions between $m_j$ and $m_l$, we force message's topic distributions $V_j$ and $V_l$ close to each other as

$$\mathcal{L}_4 = \min \sum_{j=1}^{N} \sum_{l \in G(v)} sim(j,l) \|V_j - V_l\|_F^2 \tag{10}$$

where a large value of $sim(j,l)$ indicates that message's semantic distances $V_j$ and $V_l$ should be very close, while a small value of $sim(j,l)$ indicates that the distance of message's semantic relationships $V_j$ and $V_l$ could be large.

**Context-Aware Mention Recommendation Ensemble.** Now, we design an integrated probabilistic matrix factorization model to find whom-to-mention by simultaneously considering topic relevance, mention affinity, user profile similarity and message semantic similarity, and then solve the optimization. The graphical representation of our proposed model is given in Fig. 2.

We model the conditional distribution of $U$ and $V$ over users and messages incorporating $\mathcal{L}_1, \mathcal{L}_2, \mathcal{L}_3, \mathcal{L}_4$ based on Bayesian inference as

$$P(U, V | R, W, S, \sigma_R^2, \sigma_U^2, \sigma_V^2) \propto P(R | W, S, U, V, \sigma_R^2) P(U | \sigma_U^2) P(V | \sigma_V^2)$$

$$= \prod_{i=1}^{M} \prod_{j=1}^{N} [\mathcal{N}(R_{ij} | g(W_{ij} U_i^T V_j + S_{ij} U_i^T V_j), \sigma_R^2)]^{I_{ij}}$$

$$\times \prod_{i=1}^{M} \prod_{f=1}^{H(u)} [\mathcal{N}(U_i | U_f, \sigma_U^2)] \times \prod_{j=1}^{N} \prod_{l=1}^{G(v)} [\mathcal{N}(V_j | V_l, \sigma_V^2)] \tag{11}$$

$$\times \prod_{i=1}^{M} \mathcal{N}(U_i | 0, \sigma_U^2 \mathbf{I}) \times \prod_{j=1}^{N} \mathcal{N}(V_j | 0, \sigma_V^2 \mathbf{I})$$

Maximizing the log-posterior distribution with respect to $U$, $V$, $W$ and $S$ is equivalent to minimizing the sum-of-of-squared errors function with quadratic regularization terms:

$$\mathcal{L} = \sum_{i=1}^{M} \sum_{j=1}^{N} I_{ij} \left\| R_{ij} - g(\alpha W_{ij} U_i^T V_j + (1-\alpha) S_{ij} U_i^T V_j) \right\|_F^2$$

$$+ \beta \sum_{i=1}^{M} \sum_{f \in H(u)} sim(i,f) \|U_i - U_f\|_F^2 \tag{12}$$

$$+ \gamma \sum_{j=1}^{N} \sum_{l \in G(v)} sim(j,l) \|V_j - V_l\|_F^2$$

$$+ \lambda (\|U\|_F^2 + \|V\|_F^2)$$

where $\beta = \frac{\sigma_R^2}{\sigma_U^2}$, $\gamma = \frac{\sigma_R^2}{\sigma_V^2}$. To simplify the model, we set $\beta = \gamma$ in the experiments.

We perform stochastic gradient descent approach to find the local minimum of Eq. (12) on feature vectors $U_i$ and $V_j$ as

$$\frac{\partial \mathcal{L}}{\partial U_i} = \sum_{j=1}^{N} I_{ij} g'(\alpha W_{ij} U_i^T V_j + (1 - \alpha) S_{ij} U_i^T V_j)$$

$$\times (g((\alpha W_{ij} + (1 - \alpha) S_{ij}) U_i^T V_j) - R_{ij}) \times (\alpha W_{ij} V_j + (1 - \alpha) S_{ij} V_j)$$

$$+ \beta \sum_{f \in H(u)} sim(i, f) \times (U_i - U_f) + \lambda U_i \tag{13}$$

$$\frac{\partial \mathcal{L}}{\partial V_j} = \sum_{i=1}^{M} I_{ij} g'(\alpha W_{ij} U_i^T V_j + (1 - \alpha) S_{ij} U_i^T V_j)$$

$$\times (g((\alpha W_{ij} + (1 - \alpha) S_{ij}) U_i^T V_j) - R_{ij}) \times (\alpha W_{ij} U_i + (1 - \alpha) S_{ij} U_i)$$

$$+ \gamma \sum_{l \in G(v)} sim(j, l) \times (V_j - V_l) + \lambda V_j \tag{14}$$

where $g'(x) = exp(x)/(1 + exp(x))^2$ is the derivative of logistic function.

**Table 1.** Statistics of the dataset.

| Dataset | #Users | #Messages | #Relations | #Mentions | Sparseness |
|---------|--------|-----------|------------|-----------|------------|
| Weibo | 11,925 | 208,274 | 5,368,253 | 2,382,052 | 0.1% |

# 4 Experiments and Analysis

## 4.1 Dataset and Setup

Weibo is one of the most popular social network platforms in China, and allows a user to mention other users in a message. In this paper, we use a publicly available Weibo dataset [29], which consists of user profile, message and the snapshot of network structure, etc. The user profile contains the number of friends, followers and messages, etc. The message contain the message content, the number of retweetings and comments, etc. The network has following relationship among users. Table 1 summarizes the detailed information of the used dataset.

We use Precision, Recall, and F-Score to evaluate the experimental results, and employ Hits@3 and Hits@5 to represent the percentage of correct results recommended from the top results. Moreover, we also use the Mean Reciprocal Rank (MRR) metrics to evaluate the rank of the recommended results.

## 4.2 Baseline Methods

We compare the proposed model with the following state-of-the-art baseline methods on the dataset.

- **Random Guess (RG):** RG randomly recommend the candidate users for each message. These candidates are chose from the friends by ordering the number of followers.
- **Majority Guess (MG):** This strategy is a simple but powerful method that the candidate user mentioned with a higher frequency by the author would have a higher recommendation probability.
- **PMF:** This method only uses user-message mention matrix for the mention recommendations based on the observed entries [18].
- **PMPR:** This model considers the mention recommendation as a probabilistic ranking problem to find the maximal possibility candidate by using probabilistic factor graph model in the heterogeneous social network [10].
- **CAR:** CAR uses a ranking support vector machine model by considering content, social, location and time based features to recommend the mentioned target users [21].
- **AU-HMNN:** The model introduces end-to-end memory network architecture by incorporating the textual information of query tweets and the history interests of the author and candidate users [6].

For our model, we empirically set $\alpha = 0.6$, $\beta = \gamma = \lambda = 0.01$, the number of latent factors is 100 and the number of iterations is 100, the number of clusters on users and messages are set to 40 and 80, respectively. The dimension of topic model is set to 100, and the number of iterations is set to 1000.

**Table 2.** Comparison of mention results with Precision, Recall, F-Score, MRR, Hits@3 and Hits@5 metrics on the dataset.

| Category | Method | Precision | Recall | F-Score | MRR | Hits@3 | Hits@5 |
|----------|--------|-----------|--------|---------|-----|--------|--------|
| Baselines | RG | 0.353 | 0.369 | 0.361 | 0.462 | 0.459 | 0.468 |
| | MG | 0.539 | 0.564 | 0.551 | 0.537 | 0.541 | 0.554 |
| | PMF | 0.644 | 0.648 | 0.646 | 0.643 | 0.658 | 0.669 |
| | PMPR | 0.708 | 0.716 | 0.712 | 0.717 | 0.735 | 0.767 |
| | CAR | 0.736 | 0.756 | 0.746 | 0.744 | 0.753 | 0.782 |
| | AU-HMNN | 0.774 | 0.784 | 0.779 | 0.753 | 0.775 | 0.786 |
| | CMPMF | **0.802** | **0.823** | **0.812** | **0.818** | **0.797** | **0.807** |
| Variants | TR-CMPMF | 0.723 | 0.702 | 0.712 | 0.728 | 0.735 | 0.747 |
| | UA-CMPMF | 0.733 | 0.716 | 0.724 | 0.738 | 0.751 | 0.762 |
| | PS-CMPMF | 0.682 | 0.677 | 0.679 | 0.693 | 0.729 | 0.736 |
| | SS-CMPMF | 0.656 | 0.643 | 0.649 | 0.662 | 0.678 | 0.715 |

## 4.3    Performance and Analysis

Table 2 with baselines category shows the comparisons of the proposed method with the state-of-the-art methods on the dataset. From the results, we can draw the following observations: (1) Our proposed model consistently achieves better performance than other baseline methods on the dataset, which indicates

the discovered mention contextual factors are effective for mention recommendation on social network; (2) MG can yield better results than RG, indicating
that the candidate user mentioned with a higher frequency by the publisher in
the past shares a good relationship and would have a higher mention probability in the future; (3) PMF performs even better than RG and MG, which
demonstrates that the framework of probabilistic matrix factorization is more
suited for the mention task; (4) PMPR and CAR significantly better than PMF.
The experimental results show the fact that incorporating mention contextual
factors into the learning algorithm indeed improve the performance of mention
recommendation; (5) AU-HMNN based on deep neural networks outperforms
most of baselines methods, which illustrates that neural network-based model
offers more benefit for mention recommendation. Particularly, the best results
of our proposed model for MRR and Hits@5 are relatively greater performance.
Hence, by incorporating topic relevance, mention affinity, user profile similarity
and message semantic similarity, our proposed model indeed performs well on
the mention recommendation task. Taken together, these results suggest that
our proposed model can achieve the best performance by considering mention
contextual factors under probabilistic matrix factorization.

We implement the different variants of our proposed model to demonstrate
the effectiveness of our proposed algorithm in Table 2 with variants category:
(1) TR-CMPMF only considers the topic relevance feature of messages and the
candidate users; (2) UA-CMPMF only incorporates the mention affinity feature
between publishers and the candidate users; (3) PS-CMPMF only introduces
the profile similarity feature among the candidate users; (4) SS-CMPMF only
models the semantic similarity feature among messages. From the comparison
results, we can conclude that UA-CMPMF achieves a relatively better result
than other three variants of our proposed model. The comparison of TR-CMPMF
and UA-CMPMF shows that the mention affinity strength of both users is more
important factor than the matching relevance of topic interests while making
decision for whom to mention. TR-CMPMF has larger boost than PS-CMPMF
and SS-CMPMF in term of F-Score, indicating that the topic relevance is still
an important consideration factor for mention recommendation. In contrast to
the results of SS-CMPMF, PS-CMPMF could generate better results, indicating
that the user profile information is an important consideration. The further
explanation of the phenomenon is that the main purpose of mention is to expand
the scale of information diffusion, thus the user profile information are more
taken into account than semantic information information. In summary, from
the results of TR-CMPMF, UA-CMPMF, PS-CMPMF and SS-CMPMF, we
can observe that treating the topic relevance and mention affinity as the primary
domain of factors and the profile similarity and semantic similarity as the second
domain of factors can significantly improve the performance.

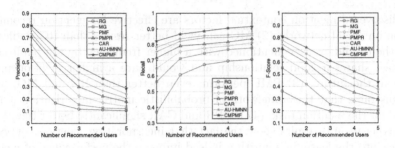

**Fig. 3.** Precision, Recall and F-score with different number of recommended users.

## 4.4  Effect of Mention Count

Figure 3 shows our proposed model and the baseline models with different numbers of recommended users, varying from 1 to 5. From the figure, we can see that (1) our proposed model achieves consistently better performance than the other methods with different number of recommendations; (2) with the number of recommended users increasing, Precision decreases and Recall increases gradually, indicating while the number of mentions in a message are neither too much nor too little, the recommendation achieves the best performance. It is a reasonable explanation that each message to be mentioned with a small number of users (e.g., one to three persons) would be viewed as a intimate chat with close friends. Otherwise, it may lead to potential mention overload and be considered as a spam when a lot of users are mentioned. Moreover, we observe that the best performance of mention recommendation is obtained in term of F-Score when recommending the top one user. It is also noticeable that our proposed method is significantly better than the state-of-the-art methods.

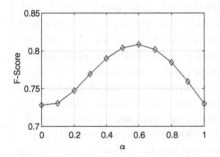

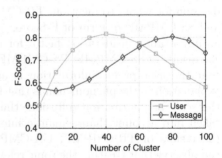

**Fig. 4.** Performance of our model under different values of $\alpha$.

**Fig. 5.** Performance of our model under different cluster on user and message.

**Table 3.** Performances of our proposed model with different number of topics

| Number of topics | Precision | Recall | F-Score |
|---|---|---|---|
| 10 | 0.572 | 0.773 | 0.659 |
| 50 | 0.703 | 0.753 | 0.723 |
| 100 | **0.802** | **0.823** | **0.812** |
| 500 | 0.722 | 0.743 | 0.732 |

### 4.5   Effect of Parameters

In this paper, the proposed model contains several critical parameters. Here, we analyze the effect of these parameters from the following aspects.

**Impact of $\alpha$.** The parameter $\alpha$ is a tunable weight to balance the strength of topic relevance and mention affinity. Figure 4 plots the performance of our proposed model with various values of $\alpha$. From the figure, we observe that the plot first gradually rises and then drops gradually while the values of $\alpha$ increasing. In particular, $\alpha = 0$ indicates that our model only incorporates mention affinity feature, and $\alpha = 1$ indicates that our method only considers topic relevance. It is clear that the best performance is achieved when $\alpha$ is 0.6.

**Number of Topic.** Table 3 shows how the topic latent features affect the mention performance. From the results, we can see that our proposed model achieves better performance as the number of topics increasing. Clearly, the optimal value of topic latent features can be observed at 100 dimensions.

**Number of Clustering.** We also try to investigate how number of clusters on users and messages affect the performance of mention recommendation. Evaluation experiments are recorded at different points as shown in Fig. 5. From the result, we can see that (1) the dataset have a optimal value for number of clusters on users and messages as the number of clusters enlarging; (2) the best number of user and message clusters is achieved around 40 and 80, respectively. It is reasonable in reality that users with the similar preferences and daily habits are likely to be friends and form a community, and messages with the similar semantic hold the same distribution.

## 5   Conclusion

In this paper, we propose a novel mention recommendation model by incorporating topic relevance, mention affinity, user profile similarity and message semantic similarity. We use topic relevance to learn how the candidate users are interested in the message, and measure the strength of mention affinity based on mention histories, and quantify the distance of user profile and message semantic similarities based on homophily influence. We use these mention contextual factors to constrain objective function under the framework of probabilistic matrix factorization for mention recommendation task. To demonstrate the effective of our model, we construct extensive experiments. The experimental results reveal that our proposed method can outperform the state-of-the-art baseline methods.

**Acknowledgments.** This work is supported by Natural Science Foundation of China (No. 61702508, No. 61802404), and Strategic Priority Research Program of CAS (XDC02040100, XDC02030200). This work is also partially supported by Key Laboratory of Network Assessment Technology, Chinese Academy of Sciences and Beijing Key Laboratory of Network security and Protection Technology.

# References

1. Abdullah, N.A., Nishioka, D., Tanaka, Y., Murayama, Y.: User's action and decision making of retweet messages towards reducing misinformation spread during disaster. J. Inf. Process. **23**(1), 31–40 (2015)
2. Bakshy, E., Hofman, J.M., Mason, W.A., Watts, D.J.: Everyone's an influencer: quantifying influence on Twitter. In: WSDM, pp. 65–74. ACM (2011)
3. Bao, P., Shen, H.-W., Huang, J., Chen, H.: Mention effect in information diffusion on a micro-blogging network. PloS One **13**(3), e0194192 (2018)
4. Ding, Z., et al.: Mentioning the optimal users in the appropriate time on Twitter. In: Li, F., Shim, K., Zheng, K., Liu, G. (eds.) APWeb 2016. LNCS, vol. 9932, pp. 464–468. Springer, Cham (2016). https://doi.org/10.1007/978-3-319-45817-5_47
5. Gong, Y., Zhang, Q., Sun, X., Huang, X.: Who will you@? In: CIKM, pp. 533–542. ACM (2015)
6. Huang, H., Zhang, Q., Huang, X., et al.: Mention recommendation for Twitter with end-to-end memory network. In: IJCAI, pp. 1872–1878 (2017)
7. Jamali, M., Ester, M.: A matrix factorization technique with trust propagation for recommendation in social networks. In: RecSys, pp. 135–142 (2010)
8. Jiang, B., Sha, Y., Wang, L.: Predicting user mention behavior in social networks. In: Li, J., Ji, H., Zhao, D., Feng, Y. (eds.) NLPCC 2015. LNCS (LNAI), vol. 9362, pp. 146–158. Springer, Cham (2015). https://doi.org/10.1007/978-3-319-25207-0_13
9. Jiang, M., et al.: Social contextual recommendation. In: CIKM, pp. 45–54 (2012)
10. Li, Q., Song, D., Liao, L., Liu, L.: Personalized mention probabilistic ranking – recommendation on mention behavior of heterogeneous social network. In: Xiao, X., Zhang, Z. (eds.) WAIM 2015. LNCS, vol. 9391, pp. 41–52. Springer, Cham (2015). https://doi.org/10.1007/978-3-319-23531-8_4
11. Ma, H., Yang, H., Lyu, M.R., King, I.: SoRec: social recommendation using probabilistic matrix factorization. Comput. Intell. **28**(3), 289–328 (2008)
12. Ma, H., King, I., Lyu, M.R.: Learning to recommend with social trust ensemble. In: SIGIR, pp. 203–210 (2009)
13. Ma, H., Zhou, D., Liu, C., Lyu, M.R., King, I.: Recommender systems with social regularization. In: WSDM, pp. 287–296. ACM (2011)
14. Ma, R., Zhang, Q., Wang, J., Cui, L., Huang, X.: Mention recommendation for multimodal microblog with cross-attention memory network. In: SIGIR, pp. 195–204. ACM (2018)
15. Metaxas, P., Mustafaraj, E., Wong, K., Zeng, L., O'Keefe, M., Finn, S.: What do retweets indicate? Results from user survey and meta-review of research. In: ICWSM (2015)

16. Michelson, M., Macskassy, S.A.: Discovering users' topics of interest on Twitter: a first look. In: AND, pp. 73–80. ACM (2010)
17. Pramanik, S., et al.: On the role of mentions on tweet virality. In: DSAA, pp. 204–213. IEEE (2016)
18. Salakhutdinov, R., Mnih, A.: Probabilistic matrix factorization. In: NIPS, pp. 1257–1264 (2007)
19. Tang, J., Gao, H., Liu, H.: mTrust: discerning multi-faceted trust in a connected world. In: WSDM, pp. 93–102. ACM (2012)
20. Tang, J., et al.: Recommendation with social dimensions. In: AAAI, pp. 251–257 (2016)
21. Tang, L., Ni, Z., Xiong, H., Zhu, H.: Locating targets through mention in twitter. World Wide Web **18**(4), 1019–1049 (2015)
22. Wang, B., et al.: Whom to mention: expand the diffusion of tweets by@ recommendation on micro-blogging systems. In: WWW, pp. 1331–1340. ACM (2013)
23. Wang, X., Hoi, S.C.H., Ester, M., Bu, J., Chen, C.: Learning personalized preference of strong and weak ties for social recommendation. In: WWW, pp. 1601–1610 (2017)
24. Wang, X., Lu, W., Ester, M., Wang, C., Chen, C.: Social recommendation with strong and weak ties. In: CIKM, pp. 5–14. ACM (2016)
25. Wu, S., Hofman, J.M., Mason, W.A., Watts, D.J.: Who says what to whom on Twitter. In: WWW, pp. 705–714. ACM (2011)
26. Yan, X., Guo, J., Lan, Y., Cheng, X.: A biterm topic model for short texts. In: WWW, pp. 1445–1456. ACM (2013)
27. Yang, J., Counts, S.: Predicting the speed, scale, and range of information diffusion in Twitter. In: ICWSM, vol. 10, pp. 355–358 (2010)
28. Ye, S., Wu, S.F.: Measuring message propagation and social influence on Twitter.com. In: Bolc, L., Makowski, M., Wierzbicki, A. (eds.) SocInfo 2010. LNCS, vol. 6430, pp. 216–231. Springer, Heidelberg (2010). https://doi.org/10.1007/978-3-642-16567-2_16
29. Zhang, J., Tang, J., Li, J., Liu, Y., Xing, C.: Who influenced you? Predicting retweet via social influence locality. TKDD **9**(3), 25 (2015)
30. Zhou, G., Yu, L., Zhang, C.-X., Liu, C., Zhang, Z.-K., Zhang, J.: A novel approach for generating personalized mention list on micro-blogging system. In: ICDMW, pp. 1368–1374. IEEE (2015)

# Synchronization Under Control in Complex Networks for a Panic Model

Guillaume Cantin[(✉)][ID], Nathalie Verdière[ID], and Valentina Lanza[ID]

Laboratoire de Mathématiques Appliquées du Havre, Normandie Univ,
FR CNRS 3335, ISCN, 76600 Le Havre, France
guillaumecantin@mail.com

**Abstract.** After a sudden catastrophic event occurring in a population of individuals, panic can spread, persist and become more problematic than the catastrophe itself. In this paper, we explore through a computational approach the possibility to control the panic level in complex networks built with a recent behavioral model. After stating a rigorous theoretical framework, we propose a numerical investigation in order to establish the effect of the topology of the network on this control process, with randomly generated networks, and we compare the panic level for two distinct topology sets on a given network.

**Keywords:** Optimal control · Numerical computation ·
Dynamical system · Complex network · Synchronization · Panic

## 1 Introduction

The aim of this paper is to explore the possibility to control the panic spreading within a population of individuals facing a catastrophic event, using a recent mathematical modeling called the Panic-Control-Reflex system (PCR system) [5,9]. This modeling is given by a set of ordinary differential equations and reproduces the behavioral process from reflex to control behavior, with the eventuality to transit through panic, and possibly to exhibit a persistence of panic. The geographical background of the areas impacted by catastrophic events naturally leads us to consider complex networks of PCR systems [4,6], that is, geographical networks whose nodes are coupled with multiple instances of a PCR system, with connections between those nodes, corresponding to physical displacements. In [4], it is proved that the evacuation of high risk zones towards refuge zones is a necessary and sufficient condition for the whole population in the network to return to a daily behavior, and to avoid a persistence of panic. But this necessary evacuation can be awkward in some particular places, or even impossible.

This work has been supported by the French government, through the National Research Agency (ANR) under the Societal Challenge 9 "Freedom and security of Europe, its citizens and resident" with the reference number ANR-17-CE39-0008, and the UCA-JEDI Investments in the Future project managed by the National Research Agency (ANR) with the reference number ANR-15-IDEX-01.

© Springer Nature Switzerland AG 2019
J. M. F. Rodrigues et al. (Eds.): ICCS 2019, LNCS 11537, pp. 262–275, 2019.
https://doi.org/10.1007/978-3-030-22741-8_19

Thus it is natural to ask if it is possible to reach a synchronization state in the network, which would guaranty that panic vanishes. This question is of great interest, since it is related to the more general problem of controlling and decision making in complex systems [1,3]. To this end, we consider the superposition of two controls, exerted on each node and on each edge in the network. Here, we propose to investigate with a computational approach the effect of the topology on the control process. In the first part, we show how to construct a complex network of PCR systems, and illustrate by a computation of randomly generated networks the risk of panic persistence. In the second part, we set the general control problem together with its performance criterion, and finally, we present a comparison of two computations obtained for two distinct topology sets of a given network.

## 2 Non Identical PCR Networks

In this section, we briefly present the Panic-Control-Reflex system (PCR), show how to construct a complex network of PCR systems, and illustrate by a computation of 800 randomly generated networks the risk of panic spreading.

### 2.1 Panic-Control-Reflex System

The Panic-Control-Reflex system (PCR system) is a mathematical model for human behaviors during catastrophic events, developed with the collaboration of geographers in order to better understand, predict and control the behavioral reactions of individuals facing a brutal disaster [5,9]. It is given by the following system of ordinary differential equations

$$\begin{cases} \dot{r} = \gamma(t)q(r_m - r) - Br + f(r,c)rc + g(r,p)rp \\ \dot{c} = B_1 r - C_2 c + C_1 p - f(r,c)rc + h(c,p)cp - \varphi(t)c(b_m - b) \\ \dot{p} = B_2 r + C_2 c - C_1 p - g(r,p)rp - h(c,p)cp \\ \dot{q} = -\gamma(t)q(r_m - r) \\ \dot{b} = +\varphi(t)c(b_m - b), \end{cases} \tag{1}$$

where the unknowns $r$, $c$, $p$, $q$, $b$ are real-valued functions defined on $\mathbb{R}$, which model the densities of individuals in *reflex, control, panic, daily* and *back to daily* behaviors respectively. The parameters $B_1 > 0$, $B_2 > 0$, $B = B_1 + B_2$, $C_1 \geq 0$, $C_2 \geq 0$, $r_m > 0$, and $b_m > 0$ are real coefficients, $\gamma$, $\varphi$ are smooth functions of $t$ with positive values, $f$, $g$, $h$ smooth functions defined on $\mathbb{R}^2$ with values in $\mathbb{R}$.

When the catastrophic event occurs, individuals are brought to the reflex behavior; this evolution is modeled by the non-linear term $\gamma(t)q(r_m - r)$, in which $\gamma(t)$ corresponds to the impact of the catastrophe. Next, individuals are subject to a behavioral evolution towards control behavior or panic; this evolution is modeled by the linear terms $B_1 r$ and $B_2 r$. Additionally, contagion phenomena can act in parallel between the 3 main behavioral subgroups (reflex, control behavior, panic); those contagion phenomena are modeled by the non-linear terms $f(r, c)rc$, $g(r, p)rp$ and $h(c, p)cp$. We emphasize that the functions $f$, $g$ and $h$ have been designed to change their signs according to the values of the proportions $\frac{r}{c}$, $\frac{r}{p}$ and $\frac{c}{p}$ [5,9]. For instance, if the density of individuals in panic is widely greater than the density of individuals in control behavior, then $h(c, p) < 0$, which means that the contagion brings individuals in control behavior to imitate individuals in panic. In the mean time, evolution between panic and control behavior is modeled by the linear terms $C_1p$, $C_2c$. Finally, the return to daily behavior operates from the control behavior; it is modeled by the non-linear term $\varphi(t)c(b_m - b)$.

*Remark 1.* The PCR system has been considered for simulations of concrete scenarios of catastrophic events, thanks to a narrow collaboration with geographers and psychologists. The example of an earthquake in Japan is studied in [9], and the risk of tsunami on the Mediterranean coast is presented in [6]. Catastrophic events of industrial origin are also of great interest.

The parameter $r_m$ models the maximum capacity of individuals which can be in reflex behavior. Without loss of generality, we will set $r_m = 1$ in the rest of the paper. The parameter $b_m$ models the maximum capacity of individuals which can return to the daily behavior. We assume that this maximum capacity coincides with the total population $\Lambda = r + c + p + q + b$ involved in the catastrophic event, thus we can reduce system (1) to a 4 equations system

$$\dot{x} = \psi(t, x), \tag{2}$$

where $x = (r, c, p, q)^T$ and

$$\psi(t, x) = \begin{pmatrix} \gamma(t)q(1 - r) - Br + f(r, c)rc + g(r, p)rp \\ B_1r - C_2c + C_1p - f(r, c)rc + h(c, p)cp - \varphi(t)c(r + c + p + q) \\ B_2r + C_2c - C_1p - g(r, p)rp - h(c, p)cp \\ -\gamma(t)q(1 - r) \end{pmatrix}.$$

The following Theorem summaries its dynamics, and highlights the decisive role of the parameter $C_1$ which models the evolution from panic to control behavior. The proof is detailed in [5].

**Theorem 1.** *For any initial condition $x_0 \in (\mathbb{R}^+)^4$, the initial value problem*

$$\begin{cases} \dot{x} = \psi(t, x), & t > 0, \\ x(0) = x_0, \end{cases}$$

*admits a unique global solution whose components are non-negative and bounded. If $C_1 > 0$, then the trivial equilibrium $0 \in \mathbb{R}^4$ is the only equilibrium, and it is globally asymptotically stable. If $C_1 = 0$, then the solution of system (2) starting from any initial condition $(r_0, c_0, p_0, q_0)$ such that $r_0 + c_0 + p_0 + q_0 > 0$ presents a persistence of panic, that is*

$$\lim_{t \to +\infty} p(t) = \bar{p} > 0.$$

## 2.2   Complex Networks of Non-identical PCR Systems

When considering the geographical relief of the zone impacted by the catastrophic event, it is natural to improve the previous modeling by a spatial modeling. One way is to construct a complex network whose nodes are coupled with multiple instances of the PCR system [4,6]. Let us consider a simple graph $\mathscr{G} = (\mathscr{V}, \mathscr{E})$ made with a finite set $\mathscr{V} = \{1, \ldots, n\}$ of $n$ vertices, where $n$ is a positive integer, and a finite set $\mathscr{E} = \{e_1, \ldots, e_k\}$ of $k$ weighted edges, with non-negative weights $\varepsilon_1, \ldots, \varepsilon_k$. For each integer $l \in \{1, \ldots, k\}$, there exists a unique pair of vertices $(i, j)$ such that $e_l$ connects vertex $i$ towards vertex $j$. We set $\varepsilon = (\varepsilon_1, \ldots, \varepsilon_k) \in (\mathbb{R}^+)^k$, and introduce the matrix of connectivity $L(\varepsilon)$ of order $n$, whose off-diagonal coefficients are given by

$$
L_{ji}(\varepsilon) = \begin{cases} \varepsilon_l \text{ if } e_l = (i, j) \in \mathscr{E}, \\ 0 \text{ else}, \end{cases}
$$

and whose diagonal coefficients satisfy

$$
L_{ii}(\varepsilon) = - \sum_{j \neq i} L_{ji}(\varepsilon).
$$

Next we couple each node in the graph with an instance of the PCR system (2). Thus we set

$$
x_i = (r_i, c_i, p_i, q_i)^T, \quad X = (x_1, \ldots, x_n)^T,
$$
$$
H = \operatorname{diag}\{1, 1, 1, 0\}, \quad HX = (Hx_1, \ldots, Hx_n)^T. \tag{3}
$$

The definition of the matrix $H$ means that individuals in daily behavior $q$ are not concerned with migrations in the network. We allow the different instances of system (2) to admit different values of parameters, and we will especially focus on the effect of coupling PCR systems with different values of the parameter $C_1$, identified previously as a bifurcation parameter.

**Definition 1.** *We will call* node of type (1) *a node coupled with an instance of the PCR system such that $C_1 = 0$, and* node of type (2) *a node coupled with an instance of the PCR system such that $C_1 > 0$.*

A PCR network is given by

$$
\dot{X} = \Psi(t, X) + L(\varepsilon) HX, \tag{4}
$$

where $\Psi(t, X) = \left(\psi^{(1)}(t, x_1), \ldots, \psi^{(n)}(t, x_n)\right)^T$. The above index in $\psi^{(i)}(t, x_i)$, $1 \leq i \leq n$, indicates that the values of parameter $C_1$ can differ from one node in the network to another. The next theorem, presented in [4], establishes a necessary and sufficient condition for the solution of the PCR network (4) to converge to the trivial equilibrium, which corresponds to a global return of all individuals to the daily behavior. It is also a condition for *synchronization* in the network, since every node exhibits the same asymptotic dynamics under the considered assumptions.

**Theorem 2.** *For any initial condition* $X_0 \in (\mathbb{R}^+)^{4n}$, *the network problem*

$$\begin{cases} \dot{X} = \Psi(t, X) + L(\varepsilon)\,HX, \quad t > 0, \\ X(0) = X_0, \end{cases}$$

*admits a unique solution whose components are non negative. The trivial equilibrium* $0 \in \mathbb{R}^{4n}$ *is the only equilibrium if and only if every node of type (1) is connected to at least one node of type (2) by an oriented chain. It that case, the trivial equilibrium is globally asymptotically stable.*

## 2.3 Computation of the Panic Level for Randomly Generated Networks

In this section, we aim to illustrate, deepen and qualify Theorem 2 with a computational approach. Thus we have generated 800 PCR networks, built with 15 nodes of type (1), 15 nodes of type (2), and $N_e$ randomly chosen edges ($0 \leq N_e \leq 150$). This random generation can yield a great variety of topology disposals (see Fig. 1). The number of isolated nodes of type (1) is computed by running a path-finding algorithm implemented in the `networkx` library [2] of the `python` language. Meanwhile, the panic level $\bar{P}$ in the whole network is computed by integrating the system of ordinary differential equations (4) on a finite time interval $[0, 60]$, using a Runge-Kutta scheme of order 4 (each PCR network corresponds to a system of 120 equations).

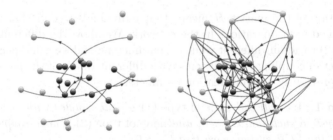

**Fig. 1.** Randomly generated PCR networks, built with 15 nodes of type (1) (depicted in red), 15 nodes of type (2) (depicted in green), and $N_e$ randomly chosen edges (with $0 \leq N_e \leq 150$). Left: weakly dense topology. Right: highly dense topology. (Color figure online)

The numerical computation has been performed on the server of the Laboratory of Applied Mathematics at University of Le Havre Normandie, in a GNU/Linux environment. The values of the parameters are $B_1 = B_2 = 0.5$,

$C_1 = 0.3$, $C_2 = 0.2$, and the imitation functions $f$, $g$, $h$ are given by

$$f(r, c) = -\alpha_1 \xi \left( \frac{r}{c + \nu_0} \right) + \alpha_2 \xi \left( \frac{c}{r + \nu_0} \right),$$

$$g(r, p) = -\delta_1 \xi \left( \frac{r}{p + \nu_0} \right) + \delta_2 \xi \left( \frac{p}{r + \nu_0} \right),$$

$$h(c, p) = -\mu_1 \xi \left( \frac{c}{p + \nu_0} \right) + \mu_2 \xi \left( \frac{p}{c + \nu_0} \right),$$

with $\alpha_i = \delta_i = \mu_i = 0.1$, $i \in \{1, 2\}$, $\nu_0 = 10^{-2}$ and

$$\xi(s) = \begin{cases} 1 & \text{if } s < 0, \\ 0 & \text{if } s > 1, \\ \frac{1}{2} + \frac{1}{2} \cos (\pi s) & \text{else.} \end{cases}$$

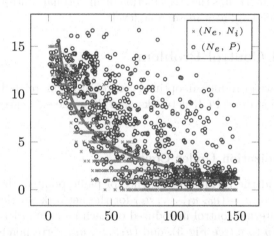

**Fig. 2.** Numerical results for the computation of 800 randomly generated PCR networks. For each randomly generated network, $N_e$ denotes the number of edges, $N_i$ the number of isolated nodes of type (1), and $\bar{P}$ the panic level in the network after a finite time. The red crosses have coordinates $(N_e, N_i)$, and the blue circles have coordinates $(N_e, \bar{P})$. The gray line corresponds to the approximation of the cloud of blue circles by an heuristic inverse power law of the form $\bar{P} = \frac{k_1}{N_e^\nu} - k_2$. (Color figure online)

The results are depicted in Fig. 2. For each randomly generated network, $N_e$ denotes the number of edges, $N_i$ the number of isolated nodes of type (1), and $\bar{P}$ the panic level in the network after a finite time $T$, defined by

$$\bar{P} = \sum_{i=1}^{n} p_i(T).$$

The red crosses have coordinates $(N_e, N_i)$, and the blue circles have coordinates $(N_e, \bar{P})$. Obviously, 15 edges could be sufficient to evacuate 15 nodes of type (1) towards nodes of type (2). However, we remark that the number of isolated nodes of type (1) can be relatively high, even with a dense topology. For example, some of the generated networks admit 70 edges and 5 isolated nodes of type (1). Furthermore, the panic level can remain high after a large time, even if the number of evacuated nodes of type (1) is law, which means that the asymptotic result of Theorem 2 has to be qualified. In other words, a dense topology is not an absolute warranty for the network to converge to the trivial equilibrium that corresponds to the return of all individuals to a daily behavior. Finally, the cloud of blue circles can be approximated by an heuristic inverse power law (see gray curve in Fig. 2) of the form

$$\bar{P} = \frac{k_1}{N_e^\nu} - k_2,$$

with positive coefficients $k_1$, $k_2$, $\nu$, which can be used for prediction. This numerical computation motivates the introduction of an optimal control process in order to limit the panic level at a reasonable level.

## 3 Optimal Control Problem

In this section, we introduce an optimal control problem, related to the expected synchronization state of PCR networks, and show the existence of a solution to that problem.

### 3.1 Synchronization Under Control

In order to avoid the possible persistence of panic pointed above, we consider a multiple control $u = (u_0, u_1, \ldots, u_k)$ for the network problem (4), in which $u_0$ models an internal control introduced on each node in order to facilitate the evolution from $p$ to $c$ (see Fig. 3), and $(u_1, \ldots, u_k)$ corresponds to an external control exerted in order to increase the coupling strength along each edge in the network. Thus we consider the following general control problem

$$\dot{X} = \tilde{\Psi}(t, X, u) + \tilde{L}(\varepsilon, u)\, HX, \tag{5}$$

where $\tilde{\Psi}(t, X, u) = \left(\tilde{\psi}(t, x_1, u), \ldots, \tilde{\psi}(t, x_n, u)\right)^T$, with

$$\tilde{\psi}(t, x, u) = \begin{pmatrix} -Br + \gamma(t)q(1-r) + f(r,c)rc + g(r,p)rp \\ B_1 r - C_2 c + (C_1 + u_0)p - f(r,c)rc + h(c,p)cp - \varphi(t)c\theta \\ B_2 r + C_2 c - (C_1 + u_0)p - g(r,p)rp - h(c,p)cp \\ -\gamma(t)q(1-r) \end{pmatrix},$$

where $\theta = r + c + p + q$ and the matrix $\tilde{L}(\varepsilon, u)$ is defined by

$$\tilde{L}(\varepsilon, u) = L(\varepsilon_1 + u_1, \ldots, \varepsilon_k + u_k). \tag{6}$$

The internal control $u_0$ appears in the system through the term $(C_1 + u_0)p$, whereas the external controls $(u_1, \ldots, u_k)$ appear through the coupling terms $(\varepsilon_i + u_i)r_i$, $(\varepsilon_i + u_i)c_i$, $(\varepsilon_i + u_i)p_i$, $1 \le i \le k$.

*Remark 2.* For the internal control $u_0$, awareness campaigns can be organized in the areas for which a high potential of catastrophic event is identified; those campaigns can be integrated to the educational programs to better prepare individuals to the known risks. External controls $u_1, \ldots, u_k$ can be made through rescue services, in order to clear some hindered avenue, or to repair any urban installation. We emphasize that it is a work in progress, in collaboration with geographers and psychologists, to establish an exhaustive list of possible actions in concordance with the mathematical control functions $(u_0, u_1, \ldots, u_k)$.

The aim of introducing the control $u = (u_0, u_1, \ldots, u_k)$ is to limit panic at a reasonable level in complex networks for which we can predict, using Theorem 2, a persistence of panic on nodes of type (1). However, we will see below that the solution of the control problem satisfies $q_i(t) > 0$ for all $t > 0$, $1 \le i \le n$, and for each initial condition such that $q_i(0) > 0$, $1 \le i \le n$ (see Eq. (10)). This demonstrates that the trivial equilibrium $0 \in \mathbb{R}^{4n}$ cannot be reached in a finite time, when starting from such initial conditions. A more pragmatic goal would be to reach a neighborhood $\mathcal{N}$ of the trivial equilibrium. For instance, we can look for a multiple control so that the panic level is limited under a given proportion of the total population after a finite time.

## 3.2  Performance Criterion

In what follows, we denote by $\mathcal{U}$ the set of admissible control functions, composed with Lebesgue-integrable functions $u = (u_0, u_1, \ldots, u_k)$ for which there exists $T > 0$ such that $u$ is defined on $[0, T]$ with values in $K = [0, 1]^{k+1}$. Note that $T$ may depend on $u$. Let $u = (u_0, u_1, \ldots, u_k) \in \mathcal{U}$ denote an admissible control for the general control problem (5). Applying this multiple control, we aim to reach a neighborhood $\mathcal{N}$ of the trivial equilibrium $0 \in \mathbb{R}^{4n}$ which corresponds to a synchronization state of the network. Additionally, we would like to minimize on $\mathcal{U}$ the performance index:

$$J(X_0, u, T) = \int_0^T \left[ \sum_{i=1}^n p_i^2(t) + u_0^2(t) + \sum_{l=1}^k u_l^2(t) \right] dt, \tag{7}$$

which models the wish to limit the level of panic during the control process, while mobilizing the less rescue services to operate during the catastrophic event.

Finally, we can state the optimal control problem for the complex network of non-identical PCR systems. The problem is to find a pair $(X, u)$ defined on some interval $[0, T]$, such that $u \in \mathcal{U}$ and

$$\begin{cases} \dot{X} = \tilde{\Psi}(t, X, u) + \tilde{L}(\varepsilon, u) HX, & t > 0, \\ X(0) = X_0, \quad X(T) \in \mathcal{N}, \\ \min_{u \in \mathcal{U}} J(X_0, u, T), \end{cases} \tag{8}$$

where $X_0$ is a given initial datum in $(\mathbb{R}^+)^{4n}$, and $\mathcal{N}$ denotes a neighborhood of the trivial equilibrium $0 \in \mathbb{R}^{4n}$.

## 3.3   Existence of an Optimal Control

**Theorem 3.** *Let $u = (u_0, u_1, \ldots, u_k)$ denote a multiple control with non-negative values. We assume that $u$ is continuous in $t$ and bounded. Then for any $X_0 \in (\mathbb{R}^+)^{4n}$, the initial value problem*

$$\begin{cases} \dot{X} = \tilde{\Psi}(t, X, u) + \tilde{L}(\varepsilon, u) H X, & t > 0, \\ X(0) = X_0, \end{cases} \tag{9}$$

*admits a unique global solution whose components are non-negative and bounded. Furthermore, the optimal control problem (8) for the complex network of PCR systems admits a solution $(X_0, u^*)$, with $u^* \in \mathcal{U}$ minimizing (7).*

*Proof.* Given any initial condition $X_0 \in (\mathbb{R}^+)^{4n}$, existence and uniqueness of a local in time solution $X(t, X_0)$ to problem (9), defined on some interval $[0, \tau]$ with $\tau > 0$, are a straightforward consequence of Cauchy-Lipschitz Theorem [8].

*Non-negativity.* In order to prove the non-negativity property, we introduce a modified problem as follows. For $\hat{x} = (\hat{r}, \hat{c}, \hat{p}, \hat{q}) \in \mathbb{R}^4$, define

$$\hat{\psi}(\hat{x}, u) = \left( \hat{\psi}_1(\hat{x}, u), \hat{\psi}_2(\hat{x}, u), \hat{\psi}_3(\hat{x}, u), \hat{\psi}_4(\hat{x}, u) \right)^T \text{ with}$$

$$\hat{\psi}_1(\hat{x}, u) = +\gamma \hat{q}(1 - \hat{r}) - B\hat{r} + f(\hat{r}, \hat{c})\hat{r}\hat{c} + g(\hat{r}, \hat{p})\hat{r}\hat{p}$$

$$\hat{\psi}_2(\hat{x}, u) = B_1\hat{r} - C_2\hat{c} + (C_1 + u_0)|\hat{p}| - f(\hat{r}, \hat{c})\hat{r}\hat{c} + h(\hat{c}, \hat{p})\hat{c}\hat{p} - \varphi\hat{c}\hat{\theta}$$

$$\hat{\psi}_3(\hat{x}, u) = B_2\hat{r} + C_2\hat{c} - (C_1 + u_0)\hat{p} - g(\hat{r}, \hat{p})\hat{r}\hat{p} - h(\hat{c}, \hat{p})\hat{c}\hat{p}$$

$$\hat{\psi}_4(\hat{x}, u) = -\gamma \hat{q}(1 - \hat{r}),$$

where $\hat{\theta} = \hat{r} + \hat{c} + \hat{p} + \hat{q}$ (we omit the dependence in $t$ in order to lighten our notations). Next, for $\hat{X} = (\hat{x}_1, \ldots, \hat{x}_n)$, consider the modified network problem

$$\dot{\hat{x}}_i = \hat{\psi}(\hat{x}_i, u) + \sum_{\substack{k=1 \\ k \neq i}}^{n} \tilde{L}_{ik} H |\hat{x}_k| - \sum_{\substack{k=1 \\ k \neq i}}^{n} \tilde{L}_{ik} H \hat{x}_i, \quad 1 \leq i \leq n,$$

with the notation $|\hat{x}| = \left( |\hat{r}|, |\hat{c}|, |\hat{p}|, |\hat{q}| \right)$, and $\tilde{L}_{ik}$ denoting the coefficient of index $(i, k)$ in the matrix $\tilde{L} = \tilde{L}(\varepsilon, u)$. For the same initial condition $X_0$ as in the non-modified problem, existence and uniqueness of a local in time solution $\hat{X}(t, X_0)$ defined on some interval $[0, \hat{\tau}]$ with $\hat{\tau} > 0$, are also obtained by Cauchy-Lipschitz Theorem.

Now, we recall that $H = \text{diag}\{1, 1, 1, 0\}$ (see Eq. (3)), which means that the $\hat{q}_i$ components, $1 \leq i \leq n$, are not coupled. It follows that

$$\hat{q}_i(t) = \hat{q}_i(0)e^{-\int_0^t \gamma(s)(1 - \hat{r}_i(s))ds}, \quad t \in [0, \hat{\tau}], \quad 1 \leq i \leq n, \tag{10}$$

which implies $\hat{q}_i(t) \geq 0$ for all $t \in [0, \hat{\tau}]$ and $1 \leq i \leq n$, since $\hat{q}_i(0) \geq 0$. Let us next examine the non-negativity of other components $\hat{r}_i$, $\hat{c}_i$ and $\hat{p}_i$, $1 \leq i \leq n$. We employ a truncation method presented in [12]. Define the real-valued function $\chi$ on $\mathbb{R}$ by

$$\chi(s) = \begin{cases} 0 & \text{if } s > 0, \\ \frac{1}{2}s^2 & \text{if } s \leq 0. \end{cases}$$

The function $\chi$ is of class $\mathscr{C}^1$ on $\mathbb{R}$, with $\chi'(s) = 0$ if $s > 0$, $\chi'(s) = s$ if $s \leq 0$. Furthermore, it enjoys the properties

$$\chi(s) \geq 0, \quad \chi'(s) \leq 0, \quad 0 \leq \chi'(s)\,s = 2\chi(s), \quad \forall s \in \mathbb{R}. \tag{11}$$

Next we introduce for each $i$ such that $1 \leq i \leq n$ the function $\rho_i$ defined by

$$\rho_i(t) = \chi\big(\hat{r}_i(t)\big), \quad t \in [0, \hat{\tau}].$$

We easily prove that $\rho_i(t) = 0$ for all $t \in [0, \hat{\tau}]$, thus $\hat{r}_i(t) \geq 0$ for all $t \in [0, \hat{\tau}]$. Applying the same method leads to $\hat{c}_i(t) \geq 0$ and $\hat{p}_i(t) \geq 0$ for all $t \in [0, \hat{\tau}]$.

Hence, the components of the solution $\hat{X}(t, X_0)$ of the modified problem are non-negative, so $\hat{X}(t, X_0)$ is also a solution of the initial non-modified problem on $[0, \hat{\tau}]$. By uniqueness, we have $\hat{X}(t, X_0) = X(t, X_0)$ on $[0, \tau] \cap [0, \hat{\tau}]$. Finally, it is seen that $\tau = \hat{\tau}$, thus we have proved the non-negativity of the components of $X(t, X_0)$ on $[0, \tau]$.

*Boundedness.* In order to prove the boundedness of the solution, we introduce the function $\theta$ defined by

$$\theta(t) = \sum_{i=1}^{n} \big[r_i(t) + c_i(t) + p_i(t) + q_i(t)\big], \quad t \in [0, \tau].$$

After basic computations, we prove that $\dot{\theta}(t) \leq 0$ for all $t \in [0, \tau]$. This implies that $\theta(t) \leq \theta(0)$ for all $t > 0$, which guarantees the boundedness of the solution.

*Optimal Control Problem.* The existence of an optimal control follows from Theorem III.4.1 in [7], since $\tilde{\Psi}$ and $\tilde{L}$ are linear in $u$, $K$ is convex, and $J$ is defined by integrating a convex function.  $\square$

## 4   Numerical Computation of an Optimal Control

We end our paper with two numerical computations of optimal controls for two distinct topology disposals of a given PCR network of 8 nodes, composed with 4 nodes of type (1) and 4 nodes of type (2). For each network, the values of the parameters are $B_1 = B_2 = 0.3$, $C_1 = 0.1$, $C_2 = 0.2$, $\alpha_i = \delta_i = \mu_i = 0.01$, and the neighborhood $\mathcal{N}$ of the trivial equilibrium which is aimed to be reached after a finite time is defined by $\mathcal{N} = [0, 10^{-1}]^{32}$. The values of other parameters are unchanged. This second computation has also been performed on the server of the Laboratory of Applied Mathematics at University of Le Havre Normandie, using the free and open-source optimal control software BOCOP developed at the INRIA [11].

*Remark 3.* The two topology disposals presented in this section are inspired by the shape of geographical networks, used for the simulation of panic spreading in the case of a tsunami of low intensity on the Mediterranean coast [6]: nodes of type (1) correspond to exposed beaches, while nodes of type (2) model refuge zones in the city center.

## 4.1 Symmetric Topology

First, we examine a symmetric topology shown in Fig. 3. Nodes 7 and 8 are not evacuated towards any node of type (2), thus we can predict, by virtue of Theorem 2, a persistence of panic on this network. We apply a multiple control $(u_0, u_1, \ldots, u_8)$ on this network, and compute an optimal control with respect to the performance criterion $J$ given by (7). The results of the computation are presented in Fig. 4.

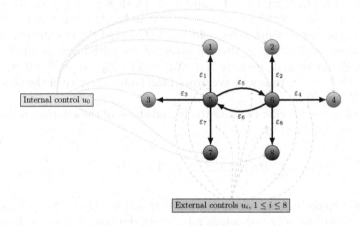

**Fig. 3.** Symmetric topology for an 8 nodes PCR network, composed with 4 nodes of type (1) (depicted in red) and 4 nodes of type (2) (depicted in green). Nodes 7 and 8 are not evacuated towards any node of type (2). (Color figure online)

In this first case, the value of the performance criterion is $J_{min} \simeq 4.1734$, and the final time is $T_f \simeq 14.11$. We observe that the main control that has to be exerted is the internal control $u_0$. This internal control favors the behavioral evolution from panic $p$ to control behavior $c$ on each node. Meanwhile, the controls $u_i$, $1 \leq i \leq 4$, exerted along the edges $e_i$, $1 \leq i \leq 4$, facilitate the evacuation of individuals of nodes 5 and 6. Finally, the controls $u_i$, $i \in \{7, 8\}$, exerted along the edges $e_i$, $i \in \{7, 8\}$, are of very low intensity. Roughly speaking, it seems unnecessary to displace individuals from a node of type (1) towards another node of type (1), if the latter is not itself evacuated towards any node of type (2).

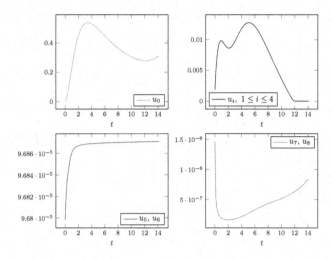

**Fig. 4.** Numerical results for the computation of an optimal control for the 8 nodes PCR network shown in Fig. 3.

## 4.2  Asymmetric Topology

Next, we consider an asymmetric topology presented in Fig. 5, motivated by the well-known symmetry-breaking effect [10]. This topology is of great interest, since it exhibits the cohabitation of a cycle, given by the path $2-5-6-8-4-2$, and a tree, centered at node 5, with leaves 1, 3, 6, 7. With this topology, node 7 is the only non-evacuated node of type (1).

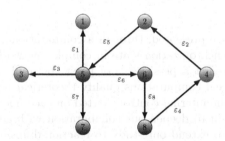

**Fig. 5.** Asymmetric topology for an 8 nodes PCR network, composed with 4 nodes of type (1) (depicted in red) and 4 nodes of type (2) (depicted in green). Node 7 is not evacuated towards any node of type (2). (Color figure online)

The numerical results are shown in Fig. 6. In this second case, the value of the performance criterion is $J_{min} \simeq 4.2187$, and the final time is $T_f \simeq 15.17$. Once again, we observe that the main control that has to be exerted is the internal control $u_0$. The controls $u_1$ and $u_3$ are equal to each other. Surprisingly, the control $u_6$ exerted between the nodes 5 and 6 which are both of type (1) is of

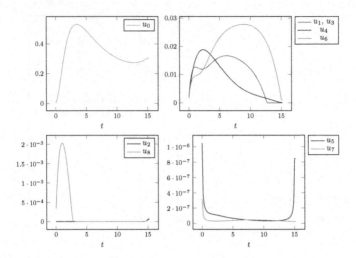

**Fig. 6.** Numerical results for the computation of an optimal control for the 8 nodes PCR network shown in Fig. 5.

greater intensity than in the first case, which seems to be a consequence of the effect of the cycle $2 - 5 - 6 - 8 - 4 - 2$. Finally, the control $u_5$ exerted along the edge $e_5$ admits very low values, since it can worsen the panic level to displace individuals from the node 2 of type (2) towards the node 5 of type (1); this low value of the control $u_5$ seems to "cut" the cycle $2 - 5 - 6 - 8 - 4 - 2$ between nodes 2 and 5.

## 5    Conclusion

In this paper, we have presented, through a computational approach, the possibility to reach a synchronization state in complex networks of PCR systems, for which we can predict a persistence of panic, by applying an optimal control process. Numerical computations qualify theoretical results, and highlight the decisive role of the internal control exerted on each node of the network. In a future work, we aim to deepen our collaboration with geographers and psychologists, in order to extend our study to reaction-diffusion networks of PCR systems, by taking into account the effect of a local diffusion of panic by random walk.

## References

1. Arenas, A., Díaz-Guilera, A., Kurths, J., Moreno, Y., Zhou, C.: Synchronization in complex networks. Phys. Rep. **469**(3), 93–153 (2008)
2. Aric, A., Hagberg, D., Schult, A., Pieter, J.: Exploring network structure, dynamics, and function using networkX. In: Proceedings of the 7th Python in Science Conference (2008)

3. Bruzzone, A.: Perspectives of modeling & applied simulation: modeling, algorithms and simulations: advances and novel researches for problem-solving and decision-making in complex, multi-scale and multi-domain system. J. Comput. Sci. **10**, 63–65 (2015)
4. Cantin, G.: Non identical coupled networks with a geographical model for human behaviors during catastrophic events. Int. J. Bifurcat. Chaos **27**(14), 1750213 (2017)
5. Cantin, G., et al.: Mathematical modeling of human behaviors during catastrophic events: stability and bifurcations. Int. J. Bifurcat. Chaos **26**(10), 1630025 (2016)
6. Cantin, G., et al.: Control of panic behavior in a non identical network coupled with a geographical model. In: PhysCon 2017, pp. 1–6. University, Firenze (2017)
7. Fleming, W., Rishel, R.: Deterministic and Stochastic Optimal Control, vol. 1. Springer, Berlin (2012)
8. Perko, L.: Differential Equations and Dynamical Systems, 3rd edn. Springer, New York (2001)
9. Provitolo, D., et al.: Les comportements humains en situation de catastrophe : de l'observation à la modélisation conceptuelle et mathématique. Cybergeo: Eur. J. Geogr. **735** (2015)
10. Stewart, I., Golubitsky, M., Pivato, M.: Symmetry groupoids and patterns of synchrony in coupled cell networks. SIAM J. Appl. Dyn. Syst. **2**(4), 609–646 (2003)
11. Team Commands, Inria Saclay: Bocop: an open source toolbox for optimal control (2017). http://bocop.org
12. Yagi, A.: Abstract Parabolic Evolution Equations and Their Applications. Springer, Heidelberg (2009). https://doi.org/10.1007/978-3-642-04631-5

# Personalized Ranking in Dynamic Graphs Using Nonbacktracking Walks

Eisha Nathan$^{(\boxtimes)}$, Geoffrey Sanders, and Van Emden Henson

Lawrence Livermore National Laboratory, Livermore, CA 94550, USA
{nathan4,sanders29,henson5}@llnl.gov

**Abstract.** Centrality has long been studied as a method of identifying node importance in networks. In this paper we study a variant of several walk-based centrality metrics based on the notion of a nonbacktracking walk, where the pattern $i \rightarrow j \rightarrow i$ is forbidden in the walk. Specifically, we focus our analysis on dynamic graphs, where the underlying data stream the network is drawn from is constantly changing. Efficient algorithms for calculating nonbactracking walk centrality scores in static and dynamic graphs are provided and experiments on graphs with several million vertices and edges are conducted. For the static algorithm, comparisons to a traditional linear algebraic method of calculating scores show that our algorithm produces scores of high accuracy within a theoretically guaranteed bound. Comparisons of our dynamic algorithm to the static show speedups of several orders of magnitude as well as a significant reduction in space required.

**Keywords:** Non-backtracking walks · Dynamic graphs · Centrality

## 1 Introduction

Calculating node rankings is a commonly studied problem in graph analysis, typically done to identify the "most important" vertices in a network. Several centrality metrics are calculated by quantifying traversals around a network, or by counting walks, where a walk in a graph is a sequence of vertices that allows for both vertices and edges to repeat. More recently, several authors have presented the notion that all walks are not created equally and more importance ought to be given to walks that do not *backtrack* on themselves (walks that visit a particular vertex $i$, visit its neighbor $j$, then immediately backtrack to vertex $i$) [1]. In the case of networks modeling disease spread, a walk that backtracks upon itself provides little to no useful information about the spread of disease. Backtracking walks in information diffusion networks do not allow the user to glean any new information. Furthermore, [2,3] noted that localization effects can

This work was performed under the auspices of the U.S. Department of Energy by Lawrence Livermore National Laboratory under Contract DE-AC52-07NA27344 with release number LLNL-CONF-765259.

be avoided by studying these nonbacktracking walks. In this paper, we study nonbacktracking walks and the associated centrality scores and propose a new algorithm for computing them. We additionally extend our analysis to dynamic graphs, where data is allowed to change over time, creating new relationships in the network. When studying analytics on dynamic graphs, it is important to have measures that can update efficiently given just the changes made to the graph from the previous timestep. This avoids a full recomputation every time the underlying graph is changed and avoids unnecessary computation time. The contributions of this paper can thus be summarized as:

- A new algorithm for approximating personalized nonbacktracking centrality scores in **static** graphs
- Theoretical and empirical proof that our static algorithm produces approximate values close to exact nonbacktracking centrality scores
- A new algorithm for approximating nonbacktracking centrality scores in **dynamic** graphs
- Evidence that our dynamic algorithm is more efficient than a static recomputation for real datasets

The rest of the paper is organized as follows. Section 2 presents the necessary notation used to understand the problem and a brief overview of some related work. Section 3 presents both the static and dynamic algorithms and Sect. 4 presents our results. In Sect. 5 we conclude.

## 2    Background

### 2.1    Terminology

Let $G = (V, E)$ be a graph, where $V$ is the set of $n$ vertices and $E$ the set of $m$ edges. $A$ is the $n \times n$ adjacency matrix of a graph where $A_{ij} = 1$ if there is an edge between vertices $i$ and $j$ in the graph, and 0 otherwise. Although all the work presented in this paper can be applied to both undirected and directed graphs, here we focus on undirected graphs so $\forall e = (i, j) \in E$, $A_{ij} = A_{ji} = 1$. Dynamic graphs represent data that is changing over time and can be modeled as snapshots of the current state of the data at different points in time, or a sequence of static graphs. Dynamic graphs can be thought of as graph data that has timestamps associated with it. Denote the current snapshot at time $t$ of the dynamic graph $G$ and its corresponding adjacency matrix $A$ as $G_t = (V_t, E_t)$ and $A_t$ respectively. The difference in subsequent snapshots of the graph at different points in time $t$ and $t + 1$ can be written as $\Delta A = A_{t+1} - A_t$, where $\Delta A$ represents the change to the graph at time $t$. If an edge $(i, j)$ is inserted into the graph at time $t$, then $\Delta A_{ij} = 1$. A walk of length $k$ in a graph is a series of connected vertices $v_1, v_2, \cdots, v_k$, where vertices are allowed to repeat. Powers of the adjacency matrix are used to count walks of different lengths where $A_{ij}^k$ is the number of walks of length $k$ from vertex $i$ to $j$ [4]. A nonbacktracking walk (NBTW) is defined as a walk which does not backtrack upon itself, meaning it contains no vertex sequences of the form $i \rightarrow j \rightarrow i$. Let $N(i)$ denote the set of neighbors of vertex $i$. For a particular NBTW $w$ that ends with the sequence of vertices $\cdots, j, i$, let $\widetilde{N}_j(i) = N(i) \backslash j$, meaning the set of neighbors of vertex $i$ without the vertex the NBTW $w$ came from (vertex $j$).

## 2.2  Related Work

Several walk-based centrality metrics are calculated as functions of the adjacency matrix [5]. Katz Centrality, for example, weights walks starting at each vertex in the graph of different lengths by increasing powers of some parameter $\alpha$, where longer walks are given less importance [6]. The parameter $\alpha$ must fall somewhere in the range $(0, 1/\|A\|_2)$, where $\|A\|_2$ is the 2-norm of $A$. As $\alpha$ reaches its upper limit, however, the centrality scores correspond to eigenvector centrality [7]. A subset of these walk-based centrality metrics and their generalized equation is given in Table 1.

**Table 1.** Several walk-based centralities as functions of the adjacency matrix

| Centrality metric | Generalized equation |
| --- | --- |
| Katz centrality [6] | $\sum_{k=0}^{\infty} \alpha^k A^k$ |
| PageRank [8] | |
| Eigenvector centrality [7] | $Ax = \lambda x$ |
| Exponential centrality [9] | $\sum_{k=0}^{\infty} \frac{A^k}{k!}$ |
| Subgraph centrality [10] | |
| Total communicability [11] | |

The similarity amongst all these walk-based centrality metrics stems from the fact that they weight all walks of the same length equally. For example, a walk of length 4 between vertices $0 \rightarrow 1 \rightarrow 0 \rightarrow 1 \rightarrow 0$ is given the same weight as a walk of length 4 between vertices $0 \rightarrow 1 \rightarrow 2 \rightarrow 3 \rightarrow 4$. Therefore, a new measure of centrality was proposed in [1], based on the concept of a *nonbacktracking walk*. Nonbacktracking walk centrality scores are computed by counting NBTWs in graphs and weighting longer ones by successive powers of some parameter $\alpha \in (0, 1)$. In this paper we calculate personalized centrality scores (w.r.t. seed vertices of interest) using NBTWs by counting NBTWs originating at some seed vertex and ending at all other vertices in the graph.

When NBTW-centrality was first introduced in [1], the authors presented a linear algebraic formulation for calculating the centrality scores based off of a *deformed graph Laplacian*. This was later extended to analysis on directed networks in [12] and in [2] a nonbacktracking variant of eigenvector centrality was introduced. The infinite sum in Theorem 1 converges to the exact solution for the scores $\mathbf{x}^*$, and if the vector $\mathbf{1}$ is replaced with $\mathbf{e}_i$ we obtain the personalized scores w.r.t. a seed vertex $i$ instead of the global scores. Here, $D$ is the associated diagonal degree matrix of the adjacency matrix $A$.

**Theorem 1.** *For $P_k = AP_{k-1} + (I - D)P_{k-2}$, where $D$ is the diagonal degree matrix of $A$, $P_0 = I$, $P_1 = A$, and $P_2 = A^2 + (I - D)$, $\mathbf{x}^* = \sum_{k=1}^{\infty} \alpha^k P_k \mathbf{1}$ [1].*

This sum converges to the linear system in Eq. 1.

$$(I - \alpha A + \alpha^2(D - I))\mathbf{x}^* = (1 - \alpha^2)\mathbf{e}_i \tag{1}$$

However, for large graphs, this linear system is computationally intensive to solve [13] and for personalized scores, the scores of vertices far away from the seed are often negligible. Therefore, it is desirable to have an alternate method to calculate these centrality scores and in this work we present one such alternate algorithm by directly tabulating walks up to a certain length. Since walks (and NBTWs) in graphs can be infinitely long, if we are counting walks manually (without using linear algebra), we can approximate the corresponding centrality metric by counting walks up to a certain length. We use the notation developed in Theorem 1 as the foundation for our algorithm. Furthermore, for dynamic updates, the linear algebraic approach of solving a linear system has several disadvantages. The linear algebraic approach will take at least several matrix-vector multiplications (on the order of $\mathcal{O}(m)$ to converge to a new solution every time the graph is updated). If the update to the graph is only a few edge insertions, this much computation is unnecessary and is avoided by our approach. Therefore, temporal fidelity is limited.

Apart from ranking, nonbacktracking walks have been studied in the context of community detection as well. Community detection is the task of identifying groups of vertices more closely related to each other than to the rest of the network [14]. In [15], a nonbacktracking variant of spectral clustering was proposed. By using the nonbacktracking matrix $B$, a $2\,\mathrm{m} \times 2\,\mathrm{m}$ matrix with entries $B_{(u \to v),(w \to x)} = 1$ if $v = w$ and $u \neq x$ and 0 otherwise, the authors show that performance of spectral algorithms in sparse networks fares better compared to using other commonly used linear operators. They show that the spectrum of this operator maintains a stronger separation between the eigenvalues relevant to showing community structure and the rest of the eigenvalues than other matrices that are more frequently used. Results are optimal for stochastic block model graphs, synthetic networks that have associated ground truth for community structure.

## 3    Algorithms

### 3.1    Approximation Theory

The logic behind our static algorithm is as follows: if we are interested in NBTW-centrality scores w.r.t. seed vertex *seed*, we count all NBTWs originating from *seed* up to some maximum walk length $k$. Tabulating walks in this manner inherently introduces error into the final solution since we do not count walks up to infinite lengths in the network. Let $\mathbf{x}^*$ be the solution to the linear system (the exact NBTW-centrality scores) and $\mathbf{x}_k$ be the approximation from our algorithm by counting up to length $k$ NBTWs. We can bound the error between $\mathbf{x}^*$ and $\mathbf{x}_k$ as in Theorem 2. Using the notation from Theorem 1, our approximation $\mathbf{x}_k$ can mathematically be written as $\mathbf{x}_k = \sum_{r=1}^{k} \alpha^r P_r \mathbf{1}$.

**Theorem 2.** *The error between the exact solution and the $k$th approximation can be bounded by $\|\mathbf{x}^* - \mathbf{x}_k\|_2 \leq \epsilon_k$ where $\epsilon_k = \frac{(\alpha\phi\|A\|)^{k+1}}{1-(\alpha\phi\|A\|)}$ for $(\alpha\phi\|A\|) < 1$, where $\phi = \frac{1+\sqrt{5}}{2}$.*

*Proof.* We first bound the norm of $P_k$ for any $k$:

- Let $\rho_k = \|P_k\|_2$
- Using the recursive formula for $P_k$ from Theorem 1, we have $\rho_k \leq \|A\|\rho_{k-1} + \|I - D\|\rho_{k-2}$
- This leads to a closed-form solution for $\rho_k$:
  - Let $\rho_k = a^k$ where $a$ is the root to characteristic polynomial $a^2 - \|A\|a - \|I - D\| = 0$
  - $a = \frac{1}{2}(\|A\| \pm \sqrt{\|A\|^2 + 4\|D - I\|} \leq \frac{1+\sqrt{5}}{2}\|A\|$

$$\Rightarrow \rho_k = (\phi\|A\|)^k$$

$$\|\mathbf{x}^* - \mathbf{x}_k\|_2 = \|\sum_{r=1}^{\infty}\alpha^r P_r - \sum_{r=1}^{k}\alpha^r P_r\|_2 = \|\sum_{r=k+1}^{\infty}\alpha^r P_r\|_2$$

$$\leq \sum_{r=k+1}^{\infty}\alpha^r\|P_r\|_2 = \sum_{r=k+1}^{\infty}\alpha^r(\phi\|A\|)^r = \sum_{r=0}^{\infty}(\alpha\phi\|A\|)^{r+k+1}$$

$$\leq (\alpha\phi\|A\|)^{k+1}\sum_{r=0}^{\infty}(\alpha\phi\|A\|)^r = \frac{(\alpha\phi\|A\|)^{k+1}}{1-(\alpha\phi\|A\|)} =: \epsilon_k$$

$\square$

Results shown in Sect. 4 demonstrate that our method produces centrality scores at least within $\epsilon_k$ of the exact solution as we count up to length $k$ NBTWs.

### 3.2  Static Algorithm

For a graph with $n$ vertices, we maintain an $n \times k$ array *walks* where *walks[i][j]* represents the number of nonbacktracking walks from *seed* to vertex $i$ of length $j$. The effect of a walk from *seed* to one of its direct neighbors can be propagated throughout the network, where we only advance the walk to a vertex if we don't backtrack. Since walks are required to be nonbacktracking, at step $r$ we need to keep track of the vertex that was visited at step $r - 1$.

We use a priority queue [16] to keep track how many NBTWs of different lengths exist in the network at any given point. The priority queue is filled with 4-tuple elements of the type $(prev, curr, k, num\_walks)$, where an element $u$ in the priority queue means we are currently processing $u.num\_walks$ of length $u.k$ ending at $u.curr$ that came from $u.prev$. The queue is prioritized by $k$, meaning elements with higher values of $k$ are processed first, searching in a manner consistent with depth-first search, starting from the seed set. If the queue is not prioritized by $k$ and the algorithm is allowed to search in a breadth-first manner (starting from the seed, then adding all immediate neighbors of the seed, then neighbors one step out), the size of the queue would grow exponentially and become impractical to use for very large graphs memory-wise.

Figure 1 gives an example of our static algorithm on a toy network. The example graph is shown in Fig. 1a with a seed vertex 0 outlined in green. Figure 1c shows the progression of the priority queue as we process elements. Walks that are terminated (either due to reaching the maximum length or to the lack of neighbors) are depicted by an 'X.' Since the seed vertex 0 has three distinct neighbors (vertices 1, 2, and 3) we initialize the priority queue with their respective elements (the three elements under the $k = 1$ heading). Each of the elements is processed: since vertex 1 has no neighbors that don't involve backtracking, the element $(0, 1, 1, 1)$ is removed from the priority queue and no new elements are added (depicted in row A). The element $(0, 2, 1, 1)$ at the start of row B indicates we have 1 NBTW from vertex 0 to 2. We follow the progression of this NBTW through the priority queue shown by the blue elements. The corresponding NBTW counts is also shown in blue in Fig. 1b. A similar progression is shown for the walk starting from vertex 0 to 3 in the red elements (starting at row C in the priority queue). Note that elements are processed in priority order according to the value of $k$, to ensure a depth first traversal (row A first, then row B, then row C).

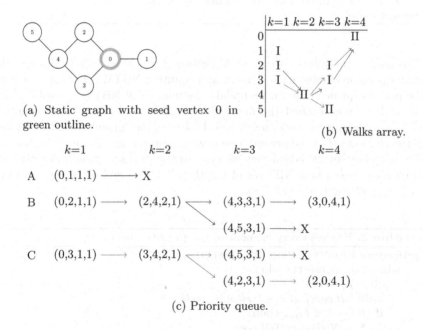

(a) Static graph with seed vertex 0 in green outline.

(b) Walks array.

(c) Priority queue.

**Fig. 1.** Example of STATIC_NBTW. Propagation of different walks is shown in different colors. For a seed vertex of 0, we propagate walks from neighbors vertex 1, 2, and 3 throughout the network. (Color figure online)

Algorithm 1 gives the overall static algorithm for counting all the NBTWs up to length $k\_max$ in the network originating at the *seed* vertex. By the definition of a NBTW, the only vertices that will have a NBTW of length 1 from *seed* are the its direct neighbors. Line 2 initializes the priority queue with the seed's neighbors obtained in Line 3. For each of the *seed*'s neighbors $nbr$, there is one NBTW of length 1 from the *seed* to $nbr$ (Lines 4–6). For each of these NBTWs, their effect is then propagated throughout the rest of the network. A new element is created to be inserted into the priority queue in Line 7, indicating that we have one walk of length 1 from *seed* to $nbr$.

---

**Algorithm 1.** Count NBTWs up to length $k\_max$ from a seed vertex *seed* in a static graph.

---

```
 1: procedure STATIC_NBTW(seed, k_max)
 2:     PriorityQueue *pq = NEW PriorityQueue
 3:     Nbrs = N(seed)
 4:     for nbr in Nbrs do
 5:         num_walks = 1
 6:         walks[nbr][1] = num_walks
 7:         new_elt = (seed, nbr, 1, num_walks)
 8:         pq→INSERT(new_elt)
 9:     walks = EVALUATE_PRIORITY_QUEUE(pq, k_max)
10: return walks
```

---

The main computation occurs in Algorithm 2, where we iterate through the priority queue processing each element and counting NBTWs. For each element in the priority queue (Line 2), we update the number of NBTWs possible (Line 4). If we have not reached the maximum length of NBTWs we are counting (Line 5), we examine the set of neighbors (Line 6) of the current vertex associated with the element $elt$ of the priority queue we are processing. For each vertex $nbr$ in this neighbor set we add a new element to the priority queue indicating we have $elt.num\_walks$ new NBTWs of length $(elt.k + 1)$ ending in the sequence $\cdots, elt.prev, elt.curr, nbr$ (Line 8).

---

**Algorithm 2.** Process every element in the priority queue.

---

```
1: procedure EVALUATE_PRIORITY_QUEUE(pq, k_max)
2:     while !pq→IS_EMPTY() do
3:         elt = pq→POP()
4:         walks[elt.curr][elt.k]+ = elt.num_walks
5:         if elt.k + 1 < k_max then
6:             S = N(elt.curr)\elt.prev
7:             for nbr ∈ Nbrs do
8:                 new_elt = (elt.curr, nbr, elt.k + 1, elt.num_walks)
9:                 pq→INSERT(new_elt)
        return walks
```

## 3.3   Dynamic Algorithm

For dynamic graphs, a naive implementation to obtain updated NBTW counts after changes to the graph occur would recompute from scratch the number of NBTWs from *seed*. However, as the graph grows larger, this naive static recomputation becomes increasingly computationally intensive. By exploiting the locality of edge insertions we can develop a more efficient dynamic algorithm that only updates NBTW counts relevant to new edges inserted into the graph. We consider the case of inserting a single edge $e = (src, dest)$.

Figure 2 gives an example of our dynamic algorithm using the same toy network as earlier. Consider the effect of adding a single edge $e = (2, 5)$ (shown in red in Fig. 2a). Here we do not show the priority queue at each stage, but rather the effect of updating NBTW counts in the *walks* array. Figure 2b gives the initial NBTW counts for the network before adding edge $e$. After inserting the edge between vertices 2 and 5, there are three NBTWs and their counts to update: (1) the NBTW of length 1 starting at vertex 2, (2) the NBTW of length 3 starting at vertex 2, and (3) two NBTWs of length 3 starting at vertex 5. These edge propagations are given in Figs. 2c, d and e respectively.

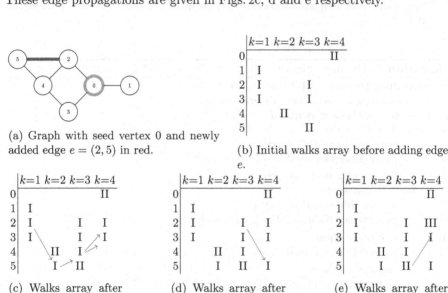

(a) Graph with seed vertex 0 and newly added edge $e = (2, 5)$ in red.

(b) Initial walks array before adding edge $e$.

(c) Walks array after propagating NBTW of length $k=1$ ending at vertex 2.

(d) Walks array after propagating NBTW of length $k=3$ ending at vertex 2.

(e) Walks array after propagating NBTWs of length $k=3$ ending at vertex 5.

**Fig. 2.** Example of DYNAMIC_NBTW. After adding edge $e$ between vertices 2 and 5 we show the steps of the dynamic algorithm to update NBTW counts taking into consideration the new edge (Color figure online)

Our dynamic algorithm is given in DYNAMIC_NBTW in Algorithm 3. All current NBTWs ending in *src* need to be updated since we can now visit *dest* from *src* by traversing the newly added edge. To identify which NBTWs need to be examined, we first find all the values of $k$ where $walks[src][k]$ is nonzero in Line 3 (obtained in the array $k\_src$). If $walks[src][k] > 0$, then there are a nonzero number of NBTWs of length $k$ that end in *src* and we need to propagate these using the newly added edge $e$. The corresponding element is added to the priority queue in Line 6. This same procedure is repeated for the *dest* vertex in Lines 7–10. Lines 11–14 take care of the edge case when either *src* or *dest* is the seed vertex. In this case we need to perform a full propagation from the start similar to the static algorithm. Since we are not recalculating all the counts of NBTWs for all the vertices from *seed*, and are only examining the effect of a single edge, we only add elements to the priority queue corresponding to walks that use the newly added edge. Therefore, this dynamic approach will be significantly faster than a naive static recomputation every time the graph is changed and we see this in Sect. 4.

Both STATIC_NBTW and DYNAMIC_NBTW return an $n \times k$ array *walks* that can then be used to calculate the centrality scores. This procedure is given in Algorithm 4 where we obtain the centrality value for vertex $i$ by weighting NBTWs of different lengths by successive powers of some user-chosen parameter $\alpha \in (0, 1)$.

---

**Algorithm 3.** Dynamic algorithm for calculating NBTW centrality scores.

```
 1: procedure DYNAMIC_NBTW(e = (src, dest))
 2:     PriorityQueue *pq = NEW PriorityQueue
 3:     k_src = walks[src].nonzero
 4:     for k in k_src do
 5:         num_walks = walks[src][k]
 6:         pq→INSERT((src, dest, k + 1, num_walks))
 7:     k_dest = walks[dest].nonzero
 8:     for k in k_dest do
 9:         num_walks = walks[dest][k]
10:         pq→INSERT((dest, src, k + 1, num_walks))
11:     if seed is src then
12:         pq→INSERT((src, dest, 1, 1))
13:     if seed is dest then
14:         pq→INSERT((dest, src, 1, 1))
15:     walks = EVALUATE_PRIORITY_QUEUE(pq, k_max)
16: return walks
```

---

**Algorithm 4.** Calculate NBTW-centrality scores from walk counts.

```
 1: procedure CALCULATE_SCORES(walks, α)
 2:     x = n × 1 array initialized to 0
 3:     for i = 1 : n do
 4:         for j = 1 : k do
 5:             x[i] += α^j · walks[i][k]
         return x
```

# 4    Results

We evaluate STATIC_NBTW and DYNAMIC_NBTW on five real-world graphs drawn from the KONECT collection [17]. Graph information is given in Table 2. For all results, five vertices from each graph are chosen randomly as seed vertices and results shown are averaged over these five seeds. For the dynamic experiments, we use temporal datasets to simulate dynamic graphs, meaning the edges already have associated timestamps. For our dynamic algorithm, we initialize the graph with half the edges and then insert the remaining edges in different batch sizes in timestamped order. We test batch sizes of 1, 10, 100, and 1000. A batch size of $b$ means at each time point we insert $b$ edges and run both the dynamic and static algorithms for comparison purposes. As previously discussed, many real graphs are small-world networks [18], meaning the graph diameter is on the order of $\mathcal{O}(\log(n))$. We set $k = \lceil \log(n) \rceil$, so by counting walks up to length $\approx \log(n)$, we can reach most vertices in the graph. The code was implemented in C++.

**Table 2.** Real graphs used in experiments

| Graph | $|V|$ | $|E|$ | Graph | $|V|$ | $|E|$ |
|---|---|---|---|---|---|
| karate | 34 | 78 | digg | 279,630 | 1,731,653 |
| lesmis | 77 | 254 | wiki-french | 1,420,367 | 4,641,928 |
| copperfield | 112 | 425 | wiki-english | 2,987,535 | 24,981,163 |
| slashdot | 50,835 | 140,451 | youtube | 3,223,585 | 9,375,374 |

## 4.1    Static Algorithm Results

For our static algorithm we present comparisons to a conventional linear algebraic method of solving the system in Eq. 1 discussed in Sect. 2. The goal here is to ensure our algorithm returns similar quality scores to a traditional linear algebraic computation of centrality scores. We measure error as the 2-norm difference between the two vectors as $error = \|\mathbf{x}^* - \mathbf{x}_k\|_2$. Results are shown for the three smallest graphs (COPPERFIELD, KARATE, LESMIS) to ensure we can solve the linear system accurately to present accurate comparisons.

Figure 3 plots the error (on the y-axis) for different values of $k$ (on the x-axis) for the three smallest graphs. Each color corresponds to a specific graph and the dotted lines indicate the theoretically guaranteed error $\epsilon_k$ from Theorem 2 and the solid lines with square markers plot the obtained error from our approximation algorithm. The first trend to note is the most intuitive: as we include counts of longer NBTWs in the calculations of the scores, the error between our approximation and the exact scores decreases. Furthermore, our actual obtained error is always below that of the theoretically guaranteed error (usually by several orders of magnitude), showing that our approximation algorithm produces good quality scores compared to the exact linear algebraic scores.

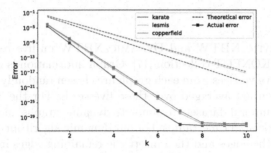

**Fig. 3.** Theoretical and actual error between exact NBTW-centrality scores $\mathbf{x}^*$ and our approximation $\mathbf{x}_k$.

## 4.2   Dynamic Algorithm Results

Our dynamic algorithm produces the same NBTW counts as our static algorithm (and therefore, the same scores), so we only examine the performance of our dynamic algorithm w.r.t. speedup in execution time compared to the static algorithm. Let $T_S$ be the time taken by our static algorithm to compute the NBTW-based centrality scores for a particular graph and $T_D$ be the time taken by our dynamic algorithm. We calculate the speedup in time as $speedup = \frac{T_S}{T_D}$. Higher values of the speedup indicate our dynamic algorithm has significant performance improvement compared to our static.

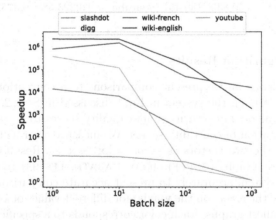

**Fig. 4.** Speedup versus batch size for real graphs.

Figure 4 plots the speedups for the five largest real graphs (on the y-axis) versus the batch size (on the x-axis). In all cases even the minimum speedup obtained is above 1×. We see the greatest speedup for smaller batch sizes of 1 and 10, indicating that our method is most beneficial for low latency applications with small number of data changes. The average speedup obtained decreases for larger batch sizes. This is due to the fact that as the batch size grows larger, the amount of time needed to process the updates grows because all endpoints of all newly added edges must be taken into account. Essentially, new NBTWs must be propagated from all the endpoints of the newly added edges. However, our dynamic algorithm still on average is able to obtain several orders of magnitude in speedup over the static recomputation. In very large graphs of millions of vertices where a static recomputation is computationally infeasible given edge updates to a graph, our dynamic algorithm offers significant savings because it just targets a localized portion of the graph where the edge has been added. Additionally, in applications where the entire graph is not able to fit in memory, our dynamic algorithm only needs to access portions of the graph directly affected by edge updates. However, since the overall trend shows decreasing speedup as the batch size is increased, this tells us that there is a batch size large enough at which it is computationally more efficient to recompute from scratch using the static algorithm.

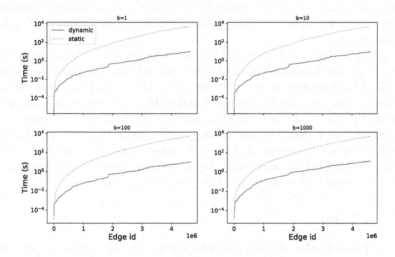

**Fig. 5.** Speedup aggregated over time for SLASHDOT graph. (Color figure online)

Figure 5 plots the speedup over time for the WIKI-FRENCH graph. As edges are inserted into the graph and both algorithms are run, we plot the aggregated time taken in seconds for the dynamic (solid blue line) and the static (dotted orange line) algorithms. The y-value plotted at any given edge id $m$ is the time taken by either the static or dynamic to process all edges up to and including edge $m$. Note that the y-axis is on a log scale. For this experiment, instead of starting with half the number of edges in the graph, we start with an empty graph for both the static and dynamic algorithms. Although the times taken by both algorithm are initially the same, over time we see that the time taken by our dynamic algorithm is several orders of magnitude lower than the time taken by our static algorithm. While there is a spike in the time taken both

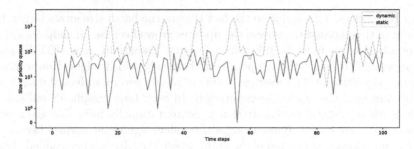

**Fig. 6.** Size of priority queue for static and dynamic algorithms for the SLASHDOT graph for a batch size of 100. (Color figure online)

algorithms initially, the aggregated times for the dynamic algorithm increase linearly (indicating processing each batch of edges at each time point takes more or less the same time), unlike those from the static algorithm. This indicates that our method of only examining places in the graph that are directly affected by the edge updates results in a highly efficient computation of the NBTW-based centrality scores.

Finally, Fig. 6 plots the sizes of the priority queues in both algorithms for the SLASHDOT graph. We sample at 100 evenly spaced time points throughout the duration of both algorithms and plot the size of the priority queue at that time point for the dynamic (solid orange line) and static (dotted blue line) algorithms. The size of the priority queue in the dynamic algorithm consistently remains orders of magnitude lower than the size of the priority queue in the static algorithm. The periodic spikes in the size of the priority queue in the static algorithm can be attributed to the periods after processing an element where we add all the neighbors of the currently processed vertex to the priority queue.

## 5 Conclusions

This paper presented a new algorithm for computing the values of personalized nonbacktracking walk-based centrality scores of the vertices in both static and dynamic graphs. The algorithm returns approximations of scores by counting NBTWs up to a certain length starting at a given seed vertex. In past literature, these centrality values have been computed using a linear algebraic formulation and only on static graphs. Our algorithm agglomeratively counts NBTWs in graphs to obtain the corresponding centrality scores and for static graphs the results presented indicate that our method obtains good quality approximations of the scores compared to a linear algebraic computation. For dynamic graphs, our algorithm is able to avoid a full static recomputation and efficiently computes updated scores, given edge updates to the graph. Our dynamic algorithm returns exactly the same scores as the static algorithm, meaning we have no approximation error. Furthermore, our dynamic algorithm is several orders of magnitude faster than the static algorithm, indicating our approach has large performance benefits.

# References

1. Grindrod, P., Higha, D.J., Noferini, V.: The deformed graph laplacian and its applications to network centrality analysis. SIAM J. Matrix Anal. Appl. **39**(1), 310–341 (2018)
2. Martin, T., Zhang, X., Newman, M.E.J.: Localization and centrality in networks. Phys. Rev. E **90**(5), 052808 (2014)
3. Kawamoto, T.: Localized eigenvectors of the non-backtracking matrix. J. Stat. Mech.: Theory Exp. **2016**(2), 023404 (2016)
4. Cvetkovic, D., Cvetković, D.M., Rowlinson, P., Simic, S.: Eigenspaces of Graphs, vol. 66. Cambridge University Press, New York (1997)
5. Benzi, M., Estrada, E., Klymko, C.: Ranking hubs and authorities using matrix functions. Linear Algebra Appl. **438**(5), 2447–2474 (2013)
6. Katz, L.: A new status index derived from sociometric analysis. Psychometrika **18**(1), 39–43 (1953)
7. Bonacich, P.: Some unique properties of eigenvector centrality. Soc. Netw. **29**(4), 555–564 (2007)
8. Gleich, D.F.: Pagerank beyond the web. SIAM Rev. **57**(3), 321–363 (2015)
9. Aprahamian, M., Higham, D.J., Higham, N.J.: Matching exponential-based and resolvent-based centrality measures. J. Complex Netw. **4**, 157–176 (2016)
10. Estrada, E., Rodriguez-Velazquez, J.A.: Subgraph centrality in complex net. Phys. Rev. E **71**(5), 056103 (2005)
11. Benzi, M., Klymko, C.: Total communicability as a centrality measure. J. Complex Netw. **1**(2), 124–149 (2013)
12. Arrigo, F., Grindrod, P., Higham, D.J., Noferini, V.: Non-backtracking walk centrality for directed networks. J. Complex Netw. **6**(1), 54–78 (2017)
13. Brandes, U., Pich, C.: Centrality estimation in large networks. Int. J. Bifurcat. Chaos **17**(07), 2303–2318 (2007)
14. Girvan, M., Newman, M.E.J.: Community structure in social and biological networks. Proc. Nat. Acad. Sci. **99**(12), 7821–7826 (2002)
15. Krzakala, F.: Spectral redemption in clustering sparse networks. Proc. Nat. Acad. Sci. **110**(52), 20935–20940 (2013)
16. Cormen, T.H., Leiserson, C.E., Rivest, R.L., Stein, C.: Introduction to Algorithms. MIT Press, Cambridge (2009)
17. Kunegis, J.: Konect: the Koblenz network collection. In: Proceedings of the 22nd International Conference on World Wide Web, pp. 1343–1350. ACM (2013)
18. Albert, R., Jeong, H., Barabási, A.-L.: Internet: diameter of the world-wide web. Nature **401**(6749), 130–131 (1999)

# An Agent-Based Model for Emergent Opponent Behavior

Koen van der Zwet[1,2,4(✉)], Ana Isabel Barros[2,4], Tom M. van Engers[1,2,3], and Bob van der Vecht[4]

[1] Science Faculty, University of Amsterdam, Amsterdam, The Netherlands
[2] Institute for Advanced Study, Amsterdam, The Netherlands
[3] Leibniz Center, University of Amsterdam, Amsterdam, The Netherlands
[4] TNO Defence, Safety and Security, The Hague, The Netherlands
koen.vanderzwet@tno.nl

**Abstract.** Organized crime, insurgency and terrorist organizations have a large and undermining impact on societies. This highlights the urgency to better understand the complex dynamics of these individuals and organizations in order to timely detect critical social phase transitions that form a risk for society. In this paper we introduce a new multi-level modelling approach that integrates insights from complex systems, criminology, psychology, and organizational studies with agent-based modelling. We use a bottom-up approach to model the active and adaptive reactions by individuals to the society, the economic situation and law enforcement activity. This approach enables analyzing the behavioral transitions of individuals and associated micro processes, and the emergent networks and organizations influenced by events at meso- and macro-level. At a meso-level it provides an experimentation analysis modelling platform of the development of opponent organization subject to the competitive characteristics of the environment and possible interventions by law enforcement. While our model is theoretically founded on findings in literature and empirical validation is still work in progress, our current model already enables a better understanding of the mechanism leading to social transitions at the macro-level. The potential of this approach is illustrated with computational results.

**Keywords:** Opponent behavior · Opponent networks ·
Multidisciplinary · Complex adaptive systems · Agent-based modelling

## 1 Introduction

Terrorists, insurgents and criminals are typical examples of opponents or opponent organizations, as their actions may destabilize societies and endanger democracy and peace [28]. Efforts to intervene and control the behavior of these groups can also cause undesired effects like retaliation, escalation and displacement [39]. Therefore, understanding the dynamics of opponent organizations (e.g. growth, decline, merging, splitting) is essential in order to maintain a stable

© Springer Nature Switzerland AG 2019
J. M. F. Rodrigues et al. (Eds.): ICCS 2019, LNCS 11537, pp. 290–303, 2019.
https://doi.org/10.1007/978-3-030-22741-8_21

society [3,39]. This dynamic behavior can be seen as a multi scale phenomenon emerging from multifaceted individual and societal interactions [37]. In fact the complex interactions between the different systems such as opponent organizations, law enforcement agencies and the society [39] yield the basic elements of complex adaptive systems (CAS): self-organizing, emergence, feedback loops, adaptive realignment and non-linearity [12,23]. On a micro-level, *individual people* interact and act relatively autonomous. Nonetheless, different relationships, such as kinship, social, cooperation and financial, connect the individuals. [5]. These networks enable communication and autonomous cooperation, which yield self-organizing behavior of the individuals (at the micro-level). These local interactions yield emerging organizational behavior (meso-level). For instance, cooperative actions by individuals aim to generate synergy and establish a competitive edge over rivals. At a meso-level, these seemingly in-articulated individual actions, yield a structure that enable opponent organizations to execute violent actions and/or access to financial and other resources. As the opponent organizations are embedded in the society (macro-level), observations of events by individuals will influence the intrinsic, self-organizing behavior of individuals and groups. The economic situation or the intensity of law enforcement actions can yield positive and negative feedback loops. For instance, a bad economic situation can trigger opponent behavior as the scarcity of economic opportunities can nudge people to capitalize opponent opportunities (opportunities that enable opponent behavior are referred to as opponent opportunities). On the other hand, an increase of law enforcement activities is a typical example of a negative feedback loop as it reduces the attractiveness of opponent behavior [10]. These negative feedback loops constrain the growth of opponent groups within the society [8]. Understanding these dynamics is essential in order to grasp the emergent opponent behavior.

In this paper we will explore how a complex systems perspective can provide new insights into the behavioral dynamics of opponent organizations. We introduce a modelling approach extending the above insights from complex systems, by integrating criminology, psychology and organizational studies within an agent-based modelling (ABM) approach. Instead of modelling organizations as a whole, we use a bottom-up approach wherein agents represent individuals with active and adaptive reactions and act driven by self-interest. The resulting model enables at a micro-level the analysis of evolution of individuals and their networks and their potential to form opponent organizations. At a meso-level it provides an experimentation analysis of the development of opponent organizations subject to the dynamics of the environment and possible interventions. Finally, it enables a better understanding of the mechanisms leading to social transitions at the macro-level.

In the next section we will provide an overview of related work. In Sect. 3 we introduce our modeling approach, while in Sect. 4 we describe in more detail the agent-based model. Computational results can be found in Sect. 5. Finally in Sect. 6 we discuss the research results and implications for future work.

## 2    Related Work

Within complex systems theory, behavioral dynamics by individuals and groups are described as social phase transitions [25]. A social phase transition could be, for instance, a dramatic change in the way groups organize themselves. These transitions, similar as phase transitions in physics, are triggered by either slow processes or small events, which breach the resilience of individuals or groups [42] or alter their motivations [43]. Current studies show promising results using ABM to analyze how interactions and networks of individuals cause social transitions at multiple levels of a system [14,38]. These efforts, however, describe dynamics of a population in general not specifically focused on opponent behavior. Specific studies are focused on the resilience of criminal [10] and terrorist networks [6, 24,27] using simulation to identify vulnerabilities of these networks. However, the authors of these papers are unable to account for the impact of complex behavior as they perceive a closed environment with a preexisting network [40].

ABM is a relatively novel method to study social dynamics in environments of conflict. In 2002, Epstein introduced ABM to study the emergence of civil violence resulting from political grievance [14]. This model incorporated hetero- geneous agents (civilians and peacekeepers) with specific decision rules. Within the model, the civilian agents interact and decide upon a rebellion based on the legitimacy of the authority. This approach demonstrated the ability to study complex social dynamics through simulation. Cioffi-Revilla et al. provided an ABM with an extended population and government model to analyse the impact of different governance strategies on the potential onset of civil unrest [7]. Moon and Carley extended upon this approach, as they presented a multi-agent model with social and geospatial dimensions [32]. This approach introduced multilay- ered network models of social and resources dependencies between individuals in covert networks. This methodology enabled analysis of organizational struc- tures and identification of individuals with a critical position for the functioning of these networks. The modelling approach proposed in this paper builds on these efforts by integrating fine-grained micro mechanisms and form hypothe- ses on the causal relationships that cause the emergent phenomena of opponent organization.

## 3    Emergent Opponent Behavior

Once opponent behavior, which conflicts with norms and rules of the society, emerges we experience changes at different scales. At a micro-level the change of individual behavior from social accepted to opponent and vice versa can be observed. At a higher abstraction level, groups of different sizes and with var- ious activities emerge. Rational choice theory offers an economic approach to explain how individuals rationally select their actions [9] based on a trade-off of expected benefits and costs. Depending on the intrinsic motivation of the indi- vidual, benefits can be associated to acquiring tangible assets such as money, but also non-tangible assets like increase of respect or the spread of fear [13].

## 3.1  Opponent Opinions and Actions

Psychological research of radicalization identify individual, group and mass mechanisms leading to justification and eventually encouragement of opponent behavior [29]. Empirical research has described two separate processes that lead to polarization between groups: developing opinions and enact opponent activities [30]. According to Pruyt et al. [35] and McCauley et al. [30] individuals navigate back and forth through states which can be described as neutral, sympathizing, justifying to moral obligation with regard to opinion and inactive, activist, radical, and terrorist with regard to activities [30,35]. These individual attitudes are influenced by events at meso and macro-level [29]. At meso-level the growth of like-minded groups increases the perception of security by its members and creates obstacles for law enforcement interventions [18]. According to Ganor et al. [18] the extent of public support for an organization is an important factor to ensure the existence of the organization. To an extent, repercussions by law enforcement towards radical actions can create a breeding ground for justification of radical opinions. This can lead to an increase of individuals with radical opinions and stimulating polarizing developments. This phenom at the macro-level can be seen as a *boomerang* effect [36]. Research focused on deradicalization and disengagement reveal the importance of social and economic factors to trigger processes reverse processes [43].

## 3.2  Characteristics of Individuals and Their Relationships

At micro-level, individual and cooperative activities are observable. Bichler et al. [17] focus on the networks structures of opponent organizations and demonstrated the importance of human and social capital for successful opponent individuals and opponent networks. Human capital consists of the information, skills and resources possessed by individuals that enable capabilities. Social capital are the *social ties* between individuals that enable them to contact others, share information and initiate cooperation. Carley [6] and Bright et al. [5] emphasize the existence of multilayered networks of social relationships to enable exchange of both tangible and non-tangible assets.

Our rational agent model incorporates the psychological and social dimension. Figure 1 illustrates the adaptive individual reasoning in an opponent environment. Individuals perceive opportunities and threats by observing events in their social network and environment. Subsequently the individual (de)radicalizes based upon these opportunities, threats, their social interactions and individual success. This influences the radical opinion state of the individual, which is an aggregated factor that represents the attitude towards possible activities. Simultaneously this influences the perceived benefits and costs and thus the expected utility of possible actions. According to the rational choice theory, individuals select the actions with the highest utility. As individuals might have different radical states the perceived utility of their actions will be different.

Self-organization causes that these emergent processes and components are becoming more organized, which yields interdependencies that constrain the

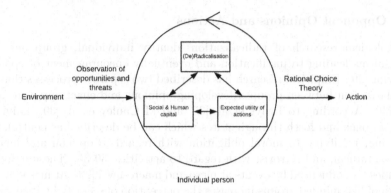

**Fig. 1.** Adaptive individual behavior

autonomy and controllability of the individual behavior [22,34]. Furthermore, rational choice theory dependent on available information on opportunities and threats and influenced by the development of social ties and attitude enables us to model decisions that cause the social phase transitions, which we observe at multiple levels of our society [25,42].

### 3.3 Dynamics of "Collective" Emergent Behavior

The mechanisms underlying collective behavior of opponent individuals are illustrated in Fig. 2. At the micro-level, individuals may cooperate and/or compete proactively or as a reaction to the environment yielding observable emergent organizational opponent behavior. Similarly to the rational decision making at micro-level, organizations also have specific goals, which should be achieved in the most effective way. As such we will use concepts of the field of organization theory, which focus on the deliberate and emergent strategies of cooperation, to analyze effective practises by opponent organizations [31].

The literature on opponent organizational theory is vast. In particular Framis [16] and Ligon et al. [26] describe a continuum of organizational sophistication. Respectively, this continuum from mechanistic to organic is characterized by a more hierarchical structure and a predictable design with a higher degree of formal rules and decision making, to a flatter structure and unpredictable design of cooperation described as a flexible network.

Whereas a lot of similarities can be found between common and opponent organizations, criminal, terrorist, and insurgent organizations are fundamentally differentiated due to the essential need for secrecy of operations [33]. Intensifying cooperation by opponents increases the chance of infiltration by law enforcement agencies or leakage of information [13]. The tension between the efficiency and security in this manner yield a trade-off while optimizing the effectiveness of opponent organizations.

Furthermore, opponent organizations often compete and become rivals [39]. Rios [36] describes a self-reinforcing equilibrium between rivalry by opponent

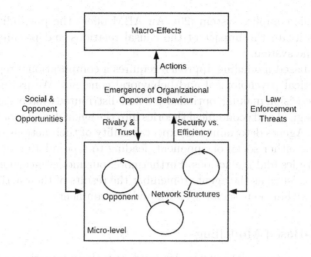

**Fig. 2.** Adaptive emergent opponent behavior

organizations, violence and law enforcement activities. Competition on illegal markets is unstable as they lack formal mechanisms, rules and institutes to cope with disputes, forcing participants to rely on trust [41]. Removing individuals and their relationships from illicit markets will create new power vacuums and a bigger unbalance within the market. Once organizational forms on illicit markets change from mechanic to organic, opponents become more individualistic and adaptive [36]. These dynamics increase competition and rivalry between the organized opponent groups.

The above drivers of organizational opponent behavior play an important role on top of the individual adaptation and self-organization, which form the basis of our approach. In particular, the emergence of different opponent organizational structures depends on the feedback mechanisms of opportunity, threat and competition at micro and meso-level.

## 4   An Agent-Based Model for Emergent Opponent Behavior

The covertness of illicit operations hinders the possibility of extensive and detailed empirical research. Moreover, experimenting with interventions is both undesirable as it will disclose the strategy of the law enforcer as well as it hinders testing alternative intervention strategies. This is one of the motivations we have for promoting an in-silico agent-based simulation environment that allows us to conduct scenario-based experiments. These simulations provide insights into the nature of the underlying complex mechanisms and create a better understanding of the evolution of individual and organizational opponent behavior. Deriving outcomes of emergent behavior is impossible by a strict mathematical approach due to the amount of interactions, activity, decision rules, states, and variables etc. within this complex system [20]. An ABM offers the possibility to experiment and estimate the impact of the causal relations and parameters on the behavior of the system.

An agent-based modelling approach requires a computational representation of the theoretical psychological and behavioral concepts. We use paradigms of autonomous decision making, optimization and distributed systems and translate some psychological and economical theories into the agent-based model presented in this paper. Agents determine the expected utility of their actions to steer their behavior being either social or opponent, leading to a population of agents with different attitudes and (re)actions. Furthermore, our model incorporates typical ABM characteristics as the agents remember the results of their actions, observe and interact within their social network and environment.

## 4.1    Agent-Based Modelling

The purpose of our model is to understand how the interactions of opponents (individuals) under different conditions yield emergent behavior. Our simulation starts with a population of agents with a neutral attitude towards either social or opponent actions, which both 25 can yield a reward. These agents have to pay a certain amount of tax to their environment each time step, which they can earn through either social or opponent activities. The economic context of the environment has a scarce amount of social and opponent opportunities available for the agents to earn money to pay their respective taxes. Actions by law enforcement agencies (governance) pose a threat to opponent activity as they can disrupt opponent activities. This scarcity of opportunities compel competitive behavior by the individuals. At micro-level, the individual opponents operate by self-interest, autonomously and utility driven to mimic the rational decision making of opponents. To monitor the evolution of the system, different metrics have been identified from literature [4]: opponent density, ratio between opponent organizations and individuals, the average opponent organization size, the density of the cooperation by opponents and the effectiveness of opponent activity [33, 39].

Our model was implemented in NetLogo according the ODD-protocol to ensure the model is comprehensive and reproducible [20, 21]. The model consists of a population of agents and a contextual environment. The *environment* represents a two dimensional socio-spatial structure which mimics the complex mechanisms of the systems of society, economy and governments in reality. The environment yields adversity and competition between the agents. Law enforcement activities and the environment economic situation create challenges to the agents. Competition is forced by the scarce amount of resources that agents need to conduct either socially accepted or opponent behavior. The *agents* are embedded in underlying *networks*, which feature the functions of observation, communication and cooperation. These networks mimic relationships of people in reality [5]. The agents in our model are constructed according a BDI-framework to enable adaptive and reactive agent responses [19]. Movement and modification of networks enable agents to compete for an improved position. The agents are able to observe and conduct social and opponent activities individually or cooperatively with their associates.

## 4.2   Radicalization and Rational Choice

Various motives and attitudes were distilled from literature and modeled as agent states: *radical opinion state, desire state, intention state* [30,43]. The radical opinion state represents the attitude of individuals towards opponent behavior. The desire state determines whether the agent aims to exploit social or opponent opportunity. The intention state determines how the agent aims to fulfill this desire, which in case of opponent behavior can be either individual or in a network or organization. Depending on the success of their activities and available opportunity, agents are rewarded for their actions. Agents can initiate bilateral cooperation to create synergy between two agents and increase their individual and collective effectiveness. In order to perform advanced forms of opponent activities, agents can initiate an *organization* to create additional synergy and extent cooperation.

A law enforcement agent is modelled to conduct *law enforcement activities*. These activities focus on specific agents that conduct opponent behavior and attempt to disrupt these activities. At each simulation time step, the modelled law enforcement agencies detect the opponent activities at a certain rate. They decide on the intensity of their counter actions and choose an action type: *direct countering* or *infiltration*. With a direct countering action, the most effective opponents will be countered. Successful counter actions will drop the reward of the opponent activity to zero. An infiltration action aims at uncovering opponents by infiltrating in the cooperation and communication networks of the initial known opponents. Subsequently law enforcement agents attempt to disrupt the opponent activities of one of the opponents detected by this infiltration.

The (de)radicalization process of individuals is modelled by incorporating three feedback mechanisms repeatedly found by empirical research [1,29,43]. The first mechanism to change the radical opinion state of an individual is caused by engagement and disengagement in activities [29,43]. Whenever individuals conduct either social or opponent actions, their satisfaction $(d_n)$ influenced by the reward of these actions, changes their attitude. For example, when an individual conducts a successful opponent activity their radical state increases and vice versa. This creates a positive or negative feedback. The second mechanism is found in studies towards social influence in radical groups [1]. The social network enables individuals to spread opinions and ideas, that ultimately drive individuals to create groups that think alike. The third mechanism is caused by the interaction between the individual, the society and the government [15,18]. This mechanism, as outlined previously, causes that government actions against radical groups create a backlash in the deradicalization process of individuals with a radical attitude [18].

The radical opinion state is bounded by extreme values 0 (social) and 1 (radical). The radical opinion state of an agent on a given time $(r_{i,t})$ is the result of its opinion state in the previous time step $(r_{i,t-1})$ influenced by the satisfaction about the reward by activities $(d_n)$, the average radical opinion level $(r_{t-1})$ of the communication network $(G)$ of the agent and attitude towards law enforcement activities. By multiplying all these factors, the radical feedback factor $(s_{i,t})$ is obtained. Depending on the social $(K)$ and communication $(G)$ network of the agent, the radical opinion state of the agent is determined by the formula:

$$r_{i,t}(K, G, d_n) = \min\left(\max\left((r_{i,t-1} + s_{i,t}), 0\right), 1\right) \tag{1}$$

The radical opinion state effects the *expected costs* of opponent activities. The expected utility $(U_i)$ of an opponent activity $x$ by agent $i$ $(x_i)$ is given by the expected benefits $(y_i)$ and costs $(c_i)$. These benefits and costs are dependent on the cooperation network $(H)$ and organization network $(O)$ of the agent, which the agent can modify by negotiation with other agents in its communication network $(G)$. The agents deliberately add or deduct cooperation links to control the density of their network, in order to increase the amount of synergy or to cope with expected law enforcement threat [11,13]. The formula of rational choice behavior for opponent behavior is given by:

$$\max U_i(H, x_i) = y_i(G, H, O, x_i) - c_i(G, H, O, x_i) \tag{2}$$

The *expected benefits* of opponent activities $(y_i)$ are based on the individual activity $x_i$, its chance of success $(e_{d_c})$, and activity of others $(x_j)$ and the added synergy from cooperation $(b_c)$ and added synergy from organization $(b_j)$ in case the agents in the cooperation network of the agent $(H)$ cooperate in one of both ways. In order to cooperate, agents need communication, such that $H \subseteq G$, with n amount of people in the respective networks. The level of crime activity of the cooperator equals $x_j$. Cooperation links are indicated by $h_{ij} = 1$, and are undirected.

$$y_i(G, H, x_i) = x_i e_{d_c}\left(1 + b_c * \sum_{j=1}^{n} h_{ij} b_j x_j e_{d_c}\right) \tag{3}$$

The *expected costs* of opponent activities $(c_i)$ are based on the individual activity $x_i$ and radical state $(r_i)$. The vulnerability of the cooperation network $(H)$ depends on the amount of law enforcement activity $(d_{la})$, their focus $(f_{la})$ and success rate $(e_{la})$. $d_{la}$ and $e_{la}$ are combined by multiplication yielding parameter $s$. The amount of law enforcement disruption attempts is determined by the law enforcement rate. The law enforcement focus will prioritise specific counter activities. The focus on effectiveness results in targeting the known opponents with the highest reward, which can be those who have the highest amount of connections, those with the most cooperation links or those at top of an organization. The infiltration focus will select one of the cooperation links of each of the potential targets. This causes an exponential risk for cooperation links as cooperation links are targeted by both policies. Fellow members of a organization $(\sum_{o=1}^{n} x_o)$ pose an additional risk, as they attract attention from law enforcement and create an additional security breach in the network. As agents are aware of these potential counter measures, the following equation to calculate the potential costs is used:

$$c_i(G, H, O, x_i, r_{i,t}) = \frac{0.5}{r_{i,t}}(x_i + s(p(\sum_{j=1}^{n} h_{ij}x_j + x_i + \sum_{o=1}^{n} x_o) + q(\sum_{j=1}^{n} h_{ij})^2)) \quad (4)$$

A radical opinion state above 0.5 corresponds with justification of opponent behavior and will discount the envisioned costs of opponent activities. Once an opponent group grows, law enforcement has to focus on a larger group, which will discount the costs of the individual [15]. An optimum in the network will exist if $\frac{50}{r} + s > b_c$, as increase of costs by the addition of cooperation links will exceed the benefits. The parameters for optimization of the networks, other than $\sum_{j=1}^{n} h_{ij}$ are updated in the belief stage of the BDI-agent framework. If beneficiary, an agent will attempt to add cooperation links by negotiation with neighbours. Other agents use an equal procedure in order to consider whether addition of cooperation links is beneficial.

## 5   Model Analysis and Experiments

Social based computational models require validation to estimate the value of the model output [2]. We both verified and validated our agent-based model in order to analyze the accuracy and applicability of the computational model [20]. Predictive ABM requires *real-world data validation* to test whether the model output can be generalized to situations in reality. The intended application of our current computation model is to explore the effects of complex adaptive system mechanisms that underlie emergent opponent behavior. Validation against empirical data is still work in progress, that will allow us to test the predictive power at a later stage. Nevertheless our model demonstrates implications of the mechanisms that underlie emergent opponent behavior. Therefore a validation process for our computational model was conducted at the dimensions of *internal validity* and *methodological validity* [2]. The estimated validity concerns the value of input parameters to construct experiments, model concepts to explore complexities and sensitivity of the input parameters upon the output values. As a result, the model should be applicable and interpretable. The initialization of our experiments demonstrates the power to explore behavior under different scenarios and study the influence of environmental context to possible emergence of opponent groups.

The experimental design includes scenarios which varies the opportunity for social and opponent behavior, the intensity of the law enforcement activities and vary the focus of these activities. Using our simulations the following metrics were collected: amount of active opponents, opponent cooperation density, amount of opponent organizations and the collective effectiveness of opponent activity by individuals. Most interestingly we were able to distinct some expected and unexpected behavioral transitions in the system based upon our model.

The simulation results indicated social opportunity as the most important factor to influence the attitude towards opponent behavior (Fig. 3). The output

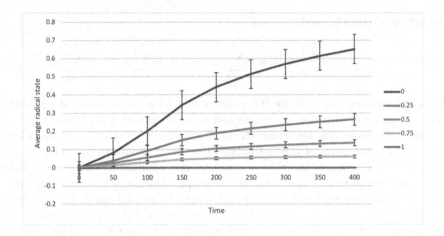

**Fig. 3.** Average radical state of agents by various amounts of social opportunity

of the model describes a social phase transition of the system from a society with a low breeding ground for opponent activities towards a society with a high probability of justification of opponent behavior. This transition can be explained by the feedback mechanisms for individual satisfaction and social influence. However, the average radical attitude remained stable in scenarios with different amounts of opponent opportunity. Due to the scarcity of social opportunity, the agents are unable to satisfy themselves conducting social activities. Thus individuals will not deradicalize by government efforts to decrease the amount of opponent opportunity.

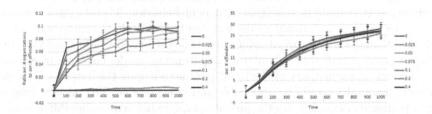

**Fig. 4.** Simulation results of random network initialization of 200 agents. Various intensities of law enforcement activity cause different organizational behavior; while the average amount of individuals that yield opponent behavior remains stable in different scenarios, there appears to be a transition in their collective behavior.

The metrics of the amount opponent organizations and opponent cooperation density also show interesting results. Notably the amount of opponent organizations decline as the rate of law enforcement activity to opponent activity increases from 0.1 to 0.2 (Fig. 4), while the amount of opponents remain stable. They either operate in an organization or, as law enforcement activities threaten their operations, they operate in a loose network to deal with the increased scrutiny. This indicates that the organization process which determines the amount of cooperation and organization form might be constrained and subjective to a threshold set by the context in which it takes place [22,34]. The spikes and slope in the diagram indicate quick and gradual social phase transitions by agents in organizational manners, which shift from mechanic to organic and vice versa. These organizational manners are influenced by the social network and radical opinion state of the agents and subject to the opportunities and threats posed by the environment. While opponent behavior is subjective to these factors, the experiment demonstrates that the behavior remains relatively unpredictable and resilient to changes by the law enforcement agencies.

Additionally, the ratio of organizations and their size were evaluated under scenarios with different strategies by law enforcement agencies, which either target the most effective opponents or infiltrate networks. The increased amount of organizations compared to loose networks shows the capability of individuals to adapt to the law enforcement infiltration strategy. Although the emergent behavior changes, the average effectiveness of opponents remained relatively stable. This demonstrates the resilience of opponents to disruptive strategies and complexity of analyzing and mitigating opponent behavior.

## 6   Discussion and Future Work

The emergent behavior of opponent organizations is a challenging topic in academic research and practice. The covertness of opponent activities and the fact that one cannot experiment with interventions in actual social systems without impacting that system have led us to develop and use in silico experiments that allow us to test the effectiveness of alternative intervention strategies and comparing the results. Our computational model provides an opportunity to experiment and uncover potential effects of governmental behavior when intervening in a social system with opponents. The ability to reveal complexities regarding opponent behavior using an interdisciplinary approach, and the computational agent-based modelling approach allowing us to study mechanisms relevant to subversive organization, such as competition and law enforcement are the main contributions of our developed methodology.

Our proposed approach provides tools to model the fine-grained evolution from the level of an individual. For future research we aim to extend the current components of our agent-based modelling efforts. We intend to include additional attributes of learning, psychological models and advanced game theoretic based behavior in a next version of our model. We then could for instance be able to experiment with scenarios where agents aim to minimize the utility of their competitors rather than maximize their personal utility or memorize past events and attempt to learn from the evolvement of past time interactions.

# References

1. Barsade, S.G.: The ripple effect: emotional contagion and its influence on group behavior. Adm. Sci. Q. **47**(4), 644–675 (2002)
2. Bharathy, G.K., Silverman, B.: Validating agent based social systems models. In: Proceedings of the 2010 Winter Simulation Conference (WSC), pp. 441–453. IEEE (2010)
3. Bohorquez, J.C., Gourley, S., Dixon, A.R., Spagat, M., Johnson, N.F.: Common ecology quantifies human insurgency. Nature **462**(7275), 911 (2009)
4. Bright, D.A., Delaney, J.J.: Evolution of a drug trafficking network: mapping changes in network structure and function across time. Glob. Crime **14**(2–3), 238–260 (2013)
5. Bright, D.A., Greenhill, C., Ritter, A., Morselli, C.: Networks within networks: using multiple link types to examine network structure and identify key actors in a drug trafficking operation. Glob. Crime **16**(3), 219–237 (2015)
6. Carley, K.M., Dombroski, M., Tsvetovat, M., Reminga, J., Kamneva, N., et al.: Destabilizing dynamic covert networks. In: Proceedings of the 8th international Command and Control Research and Technology Symposium (2003)
7. Cioffi-Revilla, C., Rouleau, M.: Mason rebeland: an agent-based model of politics, environment, and insurgency. Int. Stud. Rev. **12**(1), 31–52 (2010)
8. Clauset, A., Gleditsch, K.S.: The developmental dynamics of terrorist organizations. PloS one **7**(11), e48633 (2012)
9. Cornish, D.B., Clarke, R.V.: Understanding crime displacement: an application of rational choice theory. Criminology **25**(4), 933–948 (1987)
10. Duijn, P.: Detecting and disrupting criminal networks: a data driven approach (2016)
11. Easton, S.T., Karaivanov, A.K.: Understanding optimal criminal networks. Glob. Crime **10**(1–2), 41–65 (2009)
12. Eidelson, R.J.: Complex adaptive systems in the behavioral and social sciences. Rev. Gen. Psychol. **1**(1), 42 (1997)
13. Enders, W., Su, X.: Rational terrorists and optimal network structure. J. Conflict Resolut. **51**(1), 33–57 (2007)
14. Epstein, J.M.: Modeling civil violence: an agent-based computational approach. Proc. Nat. Acad. Sci. **99**(suppl 3), 7243–7250 (2002)
15. Faria, J.R., Arce, D.: Counterterrorism and its impact on terror support and recruitment: accounting for backlash. Def. Peace Econ. **23**(5), 431–445 (2012)
16. Framis, A.G.S.: Illegal networks or criminal organizations: power, roles and facilitators in four cocaine trafficking structures (2011)
17. Bichler, G., Malm, A., Cooper, T.: Drug supply networks: a systematic review of the organizational structure of illicit drug trade. Crime Sci. **6**(1), 2 (2017)
18. Ganor, B.: Terrorist organization typologies and the probability of a boomerang effect. Stud. Confl. Terror. **31**(4), 269–283 (2008). https://doi.org/10.1080/10576100801925208
19. Georgeff, M.P., Lansky, A.L.: Reactive reasoning and planning. In: AAAI, vol. 87, pp. 677–682 (1987)
20. Gilbert, N.: Agent-based models (2008). https://doi.org/10.4135/9781412983259
21. Grimm, V., Berger, U., DeAngelis, D.L., Polhill, J.G., Giske, J., Railsback, S.F.: The odd protocol: a review and first update. Ecol. Model. **221**(23), 2760–2768 (2010)

22. Holland, J.H.: Hidden Orderhow Adaptation Builds Complexity. Reading, Addison-Wesley (1995). No. 003.7 H6
23. Johnson, N.: Simply Complexity: A Clear Guide to Complexity Theory. Oneworld Publications, Oxford (2009)
24. Keller, J.P., Desouza, K.C., Lin, Y.: Dismantling terrorist networks: evaluating strategic options using agent-based modeling. Technol. Forecast. Soc. Chang. **77**(7), 1014–1036 (2010)
25. Levy, M.: Social phase transitions. J. Econ. Behav. Organ. **57**(1), 71–87 (2005)
26. Ligon, G.S., Simi, P., Harms, M., Harris, D.J.: Putting the "o" in veos: What makes an organization? Dyn. Asymmetric Confl. **6**(1–3), 110–134 (2013)
27. MacKerrow, E.P.: Understanding why-dissecting radical islamist terrorism with agent-based simulation. Los Alamos Sci. **28**, 184–191 (2003)
28. Makarenko, T.: The crime-terror continuum: tracing the interplay between transnational organised crime and terrorism. Glob. Crime **6**(1), 129–145 (2004)
29. McCauley, C., Moskalenko, S.: Mechanisms of political radicalization: pathways toward terrorism. Terror. Polit. Violence **20**(3), 415–433 (2008)
30. McCauley, C., Moskalenko, S.: Understanding political radicalization: the two-pyramids model. Am. Psychol. **72**(3), 205 (2017)
31. Mintzberg, H., Waters, J.A.: Of strategies, deliberate and emergent. Strateg. Manag. J. **6**(3), 257–272 (1985)
32. Moon, I.C., Carley, K.M.: Modeling and simulating terrorist networks in social and geospatial dimensions. IEEE Intell. Syst. **22**(5), 40–49 (2007)
33. Morselli, C., Giguère, C., Petit, K.: The efficiency/security trade-off in criminal networks. Soc. Netw. **29**(1), 143–153 (2007)
34. Prokopenko, M., Boschetti, F., Ryan, A.J.: An information-theoretic primer on complexity, self-organization, and emergence. Complexity **15**(1), 11–28 (2009)
35. Pruyt, E., Kwakkel, J.H.: Radicalization under deep uncertainty: a multi-model exploration of activism, extremism, and terrorism. Syst. Dyn. Rev. **30**(1–2), 1–28 (2014)
36. Ríos, V.: Why did mexico become so violent? a self-reinforcing violent equilibrium caused by competition and enforcement. Trends Organ. Crime **16**(2), 138–155 (2013)
37. Taylor, S.: Crime and Criminality: A Multidisciplinary Approach. Routledge (2015)
38. Singh, K., Sajjad, M., Ahn, C.W.: Towards full scale population dynamics modelling with an agent based and micro-simulation based framework. In: 2015 17th International Conference on Advanced Communication Technology (ICACT), pp. 495–501. IEEE (2015)
39. Spapens, T.: Macro networks, collectives, and business processes: an integrated approach to organized crime. Eur. J. Crime Crim. L. Crim. Just. **18**, 185 (2010)
40. Vecht, B., Barros, A., Boltjes, B., Keijser, B., de Reus, N.: A multi-methodology framework for modelling opponent organisations in the operational context. In: Proceedings 11th NATO Operations Research & Analysis Conference, pp. 6.21–6.2.20. NATO (2017)
41. Von Lampe, K., Ole Johansen, P.: Organized crime and trust: on the conceptualization and empirical relevance of trust in the context of criminal networks. Glob. Crime **6**(2), 159–184 (2004)
42. Walker, J.T.: Advancing science and research in criminal justice/criminology: complex systems theory and non-linear analyses. Justice Q. **24**(4), 555–581 (2007)
43. Windisch, S., Simi, P., Ligon, G.S., McNeel, H.: Disengagement from ideologically-based and violent organizations: a systematic review of the literature. J. Deradicalization (9), 1–38 (2016)

# Fine-Grained Label Learning via Siamese Network for Cross-modal Information Retrieval

Yiming Xu[1,3], Jing Yu[1,2(✉)], Jingjing Guo[1,2], Yue Hu[1(✉)], and Jianlong Tan[1]

[1] Institute of Information Engineering, Chinese Academy of Sciences, Beijing, China
yiming.tsui@gmail.com,{guojingjing,huyue,tanjianlong}@iie.ac.cn
[2] School of Cyber Security, University of Chinese Academy of Sciences,
Beijing, China
yujing02@iie.ac.cn
[3] School of Computer and Information Engineering, Henan University,
Kaifeng, China

**Abstract.** Cross-modal information retrieval aims to search for semantically relevant data from various modalities when given a query from one modality. For text-image retrieval, a common solution is to map texts and images into a common semantic space and measure their similarity directly. Both the positive and negative examples are used for common semantic space learning. Existing work treats the positive/negative text-image pairs as equally positive/negative. However, we observe that many positive examples resemble the negative ones in some degrees and vice versa. These "hard examples" are challenging for existing models. In this paper, we aim to assign fine-grained labels for the examples to capture the degrees of "hardness", thus enhancing cross-modal correlation learning. Specifically, we propose a siamese network on both the positive and negative examples to obtain their semantic similarities. For each positive/negative example, we use the text description of the image in the example to calculate its similarity with the text in the example. Based on these similarities, we assign fine-grained labels to both the positives and negatives and introduce these labels to a pairwise similarity loss function. The loss function benefits from the labels to increase the influence of hard examples on the similarity learning while maximizing the similarity of relevant text-image pairs and minimizing the similarity of irrelevant pairs. We conduct extensive experiments on the English Wikipedia, Chinese Wikipedia, and TVGraz datasets. Compared with state-of-the-art models, our model achieves significant improvement on the retrieval performance by incorporating with fine-grained labels.

**Keywords:** Fine-grained labeling · Siamese network ·
Graph convolutional network · Hard examples ·
Cross-modal information retrieval

This work is supported by the National Key Research and Development Program (Grant No. 2017YFB0803301) and the Innovation Program for College Students by Chinese Academy of Sciences (Grant No. Y8YY261101).

J. M. F. Rodrigues et al. (Eds.): ICCS 2019, LNCS 11537, pp. 304–317, 2019.
https://doi.org/10.1007/978-3-030-22741-8_22

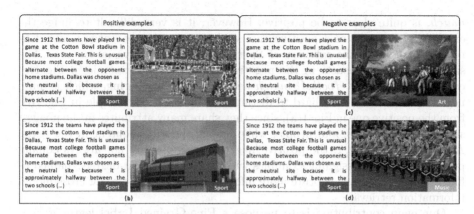

**Fig. 1.** Illustration of the examples of the positive/negative text-image pairs for cross-modal correlation learning. They are not equally positive/negative.

# 1 Introduction

In recent decades, online heterogeneous data of different modalities, such as text, image, video and audio, have been accumulated in huge quantity. There is a great need to find semantically relevant information in various modalities. Traditional single-modal search engines retrieve data of the same modality as the query, such as text-based retrieval or content-based image retrieval. With the development of natural language processing (NLP) and computer vision (CV), cross-modal information retrieval (CMIR), which supports retrieval across multi-modal data, brings increasing attention to break the boundary of natural language and vision. The main challenge of CMIR is to bridge the "heterogenous gap" and measure the semantic similarity between different modalities.

The typical solution for CMIR is to learn a common semantic space and project the features of each modality to this space for similarity measurement. The basic paradigm is based on statistical correlation analysis, which learns linear projection matrices by maximizing the statistical correlations of the positive cross-modal pairs. With the advance of deep learning, deep neural network is used for common space learning, which shows great ability of learning the non-linear correlations across modalities. Typically, two subnetworks model positive and negative text-image pairs simultaneously and optimize their common representations by matched and unmatched constrains. Though the progress in the cross-modal retrieval performance, most of existing algorithms treat positive/negative examples as equally positive/negative and ignore their difference in the degrees of "positivity/negativity".

In this paper, we focus on cross-modal information retrieval by exploring fine-grained labels. We observe that many positive examples resemble the negative ones in some degrees and vice versa. To exemplify the above issue, we give an example using the text-image pairs about *Sport* in Fig. 1. The positive text-image pair in Fig. 1(a), which contains an image and a text describing *football match*, is quite clear to be positive. However, it is very difficult to judge the example in Fig. 1(b) as a positive one, since the *gym* in the image is not definitely belonging to the category of *Sport* according to the visual features. The similar observation exists in the negative examples. The example in Fig. 1(c) consists a text about *football match* and an image of painting, which is quite negative. In contrast, it's hard to distinguish the negative example in Fig. 1(d) from the positive one in Fig. 1(a), since they share many attributes in the images, such as grasses, flags, and people, while having the same content in the texts. As illustrated in Fig. 1, the positive/negative examples are not always equally positive/negative. Assigning examples with labels in different degrees, named as fine-grained labels, will capture more informative characteristics for cross-modal information retrieval.

Our main contribution is to propose a **Fine-Grained Label** learning approach (**FGLab** for short) for cross-modal information retrieval. Our approach first leverage the text description of the image in each text-image example to represent the semantics of the image. Then we propose a siamese network on both the positive and negative examples to obtain their semantic similarities and assign the fine-grained labels accordingly. Finally, we incorporate these labels to a pairwise similarity loss function, which enables the model to pay more attention on the hard examples while maximizing the similarity of positive examples and minimizing the similarity of negative examples. Our proposed approach could be easily applied to other existing models and provide more informative cues for cross-modal correlation learning.

## 2    Related Work

**Cross-modal Information Retrieval.** The mainstream solution for CMIR is to project data from different modalities into a common semantic space where the similarity of different modalities data can be measured directly. Traditional statistical correlation analysis methods like Canonical correlation analysis (CCA) [4,5] learn a common semantic space to maximize the pairwise correlation between two sets of heterogeneous data. With the development of deep learning, the DNN-based CMIR methods have attracted much attention due to its strong learning ability. This method usually constructs two subnets to extract the features of different modal data, and the inputs of different media types learn the common semantic space through the shared layer [12,19,23]. In this work, we follow this two-path deep model to learn the common semantic space of the text and image.

**Study on Hard Examples.** Hard examples analysis can help classification tasks and retrieval tasks obtain better results [1,3,11,24]. Dalal *et al.* [1] construct a fixed set of an initial negative set, train the preliminary detector by using the original dataset, then use the preliminary detector to search the hard examples in the initial negative set. After that, add the hard examples to the original dataset, which was used to re-train the detector. Although the Dalal *et al.* method improves the accuracy of the detector, the expanded data set creates more memory and time overhead, which makes the training efficiency of the model declining. In order to solve this problem, Felzenszwalb *et al.* [3] proposed a method of deleting simple examples with high classification accuracy while adding hard examples, and effectively controlling the size of the data set while further improving the accuracy. However, both of the above methods require repeated training of the entire data set to update the model. In order to improve this method, Shrivastava *et al.* [17] proposed an online hard examples mining method. In the training, the negative examples in each batch are sorted, and the k hardest examples are selected for the back propagation to update the model. The above works focus on the mining of hard negative examples. On this basis, Jin *et al.* [6] proposed the view of hard positive, and mining the hard negative and hard positive examples in the video target detection task, which further improved the training efficiency.

These methods focus on updating the model with hard examples, which effectively improves the accuracy of the model. However, different hard examples are not equally "hard", in order to capture more informative cues from the hard examples, Ma *et al.* [10] assign fine-grained labels to hard negative videos based on the hard degree of negative examples, and obtain better detection results in the complex event detection task in the video. Recently, hard examples have begun to receive attention in cross-modal information retrieval. Faghri *et al.* [2] use the hard negative examples in cross-modal information retrieval to converge the model to better results, which significantly improves the retrieval performance. Different for their work, in this paper, we extend the idea of hard negatives to positive examples and assign fine-grained labels to the different degree of hard examples. For text-image pairs in cross-modal information retrieval, we extract the original text information of the image to calculate the similarity between the original text and the text in the examples. Then we assign fine-grained labels to the examples. We modify the loss function of the distance metric learning and add the fine-tuning effect of the fine-grained label, which increases the influence of the hard positive and negative examples.

## 3  Methodology

Our model consists of two stages, as shown in Fig. 2. The first stage is a fine-grained scoring model based on the similarity degree of text, and the second stage is a cross-modal information retrieval model. In the first stage, the main objective is measuring the correlation between the text in text-image pairs and the text description of images. Compared with the image, the text description often contains richer and more specific information. Therefore, we select the corresponding text description of the image in the example. We use three different feature extraction methods to modeling text, including GCN, Multi-head attention, and Text-CNN and assigning a fine-grained label to the example according to the similarity degree. In the second stage, we use the fine-grained label to adjust the cross-modal information retrieval model.

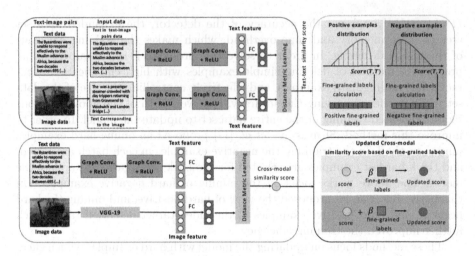

**Fig. 2.** The proposed model overview is divided into two parts. The top part is the fine-grained label learning based on text similarity evaluation, which includes text feature extraction (red box) and fine-grained labels assign (yellow box). The bottom part is the cross-modal information retrieval model, fine-grained labels play an important role in the model update. (Color figure online)

### 3.1 Fine-Grained Label Learning

In the first stage, the main objective is to measure the difficulty degree of the text-image pairs and generate fine-grained labels for them. We evaluate the difficulty degree of text-image pairs by the similarity degree of text-image pairs. For positive examples, the smaller the similarity score of text-image pairs, the greater the difficulty degree. For negative examples, the greater the similarity score of text-image pairs, the greater the difficulty degree. Compared with the image, the text usually contains richer and more specific information. Therefore, we calculate the similarity of text-image pairs by selecting the original text description of the image and the text in the example. We use three different feature extraction methods to modeling text and assigning a fine-grained label to the example.

We design a dual-path neural network to extract the text features and text features and learn the potential common semantic space. Define the original dataset as $D = \{(T_i^D, I_i^D)\}_{i=1}^N$, which contains of C classes. Where $T_i^D$ represents the $i_{th}$ text in the original dataset, $I_i^D$ represents the $i_{th}$ image, and $(T_i^D, I_i^D)$ represents the $i_{th}$ text-image pair in the original dataset D. $T_i^D$ is the original text description of the image $I_i^D$, which have the same semantic and belong to the same class.

Follow the previous work [25], we construct a positive examples dataset $P = \{(T_j^P, I_j^P)\}_{j=1}^M$ and a negative examples $E = \{(T_k^E, I_k^E)\}_{k=1}^K$ dataset based on the original training data $D$. Specifically, we randomly select $T_i^D$ and $I_i^D$ of the same class from $D$ to constitute the positive examples dataset $P$, where $M$ represents the number of text-image pairs in $P$. Similarly, for $E$, we randomly select $T_x^D$ and $I_k^D$ which do not belong to the same class from $D$, where $K$ represents the number of text-image pairs in $N$. To ensure the same number of positive and negative examples for the model training, we set $M = K$.

We focus on the learning difficulty of different text-image pairs. For positive examples, the closer the semantics of text and image, the lower the learning difficulty, and the larger the semantic difference, the higher the learning difficulty. For negative examples, the closer the semantics of text and image, the higher the learning difficulty, and the larger the semantic difference, the lower the learning difficulty. Compared with the image, the original text description contains richer and more specific information, which can express high-level semantics so that we will learn the difficulty of positive and negative examples through the semantic similarity between texts. Specifically, for all image $I_j^P$ or $I_k^E$, we extract the original text descriptions $T_j^D$ or $T_k^D$ corresponding to the image in D, forming a positive text-text pair $(T_j^P, T_j^D)$ or the negative text-text pairs $(T_k^E, T_i^D)$. We use an end-to-end dual-path neural network to learn the similarity between the texts. The output of the network is text-text similarity score $S$, where $S^{Pos}(T_j^P, T_j^D)$ indicates the similarity score of the positive example, $S^{Neg}(T_k^E, T_k^D)$ indicates the similarity score of the negative example. The loss function is formal as:

$$Loss = (\sigma_{T-T}^{2+} + \sigma_{T-T}^{2-}) + \lambda max(0, m - (\mu_{T-T}^+ - \mu_{T-T}^-)) \qquad (1)$$

Where $\mu_{T-T}^+$ and $\sigma_{T-T}^{2+}$ as the mean and variance of the associated text pairs, and $\mu_{T-T}^-$ and $\sigma_{T-T}^{2-}$ denote the mean and variance of the unrelated text pairs.

Given positive examples dataset $P = \{(T_j^P, I_j^P)\}_{j=1}^M$, we can obtain the positive examples similarity score dataset $C^{Pos}$. Similarly, given negative examples dataset $E = \{(T_k^E, I_k^E)\}_{k=1}^K$, we can obtain the negative examples similarity score dataset $C^{Neg}$. We assign fine-grained labels $L$ for positive examples and negative examples. The fine-grained labels are allocated from 0 and 1. Given a text-text similarity score $S_i$, the fine-grained label $L_i$ is defined as follows:

$$L_i^{Pos}(T_i^P, I_i^P) = 1 - \frac{S_i(T_i^P, T_j^D) - S_{min}^{Pos}}{S_{max}^{Pos} - S_{min}^{Pos}} \qquad (2)$$

$$L_i^{Neg}(T_i^E, I_i^E) = \frac{S_i(T_i^E, T_j^D) - S_{min}^{Neg}}{S_{max}^{Neg} - S_{min}^{Neg}} \qquad (3)$$

where $L_i$ denotes the fine-grained label of $i_{th}$ text-image pair, $S_{max}^{Pos}$ and $S_{min}^{Pos}$ denotes the maximum and the minimum similarity score in positive examples similarity score dataset $C^{Pos}$, $S_{max}^{Neg}$ and $S_{min}^{Neg}$ denotes the maximum and the minimum similarity score in $C^{Neg}$.

## 3.2   Cross-modal Information Retrieval Model

In this stage, we build a dual-path neural network to extract text and image features and learn the potential common semantic space. Then we attain the similarity of text and image by metric learning. Fine-grained labels are used in the model training process to update the similarity score and to further adjust the loss function. Our model makes hard examples have a greater impact in the model training.

**Text and Image Feature Extraction.** The GCN model has a strong ability to learn the local and fixed features of the graph and has been successfully used for text classification [9]. In recent research [25], GCN has shown to have a strong capability for text semantic modeling and text categorization. In our model, the text GCN contains two convolution layers each followed by a ReLU. Then, we set up a fully connected layer to map the text features to the common latent semantic space. Given a text $T$, the text feature $v_T$ can be extracted by the text GCN model $H_T(\cdot)$, which is defined as: $v_T = H_T(T)$.

For image modeling, we use the pre-trained VGG-19 [18] as the basic model to obtain image features. Given a $224 \times 224$ image, a 4096-dimensional feature vector is generated from the FC7 layer, which is the penultimate fully connected layer in VGG-19. Next, a fully connected layer map the image to the common semantic space. Given a image $I$, image vector $v_I$ is extracted by the VGG-19 model $H_I(\cdot)$, which is defined as: $v_I = H_I(I)$.

**Object Function.** In our model, we set up two paths to obtain the text features $v_T$ and images features $v_I$. Then we use element-wise product to attain the correlation between the text features and the image features, and a fully connected layer is followed to obtain the similarity score. We use the same loss function with Sect. 3.1, which aims to reduce the proportion of false positives and false negatives, as shown in Fig. 3(a). The left curve represents the distribution of the matched text-image pairs where $\mu^-$ denotes the mean and $\sigma^{2-}$ demotes the variance. The right curve represents the distribution of non-matching text-image pairs where $\mu^+$ denotes the mean and $\sigma^{2+}$ demotes the variance. The objective function is to maximize $\mu^+$ and minimize $\mu^-$, $\sigma^{2+}$ and $\sigma^{2-}$. In our work, our goal is further reducing the proportion of false positives and false negatives by enhancing the impact of the hard examples in the shadows to make the model converge to better results, which is shown in Fig. 3(b).

In the training process, the distribution of a few of false positives and false negatives in the shaded portion is updated to a more erroneous degree by the fine-grained label, so that the influence of these hard examples is increased, and the model obtains a better descending gradient. Given text features $v_T$ and images features $v_I$, the similarity of the matched and non-matched text-image pairs is defined as $Y(T, I)$, we use fine-grained labels to update the similarity score of text-image pairs, which is formulated as follow:

$$\widetilde{Y}^{Pos}(T^P, I^P) = Y(T^P, I^P) - \beta L^{Pos}(T^P, I^P) \tag{4}$$

$$\widetilde{Y}^{Neg}(T^E, I^E) = Y(T^E, I^E) - \beta L^{Neg}(T^E, I^E) \tag{5}$$

Where $\widetilde{Y}^{Pos}$ and $\widetilde{Y}^{Neg}$ represents the similarity score after the fine-grained label update for positive and negative examples. $\beta$ is a hyperparameter that adjusts

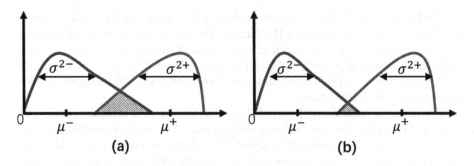

**Fig. 3.** (a) The original loss function and (b) the upgraded loss function by our fine-grained labels.

the effect of fine-grained labels on similarity scores. The value of $\beta$ is related to the range of $Y$. The loss function is defined as follows:

$$Loss = (\sigma^{2+}_{T-I}\sigma^{2-}_{T-I}) + \lambda max(0, m - (\mu^{+}_{T-I} - \mu^{-}_{T-I}))$$ (6)

$$\mu^{+}_{T-I} = \sum_{n=1}^{Q_1} \frac{\widetilde{Y}^{Pos}}{Q_1}, \qquad \sigma^{2+}_{T-I} = \sum_{n=1}^{Q_1} \frac{\widetilde{Y}^{Pos} - \mu^{+}_{T-I}}{Q_1}$$

$$\mu^{-}_{T-I} = \sum_{n=1}^{Q_2} \frac{\widetilde{Y}^{Neg}}{Q_2}, \qquad \sigma^{2-}_{T-I} = \sum_{n=1}^{Q_2} \frac{\widetilde{Y}^{Neg} - \mu^{-}_{T-I}}{Q_2}$$

Where $\mu^{+}_{T-I}$ and $\sigma^{2+}_{T-I}$ denote the mean and variance of the matched text and image, and $\mu^{-}_{T-I}$ and $\sigma^{2-}_{T-I}$ are the mean and variance of the non-matched text and image. $\lambda$ can adjust the ratio of the mean, and $m$ controls the upper limit between the average of the matching and non-matching similarities.

## 4    Experiments

### 4.1    Datasets and Evaluation

We accomplish the general cross-modal information retrieval tasks: image-query-texts and text-query-images. We evaluate our model on three benchmark datasets, i.e. English Wikipedia, TVGraz, and Chinese Wikipedia. Each dataset contains a set of text-image pairs, where the texts are long descriptions with rich content instead of tags or captions.

**English Wikipedia.** The English Wikipedia dataset was divided into 10 categories, containing 2866 image-text pairs. We selected 2173 pairs for training and 693 pairs for testing. Each text was represented by a graph containing 10055 vertices and each image by a 4096-dimensional vector representation of the last layer of the fully connected layer of the VGG-19 model [18].

**TVGraz.** The TVGraz dataset contained 10 categories from the Caltech-256 [13] dataset, stored in an URL format. We selected more than 10 words of text from 2592 web pages, which comprised 2360 image-text pairs; they were randomly divided into 1885 for training and 475 pairs for testing. Each text was represented by a graph containing 8172 vertices, and each image by a 4096-dimensional VGG-19 feature.

**Chinese Wikipedia.** Chinese Wikipedia dataset (Ch-Wiki for short) The Chinese Wikipedia dataset [14] was divided into 9 categories, containing 3103 image-text pairs. We randomly selected 2482 pairs for training and 621 pairs for testing. Each text was represented by a graph with 9613 vertices, and each image by a 4096-dimensional VGG-19 feature.

Mean Average Precision (MAP) is used to evaluate our model. MAP is the mean of average precision (AP) for all queries. AP is defined as:

$$AP = \frac{1}{R} \sum_{k=1}^{n} \frac{R_k}{k} \times rel_k \tag{7}$$

where $R$ represents the number of relevant retrieved results. $R_k$ represents the top $k$ results. $n$ represents the number of all the retrieved results. $rel_k = 1$ indicates that the $k_{th}$ result is related, and otherwise 0.

## 4.2 Implementation Details

Our model consists of two stages, i.e. fine-grained labeling model and cross-modal information retrieval model. We follow the strategy in [25] to randomly select 40,000 positive examples and 40,000 negative examples from the training set in all the datasets. We set the learning rate 0.001 with an Adam optimization, and 50 epochs for training. The regularization is 0.005. $m$ and $\lambda$ in the loss function are set as 0.8 and 0.35, respectively. In the fine-grained labeling model, the dimension of text features from the two paths is reduced to 1024 after the fully connected layer. Similarly, in the cross-modal information retrieval model, the dimension of both image features and the text features are reduced to 1024 before feeding to the feature fusion module.

## 4.3 Comparison with State-of-the-Art Methods

We compare the performance of our proposed FGLab with state-of-the-art models, including CCA [16], LCFS [22], ml-CCA [15], LGCFL [7], AUSL [26], JFSSL [21], GIN [25], TTI [14], CM [13], SM [13], SCM, TCM, w-TCM, c-TCM. All the models are well cited in the literature. GIN also serves as the baseline model, which has the same architecture as our cross-modal retrieval model without fine-grained labels in the loss function. We compare the MAP scores with their publicly reported results in Table 1.

From the performance in Table 1, we observe that our model is superior to all the other models for the text queries on the three datasets. Compared with the second best method GIN (baseline), the performance of our method in MAP is

**Table 1.** MAP score comparison of text-image retrieval on three datasets.

| Dataset | Model | Retrieval performance | | |
|---------|-------|-------------|------------|---------|
|         |       | Text query | Image query | Average |
| Eng-Wiki | CCA | 0.187 | 0.216 | 0.201 |
|          | ml-CCA | 0.287 | 0.352 | 0.312 |
|          | LCFS | 0.231 | 0.297 | 0.264 |
|          | LGCFL | 0.316 | 0.377 | 0.312 |
|          | AUSL | 0.332 | 0.396 | 0.364 |
|          | JFSSL | 0.410 | **0.467** | 0.438 |
|          | GIN | 0.767 | 0.452 | 0.609 |
|          | FGLab (ours) | **0.837** | 0.457 | **0.647** |
| TVGraz | TTI | 0.153 | 0.216 | 0.184 |
|        | CM | 0.450 | 0.460 | 0.4550 |
|        | SM | 0.585 | 0.619 | 0.602 |
|        | SCM | 0.696 | 0.693 | 0.694 |
|        | TCM | 0.706 | 0.694 | 0.695 |
|        | GIN | 0.719 | 0.818 | 0.769 |
|        | FGLab (ours) | **0.763** | **0.833** | **0.798** |
| Ch-Wiki | w-TCM | 0.298 | 0.241 | 0.269 |
|         | c-TCM | 0.317 | 0.310 | 0.313 |
|         | GIN | 0.384 | 0.334 | 0.359 |
|         | FGLab (ours) | **0.517** | **0.390** | **0.457** |

increased by about 7%, 5%, and 13% on the Eng-Wiki, TVGraz, and Ch-Wiki, respectively. For the image query, FGLab outperforms state-of-the-art models by 1.5% and 5.6% on the TVGraz and Ch-Wiki, respectively. The results on the Eng-Wiki dataset is slightly inferior to the second best model JFSSL. Compare with the image query, the improvement on the text query is more remarkable. It's because that when the fine-grained labels enable the model to pay more attention on the "hard" examples, the parameter tuning of the model has bias on the text modeling path. Specifically, the parameters in the text modeling path are tuned more greater than these parameters in the image modeling path, which enhances the generalization ability of the text representations obviously compared to the image. For the average performance, our model is superior to all the other models. Compared with GIN, the MAP scores is increased by about 4%, 3%, and 10% on the Eng-Wiki, TVGraz, and Ch-Wiki, respectively. It proves that, by incorporated with the "hardness" information of the training examples by fine-grained labels, existing model (GIN) can achieve great improvements on the cross-modal retrieval performance. Meanwhile, the remarkable improvements indicate that our proposed FGLab is able to capture the informative clues of

**Table 2.** MAP score comparison of FGLab models with different kinds of fine-grained labels on the Eng-wiki dataset.

| Model | Text feature | Text query | Image query | Average |
|---|---|---|---|---|
| FGLab (main model) | GCN | **0.837** | **0.457** | **0.647** |
| FGLab-Att | Multi-head attention | 0.806 | 0.438 | 0.622 |
| FGLab-CNN | Text-CNN | 0.7501 | 0.437 | 0.5931 |

fine-grained labels and effectively affect the cross-modal correlation learning to focus on "hard" examples.

### 4.4 The Influence of Text Features on Fine-Grained Labeling

Besides our proposed fine-grained labeling model based on GCN, we also implement another two baseline models based on different text features to evaluate their influence on the fine-grained labeling as well as the cross-modal retrieval performance. The other two models are respectively utilize the multi-head attention [20] (i.e. the encoder in transformer) and Text-CNN [8] instead of GCN in the fine-grained labeling model, with other structures unchanged. The two models are respectively named as FGLab-Att and FGLab-CNN for short.

Figure 4 shows the samples of the fine-grained labels obtained by the three models on the Eng-Wiki dataset. The bar graphs on the top or bottom of each image show the fine-grained labels of the three models using different text features for each positive or negative example. The text is identical for all the examples while the images are different. Because of the space limitation, we only show a part of the whole text, which introduces a war happened on the sea. For clear semantic comparisons, we show the corresponding image of the text on the middle left. For the three positive examples, the images in the text-image pairs from left to right contains *map, bird's-eye view of ships*, and the *close shot of a warship*, respectively. Intuitively, the semantic correspondence of the three images with the text increases from left to right while the "hardness" of the three positive examples decrease accordingly. The fine-grained labels obtained by GCN features are identical to human judgement. The results of multi-head attention and Text-CNN are not satisfied to some extent. Similar observations exist for the negative examples and other examples not shown in the paper. Therefore, from the qualitative analysis of the fine-grained labels, the labeling model based on GCN could obtain more accurate labels by human evaluation.

To further prove the effect of fine-grained labels by different text features, we train the retrieval models based on the aforementioned three kinds of labels on the Eng-Wiki dataset. The retrieval results is given in Table 2. It's obvious that FGLab with GCN achieves the highest MAP scores compared the other two models. The performance of FGLab-Att is slightly lower than that of FGLab while FGLab-CNN has the worst performance. The retrieval results of the three models are identical with their performance in fine-grained labeling.

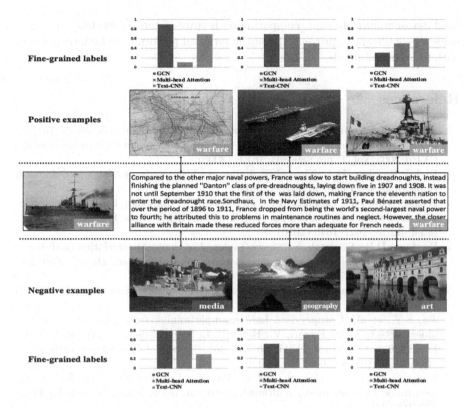

**Fig. 4.** Samples of the fine-grained labels obtained by three models on the Eng-Wiki dataset. For easy evaluation of the label quality, we show the positive examples and negative examples containing the same text (middle) but different images (top and down, respectively). For clear semantic comparisons, we show the corresponding image of the text on the middle left. The bar graphs on the top/bottom of each image show the fine-grained labels of the three models using different text features.

Both Multi-head attention approach and Text-CNN approach obtains some unreasonable labels, which degrades the retrieval performance.

## 5   Conclusion

In the paper, we propose a Fine-Grained Label learning approach for cross-modal information retrieval. We design a siamese network to learn fine-grained labels for both the positive and negative examples to capture the degrees of hardness, thus enhancing cross-modal correlation learning. We introduce these labels to a rank-based pairwise similarity loss function. The loss function benefits from the labels to increase the influence of hard examples on the similarity learning while maximizing the similarity of relevant text-image pairs and minimizing the similarity of irrelevant pairs. The experimental results on three widely used datasets indicate that, comparing with state-of-the-art models, our model achieves significant improvements on the retrieval performance by incorporating with fine-grained labels.

# References

1. Dalal, N., Triggs, B.: Histograms of oriented gradients for human detection. In: 2005 IEEE Computer Society Conference on Computer Vision and Pattern Recognition, CVPR 2005, vol. 1, pp. 886–893. IEEE (2005)
2. Faghri, F., Fleet, D.J., Kiros, J.R., Fidler, S.: VSE++: improving visual-semantic embeddings with hard negatives (2017)
3. Felzenszwalb, P.F., Girshick, R.B., McAllester, D., Ramanan, D.: Object detection with discriminatively trained part-based models. IEEE Trans. Pattern Anal. Mach. Intell. **32**(9), 1627–1645 (2010)
4. Hardoon, D.R., Szedmak, S., Shawe-Taylor, J.: Canonical correlation analysis: an overview with application to learning methods. Neural Comput. **16**(12), 2639–2664 (2004)
5. Hotelling, H.: Relations between two sets of variates. Biometrika **28**(3/4), 321–377 (1936)
6. Jin, S.Y., et al.: Unsupervised hard example mining from videos for improved object detection. In: Ferrari, V., Hebert, M., Sminchisescu, C., Weiss, Y. (eds.) ECCV 2018. LNCS, vol. 11217, pp. 316–333. Springer, Cham (2018). https://doi.org/10.1007/978-3-030-01261-8_19
7. Kang, C., Xiang, S., Liao, S., Xu, C., Pan, C.: Learning consistent feature representation for cross-modal multimedia retrieval. IEEE Trans. Multimedia **17**(3), 370–381 (2015)
8. Kim, Y.: Convolutional neural networks for sentence classification. arXiv preprint arXiv:1408.5882 (2014)
9. Kipf, T.N., Welling, M.: Semi-supervised classification with graph convolutional networks. arXiv preprint arXiv:1609.02907 (2016)
10. Ma, Z., Chang, X., Yang, Y., Sebe, N., Hauptmann, A.G.: The many shades of negativity. IEEE Trans. Multimedia **19**(7), 1558–1568 (2017)
11. Malisiewicz, T., Gupta, A., Efros, A.A.: Ensemble of exemplar-SVMs for object detection and beyond. In: 2011 IEEE International Conference on Computer Vision (ICCV), pp. 89–96. IEEE (2011)
12. Ngiam, J., Khosla, A., Kim, M., Nam, J., Lee, H., Ng, A.Y.: Multimodal deep learning. In: Proceedings of the 28th International Conference on Machine Learning (ICML-11), pp. 689–696 (2011)
13. Pereira, J.C., et al.: On the role of correlation and abstraction in cross-modal multimedia retrieval. IEEE Trans. Pattern Anal. Mach. Intell. **36**(3), 521–535 (2014)
14. Qin, Z., Yu, J., Cong, Y., Wan, T.: Topic correlation model for cross-modal multimedia information retrieval. Pattern Anal. Appl. **19**(4), 1007–1022 (2016)
15. Ranjan, V., Rasiwasia, N., Jawahar, C.: Multi-label cross-modal retrieval. In: Proceedings of the IEEE International Conference on Computer Vision, pp. 4094–4102 (2015)
16. Rasiwasia, N., et al.: A new approach to cross-modal multimedia retrieval. In: Proceedings of the 18th ACM International Conference on Multimedia, pp. 251–260. ACM (2010)

17. Shrivastava, A., Gupta, A., Girshick, R.: Training region-based object detectors with online hard example mining. In: Proceedings of the IEEE Conference on Computer Vision and Pattern Recognition, pp. 761–769 (2016)
18. Simonyan, K., Zisserman, A.: Very deep convolutional networks for large-scale image recognition. arXiv preprint arXiv:1409.1556 (2014)
19. Srivastava, N., Salakhutdinov, R.R.: Multimodal learning with deep boltzmann machines. In: Advances in Neural Information Processing Systems, pp. 2222–2230 (2012)
20. Vaswani, A., et al.: Attention is all you need. In: Advances in Neural Information Processing Systems, pp. 5998–6008 (2017)
21. Wang, K., He, R., Wang, L., Wang, W., Tan, T.: Joint feature selection and subspace learning for cross-modal retrieval. IEEE Trans. Pattern Anal. Mach. Intell. **38**(10), 2010–2023 (2016)
22. Wang, K., He, R., Wang, W., Wang, L., Tan, T.: Learning coupled feature spaces for cross-modal matching. In: Proceedings of the IEEE International Conference on Computer Vision, pp. 2088–2095 (2013)
23. Wang, L., Li, Y., Lazebnik, S.: Learning deep structure-preserving image-text embeddings. In: Proceedings of the IEEE Conference on Computer Vision and Pattern Recognition, pp. 5005–5013 (2016)
24. Wu, C.Y., Manmatha, R., Smola, A.J., Krahenbuhl, P.: Sampling matters in deep embedding learning. In: Proceedings of the IEEE International Conference on Computer Vision, pp. 2840–2848 (2017)
25. Yu, J., et al.: Modeling text with graph convolutional network for cross-modal information retrieval. In: Hong, R., Cheng, W.-H., Yamasaki, T., Wang, M., Ngo, C.-W. (eds.) PCM 2018. LNCS, vol. 11164, pp. 223–234. Springer, Cham (2018). https://doi.org/10.1007/978-3-030-00776-8_21
26. Zhang, L., Ma, B., He, J., Li, G., Huang, Q., Tian, Q.: Adaptively unified semi-supervised learning for cross-modal retrieval. In: International Conference on Artificial Intelligence, pp. 3406–3412 (2017)

# MeshTrust: A CDN-Centric Trust Model for Reputation Management on Video Traffic

Xiang Tian[1,2,3], Yujia Zhu[2,3]($\boxtimes$), Zhao Li[1,2,3], Chao Zheng[2,3], Qingyun Liu[2,3], and Yong Sun[2,3]

[1] School of Cyber Security, University of Chinese Academy of Sciences, Beijing, China
[2] National Engineering Laboratory for Information Security Technologies, Beijing, China
[3] Institute of Information Engineering, Chinese Academy of Sciences, Beijing, China
{tianxiang,zhuyujia,lizhao,zhengchao,liuqingyun,sunyong}@iie.ac.cn

**Abstract.** Video applications today are more often deploying content delivery networks (CDNs) for content delivery. However, by decoupling the owner of the content and the organization serving it, CDNs could be abused by attackers to commit network crimes. Traditional flow-level measurements for generating reputation of IPs and domain names for video applications are insufficient. In this paper, we present *MeshTrust*, a novel approach that assessing reputation of service providers on video traffic automatically. We tackle the challenge from two aspects: the multitenancy structure representation and CDN-centric trust model. First, by mining behavioral and semantic characteristics, a *Mesh Graph* consisting of video websites, CDN nodes and their relations is constructed. Second, we introduce a novel CDN-centric trust model which transforms *Mesh Graph* into *Trust Graph* based on extended network embedding methods. Based on the labeled nodes in *Trust Graph*, a reputation score can be easily calculated and applied to real-time reputation management on video traffic. Our experiments show that *MeshTrust* can differentiate normal and illegal video websites with accuracy approximately 95% in a real cloud environment.

**Keywords:** CDN · Reputation management · Network embedding · DNS · Trust model · Video traffic analysis

## 1 Introduction

In the past few years, the Internet has witnessed an explosion of video streaming applications. The explosive growth of video websites and traffic is largely due to the popularity of content delivery networks (CDNs).

Besides being used for obvious benign purposes, video websites are also popular for malicious or illegal use [4]. e.g., websites are increasingly playing a role

© Springer Nature Switzerland AG 2019
J. M. F. Rodrigues et al. (Eds.): ICCS 2019, LNCS 11537, pp. 318–331, 2019.
https://doi.org/10.1007/978-3-030-22741-8_23

for the management of disseminating illegal content. Going beyond being popular among adolescents, video content has now evolved into a much-debated public concern because of excessive or maladaptive use. For decades, the public have been consistently concerned about the potentially harmful influences of exposure to pornographic and violent content, being targeted for harassment, cyber-bullying, sexual solicitation, and Internet addiction [12]. Therefore, it is of great significance to label the reputation for video services.

The traditional mechanisms for generating trust and protecting content security by finding the IPs and domain names attached to the website, using the historical data in traffic log cannot be used for tangled networks [2]. This is because the deployment of a horizontally scalable website tending to use public infrastructure, especially video websites distribute video to CDNs. The spreading of video content by CDN service have been brought new challenges to website reputation evaluation problems: (i) dynamic changes of IPs and domains, and (ii) the relationship between website and domains is no longer a one-to-one relationship. But a multi-tenant, multi-CDN graph structure is formed by CDN as a common software service [8].

However, the problem still exits and we are trying to identify the website reputation through the reputation of rented CDN nodes. Our intuition is that the CDN nodes which hosting similar content or providing similar services are likely to be in a homophilic state. In this work, we propose *CCTrust*, a model for evaluating reputation of CDN nodes, and *MeshTrust*, a mechanism that is able to build a CDN-centric model for reputation management. First, we need measurement video websites and CDNs from distributed vantage points in order to characterize lease relationship between them. Second, realizing the reputation of CDN nodes based on network embedding. Third, *MeshTrust* can assign appropriate reputation score of video websites.

This paper provides the following contribution:

(1) We consider the real situation of multi-CDN. Based on this, an automatic detection mechanism, *MeshTrust* is proposed, which solving the problem of reputation evaluation in multi-tenant scenario. To the best of our knowledge, we are the first to detect illegal or malicious websites in cloud scenes using CDN trust model.

(2) We design a novel model, *CCTrust*, constructing *Mesh Graph* from network video traffic and transforming the heterogeneous graph into homogeneous *Trust Graph*.

(3) We extend graph embedding algorithm, transforming the sparse graph from high dimension to low dimension to form a labeled *Trust Graph*. On this basis, a reputation score can be easily calculated and applied to real-time reputation management.

(4) We verify the effectiveness of our mechanism in a real environment. The results show that *MeshTrust* can evaluate websites reputation with accuracy 95%. Meanwhile, our mechanism is lightweight and can be applied to real-time detection.

The rest of the paper is organized as follows. Section 2 presents problem and preliminary. Section 3 presents our *CCTrust* Model. In Sect. 4, we provide details of our mechanism. Experiments on real world datasets are shown in Sect. 5. We comment on related works in Sect. 6. Section 7 gives a conclusion.

## 2   Problem and Preliminary

In this paper, we aim at assessing reputation of video websites by identifying CDN nodes behind specific websites, rather than directly classifying the website's IPs and domains. Before presenting our framework, we introduce multi-CDN. And our problem statement is given.

### 2.1   Multi-CDN

To further ensure the availability of content, customers may use multi-CDN deployments. Other reasons to utilize more than one CDN provider may be cost efficiency, e.g., different prices to serve content at different times or due to traffic volume contracts [8]. Video websites deploy multi-CDN, which mixed multiple CDNs to satisfy various requirements. In an attempt to minimize single points of failure and eke out even more performance advantages many video websites have started to deploy multi-CDN for higher availability, better performance, increased capacity, and better security. Maybe one CDN is suitable for video content while another is suited to hosting image or picture content. By tapping into both, the content owner is getting a more global friendly solution.

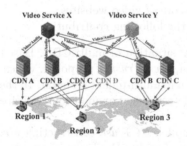

**Fig. 1.** Multi-CDN

In Fig. 1, CDN provider A, B, C, D host content for video service X and Y. For better performance and security in disparate geographic region, CDN providers host different types of resources. Multi-CDN results in a complicated subscription relationships for video websites. Instead of helping users to choose CDN providers, we utilize the complex relationship between websites and CDNs to build a graph. This is also an innovation of our work.

## 2.2   Problem Statement

*MeshTrust* is a novel mechanism for reputation management on video traffic. We focus on multi-tenancy structure in our trust model. First *MeshTrust* builds a *Mesh Graph* consisting of video websites, CDN nodes and their relations by mining behavioral and semantic characteristics from large-scale passive video traffic. Second, *MeshTrust* assesses reputation of video websites by a novel CDN-centric trust model (*CCTrust*). *CCTrust* transforms *Mesh Graph* into *Trust Graph* and extends network embedding methods to learn a mapping function. *CCTrust* outputs the reputation evaluating results of CDN nodes, which is an input of online assessment for websites reputation.

In this paper, we analyze the CDN nodes from the perspective of domains and IPs. We refer to the first sub-domain after the Top-Level Domain (TLD) as Second Level Domain (2LD). It generally refers to the organization that owns the domain name. Fully Qualified Domain Name (FQDN) is the domain name complete with all the labels that unambiguously identify a resource [3].

A Graph is $G = (V, E)$, where $v \in V$ is a node and $e \in E$ is an edge. $G$ is associated with a node type mapping function $f_v : V \to \tau^v$ and an edge type mapping function $f_e : E \to \tau^e$. $\tau^v$ and $\tau^e$ denote the set of node types and edge types, respectively [5].

**Definition 1. *Mesh Graph*.** *Mesh Graph is an interaction information graph, which is defined as $G_m = (W, C, R, \lambda)$, where $W = \{w_1, ..., w_n\}$ represents n websites nodes, $C = \{c_1, ..., c_m\}$ represents m CDN nodes and $R = r_{ij}$ represents edges from W to C, edges means lease relationship. If there exists an edge from $w_i$ to $c_j$, $r_{ij} = 1$. Otherwise, $r_{ij} = 0$. $\lambda$ is the representation of CDN nodes. In our work, $\lambda = 2LD, FQDN, IP$.*

**Definition 2. *Trust Graph*.** *Trust Graph $G_t = (C, E, W, \lambda, L)$ is an undirected graph where C is a node set which is the same in Mesh Graph, the node represents the CDN nodes. E is defined as an edge set, the edge indicates that there are leased relation between CDN nodes with the same video websites. W is the edge weight corresponding to the node similarity. $\lambda$ is the CDN nodes representation. L is a label set of all CDN nodes. The purpose of Trust Graph with network embedding is finding L.*

**Definition 3. *Reputation Score*.** *The reputation score enable dynamic domain name blacklists to counter illegal or malicious website much more effectively [1]. Given a specific video website v, $Score_v$ is reputation scores for this video website. $Score_c$ is reputation scores for a given CDN node c.*

Our main goal is to construct a trust graph for reputation management on video traffic, which can be applied to effectively differentiate normal and illegal applications and activities. Thus the goal of the evaluation reputation problem can be expressed as how to determine $L$ in *Trust Graph* $G_t = (C, E, W, \lambda, L)$.

# 3    The CCTrust Model

We consider identification method of the CDN nodes reputation, based on video website content hosted on these nodes. Combining the historical data in traffic, *CCTrust* model aims to solving the credibility evaluation of CDN nodes. Starting from the *Mesh Graph* of video websites, our model: ($i$) transforms *Mesh Graph* to *Trust Graph*, ($ii$) achieves classification of CDN nodes through applying the network embedding algorithm which could extract the structure feature of a graph, and ($iii$) evaluates the reputation of CDN nodes (Fig. 2).

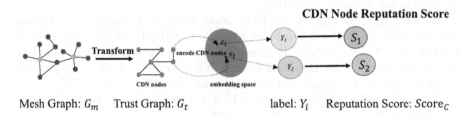

Mesh Graph: $G_m$    Trust Graph: $G_t$        label: $Y_i$    Reputation Score: $Score_c$

**Fig. 2.** CCTrust model overview

## 3.1    Graph Transformation

*Mesh Graph* $G_m = (W, C, R, \lambda)$ is a heterogeneous graph, which $|\tau^v| = 2$ and $|\tau^e| = 1$. All nodes belong to two types, while all edges belong to a single type. Given an *Mesh Graph*, *CCTrust* transform the Graph to *Trust Graph* $G_t = (C, E, W, \lambda, L)$. *Trust Graph* is a homogeneous graph, which $|\tau^v| = |\tau^e| = 1$. All nodes and edges in $G_t$ belong to one single type.

## 3.2    Network Embedding and Classification

Considering the sparsity of the *Trust Graph*, *CCTrust* choose network embedding to finding $L$. **Network embedding** is an important method for learning the low-dimensional representation of vertices in a network. It transforms network information into low-dimensional dense real vectors and is used for input of existing machine learning algorithms to capture and retain the network structure. GCN (Graph Convolutional Networks) pertain to deep learning based graph embedding without random walk paths [5]. The first GCN introduced for learning representations at the node level was in [10], where they utilized GCNs for the semi-supervised node classification problem.

Reputation of nodes based network embedding learns a mapping function $f : C \to \mathbb{R}^d$, where $d \ll |W|$. The mapping function should preserve the graph structure information. After that, all CDN nodes can be classified in a low-dimensional latent space. The input data of the network embedding is *Trust Graph* $G_t = (C, E, W, \lambda, L)$. The output result is the node feature matrix with $N \times F$ dimensions, $F$ is the dimension to feature vector. Our intuition is that once two CDN nodes serve the same website, the two nodes are similar. Based on the fact, we using **Jaccard Distance** to determine the node similarity. In addition, considering the sparsity of the graph, we apply add-one smoothing method to smooth similarity. The similarity between two CDN nodes $c_i$ and $c_j$ can be calculated as:

$$sim_{ij} = \begin{cases} \frac{|V_i \cap V_j|}{|V_i \cup V_j| + n} & |V_i \cap V_j| = \varnothing \\ \frac{|V_i \cap V_j| + (n-1)}{|V_i \cup V_j| + n} & |V_i \cap V_j| = |V_i \cup V_j| \\ \frac{|V_i \cap V_j|}{|V_i \cup V_j|} & otherwise \end{cases} \tag{1}$$

*CCTrust* classifies CDN nodes, determining what is the reputation score of the nodes. The node feature matrix and node label matrix as the input data of GCN. We produce the node label matrix using the label of video website hosted by CDN node. Intuitively, relations between the two nodes directly connected is equivalent to the adjacency matrix of the original network modeling. However, the relationships tend to be very sparse in the network. It is necessary to further depict the global and local similarity to consider nodes which is not directly connected, whereas have respectable common neighbor nodes or same structure. Thus GCN is used for evaluating the CDN nodes reputation.

A two-layer GCN is adapted to CDN node classification on *Trust Graph* with a symmetric adjacency matrix $A_C$ (weighted). The model applies a softmax classifier on the output features:

$$Z = softmax(\hat{A}_C ReLU(\hat{A}_C X W_C^{(0)}) W_C^{(1)}) \tag{2}$$

where $\hat{A}_C = \tilde{D}_C^{-\frac{1}{2}} \tilde{A}_C \tilde{D}_C^{-\frac{1}{2}}$. The loss function is defined as the cross-entropy error over all labeled examples:

$$\zeta = \sum_{c \in y_C} \sum_{l=1}^{|L|} Y_{cl} ln Z_{cl} \tag{3}$$

where $y_C$ is the set of node indices that have labels.

### 3.3 CDN Nodes Reputation Evaluation

In order to evaluate the reputation of CDN nodes based on classified results. Given a CDN node $c_i$, the probability of each label represented by $p_{il}$. According to the predict type, *CCTrust* cacultes the reputation of a specific CDN node $c_i$. The sum total of the probability of each category label is the reputation of the CDN node.

$$Score_{c_i} = \sum_{l=1}^{|L|} p_{il} \tag{4}$$

# 4   The MeshTrust Mechanism

In this section we describe how *MeshTrust* works. Figure 3 shows a high-level overview of this mechanism. *MeshTrust* is composed of off-line and on-line procedures: (*i*) **off-line:** identifying CDN nodes and websites, measurement relationship between them, constructing *Mesh Graph*, and evaluating reputation of CDN nodes by *CCTrust* model. (*ii*) **on-line:** identifying CDN nodes of a specific website, and evaluating the website reputation combined the CDN node reputation results (Fig. 3).

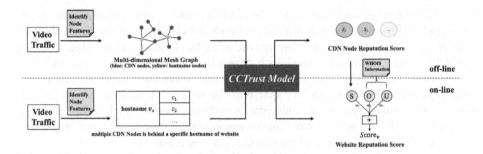

**Fig. 3.** MeshTrust mechanism overview

## 4.1   Mining Nodes from Passive Video Traffic

Research about CDN measurement mostly starting from identifying CDN. The main method of identifying CDN nodes: (*i*) **based on CNAME:** Satellite exploits the reverse PTR records and the WHOIS organization to identify CDN nodes [15]. Singh performs a reverse DNS lookup on the server IP address, develop regular expressions to identify specific CDNs based on the returned hostname [16]. (*ii*) **based on HTTP header:** McDonald identifies CDN population by HTTP hearder [14]. Based on these studies, we use CNAME and HTTP headers to identify CDN nodes.

While there are a large number of CDN domains in passive traffic, we could mine more CDN subdomains corresponding to providers. Monitoring the traffic passively, then parsing the DNS and HTTP traffic. The candidate CDN domains are obtained based on the multi-features identification whether the video resource is hosted on CDN node. Multi-features refer to behavioral and semantic characteristics: (*i*) **behavioral features:** the same content resource corresponds to multiple server-side IPs, (*ii*) **semantic features:** CNAME has CDN keywords or contains the name of the CDN providers.

For the candidate CDN domains, it is necessary to verify the validity of the domain and provider pair through the WHOIS information. The pair determines whether the domain belongs to the CDN provider. For domains that have blank WHOIS information, we can utilize the search engine searching for information. According to match whether the provider name appears in the search result to verify the validity. Finally, we obtain a CDN identification feature library which including CDN provider domain name set, CDN domain general feature words (e.g., cachecdn), and HTTP header (e.g., a special header: "cdn cache server" arises in Server field means it is ChinaNetCenter CDN nodes).

We monitor DNS and HTTP video traffic passively. Eventually, we analyze the relationship between the video website and the CDN nodes.

## 4.2   Mesh Graph Construction

To better utilize content-multihoming, multi-CDN enable content providers to realize custom and dynamic routing policies to direct traffic to the different CDNs hosting their content [8]. The relationship of hosting is intricate. We construct the *Mesh Graph* of video website and CDN nodes, based on multi-domain integration. In a word, our mechanism builds the *Mesh Graph* automatically by merging CDN nodes from the dimensions of providers (such as Akamai), 2LD (such as akamai.net), FQDN (such as cdn.cdn-baidu.net), and IP.

## 4.3   CCTrust Model Construction

The main tasks of *CCTrust* model in reputation evaluation mechanism: transform *Mesh Graph* to *Trust Graph*, identify and evaluate the reputation score of CDN nodes. *CCTrust* transforms the heterogeneous *Mesh Graph* to homogeneous *Trust Graph* which only consist of CDN nodes firstly. The reputation result $Score_c$ based on classifying the CDN nodes in the new low-dimensional space to gain the probability that each node belongs to each category label (e.g., benign, porn, malicious).

## 4.4   Websites Reputation Evaluation

When a specific website occurs in video passive traffic, *MeshTrust* identifying the CDN nodes provide service for the website. In addition, we consider two other factors: WHOIS organization, the update frequency of WHOIS information, which influence the degree of trust on. WHOIS organization is represented by O. The update frequency is represented by U. Our key insight in these reputation factors is that as benign domain names: ($i$) updates WHOIS information more often, while most illegal or malicious domain names almost change WHOIS data rarely, even never change, ($ii$) always has a clear and specific organization information, while illegal domains don't have. The reputation score $Score_v$ of the video website can be calculated as follows, m is the number of domains belonging to the website $v$, n is the number of CDN nodes belonging to the domain $j$:

$$Score_v = w_1 \sum_{j=1}^{m} \sum_{i=1}^{n} c_j Score_{c_{ij}} + w_2 \sum_{j=1}^{m} o_j + w_3 \sum_{j=1}^{m} u_j \tag{5}$$

After determining the reputation of these CDN nodes. WHOIS organization and update frequency are supplementary, together calculating the reputation score $Score_v$.

# 5    Experimental Results

In order to evaluate the availability of *MeshTrust* for identifying illegal video websites, we verify the evaluation effectiveness on the reputation of CDN nodes and websites. *MeshTrust* (*i*) surveys and maps the dependency relation of video websites and CDNs, (*ii*) building *CCTrust* model, and (*iii*) evaluating the reputation of the video websites.

## 5.1    Input Data

We adopt the observation configuration of passive traffic, and verify the effect of our mechanism according to the known website category. We judge whether it is video traffic by file suffix. We sniffed 24-h long traffic in February 28, 2018 and June 23, 2018 within a large ISP. We focus on 2,000 video websites, which contain the Baidu rank[1] of China video websites TOP 700, Alexa rank[2] of other countries video websites TOP 500 (we remove the websites from the list of publicly available illegal websites), and 800 porn video websites which are produced from public blacklist. We extract the traffic which is relevant to these specific websites by matching the domain names. We labeled the websites derived from Baidu rank and Alexa rank are benign, these sites continue to appear in the ranking after a long period of observation.

## 5.2    Characterizing CDNs of Video Websites

The CDN service lease relationship is measured from the dimension of geographical location. Besides passive video traffic, we sniff traffic with measurement points covering six countries: China, American, Australia, Japan, South Korea and Singapore. At each measurement point, our mechanism automatically simulated user behaviour to visit video URL. The measurement of CDN service shows that there are 388 mainstream benign video websites rent CDNs. Meanwhile, the number of porn websites which exist lease relationship with CDNs is 66. We focus on analyzing the measurement results of video websites which rent CDNs in China.

**Measurement the Tenancy of Chinese Website.** In the measurement result of Chinese website, the top 100 video websites all use CDN, and 33% of video websites rely on more than one CDN provider. About 10% of video websites rely on three or more CDN, e.g., iQIYI mainly use self-built CDN in China, while using Akamai CDN in American and Australia primarily. The video resource of Youku is hosted by AliCloud, but the picture resources are mostly hosted by Akamai and ChinaNetCenter. These further indicates that video websites tend to rely on the parallel mode of multi-CDN providers for distributing service.

---

[1] http://top.chinaz.com/hangye.

[2] https://www.alexa.com/topsites.

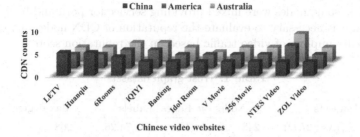

**Fig. 4.** Popularity of multi-CDN

## 5.3 Evaluating Website Reputation Based on CCTrust Model

To evaluate the accuracy of identifying result and prove the effectiveness of *MeshTrust*. We perform *CCTrust* to model the CDN nodes in *Mesh Graph*. Consequently, we evaluate the reputation of video websites. *Mesh Graph* has been constructing completely in the previous step of the mechanism.

The input data of *CCTrust* model is *Mesh Graph*, which contains CDN and website nodes. Firstly, we produce the input data of evaluating CDN nodes reputation by transforming *Mesh Graph* to *Trust Graph*.

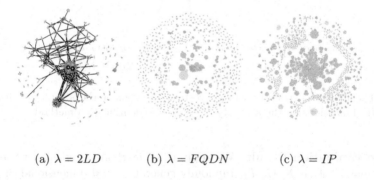

(a) $\lambda = 2LD$          (b) $\lambda = FQDN$          (c) $\lambda = IP$

**Fig. 5.** Mesh Graph of Video Websites: the porn or benign websites with orange, yellow respectively, and CDN nodes with gray (Color figure online)

*Mesh Graph.* We construct the *Mesh Graph* form three dimensions: $(i)$ $\lambda = 2LD$, we construct the video website *Mesh Graph* $G_{m_1}$, and label the websites porn or benign. There are 248 CDN nodes which belongs to 168 CDN providers. $(ii)$ $\lambda = FQDN$, analogously, we build $G_{m_2}$, which contains 1580 grey FQDN nodes. $(iii)$ $\lambda = IP$, $G_{m_3}$ contains 4408 CDN IP nodes.

In Fig. 5, we can visually observe that some CDN nodes provide services primarily for porn websites, and the platforms of these nodes have poor self-regulation ability. While some provide services only for non-porn websites, and the platforms of these nodes are likely to have strong self-regulation ability.

In addition, some nodes were mixed providing service for porn and benign websites. And it is necessary to evaluate the reputation of CDN nodes and conduct hierarchical sampling of video traffic based on the reputation scores.

**Table 1.** Trust graph datasets

| Datasets | # Nodes | # Edges | Labels | Max degree |
|---|---|---|---|---|
| $G_{t_1}(\lambda = 2LD)$ | 248 | 1,316 | 148;74;26 | 208 |
| $G_{t_2}(\lambda = FQDN)$ | 1,604 | 23,282 | 1179;411;14 | 152 |
| $G_{t_3}(\lambda = IP)$ | 4,408 | 298,336 | 3157;1140;111 | 65 |

*Trust Graph.* Three *Mesh Graph* transform to three corresponding *Trust Graph*: (*i*) $G_{t_1}$: $\lambda = 2LD$, including 148 benign, 74 porn and 26 CDN nodes serving the porn and benign website, a total of 248 CDN nodes, and *porn : benign* = 1 : 2. (*ii*) $G_{t_2}$: $\lambda = FQDN$, *porn : benign* = 1 : 3. (*iii*) $G_{t_3}$: $\lambda = IP$, *porn : benign* = 1 : 3 (Table 1).

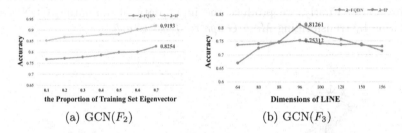

(a) GCN($F_2$)          (b) GCN($F_3$)

**Fig. 6.** The influence of training set eigenvector proportion on the accuracy of GCN($F_2$) and LINE dimensions on the accuracy of GCN($F_3$) identification method

**Off-line Evaluation Result.** We apply three methods to construct the node feature matrix of GCN: (*i*) $F_1$: randomly generate $n \times n$ dimensional diagonal matrix, n denotes the number of nodes, (*ii*) $F_2$: produce website feature matrix, that the feature vector of each node is formed with the presence of all the websites. (*iii*) $F_3$: obtain the output node feature matrix of network embedding, such as LINE [17]. LINE is a algorithm fall into edge reconstruction based optimization. It means the edges established based on node embedding should be as similar to those in the input graph as possible [5]. As shown in Fig. 6(a), even when the percentage of training set is 10%, the accuracy of GCN($F_2$) method is better than 77%. Therefore, our model can be used to identification CDN nodes on video traffic by training fewer samples. When the dimension of LINE is 96, GCN($F_3$) could achieve the higher accuracy in Fig. 6(b). The classification result indicates that the optimum solution obtained by using $F_2$ matrix. The optimal accuracy could achieve **82.54%** ($G_{t_2}$) and **91.93%** ($G_{t_3}$).

**Table 2.** Evaluating the website reputation results

| Datasets | # Website | Benign | Illegal | # Max CDN nodes | Accuracy |
|----------|-----------|--------|---------|-----------------|----------|
| $W_1$ | 968 | 385 | 583 | 87 | **95.25%** |
| $W_2$ | 1,287 | 380 | 907 | 214 | **92.23%** |

**On-line Evaluation Result.** We evaluate our model on the datasets, and the result is illustrated in Table 2. We monitor traffic passively to acquire CDN nodes behind specific websites. We collect $W_1$ and $W_2$ dataset to validate the reputation on account of the reputation result of $G_{t_2}$ and $G_{t_3}$ respectively. CCTrust performs reputation evaluation on account of $W_1$ and $W_2$. The results show that MeshTrust could evaluate the reputation of websites, identify the illegal and benign websites with accuracy **95.25%** ($W_1$) and **92.23%** ($W_2$) approximately when $w_1 = 0.7$, $w_2 = 0.15$, $w_3 = 0.15$.

# 6   Related Work

Several prior studies have investigated the identification of malicious domains and reputation models for websites. These relevant research based on characteristics: content characteristics (e.g., tag information of HTML pages), URL information (e.g., URL vocabulary information, URL length), DNS logs (e.g., WHOIS information, AS numbers).

**Based-On Content.** Canali used the abnormal content and abnormal structure of the web page as a sign to judge the maliciousness of the domains [6]. Such methods require more computing resources and network bandwidth, and generate a larger time overhead. The time required to analyze a web page also depends on the network latency and the complexity of the web content.

**Based-On URL Information.** Starting from the URL structure information, Le improved the accuracy of malicious domain recognition [11]. Li based on the network topology relationship with the PageRank algorithm, identifying malicious domains, which makes the false detection rate control within 2% [13]. Due to the widespread practice of HTTPS, the URL information cannot be obtained. Additionally, with the feature scale grows linearly, the characteristic set expanding.

**Based-On DNS Log.** Antonakakis proposed Notos, which handles DNS query responses from passive DNS traffic and extracts a set of 41 features from the observed FQDN and IP [1]. Chiba proposed Domain-Profiler to actively collect DNS logs, analyze time-change patterns, and predict whether a given domain will be used for malicious purposes [7]. In line with the global association diagram between domain and IP, Khalil proposed a path-based mechanism to derive the malicious score of each domain [9]. The above systems almost utilize the characteristics of the domain to identify the type of domains and further evaluate the reputation of the websites. Due to DNS encryption and cross-use of IPs and domains, DNS traffic is no longer suitable for detection.

Due to the web tangle: ($i$) multiple services and resources co-located on the same CDN ($ii$) more and more content providers employ multiple CDNs to serve the same content to reduce costs or to select CDNs by optimal performance, these methods above are insufficient. As far as we know, we are the first to use the CDN trust graph to solve the problem of reputation assessment for video applications in a multi-tenant scenarios.

# 7    Conclusion

In this paper, we propose *MeshTrust* based on *CCTrust* Model. *MeshTrust* evaluates video website reputation by *CCTrust* rather than directly using the website IPs or domains. Our mechanism is able to evaluate the websites reputation with accuracy approximately 95%. Then within the evaluation result, the back-end priority processes the associated traffic of less reputable domains. Treating video traffic differently which can greatly improve the processing content censorship performance of video traffic.

The passive traffic exploited in our work is sniffed from only one ISP with a limited range of traffic coverage. Therefore, the measured dependence of video websites and CDNs is not comprehensive, resulting in a smaller coverage of the discovered CDN nodes. We evaluate the reputation of the nodes over static data acquired in a single time point, which can be seen as a snapshot of network. Therefore, the *Mesh Graph* is static. In the future, we would deploy *MeshTrust* in a larger network environment.

**Acknowledgments.** We would like to thank hard work of MESA TEAM (www.mesalab.cn). This work was supported by National Key R&D Program 2016 (Grant No. 2016YFB0801300); the Strategic Priority Research Program of the Chinese Academy of Sciences (Grant No. XDC02030600). The corresponding author is Yujia Zhu.

# References

1. Antonakakis, M., Perdisci, R., Dagon, D., Lee, W., Feamster, N.: Building a dynamic reputation system for DNS. In: USENIX Security Symposium, pp. 273–290 (2010)
2. Berger, A., D'Alconzo, A., Gansterer, W.N., Pescapé, A.: Mining agile DNS traffic using graph analysis for cybercrime detection. Comput. Netw. **100**, 28–44 (2016)
3. Bermudez, I.N., Mellia, M., Munafo, M.M., Keralapura, R., Nucci, A.: DNS to the rescue: discerning content and services in a tangled web. In: Proceedings of the 2012 Internet Measurement Conference, pp. 413–426. ACM (2012)
4. Bilge, L., Kirda, E., Kruegel, C., Balduzzi, M.: EXPOSURE: finding malicious domains using passive DNS analysis. In: Ndss (2011)
5. Cai, H., Zheng, V.W., Chang, K.: A comprehensive survey of graph embedding: problems, techniques and applications. IEEE Trans. Knowl. Data Eng. **30**, 1616–1637 (2018)

6. Canali, D., Cova, M., Vigna, G., Kruegel, C.: Prophiler: a fast filter for the large-scale detection of malicious web pages. In: Proceedings of the 20th International Conference on World Wide Web, pp. 197–206. ACM (2011)
7. Chiba, D., et al.: DomainProfiler: discovering domain names abused in future. In: 2016 46th Annual IEEE/IFIP International Conference on Dependable Systems and Networks (DSN), pp. 491–502. IEEE (2016)
8. Hohlfeld, O., Rüth, J., Wolsing, K., Zimmermann, T.: Characterizing a meta-CDN. In: Beverly, R., Smaragdakis, G., Feldmann, A. (eds.) PAM 2018. LNCS, vol. 10771, pp. 114–128. Springer, Cham (2018). https://doi.org/10.1007/978-3-319-76481-8_9
9. Khalil, I., Yu, T., Guan, B.: Discovering malicious domains through passive DNS data graph analysis. In: Proceedings of the 11th ACM on Asia Conference on Computer and Communications Security, pp. 663–674. ACM (2016)
10. Kipf, T.N., Welling, M.: Semi-supervised classification with graph convolutional networks. arXiv preprint arXiv:1609.02907 (2016)
11. Le, A., Markopoulou, A., Faloutsos, M.: PhishDef: URL names say it all. In: 2011 Proceedings IEEE INFOCOM, pp. 191–195. IEEE (2011)
12. Leung, L.: Predicting internet risks: a longitudinal panel study of gratifications-sought, internet addiction symptoms, and social media use among children and adolescents. Health Psychol. Behav. Med. Open Access J. **2**(1), 424–439 (2014)
13. Li, Z., Alrwais, S., Xie, Y., Yu, F., Wang, X.: Finding the linchpins of the dark web: a study on topologically dedicated hosts on malicious web infrastructures. In: 2013 IEEE Symposium on Security and Privacy (SP), pp. 112–126. IEEE (2013)
14. McDonald, A., et al.: 403 forbidden: a global view of CDN geoblocking. In: Proceedings of the Internet Measurement Conference 2018, pp. 218–230. ACM (2018)
15. Scott, W., Anderson, T.E., Kohno, T., Krishnamurthy, A.: Satellite: joint analysis of CDNS and network-level interference. In: USENIX Annual Technical Conference, pp. 195–208 (2016)
16. Singh, R., Dunna, A., Gill, P.: Characterizing the deployment and performance of multi-CDNS. In: Proceedings of the Internet Measurement Conference 2018, pp. 168–174. ACM (2018)
17. Tang, J., Qu, M., Wang, M., Zhang, M., Yan, J., Mei, Q.: Line: large-scale information network embedding. In: Proceedings of the 24th International Conference on World Wide Web, pp. 1067–1077. International World Wide Web Conferences Steering Committee (2015)

# Optimizing Spatial Accessibility
# of Company Branches Network
# with Constraints

Oleg Zaikin[1]([envelope]) [ORCID], Ivan Derevitskii[1] [ORCID], Klavdiya Bochenina[1] [ORCID],
and Janusz Holyst[1,2] [ORCID]

[1] ITMO University, Saint Petersburg, Russia
zaikin.icc@gmail.com, kbochenina@mail.ifmo.ru
[2] Faculty of Physics, Warsaw University of Technology, Warsaw, Poland

**Abstract.** The ability of customer data collection in enterprise corporate information systems leads to the emergence of customer-centric algorithms and approaches. In this study, we consider the problem of choosing a candidate branch for closing based on the overall expected level of dissatisfaction of company customers with the location of remaining branches. To measure the availability of branches for individuals, we extract points of interests from the traces of visits using the clustering algorithm to find centers of interests. The following questions were further considered: (i) to which extent does spatial accessibility influence the choice of company branches by the customers? (ii) which algorithm provides better trade-off between accuracy and computational complexity? These questions were studied in application to a bank branches network. In particular, data and domain restrictions from our bank-partner (one of the largest regional banks in Russia) were used. The results show that: (i) spatial accessibility significantly influences customers' choice (65%–75% of customers choose one of the top 5 branches by accessibility after closing a branch), (ii) the proposed greedy algorithm provides on optimal solution in almost all of cases, (iii) output of the greedy algorithm may be further improved with a local search algorithm, (iv) instance of a problem with several dozens of branches and up to million customers may be solved with near-optimal quality in dozens of seconds.

**Keywords:** Branch network optimization · Location ·
Spatial accessibility · Banking · Black-box optimization

## 1 Introduction

Networks of company branches operating within the city arise in different domains like retail industry, banking and finance, car manufacturing and many other. Different units, e.g., shops or bank branches, may be located in different areas, serve different number of customers and may have different efficiency of functioning related to actions of line staff and managers. In this regard, modification of the network of branches (closing or opening units) is considered as

© Springer Nature Switzerland AG 2019
J. M. F. Rodrigues et al. (Eds.): ICCS 2019, LNCS 11537, pp. 332–345, 2019.
https://doi.org/10.1007/978-3-030-22741-8_24

multi-criteria optimization problem with at least two criteria: (i) profitability of a branch (taking into account cost of maintenance, remuneration, rental charges etc.), (ii) Quality-of-Service (QoS) requirements (related to the business standards of customer care).

Collection of massive data sets about customer behavior makes possible to conclude about implicit factors of customer satisfaction (QoS criteria). For the set of branches distributed within the city and having identical functionality, one of the determinant factors of customer satisfaction is spatial availability of the branches. To decide if a certain branch is located conveniently for a certain customer, one needs to know the areas of his or her frequent visits. After that, we can optimize branches network to achieve better spatial availability for the population of customers or to reduce the effect of closing the branches.

This study is focused on the latter case when a decision maker is aimed to close a fixed number of branches to reduce operational costs of a network. As a baseline business case, we consider a problem of optimizing bank branches network under constraint on maximum desirable number of units after modification of the network. We assume that each customer is assigned with a number of points of interest, and available branches may be located in the neighborhood of each of these points. We formulate a problem of optimizing spatial availability of branches networks and propose several greedy algorithms to solve it. To show that spatial availability is related to actual choice of branches by customers, we make a prediction on preferable set of branches for each customer according to her points of interests, and compare the predictions with the actual places of visits after closing the branch. As the research is conducted in partnership with one of the largest regional banks in Russian Federation, for the experimental study we use the data about visits and transactions of more than 800 000 customers for 2 years.

The outline of the paper is as follows. In the next section, related works are discussed. In Sect. 3 a formulation of the problem is given. In Sect. 4 algorithms for solving the considered problem are proposed. Section 5 contains a description of the data set and the results of computational experiments. In the last section, conclusions are drawn.

## 2   Related Work

The problem of branch location evaluation has been widely studied. In [12], this problem in application for shopping centers is solved. The following characteristics are taken into account: diversity of the tenant inside the shopping center, retail agglomeration near the shopping center, distance to metro stations, and distance between consumers and shopping center. To calculate the last two characteristics, the Euclidean distance is used. The problem of branch network optimization was studied in [14]. In that paper, a Lagrangian relaxation optimization method is proposed to optimize location of locomotive maintenance shops. In [10], a multi-agent model for optimizing supermarkets location is proposed. Using the Particle Swarm Optimization heuristic, the model iteratively determines a place maximizing the sales volume of the new supermarket.

The problem of branch location evaluation in application to a bank network is studied in a number of papers. In [2] the problem is solved by an integrated method, based on analytic hierarchy process, geographic information system and maximal covering location problem. In [13], this problem is formulated and studied as a Markov decision process. In [6] the problem was studied using demand–covering models which determine the locations that achieve the maximum capture of the demand.

The problem of optimizing a bank branches network has been also studied recently. In [9], the problem is solved for the branch network of a German retail bank. A decision support system us used for this purpose. In [3] the problem was studied by a tabu search optimization algorithm [5] in application to a Turkish bank's branches network. In [7] the problem was solved by the following algorithm. A related linear problem is solved first, then the obtained solution is iteratively improved by a local search algorithm. In [11], three heuristic algorithms are suggested to solve the problem: greedy interchange, tabu search and Lagrangian relaxation approximation. A bank's branches network in a large-size town Amherst, USA was analyzed by these algorithms.

In this study, due to the availability of digital traces of customers of our industrial partner, we present new statement of the problem of optimizing branch location which uses zones of interests of individuals instead of city-wide information about popularity of different areas. As a result, we formulate target function as spatial availability estimated directly from points of interests of different customers. Thus, the present study is a step forward in branch network optimization and, in particular, in a bank branches network optimization.

## 3    Problem Statement

Suppose we have a company with branches network of size $M$ within the city, and $b_1, ..., b_k, ..., b_M$ – branches represented with pairs of geo coordinates. This company has $N$ associated customers $c_1, ..., c_i, ..., c_N$. Customers are described with the following attributes: (i) place of living $L$ (geo coordinates); (ii) place of work $W$ (geo coordinates); (iii) ordered sequence of pairs $\langle date, branch \rangle$ of visits to branches; (iv) ordered sequence of pairs $\langle date, P_{ij} \rangle$ of visits to different locations within the city where $P_{ij}$ is $j$-th known location for $i$-th customer. If the considered company is a bank, then (iv) may be extracted, from transactional data, because transactions have the field 'Address', and one can extract from this field geo coordinates of a location where purchase was made. In this case, (iv) for a fixed date is a daily path of payments of a user within the city.

Let's assume that a company decided to close $K$ branches of $M$ (e.g. for optimizing the expenses). The problem is to find $K$ of $M$ branches which closing leads to the smallest reduction in spatial availability of branches network for the customers.

In the simplest case, all branches for customer $i$ can be divided into two types, *accessible* and *non-accessible*. By the accessible branch for a customer $i$, we mean a branch that is within the $\delta$-neighborhood of **any** of points $S = \{H, L, P_{ij}\}$.

Here $\delta$ may be chosen as a maximum walking distance still comfortable for a customer, e.g. 1 km.

Let's introduce the **accessibility coefficient** $a_{ik}$ as a measure of accessibility of a branch $k$ for a customer $i$. This coefficient may be defined in several ways as follows:

(i) $a_{ik} = \mathbb{I}_\delta(c_i, b_k)$ where $\mathbb{I}_\delta(c_i, b_k)$ is a binary variable. If its value is 0, it denotes that $k$-th branch is not $\delta$-accessible for customer $i$, and 1 denotes opposite situation;

(ii) $a_{ik} = \left(\frac{\mathbb{I}_\delta(c_i, b_k)}{\mathbb{B}_\delta(c_i)}\right)^\gamma$ for $\mathbb{B}_\delta(c_i) = \sum_k \mathbb{I}_\delta(c_i, b_k) > 0$, and 0 otherwise ($\gamma > 0$). The meaning of this expression is as follows. When $b_k$ is not accessible for customer $i$ (or customer $i$ does not have any accessible branch in $\delta$-neighborhood), the accessibility coefficient is equal to zero. If $b_k$ is accessible, $a_{ik}$ is the greater, the smaller is the number of accessible branches for customer $i$, $\mathbb{B}_\delta(c_i)$. The logic behind that is that customers with a higher number of accessible branches should influence the decision about closing one of them to a smaller extent than customers for which the accessible branch is unique. Here, coefficient $\gamma$ allows to tune the extent to which customers with increasing number of accessible branches will influence the results.

(iii) $a_{ik} = e^{-\lambda \cdot d_{min}(c_i, b_k)}, \lambda > 0$, where $d_{min}(c_i, b_k) = \min_j d(P_{ij}, b_k)$—minimum distance to branch $b_k$ from any of the points of interests of customer $i$. In such a case, accessibility coefficient exponentially decreases with an increase of a distance of branch $b_k$ from the nearest of points of customer $i$.

Further, we create an **accessibility matrix** $A = \{a_{ik}\}$, in which the sum of elements for $i$-th row ($i$-th customer) means *the potential coverage of a customer*, that is, a large value of the sum corresponds to the case when a customer has a lot of appropriate branches; and a small value of the sum denotes critical customers. For the case (ii), this value for the customers with at least one accessible branch, is equal to 1. The sum by columns for all three cases means *the potential popularity of the branch*.

Considering this matrix, we may also come to conclusion that if $\mathbb{B}_\delta(c_i) > K$ that customer $c_i$ will be satisfied after elimination of each $K$ columns, and then, to exclude them from consideration.

The optimization problem can be stated as follows: to find $K$ columns of accessibility matrix $A$ which deletion will minimize the number of customers with $\mathbb{B}_\delta(c_i) = 0$ in the case (i) or minimize the decrease of $\sum_i a_{ik}$ in the cases (ii) and (iii). In other words, we would like to reduce the number of branches keeping the number of potentially satisfied customers as large as it is possible. Further we shall focus on the case (i) and by *potentially satisfied customer*, we mean customer having at least one branch in the $\delta$-neighborhood of any of his or her visiting points.

# 4  Proposed Algorithms

In order to solve the problem considered in the previous section, algorithms of three types were developed: (i) a brute-force algorithm, that guarantees to find exact solutions, but work very slowly; (ii) greedy algorithms, which are able to quickly find approximate solutions; (iii) a local search algorithm that can be used to find an approximate solution or to improve a solution found by a greedy algorithm. All these algorithms are described below. Also, in Subsect. 4.4 several examples of the accessibility matrices and results obtained by the proposed algorithms on them are shown.

## 4.1  Brute-Force Algorithm

The optimal solution of the considered problem can be found by the brute-force approach. There are two possible variants. According to the first one, $\binom{M}{M-K}$ different combinations of choice of $M - K$ 'good' branches from $M$ branches are generated, then for each of them the number of potentially satisfied customers is calculated, and the optimal combination (with the greatest number of satisfied customers) of $M - K$ branches is chosen. According to the second variant, $K$ 'bad' branches should be found, so $\binom{M}{K}$ combinations are generated, then for each of them the number of potentially unsatisfied customers is calculated, and the optimal combination (with the smallest number of unsatisfied customers) of $K$ branches is chosen.

In practice, $M \gg K$, and also $K$ is quite small (as one wants to close just several branches), so in the following computational experiments the second variant (choosing $K$ 'bad' branches) is used. Since $\binom{M}{K} = O(M^K)$, the algorithm works in polynomial time. The pseudo-code is shown in Algorithm 1.

---

**Algorithm 1.** Brute-force algorithm

---

**Input:** $M$ bank branches, the number $K$ of branches to close,
            accessibility matrix $A$
**Output:** Set $S_{out}$ of $K$ bank branches recommended for closing
1  Generate $\binom{M}{K}$ different combinations of $K$ bank branches
2  For each combination calculate the number of unsatisfied customers
3  $S_{out} \leftarrow$ the combination with the lowest number of unsatisfied customers
4  **return** $S_{out}$

---

## 4.2  Greedy Algorithms

Despite the fact, that the proposed brute-force algorithm works in polynomial time, on some hard instances of the considered problem it cannot be executed in reasonable time. A greedy algorithm can be used instead to quickly find an approximate solution, which can be similar to the optimal one (but it is not guaranteed).

The following greedy algorithm is based on closing $K$ branches with the smallest number of unsatisfied customers one by one. Further this algorithm is denoted as GreedyLP (LP stands for Lowest Popularity). The pseudo-code of this algorithm is shown in Algorithm 2.

---

**Algorithm 2.** GreedyLP algorithm

---

**Input:** $M$ bank branches, the number $K$ of branches to close,
accessibility matrix $A$
**Output:** Set $S_{out}$ of $K$ bank branches recommended for closing
1  $S_{out} \leftarrow \emptyset$
2  **for** $step = 1 \dots K$ **do**
3  | Choose branch $b^*$ with the smallest number of unsatisfied customers
4  | $S_{out} \leftarrow b^*$
5  | Remove $b^*$ from $A$
6  | Recalculate the number of unsatisfied customers of remaining branches
7  **return** $S_{out}$

---

The following greedy algorithm is based on keeping branches with the greatest number of satisfied customers. Further this algorithm is denoted as GreedyHP (HP stands for Highest Popularity). The pseudo-code of this algorithm is shown in Algorithm 3.

---

**Algorithm 3.** GreedyHP algorithm

---

**Input:** $M$ bank branches, the number $K$ of branches to close,
accessibility matrix $A$
**Output:** Set $S_{out}$ of $K$ bank branches recommended for closing
1  $S_{out} \leftarrow$ all $M$ branches
2  **for** $step = 1 \dots M - K$ **do**
3  | Choose branch $b^*$ with the greatest number of satisfied customers
4  | Remove $b^*$ from $S_{out}$
5  | Delete from $A$ all rows with $a_{ib^*} = 1$ and column $b^*$
6  | Recalculate the number of unsatisfied customers of remaining branches
7  **return** $S_{out}$

---

Both proposed greedy algorithms have linear time complexity. If $K < \frac{M}{2}$, GreedyLP requires less iterations than GreedyHP.

## 4.3  Simple Hill Climbing Algorithm

To find an approximate solution, it is also possible to use local search heuristics. A set of $K$ branches corresponds to a Boolean vector of size $M$, where $i - th$ component is 1 if the set contains branch number $i$ and 0 otherwise. Such a vector, in turn, corresponds to a point in a search space $S$. This search space consists of all different Boolean vectors of size $M$ and with exactly $K$ 1s, so its size is $\binom{M}{K}$. For an arbitrary point $\chi \in S$, a neighborhood $Nh(\chi)$ of radius $R$ is defined as a set of such points $\chi'$, $\chi' \in S$, that $d_H(\chi, \chi') = R$, where $d_H(\chi, \chi')$ stands for Hamming distance between $\chi$ and $\chi'$.

Let's consider an objective function which operates in this search space. Given a point from the search space, it calculates the number of customers that would be potentially unsatisfied after the closure of all corresponding branches. Any discrete black-box optimization algorithm can be employed to minimize it, e.g., the genetic algorithm [1] or the tabu search [5]. As an optimization algorithm, the simple hill climbing algorithm [8] was chosen.

Simple hill climbing starts from a given point, in the role of which it is possible to use a random point from the search space. Then the algorithm checks points from the neighborhood of the given point. If a better point (i.e. a point with lower value of the objective function) is found, then this new point is considered as a new best point, and checking of its neighborhood is started. If all points from a current neighborhood are worse than a current best point, then it means that a local minimum is reached. The pseudo-code of the proposed algorithm is shown in Algorithm 4. Here $\chi_{best}$ is the point with the best found value of the objective function.

---

**Algorithm 4.** Simple hill climbing algorithm

---

**Input**: Set of $M$ bank branches, the number $K$ of branches to close, accessibility matrix $A$

**Output**: Set $S_{out}$ of $K$ bank branches recommended for closing

1  $\chi_{best} \leftarrow$ randomly chosen point of size $M$ and weight $K$
2  **repeat**
3  | NewOptimum $\leftarrow$ false
4  | **for each** $\chi \in$ GetNeigborhood($\chi_{best}$) **do**
5  | | **if** $ObjFunction(\chi, A) < ObjFunction(\chi_{best}, A)$ **then**
6  | | | $\chi_{best} \leftarrow \chi$
7  | | | NewOptimum $\leftarrow$ true
8  | | | Break
9  **until** NewOptimum = false **or** TimeExceeded()
10  $S_{out} \leftarrow$ branches for which $\chi_{best}$ contains 1s
11  **return** $S_{out}$

---

In the proposed algorithm, checking a neighborhood of a given point has a linear time complexity. The amount of such checks cannot be predicted, but in practice for the considered problem it is usually less than 10.

## 4.4   Examples

Let's consider two instances of the proposed problem. In the first one, $M = 4, K = 2$, the accessibility matrix is $A_1$, see Table 1. In the second one, $M = 3, K = 1$, the accessibility matrix is $A_2$.

GreedyLP finds a non-optimal solution for the first instance. It removes branch 1 on step 1 and branch 2 on step 2. As a result, one customer which has only one accessible branch is unsatisfied. As for GreedyHP, it keeps branch 1 on step 1 and branch 4 on step 2, and as a result all customers are satisfied. However, GreedyHP finds a non-optimal solution for the second instance.

**Table 1.** Examples of the accessibility matrices

$$
A_1 = \left(\begin{array}{cccc|c}
1 & 1 & 1 & 0 & 3 \\
1 & 1 & 1 & 0 & 3 \\
1 & 1 & 1 & 0 & 3 \\
1 & 1 & 0 & 1 & 3 \\
0 & 0 & 0 & 1 & 1 \\
\hline
4 & 4 & 3 & 2 &
\end{array}\right)
\qquad
A_2 = \left(\begin{array}{ccc|c}
1 & 0 & 0 & 1 \\
1 & 0 & 1 & 2 \\
1 & 0 & 1 & 2 \\
1 & 0 & 1 & 2 \\
0 & 1 & 1 & 2 \\
0 & 1 & 1 & 2 \\
0 & 1 & 0 & 1 \\
0 & 1 & 0 & 1 \\
\hline
4 & 4 & 5 &
\end{array}\right)
$$

GreedyHP keeps branch 3 at step 1, and branch 2 at step 2, leaving customer 1 unsatisfied. However, if branches 1 and 2 are kept, all customers will be satisfied.

As for simple hill climbing, it finds the optimal solution for both considered instances if it starts from a solution found by any proposed greedy algorithm.

## 5   Computational Experiments

In this section, a data set used for computational experiments is described first. Then the influence of the spatial accessibility on customers choice of bank branches is studied. Finally, the experimental study of the algorithms, proposed in Sect. 4 is presented.

### 5.1   Data Set Description

We consider a data set from one of the largest regional banks in Russia. This data set contains data about 58 bank branches, 844 864 customers and 3 132 296 customers' visits. Customers visits in one year (from September 2017 to September 2018) were taken into account in this study. Note, that only 51 out of 58 bank branches were active at the moment of this study. In Fig. 1a the distribution of bank branches by the amount of customers visits is presented. In Fig. 1b the distribution of bank branches by customers visits per month is shown. Seasonal effects can be seen: the activity during the summer is decreasing, while March and December are highly popular months.

According to Sect. 3, three types of points of interests (POIs) are used for each customer. The first two types (home address and work address) were provided by the industrial partner. POIs of the third type (places of visits) were taken from history of debit card transactions (the text field "Address" was used). This allows us to supplement information about home and work POIs which a customer reports to a bank because this information may become outdated.

To transform address data into pairs of geo coordinates, we use Open-StreetMap open database. OpenStreetMap is a map service, in which data are updated by its users in a collaborative manner. Using OpenStreetMap, we have

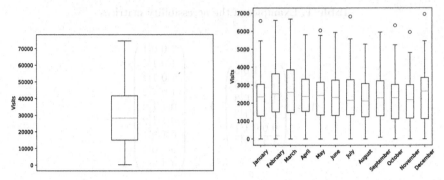

(a) Distribution of bank branches by the amount of customers visits

(b) Distribution of bank branches by customers visits per month is shown

**Fig. 1.** Customers visits to bank branches

compiled a database of addresses of St. Petersburg. This database matches text representation of addresses to their coordinates. Then, we were able to identify geo coordinates of customers' POI.

POI extracted from transactions may contain outliers—points that were visited only once and that are off the usual routes of a customer. As our goal is to identify the frequently visited locations, we propose the following approach of getting POI from traces of visits to locations for a given customer $c_i$:

1. Collect a set of initial POI $\{l_{iv}\}, v = 1 \ldots V_i$ where $V_i$ is a number of distinct locations of a customer $i$, $l_{iv}$ is a location of $v$-th point.
2. Group a set of locations $\{l_{iv}\}$ into $\hat{V}_i$ clusters.
3. Remove clusters which contain a single point to get $V_i^*$ clusters.
4. Find centers of $V_i^*$ clusters and replace initial set of POI with $V_i^*$ points which stand for the centers of payment interests of customer $c_i$.

We used this algorithm to get POI from transactional data. For step 2, DBSCAN clustering method was used [4].

## 5.2 On the Importance of Spatial Accessibility

We tested the hypothesis that the spatial accessibility of bank branches is important for customers. Two bank branches closed in Spring of 2018 were chosen for this purpose. Hereinafter they are called Branch 1 and Branch 2. For each of these branches, customers visiting it at least twice during the last 6 months before the closure were considered. It turned out, that for Branch 1 there were 964 such customers, while for Branch 2 there were 1483 of them. It was counted, how many customers visited other bank branches during 6 months after the closure. For Branch 1, there were 252 such customers, for Branch 2 – 409 customers. Then it was analyzed how many of these customers visited the most accessible active branches. The results are presented in Table 2. Here 'Top-i' stands for the amount of customers visited at least one of the $i$ most accessible active branches (in accordance with case (iii), see Sect. 3). For instance, value 109 of Top-2 for Branch 1 means, that after the closure of Branch 1, 109 out of 252 of its customers visited one of two most accessible branches. Note, that each customer has its own list of the most accessible branches.

**Table 2.** The comparison of bank branches spatial accessibility and real visits for Branch 1 and Branch 2 customers

| Branch 1 | | |
| --- | --- | --- |
| Top accessible branches | Customers (out of 252) | Percentage of customers |
| Top-1 | 68 | 27% |
| Top-2 | 109 | 43% |
| Top-3 | 134 | 53% |
| Top-4 | 149 | 59% |
| Top-5 | 163 | 65% |
| Branch 2 | | |
| Top accessible branches | Customers (out of 409) | Percentage of customers |
| Top-1 | 142 | 35% |
| Top-2 | 210 | 51% |
| Top-3 | 262 | 64% |
| Top-4 | 291 | 71% |
| Top-5 | 306 | 75% |

It turned out, that 65%–75% of customers chose one of top 5 branches by accessibility (out of 51) after the closure. Thus, it is can be concluded, that the hypothesis on spatial accessibility influence was confirmed.

## 5.3 Experimental Study of the Proposed Algorithms

The algorithms described in Sect. 4 were implemented in a form of a sequential C++ program. While in almost all cases $K$ was significantly smaller than $M$, only one greedy algorithm, GreedyLP, was used. It was done because GreedyLP suits better for small $K$ than GreedyHP. Hereinafter 'GreedyLP+SHC' stands for simple hill climbing which starts from a point found by the GreedyLP algorithm. In simple hill climbing, radius $R$ was equal to 2 (see Subsect. 4.3).

Computational experiments were held on a computer equipped with the 6-core processor Intel i7-3930K and 16 GB of RAM. Two series of experiments were conducted. In the first one, all 51 branches were available for closing (i.e. $M$ was equal to 51), $K$ (the amount of branches for closing) was varied from 1 to 10. The obtained results are presented in Table 3. Here 'calc.' stands for the number of the objective function calculations, while 'unsat.' stands for the number of unsatisfied customers in the solution found by the corresponding algorithm. Hereinafter the best found solutions are marked with bold, while solutions which are worse than the best ones, are marked with Italic. It should be noted, that the brute-force algorithm was not launched for $K = 5, .., 10$ because the corresponding experiments could not be finished in reasonable time. For instance, for $K = 5$ it would take about 10 days.

**Table 3.** Results for $M = 51, K = 1, .., 10$, time in seconds

| $K$ | Brute-force | | | GreedyLP | | | GreedyLP+SHC | | |
|---|---|---|---|---|---|---|---|---|---|
| | calc. | time | unsat. | calc. | time | unsat. | calc. | time | unsat. |
| 1 | 51 | 2 | **17** | 51 | 2 | **17** | 51 | 2 | **17** |
| 2 | 1275 | 118 | **520** | 101 | 39 | **520** | 199 | 46 | **520** |
| 3 | 20825 | 3533 | **1216** | 150 | 61 | **1216** | 294 | 71 | **1216** |
| 4 | 249900 | 66710 | **1972** | 198 | 81 | **1972** | 386 | 99 | **1972** |
| 5 | 2349060 | - | - | 245 | 98 | **2775** | 475 | 133 | **2775** |
| 6 | 18009460 | - | - | 291 | 115 | *3984* | 831 | 225 | **3811** |
| 7 | 115775100 | - | - | 336 | 141 | *5052* | 952 | 285 | **4898** |
| 8 | 636763050 | - | - | 380 | 154 | *6166* | 1068 | 370 | **6082** |
| 9 | 3042312350 | - | - | 423 | 176 | **7375** | 801 | 323 | **7375** |
| 10 | 12777711870 | - | - | 465 | 190 | **9003** | 875 | 380 | **9003** |

According to Table 3, for $K = 1, ..4$ all three algorithms found optimal solutions. As for $K = 5, ..10$, the best solutions are unknown, because, as it was mentioned above, the brute-force algorithm was not launched in these cases. Nevertheless, for $K = 6, 7, 8$ GreedyLP+SHC found better solutions than GreedyLP. It is not guaranteed that solutions, found by GreedyLP+SHC, are optimal. However, these solutions are local minima (in the sense described in Subsect. 4.3). It means that each of them can not be improved by replacing any branch from the corresponding set by any other possible branch. Note that GreedyLP+SHC takes a little more time compared to GreedyLP. However, it takes reasonable time even for the most difficult instances of the problem. Note, the in all cases GreedyLP+SHC required from 1 up to 3 neighborhood checks (see Subsect. 4.3) to reach a local minimum. These results show, that a solution found by GreedyLP is a very good start point for GreedyLP+SHC.

The results from Table 3 are also shown in Fig. 2. In Fig. 2a the solving time in seconds are shown on a logarithmic scale, while in Fig. 2b the objective function values are shown on a linear scale.

In the second series of experiments, a constraint was added. Only 17 out of 51 branches were available for closing (i.e. $M$ was equal to 17). These branches were chosen because other 34 branches cannot be closed in accordance with the bank's strategy. $K$ was varied from 1 to 10. Note, that in the accessibility matrix all 51 branches were presented, so the whole bank branches network was taken into account during the objective function calculations. The obtained results are presented in Table 4. The brute-force algorithm managed to finish its work within

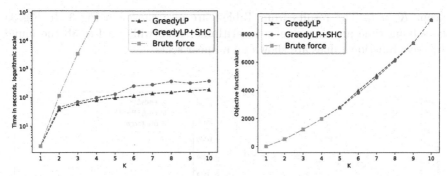

(a) Solving time in seconds (logarithmic (b) Objective functions values (linear scale)
scale)

**Fig. 2.** Results for $M = 51, K = 1, .., 10$

**Table 4.** Results for $M = 17, K = 1, .., 10$, time in seconds

| $k$ | Brute-force | | | GreedyLP | | | GreedyLP+SHC | | |
|---|---|---|---|---|---|---|---|---|---|
| | calc. | time | unsat. | calc. | time | unsat. | calc. | time | unsat. |
| 1 | 17 | 1 | **498** | 17 | 1 | **498** | 17 | 1 | **498** |
| 2 | 136 | 9 | **1265** | 33 | 8 | **1265** | 63 | 9 | **1265** |
| 3 | 680 | 86 | **2722** | 48 | 52 | **2722** | 42 | 57 | **2722** |
| 4 | 2380 | 476 | **4224** | 62 | 70 | **4224** | 114 | 78 | **4224** |
| 5 | 6188 | 1785 | **5880** | 75 | 88 | *5952* | 195 | 111 | **5880** |
| 6 | 12376 | 4759 | **7614** | 87 | 103 | **7614** | 153 | 127 | **7614** |
| 7 | 19448 | 17524 | **10064** | 98 | 118 | **10064** | 168 | 143 | **10064** |
| 8 | 24310 | 18197 | **12590** | 108 | 139 | **12590** | 180 | 168 | **12590** |
| 9 | 24310 | 18287 | **15423** | 117 | 151 | **15423** | 189 | 198 | **15423** |
| 10 | 19448 | 17519 | **18429** | 125 | 169 | **18429** | 195 | 219 | **18429** |

the runtime limit of 1 day for all $K$, so it was possible to compare solutions found
by GreedyLP and GreedyLP+SHC with optimal solutions in all cases.

It turned out, that the branches, recommended for closing in accordance
with the first experiment (without any constraint) significantly differ from the
ones recommended in accordance with the second experiment. In particular, for
$K = 1$ in the first experiment the closure of the recommended branch will lead
to 17 unsatisfied customers (see Table 3), while in the second experiment the
closure of the recommended branch will lead to 498 unsatisfied customers. It
means that the employed constraint is very strong and some branches which
suit well for closing from the spatial accessibility point of view, cannot be closed
due to some additional reasons.

According to Table 4, both Brute-force and GreedyLP+SHC found optimal
solutions in all cases. As for GreedyLP, it found optimal solutions in all cases

except $K = 5$. The results from Table 4 are also shown in Fig. 3. In Fig. 3a the solving time are shown on a logarithmic scale, while in Fig. 3b the found objective function values are shown on a linear scale.

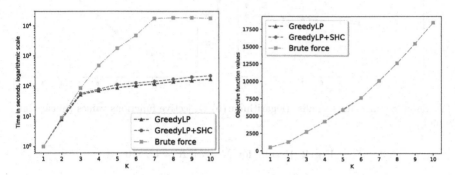

(a) Solving time in seconds (logarithmic scale)

(b) Objective functions values (linear scale)

**Fig. 3.** Results for $M = 17, K = 1, .., 10$

The obtained results show, that GreedyLP suits very well for the considered problem. It works fast and in almost all cases finds an optimal solution. GreedyLP+SHC, being a little slower than GreedyLP, finds optimal solutions in all practical cases.

## 6    Conclusions

In this paper, the problem of company branches location optimization was considered. It was considered as a problem of minimizing the number of potentially dissatisfied customers according to the distances between points of individuals and branches. According to the conducted experiments, a greedy algorithm found exact solutions of the considered problem in almost all cases. A local search algorithm, which starts from a solution found by the greedy algorithm, managed to find exact solutions in all cases, where it was possible to verify by the brute-force algorithm. Meanwhile, both suggested algorithms work much faster than the brute-force algorithm. We have also shown that in the case of constraints related to the selection of branches for closing, the number of unsatisfied customers can increase significantly. In the future, we are planning to apply the suggested approach to other data sets.

**Acknowledgments.** This research is financially supported by the Russian Science Foundation, Agreement 17-71-30029 with co-financing of Bank Saint Petersburg.

# References

1. Abualigah, L., Hanandeh, E.: Applying genetic algorithms to information retrieval using vector space model. Int. J. Comput. Sci. Eng. Appl. **5**, 19–28 (2015). https://doi.org/10.5121/ijcsea.2015.5102
2. Allahi, S., Mobin, M., Vafadarnikjoo, A., Salmon, C.: An integrated AHP-GIS-MCLP method to locate bank branches. In: Proceedings of Industrial and Systems Engineering Research Conference, July 2015
3. Basar, A., Kabak, O., Topcu, Y.I.: A tabu search algorithm for multi-period bank branch location problem: a case study in a Turkish bank. Scientia Iranica (2018). https://doi.org/10.24200/sci.2018.20493
4. Ester, M., Kriegel, H.P., Sander, J., Xu, X.: A density-based algorithm for discovering clusters a density-based algorithm for discovering clusters in large spatial databases with noise. In: Proceedings of the Second International Conference on Knowledge Discovery and Data Mining, KDD 1996, pp. 226–231. AAAI Press (1996)
5. Glover, F.: Future paths for integer programming and links to artificial intelligence. Comput. Oper. Res. **13**(5), 533–549 (1986). https://doi.org/10.1016/0305-0548(86)90048-1
6. Miliotis, P., Dimopoulou, M., Giannikos, I.: A hierarchical location model for locating bank branches in a competitive environment. Int. Trans. Oper. Res. **9**, 549–565 (2002). https://doi.org/10.1111/1475-3995.00373
7. Monteiro, M.S.R., Fontes, D.B.M.M.: Locating and sizing bank-branches by opening, closing or maintaining facilities. In: Haasis, H.D., Kopfer, H., Schönberger, J. (eds.) Operations Research Proceedings 2005, pp. 303–308. Springer, Heidelberg (2006). https://doi.org/10.1007/3-540-32539-5_48
8. Russell, S., Norvig, P.: Artificial Intelligence: A Modern Approach, 3rd edn. Prentice Hall, Englewood cliffs (2009)
9. Schneider, S., Seifert, F., Sunyaev, A.: Market potential analysis and branch network planning: application in a german retail bank. In: 2014 47th Hawaii International Conference on System Sciences, pp. 1122–1131, January 2014. https://doi.org/10.1109/HICSS.2014.145
10. Thiel, D., Hovelaque, V., Pham, D.N.: A multi-agent model for optimizing supermarkets location in emerging countries. In: 2012 IEEE 13th International Symposium on Computational Intelligence and Informatics (CINTI), pp. 395–399 (2012). https://doi.org/10.1109/CINTI.2012.6496798
11. Wang, Q., Batta, R., Bhadury, J., Rump, C.M.: Budget constrained location problem with opening and closing of facilities. Comput. Oper. Res. **30**(13), 2047–2069 (2003). https://doi.org/10.1016/S0305-0548(02)00123-5
12. Wu, S.S., Kuang, H., Lo, S.M.: Modeling shopping center location choice: Shopper preference-based competitive location model. J. Urban Plan. Dev. **145**, March 2019. https://doi.org/10.1061/(ASCE)UP.1943-5444.0000482
13. Xia, L., Xie, M., Yin, W., Dong, J., Shao, J.: Markov decision processes formulation for stochastic and dynamic bank branches location problems. In: Proceedings of 2008 IEEE International Conference on Service Operations and Logistics, and Informatics, IEEE/SOLI 2008, vol. 1, pp. 419–424, October 2008. https://doi.org/10.1109/SOLI.2008.4686432
14. Xie, W., Ouyang, Y., Somani, K.: Optimizing location and capacity for multiple types of locomotive maintenance shops. Comp.-Aided Civ. Infrastruct. Eng. **31**(3), 163–175 (2016). https://doi.org/10.1111/mice.12114

# Track of Advances in High-Performance Computational Earth Sciences: Applications and Frameworks

# A Fast 3D Finite-Element Solver for Large-Scale Seismic Soil Liquefaction Analysis

Ryota Kusakabe[1]([⊠]), Kohei Fujita[1], Tsuyoshi Ichimura[1], Muneo Hori[2], and Lalith Wijerathne[1]

[1] The University of Tokyo, Bunkyo-ku, Tokyo, Japan
{ryota-k,fujita,ichimura,lalith}@eri.u-tokyo.ac.jp
[2] Japan Agency for Marine-Earth Science and Technology,
Yokosuka-shi, Kanagawa, Japan
horimune@jamstec.go.jp

**Abstract.** The accumulation of spatial data and development of computer architectures and computational techniques raise expectations for large-scale soil liquefaction simulations using highly detailed three-dimensional (3D) soil-structure models; however, the associated large computational cost remains the major obstacle to realizing this in practice. In this study, we increased the speed of large-scale 3D soil liquefaction simulation on computers with many-core wide SIMD architectures. A previous study overcame the large computational cost by expanding a method for large-scale seismic response analysis for application in soil liquefaction analysis; however, that algorithm did not assume the heterogeneity of the soil liquefaction problem, resulting in a load imbalance among CPU cores in parallel computations and limiting performance. Here we proposed a load-balancing method suitable for soil liquefaction analysis. We developed an efficient algorithm that considers the physical characteristics of soil liquefaction phenomena in order to increase the speed of solving the target linear system. The proposed method achieved a 26-fold increase in speed over the previous study. Soil liquefaction simulations were performed using large-scale 3D models with up to 3.5 billion degrees-of-freedom on an Intel Xeon Phi (Knights Landing)-based supercomputer system (Oakforest-PACS).

**Keywords:** Soil liquefaction · Fast and scalable solver ·
Large-scale analysis · Finite-element method

## 1 Introduction

Soil liquefaction induced by earthquakes has caused various kinds of damage, including ground settlement, lateral flow, and tilting and destruction of buildings. When the Great East Japan Earthquake occurred in 2011, soil liquefaction was induced in a large area ranging from Kanto to Tohoku and resulted in sand

© Springer Nature Switzerland AG 2019
J. M. F. Rodrigues et al. (Eds.): ICCS 2019, LNCS 11537, pp. 349–362, 2019.
https://doi.org/10.1007/978-3-030-22741-8_25

boiling and the uplift of manholes among other damage [11,13]. To reduce this type of damage, various methods for soil liquefaction analyses have been developed. The recent accumulation of spatial data and development of computer architectures and computational techniques have enabled larger-scale physical simulations, thereby raising the expectation for realization of large-scale soil liquefaction simulations using highly detailed three-dimensional (3D) soil-structure models. Such simulations will contribute to the mitigation of damage by soil liquefaction, as well as soil liquefaction itself. However, these numerical analyses using large-scale 3D models require huge computational requirements, with most soil liquefaction analyses performed either on two-dimensional (2D) models under plane-strain conditions or on small-scale 3D models. The aim of this study was to increase the speed of large-scale 3D soil liquefaction simulation on computers with many-core wide SIMD architectures, which represent a type of architecture used for the recent development of exascale supercomputers.

For seismic wave-propagation simulations in urban areas not limited to soil liquefaction analyses, parallel 3D finite-element solvers suitable for massively parallel computers have been developed and used to analyze the actual damage. A finalist for the Gordon Bell Prize in SC14 [5] (hereafter, referred to as the SC14 solver) enabled a large-scale earthquake simulation by developing a fast algorithm for solving the huge linear system obtained by discretizing the target physical problem, with the solver algorithm designed based on analyzing the characteristics of the discretized linear system of equations. Such an algorithm can be effective for seismic soil liquefaction simulations. Our previous study [9] expanded the algorithm used for the SC14 solver for use in soil liquefaction simulation to enable large-scale 3D soil liquefaction analysis; however, although the target problem of the SC14 solver was relatively homogeneous, soil liquefaction analysis is a heterogeneous problem, where different physical problems are solved in different soil layers [i.e., non-liquefiable (linear) layers and liquefiable (nonlinear) layers]. Therefore, our previous solver [9], which used the algorithm used for the SC14 solver, displayed low parallel efficiency at each parallelization level of the SIMD lane, thread, and process when executed on a computer with many-core wide SIMD architecture. In the present study, we developed methods to improve the load balance in each parallelization level to overcome this inefficiency and to potentially increase the speed of soil liquefaction analysis and enable larger-scale simulations.

In addition to the development of algorithms appropriate for the discretized linear system, our research group proposed algorithms to reduce the computational cost by considering the characteristics of the target physical problem underlying the linear system (a finalist for the Gordon Bell Prize in SC18 [6]; hereafter referred to as the SC18 solver). In this study, we applied this idea to soil liquefaction analysis. Considering the locality of soil liquefaction phenomena, we partially approximated the heavy computation in the nonlinear layer by using light-weight computation, which is expected to reduce analysis cost and time.

We compared the developed solver with our previous solver [9] and performed large-scale soil liquefaction simulations using 3D ground models with up to 3.5 billion degrees-of-freedom (DOF) on an Intel Xeon Phi (Knights Landing)-based supercomputer system (Oakforest-PACS [8]; OFP).

## 2  Target Problem

Equation (1) is the target linear system of soil liquefaction analysis under undrained conditions. This equation is discretized spatially using the finite-element method and temporally using the Newmark-$\beta$ method ($\beta = 0.25$).

$$A\delta u^{(n)} = b, \tag{1}$$

where

$$A = \frac{4}{dt^2}M + \frac{2}{dt}C^{(n-1)} + K^{(n-1)},$$

$$b = f^{(n)} - q^{(n-1)} + C^{(n-1)}v^{(n-1)} + M\left(a^{(n-1)} + \frac{4}{dt}v^{(n-1)}\right).$$

Here, $M$, $C$, and $K$ are respectively the consistent mass, Rayleigh damping, and stiffness matrices, $f$, $q$, $\delta u$, $v$, and $a$ are the external force, inner force, displacement increment, velocity, and acceleration vectors, respectively, $dt$ is the time increment, and $*^{(n)}$ is the variable $*$ at the $n$-th time step. To carry out soil liquefaction analysis, we perform the following at each time step.

1. Read the external force $f$, and calculate the coefficient matrix $A$, and right-hand side vector $b$.
2. Solve the linear system (1) and obtain the displacement increment $\delta u$.
3. Update the displacement $u$, velocity $v$, and acceleration $a$, using $\delta u$.
4. Update the stiffness $K$, and inner force $q$ [Eqs. (2) and (3)].

$$K^{(n)} = \sum_e \int_{V_e} B^T\left(D_s^{(n)} + D_f\right)B\ dV, \tag{2}$$

$$q^{(n)} = \sum_e \int_{V_e} B^T\left(\sigma'^{(n)} - p^{(n)}m\right)\ dV, \tag{3}$$

where, $B$ is the matrix used to convert the displacement into the strain, $D_s$ and $D_f$ are the elasto-plastic matrices of the soil and pore water, respectively, $\sigma'$ is the effective stress of the soil, $p$ is the excess pore water pressure and $m = \{1\ 1\ 1\ 0\ 0\ 0\}^T$. $\int_e *dV$ is the volume integration of the variable $*$ in the $e$-th finite element. $D_s$, $D_f$, $\sigma'$, and $p$ are defined by the constitutive law [3]. For detailed information, consult [3] and a 2D constitutive law [4] which [3] is based on.

## 3  Previous Method

Solving the linear system (1) and updating the stiffness $K$, and inner force $q$ occupy a considerable amount of the calculation time in liquefaction analysis. Solving the linear system (1) requires a huge computational cost, especially in large-scale 3D analysis, which has been the major obstacle to realizing large-scale

3D soil liquefaction analysis. Our previous study [9] overcame the computational cost by applying the algorithm of the SC14 solver for large-scale seismic response analysis to large-scale soil liquefaction analysis. In this section, we explain the algorithm associated with the SC14 solver and the problems in our previous study [9] about implementing the algorithm of the SC14 solver to soil liquefaction analysis.

## 3.1    Finite-Element Solver

The SC14 solver is a fast and scalable method that uses unstructured tetrahedral second-order elements developed for solving the target linear system involved in seismic response analysis without soil liquefaction. It is based on the adaptive conjugate gradient (CG) method [2] and uses hybrid parallelization with message passing interface (MPI) and OpenMP.

The conventional preconditioned CG (PCG) method improves the convergence characteristics by preconditioning $z = M^{-1}r$, where $M$ is a preconditioning matrix that mimics the coefficient matrix $A$. Preconditioning in the adaptive CG method [2] is to *roughly* solve $Az = r$ instead of $z = M^{-1}r$. The preconditioning matrix $M$ in this method is supposed to be $M \approx A$, which describes it as having high preconditioning performance. The PCG method with the $3 \times 3$ block Jacobi preconditioner is used to solve $Az = r$ as the preconditioning in order to achieve parallel efficiency.

Increases in speed are limited when merely solving $Az = r$ during preconditioning. The SC14 solver reduces the computational cost by using a first-order tetrahedral-element model generated as the geometric multigrid of the second-order tetrahedral-element model of the target problem. First, $A_c z_c = r_c$ is solved on the first-order element model, which requires much less computational cost than on the second-order element model. Using the solution of the first-order element model as the initial solution, $Az = r$ is then solved on the second-order element model, thereby significantly reducing the amount of computation and communication. The PCG method is used twice in preconditioning on the first- and second-order element models. Hereafter, the former CG is referred to as an inner coarse CG, and the latter is referred to as an inner fine CG. The main CG that uses the two inner CGs for preconditioning is called the outer CG. To further reduce the computational cost, the preconditioning step, which does not require high computational accuracy, is computed in single precision, whereas the other calculation is performed in double precision.

The SC14 solver uses the element-by-element (EBE) method [12] to efficiently compute matrix-vector multiplication, which is the heaviest computation in the CG method. In conventional matrix-vector multiplication calculations, the global coefficient matrix, $A = \sum_e A_e$, is calculated, followed by multiplication of the matrix $A$ by the vector $x$ to obtain the matrix-vector multiplication $Ax$. In the EBE method, the global matrix $A$ is not calculated. The element-wise matrix-vector multiplications $A_e x_e$ are calculated on the fly and summed to obtain the global matrix-vector multiplication $Ax = \sum_e (A_e x_e)$, where $x$ is the component of $x$ for the $e$-th element. This method reduces the cost to read and write the global coefficient matrix, as well as the memory usage necessary to store it. Therefore matrix-vector multiplication using the EBE method requires less memory access and communication, which shortens computation time.

## 3.2   Implementation of Our Previous Solver and Its Problems

Our previous study [9] overcame the huge computational cost involved with solving the target linear system (1) in order to allow large-scale soil liquefaction analysis using the method of the SC14 solver, which is explained in the previous section. However, the heterogeneity of soil liquefaction analysis causes load imbalance in parallel computation, which results in low parallel efficiency and long computation times. This is because the target problem for the SC14 solver is seismic response analysis, which is relatively homogeneous; therefore, the algorithm does not assume heterogeneity. The load imbalance occurs during calculation of matrix-vector multiplication using the EBE method and update of the stress and elasto-plastic matrix according to the constitutive law. These two computations have the highest computational costs; therefore, the resulting load imbalance potentially results in a worse time to solution.

The difference in the number of integration points in different elements causes the load imbalance in the calculation of matrix-vector multiplication. When calculating the element-wise coefficient matrix $A_e$ during matrix-vector multiplication using the EBE method, the element-wise stiffness matrix $K_e = \int_{V_e} B^{\mathrm{T}}(D_s + D_f)B\mathrm{d}V$, is calculated on the fly. $B$ is a linear function, and the elasto-plastic matrix in the linear layer is constant, suggesting that the integrand in a linear element is second order, and computing the volume integration needs only four integration points. On the other hand, the elasto-plastic matrix in the nonlinear layer spatially varies. Because soil liquefaction occurs locally, the elasto-plastic matrix should be assumed to vary in an element to accurately analyze the phenomenon. Therefore, the integrand in a nonlinear element is more than third order, and more than four integration points are needed. In this study, five integration points were used, enabling calculation of the integration of up to a third-order function.

The difference in the constitutive law at different layers causes load imbalance in the calculation of the effective stress $\sigma'$, and the elasto-plastic matrix $D_s$, of soil. In the linear layer, $D_s$ is constant and unnecessary to update, and $\sigma'$ is calculated with very low computational cost, as $\sigma' = D_s\varepsilon$. Therefore, it is not explicitly calculated and, thus, computed on the fly when $q$ is computed using Eq. (3). In the nonlinear elements, $\sigma'$ and $D_s$ are updated at every time step in the five integration points per element. This requires considerable computation, because calculations for 300 one-dimensional springs are required at each integration point for the implementation in the present study.

The SC14 solver achieves load balance by distributing the same number of elements to each CPU core and by assuming that the computation amount in each element is highly similar. However, for soil liquefaction analysis, the computation amount in each element differs; therefore, load imbalance among processes and threads occurred in our previous solver [9], which used the original version

| As is | Reordering |
|---|---|
| do ie = 1, ne<br>  if(ie-th element is linear element)<br>    [calc. for a linear element]<br>  else #nonlinear<br>    [calc. for a nonlinear element]<br>  endif<br>end do | [Reordering of the elements]<br>do ie = 1, ne_linear<br>  [calc. for a linear element]<br>enddo<br>do ie = ne_linear+1, ne<br>  [calc. for a nonlinear element]<br>end do |

**Fig. 1.** Element reordering to improve thread load balance and exploit automatic vectorization by the compiler.

of the algorithm of the SC14 solver. Another problem in speeding up on our target architecture of many-core wide SIMD is that it is diffcult to exploit a wide SIMD in heterogeneous calculations.

## 4   Developed Method

In this section, we propose methods for improving the load balance among SIMD lanes, among OpenMP threads, and among MPI processes in order to overcome the problem of load imbalance experienced in our previous study [9]. Additionally, we developed a physics-aware algorithm to solve the target linear system (1) in order to achieve a further increase in speed.

### 4.1   Improving Load Balance

First, we reordered the elements to improve load balance among OpenMP threads in each process and to exploit efficient use of wide SIMD arithmetic units. In a conventional implementation, variables for both linear and nonlinear elements are stored in arrays in mixed order. We implemented element reordering to enable storing variables for linear elements in the first part of arrays and variables for nonlinear elements in the remainder of the arrays. In this implementation, the computation for the linear and nonlinear elements is separated (Fig. 1) and OpenMP parallelized independently, thereby improving thread load balance. Additionally, separating the computation of the linear and nonlinear elements is suitable for automatic vectorization by the compiler, which will increase the speed of simulation, especially on our target computers of manycore architecture with wide SIMD.

Second, we revised the model-partitioning method in order to improve load balance among MPI processes. The conventional method partitions the model so that the number of elements assigned to each process is almost the same. This causes process load imbalance because the ratio of the number of linear and nonlinear elements could be different. We propose that the linear and nonlinear layers be independently partitioned into the number of MPI processes and that each process be assigned to a domain comprising one partition of the linear layer and one partition of the nonlinear layer (Fig. 2). This method improves load

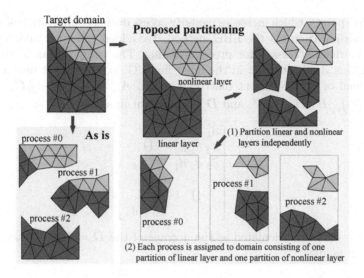

**Fig. 2.** Overview of the proposed partitioning of the simulation domain.

balance among the processes because each process works on the same number of linear and nonlinear elements.

## 4.2 Physics-Aware Preconditioner

We developed a preconditioner that considers the physical characteristics of soil liquefaction based on the preconditioner of the SC14 solver in order to achieve additional increases in speed. The SC18 solver was developed based on the SC14 solver and used the preconditioner based on the physical characteristics of the target problem. The target problem of the SC18 solver includes a domain involving a high degree of stiffness, such as concrete structures. The convergence characteristics of the CG method are extremely poor around these structures. Another inner CG was added between the inner coarse and fine CGs for pre-conditioning. The only domain having poor convergence characteristics, which is detected by an AI before the analysis, is solved in the additional inner CG, which efficiently improved the convergence characteristics. The physics-aware preconditioner introduced in the present study uses the same concept involving the addition of another inner CG (called the inner middle CG) between the inner coarse and fine CGs. Note that the inner middle CG in this study exploits the locality of soil liquefaction phenomena and differs from the additional inner CG in the SC18 solver.

The target problem of the SC14 solver has a relatively small spatial variation in soil-physical properties. The soil elasto-plastic matrix $\boldsymbol{D}_s$, is assumed to be constant not only in a linear element but also in a nonlinear element. On the other hand, for soil liquefaction analysis implemented in this study, the volume integration in nonlinear elements is calculated using $\boldsymbol{D}_s$ at the five integration points, which increases memory access during computation of matrix-vector multiplication via the EBE method. We added the inner middle CG to reduce the data necessary for preconditioning. The computation in the inner middle CG is similar to that in the inner fine CG, except that it uses a matrix $A_m$, instead of the coefficient matrix $A$. Here, $A_m = \frac{4}{dt^2}M + \frac{2}{dt}C_m + K_m$, $K_m = \sum_e \int_{V_e} B^T D_m B \, dV$, and $D_m$ is constant in an element:

$$
D_m = \begin{bmatrix}
d_1 & d_2 & d_2 & & & \\
d_2 & d_1 & d_2 & & \mathbf{0} & \\
d_2 & d_2 & d_1 & & & \\
& & & d_3 & & \\
& \mathbf{0} & & & d_4 & \\
& & & & & d_5
\end{bmatrix}. \tag{4}
$$

$d_i$ $(i = 1, \cdots, 5)$ are calculated as the *average* of the $D = D_s + D_f$ at the five integration points[1]:

$$
d_1 = \frac{1}{15} \sum_{j=1}^{5} \left( D_{11}^j + D_{22}^j + D_{33}^j \right), \quad d_2 = \frac{1}{15} \sum_{j=1}^{5} \left( D_{12}^j + D_{23}^j + D_{31}^j \right),
$$

$$
d_3 = \frac{1}{5} \sum_{j=1}^{5} D_{44}^j, \quad d_4 = \frac{1}{5} \sum_{j=1}^{5} D_{55}^j, \quad d_5 = \frac{1}{5} \sum_{j=1}^{5} D_{66}^j, \tag{5}
$$

where $D^j$ is the elasto-plastic matrix at the $j$-th integration point. This reduces the memory access necessary during the computation. Additionally, the integration in the calculation of $K_m$ uses only four integration points, because $D_m$ is constant, and requires a smaller amount of computation than that in the inner fine CG, which uses five integration points.

Because soil liquefaction occurs locally, there are only a few elements in which the physical property varies greatly and in the most elements, $D^j \approx D_m$ $(j = 1, \cdots, 5)$. Therefore, the computation using the inner fine CG can be partially replaced by the computation using the inner middle CG, which has a lower computational cost.

Algorithm A summarizes the proposed method.

---

[1] There are several ways to average $D$, with the optimal method depending upon the problem. Non-zero components of $D_f$ are only present in the upper-left 3 × 3 components and correspond to the bulk modulus of pore water, which is constant and 10-fold greater than that of soil. Therefore, the upper-left 3 × 3 components of $D$ are only slightly influenced by soil heterogeneity and are supposed to be approximated using two parameters: $d_1$ and $d_2$. On the other hand, a shear wave has a dominant influence on seismic response analysis, and the soil is supposed to show its heterogeneity, such that only shear stiffness in the vibration direction is reduced. Therefore, the lower-right diagonal three components, which correspond to the shear stiffness in different directions are approximated using different parameters: $d_3, d_4$, and $d_5$. The zero components of $D_m$ could be non-zero in $D$ but are sufficiently small as compared with $d_i$ $(i = 1, \cdots, 5)$. Approximating these small components by zero efficiently reduces the amount of data in $D_m$.

**Algorithm A.** The outline of the proposed method. $\bar{*}$ indicates that the variable $*$ is in single precision. $*_c$ indicates that the variable $*$ is associated with the first-order model. $\bar{P}$ is the mapping matrix from a variable for the first-order model to the equivalent variable for the second-order model. $nt$ is the number of time steps.

| | |
|---|---|
| 1 | [Model partitioning] |
| 2 | [Element reordering] |
| 3 | Calculate the initial values of $u$, $q$, $K$ by self-weight analysis. |
| 4 | $v, a \Leftarrow 0$ |
| 5 | **do** $it = 1, nt$ |
| 6 | Read the external force $f$. |
| 7 | $b \Leftarrow f - q + Cv + M\left(a + \frac{4}{dt}v\right)$ |
| 8 | $A \Leftarrow \frac{4}{dt^2}M + \frac{2}{dt}C + K$ |
| 9 | Set the initial values of $\delta u$ and other variables for the CG method. |
| 10 | $r \Leftarrow b - A\delta u$ |
| 11 | (Outer CG starts: solve $A\delta u = b$) |
| 12 | **while** not converged **do** |
| 13 | (Preconditioning starts) |
| 14 | $\bar{r} \Leftarrow r$ |
| 15 | $\bar{r}_c \Leftarrow \bar{P}^T\bar{r}$ |
| 16 | solve $\bar{A}_c\bar{z}_c = \bar{r}_c$   by PCG method (inner coarse CG) on first-order-element mesh. |
| 17 | $\bar{z}_0 \Leftarrow \bar{P}\bar{z}_c$ |
| 18 | solve $\bar{A}_m\bar{z} = \bar{r}$   by PCG method (inner middle CG) with initial solution $\bar{z}_0$, on second-order-element mesh. |
| 19 | $\bar{z}_0 \Leftarrow \bar{z}$ |
| 20 | solve $\bar{A}\bar{z} = \bar{r}$   by PCG method (inner fine CG) with initial solution $\bar{z}_0$, on second-order-element mesh. |
| 21 | $z \Leftarrow \bar{z}$ |
| 22 | (Preconditioning ends) |
| 23 | Update $\delta u$ and $r (= b - A\delta u)$, using $z$ |
| 24 | **end while** |
| 25 | (Outer CG ends: displacement increment $\delta u$ is obtained.) |
| 26 | Update $u$, $v$ and $a$ by Newmark-$\beta$ Method, using $\delta u$. |
| 27 | Update $\sigma'$ and $p$ by constitutive law, using $u$ and $\delta u$ |
| 28 | $Q \Leftarrow \sum_e \int_{V_e} B^T \left(\sigma'^{(n)} - pm\right) \, dV$ |
| 29 | Update $D_s$ by constitutive law, using $u$ and $\delta u$ |
| 30 | $K \Leftarrow \sum_e \int_{V_e} B^T (D_s + D_f) B \, dV$ |
| 31 | **end do** |

## 5   Performance Measurement

We performed large-scale soil liquefaction simulations using 3D ground models that mimic actual ground structures in order to demonstrate the effectiveness of the proposed method. The models used in the simulations were generated by replicating the model shown in Fig. 3 in the x and y directions. Table 1 shows

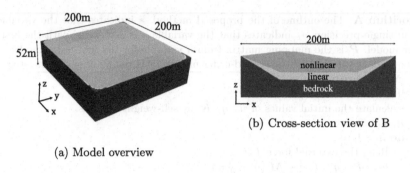

(a) Model overview

(b) Cross-section view of B

**Fig. 3.** Model configuration

**Table 1.** Soil profile properties. $\rho$: density, $V_p, V_s$: velocity of primary and secondary wave, $h$: damping ratio.

|         | $\rho[\mathrm{kg/m^3}]$ | $V_p[\mathrm{m/s}]$ | $V_s[\mathrm{m/s}]$ | $h$    | Constitutive law |
|---------|-------------------------|---------------------|---------------------|--------|------------------|
| Layer1  | 1500                    | —                   | —                   | 5%     | Nonlinear        |
| Layer2  | 1800                    | 1380                | 255                 | 5%     | Linear           |
| Bedrock | 1900                    | 1770                | 490                 | 0.5%   | Linear           |

the physical properties of the soil. The parameters for soil liquefaction were the same as those used previously [9]. The groundwater level was assumed to be 2-m below the surface of the model. We used the seismic wave observed during the Hyogo-ken Nambu earthquake in 1995 [7], which induced soil liquefaction in large areas, including reclaimed land along the coast, causing significant damage to buildings [1,10]. The resolution of the models was decided based on a previous study [9], where the maximum element size was 2 m in the nonlinear layer, 20 m in the linear layer, and 40 m in the bedrock, and the time increment was 0.001 s.

All performance measurements were undertaken on Oakforest-PACS [8] (OFP), a computer with many-core wide SIMD architecture. The OFP is a supercomputer introduced by the Joint Center for Advanced High Performance Computing, Japan. It has 8,208 compute nodes, each of which comprises a 68-core Intel Xeon Phi processor 7250 (Knights Landing) CPU, 16GB-MCDRAM memory, and six 16GB-DDR4-2400 memory chips. The network between compute nodes is interconnected by an Intel Omni-Path architecture, and each core has two 512-bit vector units and supports AVX-512 SIMD instructions.

## 5.1   Time to Solution

First, time to solution was compared with our previous study [9]. Table 2 shows the simulation cases. Asis is the implementation in our previous study [9]. Asis is an incomplete implementation that does not implement OpenMP parallelization in the calculation of the constitutive law. In the present study, asisOMP, which implements OpenMP parallelization for calculation of the constitutive law, was

**Table 2.** Simulation cases: *OMP(constitutive law)* describes OpenMP parallelization in the calculation of the constitutive law; *Thread load balance/SIMD* describes load balancing among OpenMP threads in an MPI process and SIMD vectorization by reordering elements; *Process load balance* describes load balancing among MPI processes by independently partitioning the linear and nonlinear domains; and *Physics-aware preconditioner* describes preconditioning using the inner coarse, middle, and fine CGs. "+" indicates that the feature is implemented, and "–" indicates that it is not implemented.

|  | asis | asisOMP | case1 | case2 | case3 |
|---|---|---|---|---|---|
| OMP(constitutive law) | - | + | + | + | + |
| Thread load balance/SIMD | - | - | + | + | + |
| Process load balance | - | - | - | + | + |
| Physics-aware preconditioner | - | - | - | - | + |

used to measure the reference performance. Case1 implements element reordering in order to improve the thread load balance and exploit the wide SIMD. Case2 implements revised model partitioning to improve the process load balance in addition to case1 implementation. Case3 implements the physics-aware preconditioner to increase the speed of solving the linear system (1) in addition to case2 implementation. We used a 54,725,427-DOF model generated by replicating the model shown in Fig. 3 in the x and y directions. Soil liquefaction analysis for 7,500 time steps was performed with 512 MPI processes × eight OpenMP threads per one MPI process = 4,096 CPU cores (64 OFP nodes). Figure 4a shows the time to solution. Case1 compared with asisOMP achieved a 2.03-fold increase in speed in solving the linear system and a 6.78-fold increase in the speed of calculating the constitutive law. This result showed that thread load balancing and SIMD vectorization successfully reduced the computation time. Specifically, calculation of the constitutive law exploited the wide SIMD based on the SIMD width of the OFP at eight. Case2 compared with case1 achieved a 1.12-fold increase in speed in solving the linear system and a 1.89-fold increase in the speed of calculating the constitutive law, with calculation of the constitutive law showing greater performance improvement, because the difference in computational cost between the linear and nonlinear layer was larger. Case3 compared with case2 achieved a 13% increase in the speed of solving the linear system. Figure 4b shows the total number of iterations in the outer and inner CGs and the analysis times for case2 and case3. The number of iterations in the inner fine CG in case2 and the summation of the number of iterations in the inner middle and fine CGs in case3 were similar. Case3 reduced computation time by using the inner middle CG accompanied by a lower computational cost. Case3 achieved a 26-fold increase in speed over asis and a 4.85-fold increase in speed over asisOMP, thereby demonstrating the effectiveness of the proposed method.

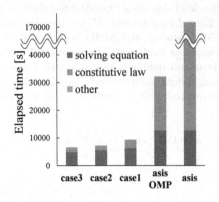
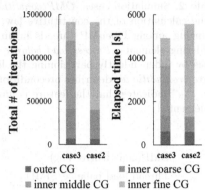

(a) Time to solution of the proposed method. Elapsed time to analyze 7,500 time steps.

(b) Number of iterations and the elapsed time in the outer CG and inner CGs. Results represent analyses of 7,500 time steps.

**Fig. 4.** Measurement of the performance of the proposed method.

**Table 3.** Weak-scale configuration.

| # of OFP nodes | # of MPI procs. | DOF | DOF per process | # of elements |
|---:|---:|---:|---:|---:|
| 64 | 512 | 109,425,213 | 213,721 | 26,810,034 |
| 128 | 1,024 | 218,798,889 | 213,671 | 53,620,068 |
| 256 | 2,048 | 437,546,541 | 213,646 | 107,240,212 |
| 512 | 4,096 | 874,990,053 | 213,621 | 214,480,424 |
| 1,024 | 8,192 | 1,749,877,677 | 213,608 | 428,961,000 |
| 2,046 | 16,368 | 3,499,549,341 | 213,804 | 857,922,000 |

## 5.2 Scalability

We then measured the weak scalabilities of case3 and asisOMP. The model shown in Fig. 3 was replicated in the x and y directions in order to generate models, and Table 3 shows the model configuration.

100 time steps of soil liquefaction analyses were performed using 512 to 16,368 MPI processes while maintaining a constant DOF per MPI process. The number of OpenMP threads per MPI process was eight. In the analysis using 16,368 MPI processes, up to 29.2 GiB per OFP node was used out of 16GB-MCDRAM memory and 96GB-DDR memory. Results are shown in Fig. 5.

For all problem sizes, case3 showed an about 5-fold increase in speed relative to asisOMP. The largest-scale problem used 2,046 of the 8,208 OFP compute nodes, indicating that the proposed solver showed a high level of performance using 25% of the OFP. The weak scalability of case3 from 512 MPI processes to 16,368 MPI processes was 75 %. This can be considered a high level of

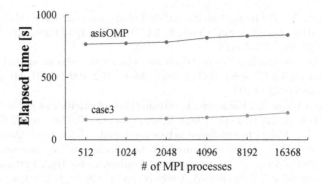

**Fig. 5.** Weak scalability

performance based on the complexities of the target problem. The finite-element method using unstructured elements is complicated, making it difficult to achieve a high degree of scalability. Soil liquefaction analysis involves additional complexities in regard to load balancing, and the high memory usage for calculating the constitutive law limits the problem scale per MPI process, resulting in a relatively larger percentage of communications.

## 6   Conclusion

In this study, we developed an efficient algorithm for soil liquefaction analysis based on that reported previously [9]. Additionally, we developed a preconditioner that reduced the amount of computation and memory access by considering the locality of soil liquefaction phenomena. The proposed method achieved a 26-fold increase in speed as compared with that demonstrated previously [9]. We performed soil liquefaction simulations using large-scale 3D models with up to 3.5 billion DOFs, which is 1,000-fold larger than the 3.4-million-DOF model computed in the previous study [9]. These results demonstrated that the proposed method is fast and scalable.

The load-balancing method proposed in this study can be applied not only to soil liquefaction analysis but also to other heterogeneous problems, such as other multi-physics coupled analyses. The physics-aware preconditioner developed in this study achieved increases in speed by considering the physical characteristics of soil liquefaction phenomena, implying that smarter algorithms for solving mathematical problems can be developed by considering the characteristics of the underlying physical phenomenon, which the SC18 solver [6] demonstrated for a different physical problem.

## References

1. Akai, K.: Geotechnical reconnaissance of the effects of the January 17, 1995, Hyogoken-Nanbu earthquake. University of California at Berkeley, Japan. Earthquake Engineering Research Center (1995)

2. Golub, G.H., Ye, Q.: Inexact preconditioned conjugate gradient method with inner-outer iteration. SIAM J. Sci. Comput. **21**(4), 1305–1320 (1999). https://doi.org/10.1137/S1064827597323415

3. Iai, S.: Three dimensional formulation and objectivity of a strain splace multiple mechanism model for sand. Soils Found. **33**(1), 192–199 (1993). https://doi.org/10.3208/sandf1972.33.192

4. Iai, S., Matsunaga, Y., Kameoka, T.: Strain space plasticity model for cyclic mobility. Soils Found. **32**(2), 1–15 (1992). https://doi.org/10.3208/sandf1972.32.2_1

5. Ichimura, T., et al.: Physics-based urban earthquake simulation enhanced by 10.7 BlnDOF 30 k time-step unstructured FE non-linear seismic wave simulation. In: SC 2014: Proceedings of the International Conference for High Performance Computing, Networking, Storage and Analysis, pp. 15–26 (2014). https://doi.org/10.1109/SC.2014.7

6. Ichimura, T., et al.: A fast scalable implicit solver for nonlinear time-evolution earthquake city problem on low-ordered unstructured finite elements with artificial intelligence and transprecision computing. In: Proceedings of the International Conference for High Performance Computing, Networking, Storage, and Analysis, SC 2018, pp. 49:1–49:11 (2018). http://dl.acm.org/citation.cfm?id=3291656.3291722

7. JMA observatory, Japan Meteorological Agency: Strong ground motion of the southern Hyogo prefecture earthquake in 1995 observed at Kobe. http://www.data.jma.go.jp/svd/eqev/data/kyoshin/jishin/hyogo_nanbu/dat/H1171931.csv. Accessed 3 Jan 2019

8. Joint Center for Advanced High Performance Computing: Basic specification of oakforest-pacs. http://jcahpc.jp/files/OFP-basic.pdf

9. Kusakabe, R., Ichimura, T., Fujita, K., Hori, M., Wijerathne, L.: A finite element analysis method for simulating seismic soil liquefaction based on a large-scale 3D soil structure model. Dyn. Earthq. Eng. **123**, 64–74 (2019)

10. Okimura, T., Takada, S., Koid, T.H.: Outline of the great Hanshin earthquake, Japan 1995. Nat. Hazards **14**(1), 39–71 (1996). https://doi.org/10.1007/BF00229911

11. Towhata, I., et al.: Liquefaction in the Kanto region during the 2011 off the pacific coast of Tohoku earthquake. Soils Found. **54**(4), 859–873 (2014). https://doi.org/10.1016/j.sandf.2014.06.016

12. Winget, J.M., Hughes, T.J.R.: Solution algorithms for nonlinear transient heat conduction analysis employing element-by-element iterative strategies. Comput. Methods Appl. Mech. Eng. **52**(1–3), 711–815 (1985). https://doi.org/10.1016/0045-7825(85)90015-5

13. Yamaguchi, A., Mori, T., Kazama, M., Yoshida, N.: Liquefaction in Tohoku district during the 2011 off the pacific coast of Tohoku earthquake. Soils Found. **52**(5), 811–829 (2012). https://doi.org/10.1016/j.sandf.2012.11.005

# Performance Evaluation of Tsunami Inundation Simulation on SX-Aurora TSUBASA

Akihiro Musa[1,2(✉)], Takashi Abe[1], Takumi Kishitani[1], Takuya Inoue[1,3],
Masayuki Sato[1], Kazuhiko Komatsu[1], Yoichi Murashima[1,3],
Shunichi Koshimura[1], and Hiroaki Kobayashi[1]

[1] Tohoku University, Sendai, Miyagi 908-8578, Japan
{musa,masa,komatsu,koba}@tohoku.ac.jp
{t-abe,koshimura}@irides.tohoku.ac.jp,
takumi.kishitani.p6@dc.tohoku.ac.jp
[2] NEC Corporation, Minato-ku, Tokyo 108-8001, Japan
[3] Kokusai Kogyo Co., LTD., Fuchu-shi, Tokyo 183-0057, Japan
{takuya_inoue,yoichi_murashima}@kk-grp.jp

**Abstract.** As tsunamis may cause damage in wide area, it is difficult to immediately understand the whole damage. To quickly estimate the damages of and respond to the disaster, we have developed a real-time tsunami inundation forecast system that utilizes the vector supercomputer SX-ACE for simulating tsunami inundation phenomena. The forecast system can complete a tsunami inundation and damage forecast for the southwestern part of the Pacific coast of Japan at the level of a 30-m grid size in less than 30 min. The forecast system requires higher-performance supercomputers to increase resolutions and expand forecast areas. In this paper, we compare the performance of the tsunami inundation simulation on SX-Aurora TSUBASA, which is a new vector supercomputer released in 2018, with those on Xeon Gold and SX-ACE. We clarify that SX-Aurora TSUBASA achieves the highest performance among the three systems and has a high potential for increasing resolutions as well as expanding forecast areas.

**Keywords:** System performance · Supercomputer · Tsunami simulation

## 1 Introduction

Tsunamis are a natural disaster that may cause many fatalities and extensive damage of infrastructures in coastal areas. In 2011, the 2011 Tohoku earthquake tsunami, which was generated by a 9.0-magnitude earthquake, struck the northeastern part of the Pacific coast of Japan. Its inundated area ranged up to 2000 km along the Japanese coastlines and spread out over 500 $km^2$ [4,13]. The number of fatalities was more than 19,000, and many houses and infrastructures

© Springer Nature Switzerland AG 2019
J. M. F. Rodrigues et al. (Eds.): ICCS 2019, LNCS 11537, pp. 363–376, 2019.
https://doi.org/10.1007/978-3-030-22741-8_26

were destroyed [4,18]. Since the damaged area was widespread, the Japanese government was not able to grasp the entire picture of the damage shortly after the tsunami. Under this situation, it was difficult to promptly cope with the disaster [11]. Therefore, a forecast system that can quickly estimate the damage of a tsunami inundation after an earthquake is required worldwide in order to provide an early understanding of damage situations.

We have developed a real-time tsunami inundation forecast system using the vector supercomputer SX-ACE [15] to quickly estimate tsunami inundations and infrastructure damages [9]. The forecast system is automatically activated by a large earthquake that may generate a tsunami, simulates the tsunami inundation, and finally visualizes the results of the damage estimations on a Web site. The total processing time is within 20 min for a coastal city region at the resolution of a 10-m grid size [7,15]. The forecast system has also been implemented as the disaster information system of the Cabinet Office of Japan, however, the forecast area is needed to been expanded from a city region to the southwestern part of the Pacific coast of Japan, resulting in the coastal length of 5,500 km. Therefore, in the implementation for the Cabinet Office, the grid size and the total processing time are changed from 10 m to 30 m and from 20 min to 30 min due to the processing time of the SX-ACE system. In addition, since the entire Pacific coast of Japan has the high probability of damages by large tsunamis, it is necessary to expand the forecast areas. Moreover, the resolution of a 10-m grid size is required for accurate forecasting. Therefore, we need performance improvements of the tsunami inundation simulation in order to improve disaster prevention and mitigation by expanding forecast areas and increasing resolutions.

Modern supercomputers consist of multi-nodes, each of which is equipped with multi-core processors. These processors have recently employed vector (SIMD) instructions to further exploit loop-level parallelism. Intel Corporation released a Xeon Phi processor (*Knights Landing*) in 2016 and a Xeon Gold (*Skylake*) in 2017, and NEC Corporation developed an SX-Aurora TSUBASA processor in 2018. We evaluated the performance of our tsunami inundation on the Xeon Phi processor, and found that Xeon Phi needs optimizations for vectorization to achieve a high performance [14]. However, since the performance of Xeon Phi did not exceed that of SX-ACE, the forecast system cannot increase resolutions and expand forecast areas by Xeon Phi. In this paper, we discuss the performance of the tsunami inundation simulation on the latest vector supercomputer SX-Aurora TSUBASA [20] and Xeon processor (*Skylake*) and the possibilities to increase the forecast resolutions and areas.

The rest of this paper is organized as follows. In Sect. 2, we present related work on performance evaluations using supercomputers. Section 3 explains the tsunami inundation simulation. Section 4 describes the experimental setup of the performance evaluation, which includes an overview of SX-Aurora TSUBASA. Section 5 presents the results of the performance evaluations. We conclude this paper in Sect. 6 with a brief summary and mention of future work.

## 2    Related Work

A real-time tsunami forecast system requires reduction in execution time in order to provide adequate disaster prevention and mitigation. To this end, there have been studies on reducing the execution time of tsunami simulations using high-performance computing. Christgau et al. [3] evaluated the performance of Intel Xeon Phi (*Knights Corner*), Xeon (*Ivy Bridge*), and NVIDIA Tesla K40m GPGPU (general-purpose computing on graphics processing units) using the tsunami simulation EasyWave, which is a TUNAMI-F1 model [6]. They optimized the program to increase the reusability of cache data. Specifically, the computer domain was divided into smaller blocks that fit into the last level cache on Xeon and Xeon Phi. The program was also vectorized by the Intel compiler. They parallelized it using OpenMP on Xeon and MPI on XeonPhi, and the program has a CUDA version for GPGPU. They compared the performances of the program with an $18{,}001 \times 16{,}301$ grid on Xeon, Xeon Phi, and Tesla K40m. The performance of the single Xeon processor (10 cores) was 1.23 and 1.8 times higher than those of the single Xeon Phi processor (61 cores) and the single Tesla K40m GPGPU (2880 cores), respectively. They found that the tsunami simulation EasyWave was memory intensive and required the load reduction of the memory to achieve higher performance, and the vector processing on Xeon and Xeon Phi was useful for accelerating the tsunami simulation. We will discuss the potential of vector processing and high memory bandwidth by comparing the performances of vector supercomputers, SX-Aurora TSUBASA and SX-ACE, as well as Xeon Gold in this paper.

Nagasu et al. [16] developed a field-programmable gate array (FPGA)-based tsunami simulation system using the method of splitting tsunami (MOST) model, which solves shallow water equations for ocean-wide tsunami propagation. The machine contained an ALTERA Arria 10 FPGA, two DDR3 PC12800 SDRAMs, and a PCI-Express Gen2 interface. They evaluated the performance and power consumption of the tsunami propagation simulation with a $2581 \times 2879$ grid in the Pacific Ocean on the FPGA system, NVIDIA Tesla K20c, and AMD Radeon R9 280X GPGPU. The sustained performances were 383.0 Gflop/s on the FPGA system, 44.6 Gflop/s on Tesla K20c and 219.2 Gflop/s on Radeon R9 280X, respectively. The performance per power on the system was 17.2 and 7.1 times higher than those on Tesla K20c and Radeon R9 280x, respectively. The research demonstrated the significance of power efficiency. However, their resolutions were several kilometers and the tsunami inundation behavior was not simulated. This paper conducts the tsunami inundation simulation with higher resolutions that is necessary for the forecast system.

Arce Acuña et al. [1] developed a non-linear spherical shallow water model to compute tsunami propagation across an ocean on GPGPU. The model utilized the tree-based mesh refinement method to generate a computational mesh and was used to replicate the 2004 Indonesia tsunami off the Indian Ocean ($1{,}500\,\text{km} \times 500\,\text{km}$) with grid sizes from $1852\,\text{m}$ to $50\,\text{m}$. The execution time for ten hours of tsunami behaviors was $8.19\,\text{min}$ on eight NVIDIA Tesla P100 GPGPUs. These results indicated that their model had a potential for real-time tsunami forecasting using GPGPU. However, the area of the 50-m grid size was a city region. Meanwhile, we will evaluate the performance of our tsunami inundation simulation using the 10-m grid size along coastlines.

## 3    Tsunami Inundation Simulation

Our tsunami inundation simulation is based on the TUNAMI model [10] that solves non-linear shallow water equations with the staggered leap-frog finite difference method. The governing equations are

$$\frac{\partial \eta}{\partial t} + \frac{\partial M}{\partial x} + \frac{\partial N}{\partial y} = 0, \tag{1}$$

$$\frac{\partial M}{\partial t} + \frac{\partial}{\partial x}\left(\frac{M^2}{D}\right) + \frac{\partial}{\partial y}\left(\frac{MN}{D}\right) + gD\frac{\partial \eta}{\partial x}$$
$$+ \frac{gn^2}{D^{7/3}}M\sqrt{M^2 + N^2} = 0, \tag{2}$$

$$\frac{\partial N}{\partial t} + \frac{\partial}{\partial x}\left(\frac{MN}{D}\right) + \frac{\partial}{\partial y}\left(\frac{N^2}{D}\right) + gD\frac{\partial \eta}{\partial y}$$
$$+ \frac{gn^2}{D^{7/3}}N\sqrt{M^2 + N^2} = 0, \tag{3}$$

where $\eta$ is the vertical displacement of the water surface, $D$ is the total water depth, $g$ is the gravitational acceleration, $n$ is Manning's roughness coefficient, $x$ and $y$ are the horizontal directions. $M$ and $N$ are the discharge fluxes in the x and y directions, respectively. $M$ and $N$ are given by

$$M = \int_{-h}^{\eta} u dz = \bar{u}(\eta + h), \tag{4}$$

$$N = \int_{-h}^{\eta} v dz = \bar{v}(\eta + h), \tag{5}$$

where $h$ is the still water depth. $u$ and $v$ are the water velocities in the x and y directions, respectively. $\bar{u}$ and $\bar{v}$ are the average water velocities in the x and y directions, respectively. The tsunami source is calculated using Okada's equation [19], and the damage estimation is based on a study by Koshimura et al. [10].

The tsunami inundation simulation leverages a nested grid system to simulate the multiscale nature of a tsunami. The computational domain is a polygonal domain [5], which can reduce the number of computations in the simulation [14]. Also, the system of the Cabinet Office simulates the coastal length of 5,500 km, and the simulated area is divided into fifteen subareas, and each subarea is individually simulated because the geographical coordinates of the subareas are different and a different tidal level is used for each region.

## 4    Experimental Setup for Performance Evaluation

We evaluate the performance of the tsunami inundation simulation by using SX-Aurora TSUBASA, SX-ACE, and Xeon Gold. Table 1 lists the specifications of each system. Here, SP and DP indicate single precision and double precision floating-point operations, respectively.

**Table 1.** Specifications of systems used in the evaluation

| Clock freq. (GHz) | | SX-Aurora TSUBASA | SX-ACE | Xeon Gold |
|---|---|---|---|---|
| | | 1.4 | 1.0 | 2.6 |
| Core | Performance (Gflop/s) | 537.6 (SP) | 64 (SP/DP) | 147.2 (SP) |
| | | 268.8 (DP) | | 73.6 (DP) |
| | Memory bandwidth (GB/s) | 358 | 256 | N/A |
| | B/F | 0.67 (SP) | 4 (SP/DP) | N/A |
| | | 1.33 (DP) | | |
| Processor | No. of cores | 8 | 4 | 12 |
| | Performance (Tflop/s) | 4.3 (SP) | 0.256 (SP/DP) | 1.77 (SP) |
| | | 2.15 (DP) | | 0.88 (DP) |
| | LLC (MB) | 16 (shared) | 1 (private) | 19.5 |
| | LLC bandwidth (GB/s) | 2854 | 1000 | N/A |
| | B/F of LLC | 0.67 (SP) | 4 (SP/DP) | N/A |
| | | 1.33 (DP) | | |
| | Memory (GB) | 48 | 64 | 96 |
| | Memory bandwidth (GB/s) | 1228 | 256 | 128 |
| | B/F of memory | 0.28 (SP) | 1 (SP/DP) | 0.07 (SP) |
| | | 0.56 (DP) | | 0.14 (DP) |
| Node | No. of processors | 8 | 1 | 2 |
| | Memory (GB) | 384 | 64 | 192 |
| | Memory bandwidth (GB/s) | 9760 | 256 | 256 |
| | Network (GB/s) | 25 | 8 | 12.5 |

## 4.1   SX-Aurora TSUBASA

SX-Aurora TSUBASA is a new vector supercomputer released in 2018 [8,20]. It consists of one or more card-type vector engines (VEs) and a vector host (VH). The VE is the main part of SX-Aurora TSUBASA and contains a vector processor, a 16 MB shared last-level cache (LLC), and six HBM2 memory modules, as shown in Fig. 1. The vector processor has eight vector cores and two types of the clock frequency: 1.6 GHz and 1.4 GHz. In this evaluation, we use the VE of 1.4 GHz. The peak performances of single precision floating-point operations on a vector core and a vector processor are 537.6 Gflop/s and 4.3 Tflop/s, respectively. The memory bandwidths are 358 GB/s per vector core and 1.22 TB/s per vector processor. The ratios of the memory bandwidth to the floating-point operation per second, so called the Bytes/Flop ratio (B/F), on the vector core and the vector processor are 0.67 and 0.28, respectively. The VH is a standard x86 Linux server and executes an operating system (OS) of the VE. While the conventional SX series runs an OS on a vector processor, SX-Aurora TSUBASA

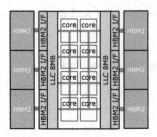   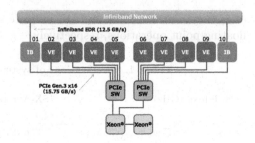

**Fig. 1.** Block diagram of a vector engine     **Fig. 2.** Block diagram of SX-Aurora TSUB-ASA system

offloads OS-related processing on the VH since the VH is suitable for scalar processing such as the OS-related processing. Figure 2 shows the maximum configuration of one VH that contains eight VEs, two Infiniband interfaces, and two Xeon processors. Moreover, SX-Aurora TSUBASA can compose a large system by connecting the VHs via the Infiniband switch.

The SX-Aurora TSUBASA system supports Red Hat Enterprise Linux and CentOS as the OS, and the VH provides OS functions such as process scheduling and handling of system calls invoked by the application running on the VEs. The development environment is NEC MPI and NEC Software Development Kit for Vector Engine, which contains NEC Fortran compiler, NEC C/C++ compiler, NEC Numeric Library Collection, NEC Parallel Debugger, and NEC Ftrace Viewer. The Fortran and C/C++ compilers support functions of automatic optimization, automatic vectorization, automatic parallelization, and OpenMP. NEC MPI is implemented in accordance with MPI-3.1.

### 4.2 SX-ACE

SX-ACE, which is also a vector supercomputer, was released in 2014. It consists of one vector processor and a 64 GB memory [12]. The vector processor has four vector cores, each of which is comprised of a scalar processing unit, a vector processing unit, and an on-chip private cache with 1 MB. The peak performances on a vector core and a vector processor are 64 Gflop/s and 256 Gflop/s, respectively. The memory bandwidth is 256 GB/s per vector core and per vector processor. The B/F ratios of a vector core and a vector processor are 4.0 and 1.0, respectively. The SX-ACE system is composed of 512 nodes connected via a custom interconnect network (called IXS) by an 8 GB/s bandwidth.

The OS of the SX-ACE is SUPER-UX, which is based on the UNIX System V with extensions for performance and functionality. The FORTRAN compiler for SX-ACE, FORTRAN90/SX, has functions of automatic optimization, automatic vectorization, automatic parallelization and OpenMP. The MPI Library, MPI2/SX, implements the Message Passing Interface specifications MPI-1.3 and MPI-2.1.

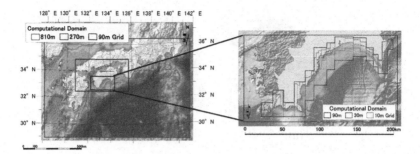

**Fig. 3.** Computational domains for Kochi prefecture

**Table 2.** Computational domain

| Domain | Grid size (m) | Number of grid points |
|--------|---------------|-----------------------|
| 1 | 810 | $1.6 \times 10^6$ |
| 2 | 270 | $1.7 \times 10^6$ |
| 3 | 90 | $2.9 \times 10^6$ |
| 4 | 30 | $6.0 \times 10^6$ |
| 5 | 10 | $22.8 \times 10^6$ |
| Total | | $3.5 \times 10^7$ |

### 4.3  Xeon Gold

The Xeon Gold system consists of two Intel Xeon Gold 6126 processors (released in 2017) with a 192 GB memory and 128 GB/s memory bandwidth. The processor contains 12 cores and a shared last-level cache of 19.5 MB. The peak performances of single precision floating-point operations on a core and a processor are 147.2 Gflop/s and 1766.4 Gflop/s, respectively. The B/F of a processor is 0.07. Each node is connected with an InfiniBand EDR at 12.5 GB/s.

The OS of the Xeon Gold system is Red Hat Enterprise Linux, and the programming environment is Intel Parallel Studio XE Cluster Edition, which contains Fortran and C/C++ compilers and an MPI library. The Fortran compiler has the functions of automatic optimization, automatic vectorization, automatic parallelization, and OpenMP.

### 4.4  Evaluated Model and Compiler Options

We evaluate the performances of a tsunami inundation simulation in the Kochi prefecture region in Japan with a 10-m grid size, as shown in Fig. 3. This coastal length is about 720 km, and this region is one of the fifteen subareas in the system of Cabinet Office of Japan. Table 2 lists the specifications of the model grid sizes and the number of grid points in each domains. *Domains 4 and 5* refer to the polygonal domains. The simulation case is the tsunami generated

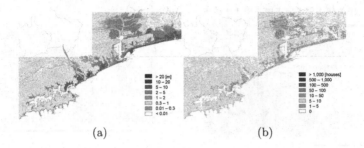

(a)                                              (b)

**Fig. 4.** Simulation results. *a* Maximum tsunami inundation depth. *b* Structural damage (houses)

**Table 3.** Fortran compiler's options and directives on each system

| System | Version | Options | Directives |
|---|---|---|---|
| SX-Aurora TSUBASA | Fortran compiler for Vector Engine 1.5.7 | -O3 -msched-block | - |
| | | -finline-functions | |
| SX-ACE | FORTRAN 90/SX Rev.535 | -Chopt -pi | - |
| Xeon Gold | Intel Fortran 18.0.3.222 | -O3 -ipo | Vector always |
| | | -xCORE-AVX512 | |

by the 9.0-magnitude Nankai Trough earthquake [2], and tsunami inundation behaviors are calculated with the time interval of 0.2 s for six hours after the earthquake. Here, the simulation is executed with a single precision floating-point operation. Figure 4 shows the simulation results of the maximum tsunami inundation depth and structural damage. In the traditional tsunami forecast, the items of prediction information were tsunami arrival times and tsunami heights on coastlines. But our forecast system can estimate tsunami inundations and infrastructure damages of coastal land areas.

Table 3 lists Fortran compiler's versions, options, and directives on each system. The options, *-Chopt*, *-O3*, and *-msched-block* are high-level optimization options, and *-finline-functions*, *-pi*, and *-ipo* are inline options. As Xeon Gold is unable to vectorize two do-loops, which are time-consuming do-loops, just with the compiler options, we insert the *vector always* directive into the simulation code. *-xCORE-AVX512* also creates the object code including the AVX512 operations.

## 5    Performance Results

### 5.1    Single Core Performance

Figure 5 shows the execution times of a single core on SX-Aurora TSUBASA, SX-ACE and Xeon Gold. The CPU time is the running time of the CPU. The memory time indicates the stall time of the CPU due to data loads from the memory. *Directive* and *Non-directive* indicate the cases with and without the vectorization directive on Xeon Gold, respectively. The vectorization directive *vector always* can vectorize two time-consuming do-loops, which take up about 80 % of the total execution time even when the loops are vectorized. The CPU time of the *directive* case is less than half that of the *non-directive* case. This result shows that the vector processing is key in the achievement of a high performance on the Xeon Gold processor.

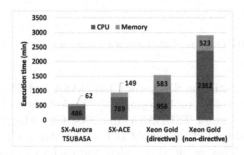

**Fig. 5.** Execution times on a single core. *CPU*: running time of CPU, *Memory*: Stall time of CPU due to data loads

**Table 4.** Code B/F and actual B/F on SX-ACE and SX-Aurora TSUBASA

| Code B/F | Actual B/F (SX-Aurora TSUBASA) | Actual B/F (SX-ACE) |
|----------|-------------------------------|---------------------|
| 1.8      | 0.48                          | 0.66                |

The execution times of both the CPU and memory on SX-Aurora TSUBASA is the shortest among all the systems. SX-Aurora TSUBASA can calculate the tsunami inundation simulation in 58% and 36% of the execution times of SX-ACE and Xeon Gold (directive), respectively. The memory time on SX-Aurora TSUBASA is 40% and 11% of those on SX-ACE and Xeon Gold, respectively. This is because SX-Aurora TSUBASA has the highest LLC bandwidth among all the systems. Moreover, the memory accesses on SX-Aurora TSUBASA can be hidden by vector arithmetic operations since SX-Aurora TSUBASA has long vector pipelines whose vector length is 256. SX-ACE can also hide them and has 61% of the execution time of Xeon Gold (directive). The memory time on SX-ACE is 25% of that on Xeon Gold (directive).

To analyze the performance using the B/F ratios, we use two metrics: a code B/F and an actual B/F. The code B/F is defined as the necessary data in bytes divided by the floating-point operations in a code. The numbers of memory operations and floating-point operations are counted from the simulation code. Since a processor contains a cache memory, the amount of data from a memory is smaller than that of the necessary data. Thus, we utilize the actual B/F calculated from the numbers of block memory accesses and the floating-point operations in the actual behavior of a processor when the simulation is executed.

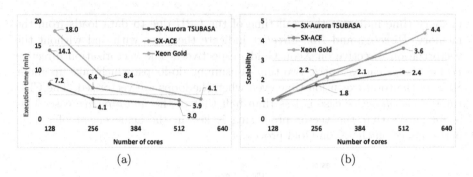

**Fig. 6.** Multi core performance. *a* Execution times. *b* Scalability

These execution statistics on SX-Aurora TSUBASA and SX-ACE are obtained by using the NEC profiler tool *proginf* [17]. Table 4 shows the code B/F and actual B/F on SX-Aurora TSUBASA and SX-ACE. The code B/F of 1.8 is reduced to 0.48 and 0.66 of the actual B/F on SX-Aurora TSUBASA and SX-ACE, respectively. This result shows that the cache memory can reduce the memory loads of the tsunami inundation simulation on these processors.

The B/F of a vector core on SX-Aurora TSUBASA is 0.67, which is smaller than the code B/F. This indicates that the tsunami inundation simulation can be considered a memory-bandwidth-limited program for SX-Aurora TSUBASA. In addition, since the actual B/F of 0.48 is smaller than the B/F of 0.67, the bandwidth between the LLC and the memory should be sufficient. Therefore, the bandwidth between the vector core and the LLC (LLC bandwidth) poses a performance bottleneck, and the performance of the simulation is limited by the performance of the LLC. Although the LLC bandwidth becomes a bottleneck on SX-Aurora TSUBASA, the execution time is the shortest among all the systems. This is because the bandwidth of the LCC on SX-Aurora TSUBASA is the highest.

On SX-ACE, the B/F of the vector core is 4.0, which is greater than the code B/F. Therefore, the tsunami inundation simulation can be considered a computation-bound program for SX-ACE and does not require the full memory bandwidth. Although the peak performance of the core is a mere 64 Gflop/s, the CPU time is shorter than that of Xeon Gold (directive). Here, the calculation performance is 12.8 Gflop/s and the computational efficiency reaches 20%. This is because the B/F of the vector core on SX-ACE is high and the vector core can efficiently execute the simulation.

In the case of Xeon Gold, as the bandwidth of core is not officially announced, we assume that one core can utilize the full memory bandwidth of 128 GB/s. Its B/F of the core is 0.87 and smaller than the code B/F. Therefore, the tsunami inundation simulation can be considered a memory-bandwidth-limited program for Xeon Gold, and the bandwidth of the core or memory causes a performance bottleneck. As the memory system of Xeon has the lowest performance among three systems, the execution time on Xeon becomes the longest.

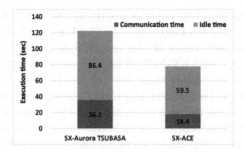

**Fig. 7.** MPI communication and idle times with 512 cores on SX-Aurora TSUBASA and SX-ACE

From these evaluations and analyses, we clarify that SX-Aurora TSUBASA has the shortest execution time among all the systems due to its high LLC bandwidth. This performance has a possibility to improve the functions of the forecast system.

## 5.2 Multi-core Performance

We evaluate the parallel performances of the tsunami inundation simulation using 512 cores of SX-Aurora TSUBASA and SX-ACE, and 576 cores of Xeon Gold. The peak system performances of single precision floating-point operations on the SX-Aurora TSUBASA, SX-ACE, and Xeon Gold systems are 275.3 Tflop/s, 32.8 Tflop/s, and 84.8 Tflop/s, respectively. Figure 6a shows the execution times of the tsunami inundation simulation. The execution time of SX-Aurora TSUBASA is the shortest time among all the systems due to the highest-core sustained performance. Specifically, it takes three minutes on 512 cores, which means 64 VEs and eight nodes. Figure 6b depicts the scalability of each system. SX-ACE and Xeon Gold achieve super-scalability when the numbers of cores are 256 on SX-ACE, and 288 and 576 on Xeon Gold. This is due to a load imbalance of the tsunami inundation simulation that utilizes the polygonal domain. The polygonal domain has the characteristic of variation in the number of grid points in each MPI process. Specifically, since the variation in the number of grid points per MPI process at small parallel number is larger than that at large parallel number, the load imbalance at small parallel number is large. Therefore, the performances at 128 and 144 cores are low and the super-scalability is generated. Figure 6b also shows that the scalability on SX-Aurora TSUBASA is the lowest among all the systems. The performance of 512 cores on SX-Aurora TSUBASA is 2.4 times higher than that on 128 cores. Meanwhile, the 512 cores on SX-ACE has a 3.6 times higher performance of 128 cores, and the performances of 576 cores on Xeon Gold is 4.4 times higher than that of 144 cores. This is because the network bandwidth per core between the nodes on SX-Aurora TSUBASA is lower than those on SX-ACE and Xeon Gold.

The peak network bandwidths per core are 0.4 GB/s on SX-Aurora TSUBASA, 2 GB/s on SX-ACE, and 0.5 GB/s on Xeon Gold. Figure 7 shows the MPI communication times and idle times of communications in the cases of 512 cores on SX-Aurora TSUBASA and SX-ACE. The communication time on SX-Aurora TSUBASA is 1.96 times longer than that on SX-ACE, and the idle time with SX-Aurora TSUBASA has a longer idle time than with SX-ACE.

We set a criterion of the execution time of the tsunami inundation simulation for the forecast system as five minutes in order to terminate the total processing from system start-up to visualization of the results within 30 min. In this evaluation, SX-Aurora TSUBASA, SX-ACE, and Xeon Gold achieve this time on 200 cores (25 processors), 384 cores (96 processors), and 492 cores (41 processors), respectively. In the case of the system used by the Cabinet Office of Japan, the southwestern part of the Pacific coast of Japan is now simulated with the 30-m grid size. When the grid size is the 10-m size, the number of grid points increases by a factor of 9, and the time step is supposed to be one third of the 30-m size case, thus the overall calculation amount is 27 times larger than that of this evaluation case. Moreover, when the forecast system covers the entire Pacific coast of Japan, we assume that length of the coast becomes triple, and thus the calculation amount is also about 81 times larger than that of this evaluation case. Table 5 lists the number of processors required for the forecast system with the 10-m grid size within five minutes. SX-Aurora TSUBASA can build the system by using the smallest number of processors. The numbers of nodes are 85 and 254 nodes in the cases of the southwestern part of the Pacific coast and entire Pacific coast of Japan, respectively. The forecast system for the entire Pacific coast of Japan with the 10-m grid size is feasible by SX-Aurora TSUBASA. Therefore, SX-Aurora TSUBASA has a high potential for increasing resolution and expanding forecast areas for the forecast system.

Table 5. Number of processors required for the forecast system

| Case | SX-Aurora TSUBASA | SX-ACE | Xeon Gold |
|---|---|---|---|
| Kochi prefecture | 25 | 96 | 41 |
| Southwestern part of the Pacific coast of Japan | 675 | 2,592 | 1,107 |
| Entire Pacific coast of Japan | 2,025 | 7,776 | 3,321 |

## 6   Summary

We have developed a real-time tsunami inundation forecast system in order to provide an early understanding of tsunami damage situations. Our system has been operated as a Japanese disaster system on the southwestern part of the Pacific coast of Japan. In order to increase the resolutions of simulation and expand the forecast areas, higher-performance supercomputers are required. In this paper, we evaluate the performance of our tsunami inundation simulation using SX-Aurora TUBASA, SX-ACE, and Xeon Gold. SX-Aurora TSUBASA has the highest sustained performance among all the systems. Specifically, the core performance of SX-Aurora TSUBASA is 1.7 and 2.8 times higher than those of SX-ACE and Xeon Gold, respectively. SX-Aurora TSUBASA can simulate a tsunami inundation in Kochi prefecture with the 10-m grid within five minutes using 200 cores. In the same scenario, SX-ACE and Xeon Gold need 384 cores and 492 cores, respectively. We found that SX-Aurora TSUBASA has the high potential to simulate the entire Pacific coast of Japan with the required time and resolution constrains. However, the performance of SX-Aurora TSUBASA is limited by the LLC bandwidth. The network bandwidth on SX-Aurora TSUBASA is also lower than those on SX-ACE and Xeon Gold.

In future work, we plan to redevaluate the performance of the multi-core environment by changing the domain decomposition for enhancement of the parallel performance in the polygonal domain method [5]. We will also implement the tsunami inundation simulation in general-purpose computing on graphics processing units (GPGPU) and investigate the performance characteristics of the GPGPUs.

**Acknowledgments.** The authors would like to thank Mr. Takayuki Suzuki and Mr. Atsushi Tanobe of Kokusai Kogyo Co., LTD. for developing the tsunami model. This study was partially supported by JSPS KAKENHI Grant Numbers 16K12845, 17H06108, 18K11322, JST CREST Grant Number JP-MJCR1411, Joint Usage/Research Center for Interdisciplinary Large-scale Information Infrastructures in Japan (Project ID: jh180040-NAH), and Ministry of Education, Culture, Sports, Science and Technology of Japan (Next Generation High-Performance Computing Infrastructures and Applications R&D Program). In this work, we used SX-ACE of the Cyberscience Center at Tohoku University, and SX-ACE and Octopus of the Cybermedia Center at the Osaka University.

# References

1. Arce Acuña, M., Aoki, T.: Tree-based mesh-refinement GPU-accelerated tsunami simulator for real-time operation. Nat. Hazards Earth Syst. Sci. **18**(9), 2561–2602 (2018). https://doi.org/10.5194/nhess-18-2561-2018. https://www.nat-hazards-earth-syst-sci.net/18/2561/2018/

2. Cabinet Office: The investigative commmmission on massive earthquake models in nankai trough (2012). http://www.bousai.go.jp/jishin/nankai/model/data_teikyou.html. (In Japanese)

3. Christgau, S., Spazier, J., Schnor, B.: A performance and scalability analysis of the tsunami simulation easywave for difference multi-core architectures and programming models. Technical report, Institute for Computer Science, University of Potsdam (2015). http://www.cs.uni-potsdam.de/bs/research/docs/techreports/2015/css15.pdf

4. Imamura, F., Anawat, S.: Damage due to the 2011 Tohoku earthquake tsunami and its lessons for future mitigation. In: Proceedings of the International Symposium on Engineering Lessons Learned from the 2011 Great East Japan Earthquake, Tokyo, Japan, March 2012

5. Inoue, T., Abe, T., Koshimura, S., Musa, A., Murashima, Y., Kobayashi, H.: Development and validation of a tsunami numerical model with the polygonally nested grid system and its MPI-parallelization for real-time tsunami inundation forecast on a regional scale. J. Disaster Res. **14**(3), 416–434 (2019). https://doi.org/10.20965/jdr.2019.p0416

6. IUGG/IOC Time Project: IUGG/IOC TIME PROJECT: Numerical method of Tsunami simulation with the leap-frog scheme. UNESCO (1997)

7. Kobayashi, H.: A case study of urgent computing on SX-ACE: design and development of a real-time tsunami inundation analysis system for disaster prevention and mitigation. In: Resch, M., et al. (eds.) Sustained Simulation Performance 2016, pp. 131–138. Springer, Cham (2016). https://doi.org/10.1007/978-3-319-46735-1_11

8. Komatsu, K., et al.: Performance evaluation of a vector supercomputer SX-Aurora TSUBASA. In: The International Conference on High Performance Computing, Networking, Storage and Analysis (SC 2018), pp. 54:1–54:12. IEEE Press, Dallas, USA, November 2018. http://dl.acm.org/citation.cfm?id=3291656.3291728

9. Koshimura, S.: Establishing the advanced disaster reduction management system by fusion of real-time disaster simulation and big data assimilation. J. Disaster Res. **11**(2), 164–174 (2016)

10. Koshimura, S., Oie, T., Yanagisawa, H., Imamura, F.: Developing fragility functions for tsunami damage estimation using numerical model and post-tsunami data from Banda Aceh, Indonesia. Coast. Eng. J. JSCE **51**(3), 243–273 (2009)

11. Koshiyama, K.: Characteristics of emergency responcy at the great east Japan earthquake. In: 5th International Disaster and Risk Conference Davos 2014 [poster]. Davos, Switzland, August 2014. https://idrc.info/fileadmin/user_upload/idrc/documents/IDRC14_PosterCollection.pdf

12. Momose, S.: SX-ACE, brand-new vector supercomputer for higher sustained performance I. In: Resch, M., et al. (eds.) Sustained Simulation Performance 2014, pp. 57–67. Springer, Cham (2014). https://doi.org/10.1007/978-3-319-10626-7_5

13. Mori, N., Takahashi, T., Yanagisawa, H.: Survey of 2011 Tohoku earthquake tsunami inundation. Geophys. Res. Lett. **38** L00G14 (2016). https://doi.org/10.1029/2011GL049210

14. Musa, A., et al.: A real-time tsunami inundation forecast system using vector supercomputer SX-ACE. J. Disaster Res. **13**(2), 234–244 (2018). https://doi.org/10.20965/jdr.2018.p0234

15. Musa, A., et al.: Real-time tsunami inundation forecast system for tsunami disaster prevention and mitigation. J. Supercomput. **74**(7), 3093–3113 (2018). https://doi.org/10.1007/s11227-018-2363-0

16. Nagasu, K., Sane, K., Kono, F., Nakasato, N.: FPGA-based tsunami simulation: performance comparison with GPUs, and roofline model for scalability analysis. J. Parallel Distrib. Comput. **106**, 153–169 (2017). https://doi.org/10.1016/j.jpdc.2016.12.015

17. NEC Corporation: NEC Fortran 2003 Programmer's Guide (2014)

18. Okada, N., Tao, Y., Yoshio, K., Peijun, S., Hirokazu, T.: The 2011 eastern Japan great earthquake disaster: Overview and comments. Int. J. Disaster Risk Sci. **2**(1), 34–42 (2011). https://doi.org/10.1007/s13753-011-0004-9

19. Okada, Y.: Internal deformation due to shear and tensile faults in a half-space. Bull. Seism. Soc. Am. **82**(2), 1018–1040 (1992)

20. Yamada, Y., Momose, S.: Vector engine processor of NECfs brand-new supercomputer SX-Aurora TSUBASA. In: International symposium on High Performance Chips (Hot Chips), Cupertino, USA, August 2018

# Parallel Computing for Module-Based Computational Experiment

Zhuo Yao[1], Dali Wang[2(✉)] ⓘ, Danial Riccuito[2] ⓘ, Fengming Yuan[2] ⓘ,
and Chunsheng Fang[3]

[1] University of Tennessee, Knoxville, TN 37996, USA
yaoz5@vols.utk.edu
[2] Oak Ridge National Laboratory, POBox 2008, MS 6301,
Oak Ridge, TN 37831, USA
{wangd,ricciutodm,yuanfm}@ornl.gov
[3] Jilin University, Changchun 130012, Jilin, People's Republic of China
fangcs@jlu.edu.cn

**Abstract.** Large-scale scientific code plays an important role in scientific researches. In order to facilitate module and element evaluation in scientific applications, we introduce a unit testing framework and describe the demand for module-based experiment customization. We then develop a parallel version of the unit testing framework to handle long-term simulations with a large amount of data. Specifically, we apply message passing based parallelization and I/O behavior optimization to improve the performance of the unit testing framework and use profiling result to guide the parallel process implementation. Finally, we present a case study on litter decomposition experiment using a standalone module from a large-scale Earth System Model. This case study is also a good demonstration on the scalability, portability, and high-efficiency of the framework.

**Keywords:** Parallel computing · Scientific software ·
Message passing based parallelization · Profiling

## 1 Introduction

Scientific code that incorporates important domain knowledge plays an important role in answering essential questions. In order to help researchers understand and modify the scientific code using good approaches from best software development practices [4,7], we prefer a framework to visualize code architecture and individual modules, so that researchers can conveniently use modules to design specific experiments and to optimize the code base. We hope the framework to be highly portable and multi-platform compatible, therefore scientists can use

This research was funded by the U.S. Department of Energy, Office of Science, Biological and Environmental Research (Terrestrial Ecosystem Sciences), Advanced Scientific Computing Research (Interoperable Design of Extreme-scale Application Software).

J. M. F. Rodrigues et al. (Eds.): ICCS 2019, LNCS 11537, pp. 377–388, 2019.
https://doi.org/10.1007/978-3-030-22741-8_27

it on different platforms. At the same time, we would like the framework to offer concurrent data analysis interface, which decouples the analysis from the file-based I/O in order to facilitate the data analysis.

In the previous work [13], we introduced and designed a unit testing framework that isolates specific functions from complex software base and offers an in-situ data communication service [11]. This service runs by an analysis code locating remotely as the original scientific code is running. The testing driver can build a necessary environment which suits the needs of a typical experiment. With the framework, the scientists can track and manipulate variables between modules or inside modules to better meet their needs. However, the above-mentioned framework is unable to meet the requirement of many scientific applications that simulate large-scale phenomena with complex mathematical models on supercomputers. The examples of these scientific applications include large-scale models to predict climate change, air traffic control, power grids, and nuclear power plants. Therefore, improving the overall performance of the functional-unit testing platform for scientific code is significant.

## 2    Related Works

Scientific software can be bulky and complicated, it is important to analyze the crucial performance factor in order to optimize the code base. A diverse set of tools and methodology were used to identify the performance and scaling problems, including shell timers, library subroutines, profilers, and consisting of tracing analysis tools and sophisticated full-featured toolsets. For example, the shell timex reports system related information in a common format across a variety of shells. Profiling measures the frequency and duration of functions or memory and time complexity of a program through instrumenting program's source code or executable file. Unlike profiling, the tracing approach records all events of an application run with precise time stamps and many event type specific properties [2]. Performance analysis toolkits include three steps: instrumentation, measurement, and analysis. Among all popular toolkits, this paper chooses Vampir to visualize the Fortran program behavior, recorded by Score-P in open trace format.

There are some unit test frameworks available for Fortran: Fortran unit test framework (FRUIT), pFUnit, ObjecxxFTK, and FLIBS. This paper takes FRUIT as an example to describe the team's work and weakness. The FRUIT is written in FORTRAN 95, and it can test all FORTRAN features. FRUIT includes five features: assertions for different types of unit tests, function files used to write tests, summary reports of the success or failure of tests, a basket module to invoke set up and tear down functions, and driver generation. Set up and tear down functions are used to perform initialization and finalization operations to all tests within a module. However, all these tools never consider testing modules with groups of global variables. It is known that defining variables on the global scope is a bad but common practice in scientific software development. Extensive usage of global variables makes dependencies analysis difficult and complicates module loading, which in turn cause complicated module interactions.

Paper [3] introduced a Python-based tool called KGEN that extracts a part of Fortran code from an application. KGEN can extract specific statements into stand-alone kernels from large applications including MPI application, it can also provide a way to automatically generate a data stream to drive and verify the execution of the extracted kernel. The tool can deal with global variables and have parallel computation configuration, but it includes excessive time statistics and built-in libraries for kernel generation, which decrease the overall performance.

Paper [13] developed a platform which first split data and library dependencies over software modules and then drove the unit functions with the extracted data stream from original scientific code. With the verified platform, the scientific model builders can track interesting variables either in one single subroutine or among different subroutines. However, the research did not deal well with the performance issues related to long time scientific simulations.

## 3    System Design

In previous research, we focus on how to design a framework to generate unit testing and how to drive the unit testing and validate the correctness of the infrastructure. The framework adapts a serial computational model and does not consider the performance issues associated with a long period of time simulation, such as a 10-year period simulation at a half-hour timestep. Therefore, in this study, in order to make our kernel generation infrastructure more reliable and practical, we improve our design to embrace parallel computing methods.

First, we create an experiment with user-required subroutines. We extract specific unit modules with built-in logic based on user requirements. Then, we apply our sequential unit testing framework to isolate the user required modules and then validate the correctness of the framework.

Second, we design an efficient model to support large data transfer between different modules and continuous step-wise simulations. In our previous work, we divided data based on data access method into three groups: *write_only*, *read_only* and the *modify* [13]. In the parallel version, since disk I/O operation was time-costly, we switched the file-based I/O operations to memory read/write to improve the overall performance. In this paper, we used the code analyzer to analyze data flow based on module relevance. Then, we divided the variables into three groups (*In-Module* group, *Out-Module* group, and *Constant* group) by the function relevance to make sure the parallel CPU cores run during multiple time step simulation. *In-Module* variables are the variables modified by a specific module and can be directly retrieved from the end of the module. These variables appeared in the outStream data flow of the module to drive the subsequent module. *Constant* variables are the environmental setting variables whose values are fixed at the beginning of the model running; hence, they only need to be initialized at the very first time step in a multi-time step simulation. *Out-Module* variables are the input variables whose values are modified by other modules. As such, we need to retrieve them from other modules during the original scientific code running and then provide them at each time step in the unit test platform. By analyzing how these variables were used, we further tagged them with two tags: disk variables and memory variables. The *Constant* variables refer to variables whose values keep constant during the experiment process, we tag them as disk variables and only read them once. The *Out-Module* variables refer to the ones whose values are received from outside of the target modules; we tagged them as disk variables and read them at each time step. Since the *In-Module* variables refer to variables whose values are changed during the experimental process, we tagged them as memory variables and transfer them to the next time step.

Third, we designed a loop-parallel algorithm for an n-case computation illustrated in Fig. 1. First, the retrieved *Constant* variables were used to set up the experiment environment. Then, n cases were initialized with a customized requirement and MPI execution environment. Inside each case, there were *ranksize* processes. Every process $i$ read data from two storage media (except the first timestep process loads all variables from disk). One storage media was disk, which was a file written with *Out-Module* variables from the original scientific code. The other storage media was MPI message buffer, which received updated *In-Module* data from the processes at the previous timestep. The process simulated the status of timestep T ranging from 1 to user-defined F and constantly sent data to the next timestep computation on the No. *(i+1) mod ranksize* process.

## 4  Implementation

The development platform used in the study is a multi-programmatic heterogeneous federated cluster with the Red Hat Enterprise Linux (RHEL) operating system. The production unit testing platform runs on Titan which contains 18,688 physical compute nodes, each with a processor, physical memory, and a connection to the Cray custom high-speed interconnect. Each compute node houses one 16-core 2.2 GHz AMD Opteron$^{TM}$ 6274 (Interlagos) processor and 32 GB of RAM. The working procedure of the parallel unit testing framework (PUTF, as shown in Fig. 2) has five stages: user specification, dataflow generation, customized experiment generation, experiment verification, and experiment execution.

User Specification: The first step is to define which module to isolate. Generally, in the model build step, the PUTF claims which modules to extract based on user requirements; the researchers customize their experiment by designing the necessary duration of the experiment simulation period and providing initial parameters. Dataflow Generation: PUTF uses a dataflow analyzer to split data dependency between modules. The analyzer first collects constant variables, in-module variables, and out-Module variables inside the code, and then inserts all these variables declarations to the corresponding modules in the original code as an "inspector". After re-compiling and re-running the scientific code, we can

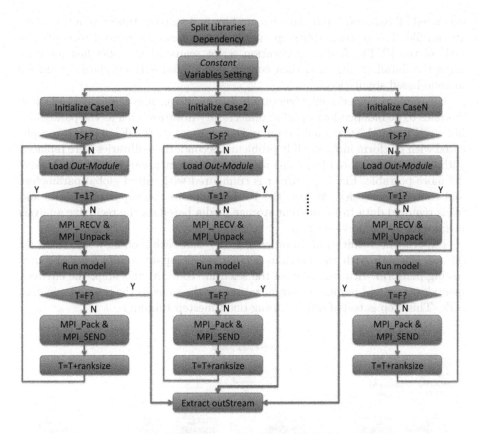

**Fig. 1.** Loop parallel method

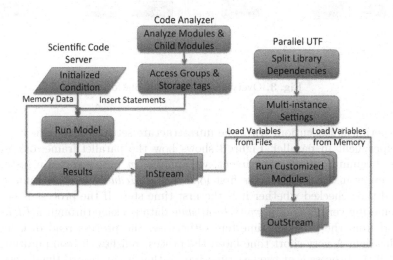

**Fig. 2.** Overview of the improved parallel infrastructure

extracted all required input data stream files and starting timestep output data stream file. The starting timestep output data stream file is used to verify the logic of the PUTF. A data generation script scans all user-specified modules using the dataflow analyzer, then collect, divide, and extracts data stream for module-based simulations.

Customized Experiment Generation: In this stage, it is necessary to isolate modules to be independent of other unnecessary libraries, such as the parallel IO library (PIO) and Networked Common Data Format (netCDF), which is complicated with platform incompatibly problems. Second, these libraries were replaced with easily implemented functions without dependent libraries, making the kernel more portable. Finally, a driver is configured with initial global parameters and constant variables. At last, the PUTF prepares the required modules and loads required data from different storage media based on the recursive analysis mentioned in the previous step.

Experiment Verification: In order to verify that each module works correctly on our platform with the previous setting, we compare results from the unit testing platform with results from the scientific code. At this step, the environmental setting and parameter initialization are the same as the original scientific code. This step is tested quickly using one timestep running.

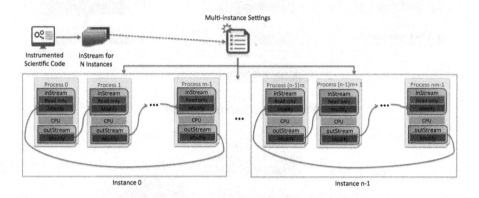

**Fig. 3.** Overview of MPI Unit platform

Experiment Execution: Once the infrastructure settings are verified, we run the experiment in parallel. Figure 3 shows how the parallel framework works. At the beginning of the experiment, we apply $n*m$ processes for n instances. Inside every instance, a process first loaded *Out-Module* variables from a disk file and then checked whether it is the first time step. If the process is not the first timestep verification, it waits *In-Module* data package through *MPI_RECV* method from the previous timestep. Otherwise, the process read disk file for initialization. A very short time later, the process finishes the computation and checks if the process is at the last timestep, so that it can record the experiment result and exit; otherwise, it sends the *In-Module* data package to process at the next timestep through *MPI_SEND* and begins to deal with calculations at the No. $i+m$ timestep. Different instances shares the same disk files but conducts different computation.

# 5  Case Study

## 5.1  Scientific Code and Module-Based Experiments

The "Accelerated Climate Model for Energy (ACME)" a fully-coupled Earth system model development and simulation project to investigate energy-relevant science using code optimized for advanced computers. Inside ACME system, the ACME Land model (ALM) is designed to understand how natural and human changes in terrestrial land surfaces will affect the climate [5]. The ALM model consists of submodels related to land biogeophysics, the hydrologic cycle, biogeochemistry, human dimensions, and ecosystem dynamics. Due to internal biogeophysics and geochemical connections, ALM simulations have to be executed with other earth system components within ACME, such as atmosphere, ocean, ice, and glaciers etc. [8].

The objective of this case study was to compare the performance of the decomposition reaction network within the ALM using data collected from the long-term intersite decomposition experiment team (LIDET). However, with more than 1800 source files and over 350,000 lines of source code, the software complexity of ALM became a barrier to rapid model improvements and validation [9,10], also it is very inconvenient to track specific modules and capture the impact of specific factors on the overall model performance.

At the center of ALM decomposition submodel is the Convergent Trophic Cascade (CTC) method. We would like to evaluate CTC using LIDET data. In a previous study [1], CTC was investigate in standalone mode without consideration of temporal variations in environmental and nutrient conditions that would occur in the full model. If we want to perform the LIDET study in ALM model directly, that may introduce unrealistic feedbacks between the simulated litter bags and vegetation growth. Therefore, we develop a model within our PUTF framework which allows the CTC submodel to operate independently while retaining the temporally varying environmental drivers calculated by ALM. Particularly, we are interested in (1) the influence of litter decomposition base rate parameters, and (2) the influence of nitrogen limitation, and the temporal variability of this limitation, on litter decomposition.

## 5.2  Experiment Setups

In the experiments, six types of leaf litter were placed in fine mesh bags at 21 sites representing a range of biomes. The mass of remaining carbon and nitrogen in this litter was measured annually over a 10-year period. To simulate the experimental conditions in the model, we first spined up the carbon and nitrogen pools using an accelerated decomposition phase for 500 years, followed by a return to normal decomposition rate for 500 years [6]. In these simulations, we used a repeating cycle of the CRU-NCEP meteorology over the years 1901–1920. Then we performed a transient simulation from the years January 1st, 1850 - October 1st, 1990, which was forced by changing atmospheric CO2 concentrations, nitrogen and aerosol deposition and land use. Globally gridded meteorological and land-surface data were used for these simulations except for plant functional type, which was replaced using site information. The model state on October 1st, 1990 represents the simulated conditions at the beginning of the experiment.

At this point, the UTF framework is used to execute a 10-year simulation. For a control simulation in which no litter is added, we run the full scientific model and save all model state variables at every timestep. These model states were then used as boundary conditions for our decomposition unit, for which only the decomposition subroutines and relevant updating codes are active. For each site, we added litter inputs to the first soil layer using the appropriate mass, quality, and C:N ratios for each of the six litter types. The decomposition unit is driven by soil moisture, temperature, and the nutrient limitation factor for decomposition from the full model. Unlike in the full version of the scientific code, there was no feedback between decomposition and the ecosystem.

## 5.3    Results and Analysis

In this section, we used a dynamic performance analysis measurement tool to help to improve the framework.

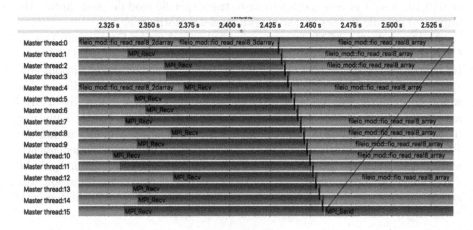

**Fig. 4.** Timeline chart

### 5.3.1    Parallel I/O

In this experiment, if we applied up to 2114 processes and 133 computing nodes in Titan, the total execution time for one site and one leaf litter's type in 4-month's simulation was 13s, the best performance among configurations. In Fig. 4, we applied 16 processes and 1 node to simulate one site and one type's 10 years's simulation. The red bar signifies the MPI functions, which include *MPI_Init*, *MPI_Recv*, *MPI_Send* and *MPI_Finalize*. The green bar between red ones represents CPU computation, the green bars after *MPI_Send* and before *MPI_Receive* are disk I/O read. Messages exchanged between different processes are depicted as black lines. Within the case, since every time step needs previous time step's data as input, the function computation is sequential, while the I/O operation is parallel. The *MPI_Recv* bar is long over time because every process is waiting for previous results.

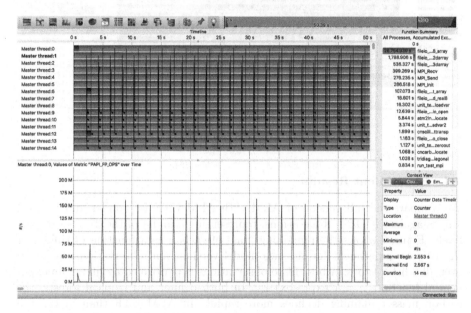

**Fig. 5.** Profiling info for improved PUTF

### 5.3.2  MPI with Parallel I/O

In Figs. 5 and 6, we applied 600 processes and 40 nodes to parallel one site and 6 types 10-years simulation experiment. The red box stands for the MPI functions, and the green boxes consist of function computation and I/O operation. The black line represents the paralleled CPU and how the MPI message goes. In this case, every type in the same site shared the same input data but computed in different ways. Therefore the function execution and the I/O operation were both parallel which improved the overall performance that can be seen from the counter in Figs. 5 and 6.

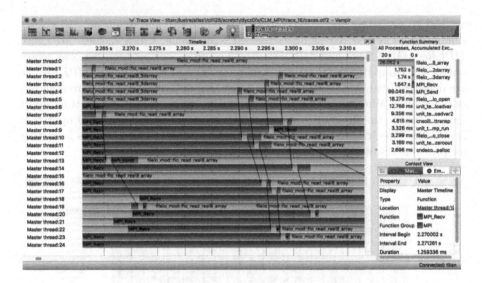

**Fig. 6.** Improved time dissection of CPU and I/O

## 5.4   Experiment Results

We compare the full version of ALM with ALM UTF for conifer and tropical forests against LIDET observations, representing 5 and 4 of the 21 sites respectively. Figure 7 shows the remaining mass of carbon as a function of time over the 10-year experiment for the two model versions and observations, averaged over all 6 litter types. A best fit to observations is performed by fitting an exponential function $y = a*\exp(-bx) + c$. In both conifer and tropical forests, the carbon mass remaining declines more rapidly in ALM UTF than in the full ALM, which is more consistent with the best fit to observations. The ALM UTF model is more consistent with the actual experimental conditions, because in the experiment the small amount of litter in the litter bag added to each plot is not large enough to induce ecosystem-scale feedbacks. However in the full ALM, the added litter effectively covers the entire land surface, causing feedbacks to vegetation growth. The additional litter unrealistically stimulates vegetation growth in the full ALM, causing more carbon fixation and increased litterfall, effectively increasing the carbon mass remaining. In both the full ALM and in the ALM UTF, the carbon mass remaining is significantly higher than that estimated by [1] when comparing the CTC submodel in ALM to the DAYCENT soil decomposition model. That analysis, while also using a functional unit approach, did not use the full model for boundary conditions and thus neglected to consider changing nutrient limitation and environmental conditions. The approach taken in ALM UTF is a useful way to perform such model-experiment intercomparisons in a consistent way while avoiding unrealistic feedback in small-scale experiments.

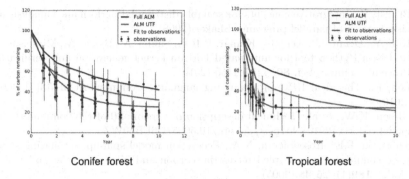

Conifer forest                                    Tropical forest

**Fig. 7.** Comparison among full version of ALM, ALM UTF and LIDET observations. Average carbon mass remaining in relation to time for leaf litter decomposed at two sites. Data are averaged across six leaf litter types for sites classified as conifer forest and tropical forest

## 6 Conclusions

Large-scale scientific code is important for scientific research. However, because of the complexity of models, it is very time-consuming to modify scientific code and to validate individual modules inside a complex modeling system. To facilitate module evaluation and validation within scientific applications, we first introduce a unit testing framework. Since scientific experiment analysis generally requires a long-term simulation with large-amount of data, we apply message passing based parallelization and I/O behavior optimization to improve the performance of the unit testing framework on parallel computing infrastructure. We also used profiling result to guide the parallel process implementation. Finally, we use a standalone moduled-based simulation, extract from a large-scale Earth System Model, to demonstrates the scalability, portability, and high-efficiency of the parallel functional unit testing framework.

**Acknowledgement.** Some part of this research is included in Yao's Ph.D. dissertation (A Kernel Generation Framework for Scientific Legacy Code [12]) with the University of Tennessee, Knoxville. TN. This research was funded by the U.S. Department of Energy, Office of Science, Biological and Environmental Research program (E3SM and TES) and Advanced Scientific Computing Research program. This research used resources of the Oak Ridge Leadership Computing Facility at the Oak Ridge National Laboratory, which is supported by the Office of Science of the U.S. Department of Energy under Contract No. DE-AC05-00OR22725.

## References

1. Bonan, G.B., Hartman, M.D., Parton, W.J., Wieder, W.R.: Evaluating litter decomposition in earth system models with long-term litterbag experiments: an example using the community land model version 4 (CLM4). Glob. Change Biol. **19**(3), 957–974 (2013)

2. Brunst, H.: Integrative concepts for scalable distributed performance analysis and visualization of parallel programs. Shaker (2008)
3. Kim, Y., Dennis, J., Kerr, C., Kumar, R.R.P., Simha, A., Baker, A., Mickelson, S.: KGEN: a Python tool for automated Fortran kernel generation and verification. Procedia Comput. Sci. **80**, 1450–1460 (2016)
4. Loh, E.: The ideal HPC programming language. Commun. ACM **53**(7), 42–47 (2010)
5. Oleson, K.W., et al.: Technical description of version 4.0 of the community land model (CLM) (2010). https://doi.org/10.5065/D6FB50WZ
6. Thornton, P.E., Rosenbloom, N.A.: Ecosystem model spin-up: estimating steady state conditions in a coupled terrestrial carbon and nitrogen cycle model. Ecol. Model. **189**(1), 25–48 (2005)
7. Vardi, M.: Science has only two legs. Commun. ACM **53**, 5 (2010)
8. Wang, D., et al.: A scientific function test framework for modular environmental model development: application to the community land model. In: Proceedings of the 2015 International Workshop on Software Engineering for High Performance Computing in Science, pp. 16–23. IEEE Press (2015)
9. Wang, D., Post, W.M., Wilson, B.E.: Climate change modeling: computational opportunities and challenges. Comput. Sci. Eng. **13**(5), 36–42 (2011)
10. Wang, D., Schuchart, J., Janjusic, T., Winkler, F., Xu, Y., Kartsaklis, C.: Toward better understanding of the community land model within the earth system modeling framework. Procedia Comput. Sci. **29**, 1515–1524 (2014)
11. Wang, D., Yuan, F., Hernandez, B., Pei, Y., Yao, C., Steed, C.: Virtual observation system for earth system model: an application to acmeland model simulations. Int. J. Adv. Comput. Sci. Appl. 8(2) (2017). https://doi.org/10.14569/IJACSA.2017.080223
12. Yao, Z.: A Kernel Generation Framework for Scientific Legacy Code. Ph.D. thesis, University of Tennessee, Knoxville (2018)
13. Yao, Z., Jia, Y., Wang, D., Steed, C., Atchley, S.: In situ data infrastructure for scientific unit testing platform. Procedia Comput. Sci. **80**, 587–598 (2016)

# Heuristic Optimization with CPU-GPU Heterogeneous Wave Computing for Estimating Three-Dimensional Inner Structure

Takuma Yamaguchi[1]([✉]), Tsuyoshi Ichimura[1], Kohei Fujita[1], Muneo Hori[2], and Lalith Wijerathne[1]

[1] Earthquake Research Institute and Department of Civil Engineering, The University of Tokyo, Bunkyo, Tokyo, Japan
{yamaguchi,ichimura,fujita,lalith}@eri.u-tokyo.ac.jp
[2] Japan Agency for Marine-Earth Science and Technology, Yokosuka, Kanagawa, Japan
horimune@jamstec.go.jp

**Abstract.** To increase the reliability of numerical simulations, it is important to use more reliable models. This study proposes a method to generate a finite element model that can reproduce observational data in a target domain. Our proposed method searches parameters to determine finite element models by combining simulated annealing and finite element wave propagation analyses. In the optimization, we utilize heterogeneous computer resources. The finite element solver, which is the computationally expensive portion, is computed rapidly using GPU computation. Simultaneously, we generate finite element models using CPU computation to overlap the computation time of model generation. We estimate the inner soil structure as an application example. The soil structure is reproduced from the observed time history of velocity on the ground surface using our developed optimizer.

**Keywords:** Heuristic optimization ·
CPU-GPU collaborative computing · CUDA ·
Finite element analysis · Conjugate gradient method

## 1 Introduction

Numerical simulations with large number of degrees of freedom are becoming feasible due to the development of computation environments. Accordingly, more reliable models are required to obtain more reliable results for the target domain with complex structures. This approach has been discussed in various fields including biomedicine [2,9], and it is also important for the numerical simulation of earthquake disasters. It is rational that we allocate resource and take countermeasures after detecting an area with potentially substantial damage.

© Springer Nature Switzerland AG 2019
J. M. F. Rodrigues et al. (Eds.): ICCS 2019, LNCS 11537, pp. 389–401, 2019.
https://doi.org/10.1007/978-3-030-22741-8_28

We can apply numerical simulation for estimating damages. The authors of [10] found that the geometry of the target domain significantly affects the distribution of displacement on the ground surface and strain in underground structures. To undertake well-suited countermeasures, three-dimensional unstructured finite element analysis is preferred, as it considers complex geometry. This analysis results in problems with large degrees of freedom because it targets large domains with high resolution. The computation mentioned above has become more attainable due to the development of the computation environment and analysis methods for CPU-based large-scale systems [5]. However, inner soil structure is not available with high resolution, which hampers the generation of finite element models. On the ground surface, [13] is used as elevation data for Japan. With the advance of sensing technology, it is possible to observe earthquake waves on many points on the ground surface. It is desirable that we generate a finite element model which can reproduce observational data on the ground surface and conduct analyses using an estimated model. On the other hand, it is difficult to measure the underground structure with high accuracy and resolution.

One of realistic ways to address this issue is introduction of an optimization method using observation data on the ground surface for a micro earthquake. If we can generate many finite element models and conduct wave propagation analyses for each model, it is possible to select a model which can reproduce available observation data most closely. Using optimized models will increase the reliability of the analyses. There are some gradient-based methods for optimization as [12] proposed for three-dimensional crustal structure optimization. These methods have the advantage that the number of trials is small; however, they may be difficult to escape from a local solution if control parameters have low sensitivity to an error function. Thus, this study focuses on heuristic methods such as simulated annealing so that we can reach the global optimal solution robustly. The expected optimization requires many forward analyses, and the challenge is an increase in the computation cost for many analyses with large number of degrees of freedom.

We use GPUs in this paper. Its hardware and development environment are rapidly evolving [8]. The computation time can be reduced by using parallel computation with many GPU cores. However, it is known that GPU computation requires the consideration of memory access and communication cost to attain better performance. This paper proposes an algorithm that combines very fast simulated annealing and wave propagation analyses and repeats generation of finite element models and the computation of the solver for estimation of inner soil structure. Some computations in our optimizer are not suitable for GPU computation. Thus, computer resources are allocated so that we can benefit further from the introduction of GPU computation. A finite element solver appropriate for GPU computation is proposed to reduce the computation time in the solver, which is the most computationally expensive part. At the same time, generation of finite element models, which requires serial operations, is computed on CPUs so that computation time for model generation can be overlapped. We confirm that the inner soil structure has a large effect on the results and that our proposed method can estimate the soil structure with sufficient accuracy for damage estimation. Our proposed optimizer is proposed in the following section. Section 3 describes the estimation of soil structure using our developed optimizer. Section 4 describes our conclusions and future prospects.

## 2 Methodology

For estimating the inner structure of the target domain, this study proposes a method to conduct many wave propagation analyses and accept the inner structure with its maximum likelihood. In this study, optimization targets the estimation of boundary surfaces of the domain that has different material properties. For simplicity and for the purposes of this study, we have assumed that the target domain has a stratified structure and that target parameters for optimization are an elevation of the boundary surface on control points which are located at regular intervals in the $x$ and $y$ directions. A boundary surface is generated in the target domain by interpolating elevation on control points using linear functions. Figure 1 depicts the scheme for optimization. In this scheme, we conduct finite element analyses for evaluation of parameters many times in very fast simulated annealing. Our optimizer is designed so that the generation of a finite element model and the finite element solver, which account for the large proportion of the whole computation time, can be computed at the same time by CPUs and GPUs, respectively. We describe the details for each part of our optimizer in the following parts.

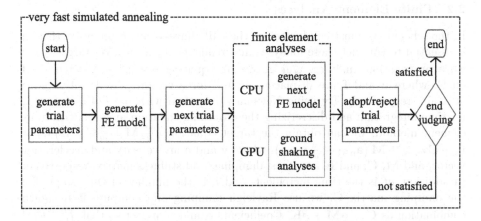

**Fig. 1.** Rough scheme for our proposed optimizer for an estimation of soil structure.

### 2.1 Very Fast Simulated Annealing

Very fast simulated annealing, which is a heuristic optimization method for problems with many control parameters [7], is applied. Simulated annealing has a parameter that corresponds to temperature, and the temperature decreases as the number of trials increases. We search and evaluate parameters in the following manner. First, trial parameters are selected randomly based on current parameters. The search parameter domain is wider when the temperature is higher. The evaluation value of trial parameters and that of previous ones are compared, and if the evaluation value is improved, parameters are always updated. Even if the evaluation value is worse, parameters are updated with a high degree of probability while the temperature is high. By repeating this procedure, this method can move out of the local optimal solution and find the global optimal solution robustly. To evaluate the parameters, finite element analysis is conducted. We assume that we have many observation points on the surface of the target domain. Our error function is defined by the time histories of displacement in the analyses and observation data on observation points. The actual error function is defined in Sect. 3.

In very fast simulated annealing, temperature at the $k$-th trial is defined using initial temperature $T_0$ and the number of control points $D$ as $T_k = T_0 \exp(-ck^{\frac{1}{D}})$, where parameter $c$ is defined by $T_0$, $D$, lowest temperature $T_f$, and the number of trials $k_f$ as $T_f = T_0 \exp(-m)$, $k_f = \exp n$, and $c = m \exp(-\frac{n}{D})$. The initial temperature, lowest temperature, and number of iterations depend on problems. A certain number of iterations are conducted for this simulation, though we can stop searching by other conditions, including acceptance frequency of new solutions. Also, we don't use re-annealing method mentioned in [7] because the cost for computing sensitivity of each parameter to the error function increases as $D$ increases.

## 2.2 Finite Element Analyses

In the scheme, we must conduct more than $10^3$ finite element analyses; thus, it is essential to conduct these analyses in a realistic timeframe. We target linear wave propagation analyses. Our governing equation is $\rho \frac{\partial^2 u}{\partial t^2} - \nabla \cdot \sigma(u) = f$ on $\Omega$, where $u$ and $f$ are displacement and force vector, $\rho$ is density, and $\sigma$ is strain, respectively. By using Newmark-$\beta$ method with $\beta = 1/4$ and $\delta = 1/2$ for time integration and discretizing the governing equation in space with finite element method, we can obtain the target equation $\left(\frac{4}{dt^2}M + \frac{2}{dt}C + K\right) u_n = f_n + Cv_{n-1} + M\left(a_{n-1} + \frac{4}{dt}v_{n-1}\right)$, where $v$ and $a$ are velocity and acceleration vector, and $M$, $C$, and $K$ are mass, damping, and stiffness matrix, respectively. In addition, $dt$ is the time increment, and $n$ is the number of time steps. For the damping matrix $C$, we use Rayleigh damping and compute it by linear combination as $C = \alpha M + \beta K$. Coefficients $\alpha$ and $\beta$ are set so that $\int_{f_{min}}^{f_{max}} (h - \frac{1}{2}(\frac{\alpha}{2\pi f} + 2\pi f\beta))^2 df$ is minimized, where $f_{max}$, $f_{min}$, and $h$ are maximum targeting frequency, minimum targeting frequency, and damping ratio. Vectors $v_n$ and $a_n$ can be described as $v_n = -v_{n-1} + \frac{2}{dt}(u_n - u_{n-1})$, $a_n = -a_{n-1} - \frac{4}{dt}v_{n-1} + \frac{4}{dt^2}(u_n - u_{n-1})$. We obtain displacement vector $u_n$ by solving the equation above and updating vectors $v_n$ and $a_n$. Computation in the finite element solver and generation of finite element models are most computationally expensive parts in our optimizer.

**Finite Element Solver.** We developed a solver based on our MPI-parallel solver using conjugate gradient method [5]. The original solver has been developed for CPU-based supercomputers. The solver combines a conjugate gradient method with adaptive preconditioning, geometric multigrid method, and mixed precision arithmetic to reduce the amount of arithmetic counts and data transfer size. In the solver, sparse matrix vector multiplication is computed by the Element-by-Element (EbE) method. It computes element matrix on-the-fly and reduces the memory access cost. Specifically, the multiplication $\mathbf{y} = \mathbf{A}\mathbf{x}$ is computed as $\mathbf{y} = \sum_{i=1}^{ne}(\mathbf{Q}^{(i)T}(\mathbf{A}^{(i)}(\mathbf{Q}^{(i)}\mathbf{x})))$, where $ne$ is the number of elements in the domain, $\mathbf{Q}^{(i)}$ is a mapping matrix from local node numbers in the $i$-th element to the global node numbers, and $\mathbf{A}^{(i)}$ is the $i$-th element matrix and satisfies $\mathbf{A} = \sum_{i=1}^{ne} \mathbf{Q}^{(i)T}\mathbf{A}^{(i)}\mathbf{Q}^{(i)}$. In this problem, $\mathbf{A}^{(i)} = \frac{4}{dt^2}\mathbf{M}^{(i)} + \frac{2}{dt}\mathbf{C}^{(i)} + \mathbf{K}^{(i)}$. The entire part of the solver is implemented in the multiple GPUs using CUDA. To exhibit higher performance using GPUs, we have to reduce the operations that are not suitable for GPU computation; thus, we modify the algorithm of the solver.

When we compute EbE kernel in GPU, we assign one thread for one element and each element adds temporal results per element into the global vector. This summation can be operated without data race conditions using atomic operation; however, many random accesses to the global vector in this scheme may decrease the performance of the kernel. To improve the performance of this part, we use shared memory as a buffer and reduce the number of data accesses to global memory. These methods are extension of the finite element solver for crustal deformation computation by [14].

In addition, we overlap computation and communication as described in [11]. In the domain of each MPI process, some elements are adjacent to domains of other MPI processes and require point-to-point communications, and others do not require these communications. First we compute elements that require data transfer among other GPUs. Next we communicate with other GPUs while we are computing elements that do not require data transfer. By following this procedure, it is possible to overlap MPI point-to-point communication in the solver.

In the conjugate gradient solver, the coefficients are derived from the result of inner product calculations so that orthogonal residual vector and A-orthogonal searching vector can be generated to those in the previous iteration, respectively. When multiple GPUs are used with MPI, calculations of these coefficients require data transfer and synchronization among MPI processes such as MPI_Allreduce. Thus, they become relatively time-consuming taking into account that other computations including vector operations and sparse matrix vector multiplication are accelerated by GPUs. In our solver, we employ the method described in [1]. This algorithm requires one MPI_Allreduce per iteration, which halves the number of MPI_Allreduce per iteration in the original conjugate gradient method. The amount of vector operation increases in this scheme. However, the reduction of calculations of coefficients is more effective for GPU-based systems.

**Generation of the Finite Element Model.** We automatically generate finite element model using the method by [3]. Its procedure utilizes OpenMP parallelization, where each thread has temporal array required to compute connectivity of elements and numbering of nodes. This enables us to compute model generation with up to $10^2$ OpenMP threads on CPU; however, we cannot apply GPU computation for this part as GPU requires more than 10,000 threads for efficient computation and memory consumption greatly increases. Generation of finite element models can account larger proportion of the whole elapsed time, which is not negligible compared to the computation time in the finite element solver. Therefore, we design our optimizer so that it is possible to generate a finite element model for the next trial on CPUs while wave propagation analysis is computed on GPUs. All of the main computation in the solver can be computed in GPUs, so we can assign only one core of CPUs for each GPU and this has little effect on the performance of the solver. Other cores in CPUs are assigned for the generation of finite element models. Program of the model generation is created separately from that of the finite element solver and we executed them asynchronously using a shell-script. Output files are shared in the file system and controlled so that they are updated in correct timing. By allocating heterogeneous computer resource as mentioned above, it is possible to overlap model generation with GPU computation.

## 3   Application Example

We use our developed optimizer to estimate soil structure. Our target domain has two layers, and we define their boundary surface. We use IBM POWER System AC922 for computation, which has two POWER9 CPUs (16 cores, 2.6 GHz) and four NVIDIA Volta V100 GPUs. We assign one CPU core to each GPU for finite element analysis with MPI and we use the remaining 28 CPU cores for the model generation with OpenMP.

**Table 1.** Material properties in target domain. $V_p$, $V_s$, and $\rho$ are primary and secondary wave velocity, and density, respectively. $h$ is the damping ratio used in the linear wave field calculation, $h_{max}$ is maximum damping ratio, and $\gamma$ is the reference strain used in the non-linear wave analyses.

|            | $V_p$(m/s) | $V_s$(m/s) | $\rho$ (kg/m$^3$) | $h$   | $h_{max}$ | $\gamma$ |
|------------|------------|------------|-------------------|-------|-----------|----------|
| Soft layer | 700        | 100        | 1500              | 0.001 | 0.23      | 0.007    |
| Hard layer | 2100       | 700        | 2100              | 0.001 | 0.001     | -        |

The target domain is 300 m × 400 m × 75 m. In this problem, material properties of the soil structure are deterministic. These properties are described in Table 1. In the model generation using [3], tetrahedral elements are generated based on a background octree-based structured grid, and according to the previous study [4], its resolution $ds$ must satisfy the condition $ds < \frac{V_s}{10 f_{max}}$ in soft layers. The frequency components below 2.5 Hz are dominant in the strain response analysis, so we set $f_{max} = 2.5$ Hz. Thus, we set $ds = 2.5$ m so that the condition above is satisfied. Elevation data at the surface are available. They are flat and we set them as $z = 0$ m. Control points are located at regular intervals in the $x$ and $y$ directions. We simplify the problem and notate the elevation of the hard layer on points $(x, y) = (100i, 100j)(i = 0\text{--}3, j = 0\text{--}4)$ as $\alpha_{ij}$ in metric units. The points $x = 0$, $x = 300$, $y = 0$, and $y = 400$ are the edges of the domain, and we assume $\alpha_{ij} = 0$ for these points. The parameters for optimization are $\alpha_{ij}(i = 1\text{--}2, j = 1\text{--}3)$. Initial parameters and reference parameters, which are true, are shown in Table 2. We assume the information from the boring survey are available at points $(x, y) = (50, 50), (150, 350), (200, 100)$. Elevations at these points are $-7.01$ m, $-5.97$ m, and $-16.3$ m, respectively, and these elevations are interpolated to make the initial boundary surface. The distributions of the boundary surface for initial and reference models are described as Fig. 2. An unstructured mesh with approximately 3,000,000 degrees of freedom is generated by using the method by [3]. Figure 3 shows one of the FE models in the analysis. Input waves for wave propagation analyses can be obtained by pulling back observed waves on the ground surface. In this paper, we assume that input waves are generated by micro earthquakes, and linear analysis can be applied. It is then possible to use Ricker wave as our input wave. We derive amplification functions from observation data and pulled back waves. Using these functions, it is possible to estimate observation data when we input Ricker wave. These operations reduce time steps for wave propagation analyses and the entire computation time. Ricker waves, represented as $(1 - 2\pi^2 f_c^2 (t - t_c)^2) \exp(-\pi^2 f_c^2 (t - t_c)^2)$, are input as $x$ and $y$ components of velocity at the bottom of models. $t$ is time in second, $f_c$ is central frequency and $t_c$ is central time and they can be set independently. For this application example, the target frequency is as much as 2.5 Hz and we set the period of each analysis to 2.56 s. Considering these settings, we set $(f_c, t_c) = (0.8, 1.2)$. Time increment of the analysis must be small enough to converge the results in the time integration. We set it to 0.01 s, which is the same as the setting in [6]; thus each wave propagation analysis requires computation for 256 times steps. We set two cases for observation points. In case 1, we allocate 35 observation points defined as $(x, y) = (50i, 50j)$ $(i = 1\text{--}5, j = 1\text{--}7)$ and in case 2, observation points are $(x, y) = (-50 + 100i, -50 + 100j)$ $(i = 1\text{--}3, j = 1\text{--}4)$ and the number of points is 12. We use an error function as follows; $Error = \frac{1}{np} \sum_{i=1}^{np} \sum_{j=1}^{3} \int_{0}^{f_{max}} |F[v_{ij}] - F[\bar{v}_{ij}]| df$, where $np$ is the number of observation points, $v$ is the velocity of the observation data, and $f_{max}$ is the maximum targeting frequency, which is 2.5 Hz in our paper. $v$ is the time history of $x$, $y$, and $z$ components of velocity on each observation point. Values with a over-line corresponds to the observation data. In addition, $F[\ ]$ corresponds to the discrete Fourier transformation. In other words, this error function is the total sum of absolute values of difference for frequency components on observation points. These settings mentioned above are the same as settings in [6].

**Table 2.** Parameters. The units of $\alpha$ are meters.

|  | $\alpha_{11}$ | $\alpha_{21}$ | $\alpha_{12}$ | $\alpha_{22}$ | $\alpha_{13}$ | $\alpha_{23}$ |
|---|---|---|---|---|---|---|
| Initial model | −9.190 | −16.260 | −6.490 | −11.660 | −4.980 | −7.050 |
| Reference model | −28.030 | −16.260 | −25.550 | −21.140 | −12.790 | −11.090 |

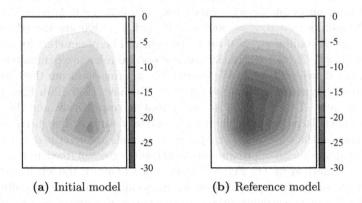

(a) Initial model          (b) Reference model

**Fig. 2.** Distribution of elevation (m) of the hard layer.

In our proposed method, we generate finite element models for the next trial and conduct wave propagation analysis at the same time. In simulated annealing, we generate next trial parameters after current trial parameters are adopted or rejected. It is desirable that we generate two models in cases that trial parameters are adopted and rejected while we are conducting wave propagation analysis; however, generation of finite element model twice takes more time than the computation in our finite element solver. Parameters in these problem settings are thought to be rejected with high probability. Thus, we generate a finite element model with prediction that trial parameters will be rejected. When trial parameters are adopted, we regenerate next finite element models for updated parameters. This regeneration has a small effect on the whole computation time. The breakdown of computation costs in our optimizer is described later in the performance evaluation part. The number of control points in very fast simulated annealing $D = 6$. Also, we set the number of trials $k_f = 1500$ and $c = 4.2975$. This $c$ satisfies that parameters which increase the value of the error function by $\Delta E$ are adopted with the probability of 80% at the initial temperature and parameters which increase the value of the error function by $\Delta E \times 10^{-5}$ are adopted with the probability of 0.1% at the lowest temperature, where $\Delta E$ is the value of error function obtained in the initial model. The history of error function is described in Fig. 4 and parameters are estimated as Table 3. Optimization of both case 1 and 2 adopted trial parameters 51 times in 1,500 trials. Trial parameters are rejected with the probability of more than 90% and we find that the generation of finite element model is mostly overlapped by the

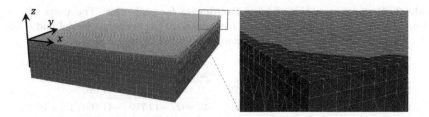

**Fig. 3.** One of finite element models in the analysis.

computation in the solver. Compared to case 2, case 1 with more observation points estimated the soil structure more accurately. Our previous study [6] used a multigrid stochastic search algorithm and optimized the same parameters in meters. The numbers of iteration were 3,000 in case 1 and 1,300 in case 2; thereby we found that parameters were efficiently optimized with higher accuracy as the number of trials by the very fast simulated annealing was 1,500. For confirmation of the optimized model, we conduct wave propagation analysis with parameters obtained in case 1. Figure 5 is the distribution of the displacement on the ground surface at time $t = 2.20$ s and Fig. 6 is the time history of the velocity on point $(x, y, z) = (150, 200, 0)$. Judging from these figures, we can confirm that the results by optimized model and reference model are consistent.

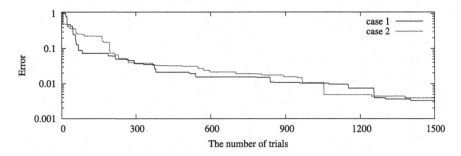

**Fig. 4.** Time history of error function. Each value is normalized by the error of the initial model.

Here we evaluate the performance of the computation in our optimization. The elapsed time for our solver part is about 18 s per trial. In [6], wave propagation analysis was computed in 263 s for a finite element model with 274,041 degrees of freedom using Intel Xeon E5-4667 v3 CPU.If this CPU-based solver is used for the analysis in this paper and we assume that computation cost increases in proportion to the number of degrees of the freedoms, estimated elapsed time will be 3,000,000/274,041=10.94 times longer and 263 s × 10.94 = 2,879 s. Therefore, our GPU-based solver has achieved about 160-fold speeding up per problem size, although it is difficult to compare the performance on different systems.

**Table 3.** Parameters obtained by the optimizer for each case. The units of $\alpha$ are meters. $RSS$ is the residual sum of squares based on the reference model and defined as $\sum_{ij}(\alpha_{ij}/\bar{\alpha}_{ij} - 1)^2$, where $\bar{\alpha}_{ij}$ are parameters of the reference model.

|  | $\alpha_{11}$ | $\alpha_{21}$ | $\alpha_{12}$ | $\alpha_{22}$ | $\alpha_{13}$ | $\alpha_{23}$ | $RSS$ |
|---|---|---|---|---|---|---|---|
| Reference | $-28.030$ | $-16.260$ | $-25.550$ | $-21.140$ | $-12.790$ | $-11.090$ | - |
| Case 1 | $-28.007$ | $-16.290$ | $-25.611$ | $-21.092$ | $-12.770$ | $-11.100$ | $1.6 \times 10^{-5}$ |
| Case 2 | $-28.119$ | $-16.133$ | $-25.484$ | $-21.182$ | $-12.849$ | $-11.086$ | $1.2 \times 10^{-4}$ |

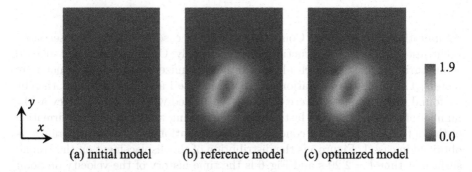

(a) initial model        (b) reference model        (c) optimized model

**Fig. 5.** Norm distribution of displacement (m) on the ground surface at $t = 2.20\,\mathrm{s}$ in the linear ground shaking analysis.

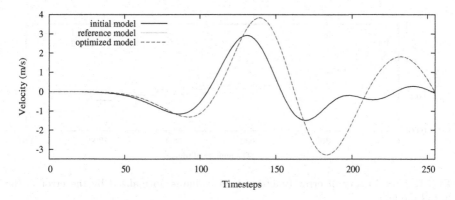

**Fig. 6.** $x$ component of the velocity at $(x, y) = (150, 200)$ on the ground surface in the linear ground shaking analysis.

Here we use peak memory bandwidth to evaluate the speeding up ratio, as general finite element analyses are memory bandwidth bound computations. Intel Xeon E5-4667 v3 CPU has 68 GB/s and four NVIDIA V100 GPUs have 900 GB/s $\times$ 4 = 3,600 GB/s of memory bandwidth. We attained higher speeding up ratio than the ratio of peak memory bandwidth; this indicates that we efficiently computed on GPUs. The optimization in case 1 was computed in 13 h 32 min. The elapsed time per trial in simulated annealing was 32 s. The breakdown of computation cost was as follows: The computation part of our solver took 18 s, other part of our solver such as I/O and data transfer from CPU to GPU before computation took 7 s, model decomposition for MPI parallelization using METIS took 3 s, postprocessing to obtain the response on the ground surface took 2 s, and other computations including Fourier transformation took 2 s. Besides, it took about 10 s for the generation of each model, which is overlapped with the computation in the solver; thus, the whole elapsed time would be 13 h 32 min + 10 s × 1500 = 17 h 42 min and increase by 30% if model generation and finite element solver were computed sequentially. Thereby we confirmed that efficient allocation of computer resources is important for this optimization. Considering our previous method took 9.4 days for parameter optimization in meters using finite element model with 1/10 of degrees of freedoms, our proposed method has achieved great reduction in computational cost.

Finally, we conduct a non-linear ground shaking analysis using the optimized model. The methods are the same as [5]. We input wave observed in the 1995 Kobe Earthquake at the Kobe Local Meteorological Office and its time increment is 0.005 s and the number of time steps is 16,384. We used the modified Ramberg-Osgood model and Masing Rule for non-linear constitutive models. We assume that a gas pipeline is buried as shown in Fig. 7(a). Figure 7(b) shows the maximum axial strain of the pipeline. We can confirm that the strain distributions obtained by our optimized model and initial model, which is derived from boring survey, are completely different. This analysis is used for screening of underground structures which will be damaged and its result shows that this optimization is important to assure the reliability of the result.

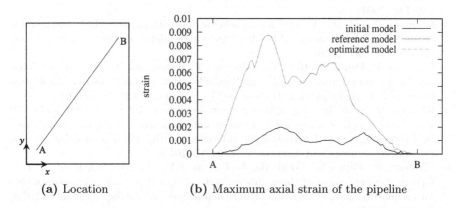

(a) Location          (b) Maximum axial strain of the pipeline

**Fig. 7.** Maximum distribution of axial strain along a buried pipeline for each model in the non-linear ground shaking analysis. The buried pipeline is located between point A $(x, y, z) = (30, 40, -1.2)$ and point B $(x, y, z) = (270, 360, -1.2)$.

## 4 Conclusion

To increase the reliability of numerical simulations, it is essential to use more reliable models. Our proposed optimizer searches for a finite element model that can reproduce observation data by combining very fast simulated annealing and finite element analyses. As an application example, we estimated soil structure using observation data with 1,500 wave propagation analyses with a finite element model with 3,000,000 degrees of freedom. The finite element solver, which accounted for the large proportion of the whole computation time, was accelerated by utilizing the GPU computation. Compared to the previous study, the elapsed time per problem size was decreased by 1/160. Generation of a finite element model was difficult to computed on GPUs. We designed our algorithm so that the computation in model generation on CPUs was overlapped by the computation in the solver on GPUs and enhanced the effect of GPU acceleration. For future prospects, more trials will be required for larger problem size or parameter searching in higher resolution, as the convergence of simulated annealing gets worse. To reduce the computation time, we must attain more speedup ratio for the solver or design a faster algorithm of our optimizer.

## References

1. Chronopoulos, A.T., Gear, C.W.: s-step iterative methods for symmetric linear systems. J. Comput. Appl. Math. **25**(2), 153–168 (1989)
2. Clement, R., Schneider, J., Brambs, H.-J., Wunderlich, A., Geiger, M., Sander, F.G.: Quasi-automatic 3D finite element model generation for individual single-rooted teeth and periodontal ligament. Comput. Methods Program. Biomed. **73**(2), 135–144 (2004)
3. Ichimura, T., et al.: An elastic/viscoelastic finite element analysis method for crustal deformation using a 3-D island-scale high-fidelity model. Geophys. J. Int. **206**(1), 114–129 (2016)
4. Ichimura, T., Fujita, K., Hori, M., Sakanoue, T., Hamanaka, R.: Three-dimensional nonlinear seismic ground response analysis of local site effects for estimating seismic behavior of buried pipelines. J. Pressure Vessel Technol. **136**(4), 041702 (2014)
5. Ichimura, T., et al.: Implicit nonlinear wave simulation with 1.08 T DOF and 0.270 T unstructured finite elements to enhance comprehensive earthquake simulation. In: Proceedings of the International Conference for High Performance Computing, Networking, Storage and Analysis, p. 4. ACM (2015)
6. Ichimura, T., Fujita, K., Yoshiyuki, A., Quinay, P.E., Hori, M., Sakanoue, T.: Performance enhancement of three-dimensional soil structure model via optimization for estimating seismic behavior of buried pipelines. J. Earthq. Tsunami **11**(05), 1750019 (2017)
7. Ingber, L.: Very fast simulated re-annealing. Math. Comput. Model. **12**(8), 967–973 (1989)
8. Kirk, D., et al.: NVIDIA CUDA software and GPU parallel computing architecture. In: ISMM, vol. 7, pp. 103–104 (2007)
9. Koch, R.M., Gross, M.H., Carls, F.R., von Büren, D.F., Fankhauser, G., Parish, Y.I.H.: Simulating facial surgery using finite element models. In: Proceedings of the 23rd Annual Conference on Computer Graphics and Interactive Techniques, pp. 421–428. ACM (1996)

10. Liang, J., Sun, S.: Site effects on seismic behavior of pipelines: a review. J. Pressure Vessel Technol. **122**(4), 469–475 (2000)
11. Micikevicius, P.: 3D finite difference computation on GPUs using CUDA. In: Proceedings of 2nd Workshop on General Purpose Processing on Graphics Processing Units, pp. 79–84. ACM (2009)
12. Quinay, P.E.B., Ichimura, T., Hori, M.: Waveform inversion for modeling three-dimensional crust structure with topographic effects. Bull. Seismol. Soc. Am. **102**(3), 1018–1029 (2012)
13. Geospatial Information Authority of Japan Tokyo ward area. 5m mesh digital elevation map. http://www.gsi.go.jp/MAP/CD-ROM/dem5m/index.htm
14. Yamaguchi, T.: Viscoelastic crustal deformation computation method with reduced random memory accesses for GPU-based computers. In: Shi, Y., et al. (eds.) ICCS 2018. LNCS, vol. 10861, pp. 31–43. Springer, Cham (2018). https://doi.org/10.1007/978-3-319-93701-4_3

# A Generic Interface for Godunov-Type Finite Volume Methods on Adaptive Triangular Meshes

Chaulio R. Ferreira and Michael Bader[(✉)]

Department of Informatics, Technical University of Munich, Munich, Germany
chaulio.ferreira@tum.de, bader@in.tum.de

**Abstract.** We present and evaluate a programming interface for high performance Godunov-type finite volume applications with the framework sam(oa)$^2$. This interface requires application developers only to provide problem-specific implementations of a set of operators, while sam(oa)$^2$ transparently manages HPC features such as memory-efficient adaptive mesh refinement, parallelism in distributed and shared memory and vectorization of Riemann solvers. We focus especially on the performance of vectorization, which can be either managed by the framework (with compiler auto-vectorization of the operator calls) or directly by the developers in the operator implementation (possibly using more advanced techniques). We demonstrate the interface's performance using two example applications based on variations of the shallow water equations. Our performance results show successful vectorization using both approaches, with similar performance. They also show that the applications developed with the new interface achieve performance comparable to analogous applications developed without the new layer of abstraction.

**Keywords:** High performance computing · Vectorization ·
Finite volume methods · Shallow water equations

## 1 Introduction

Most modern processors used in scientific computing rely on multiple levels of parallelism to deliver high performance. In this paper, we particularly focus on vectorization (i.e., SIMD parallelism) due to the recent increases in the width of the SIMD instructions provided by high-end processors. For example, latest generations of Intel Xeon architectures provide the AVX-512 instruction set, which can operate simultaneously on 8 double-precision (or 16 single-precision) values. As SIMD parallelism relies on uniform control flows, it benefits from regular data structures and loop-oriented algorithms for effective vectorization. This poses a particular challenge to applications that exploit non-regularity.

Adaptive mesh refinement (AMR) is a key strategy for large-scale applications to efficiently compute accurate solutions, as high resolution can be selectively applied in regions of interest, or where the numerical scheme requires

© Springer Nature Switzerland AG 2019
J. M. F. Rodrigues et al. (Eds.): ICCS 2019, LNCS 11537, pp. 402–416, 2019.
https://doi.org/10.1007/978-3-030-22741-8_29

higher accuracy. Managing dynamic mesh refinement and coarsening on parallel platforms, however, requires complex algorithms, which makes minimizing the time-to-solution a non-trivial task. Tree-structured approaches were proven successful to realize cell-wise adaptivity, even for large-scale applications (e.g., [6,20,23]). They allow to minimize the invested degrees of freedom, but previous work has shown that SIMD parallelism and meshing overhead push towards combining tree-structure with uniform data structures, such as introducing uniformly refined *patches* on the leaf-level [5,24]. Block-wise AMR approaches (e.g., [2,3,22]) extend refined mesh regions to uniform grid blocks – thus increasing the degrees of freedom, but allowing to stick to regular, array-based data structures. We found, however, that even then, efficient vectorization is not guaranteed, but requires careful implementation and choice of data structures [8].

In this work, we address the area of conflict between dynamic AMR and vectorization, on the development of a general Godunov-type finite volume solver in sam(oa)$^2$ [17], a simulation framework that provides memory- and cache-efficient traversals on dynamically adaptive triangular meshes and supports parallelism in distributed (using MPI) and shared memory (using OpenMP). We provide an easy-to-use and highly customizable interface that hides from the application developers (here referred to as *users*) all the complexity of the meshing algorithms and simplifies the creation of high-performance solvers for various systems of partial differential equations (PDEs). The use of patches for efficient vectorization is also transparent to the user, which is accomplished by a tailored concept of *operators*. Performance experiments confirm the effectiveness of vectorization for two shallow-water models developed with this interface, which perform comparably to analogous applications developed without the abstraction layer.

## 2    Numerical Background

We consider hyperbolic PDE systems written in the general form

$$q_t + f(q)_x + g(q)_y = \psi(q, x, y), \tag{1}$$

where $q(x, y, t)$ is a vector of unknowns and the *flux functions* $f(q)$ and $g(q)$, as well as the *source term* $\psi(q, x, y)$ are specific for each PDE. In the following, we describe two PDEs that serve to demonstrate the usability and performance of sam(oa)$^2$. Then, we discuss a numerical approach often used to solve such problems, usually known as *Godunov-type finite volume methods*.

**(Single-Layer) Shallow Water Equations.** The *shallow water equations* are depth-averaged equations that are suitable for modeling incompressible fluids in problems where the horizontal scales ($x$ and $y$ dimensions) are much larger than the vertical scale ($z$ dimension) and the vertical acceleration is negligible. As such, they are widely used for ocean modeling, since the ocean wave lengths are generally very long compared to the ocean depth [2,13].

The *single-layer shallow water equations* take the form

$$
\begin{bmatrix} h \\ hu \\ hv \end{bmatrix}_t + \begin{bmatrix} hu \\ hu^2 + \frac{1}{2}gh^2 \\ huv \end{bmatrix}_x + \begin{bmatrix} hv \\ huv \\ hv^2 + \frac{1}{2}gh^2 \end{bmatrix}_y = \begin{bmatrix} 0 \\ -ghb_x \\ -ghb_y \end{bmatrix},
\tag{2}
$$

where $h(x,y,t)$ is the fluid depth; $u(x,y,t)$ and $v(x,y,t)$ are the vertically averaged fluid velocities in the $x$ and $y$ directions, respectively; $g$ is the gravitational constant; and $b(x,y)$ is the bottom surface elevation. In oceanic applications, $b$ is usually relative to mean sea level and corresponds to submarine bathymetry where $b < 0$ and to terrain topography where $b > 0$. Here, the source term $\psi(x,y,t) = [0, -ghb_x, -ghb_y]^T$ models the effect of the varying topography, but may also include further terms, such as bottom friction and Coriolis forces.

**Two-Layer Shallow Water Equations.** Although the single-layer equations are appropriate for modeling various wave propagation phenomena, such as tsunamis and dam breaks, they lack accuracy for problems where significant vertical variations in the water column can be observed. For instance, in storm surge simulations wind stress plays a crucial role and affects more the top of the water column than the bottom [15]. The single-layer equations are not able to properly model this effect, because the water momentum gets averaged vertically.

Vertical profiles can be more accurately modeled using *multi-layer shallow water equations*, which can provide more realistic representations while keeping the computational costs relatively low. Using the simplest variation of these equations, one can model the ocean with two layers: a shallow top layer over a deeper bottom layer, allowing a more accurate representation of the wind effects. The system of *two-layer shallow water equations* can be written as

$$
\begin{bmatrix} h_1 \\ h_1u_1 \\ h_1v_1 \\ h_2 \\ h_2u_2 \\ h_2v_2 \end{bmatrix}_t + \begin{bmatrix} h_1u_1 \\ h_1u_1^2 + \frac{1}{2}gh_1^2 \\ h_1u_1v_1 \\ h_2u_2 \\ h_2u_2^2 + \frac{1}{2}gh_2^2 + rgh_2h_1 \\ h_2u_2v_2 \end{bmatrix}_x + \begin{bmatrix} h_1v_1 \\ h_1u_1v_1 \\ h_1v_1^2 + \frac{1}{2}gh_1^2 \\ h_2v_2 \\ h_2u_2v_2 \\ h_2v_2^2 + \frac{1}{2}gh_2^2 + rgh_2h_1 \end{bmatrix}_y = \begin{bmatrix} 0 \\ -(h_2)_x gh_1 - gh_1b_x \\ -(h_2)_y gh_1 - gh_1b_y \\ 0 \\ (h_2)_x rgh_1 - gh_2b_x \\ (h_2)_y rgh_1 - gh_2b_y \end{bmatrix}
\tag{3}
$$

where $h_1$, $u_1$ and $v_1$ are the quantities in the top layer and $h_2$, $u_2$ and $v_2$ are in the bottom layer; and $r \equiv \rho_1/\rho_2$ is the ratio of the densities of the fluid contained in each layer. The equations can be generalized for an arbitrary number of vertical layers [4]; however, in this work we deal only with the single- and two-layer forms discussed above, using them as example applications for our framework.

**Godunov-Type Finite Volume Methods.** To solve PDE systems as in Eqs. (1), (2) and (3), we adopt Godunov-type finite volume methods, following the numerical approach described in [12,13], which we slightly modify to match the triangular meshes in sam(oa)$^2$ (cf. [17] for details). We discretize on tree-structured adaptive triangular meshes, where the solution variables $q$ are averaged within each cell. We represent these variables as $Q_i^n$, the quantities vector in cell $C_i$ at time $t_n$. To update $C_i$, we first need to approximately solve the so-called *Riemann problem* on all of its edges, i.e., to compute the *numerical* *fluxes* $\mathcal{F}$ that reflect the quantities being transferred between $\mathcal{C}_i$ and each of its neighbors. We can then use these numerical fluxes to update $\mathcal{C}_i$, according to

$$Q_i^{n+1} = Q_i^n + \frac{\Delta t}{V_i} \sum_{j \in \mathcal{N}(\mathcal{C}_i)} \mathcal{F}(Q_i^n, Q_j^n), \tag{4}$$

where $\mathcal{N}(\mathcal{C}_i)$ is the set of neighbors of $\mathcal{C}_i$; $\mathcal{F}(Q_i^n, Q_j^n)$ is the numerical flux between cells $\mathcal{C}_i$ and $\mathcal{C}_j$; $V_i$ is the volume of $\mathcal{C}_i$; and $\Delta t$ is the time step applied (computed independently for every time step according to the CFL condition).

The *Riemann solver*, i.e., the numerical algorithm used to compute the numerical fluxes, is typically the most important and complex part of the scheme. As further discussed in Sect. 4, users need to provide a proper Riemann solver for their specific problem. In our example applications, we used solver implementations extracted from GeoClaw [2]: the augmented Riemann solver [10] for the single-layer and the solver proposed in [14] for the two-layer equations.

## 3   Patch-Based Parallel Adaptive Meshes in Sam(oa)$^2$

Sam(oa)$^2$ [17] is a simulation framework that supports the creation of finite element, finite volume and discontinuous Galerkin methods using adaptive meshes generated by recursive subdivision of triangular cells, following the newest-vertex-bisection method [18]. The cells in the resulting tree-structured mesh are organized following the order induced by the Sierpinski space-filling curve – which is obtained by a depth-first traversal of the binary refinement tree. The Sierpinski order is used to store the adaptive mesh linearly without the need to explicitly store the full refinement tree, leading to low memory requirement for mesh storage and management. The order also allows to iterate through all mesh elements in a memory- and cache-efficient way [1]. Sam(oa)$^2$ is implemented in Fortran and features a hybrid MPI+OpenMP parallelization that is also based on the Sierpinski order induced on the cells, which are divided into contiguous sections of similar size. The sections are independent units that can be assigned to different processes/threads and processed simultaneously with low communication requirements. Previous research [17] has shown that this parallelization is very efficient and scales well on up to 8,000 cores.

To better support vectorization, we modified the mesh structure to store uniformly refined *patches* in the leaves of the refinement tree, instead of single cells [7]. This allows reorganizing the simulation data into temporary arrays that effectively define lists of Riemann problems and to which vectorization is applied, following the approach proposed in [8]. In addition to enabling vectorization, the patch-based approach also leads to further performance gains due to improved memory throughput (because of the new data layout) and to lower overhead for managing the refinement tree (because its size was considerably reduced) [7].

Each patch is obtained by newest-vertex bisection of triangular cells, up to a uniform refinement depth $d$, resulting in patches with $2^d$ cells. To guarantee conforming meshes (i.e., without hanging nodes), all patches are generated with identical depth $d$ (whose value is set by the user at compilation time) and we only allow even values for $d$. Our results in Sect. 6 show that the best time-to-solution is achieved already for comparably small patches ($d = 6$ or $d = 8$).

Note that compared to [7] we recently changed the refinement strategy within patches: patch refinement now follows the adaptive tree-refinement strategy, which simplifies interpolation schemes for refinement and coarsening.

## 4    FVM Interface

To simplify the creation of new PDE solvers in sam(oa)$^2$, we designed a pro-gramming interface with which users can easily define implementation details that are particular to each system of PDEs. Using the abstraction layer of the new interface, users work with simple data structures (arrays) and application-specific algorithms, while sam(oa)$^2$ transparently manages cache-efficient grid traversals, distributed/shared-memory parallelism, adaptive mesh refinement, dynamic load balancing and vectorization. In the following, we will refer to the new programming interface as *FVM interface*, applications developed with it as *FVM applications* and subroutines provided by the users as *FVM operators*.

### 4.1    Data Structures Used in the FVM Interface

For a new FVM application, users first need to declare how many quantities should be stored for each cell. Following the approach used in GeoClaw [2], these are divided into two types: Q quantities are the actual unknowns in the system of equations and are updated at every time step following the numeri-cal scheme, as in Eq. (4); AUX quantities are other cell-specific values that are required for the simulation, but are either constant in time or updated according to a different scheme. E.g., in our FVM application for the single-layer shallow water equations, Q=(h,hu,hv) and AUX=(b).

The user specifies the number of Q and AUX variables via the preprocessor macros _FVM_Q_SIZE and _FVM_AUX_SIZE. Using these values, sam(oa)$^2$ creates 2D arrays Q and AUX for each patch to store the respective quantities for all patch cells. The arrays are stored similarly to a "structs of arrays" layout, i.e., the cell averages of any given quantity are stored contiguously, which is very important for efficient vectorization. We designed the FVM interface to transparently man-age the data in the patches, such that users do not need any knowledge of their geometry or data layout. Instead, users only need to implement simple *FVM operators* that are applied to the mesh on a cell-by-cell or edge-by-edge basis.

### 4.2    FVM Operators

In the FVM operators, users define how cells should be initialized, refined, coars-ened and updated. This includes providing a Riemann solver for the particular problem – where we anticipate that users might want to use existing solvers (e.g., from the GeoClaw package [2]). In the following we briefly describe the interface and what is expected from the most important operators. We note that due to limited space, we omit descriptions of further operators that are mostly concerned with details of the input and output formats used by sam(oa)$^2$.

InitializeCell. With this operator, users define the initial conditions and the initial mesh refinement in a cell-by-cell basis. Given the coordinates of a cell's vertices, this operator returns the initial value of all cell quantities (Q and AUX), as well as an estimate for the initial wave speed (for computing the size of the first time step), and a refinement flag: 1 if this cell should be further refined in the following initialization step, or 0 otherwise.

ComputeFluxes. This operator is responsible for solving the PDE-specific Riemann problems at all edges in the mesh and for returning the numerical fluxes at each edge, as well as the speed of the fastest wave created by the discontinuities between the cells. This is usually the most computing-intensive and time-consuming step in the simulations and where optimization efforts achieve the best results. Thus, to give users finer control of the code used at this step, especially with respect to vectorization, we provide two options to implement this operator: a *single-edge* or a *multi-edge* operator, as described in the following.

The **single-edge operator** is applied to one edge at a time, solving a single Riemann problem. In this case, the operator code provided by the users does not need to implement vectorization, because it is completely handled by the framework. Consider the example shown in Fig. 1(a): the framework loops through a list of $N$ Riemann problems extracting the input data of each problem and calling the operator with the respective data, while the operator only needs to extract the quantities from the input arrays and call the Riemann solver with the appropriate parameters. That loop is annotated with an !$OMP SIMD directive, and the operator call with a !DIR$ FORCEINLINE directive, to inform the compiler that we want the loop to be vectorized and the operator to be inlined and vectorized as well. In the operator call we use Fortran's subarrays, which introduce an overhead for their creation, but still support vectorization by the Intel Fortran Compiler. Note that whether the loop actually gets vectorized depends on the operator implementation, because it may contain complex operations or library calls that can inhibit auto-vectorization by the compiler.

In a **multi-edge operator** users can provide their own optimized code for the main loop. Instead of dealing with a single Riemann problem, the multi-edge operator takes a list of problems as input. It gives users complete control of the loop that iterates through the list, as well as of the compiler directives for vectorization. For illustration, see Fig. 1(b). The implementation of a multi-edge operator can become considerably more complex, allowing advanced users to exploit any technique for improved vectorization (intrinsic functions, etc.).

We created successfully vectorized FVM applications using single- and multi-edge operators both for the single- and two-layer shallow water equations. Section 6 will present and compare the performance achieved by each approach.

```
1 | ! OPERATOR CALL (FRAMEWORK CODE, NOT TOUCHED BY THE USER):
2 | real, dimension(N,_FVM_Q_SIZE)   :: qL, qR     ! Data from cells to
3 | real, dimension(N,_FVM_AUX_SIZE) :: auxL, auxR ! left/right of each edge
4 | real, dimension(N,2) :: normals ! Vectors that are normal to each edge
5 |
6 | !$OMP SIMD PRIVATE(waveSpeed) REDUCTION(max: maxWaveSpeed)
7 | do i=1,N ! Loop for all N Riemann problems
8 |    !DIR$ FORCEINLINE
9 |    call computeFluxesSingle(normals(i,:),qL(i,:),qR(i,:),auxL(i,:),
   |      ↪ auxR(i,:),fluxL(i,:),fluxR(i,:),waveSpeed)
10|    maxWaveSpeed = max(maxWaveSpeed, waveSpeed)
11| end do
```

```
1 | ! OPERATOR CODE, IMPLEMENTED BY THE USER:
2 | subroutine computeFluxesSingle(normal,qL,qR,auxL,auxR,fluxL,fluxR,waveSpeed)
3 |    real, dimension(2),            intent(in)  :: normal !Normal vector
4 |    real, dimension(_FVM_Q_SIZE),   intent(in)  :: qL,qR
5 |    real, dimension(_FVM_AUX_SIZE), intent(in)  :: auxL,auxR
6 |    real, dimension(_FVM_Q_SIZE),   intent(out) :: fluxL,fluxR
7 |    real,                         intent(out) :: waveSpeed
8 |    real :: hL,huL,hvL,bL, hR,huR,hvR,bR ! local variables
9 |
10|    !Extract data from input arrays to local variables
11|    hL = qL(1); huL = qL(2); hvL = qL(3); bL = auxL(1)
12|    hR = qR(1); huR = qR(2); hvR = qR(3); bR = auxR(1)
13|
14|    !The Riemann solver fills the output (fluxL, fluxR and waveSpeed):
15|    !DIR$ FORCEINLINE
16|    call RiemannSolver(normal,hL,huL,...,hvR,bR,fluxL,fluxR,waveSpeed)
17| end subroutine
```

(a) Single-edge version of the `ComputeFluxes` operator.

```
1 | ! OPERATOR CALL (FRAMEWORK CODE, NOT TOUCHED BY THE USER):
2 | ! ... (Declaration of qL, qR, auxL, auxR and normals, exactly as above)
3 |
4 | call computeFluxesMulti(normals,qL,qR,auxL,auxR,fluxL,fluxR,maxWaveSpeed)
```

```
1 | ! OPERATOR CODE, IMPLEMENTED BY THE USER:
2 | subroutine computeFluxesMulti(normals,qL,qR,auxL,auxR,fluxL,fluxR,
   |      ↪ maxWaveSpeed)
3 |    real, dim(N,2),                 intent(in)  :: normals !Normal vectors
4 |    real, dimension(N,_FVM_Q_SIZE),   intent(in)  :: qL,qR
5 |    real, dimension(N,_FVM_AUX_SIZE), intent(in)  :: auxL,auxR
6 |    real, dimension(N,_FVM_Q_SIZE),   intent(out) :: fluxL,fluxR
7 |    real,                           intent(out) :: maxWaveSpeed
8 |    real :: hL,huL,hvL,bL, hR,huR,hvR,bR, normal(2) ! local variables
9 |
10|    !$OMP SIMD REDUCTION(max: maxWaveSpeed) PRIVATE(normal,hL,huL,...,hvR,bR)
11|    do i=1,N ! Loop for all N Riemann problems
12|       !Extract data from input arrays to iteration-private variables
13|       hL = qL(i,1); huL = qL(i,2); hvL = qL(i,3); bL = auxL(i,1)
14|       hR = qR(i,1); huR = qR(i,2); hvR = qR(i,3); bR = auxR(i,1)
15|       normal = normals(i,:)
16|
17|       !DIR$ FORCEINLINE
18|       call RiemannSolver(normal,hL,huL,...,hvR,bR,fluxL,fluxR,waveSpeed)
19|       maxWaveSpeed = max(maxWaveSpeed, waveSpeed)
20|    end do
21| end subroutine
```

(b) Multi-edge version of the `ComputeFluxes` operator.

**Fig. 1.** Example implementations of the single-edge (a) and multi-edge (b) versions of the `ComputeFluxes` operator for the single-layer shallow water equations, as well as the framework codes that call them. In both we assume the existence of a subroutine called `RiemannSolver` that solves a Riemann problem.

**UpdateCell.** This operator defines how the Q quantities of each cell should be updated with the Riemann solutions. After obtaining the numerical fluxes at each edge in the mesh, sam(oa)$^2$ computes the sum of all fluxes leaving/entering each cell through its three edges. The total flux is then used to update the cell quantities, usually following Eq. (4); however, users have flexibility to modify this, and may also take care of special cases that arise for the particular problem (e.g., drying or wetting of cells with the shallow water equations). The operator also returns a refinement flag for the cell, to inform whether it should be refined (1), kept (0) or coarsened (−1) for the next time step.

**SplitCell and MergeCells.** These two operators control how adaptivity is performed on the cells. **SplitCell** takes the data from a cell $C^{in}$ as input, and outputs data for two finer cells $C_1^{out}$ and $C_2^{out}$, that result from splitting $C^{in}$. **MergeCells** takes data from two neighbor cells $C_1^{in}$ and $C_2^{in}$ as input, and returns data for a coarse cell $C^{out}$ that results from merging both input cells. These operations are often performed by simply copying or interpolating cell quantities, but again the users can customize the operator and handle special cases.

## 5 FVM Applications and Test Scenarios

We used the FVM interface to create two FVM applications that simulate tsunami wave propagation with the single- and two-layer shallow water equations discussed in Sect. 2. Here, we refer to them as FVM-SWE and FVM-SWE2L. They are based on two applications that already existed within sam(oa)$^2$ for those same systems of PDEs and were implemented directly into the framework's core (i.e., without the FVM interface). To distinguish them from the new FVM applications, we will refer to the older applications as SWE and SWE2L.

While the implementations of SWE and SWE2L are considerably more complex because their codes deal directly with the data structures and algorithms used in sam(oa)$^2$, that may be advantageous in terms of performance, because they do not have the overhead due to additional memory management performed by the FVM interface's abstraction layer. Therefore, we will use their performance as baseline for evaluating the new applications FVM-SWE and FVM-SWE2L.

Before presenting our experimental results, we first give more details about the simulation scenarios we used in our experiments in the following.

**Tohoku Tsunami 2011.** For SWE and FVM-SWE, we simulated a real tsunami event that took place near the coast of Tohoku, Japan, in 2011 – see Fig. 2. We used bathymetry data from the Northern Pacific and the Sea of Japan (GEBCO_08 Grid, version 20100927) and initial bathymetry and water displacements obtained from a simulation of the Tohoku earthquake [9]. All geographical input data was handled by the parallel I/O library ASAGI [19].

Although the entire simulation domain covers an area of 7 000 km × 4 000 km, our adaptive mesh was able to discretize it with a maximum resolution of 2852 m$^2$ in the area around the tsunami, while regions farther away can have cells with up to 5981 km$^2$. Considering all simulations, the minimum and maximum observed mesh sizes were of approx. 6.8 million to 34.2 million cells.

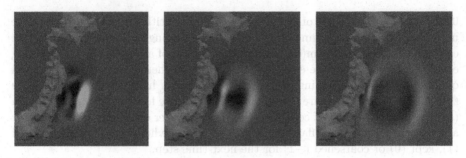

**Fig. 2.** Simulation of the tsunami in Tohoku, Japan, 2011. The pictures show the tsunami wave 10, 20 and 30 min after the earthquake, respectively.

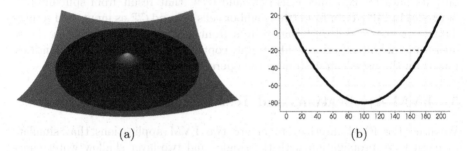

(a)                                    (b)

**Fig. 3.** Parabolic bowl-shaped lake at $t = 0$ with 3D (a) and cross-cut (b) visualizations. In (b), the thick (black) line depicts the bathymetry, while the dashed (blue) and solid (cyan) lines represent the two layers of water. (Color figure online)

The simulations were run for 1000 time steps and the measurements only include the regular time-stepping phase, i.e., they do not consider the time necessary for reading the input data and generating the initial mesh.

**Parabolic Bowl-Shaped Lake.** For SWE2L and FVM-SWE2L, we simulated waves generated by a circular hump of water propagating over a parabolic bowl-shaped bathymetry – see Fig. 3, where we show visualizations of this scenario at $t = 0$. This setup is based on an example from the GeoClaw package [2] and serves as a benchmark for the quality of numerical schemes regarding wetting/drying of layers. We used cells with sizes ranging from $2^{-27}$ to $2^{-13}$ of the computational domain, resulting in mesh sizes from around 0.8 million to 7.2 million cells. The measured times also consider 1000 time steps of the regular time-stepping phase.

## 6    Performance Results

We conducted experiments on the CoolMUC2 and CoolMUC3 cluster systems hosted at the Leibniz Supercomputing Center (LRZ). CoolMUC2 contains nodes with dual-socket Haswell systems and CoolMUC3 provides nodes with Xeon Phi "Knights Landing" (KNL) processors. An overview of the system configuration

**Table 1.** Specifications of the nodes in the experimental platforms.

| System overview | 2×Haswell | Knights landing |
|---|---|---|
| Architecture | Intel® Xeon® | Intel® Xeon Phi™ |
| Model | E5-2697v3 | 7210F |
| Cores | $2 \times 14$ | 64 (max. 256 threads) |
| Clock rate | 2.60 GHz | 1.30 GHz |
| SIMD vector width | 256-bit | 512-bit |
| Memory | 64 GB | 96 GB + 16 GB MCDRAM |
| Peak bandwidth | 136 GB/s | 102 GB/s |
| Measured bandwidth | 106 GB/s | 83 GB/s |
| Peak throughput (double) | 582 GFlop/s | 2 662 GFlops/s |
| Measured throughput (double) | 156 GFlop/s | 720 GFlops/s |

of the nodes in each system is presented in Table 1. As we focus on vectorization performance, we point out the difference in the SIMD width of these machines: the Haswells provide AVX2 instructions (256-bit), while the KNLs provide AVX-512 instructions (512-bit). As such, the benefits from vectorization are expected to be more noticeable on the KNL nodes. Table 1 lists the theoretical peak bandwidth and peak Flop/s throughput of each machine, along with measurements obtained with the STREAM benchmark [16] and with the Flop/s benchmark proposed in Chapter 2 of [11]. We use these values as estimates for the maximum performance that can be achieved in practice on those machines.

In all reported experiments we used the Intel Fortran compiler 17.0.6 and double precision arithmetic. On both systems we use only one node at a time (i.e., no MPI), but we use OpenMP threading on all available cores, i.e., 28 on the Haswells and 64 on the KNL. On the KNL we also experimented with different number of threads per core (from 1 to 4 with hyperthreading). However, here we only list the results with 2 threads per core (i.e., 128 in total), because this configuration achieved the best performance in most experiments. Also, we use the KNL in cache mode, i.e., the MCDRAM memory is used as an L3 cache.

**Simulation Performance.** We start by evaluating the performance of FVM-SWE and FVM-SWE2L with different vectorization strategies and patches with various sizes – see Fig. 4. There is a clear pattern of larger patches delivering higher performance even with vectorization turned off, which can be attributed to the improved memory throughput and the reduction of the adaptivity tree achieved by the patch-based discretizations, as mentioned in Sect. 3.

When vectorization is used ("Single-edge" and "Multi-edge"), we can observe speedups by factors of 1.1–1.6 on the Haswells and of 1.5–2.3 on the KNL, compared to the non-vectorized versions. Vectorization is clearly more effective for the single-layer applications and (as expected) on KNL processors, which agrees with recent results in literature [8]. The codes developed using single-edge operators perform only slightly slower than the multi-edge ones (up to 2%

and 5% slower on each machine), despite of the overhead introduced due to the use of Fortran subarrays in the operator calls. This not only reveals that this overhead is not so high, but also confirms that the Intel Fortran Compiler is able to efficiently handle the subarrays when vectorizing the loop.

The vectorized `FVM-SWE` and `FVM-SWE2L` implementations achieve performance very similar to their analogous applications (`SWE` and `SWE2L`) on Haswells (up to 6% slower), while on the KNL there is a noticeable difference in performance (up to 14% slower). Nevertheless, these experiments show that FVM applications can achieve performance comparable to other applications developed directly within the complex framework code, despite of the additional memory operations performed to create the interface's layer of abstraction.

**Time-to-Solution.** Although larger patches deliver higher throughput, they also lead to meshes with more cells (due to more coarse-grained adaptivity). As such, a trade-off in the patch size must be found to minimize the time-to-solution. In Fig. 5 we compare the wall time of the "Multi-edge" implementations. These results show that patches with 64–256 cells tend to minimize the time-to-solution on both machines and both applications – despite of the higher throughput, larger patches are disadvantageous, due to the considerably increased mesh size. We also point out that by combining patch-based AMR with vectorization we have been able to reduce the total execution time by factors of 2.7–4.2, compared to cell-wise adaptivity ("trivial" patches with only one cell).

**Component Analysis.** In Fig. 5 we split the execution time into two components: the "Numerical" component comprises all routines performed for the numerical time step, i.e., updating the cell quantities; and "Adaptivity" is responsible for handling the remeshing performed after every time step, i.e., refining/coarsening of cells, guaranteeing mesh conformity and updating communication structures. We can observe that, while most of the time reduction happens in the numerical component, the adaptivity component also benefits considerably, because of the reduced complexity and size of the mesh refinement tree.

**Solver Performance.** Now we evaluate the numerical routines alone, without the mesh management algorithms. We computed the data and Flop/s throughputs of the fastest run of each FVM application on each machine – see Table 2, where we also show their percentages relative to the measured peak performance of each machine. The Flop/s were measured using the library PAPI [21], and the data throughputs calculated assuming two accesses (read & write) to `Q` quantities and only one read access to `AUX` quantities for each cell updated.

Considering that the Riemann solver is much more complex than the benchmark used, the results show great utilization of the Haswell processors by the single-layer solver (42%), indicating compute-bound behavior. On the KNL the solver is also compute-bound, although it reaches only 14% of the peak throughput. We point out that the entire simulation data fits in the MCDRAM memory that is being used in cache mode, thus memory bandwidth is not an issue.

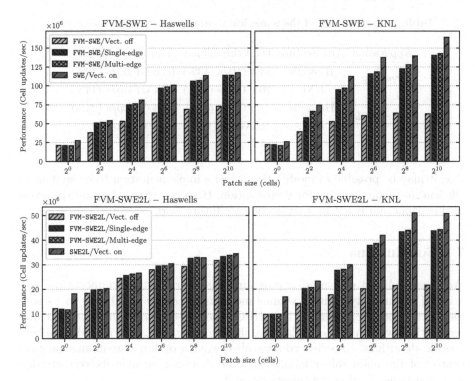

**Fig. 4.** Performance of the FVM applications and their analogous applications.

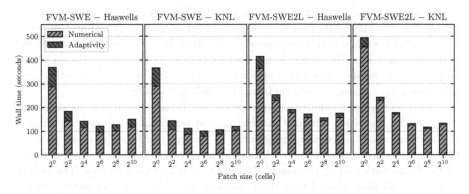

**Fig. 5.** Execution times of the FVM applications, split into components.

A similar analysis for the two-layer solver indicates fairly low performance, compared to the measured peak of each machine, both for Flop/s and data throughput. This happens because the two-layer solver is considerably more complex than the single-layer one, mainly due to several if-then-else branches necessary to handle special cases (such as dry cells/layers). The compiler converts these into masked operations, which causes vectorization overhead. This reveals

**Table 2.** Performance of the numerical routines of the FVM applications.

| Application | Architecture | Data throughput | Flop/s throughput |
|---|---|---|---|
| FVM-SWE | Haswells | 14.1 GB/s (13%) | 66.0 GFlop/s (42%) |
| | KNL | 20.9 GB/s (25%) | 97.4 GFlop/s (14%) |
| FVM-SWE2L | Haswells | 7.9 GB/s (7%) | 36.3 GFlop/s (23%) |
| | KNL | 9.5 GB/s (11%) | 44.0 GFlop/s (6%) |

that it may be possible to modify this solver's implementation to make it more efficient and more suitable for vectorization. However, that is beyond the scope of this paper and is left as a suggestion for future work.

# 7    Conclusions

We described and evaluated a programming interface that supports the creation of Godunov-type finite volume methods in sam(oa)$^2$. Thanks to its simple abstraction layer, users with no HPC expertise can create high performance applications with multiple levels of parallelism, only needing to provide problem-specific algorithms. Experienced users also have the option of assuming complete control of the main solver loop, such that they can manage its vectorization and/or attempt further optimizations on it.

The interface allows easy customization of our efficient finite volume solver to different systems of PDEs. Assuming that a Riemann solver for the specific problem is available, the user's work is reduced mainly to handling the solver calls and providing application-specific operators for initialization, refinement and coarsening of cells, which in most cases consist of trivial implementations. In particular, we developed two applications in which we could directly apply Riemann solver implementations from the package GeoClaw.

The underlying framework implements patch-based adaptive mesh refinement, which enables vectorization and at the same time reduces the simulation's computational costs considerably by applying relatively small patches. Our experiments revealed successful vectorization (both when it was managed by the framework and by the users), leading to substantial speedups. The performance results also showed that these applications achieve performance comparable to analogous applications developed directly into the complex source code of sam(oa)$^2$, with only low to moderate overhead (2–14% slower).

**Acknowledgments.** Chaulio R. Ferreira appreciates the support of CNPq, the Brazilian Council of Technological and Scientific Development (grant no. 234439/2014-9). Computing resources were provided by the Leibniz Supercomputing Center (project *pr85ri*).

# References

1. Bader, M., Böck, C., Schwaiger, J., Vigh, C.A.: Dynamically adaptive simulations with minimal memory requirement - solving the shallow water equations using sierpinski curves. SIAM J. Sci. Comput. **32**(1), 212–228 (2010)
2. Berger, M.J., George, D.L., LeVeque, R.J., Mandli, K.T.: The GeoClaw software for depth-averaged flows with adaptive refinement. Adv. Water Resour. **34**(9), 1195–1206 (2011)
3. Berger, M.J., Oliger, J.: Adaptive mesh refinement for hyperbolic partial differential equations. J. Comput. Phys. **53**(3), 484–512 (1984)
4. Bouchut, F., Zeitlin, V., et al.: A robust well-balanced scheme for multi-layer shallow water equations. Discret. Contin. Dyn. Syst.-Ser. B **13**(4), 739–758 (2010)
5. Burstedde, C., Calhoun, D., Mandli, K., Terrel, A.R.: ForestClaw: hybrid forest-of-octrees AMR for hyperbolic conservation laws. In: Parallel Computing: Accelerating Computational Science and Engineering, vol. 25, pp. 253–262 (2014)
6. Burstedde, C., Wilcox, L.C., Ghattas, O.: p4est: scalable algorithms for parallel adaptive mesh refinement on forests of octrees. SIAM J. Sci. Comput. **33**(3), 1103–1133 (2011)
7. Ferreira, C.R., Bader, M.: Load balancing and patch-based parallel adaptive mesh refinement for tsunami simulation on heterogeneous platforms using Xeon Phi coprocessors. In: Proceeding of PASC, pp. 12:1–12:12. ACM (2017)
8. Ferreira, C.R., Mandli, K.T., Bader, M.: Vectorization of Riemann solvers for the single-and multi-layer Shallow Water Equations. In: International Conference on High Performance Computing & Simulation, pp. 415–422. IEEE (2018)
9. Galvez, P., Ampuero, J.P., Dalguer, L.A., Somala, S.N., Nissen-Meyer, T.: Dynamic earthquake rupture modelled with an unstructured 3-D spectral element method applied to the 2011 M9 Tohoku earthquake. Geophys. J. Int. **198**(2), 1222–1240 (2014)
10. George, D.L.: Augmented Riemann solvers for the shallow water equations over variable topography with steady states and inundation. J. Comput. Phys. **227**(6), 3089–3113 (2008)
11. Jeffers, J., Reinders, J.: Intel Xeon Phi Coprocessor High-performance Programming. Newnes, Oxford (2013)
12. LeVeque, R.J.: Finite Volume Methods for Hyperbolic Problems. Cambridge University Press (2002). www.clawpack.org/book.html
13. LeVeque, R.J., George, D.L., Berger, M.J.: Tsunami modelling with adaptively refined finite volume methods. Acta Numer. **20**, 211–289 (2011)
14. Mandli, K.T.: A numerical method for the two layer shallow water equations with dry states. Ocean Model. **72**, 80–91 (2013)
15. Mandli, K.T., Dawson, C.N.: Adaptive mesh refinement for storm surge. Ocean Model. **75**, 36–50 (2014)
16. McCalpin, J.D.: Memory bandwidth and machine balance in current high performance computers. IEEE TCCA Newsl. **2**, 19–25 (1995)
17. Meister, O., Rahnema, K., Bader, M.: Parallel memory-efficient adaptive mesh refinement on structured triangular meshes with billions of grid cells. ACM Trans. Math. Softw. **43**(3), 19 (2016)
18. Mitchell, W.F.: Adaptive refinement for arbitrary finite-element spaces with hierarchical bases. J. Comput. Appl. Math. **36**(1), 65–78 (1991)
19. Rettenberger, S., Meister, O., Bader, M., Gabriel, A.A.: ASAGI - a parallel server for adaptive geoinformation. In: Proceedings of EASC, pp. 2:1–2:9. ACM (2016)

20. Sampath, R.S., Adavani, S.S., Sundar, H., Lashuk, I., Biros, G.: Dendro: Parallel algorithms for multigrid and AMR methods on 2:1 balanced octrees. In: Proc. 2008 ACM/IEEE Conference on Supercomputing, pp. 18:1–18:12. IEEE Press (2008)
21. Terpstra, D., Jagode, H., You, H., Dongarra, J.: Collecting performance data with PAPI-C. In: Müller, M., Resch, M., Schulz, A., Nagel, W. (eds.) Tools for High Performance Computing 2009. Springer, Heidelberg (2010). https://doi.org/10.1007/978-3-642-11261-4_11
22. Wahib, M., Maruyama, N., Aoki, T.: Daino: a high-level framework for parallel and efficient AMR on GPUs. In: Proceedings of International Conference on HPC, Networking, Storage and Analysis, pp. 53:1–53:12. IEEE Press (2016)
23. Weinzierl, T.: The Peano software - parallel, automaton-based, dynamically adaptive grid traversals. ACM Trans. Math. Softw. **45**, 14 (2019)
24. Weinzierl, T., Bader, M., Unterweger, K., Wittmann, R.: Block fusion on dynamically adaptive spacetree grids for shallow water waves. Parallel Process. Lett. **24**(3), 1441006 (2014)

# Track of Agent-Based Simulations, Adaptive Algorithms and Solvers

# Distributed Memory Parallel Implementation of Agent-Based Economic Models

Maddegedara Lalith[1(✉)], Amit Gill[2], Sebastian Poledna[3], Muneo Hori[1],
Inoue Hikaru[4], Noda Tomoyuki[4], Toda Koyo[4], and Tsuyoshi Ichimura[1]

[1] Earthquake Research Institute, The University of Tokyo, Bunkyo-ku, Tokyo, Japan
{lalith,hori,ichimura}@eri.u-tokyo.ac.jp
[2] Department of Civil Engineering, The University of Tokyo,
Bunkyo-ku, Tokyo, Japan
gill@eri.u-tokyo.ac.jp
[3] International Institute for Applied Systems Analysis, Laxenburg, Austria
poledna@iiasa.ac.at
[4] Frontier Computing Center, Fujitsu Limited, Tokyo, Japan
{inoue-hikaru,tomoyuki,toda.koyo}@jp.fujitsu.com

**Abstract.** We present a *Distributed Memory Parallel* (DMP) implementation of agent-based economic models, which facilitates large-scale simulations with millions of agents. A major obstacle in scalable DMP implementation is to distribute a balanced workload among MPI processes, while making all the topological graphs, over which the agents interact, available at a minimum communication cost. We addressed this problem by partitioning a representative employer-employee interaction graph, and all the other interaction graphs are made available at negligible communication costs by mimicking the organizations of the real-world economic entities. Cache-friendly and low-memory intensive algorithms and data structures are proposed to improve runtime and scalability, and the effectiveness of each is demonstrated. The current implementation is capable of simulating 1:1 scale models of medium-sized countries.

**Keywords:** Agent-based economic models · Large-scale simulations · MPI

## 1 Introduction

A recent trend in macroeconomics is to use Agent-based models (ABMs), that simulate the behavior of individual economic entities, often using simple decision rules. Macroeconomic ABMs relax two key assumptions at the core of the New Neoclassical Synthesis–the single, representative agent and the rational, or model-consistent, expectations hypothesis [1]. Since the financial crisis of 2007–2008, ABMs have been increasingly used [1]. In recent years several ABMs that depict entire national economies have been developed. The European Commission has in part supported this endeavor by funding large research projects like

© Springer Nature Switzerland AG 2019
J. M. F. Rodrigues et al. (Eds.): ICCS 2019, LNCS 11537, pp. 419–433, 2019.
https://doi.org/10.1007/978-3-030-22741-8_30

CRISIS[1][2,3], and EURACE[2] which is a large micro-founded macroeconomic model with regional heterogeneity [4–6]. Some of the recent ABMs aim to simulate entire national or regional economies, like the Eurozone, with comprehensive models that include each individual economic entity–each household, each firm, etc. [4,7]. Such 1:1 scale simulations with hundreds of millions of interacting agents are computationally demanding.

The lack of scalable High Performance Computing (HPC)–enhanced macroeconomic ABMs that are capable of simulating hundreds of millions of agents is a major barrier in the application of ABMs to simulate an entire national or a regional economy. A major difficulty in implementing scalable HPC extensions is the complicated interactions among agents which occur over several topological graphs. In shared-memory implementations, these interactions lead to race conditions, while in *Distributed Memory Parallel* (DMP) implementations with MPI (Message Passing Interface), each interaction generates one or more messages among MPI processes (e.g., CPU cores). A critical step in scalable DMP implementation is balanced distribution of the agents among MPI processes, while making all the interactions graphs available at least communication cost. Some parallel implementations circumvent this difficulty by making only one interaction graph available across MPI processes using non-scalable strategies [4]. Apart from such partial parallel implementations, we could not find any literature on complete parallel implementation of a macroeconomic ABM capable of simulating several million agents.

Inspired by the need of HPC-enhanced agent-based economic models, we developed a DMP code based on MPI. Poledna et al.'s [7] model is chosen as the base ABM since its features, like the credit-based market, expectations-based decisions by agents, production based on the goods purchased in previous period, etc., provide opportunities to attain higher parallel scalability. Further, their model is attractive from the application point of view: rich in level of details, parameters estimated from real data, realistic economic interactions, etc. To distribute a balanced workload, we partitioned the agents based on a representative employer-employee interaction graph, and all the other interactions were made available with a negligible amount of communications. Several methods of improving the serial performance of market interactions, which in turn reduce the load imbalances, are presented. The current implementation takes around 20 s to complete a single iteration of a 1:1 model of Austria ($\sim$10 million agents), making the simulation of a whole nation a reality. Though the current scalability is limited to several tens of MPI processes, it can be further improved by using accurate estimates of the amount of computations associated with each agent.

The rest of the paper is organized as follows. Section 2 discusses economic ABMs in a DMP perspective, and presents the challenges in scalable implementations. Section 3 explains the details of the proposed DMP implementation scheme, providing details of domain decomposition, solutions to the challenges

---

[1] FP7-ICT grant 288501, http://cordis.europa.eu/project/rcn/101350_en.html.

[2] FP6-STREP grant 035086, http://cordis.europa.eu/project/rcn/79429_en.html. See also: http://www.wiwi.uni-bielefeld.de/lehrbereiche/vwl/etace/Eurace_Unibi/.

discussed in Sect. 2, and improvements to some extensively used serial algorithms. Section 4 presents runtime statistics to demonstrate the effectiveness of solutions introduced in Sect. 3 and the parallel scalability.

For the sake of brevity, in the rest of the paper, the term *rank* or *MPI-rank* denotes an MPI process (e.g., CPU core), the term *parallel* implies distributed parallel implementation with MPI, and *message* or *communication* implies a point-to-point or collective MPI communication (e.g. `MPI_Send()` and `MPI_Recv()`, `MPI_Bcast()`, etc.).

## 2    Agent-Based Economic Models: An HPC Point of View

Though they vary in rules defining the agents' actions, most of the general economic ABMs have common agent types and interactions among them. Figure 1 shows a schematic diagram of a typical ABM. Depending on the simulated economic zone, there can be several tens to a few hundred industries, each consisting of a large number of firms. The regional economy is connected to the rest of the world through foreign buyers and sellers. The largest in number are household agents, consisting of workers, inactive households, and investors. The total number of firms is roughly 10% of the workers.

At a glance, simulating millions of agents by assigning equal-sized subsets of agents to each MPI process seems to be an ideal DMP application. However, scalable DMP implementation is a challenging task due to the complicated interactions among agents. Most of the interactions take place over either centralized graphs or dense graphs with random links. In DMP implementations, each link represents either one or two communications among ranks that possess the two agents defining the link. Unless the interactions are between random pairs of agents, all the messages between a pair of ranks can be combined and delivered in a single message. On the other hand, each interaction over random graphs requires an independent message per link, and an independent reply message if the interaction is bidirectional. The rest of this section gives the details of different interactions among the agents, and explains the difficulties in implementing a scalable HPC extension.

### 2.1    Interactions over Centralized Graphs

The interactions involving government and banks take place over centralized graphs (see Fig. 2a). Of these, uni-directional interactions are easier to parallelize, while the bi-directional interactions introduce load imbalances due to the serialization of the related decision-making of the government or bank.

**Interactions of Households, Firms, and Banks with Government.** All the workers, investors, firms, and banks pay various taxes to the government, while government pays various social benefits to the households. Tax paying is an unidirectional interaction, and can easily be parallelized by making each rank collect taxes and send them to the rank holding the government agent. On the

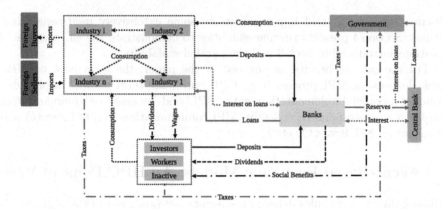

**Fig. 1.** Schematic diagram of a typical agent-based economic model. Each industry consists of a large number of firms.

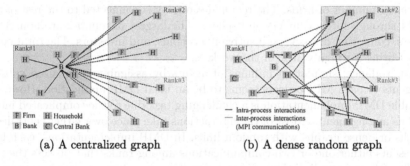

(a) A centralized graph          (b) A dense random graph

**Fig. 2.** Major sources of difficulties in shared- or distributed-memory parallel implementations. (a) hot spots like banks and government in centralized interaction graphs, (b) dense random interaction graphs like goods and job markets.

other hand, paying social benefits is bi-directional, and involves three stages: first, the income states of households are gathered using MPI_Gatherv(); next, the social benefits to be paid to each household are *sequentially* calculated by the government agent; and finally, the social benefits are scattered to respective households using MPI_Scatterv().

**Interaction of Households and Firms with Banks.** While depositing money is uni-directional, withdrawal and loan requests are bi-directional. The bi-directional interactions involve three stages, just like when the government is paying out social benefits: each MPI-rank sends households' or firms' requests to withdraw or borrow with MPI_Gatherv(); the bank sequentially processes each request; and responses are sent using MPI_Scatterv().

## 2.2    Interactions over Dense Random Graphs

Unlike the interactions over centralized graphs, the interactions of goods and job markets take place over a set of dense random graphs: the connectivity of the graphs randomly changes in each time period to mimic the randomness in the goods and job market. These random graphs make it quite hard to implement scalable DMP extensions.

## 2.3    Interactions in the Goods Market

In each period, the consumers (i.e., firms, households, foreign buyers, and government sectors) visit random sets of sellers (i.e., firms and foreign sellers) from each industry to purchase the necessary consumption and capital goods. The number of sellers visited by a consumer agent is unpredictable, and depends on its current budget and the amounts available with each seller at the time it visits the seller. As a result, each buy-sell interaction involves independent messages among the ranks that possess the buyer and the seller. Being bi-directional, each visit to a seller involves three stages; a message to the rank possessing the corresponding seller, decision-making by the seller, and a reply message to the buyer. Developing a scalable shared- or distributed-memory parallel computing code to facilitate such random interactions among millions of agents is undoubtedly challenging.

A naïve solution is to introduce a queue at each seller, into which each consumer can submit a request with an MPI point-to-point communication, with the seller sending the reply message once the request is processed. Though logically correct, obviously the computational performance is much worse, even compared to the serial version.

## 2.4    Interactions in the Labor Market

At the start of each period, firms hire a random set of unemployed workers, according to their labor requirements. When the available number of workers assigned to a rank is less than the total labor demand of the firms assigned to that rank, workers from the surrounding ranks have to be hired. Firms hire workers from a limited geographical extent to ensure real-life constraints, like travel time to work. If each firm independently seeks the required labor, it leads to a complicated situation similar to that of the goods market.

# 3    Distributed-Memory Parallel Computing Extension

This section presents the major steps involved in implementing an MPI-based DMP extension: partitioning the agents to assign balanced workloads to each rank; scalable solution for the difficulties discussed in the previous section; and some techniques for reducing the runtime of extensively called serial algorithms.

Though blocking MPI functions are used in the following text, corresponding non-blocking functions with user-defined MPI data types should be utilized to attain higher communication efficiency.

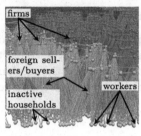

(a) Input graph to METIS              (b) 8 partitions

**Fig. 3.** Input graph with 50,000 agents, and 8 partitions obtained with METIS. For clarity, different types of agents are drawn at three different elevations.

## 3.1  Domain Decomposition

In order to distribute balanced computational workload among ranks, the agents are partitioned into mutually exclusive subsets. We use a graph representing the job market (i.e., employer-employee graph) in partitioning the agents, while taking different strategies to ensure the availability of rest of the topological graphs. The sections that follow explain how the remaining graphs, like buyer-seller, bank-customer, employer-employee, government-tax payers, foreign seller-local buyers, etc., are made available. What is desired are partitions with nearly equal sum of nodal weights (i.e., computational workload associated with agents) while minimizing the number of graph edges intersected by the partition boundaries, since each intersecting link gives rise to communications among ranks.

The graph to be partitioned is constructed by connecting each firm, $f_i$, with the closest $n_i$ worker agents, where $n_i$ is an approximate number of the labor requirement of $f_i$ in the first period. To capture the reality, each agent is placed at the respective physical location, and the closest worker agents are identified using k-d tree. In order to ensure that the graph sufficiently reflects the employer-employee relations and has insignificant effects due to the presence of inactive households, foreign sellers, and foreign buyers, these three types of agents are connected to only one or two of the closest firms with low link weights. Further, each firm is connected to a few neighboring firms, with a low link weight, to reduce the possibility of producing a disconnected graph. The nodes of the graph (i.e., the agents) are assigned weights according to the amount of computations associated with each agent type. We partitioned this graph using METIS [8]. Figures 3a and 3b show a sample input graph and the 8 partitions generated.

Investors, banks, government, and central bank are not included in the above graph partitioning. Investors are assigned according to the number of firms in each partition. Government, bank, and central bank are assigned to the master rank (e.g., rank 0) to make them interact without any communications.

## 3.2  Scalable Solutions for Interactions over Centralized Graphs

By introducing representative agents of government, and banks in each rank, it is possible to almost eliminate the associated communications and equally distribute the associated computational workloads, making the centralized interactions involving the government, and banks nearly perfectly scalable.

**Interactions with Government on Centralized Graphs.** The simple solution given in Sect. 2.1 has quite a poor scalability since a single rank has to calculate the social benefits to be paid to millions of households, and involves messages of large volumes. A scalable solution is to introduce in each rank a *local government* agent which collects tax, calculates and pays the social benefits, and finally communicates the total tax collected and the social benefits paid to the master rank which holds the government agent. This simple solution is highly scalable since the computational workload is well distributed, all the bi-directional communications are eliminated, and the numbers of messages and volume of data communicated are drastically reduced.

**Interactions with Banks on Centralized Graphs.** A scalable solution for banks (see Sect. 2.1) is to introduce into each rank *local bank branches,* that keep the accounts of all the customers in the corresponding ranks, and locally process all the deposit and withdrawal requests, thereby completely eliminating any messages and distributing the involved computations among the ranks.

When issuing loans to firms, a bank has to make sure that the total amount of the issued loans is less than a certain percentage of its total equity. That is, when processing $m^{th}$ loan in period $t$, $L(t-1) + \sum_i^m \Delta L_i < \eta E(t)$, where $L(t-1)$ is the loans granted until the end of period $t-1$, $\Delta L_i$ is the $i^{th}$ loan granted by the bank in period $t$, $E(t)$ is the equity of the bank, and $\eta$ is the maximum allowable leverage for the bank. To strictly follow this condition at each local branch in a scalable manner, either a pre-estimated upper limit of the loans issuable at each local branch should be set or a relaxed condition like $L(t-1) < \eta E(t)$ should be used. We adopted the latter option since $\Delta L \ll L(t-1)$ is always valid. Both the options allow the local branches to issue loans without any need for contacting branches in other ranks, thereby eradicating any communications.

At the end of each period, local branches send a sum of their savings, loans, etc., to the main bank located in the master rank with a single `MPI_Gather()`. This drastic reduction in the number of messages and balanced distribution of computational workload make the introduction of local branches scalable.

## 3.3  Scalable Solution for Interactions over Dense Random Graphs

Though, at a first glance, it seems impossible to implement a scalable parallel extension (see Sect. 2.2), there are simple real-world solutions for goods and labor markets: sales outlets and recruitment agencies. We visit sales outlets, like supermarkets, instead of directly buying from producers. Scalable and least

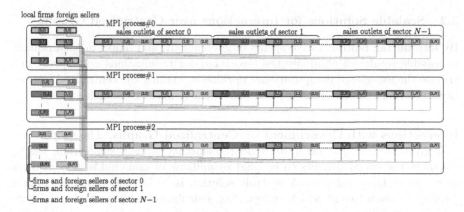

**Fig. 4.** Desired data layout of sales-outlets, and communications pattern.

compromised solutions for the labor and goods market can be implemented by mimicking these real-world solutions.

**Goods Market.** We introduced *sales-outlets*, each of which sells products of a seller (i.e., a firm or a foreign seller). Each seller has one sales-outlet in each rank. The sellers communicate the total amount to be sold to all their sales-outlets using collective communications. The sales-outlet of the seller $i$ in the rank $r$ sets its amount to be sold based on the ratio $s_i^r(t-1)/\sum_{r=0}^M s_i^r(t-1)$, where $s_i^r(t-1)$ is the amount it sold in last period, and $M$ is the total number of ranks. This requires an additional communication to gather $\sum_r s_i^r(t-1))$ to each sales-outlet. Once selling is complete, total sales and the demand of each seller are found by summing the corresponding data of its sales-outlets to master rank with MPI_Reduce(). Finally, the total sales and demands are scattered from the master rank to the corresponding parent sellers. This simple solution drastically reduces the number of MPI communications and completely eliminates any random communications, without compromising the buyer-seller interactions (i.e., buyers can use any random process within their respective rank).

While the introduction of sales-outlets solves the complicated communication problem, further planning is necessary to efficiently exchange the data among sellers and their outlets located in each rank. Figure 4 shows the desired layouts of the data of sellers and sales-outlets. The leftmost pairs of boxes indicate the sellers in each rank. The right-hand-side arrays of boxes indicate the corresponding sales-outlets. Using common MPI collective functions is not the most efficient since it involves at least $2I$ messages, where $I$ is the number of industrial sectors; $I$ calls to MPI_Allgatherv() to distribute products, and MPI_Scatterv() to collect sales information (e.g., sales and demand). This is a significant overhead [9] in simulating a country like Japan, which has 108 industrial sectors.

Some of communication overheads associated with $I$ independent collective messages can be eliminated by combining [9] all the $I$ using MPI_Alltoallw().

This more general collective function generalizes `MPI_Allgatherv()` by allowing to separately specify counts, displacements and MPI data types. Appropriately defining the MPI data types and displacements at the send and receive ends, `MPI_Alltoallw()` can be used to mimic `Allgatherw()` and `Scatterw()` operations and fetch and deliver the data in the desired format shown in Fig. 4.

In the rest of the paper, this implementation of the goods market is referred to as `buy()`, and when referring to a specific consumer it is prefixed with the name of the consumer (e.g. households' `buy()`).

**Labor Market.** As mentioned in Sect. 2.4, the labor market also poses a complicated communication problem similar to that of the goods-market. Though our partitioning scheme minimizes the number of labors to be imported/exported from/to neighboring ranks, the firms have to meet the required additional labor demands due to growth. We are exploring the advantage of mimicking recruitment agencies as a scalable solution. The recruitment agent in each rank keeps a record of which worker works where, and negotiates with the recruitment agents in neighboring ranks to allow excess workers of one rank to work in another.

### 3.4 Communication Hiding

In order to attain higher parallel scalability, almost all the communications are overlapped with computations. Due to the space limitations, which communication is hidden behind which computations is not presented. Even though the blocking MPI functions are mentioned in the above explanations, we use corresponding non-blocking MPI functions (e.g., `MPI_Ibcast()`, `MPI_Ialltoallw()`, etc.). Whenever possible, we strive to post the non-blocking messages as soon as required data are available and finalize the receive right before the data are needed at the receiving ranks. Though not explicitly measured, this communication hiding must have made significant contributions to the scalability.

### 3.5 Serial Performance Enhancements

Several attempts were made to reduce the runtime of extensively called functions and thereby reduce the load imbalance among the ranks. As shown in the next section, the `buy()` function consumes about 95% of the total runtime. The runtime of `buy()` is significantly reduced by implementing a better performing function to draw random samples, and by using a data-oriented approach.

**Draw from Random Distributions.** In most ABMs, consumer agents search for the best bargain. The consumers draw random samples from a distribution of sales-outlets of the industry from which they want to buy. The distributions are computed such that the outlets of the firms charging lower prices and/or producing larger quantities have a higher likelihood of being visited by customers.

To randomly select an outlet to visit, an agent generates a uniform random number $r$ in the range $0 < r \leq \sum_1^n p_j$, and selects the corresponding outlet $O_i$ from the distribution such that $\sum_1^{i-1} p_j < r \leq \sum_1^i p_j$, where $n$ is the total number of active outlets and $p_j$ is the likelihood of buying from active outlet $O_j$. Inclusion of the sold-out outlets in the distribution makes the consumer agents visit those pointlessly, thereby wasting a large number of CPU cycles. To eliminate this large wastage of CPU cycles, sales-outlets are removed from the distribution as soon as they are sold out. We refer this implementation of drawing from distribution as the *primitive* draw_dist() function.

The primitive draw_dist() is inefficient since deleting elements from the large arrays holding the distributions, is time-consuming. The efficiency of draw_dist() can be improved by disabling the sold-out outlets, instead of deleting. The sold-out $O_k$ can be efficiently disabled by updating the distribution with $P_j = P_j - p_k$ for $k \leq j < N_s$, where $N_s$ is the total number of sales-outlets in the industrial sector $s$, and $P_j = \sum_1^j p_l$. To skip this sold-out $O_k$, when drawing a random sales outlet from this new distribution, the smallest $j$ which satisfies $r \leq P_j$ is chosen, for a given uniform random number $r$. This *improved* draw_dist() not only eliminates deletion from large arrays, but also eliminates a large number of conditional branching (i.e., if conditions). As shown in the next section, this improved draw_dist() significantly reduces the computation time.

**Data-Oriented Design.** In the buy(), millions of consumers visit hundreds of thousands of sellers from several tens of industries, one industry at a time. Obviously, this is highly memory-bound, and the size of the consumer objects and memory access patterns significantly influence the computational performance. In our C++ implementation, a household agent consist of $12 + 2 \times I$ double-precision variables, where $I$ is the number of industries. These bloated objects makes buy() of households an excessively memory-bound computation. Further, to provide fair buying opportunities, consumer agents are made to visit outlets in random sequences. Random access of a large array with bloated objects obviously has poor cache performance, leading to further performance degeneration.

In order to reduce the memory consumption and improve the cache performance of buy(), the amount of memory involved in buy() is drastically reduced by iterating though an array $V$, into which only 6 buy() related variables are copied from the household array. Further, the cache performance is significantly improved by randomizing $V$, instead of randomly accessing its components. Though lightweight, element-wise randomization of $V$ is a memory-intensive task; hence we opt for block-wise shuffling. Since the size of $V$ is large, block-wise shuffling is sufficient to produce fair opportunities in buy(). For efficient block-wise shuffling, a modified version of the Fisher-Yates algorithm [10] is used.

## 4   Computational Performance

Based on the agent-based economic model by Sebastian et al. [7], we developed a distributed parallel code using C++11 and MPI-3.2. The code is designed according to the domain decomposition and other techniques presented in the previous section. This section accesses the effectiveness of the improvements discussed in Sect. 3.5, and the strong scalability.

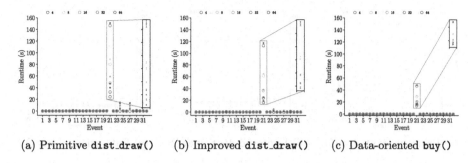

(a) Primitive dist_draw()     (b) Improved dist_draw()     (c) Data-oriented buy()

**Fig. 5.** Mean runtimes (of 20 runs) of the 32 events. Zoomed views show the standard deviations.

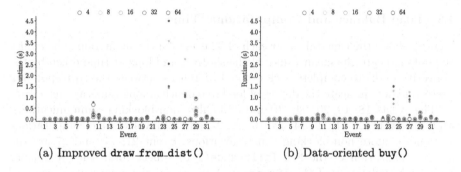

(a) Improved draw_from_dist()     (b) Data-oriented buy()

**Fig. 6.** Zoomed views of Figs. 5b and c

## 4.1  Problem Settings

All the simulations were conducted with a 1:1 scale data set pertaining to the Austrian economy: altogether 10 million agents with 62 industries, 634,019 firms, 98,270 foreign sellers, 158,505 foreign buyers, 4,130,385 inactive households, 4,267,202 workers, 634,020 investors, one bank, central government, and central bank. The population is set to increase at the rate of 0.25% per time period.

Each simulation is conducted for 20 time periods with 4, 8, 16, 32, and 64 MPI processes in the ReedBush supercomputer of the Univ. of Tokyo. Each computing node consists of Intel Xeon E5-2695 v4 (2.1 GHz with 18 cores) × 2 socket, and 256 GB memory with 153.6 GB/s bandwidth.

## 4.2  Significance of Serial Performance Enhancements

Three sets of simulations were conducted to quantitatively estimate the contributions from the serial performance enhancements presented in Sect. 3.5: primitive draw_dist(), improved draw_dist(), and data-oriented buy(). The main loop of the code comprises 32 events which include the basic events in the agent-based model, and the posting and finalizing of non-blocking messages.

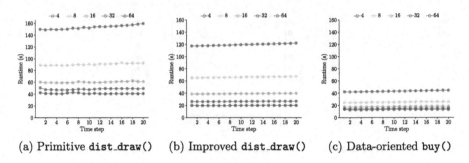

(a) Primitive dist_draw()    (b) Improved dist_draw()    (c) Data-oriented buy()

**Fig. 7.** Runtime histories for the 20 simulated time steps.

## 4.3 Load Balance and Computational Time

Figure 5 shows the time taken for each of 32 events of the main loop. The circles and dots indicate the mean values of the shortest and longest time taken by the respective ranks to complete each event. Differences between the corresponding circles and dots indicate the degree of load imbalance among ranks at each event.

The events 18, 19, 20, 24, 30, and 32 finalize non-blocking communications, while the events 2, 11, 12, 14, 15, 16, 22, 27, and 29 involve some computations and one or more non-blocking communications. Event 21, 24, and 27 are the buy(), finalization of the MPI_Iallreduce() to collect total sales to master rank, and finalization of the IScatterw() to inform the respective parents of the sales-outlets of those sales data (see Sect. 3.3); events 22 and 25 post the corresponding messages.

Comparison of event 21 of Figs. 5a and b indicates that the primitive dist_draw() introduces a significant load imbalance, and its adverse effects are visible in events 24 and 27, which finalize the non-blocking messages related to buy(). Figure 5b shows that disabling of sold-out outlets, instead of deleting them from the distribution (see Sect. 3.5), has not only eliminated a significant amount of load imbalances, but has also reduced the runtime of buy() by almost 25%. Further, it drastically reduces the imbalances in events 24 and 27.

Comparison of Figs. 5b and c shows that the lightweight data structure and cache-friendly data access introduced by the data-oriented improvements (see Sect. 3.5) reduce the runtime by nearly 50%. The zoomed views of Fig. 6, shown in Figs. 5b and c, indicate that the data-oriented improvements of buy() have further reduced the imbalances of the events 10, 24, 27, and 29, which correspond to finalization of non-blocking messages. Event 29 updates the parent firms according to the sales information delivered by event 27; hence the reduction in event 29's runtime due to improvements in buy(). Since the main loop does not have any event that synchronizes the ranks, the load imbalances in buy() are transferred to the next time step, thereby introducing load imbalance to the events 10 and 11.

Event 16 and 19 correspond to the posting and finalizing Iallgatherw() with which sellers inform respective sales-outlets of the amount to be sold

**Table 1.** Average runtime and strong scalability with improved `dist_draw()` and data-oriented `buy()`. "Dist. updated at #" indicates that disabling is called when # outlets are sold-out. "#% growth rate" is population growth per period.

| Number of MPI ranks | Dist. updated at 1 + 0.25% growth rate | | Dist. updated at 1 + 0.% growth rate | | Dist. updated at 50 + 0.25% growth rate | |
|---|---|---|---|---|---|---|
| | Runtime (s) | Strong scalability | Runtime (s) | Strong scalability | Runtime (s) | Strong scalability |
| 4 | 50.10 | | 50.0 | | 44.52 | |
| 8 | 31.97 | 78.4% | 34.16 | 73.2% | 25.97 | 85.7% |
| 16 | 22.79 | 70.1% | 22.87 | 74.7% | 18.35 | 70.7% |
| 32 | 19.12 | 59.6% | 19.19 | 59.6% | 15.23 | 60.2% |
| 64 | 16.79 | 56.9% | 16.74 | 57.3% | 13.26 | 57.4% |

(see Sect. 3.3). This is not only the most complex communication involved, but also the message carrying the largest data volume. Figure 5b and c clearly show that this, the most complex message, takes significantly less time and introduces no load imbalances, at least in relation to the imbalances introduced by `buy()`.

Figure 7 compares the runtime of the three cases considered above, for all the 20 time steps simulated. Figure 7a shows that not only is there a notable sudden variation and but also a notable increase in runtime with the time steps. Both these negative effects are induced by load imbalance in `buy()`. Though both of these effects diminish, the increase in runtime with the number of iterations is still present in Figs. 7b and c, due to the transfer of the load imbalance of `buy()` mentioned above.

## 4.4   Scalability

Table 1 shows the runtime and strong scalability[3] of the version with the improved `dist_draw()` and data-oriented `buy()` (Fig. 7c), under three different settings. In general, each setting produces a reasonable scalability up to 16 MPI ranks. As discussed, the main reason for lower scalability is the load imbalance in `buy()`. The three settings considered investigate other possible factors affecting runtime and scalability. A comparison of the $2^{nd}$ and $3^{rd}$ columns with the $4^{th}$ and $5^{th}$ shows that a gradual increase in the number of agents has a negligible impact on the runtime and scalability. On the other hand, a comparison of the $4^{th}$ and $5^{th}$ columns with the $6^{th}$ and $7^{th}$ shows that the time taken to disable the sold-out outlets by updating the cumulative probability distribution is considerable and heterogeneous among MPI ranks. Reducing the frequency of updating the probability distribution lowers total runtime by around 20%, although the scalability improvements are limited to a smaller number of ranks.

---

[3] Strong scalability, a measure of how efficiently computational resources are utilized, is defined as $(T_m/T_n)/(n/m)$, where $T_k$ is the runtime with $k$ MPI ranks and $n \geq 2m$.

It is possible to further improve the scalability by eliminating the remaining load imbalances (see Fig. 6b) induced by buy(). When partitioning, we set the nodal weights approximately according to the number of floating point operations in the rules of each type of agent. This approximate estimation does not take details, like scale of production, number of workers, etc., into account. The load imbalance in buy() can be reduced by setting the nodal weights according to the measured runtime of each agent, thereby improving the parallel scalability.

## 5    Concluding Remarks

A DMP implementation of an agent-based economic model capable of simulating medium-sized economies is presented. The major obstacle of distributing a balanced workload among MPI ranks, and facilitating all the agents' interactions at a minimum communication cost is addressed by partitioning a representative employer-employee interaction graph and mimicking real-life solutions. It is demonstrated that the runtime can be significantly reduced using a data-oriented design, and less memory-intensive, and cache-friendly algorithms. While the current implementation can simulate tens of millions of agents, larger simulations are possible through a better distribution of workload among MPI processes.

**Acknowledgments.** This work was supported by JSPS kakenhi grant 18H01675. Parts of the results are obtained using K computer at the RIKEN Center for Computational Science, and the Reedbush supercomputer at the Univ. of Tokyo.

## References

1. Haldane, A.G., Turrell, A.E.: An interdisciplinary model for macroeconomics. Oxford Rev. Econ. Pol. **34**(1–2), 219–251 (2018)
2. Assenza, T., Delli Gatti, T., Grazzini, J.: Emergentdynamics of a macroeconomic agent-based model with capital and credit. J. Econ. Dyn. Control **50**, 5–28 (2015)
3. Klimek, P., Poledna, S., Farmer, J.D., Thurner, S.: To bail-out or to bail-in? Answers from an agent-based model. J. Econ. Dyn. Control **50**, 144–154 (2015)
4. Deissenberg, C., van der Hoog, S., Dawid, H.: EURACE: a massively parallel agent-based model of the European economy. Appl. Math. Comput. **204**, 541–552 (2008)
5. Fagiolo, G., Roventini, A.: Macroeconomic policy in DSGE and agent-based models redux: new developments and challenges ahead. J. Artif. Soc. Soc. Simul. **20**(1), 1 (2017)
6. Dawid, H., Delli Gatti, D.: Agent-based macroeconomics. In: Hommes, C., LeBaron, B. (eds.) Handbook of Computational Economics, vol. 4, pp. 63–156 (2018)
7. Poledna, S., et al.: When does a disaster become a systematic event? Estimating indirect economic losses from natural disasters. arXiv:1801.09740
8. Karypis, G., Kumar, V.: A fast and highly quality multilevel scheme for partitioning irregular graphs. J. Sci. Comput. **20**(1), 359–392 (1999)

9. Balaji, P., Chan, A., Gropp, W., Thakur, R., Lusk, E.: Non-data-communication overheads in MPI: analysis on Blue Gene/P. In: Lastovetsky, A., Kechadi, T., Dongarra, J. (eds.) EuroPVM/MPI 2008. LNCS, vol. 5205, pp. 13–22. Springer, Heidelberg (2008). https://doi.org/10.1007/978-3-540-87475-1_9

10. Dursteнfeld, R.: Algorithm 235: random permutation. Commun. ACM **7**(7), 420 (1964). https://doi.org/10.1145/364520.364540

# Augmenting Multi-agent Negotiation in Interconnected Freight Transport Using Complex Networks Analysis

Alex Becheru$^{(\boxtimes)}$ and Costin Bădică

Department of Computers and Information Technology, University of Craiova,
Craiova, Romania
becheru@gmail.com, cbadica@software.ucv.ro

**Abstract.** This paper proposes the use of computational methods of Complex Networks Analysis to augment the capabilities of broker agents involved in multi agent freight transport negotiation. We have developed an experimentation environment that enabled us to obtain compelling arguments suggesting that using our proposed approach, the broker is able to apply more effective negotiation strategies for gaining longer term benefits, than those offered by the standard Iterated Contract Net negotiation approach. The proposed negotiation strategies take effect on the entire population of biding agents and are driven by market inspired purposes like for example breaking monopolies and supporting agents with diverse transportation capabilities.

**Keywords:** Complex Networks Analysis · Automated negotiation · Multi Agent Systems · Iterated Contract Net

## 1 Introduction

This paper proposes a new computational approach to endow negotiation brokers with novel and controllable instruments aiming to better impact their business environment, by providing quantitative feedback on their decisions. In general, the majority of negotiation protocols only consider short-term advantages of participants, like for example monetary benefits, by aiming to increase the profit per negotiation of the broker and/or other participant agents. However, we consider this is a limited vision as the business world is far more complex and with many facets of the longer-term benefits one might gather by engaging in negotiations.

Let us consider a well known fact in economy: monopolies are bad for an open market since the broker and the entire market environment is at the hand of a single actor. The monopoly holder establishes the monetary value and the pace of evolution, see for example the case of Standard Oil [21]. Hence, brokers/markets need to prevent the rise of monopolies. This phenomenon can be controlled either through regulatory bodies and/or through active preemptive strategies, the latter approach being considered in this paper. Since, monopoly hindering

© Springer Nature Switzerland AG 2019
J. M. F. Rodrigues et al. (Eds.): ICCS 2019, LNCS 11537, pp. 434–448, 2019.
https://doi.org/10.1007/978-3-030-22741-8_31

strategies that involve only adjustments to the monetary value are ineffective, as a monopoly will diminish the value of goods/services in order to hinder potential competition in the short term and to gain monopoly in the medium to longer term [28], a need for a broader set of actions emerges.

Hence, our motivation is to design an enhanced negotiation protocol that enables the broker to easily employ complex dynamic negotiation strategies of domain specific (transport and logistics in our case) and social inspiration. For example, a strategy for hindering monopolies might consider favouring smaller/less important transporters over larger/more important transporters when their bids are the same or marginally different. Also, sometime might be better to encourage transporters that have multiple transport capabilities in order to reduce the dependence of rather few and highly specialised transporters.

This work heavily relies on experimental evaluation of negotiations, thus we have successfully created an agent-based simulation environment by using concepts and computational methods borrowed from the fields of Complex Networks Analysis (CNA in what follows) [32] and Multi-Agent Systems (MAS in what follows) [9]. CNA is an interdisciplinary research domain inspired by Graph Theory, Statistics and Computer Science that seeks to represent and extract knowledge from complex interconnected environments where non-trivial phenomena arise. The use of MAS is justified by the requirement for developing and analysing automated negotiation models involving self-interested agents.

The results included in this paper represent a progress of our work described in previous research publications on multi-agent systems for freight transportation brokering. The initial proposal and general architecture of an agent-based system for brokering of logistics services and of semantic modelling of freight information were introduced in [16] & [17]. The details of agents' interaction protocols were defined in [19]. As per [18], the authors were able to develop and validate multiple ontologies to be used by the transportation broker for matching the cargo transportation requirements with the appropriate transport vehicles. In [20], an automated negotiation framework based on Iterated Contract Net protocol [13] was developed and experimentally analysed. The results have shown that agents using this protocol tend to behave realistically by manifesting features typically encountered in human-conducted negotiations.

The ultimate goal of our work is to develop a MAS framework for transport and logistics services – MAFTLS in what follows. The architecture of MAFTLS includes three types of actors: cargo owners, transport providers and broker. The focus of this paper is set on the broker and its capabilities to match the appropriate transport providers and transport vehicles for solving each specific transportation request. Note that transportation contracts can only be established through the broker, by employing an appropriate negotiation protocol.

The aim of this paper is to experimentally investigate if the proposed augmented negotiation protocol – Augmented Iterated Contract Net or AICNET in what follows, is able to provide the broker with the capabilities required for the dynamic selection and application of various market strategies of social inspiration in order to obtain longer-term benefits. Our approach relies on agent-based computational modelling and simulation. The novelty of our proposal is supported by the use of CNA for capturing and extracting knowledge from the negotiation environment and its further incorporation it into negotiations through an augmentation process. This enables the broker agent using AICNET to easily adapt and control its negotiation strategy for achieving higher level goals, possibly on a longer time horizon.

## 2    Background and Related Work

### 2.1    Negotiations in Freight Transport Multi Agents Systems

As early as 1998 [7] freight transportation approaches based on cooperating agents have been proposed. The aforementioned paper describes a *Holonic* MAS approach called *TeleTruck*, where transportation agents could bid for the "whole" freight to be transported or only for parts of it. In [1], Adler et al. have focused on the efficient reallocation of network capacity over time and space without seriously violating any individual user's preferences for mode, routing, departure, and/or arrival time. Neagu et al. [25], have developed an agent based solution to support human freight transport dispatchers, which was considered business-fit and therefore it was adopted by a multi-national logistics company. A recent MAS related approach to freight transportation, presented by Mes et al. [22], tackled the problem of real-time scheduling of time-sensitive full truckloads pickup-and-delivery jobs. Among the many related papers it is also noteworthy to mention a recent work by Wang et al. [34] sharing many of our research goals, that proposes a promising new *gradient knowledge* policy for matching shippers with carriers through brokering.

Note that MAS provides the perfect setting for simulating freight transport negotiation models. We argue that such models are highly interconnected due to the continuous exchange of messages between the involved agents, with the goal of establishing transport agreements. Such contracts can be reached by letting agents use a shared language and a shared set of negotiation rules that we have addressed in paper [19].

A negotiation can be defined as a complex dynamic process by which two or more parties seek a compromise to a non-trivial negotiation subject. In the context of transportation networks, large scale automated negotiations involving a large number of participant agents, fit best our model. One of first prototype mass negotiation systems was introduced by Picard [27] and it used a multi-facet analysis mechanism.

According to [14], a negotiation brings together three elements: *negotiations protocols*, *negotiation subject* and *negotiation strategies*. This paper builds upon the Iterated Contract Net as *negotiation protocol*. It allows multiple rounds of bidding before reaching an agreement among the participant agents regarding the *negotiation subjects* represented by specific freight transportation requests. Our contribution is the introduction and experimental validation of new *negotiation strategies* of the broker agent.

An overview of existing models of automated negotiation in multi-agent systems with a special focus on complex negotiations involving non-linear utility functions has been presented by Scafes et al. [30]. However, the lack of a critical mass of research papers related to our negotiation scenario can be seen. This observation is also supported by the authors of paper [5] which have formally defined and designed a conceptual software architecture of a multi-strategy negotiation agent-based system.

## 2.2 Multi-agent Framework for Transport Logistic Services

MAFTLS is a MAS framework that captures each transportation actor as an agent: cargo owner as *aCAgent*, transport provider as *aFTPAgent*, freight broker's registry as *aFBRAgent*, and freight broker as *aFBAgent*. The focus of our research is on the freight broker agent. It has a unique intermediate position between buyer and seller of transportation services. It is a self-interested agent with the goal to sell freight transport services at the highest possible value (*oval*) to cargo owners and to buy freight transport services from transport providers at the lowest possible value (*tval*). The business model of the broker implies a positive difference between *oval* − *tval* in order to support broker's operating costs and produce its profit.

For the description of MAFTLS we are using the agents' communication diagram depicted in Fig. 1. This diagram presents the workflow of MAFTLS triggered by receiving a transport request from a cargo owner. The *aCAgent* representing the requesting cargo owner forwards the request to the broker, which queries the *aFBRAgent* to obtain a list of pre-registered vehicles suited to the transport requirements of the cargo. Both the transport request and the process of determining if a vehicle is suited for specific transport requirements are using ontological demarcated information. If the matching is successful, the broker will start a one to many negotiation process with the *aFTPAgents* owning the vehicles (right-hand negotiation). If the negotiation process is successful, a winner transporter is established together with the *tval*. Based on *tval* the broker will engage in a second (one to one) negotiation with the requesting cargo owner (left-hand negotiation). If an agreement is reached with the cargo owner, then a contract between the winning transporter and the cargo owner is set. If any of the above fail then the contract is not set and all actors involved are informed accordingly. Broker strategy is based on the *list order algorithm* that is captured as a "black-box" in Fig. 1. It enables the broker to dynamically configure and adapt its strategy by possibly introducing other criteria that govern the selection of the winner transporter. We will focus on such criteria in the following sections of the paper.

As already mentioned, we adapted the Iterated Contract Net negotiation protocol for our transport scenarios [20], while also introducing basic negotiation personalities of the participant agents [20]. Experiments conducted using Java Agent Development Framework (version 4.4) [3] were successful, with a rate of less than 7% of negotiation failure and with a rather fast agreement reaching in 5.7 negotiation rounds on average. From the business point of view, the experiments were successful too, as the broker would receive an average commission of about 10% from *oval*. Results regarding the agents' personalities have shown that the negotiation protocol has favoured transport providers asking an initial low price and which are flexible during the negotiation process (Low Price Lenient personality). Also, the distribution of the *oval* prices has a Gaussian bell-like shape, as one might expect to happen in human driven negotiations.

In this paper we build our experimental evaluation on this setting, by augmenting the right-side negotiation process with new information representing social insight obtained using computational methods of Complex Networks Analysis.

### 2.3   Complex Network Analysis

The establishment of transport contracts between *aCAgents* and *aFTPAgents* enables them to interconnect into a large social network of transport stakeholders. As new transport requests are processed by the broker, a highly interconnected network emerges. Hence, we can use some of the information related to this inter-connectivity of agents to provide additional leverage for the broker during the negotiation processes. The inter-connectivity emerging between pairs of agents is called *social behaviour*, since the behaviour on one agent can influence the other agent.

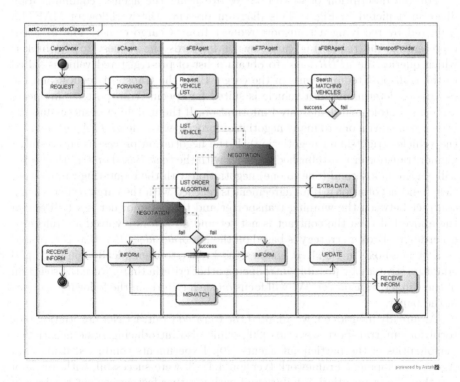

**Fig. 1.** Agents' communication diagram in MAFTLS

To the best of our knowledge we are not aware of any previous studies that augment an automated negotiation protocol with information extracted from the resulting interconnected world of the actors involved. Nevertheless, we could find some related studies, as follows. Van Doosselaere was able to infer the social rules that brought the rise of capitalism by analysing the link between commercial agreements and social processes [33]. Further more, Money has shown that social activity influences human based multilateral commercial negotiations [23]. Nan highlighted the potential impact of using social structures (networks) in conflict resolution processes through negotiation [24]. Thus, we argue there is sufficient evidence to motivate a research study into social automated negotiation, which is the purpose of this paper.

In order to better understand complex interconnected systems, the research field of *Complex Networks Analysis* has evolved from synergies of graph theory, social sciences, physics, statistics, and computer science [15]. This field of research studies overlapping non-trivial complex phenomena that can not be explained either by more "classic" approaches of lattice theory, random graphs, or statistics. CNA has matured during the last five decades and several branches have emerged, with Social Network Analysis (SNA) standing out as triggered by the rapid evolution of synergies between Computer Networks and Social Sciences [31].

SNA techniques and methods have proven their utility in many business related studies. According to [6], SNA can be used to support strategic collaborations. Greve et al. have used SNA to discover that social capital is the most important factor in productivity [11]. The authors of [8] have made a marathon in their textbook to support the utility of SNA/CNA in various areas including: game theory, auctions, bargaining, etc. In our study we are using dedicated CNA/SNA computational tools (*Gephi* [2] graphical environment and *NetworkX* [12] Python library) for analyzing social networks, as well as the Python programming language and its relevant libraries.

# 3    System Design

We have developed a simulation system using Python programming language, for experimental evaluation of our proposals, publicly available on *Github*[1] under MIT license. The following subsections introduce the conceptual model of our experiments and the supporting experimental system (ES).

## 3.1    Conceptual Model

In order to analyze social characteristics of agents involved in MAFTLS, we introduce the methods used for representing the agents' social environment. CNA defines the structures used to represent such environments as networks, while SNA often refers to them as sociograms. However, both are essentially

---

[1] https://becheru.github.io/aicnet/.

graphs (as in graph theory) augmented with specific information. Hence, we are using these terms interchangeably. Graphs have two types of constituent elements: nodes/vertices and links/edges; nodes are the portrayal of environment actors, while links are depicting relations among the actors. In our scenario we model each *aCAgent* and *aFTPAgent* by a separate node. Links are formed among nodes that have established a transport contract. Hence, we obtain a social bipartite graph of cargo owners and transport providers, where the social relationship is based on prior commercial agreements. The broker and the broker registry are not included since they are only environment artefacts, acting as match makers, with no active role in the social relationship formation and development. As a freight transport contract does not imply a leadership/direction, we use undirected graphs in our modelling. Also, as multiple contracts can be established between the same freight transporter and cargo owner, it is natural to augment our model with link weights.

As already mentioned, the broker is involved in two types of negotiation processes: one with the cargo owner (left negotiation in Fig. 1) and the other with the freight transport providers (right negotiation in Fig. 1). Since the *left negotiation* is one to one, we consider that social factors are less important here. So, for this study, we will focus only on the *right negotiation* type, where the broker is involved in one to many negotiations. Here the social characteristics of each freight transporter will be explicitly considered to differentiate between them. We are now considering possible social features for characterising freight transporters.

*Centrality measures* are indicators of the most important/influential nodes in a social graph. The *PageRank* [26] is a highly utilised centrality measure in SNA studies. It stands out as the underlying algorithm of *Google*'s search engine. The algorithm for computing this measure takes into consideration the number of links of a node (similarly with *Degree* measure) and their respective *quality*. Simply put, it is important to have as many connections as possible with nodes that are also highly connected. In our scenario, a freight transporter with high *PageRank* established many transport contracts with cargo owners that in their turn have established a significant number of contracts with other highly successful transporters. From an economic point of view, a highly rated transport provider according to *PageRank* serves the transport needs of the most active cargo owners, i.e. those cargo owners which are making most transport requests. Hence, such transport providers represent a crucial factor for the success of MAFTLS, as the revenue increases with each established transport contract.

Another measure of centrality that we are considering in our work is *Betweenness*, initially proposed by Freeman in [10] and further developed by Brandes U. in [4]. It emphasises the nodes that act as bridges between graph communities, by representing the weak ties in a social graph, as described in [29].

Note that all the metrics are able to assign a quantifiable measure to each node of the social graph, hence enabling their comparison by an appropriate ranking.

Now that we have the means to determine the social welfare of each transport provider, we must establish the goal(s) of using them. We group one or more goals into *strategies*, for a better business alignment. The broker is the agent capable of selecting and applying various strategies, according to its business and/or social interests.

It is not possible to cover all the possible strategies, even in a larger paper. Hence we will focus on few of them that we consider more relevant. We are aware that a strategy can involve more SNA metrics. However, in this work we consider only strategies that can be at least partly satisfied by a single metric. The simultaneous intertwining of multiple metrics to adhere to a specific strategy will be addressed in future works.

In the introduction we discussed that the existence of the monopoly might negatively impact a market. Hence, being able to apply a strategy to hinder/disrupt monopolies can be of great value for the broker. One such strategy may rely on increasing competitiveness by providing an advantage to transport providers that have this trait. Let us call this strategy *Competitiveness Advantage (sCA)*. As already mentioned, such transport providers can be highlighted using the *Betweenness* metrics. Hence, the transport provider with the highest *Betweenness* will be awarded contracts with *sCA*. The second strategy, let us call it *Page Rank (sPR)*, will use *PageRank* centrality measure to identify nodes that represents potential monopolies. The freight transporter with the lowest *PageRank* coefficient will be selected as winner.

Having fixed those SNA metrics that we intend to utilize as negotiation factors, we now present some details of the winner selection in negotiation protocol (we assume the reader familiarity with ICNET – a standard task allocation protocol in MAS).

The simplest solution would be to directly (i.e. in one negotiation round) select based on the metric value. However, this approach would not imply really a negotiation. Therefore, we decided to slightly update ICNET to better support the use of our metrics. This is actually a natural development of our previous results that discussed negotiation processes in MAFTLS [20], as we can use those results as comparison benchmarks.

ICNET implies that the negotiation proceeds as a series of negotiation rounds until one or more freight transport providers will agree with the broker's offer or the maximum number of iterations is reached (negotiation failure). During each round the broker proposes a monetary value in exchange for the transport service that the freight transporters can: accept – thus finishing the negotiation (negotiation success), reject but continue to the next round and reject and withdraw of the current negotiation. The accept is given when the broker bid is higher than transporter bid ceiling, while the reject is given when the broker bid is lower than the transporter bid floor. In between the bid ceiling and bid floor the transport will reject the current bid, but it will proceed to the next round of negotiation. In standard ICNET, if multiple transport providers accepted the offer in the current round then the one with lowest price would win. In our variant, AICNET, the accepting transport providers would be ranked based on a

SNA metric and the winner would be the highest/lowest ranked depending on the strategy used.

## 3.2   System Architecture

We now provide some of the details of our experimental system – ES. An experiment is organised as a series of simulation rounds. Each simulation round deals with solving of multiple transport requests. Each transport request can be satisfied by a negotiation process that usually takes several negotiation iterations, with a maximum threshold established. The negotiation protocol (ICNET or AICNET), number of experiments, rounds and iterations are defined by the user at the start of the experiment and they are fixed for the whole duration of the experiment. It is noteworthy to mention that a series of simulation rounds can be continued by another as the data is not lost between rounds, unless this is explicitly requested by the user. These experiments produce and record results and information that can be obtained by querying the ES's statistical module.

The population of participant agents is automatically generated at the start of each simulation round. Currently each cargo owner is only characterised by its ID, since cargo owners are not involved in the *right side* negotiation. The broker and the transport providers have a more complex structure since they are endowed with "personalities". The details on the personalities have been discussed in our previous paper [20]. The personalities influence the initial and reserve values (High Price or Low Price) and the level of flexibility of the agents during a negotiation (Conservative or Lenient). Transport providers change their personalities autonomously, between consecutive transport requests, as they reach specific thresholds or randomly. The broker personality is explicitly set by the user, while transporter personalities are set at the start of each experiment, by random selection from 4 available options.

As this paper is focused only on the *right-negotiation*, the simulation of a transport request issued by a cargo owner is simply reduced to: (i) randomly selecting a cargo owner as transport request issuer, and (ii) generating a random monetary values between 1 and 10000 that represents the estimated-transport-cost. In MAFTLS, the broker-estimated-cost is computed by the broker based on the details of the transport request, to determine the broker initial bid. The broker bid for each negotiation iteration is computed based on the previous bid value and the broker personality.

During each negotiation iteration, the transporters receive the transport request from the broker, and consequently they compute their own transporter-cost-estimation. In a realistic market we would expect that the transporter-cost-estimation of transporters is randomly distributed around the broker-estimated-cost. Hence, we compute the transporter-cost-estimation for each transport provider using Gaussian distribution, as follows. The *mean* is computed by adding the broker-estimated-cost with a *displacement* (a parameter set by the user for each experiment). It models the estimation error of the transporters, for the broker-estimated-cost, as in a real setting this value is private to the broker, so transporters do not know it exactly. The *standard deviation* is given as *displacement* over *deviation* (user defined scaling parameter).

After determining its private cost estimation, each transporter proceeds to compute the bid ceiling and bid floor, based on its personality. Bid floor and bid ceiling are updated in each negotiation iteration. Then the negotiation proceeds according to the negotiation protocol, until termination is reached. If the negotiation terminates with success then the social graph is updated by creating a new link between the cargo owner and the transport provider.

# 4    Experimental Results

## 4.1    Experimental Setup

ES has been developed as an object oriented program in Python which makes it very easy to use and/or adapt. The user can easily interact with ES using two classes: *Environment* and *Statistics*. The *Environment* class makes the necessary initialisation and controls the execution of the negotiation processes. The *Statistics* class is in charge of gathering data during the negotiation and presenting various statistics to the user.

ES provides the user with a wide experimentation perspective, by suitable setting of the various parameters. Because of the limited space, we focus here only on experiments with the following settings of the simulation parameters: broker personality set to Low Price Lenient, 50 transport provider agents, 1000 cargo owner agents, a threshold of maximum 12 iterations per negotiation, displacement 10 and deviation 10.

The experiments are focused on two broker strategies (*Competitiveness Advantage (sCA)* and *Page Rank (sPR)*, see Sect. 3.1), aiming to provide compelling evidence that AICNET can satisfy the goal of each strategy. Since we are restricted regarding the length of the paper, the details of using ES to run these experiments are mentioned on the GitHub page of the tool. For each parameter setting, we first run 1000 rounds of ICNET, to avoid the cold start problem for computing graph metrics, and then we run another 1000 rounds of AICNET (with ICNET in parallel), to be able to draw conclusions about the effectiveness of AICNET and compare it with ICNET. Concluding, after 2000 rounds we are able to compare the results of running ICNET (2000 rounds) with those of running 1000 rounds of ICNET followed by 1000 rounds of AICNET.

## 4.2    Results and Discussions

Table 1 presents some results obtained with ICNET and AICNET in the context of the broker strategies. Regarding negotiation related metrics, we can observe one major advantage of AICNET: the number of iterations per negotiations is slightly smaller, which translates in less waiting time for the cargo owner.

The gain per negotiation of the winning transporter is computed as the winning price over the initial estimate of the broker. Hence, from a business perspective there are no or minor differences between ICNET and AICNET, which we consider to be a major incentive for using AICNET. Some differences still arise, albeit not major, when we take into account the personalities of the winning transport providers. However, as stated before there is almost no impact on the negotiated monetary value. Our interpretation is that transporter providers that are more conservative and give higher initial prices have better chances to participate with success in AICNET than in ICNET, resulting in a slightly more inclusive negotiation protocol.

Regarding negotiation failures, both protocols act well. Although, as shown in [20], when the implementation is done in a distributed MAS framework (like Jade), some failures may arise because of communication problems between agents distributed on different machines. Based on our obtained results, we can speculate that AICNET has a slight edge over ICNET.

**Table 1.** Negotiation's results and graph related metrics.

| Metric/strategy | sCA | | sPR | |
|---|---|---|---|---|
| | ICNET | AICNET | ICNET | AICNET |
| Negotiation related metrics | | | | |
| Avg. no. of iterations per negotiation | 3.186 | 2.084 | 2.268 | 1.633 |
| Avg. gain of the transporters | 1.185 | 1.195 | 1.205 | 1.217 |
| No. wins LOW PRICE LENIENT | 1628 | 1544 | 1406 | 1292 |
| No. wins LOW PRICE CONSERVATIVE | 341 | 405 | 403 | 378 |
| No. wins HIGH PRICE LENIENT | 13 | 13 | 87 | 120 |
| No. wins HIGH PRICE CONSERVATIVE | 18 | 38 | 104 | 210 |
| No. failed negotiations | 0 | 0 | 0 | 0 |
| Graph related metrics | | | | |
| Avg weighted degree | 3.810 | 3.808 | 3.810 | 3.808 |
| Avg path length | 3.780 | 3.602 | 3.771 | 3.786 |
| Diameter | 6 | 6 | 6 | 6 |
| No. communities | 155 | 156 | 158 | 160 |

Analyzing the graph related metrics for $sPR$, they might superficially suggest that they only display minor differences, and thus supporting the idea that both graphs are in fact almost identical. However, this is not true, as clearly shown in Fig. 5. We can observe on that figure that the graph produced using AICNET is slightly more connected at the graph periphery, thus supporting the inclusion supposition. A major difference can be seen by looking at the nodes representing the transport providers (shown in purple). With ICNET, the diameter of the nodes varies more than with AICNET. Note that the diameters are proportional to the *PageRank* coefficient. Hence, we could argue that AICNET with $sPR$ has successfully hindered the rise of monopolies. This statement is further supported by the plots in Fig. 2. Both the number of wins and the *PageRank* coefficients are flatten out in AICNET. Hence *sPR* is proven to be highly effective in hindering the rise of monopolies, while having no negative impact on the monetary values and other relevant metrics.

We obtained rather similar values of graph related metrics with *sCA*. However, as shown in Fig. 4, we can observe different results than with *sPR*. The ICNET graph is by far more connected and there is clear evidence in the AICNET graph that at least two monopolies have risen, purple node at the top and the one at the right bottom part. Moreover, the distribution of the transport provider's wins and their associated *PageRank* coefficient are similar to power-law distributions, see Fig. 3. Hence, this strategy does not hinder monopolies, rather it encourages their rise. Hence, we can conclude that *sPR* is a good strategy for hindering monopolies, while on the contrary, *sCA* is facilitating them.

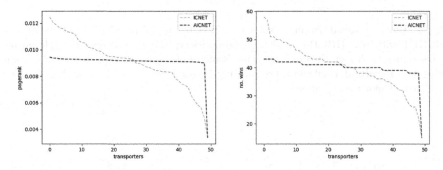

**Fig. 2.** Plots of results obtained by applying *sPR*. On the left side you can see the values of *PageRank* coefficient for each transporter, while on the right the total number of wins per each transporter is depicted.

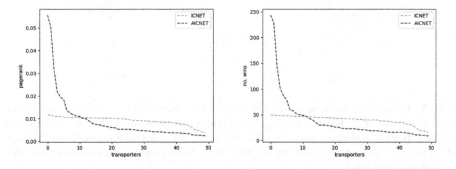

**Fig. 3.** Plots of results obtained by applying *sCA*. On the left side you can see the values of *PageRank* coefficient for each transporter, while on the right the total number of wins per each transporter is depicted.

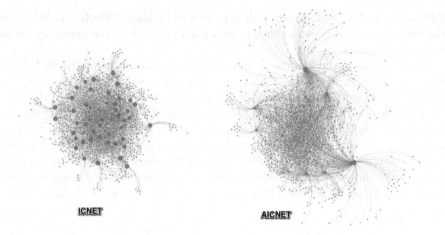

**Fig. 4.** Plots of social graphs results obtained by applying ICNET (left side) and AICNET with *sCA* (RIGHT SIDE). Green nodes represent cargo owners while purple nodes represent transport providers. The diameter of the nodes is proportional to their respective PageRank coefficient. For plotting we used *ForceAtlas 2* algorithm included in *Gephi*. (Color figure online)

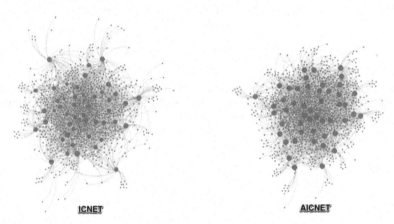

**Fig. 5.** Plots of social graphs results obtained by applying ICNET (left side) and AICNET with *sPR* (RIGHT SIDE). Green nodes represent cargo owners while purple nodes represent transport providers. The diameter of the nodes is proportional to their *PageRank* coefficient. For plotting we used *ForceAtlas 2* algorithm included in *Gephi*. (Color figure online)

## 5 Conclusions and Future Work

In this paper we proposed a new computational method based on CNA to augment the capabilities of a broker involved in multi agent freight transport negotiation. Our experiments have shown that using this approach the broker is able to apply negotiation strategies of social inspiration for gaining longer term benefits, like for example hindering monopolies and supporting agents with diverse transportation capabilities. Currently, our experimental strategies involved a single SNA metric. We plan to strengthen our results by considering strategies involving more metrics, possibly in larger scale experiments. Moreover, we plan to experimentally investigate in more detail the impact of several parameters onto the negotiation outcomes.

**Acknowledgement.** This work was supported by QFORIT programme, University of Craiova, 2017.

# References

1. Adler, J.L., Blue, V.J.: A cooperative multi-agent transportation management and route guidance system. Transp. Res. Part C: Emerg. Technol. **10**, 433–454 (2002)
2. Bastian, M., Heymann, S., Jacomy, M.: Gephi: an open source software for exploring and manipulating networks. In: Icwsm, vol. 8, pp. 361–362 (2009)
3. Bellifemine, F., Bergenti, F., Caire, G., Poggi, A.: Jade—a Java agent development framework. In: Bordini, R.H., Dastani, M., Dix, J., El Fallah Seghrouchni, A. (eds.) Multi-Agent Programming. MSASSO, vol. 15, pp. 125–147. Springer, Boston (2005). https://doi.org/10.1007/0-387-26350-0_5
4. Brandes, U.: A faster algorithm for betweenness centrality. J. Math. Sociol. **25**(2), 163–177 (2001)
5. Cao, M., Luo, X., Luo, X.R., Dai, X.: Automated negotiation for e-commerce decision making: a goal deliberated agent architecture for multi-strategy selection. Decis. Support Syst. **73**, 1–14 (2015)
6. Cross, R., Borgatti, S.P., Parker, A.: Making invisible work visible: using social network analysis to support strategic collaboration. Calif. Manag. Rev. **44**(2), 25–46 (2002)
7. Burckert, H.J., Fischer, K., Vierke, G.: Transportation scheduling with holonic MAS: the TeleTruck approach. In: proceedings of the 3rd International Conference on the Practical Application of Intelligent Agents and Multi-Agent Technology, PAAM 1998, London, England (1998)
8. Easley, D., Kleinberg, J.: Networks, Crowds, and Markets: Reasoning About a Highly Connected World. Cambridge University Press, Cambridge (2010)
9. Ferber, J., Weiss, G.: Multi-agent Systems: An Introduction to Distributed Artificial Intelligence, vol. 1. Addison-Wesley, Reading (1999)
10. Freeman, L. C.: A set of measures of centrality based on betweenness. Sociometry, 35–41, 1977
11. Greve, A., Benassi, M., Sti, A.D.: Exploring the contributions of human and social capital to productivity. Int. Rev. Sociol. **20**(1), 35–58 (2010)
12. Hagberg, A., Swart, P., S Chult, D.: Exploring network structure, dynamics, and function using NetworkX (No. LA-UR-08-05495; LA-UR-08-5495). Los Alamos National Lab. (LANL), Los Alamos, NM, USA (2008)
13. Foundation for Intelligent Physical Agents (FIPA) Iterated Contract Net protocol http://www.fipa.org/specs/fipa00030/SC00030H.pdf. Accessed 18 Jan 2019
14. Jennings, N.R., Faratin, P., Lomuscio, A.R., Parsons, S., Wooldridge, M.J., Sierra, C.: Automated negotiation: prospects, methods and challenges. Group Decis. Negot. **10**(2), 199–215 (2001)

15. Latora, V., Nicosia, V., Russo, G.: Complex Networks: Principles, Methods and Applications. Cambridge University Press, Cambridge (2017)
16. Luncean, L., Bădică, C., Bădică, A.: Agent-based system for brokering of logistics services – initial report. In: Nguyen, N.T., Attachoo, B., Trawiński, B., Somboonviwat, K. (eds.) ACIIDS 2014. LNCS (LNAI), vol. 8398, pp. 485–494. Springer, Cham (2014). https://doi.org/10.1007/978-3-319-05458-2_50
17. Luncean, L., Bădică, C.: Semantic modeling of information for freight transportation broker. In: Proceedings: 16th International Symposium on Symbolic and Numeric Algorithms for Scientific Computing (SYNASC), pp. 527–534. IEEE (2014)
18. Luncean, L., Becheru, A., Bădică, C.: Initial evaluation of an ontology for transport brokering. In: Proceedings: 19th International Conference on Computer Supported Cooperative Work in Design (CSCWD), pp. 121–126. IEEE (2015)
19. Luncean, L., Becheru, A.: Communication and interaction in a multi-agent system devised for transport brokering. In: Proceedings: 7th Balkan Conference on Informatics: Advances in ICT, pp. 51–58 (2015)
20. Luncean, L., Mocanu, A., Becheru, A. P.: Automated negotiation framework for the transport logistics service. In: Proceedings: 18th International Symposium on Symbolic and Numeric Algorithms for Scientific Computing (SYNASC), pp. 387–394. IEEE (2016)
21. McGee, J.S.: Predatory price cutting: the Standard Oil (NJ) case. J. Law Econ. **1**, 137–169 (1958)
22. Mes, M., van der Heijden, M., Schuur, P.: Interaction between intelligent agent strategies for real-time transportation planning. Central Eur. J. Oper. Res. **21**(2), 337–358 (2013)
23. Money, R.B.: International multilateral negotiations and social networks. J. Int. Bus. Stud. **29**(4), 695–710 (1998)
24. Nan, S.A.: Conflict resolution in a network society. Int. Negot. **13**(1), 111–131 (2008)
25. Neagu, N., Dorer, K., Greenwood, D., Calisti, M.: LS/ATN: reporting on a successful Agent-based solution for transport logistics optimization. In: IEEE Workshop on Distributed Intelligent Systems: Collective Intelligence and Its Applications (DIS 2006), pp. 213–218. IEEE, June 2006
26. Page, L., Brin, S., Motwani, R., Winograd, T.: The PageRank citation ranking: bringing order to the web. Stanford InfoLab (1999)
27. Picard, W.: NeSSy: enabling mass e-negotiations of complex contracts. In: Proceedings: 14th International Workshop on Database and Expert Systems Applications, pp. 829–833. IEEE (2003)
28. Posner, R.A.: Antitrust Law. University of Chicago press, Chicago (2009)
29. Sandstrom, G.M., Dunn, E.W.: Social interactions and well-being: the surprising power of weak ties. Pers. Soc. Psychol. Bull. **40**(7), 910–922 (2014)
30. Scafes, M., Badica, C.: Complex negotiations in multi-agent systems. Ann. Univ. Craiova Ser.: Autom. Comput. Electron. Mechatron. **7**(34), 53–60 (2011)
31. Scott, J.: Social Network Analysis. Sage, Thousand Oaks (2017)
32. Strogatz, S.H.: Exploring complex networks. Nature **410**(6825), 268 (2001)
33. Van Doosselaere, Q.: Commercial Agreements and Social Dynamics in Medieval Genoa. Cambridge University Press, Cambridge (2009)
34. Wang, Y., Nascimento, J.M.D., Powell, W.: Dynamic bidding for advance commitments in truckload brokerage markets. arXiv preprint arXiv:1802.08976 (2018)

# Security-Aware Distributed Job Scheduling in Cloud Computing Systems: A Game-Theoretic Cellular Automata-Based Approach

Jakub Gąsior[✉] and Franciszek Seredyński

Department of Mathematics and Natural Sciences,
Cardinal Stefan Wyszyński University, Warsaw, Poland
{j.gasior,f.seredynski}@uksw.edu.pl

**Abstract.** We consider the problem of security-aware scheduling and load balancing in Cloud Computing (CC) systems. This optimization problem we replace by a game-theoretic approach where players tend to achieve a solution by reaching a Nash equilibrium. We propose a fully distributed algorithm based on applying Iterated Spatial Prisoner's Dilemma (ISPD) game and a phenomenon of collective behavior of players participating in the game. Brokers representing users participate in the game to fulfill their own three criteria: the execution time of the submitted tasks, their execution cost and the level of provided Quality of Service (QoS). We experimentally show that in the process of the game a solution is found which provides an optimal resource utilization while users meet their applications' performance and security requirements with minimum expenditure and overhead.

**Keywords:** Collective behavior · Multi-agent systems ·
Spatial Prisoner's Dilemma Game · Second order cellular automata

## 1 Introduction

An increasing popularity of the CC paradigm [2,5] is lately accompanied by another phenomenon, namely fast developing of the Internet of Things (IoT). A large number of IoT devices will require automatic access to the resources of CC and, together with human clients, will create a huge number of potential CC users. The users sending their computational requests to CC will require fulfilling at least two criteria: predefined efficiency and predefined level of security. These criteria potentially can be fulfilled by solving in a centralized way a huge multi-criteria optimization problem with a request of full information about the system resources and users demands, what seems to be intractable for realistic problems.

Within the proposed simulated framework, we analyze a novel approach to considered multi-objective job scheduling problem aiming to optimize both aforementioned performance criteria. Firstly, we apply algorithmic game-theory to

© Springer Nature Switzerland AG 2019
J. M. F. Rodrigues et al. (Eds.): ICCS 2019, LNCS 11537, pp. 449–462, 2019.
https://doi.org/10.1007/978-3-030-22741-8_32

extend the notions of independence and selfishness of each user submitting his jobs to the cloud. We give to the users the freedom to choose the best strategy for their jobs. The proposed users' strategies and resulting job allocations are used as an input in our distributed scheduling scheme employing a non-cooperative game-theoretic approach. The whole process is realized in the two-dimensional Cellular Automata (CA) space where individual agents are mapped onto a regular square lattice. Main issues that are addressed here are: (a) incorporating the global goal of the system into the local interests of all agents participating in the scheduling game, and (b) formulation of the local interaction rules allowing to achieve those interests.

Moreover, we propose to develop a fully distributed approach based on converting an optimization problem into a game-theoretic problem where brokers representing users demands will search a solution as a Nash equilibrium. For this purpose we will use a variant of ISPD game proposed by [6] in the context of CA space. The game has many interesting features. It has built-in "soft security" mechanisms called membranes and it has potential of providing a collective behavior. Similar to [8,14], we will combine this approach with a traditional Multiobjective Genetic Algorithm (MOGA) approach in order to account for different trade-offs between QoS, computation time, and its associated costs.

The remainder of this paper is organized as follows. In Sect. 2, we describe the proposed CC system model and define the scheduling problem. Section 3 contains some background concerning CA, details of the studied ISPD game and some results of the experimental study of the game from positions of collective behavior. Section 4 presents the proposed agent-based game-theoretic scheduling scheme and its application. Section 5 demonstrates the performance metrics, the input parameters and experimental results. Finally, Sect. 6 concludes the paper.

## 2   Multi-objective Scheduling in Cloud Environment

### 2.1   Cloud Datacenter Model

The system model is an extension of the architecture introduced in [13] and further developed in our earlier paper [4]. A system consists of a set of geographically distributed Cloud nodes $M_1, M_2, ..., M_m$, which are specified by several characteristics, including their processing capacity $s_i$ in million floating point operations per second (MFLOPS), cost per hour $c_i$, memory and storage space. For the purpose of this paper, we follow the specification of Compute Optimized Virtual Machine (VM) series provided by Amazon EC2.

Users $(U_1, U_2, ..., U_n)$ submit to the system workflow application $J_k^j$ for execution. Each application is the set of $n$ tasks or jobs. Users are expected to pay appropriate fees to the Cloud provider dependent on the Service Level Agreement (SLA) requested. Job (denoted as $J_k^j$) is $j$th job produced (and owned) by user $U_k$. $J_k$ stands for the set of all jobs produced by user $U_k$, while $n_k = |J_k|$ is the number of such jobs. Each task has varied parameters defined as a tuple $<r_k^j, size_k^j, t_k^j, d_k^j>$, specifying its release dates $r_k^j \geq 0$; its size $1 \leq size_k^j \leq m_m$, that is referred to as its processor requirements or *degree of parallelism*; its work-load $t_k^j$ defined in MFLOPs and a deadline $d_k^j$.

Rigid jobs require a fixed number of processors for parallel execution: this number is not changed until the job is completed. We assume that job $J_k^j$ can only run on machine $M_i$ if $size_k^j \leq m_i$ holds, that is, we do not allow multi-site execution and co-allocation of processors from different machines. We assume a space sharing scheduling approach, therefore a parallel job $J_k^j$ is executed on exactly $size_k^j$ disjoint processors without preemptions.

In mapping jobs onto cloud resources, we have to tackle a number of security-related problems [7]. The first step is for a user to issue a Security Demand (SD) to all submitted jobs. When setting up the SD values, users should be concerned about issues related to job sensitivity, job execution environment, access control and data integration [12], etc. On the other hand, the Security Level (SL) of a machine can be attributed to the available intrusion detection mechanisms, firewalls, and anti-virus capabilities, as well as prior job execution success rates. This defense capability is evaluating the risk existing in the allocation of a submitted job to a specific machine.

Thus, a job is expected to be successfully carried out when SD and SL satisfy a security-assurance condition $(SD \leq SL)$ during the job mapping process. The SD is a real fraction in the range $[0, 1]$ with 0 representing the lowest and 1 the highest security requirement. The SL is in the same range with 0 for the most risky resource site and 1 for a risk-free or fully trusted site. Specifically, we define a *Job Failure Model* as a function of the difference between the job security demand and a resource trust (Eq. 1):

$$
P_{i,j}^{Failure} = \begin{cases} 0, & SD_j \leq SL_i, \\ 1 - exp^{-(SD_j - SL_i)}, & SD_j > SL_i. \end{cases} \tag{1}
$$

Meeting the security assurance condition $(SD_j \leq SL_i)$ for a given job-machine pair guarantees successful execution of that particular job. Such scheduling will be further called as a *Secure Job Allocation*. On the other hand, successful execution of the job assigned to machine without meeting this condition $(SD_j > SL_i)$, will be dependent on the calculated probability and further referred to as a *Risky Job Allocation*.

Our goal is to design a dynamic and fully decentralized scheduling architecture oriented to on-line CC platforms. In this paper, we make the following assumptions about the characteristics of the modeled CC system and workload scheduler:

- The model is *on-line*;
- The scheduler is *decentralized* and *clairvoyant*;
- Each resource has a different number of processing elements;
- Jobs are *parallel* and *rigid*. Job $J_k^j$ must be executed in parallel on $size_k^j$ processors of exactly one resource.
- Jobs are characterized by their release dates $(r_k^j \geq 0)$;
- *Preemption* is not allowed;

– Processors are *time-shared* between jobs. Resources are *space-shared* and *time-shared* between jobs.

Our main scheduling objective is to ensure fairness among multiple users requesting access to cloud resources by distributing user's $U_k$ jobs among the available machines. A detailed explanation of this process is provided in Sect. 4.

## 3   Collective Behavior in Iterated Spatial Prisoner's Dilemma Game

### 3.1   Cellular Automata

CA are spatially and temporally discrete computational systems originally proposed by S. Ulam and J. von Neumann in the late 1940s. Today they are a powerful tool used in computer science, mathematics and natural science to model different phenomena and develop parallel and distributed algorithms [15].

We will consider two dimensional (2D) CA which is 2D lattice consisting of $n \times n$ elementary cells. For each cell a local 2D neighborhood of a radius $r$ is defined (see, Fig. 1). At a given discrete moment of time $t$ each cell $(i, j)$ is in some state $q_{i,j}^t$, in the simplest case it is a binary number 0 or 1.

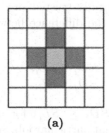          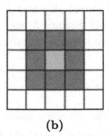

(a)                    (b)

**Fig. 1.** 2D CA of the size $5 \times 5$. Examples of a local neighborhood of the size $r = 1$ of a central cell (in light gray): (a) von Neumann neighborhood consisting of 4 neighbors, (b) Moore neighborhood consisting of 8 neighbors.

At discrete moments of time, all cells synchronously update their states according to the assigned local rules (transition functions), which depend on states of their neighborhood. A cyclic boundary condition is applied to finite size of 2D CA. If two or more different rules are assigned to CA to update cells, such CA is called non-uniform.

### 3.2   Iterated Spatial Prisoner's Dilemma Game

We consider 2D CA of the size $n \times n$. Each cell of CA has a Moore neighborhood of radius $r$, and also a rule assigned to it which depends on the state of its neighborhood.

Each cell of 2D CA will be considered as an agent (player) participating in the ISPD game [6]. Each player (a cell of CA) has two possible actions: $C$ (cooperate) and $D$ (defect). It means that at a given moment of time each cell is either in the state $C$ or the state $D$. Payoff function of the game is given in Table 1.

Table 1. Payoff function of a row player participating in ISPD game.

| Action | Cooperate (C) | Defect (D) |
|---|---|---|
| Cooperate (C) | $R = 1$ | $S = 0$ |
| Defect (D) | $T = b$ | $P = 0$ |

Each player playing a game with an opponent player in a single round (iteration) receives a payoff equal to $R$, $T$, $S$ or $P$, where $T > R > P > S$. Values of these payoffs used in this study are specified in Table 1, where $P = S = 0$, $R = 1$, and $T = b$ ($1.1 < b < 1.8$). If a player takes action $C$ and the opponent player also takes action $C$ than the player receives payoff equal to $R = 1$. However, if a player takes action $D$ and the opponent player still takes action $C$, the defecting player receives payoff equal to $T = b$. In two remaining cases, a player receives a payoff equal to 0.

In fact, each player associated with a given cell plays in a single round a game with each of his 8 neighbors and this way collects some total score. After a $q$ number of rounds (iterations of CA) each cell (agent) of CA can change its rule (strategy). We assume that considered 2D CA is non-uniform CA, and to each cell one of the following rules can be assigned: *all-C* (always cooperate), *all-D* (always defect), and *k-D* (cooperate until not more than $k$ ($0 \leq k \leq 7$) neighbors defect). The strategy *k-D* is a generalized strategy TFT, and when $k = 0$ it is exactly the strategy TFT.

A player changes its current strategy into another one comparing collected during $q$ rounds total score with scores collected by his neighbors. He selects as his new strategy the strategy of the best performing neighbor, i.e. the player whose the collected total score is the highest. This new strategy is used by a cell (player) to change its current state, and the value of the state is used in games during the next $q$ rounds.

It is worth to notice that the considered 2D CA differs from a classical CA, where rules assigned to cells do not change during evolving CA in time. A CA with a possibility of changing rules is called a second-order CA. In opposite to a classical CA, a second-order CA has potential to solve various optimization problems.

The main research issue in [6] was to study conditions of appearing of specific structures called membranes created by cells using *k-D* rules which protected cooperated players from defected players. In this study, we are interested in the collective behavior of a large team of players, in particular in conditions of emerging global cooperation in such teams.

# 4 Game-Theoretic Approach to Cloud Scheduling

This section provides a complete overview of our proposed agent-based game-theoretic distributed scheduling scheme. To develop a truly distributed multi-objective scheduling framework we propose a four-stage procedure consisting of:

- *Stage 1*: For each user $U_k$ submitting his batch of jobs $J_k$ to the cloud we assign a broker $B_k$, responsible for allocation of user's $U_k$ jobs in the system;
- *Stage 2*: A Pareto frontier is calculated for each broker $B_k$ and batch of jobs $J_k$ submitted by user $U_k$ under an assumption that all cloud resources belong exclusively to the broker $B_k$ using the MOGA approach;
- *Stage 3*: Specific job allocation solutions (*Scheduling Strategies*) from the Pareto frontier are selected according to the Pareto *Selection Policies* characterizing user's preferences (namely, *Maximum Reliability*, *Minimum Cost* and *Minimum Cost with Deadline*). These solutions will be subsequently used by a broker $B_k$ representing user's $U_k$ interests;
- *Stage 4*: Individual brokers employ previously selected *Scheduling Strategies* in the ISPD game realized in a two-dimensional CA space, trying to find a compromise global scheduling solution (Nash Equilibrium (NE) point) in their selfish attempts to obtain cloud resources.

An equilibrium is reached when none of the participating brokers is interested in changing its own task allocation strategy. This is experienced when no improvement is achieved in the individual scores and a valid global problem solution $\overline{S}$ is obtained. A more detailed explanation of this process was provided in our earlier work [3].

In this paper, we are more interested in the problem of incorporating this global goal of the system into the local interests of individual brokers. To study a possibility of emergence of a global collective behavior of players in the sense of the second class of the collective behavior classification [1,10] we will introduce a number of local mechanisms of interaction between brokers, which can be potentially spread or dismissed by the brokers during the evolution of system in the process of the iterated game and study their impact on the scheduling performance.

The first mechanism is the possibility of sharing locally profits obtained by players participating in the game. Some kind of hard local sharing was successfully used [11] in the context of LA games. Here we will be using a soft version of sharing, where a player decides to use it or not. It is assumed that each player has a tag indicating whether he wishes (on) or not (off) to share his payoff with players from the neighborhood who also wish to share their payoffs. The sharing works in such a way that if two players play a game and both wish to share their payoffs, each of them receives half of the payoff from the sum of payoffs received by both players in the game. Before starting the iterated game each player turns on its tag with a predefined probability $p_{sharing}$.

The second mechanism which will be used is a mutation of a number of the system parameters. With some predefined value of probability, a CA-based agent of the first type can change the currently assigned strategy (rule) to one of the two other strategies. Similarly, a CA-based agent of the second type can increase/decrease its probability of cooperation.

The third mechanism which can be used is a competition which is a generalization of the idea proposed in [9]. In fact, each player associated with a given cell plays in a single round a game with each of his neighbors and this way collects some total score. If the competition mechanism is in turned on, after a $q$ number of rounds (iterations) each agent compares its total payoff with total payoffs of its neighbors. If a more successful player exists in the neighborhood of a given player this player is replaced by the most successful one.

# 5   Experimental Analysis and Performance Evaluation

## 5.1   Parameter Settings

A number of experiments with the proposed system have been conducted with the following settings of parameters. A 2D array of the size of $50 \times 50$ cells (players) was used, with an initial state $C$ or $D$ (player action) set with the probability equal to 0.5. Initially, the rule $k$-$D$ was assigned (if applied) to CA cells with the probability equal to 0.7, and the remaining three rules with the probability equal to 0.1. When the competition mechanism was turned on, updating the array cells (by a winner in a local neighborhood) was conducted after $q = 1$ iterations.

## 5.2   Experiment #1. Typical Runs of ISPD Game with Variable Values of Parameter B

The purpose of this set of experiments was to observe the dynamics and conditions of appearing in the game a collective behavior of players measured by a total number of cooperating players in the game. Initially conducted experiments have shown that the ability of emergence of collective behavior depends mainly on values of parameter $b$ and only a little bit on a value of $k$. Therefore, we will show some experimental results for the same value $k = 4$ and different values of $b$.

The experiments were conducted for 2D CA of the size $n \times n = 10000$ and Fig. 2 shows results averaged over 50 runs of the game. Depending on the value of $b$ we can notice four regions of behavior resulting in different types of equilibria. When $b$ is in the range $[1.1, \ldots, 1.3]$ about 96% of players (see, Fig. 2a) decides quickly (after about 25 iterations) to cooperate (red color). Among cooperative players around 91% of players uses strategy *all-C*, (yellow color) and around 5% uses the strategy $k$-$D$, (green color). Remaining about 4% of players use strategy *all-D*.

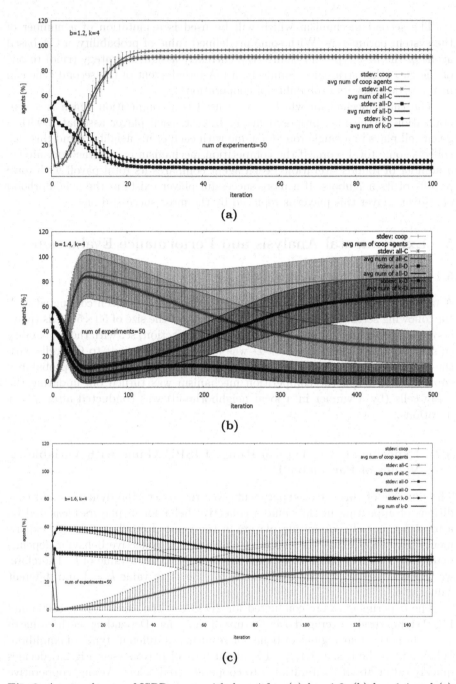

**Fig. 2.** Averaged runs of ISPD game with $k = 4$ for: (a) $b = 1.2$, (b) $b = 1.4$ and (c) $b = 1.6$. (Color figure online)

When parameter $b$ is the range $[1.4, \ldots, 1.5]$ (see, Fig. 2b) the behavior of players is more complex, and the system achieves an equilibrium after about 450 iterations. During about first 50 iterations global cooperation of the level of around 90% is reached (red color), and next it is slowly decreasing during around 200 iterations to return back to the previous level after around 450 iterations. We can observe the changing strategies structure of cooperating players.

While at the beginning of the game about 80% of coopering players use the strategy *all-C* (yellow) and about 10% of cooperating players use the strategy *k-D*, after around 450 iterations we can see a new stable proportion of cooperating players: around 65% of them use strategy *k-D* and around 25% use the strategy *all-C*. Still, a few percents of players use the strategy *all-D*. Comparing to the previous experiment we can see the increasing standard deviation of cooperation level and strategies applied by players.

When the value of $b$ is close to the value 1.6 (see, Fig. 2c) a global cooperation grows very slowly and reaches the level of around 20% (red color) after about 100 iterations, and it is supported by about 18% of players applying the strategy *all-C* (yellow color). The remaining players use either *k-D* strategy (about 41%, green color) or *all-D* (about 39%, blue color). About 2% of players applying *k-D* strategy support a global operation.

When $b \geq 1.7$ (not shown) global cooperation is not observed.

### 5.3   Experiment #2: Patterns of Agents' Strategies

For the presented above dependencies between the level of global cooperation and the parameter $b$, we can also observe accompanying spatial patterns of strategies used by players. Figure 3 (left) shows corresponding spatial pattern of players' strategies used in the game with the parameter $b = 1.1$. One can see that dominating strategy used by players is strategy *all-C* (white color) and some players use strategies *all-D* (black color) and *k-D* (tones of orange color which depends on the value of $k$). Strategies of players which use either *all-D* or *k-D* strategies create short, isolated, sometimes vertical or horizontal regular structures.

For games with values $b = 1.2$ (not shown) and $b = 1.3$ (see, Fig. 3 (right)) spatial structures of strategies are similar to the game with $b = 1.1$, but we can observe an increasing length of regular vertical and horizontal structures, which are still isolated for $b = 1.2$, but more connected with $b = 1.3$. We can see also an increasing number of strategies *k-D* which replace the strategy *all-C*.

For $b = 1.4$ (see, Fig. 4 (left)) spatial structure of strategies are created by isolated small clusters of *all-C* strategies and rarely clusters of strategies *all-D*, both surrounded by strategies *k-D*. When $b = 1.5$ (see, Fig. 4 (right)) spatial structures of strategies are created by long irregular isolated chains with two types of granularity of *k-D* strategies and *all-D* strategies. When $b = 1.6$ (see, Fig. 5 (left)) the structure of spatial strategies is similar to that observed for $b = 1.4$.

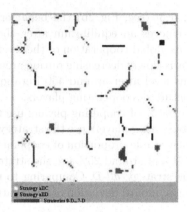

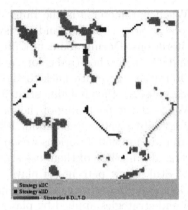

**Fig. 3.** Strategies distribution for: $b = 1.1$ (left) and $b = 1.3$ (right). (Color figure online)

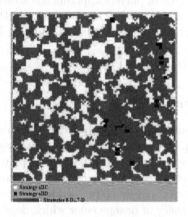

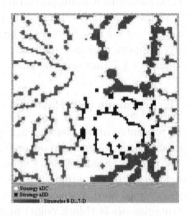

**Fig. 4.** Strategies distribution for: $b = 1.4$ (left) and $b = 1.5$ (right).

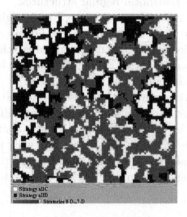

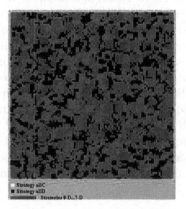

**Fig. 5.** Strategies distribution for: $b = 1.6$ (left) and $b = 1.7$ right).

However, in this case we can see appearing new small clusters of *all-D* strategies. Finally, when $b = 1.7$ (see, Fig. 5 (right)) we can notice that none of players uses *all-C* strategy but dominating strategy selected by them is the strategy $k$-$D$ with different values of $k$ and some clusters of players use the strategy *all-D*.

## 5.4 Experiment #3: A Distributed Scheduling Game

The last series of experiments incorporated our previous findings into a fully decentralized scheduler capable of determining a *Competitive Scheduling Profile* that minimizes job completion times and failure probabilities by exploiting agents' selfish needs to maximize their own payoff scores. For that purpose, 8 independent users submitted to the system a randomly generated *Bag of Tasks* containing $n_k = 1000$ job instances. Jobs were then scheduled within $m = 8$ CC nodes by independent brokering agents assigned to individual users.

Scheduling game was conducted on a rectangular CA lattice. Initial ISPD game *Actions* and *Spatial Strategies* as well as Pareto *Selection Polices* were equally and randomly distributed between the participating players. Parameters $k$ and $b$ were selected in order to maximize the emergence of global cooperation between participating players. The maximum number of game iterations has been fixed at a value of $T_{Max} = 200$ steps. 100 independent runs per configuration have been carried out to guarantee statistical significance of the results and construct an approximate Pareto frontier by gathering non-dominated solutions in all executions.

The efficiency of the analyzed job scheduling methods was measured in terms of:

- **Makespan:** the time of completion of the latest job submitted to the Cloud, defined as $C_{max} = max_k\{C^k_{max}\}$;
- **Scheduling Success Rate:** the percentage of jobs successfully completed in the Cloud;
- **Total Cost:** cost function dependent on both the requested level of SLA and a number of used CC nodes.

The simulation results are given in Fig. 6 showing the *Makespan* (Fig. 6(a)), *Scheduling Success Rate* (Fig. 6(b)) and *Total Cost* (Fig. 6(c)) during 200 iterations of ISPD game for each individual agent. These experiments aimed to analyze the number of iterations necessary to solve potential conflicts between the selfish strategies of the individual agents and converge to a global equilibrium using our proposed distributed scheduling scheme. It shows how global performance metrics improve over time due to actions taken by the brokers during our distributed resource allocation game.

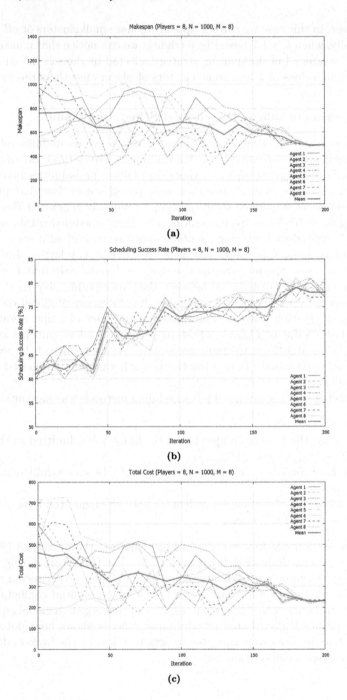

**Fig. 6.** Scheduling performance over time for a total of $n = 8000$ jobs scheduled within $m = 8$ CC nodes by 8 independent agents. (a) Makespan. (b) Scheduling success rate. (c) Total cost.

# 6    Conclusion and Future Work

We have presented a theoretical framework and an experimental software tool to study the behavior of heterogeneous multi-agent systems operating in an environment described in terms of a ISPD game. This framework was defined in order to solve global optimization tasks in a distributed way by the collective behavior of agents. The essential from this point of view was to use a concept of the second order CA and introduce some specific mechanisms of interaction between agents. A set of conducted experiments have shown that these proposed solutions are promising building blocks enabling emergence of global collective behavior in such heterogeneous multi-agent systems.

We have studied the conditions of emergence of global cooperation in large CA-based broker teams. We have shown that the phenomenon of global cooperation depends mainly on values of the parameter $b$ of payoff function reflecting a gain of a player who defects while the other players still cooperate, and to some extent on the value $k$ describing the tolerance of players for defection by their neighboring players.

We incorporated these findings into the paradigm of MOGA-based scheduling in order to use the competition among the entities involved in the scheduling process to converge towards Nash equilibrium. It allowed us to account for often contradicting interests of the clients within the CC system, without the need of any centralized control and introduced a number of desirable properties such as adaptation and self-organization. Conditions of emerging a global behavior of such systems depend on a number of parameters and these issues will be a subject of our future work.

# References

1. Brambilla, M., Ferrante, E., Birattari, M., Dorigo, M.: Swarm robotics: a review from the swarm engineering perspective. Swarm Intell. **7**(1), 1–41 (2013)
2. Buyya, R., Yeo, C.S., Venugopal, S., Broberg, J., Brandic, I.: Cloud computing and emerging IT platforms: vision, hype and reality for delivering computing as the 5th utility. Future Gener. Comput. Syst. **25**(6), 599–616 (2009)
3. Gąsior, J., Seredyński, F.: Decentralized job scheduling in the cloud based on a spatially generalized Prisoner's Dilemma game. Appl. Math. Comput. Sci. **25**(4), 737–751 (2015)
4. Gąsior, J., Seredyński, F., Tchernykh, A.: A security-driven approach to online job scheduling in IaaS cloud computing systems. In: Wyrzykowski, R., Dongarra, J., Deelman, E., Karczewski, K. (eds.) PPAM 2017. LNCS, vol. 10778, pp. 156–165. Springer, Cham (2018). https://doi.org/10.1007/978-3-319-78054-2_15
5. Jansen, W.A.: Cloud hooks: security and privacy issues in cloud computing. In: Proceedings of the 2011 44th Hawaii International Conference on System Sciences, HICSS 2011, pp. 1–10. IEEE Computer Society, Washington (2011)
6. Katsumata, Y., Ishida, Y.: On a membrane formation in a spatio-temporally generalized Prisoner's Dilemma. In: Umeo, H., Morishita, S., Nishinari, K., Komatsuzaki, T., Bandini, S. (eds.) ACRI 2008. LNCS, vol. 5191, pp. 60–66. Springer, Heidelberg (2008). https://doi.org/10.1007/978-3-540-79992-4_8

7. Kolodziej, J., et al.: Security, energy, and performance-aware resource allocation mechanisms for computational grids. Future Gener. Comput. Syst. **31**, 77–92 (2014)
8. Nesmachnow, S., Perfumo, C., Goiri, Í.: Controlling datacenter power consumption while maintaining temperature and QoS levels. In: 2014 IEEE 3rd International Conference on Cloud Networking (CloudNet), pp. 242–247, October 2014
9. Nowak, M.A., May, R.M.: Evolutionary games and spatial chaos. Nature **359**, 826 (1992)
10. Rossi, F., Bandyopadhyay, S., Wolf, M., Pavone, M.: Review of multi-agent algorithms for collective behavior: a structural taxonomy. IFAC-PapersOnLine **51**(12), 112–117 (2018). IFAC Workshop on Networked & Autonomous Air and Space Systems NAASS 2018
11. Seredynski, F.: Competitive coevolutionary multi-agent systems: the application to mapping and scheduling problems. J. Parallel Distrib. Comput. **47**(1), 39–57 (1997)
12. Song, S., Hwang, K., Kwok, Y.-K.: Risk-resilient heuristics and genetic algorithms for security-assured grid job scheduling. IEEE Trans. Comput. **55**(6), 703–719 (2006)
13. Tchernykh, A., et al.: Online bi-objective scheduling for IaaS clouds ensuring quality of service. J. Grid Comput. **14**(1), 5–22 (2016)
14. Tchernykh, A., Pecero, J.E., Barrondo, A., Schaeffer, E.: Adaptive energy efficient scheduling in Peer-to-Peer desktop grids. Future Gener. Comput. Syst. **36**, 209–220 (2014)
15. Wolfram, S.: A New Kind of Science. Wolfram Media, Champaign (2002)

# Residual Minimization for Isogeometric Analysis in Reduced and Mixed Forms

Victor M. Calo[1,2], Quanling Deng[1(✉)], Sergio Rojas[1], and Albert Romkes[3]

[1] Department of Applied Geology, Curtin University, Kent Street, Bentley,
Perth, WA 6102, Australia
{Victor.Calo,Quanling.Deng}@curtin.edu.au, srojash@gmail.com
[2] Mineral Resources, Commonwealth Scientific and Industrial Research Organisation
(CSIRO), Kensington, Perth, WA 6152, Australia
[3] Department of Mechanical Engineering, South Dakota Schoolol Mines and
Technology, 501 E. St. Joseph Street, Rapid City, SD 57701, USA
Albert.Romkes@sdsmt.edu

**Abstract.** Most variational forms of isogeometric analysis use highly-continuous basis functions for both trial and test spaces. Isogeometric analysis results in excellent discrete approximations for differential equations with regular solutions. However, we observe that high continuity for test spaces is not necessary. In this work, we present a framework which uses highly-continuous B-splines for the trial spaces and basis functions with minimal regularity and possibly lower order polynomials for the test spaces. To realize this goal, we adopt the residual minimization methodology. We pose the problem in a mixed formulation, which results in a system governing both the solution and a Riesz representation of the residual. We present various variational formulations which are variationally-stable and verify their equivalence numerically via numerical tests.

**Keywords:** Isogeometric analysis · Finite elements ·
Discontinuous Petrov-Galerkin · Mixed formulation

## 1 Introduction

Isogeometric analysis, introduced in 2005 [12,26], is a widely-used numerical method for solving partial differential equations (PDEs). This analysis unifies the finite element methods with computer-aided design tools. Within the framework of the classic Galerkin finite element methods, isogeometric analysis uses as basis functions the functions that describe the geometry in the computer-aided design (CAD) and engineering (CAE) technologies, namely, B-splines or non-uniform rational B-splines (NURBS) and their generalizations. These CAD/CAE basis functions may possess higher continuity. This smoothness improves the numerical approximations of PDEs which have highly regular solutions. Bazilevs *et al.* established the approximation, stability, and error estimates in [2].

© Springer Nature Switzerland AG 2019
J. M. F. Rodrigues et al. (Eds.): ICCS 2019, LNCS 11537, pp. 463–476, 2019.
https://doi.org/10.1007/978-3-030-22741-8_33

Cottrell *et al.* in [13] first used isogeometric analysis to study the structural vibrations and wave propagation problems. Spectral analysis shows that isogeometric elements significantly improve the accuracy of the spectral approximation when compared with classical finite elements. In [27], the authors explored the additional advantages of isogeometric analysis on the spectral approximation over finite elements. Moreover, the work [6,21,22,35] minimized the spectral approximation errors for isogeometric elements. The minimized spectral errors possess superconvergence (two extra orders) as the mesh is refined.

Galerkin isogeometric analysis uses highly-continuous B-splines or NURBS for both trial and test spaces. For PDEs with smooth solutions (high regularities), isogeometric elements based on smooth functions render solutions which have a better physical interpretation. For example, for a PDE with a solution in $H^2$, both the exact solution field and exact flux field are expected to be globally continuous. Isogeometric analysis with $C^1$ B-spline basis functions produces a globally continuous flux field while the classical finite element method does not. Therefore, for smooth problems, it makes sense to use highly-continuous B-splines for the trial spaces. However, it is not necessary to use highly-continuous B-splines for the test spaces. In the view of the work [11] which explores the cost of continuity, we expect an extra cost for solving the resulting system when we apply unnecessary high continuities for the basis functions in the test spaces. This extra cost per degree of freedom for highly-continuous discretizations was verified for direct [9,11,33] and iterative solvers [10]. Additionally, the work [24,25] shows that a reduction in the continuity of the trial and test spaces may lead to faster and more accurate solutions. Thus, we seek to develop an isogeometric analysis which uses minimal regularity for the test spaces. In this work, we establish a framework which uses highly-continuous B-splines for the trial spaces and basis functions with minimal regularity for the test spaces.

To realize this goal, we adopt the residual minimization methodology sharing with the Discontinous Petrov-Galerkin (DPG) method, the idea of a Riesz representation for the residual with respect to a stabilizing norm (c.f., [17,18,20,38]). The main idea of DPG is the use of discontinuous or broken basis functions for the test spaces within the Petrov-Galerkin framework. The method computes the optimal test function by using a trial-to-test operator. The goal is to stabilize the discrete formulation [5,30–32] without parameter tuning. The DPG method can be interpreted as a minimum residual method in which the residual is measured in the dual test norm [19]. By introducing an auxiliary unknown representing the Riesz representation of residual, the method is cast into a mixed problem. Effectively, we extend to highly continuous isogeometric discretizations the method described in [7] where standard $C^0$ finite element basis functions build the trial space while the test space uses broken polynomials. The proposed method has a feature which distinguishes us from DPG. We use highly-continuous B-splines for space that we seek for the solution, while DPG uses discontinuous basis functions. Consequently, DPG introduces additional unknowns that live on the mesh skeleton; see equation (10) or (39) in [19]. We do not introduce any additional unknowns. Instead, we use basis functions with minimal regularity in the sense that no inner products are introduced on the mesh skeleton (thus, the bilinear form only involves element integrations) while maintaining the inf-sup condition for the resulting variational formulation.

The rest of this paper is organized as follows. Section 2 describes the problem under consideration and introduces various variational formulations. Section 3 presents isogeometric formulations at the discrete level, while Sect. 4 shows numerical examples to demonstrate the performance of the formulations. Concluding remarks are given in Sect. 5.

## 2    Problem Statement

Let $\Omega \subset \mathbb{R}^d, d = 1, 2, 3$, be an open bounded domain with Lipschitz boundary $\partial\Omega$. We consider the advection-diffusion-reaction equation: Find $u$ such that

$$
\begin{aligned}
-\nabla \cdot (\kappa\nabla u - \beta u) + \gamma u &= f \quad \text{in} \quad \Omega, \\
u &= 0 \quad \text{on} \quad \partial\Omega,
\end{aligned}
\tag{1}
$$

where $\kappa$ is the diffusion coefficient, $\beta$ is the advective velocity vector, $\gamma \geq 0$ is the reaction coefficient, $u$ is the function to be found, and $f$ is a forcing function. This problem can be cast into an equivalent system of two first-order equations

$$
\begin{aligned}
q - \kappa\nabla u + \beta u &= 0 \quad \text{in} \quad \Omega, \\
-\nabla \cdot q + \gamma u &= f \quad \text{in} \quad \Omega, \\
u &= 0 \quad \text{on} \quad \partial\Omega,
\end{aligned}
\tag{2}
$$

where $q$ is an auxiliary variable standing for the flux.

### 2.1    General Setting

We present the general setting for Eqs. (1) and (2) as they have distinguishing features. Let $V$ and $W$ denote the trial and test space, resepectively. For (2) we let $(u, q)$ be the solution pair in its trial space $V^u \times V^q$ and let its corresponding test space be $W^u \times W^q$. We specify these spaces in each of the variational formulations. We denote by $\| \cdot \|_{0,\Omega}$ the $L^2$ norm. Let $H^1(\Omega)$ be the Hilbert space and $H_0^1(\Omega)$ be the space of all functions in $H^1(\Omega)$ vanishing at the boundary $\partial\Omega$. The $H^1$ seminorm is defined as $|w|_{1,\Omega} = \|\nabla w\|_{0,\Omega}$ and $H^1$ norm is defined as $\|w\|_{1,\Omega}^2 = \|w\|_{0,\Omega}^2 + |w|_{1,\Omega}^2$. Finally, $H(\text{div}, \Omega)$ consists of all square integrable vector-valued fields on $\Omega$ whose divergence is a function that is also square integrable. Similarly, the $H(\text{div}, \Omega)$ norm is defined as $\|p\|_{\text{div},\Omega}^2 = \|p\|_{0,\Omega}^2 + \|\nabla \cdot p\|_{0,\Omega}^2$.

Let $\mathcal{T}_h$ be a partition of $\Omega$ into non-overlapping mesh elements. For simplicity and the purpose of using B-splines in multiple dimensions, we assume the tensor-product structure. Let $K \in \mathcal{T}_h$ be a generic element and denote by $\partial K$ its boundary. Let $n$ be the outward unit normal vector. Let $(\cdot, \cdot)_S$ denote the $L^2(S)$ the inner product where $S$ is a $d$ or $d - 1$ dimensional domain.

At discrete level, we define the finite spaces, namely the finite subspaces of $V, W, V^u \times V^q$, and $W^u \times W^q$. Since isogeometric analysis adopts highly-continuous basis functions, we specify the corresponding finite spaces with both the polynomial order $p$ and the order $k$ of global continuity (that is, $C^k$). Let us denote by $V_{p,k}^h, W_{p,k}^h$ and $V_{p,k}^{u,h}, V_{p,k}^{q,h}, W_{p,k}^{u,h}, W_{p,k}^{q,h}$ the finite-dimensional subspaces of $V, W$ for (1) and $V^u, V^q, W^u, W^q$ for (2), respectively. In this work, we construct the test spaces $W_{q,l}^h$ and $W_{q,l}^{u,h}$ using basis functions with lower order polynomials as well as lower regularity.

## 2.2 Various Variational Formulations at Continuous Level

In this section, we present six variational formulations for both (1) and (2) following closely the six formulations described within the DPG framework in [19]. In particular, we add one more formulation which resembles the isogeometric collocation method.

We start with the abstract variational formulations of (1) and (2). The variational weak formulations of (1) can be written as: Find $u \in V$ such that

$$b(w, u) = \ell(w), \qquad \forall\, w \in W, \tag{3}$$

where $b(\cdot, \cdot)$ and $\ell(\cdot)$ are bilinear and linear forms, respectively. Similarly, the variational weak formulations of (2) can be written: Find $(u, \boldsymbol{q}) \in V^u \times V^q$ such that

$$b((w, \boldsymbol{p}), (u, \boldsymbol{q})) = \ell((w, \boldsymbol{p})), \qquad \forall\, (w, \boldsymbol{p}) \in W^u \times W^q, \tag{4}$$

where $b(\cdot, \cdot)$ and $\ell(\cdot)$ are bilinear and linear forms defined over two fields. Herein, we refer to $(u, \boldsymbol{q})$ as the solution (trial) pair while to $(w, \boldsymbol{p})$ as the weighting (test) function pair. Equations (1) and (3) are the primal forms of the PDE and the variational formulation while Eqs. (2) and (4) are their mixed forms. We keep this structure for the forms at discrete level in Sect. 3. All these forms and their associated spaces are to be specified in each particular formulation as follows.

1. Primal trivial formulation: Let $V = H_0^1(\Omega) \cap C^1(\Omega), W = L^2(\Omega)$. Find $u \in V$ satisfying (3) with

$$b(w, u) := b_1(w, u) = (w, -\nabla \cdot (\kappa \nabla u - \beta u) + \gamma u)_\Omega,$$
$$\ell(w) := \ell_1(w) = (w, f)_\Omega. \tag{5}$$

2. Primal classical (FEM) formulation: Let $V = H_0^1(\Omega), W = H_0^1(\Omega)$. Find $u \in V$ satisfying (3) with

$$b(w, u) := b_2(w, u) = (\nabla w, \kappa \nabla u - \beta u)_\Omega + (w, \gamma u)_\Omega,$$
$$\ell(w) := \ell_2(w) = (w, f)_\Omega. \tag{6}$$

3. Mixed trivial formulation I: Let $V^u = H_0^1(\Omega), V^q = H(\text{div}, \Omega), W^u = L^2(\Omega)$, $W^q = (L^2(\Omega))^d$. Find $(u, \boldsymbol{q}) \in V^u \times V^q$ satisfying (4) with

$$b((w, \boldsymbol{p}), (u, \boldsymbol{q})) := b_3((w, \boldsymbol{p}), (u, \boldsymbol{q}))$$
$$= (w, -\nabla \cdot \boldsymbol{q} + \gamma u)_\Omega + (\boldsymbol{p}, \boldsymbol{q} - \kappa \nabla u + \beta u)_\Omega, \tag{7}$$
$$\ell((w, \boldsymbol{p})) := \ell_3((w, \boldsymbol{p})) = (w, f)_\Omega + (\mathbf{0}, \boldsymbol{q})_\Omega = (w, f)_\Omega.$$

4. Mixed classical formulation II: Let $V^u = L^2(\Omega), V^q = H(\text{div}, \Omega), W^u = L^2(\Omega), W^q = H(\text{div}, \Omega)$. Find $(u, q) \in V^u \times V^q$ satisfying (4) with

$$b((w, p), (u, q)) := b_4((w, p), (u, q))$$
$$= (w, -\nabla \cdot q + \gamma u)_\Omega + (p, q + \beta u)_\Omega + (\nabla \cdot (\kappa p), u)_\Omega, \quad (8)$$
$$\ell((w, p)) := \ell_4((w, p)) = (w, f)_\Omega + (\mathbf{0}, q)_\Omega = (w, f)_\Omega.$$

5. Mixed classical formulation: Let $V^u = H_0^1(\Omega), V^q = (L^2(\Omega))^d, W^u = H_0^1(\Omega), W^q = (L^2(\Omega))^d$. Find $(u, q) \in V^u \times V^q$ satisfying (4) with

$$b((w, p), (u, q)) := b_5((w, p), (u, q))$$
$$= (\nabla w, q)_\Omega + (w, \gamma u)_\Omega + (p, q - \kappa \nabla u + \beta u)_\Omega, \quad (9)$$
$$\ell((w, p)) := \ell_5((w, p)) = (w, f)_\Omega + (\mathbf{0}, q)_\Omega = (w, f)_\Omega.$$

6. Mixed ultraweak formulation: Let $V^u = L^2(\Omega), V^q = (L^2(\Omega))^d, W^u = H_0^1(\Omega), W^q = H(\text{div}, \Omega)$. Find $(u, q) \in V^u \times V^q$ satisfying (4) with

$$b((w, p), (u, q)) := b_6((w, p), (u, q))$$
$$= (\nabla w, q)_\Omega + (w, \gamma u)_\Omega + (p, q + \beta u)_\Omega + (\nabla \cdot (\kappa p), u)_\Omega,$$
$$\ell((w, p)) := \ell_6((w, p)) = (w, f)_\Omega + (\mathbf{0}, q)_\Omega = (w, f)_\Omega.$$
$$(10)$$

Herein, the primal trivial formulation reduces to the isogeometric collocation method when applying constant test functions; see [37].

## 3   Various Isogeometric Formulations

In this section, we present the isogeometric formulations at the discrete level for both (3) and (4). We first specify the basis functions for all the finite element spaces associated with the mesh configuration $\mathcal{T}_h$. For this purpose, we use the Cox-de Boor recursion formula [15,34] on each dimension and then take tensor-product to obtain the necessary basis functions for multiple dimensions. The Cox-de Boor recursion formula generates $C^k$ and $p$-th order B-spline basis functions, where $p = 0, 1, 2, \cdots$, and $k = 0, 1, \cdots, p-1$. $C^k$ for $k = 0, 1, \cdots, p-1$ basis functions generate a finite-dimensional subspace of the $H^1(\Omega)$ and $H(\text{div}, \Omega)$. We also consider the polynomials which generate the broken space. These finite-dimensional subspaces consist of piecewise polynomials of order $p = 0, 1, 2, \cdots$ and of continuity order $k =, 0, 1, \cdots, p-1$.

The definition of the B-spline basis functions in one dimension is as follows. Let $X = \{x_0, x_1, \cdots, x_m\}$ be a knot vector with knots $x_j$, that is, a nondecreasing sequence of real numbers which are called knots. The $j$-th B-spline basis function of degree $p$, denoted as $\theta_p^j(x)$, is defined as [15,34]

$$\theta_0^j(x) = \begin{cases} 1, & \text{if } x_j \leq x < x_{j+1} \\ 0, & \text{otherwise} \end{cases}$$
$$\theta_p^j(x) = \frac{x - x_j}{x_{j+p} - x_j} \theta_{p-1}^j(x) + \frac{x_{j+p+1} - x}{x_{j+p+1} - x_{j+1}} \theta_{p-1}^{j+1}(x).$$
$$(11)$$

We then construct the finite-dimensional subspaces using the B-splines on uniform tensor-product meshes with non-repeating and repeating knots. For a $p$-th order B-spline, a repetition of $k = 0, 1, \cdots, p$ times of an internal node results in a function of $C^{p-1-k}$ continuity; see [12,34]. These B-spline basis functions characterize the finite-dimensional subspaces. For example,

$$
V_{p,k}^h = \begin{cases} S_k^p = \operatorname{span}\{\theta_j^p(x)\}_{j=1}^{N_x}, & \text{in 1D} \\ S_{k,k}^{p,p} = \operatorname{span}\{\theta_i^p(x)\theta_j^p(y)\}_{i,j=1}^{N_x,N_y}, & \text{in 2D} \\ S_{k,k,k}^{p,p,p} = \operatorname{span}\{\theta_i^p(x)\theta_j^p(y)\theta_l^p(z)\}_{i,j,l=1}^{N_x,N_y,N_z}, & \text{in 3D} \end{cases}
$$

$$
V_{p,k}^{q,h} = \begin{cases} S_{k-1}^{p-1}, & \text{in 1D} \\ S_{k-1,k-1}^{p-1,p-1} \times S_{k-1,k-1}^{p-1,p-1}, & \text{in 2D} \\ S_{k-1,k-1}^{p-1,p-1} \times S_{k-1,k-1}^{p-1,p-1} \times S_{k-1,k-1}^{p-1,p-1}, & \text{in 3D} \end{cases}
$$

(12)

where $p$ and $k$ specify the approximation order and continuity order in each dimension (they can be different in general), respectively. $N_x, N_y, N_z$ are the total numbers of basis functions in each dimension. Note that the space $V_{p,k}^{q,h}$ collapse to standard Raviart-Thomas mixed finite elements [36]; see [23, Section 5] or [3, Section 3] for more details on the construction of these spaces using B-splines.

We now adopt the mixed formulation for (3) at discrete level: Find $(u^h, \phi^h) \in V_{p,k}^h \times W_{q,l}^h$ such that

$$
\begin{aligned} g(w^h, \phi^h) + b(w^h, u^h) &= \ell(w^h), & \forall\, w^h \in W_{q,l}^h, \\ b(\phi^h, v^h) &= 0, & \forall\, v^h \in V_{p,k}^h, \end{aligned}
$$

(13)

where the spaces are chosen such that $\dim(W_{q,l}^h) \geq \dim(V_{p,k}^h)$. Herein, $\phi^h$ is the negative of the Riesz representation of the residual. The auxiliary bilinear form $g(\cdot, \cdot)$ is an inner product and it produces a Gramm matrix for the purpose of residual minimization. We define it generally as follows

$$
g(v, w) = \sum_{K \in \mathcal{T}_h} \tau_0(v, w)_K + \tau_1 h^{l_1}(\nabla v, \nabla w)_K + \tau_2 h^{l_2}(\Delta v, \Delta w)_K, \quad \forall\, v, w \in W_{q,l}^h,
$$

(14)

where $\tau_i, l_j \in \mathbb{R}, i = 0, 1, 2, j = 1, 2$ are free parameters. The default setting is $\tau_0 = 1, \tau_1 = 1, \tau_2 = 0, l_1 = 2, l_2 = 0$. Once one specifies an inner product $g(\cdot, \cdot)$, then under inf-sup assumption on the discrete bilinear formulation in (13), the approximate solution $u^h$ has a minimal error in the energy norm induced from $g(\cdot, \cdot)$; see, for example [18,19].

Similarly, we present the mixed formulation for (4) at discrete level: Find $(u^h, q^h) \in V_{p_1,k_1}^{u,h} \times V_{p_2,k_2}^{q,h}$ and $(\phi^h, \psi^h) \in \times W_{q_1,l_1}^{w,h} \times W_{q_2,l_2}^{q,h}$ such that

$$
\begin{aligned} g((w^h, p^h), (\phi^h, \psi^h)) + b((w^h, p^h), (u^h, q^h)) &= \ell((w^h, p^h)), \\ b((\phi^h, \psi^h), (v^h, r^h)) &= 0 \end{aligned}
$$

(15)

with $(w^h, p^h) \in W_{q_1,l_1}^{w,h} \times W_{q_2,l_2}^{q,h}, (v^h, r^h) \in V_{p_1,k_1}^{u,h} \times V_{p_2,k_2}^{q,h}$, where the spaces are chosen such that $\dim(W_{q_1,l_1}^{w,h}) \geq \dim(V_{p_1,k_1}^{u,h})$ and $\dim(W_{q_2,l_2}^{q,h}) \geq \dim(V_{p_2,k_2}^{q,h})$.

Herein, the continuity orders corresponding to solution $u$ and flux $q$ can be different. Potentially, their polynomial orders can also be different. Similarly, $g((\cdot,\cdot),(\cdot,\cdot))$ is an inner product for the purpose of residual minimization. We define the Gramm product generally as follows, for $(v,r),(w,p) \in W_{q_1,l_1}^{w,h} \times W_{q_2,l_2}^{q,h}$,

$$g((v,r),(w,p)) = \sum_{K \in \mathcal{T}_h} \tau_3 (v,w)_K + \tau_4 h^{\iota_3}(\nabla v, \nabla w)_K \tag{16}$$
$$+ \tau_5 (r,p)_K + \tau_6 h^{\iota_4}(\nabla \cdot r, \nabla \cdot p)_K,$$

where $\tau_i, \iota_j \in \mathbb{R}, i = 3,4,\cdots,6, j = 3,4$ are free parameters. The default setting is $\tau_3 = 1, \tau_4 = 1, \tau_5 = 1, \tau_6 = 1, \iota_3 = 2, \iota_4 = 2$.

Within these formulations, once we specify all free parameters, the bilinear and linear forms, and space settings, we have a different method. We present the following discrete variational formulations

1. Let $b(\cdot,\cdot) = b_1(\cdot,\cdot), \ell(\cdot) = \ell_1(\cdot)$ defined in (5). Let $V_{p,k}^h$ consist of B-spline basis functions of continuity $C^k, k \geq 1$ and $W_{q,l}^h$ consist of discontinuous basis functions. The *discrete primal trivial formulation* is: Find $(u^h, \phi^h) \in V_{p,k}^h \times W_{q,l}^h$ satisfying (13).
2. Let $b(\cdot,\cdot) = b_2(\cdot,\cdot), \ell(\cdot) = \ell_2(\cdot)$ defined in (6). Let $V_{p,k}^h$ and $W_{q,l}^h$ consist of B-spline basis functions of continuity at least $C^0$. The *discrete primal classical formulation* is: Find $(u^h, \phi^h) \in V_{p,k}^h \times W_{q,l}^h$ satisfying (13).
3. Let $b(\cdot,\cdot) = b_3(\cdot,\cdot), \ell(\cdot) = \ell_3(\cdot)$ defined in (7). Let the solution space consist of B-spline basis functions of continuity at least $C^0$ while the test space consist of discontinuous basis functions. The *discrete mixed trivial formulation* is: Find $(u^h, q^h) \in V_{p_1,k_1}^{u,h} \times V_{p_2,k_2}^{q,h}$ and $(\phi^h, \psi^h) \in \times W_{q_1,l_1}^{w,h} \times W_{q_2,l_2}^{q,h}$ satisfying (15).
4. Let $b(\cdot,\cdot) = b_4(\cdot,\cdot), \ell(\cdot) = \ell_4(\cdot)$ defined in (8). Let the solution space and test space for flux consist of B-spline basis functions of continuity at least $C^0$ while the test space for $u$ consist of discontinuous basis functions. The *discrete mixed classical formulation I* is: Find $(u^h, q^h) \in V_{p_1,k_1}^{u,h} \times V_{p_2,k_2}^{q,h}$ and $(\phi^h, \psi^h) \in \times W_{q_1,l_1}^{w,h} \times W_{q_2,l_2}^{q,h}$ satisfying (15).
5. Let $b(\cdot,\cdot) = b_5(\cdot,\cdot), \ell(\cdot) = \ell_5(\cdot)$ defined in (9). Let the solution space and test space for $u$ consist of B-spline basis functions of continuity at least $C^0$ while the test space for flux consist of discontinuous basis functions. The *discrete mixed classical formulation II* is: Find $(u^h, q^h) \in V_{p_1,k_1}^{u,h} \times V_{p_2,k_2}^{q,h}$ and $(\phi^h, \psi^h) \in \times W_{q_1,l_1}^{w,h} \times W_{q_2,l_2}^{q,h}$ satisfying (15).
6. Let $b(\cdot,\cdot) = b_6(\cdot,\cdot), \ell(\cdot) = \ell_6(\cdot)$ defined in (10). Let the solution and test spaces consist of B-spline basis functions of continuity at least $C^0$. The *discrete mixed ultraweak formulation* is: Find $(u^h, q^h) \in V_{p_1,k_1}^{u,h} \times V_{p_2,k_2}^{q,h}$ and $(\phi^h, \psi^h) \in \times W_{q_1,l_1}^{w,h} \times W_{q_2,l_2}^{q,h}$ satisfying (15).

The primal classical formulation can be reduced to the standard finite element and isogeometric element methods. If we set $k = 0, l = 0$ for $p \geq 1$, then we have $\phi^h = 0$ due to the orthogonal condition (second equation) in (13). Thus, the method reduces to finite element method. Similarly, if we set $k = l = 1, \cdots, p-1$

for $p \geq 2$, then we have $\phi^h = 0$ in (13) as well and the method reduces to isogeometric analysis. For all other discrete variational formulations, we may constrain the solution and test spaces in such a way that the variational forms we discuss render standard discretization techniques when the Riesz representation of the residual is identically zero.

For the scenarios where we use different trial and test spaces, we obtain a non-zero discrete representation of the residual $\Phi$, which we use as an error estimator to guide the refinements of the meshes accordingly. The error estimators are defined in the sense of $G$ which is a result of the bilinear form $g(\cdot, \cdot)$. We refer to the DPG work [17,18,20,38] for more details in this direction.

These discrete variational formulations lead to a linear matrix problem

$$\begin{bmatrix} G & B \\ B^T & 0 \end{bmatrix} \begin{bmatrix} \Phi \\ U \end{bmatrix} = \begin{bmatrix} L \\ \mathbf{0} \end{bmatrix}, \tag{17}$$

where $L$ is the corresponding forcing term arising from the linear form $\ell(\cdot)$, $\Phi, U$ are the solution pairs for the mixed formulations, $G$ represents the Gramm product in which we minimize the residual that arises from the bilinear form $g(\cdot, \cdot)$ and $B$ is the matrix arising from the bilinear form $b(\cdot, \cdot)$. We solve the first equation in (17) and substitute to the second equation in (17) to obtain

$$B^T G^{-1}(L - BU) = B^T G^{-1} L - B^T G^{-1} BU = \mathbf{0}. \tag{18}$$

Herein, $BU - L$ is the residual and (18) is a least-square type of problem.

At continuous level, the bilinear forms defined in the formulations (6)–(10) satisfy simultaneously the inf-sup condition, that is,

$$\inf_{u \neq 0} \sup_{w \neq 0} \frac{|b(w, u)|}{\|w\|_W \|u\|_V} \geq \alpha > 0, \tag{19}$$

where $\alpha$ is a constant; see [16,19]. The discrete inf-sup condition depends on the spaces. Using the coercivity of $b(\cdot, \cdot)$, the discrete primal classical formulation with $q = p, 0 \leq l \leq k$ (hence $V_{p,k}^h \subset W_{q,l}^h$) satisfies

$$\inf_{V_{p,k}^h \ni u^h \neq 0} \sup_{W_{q,l}^h \ni w^h \neq 0} \frac{|b(w^h, u^h)|}{\|w\|_W \|u\|_V} \geq \inf_{V_{p,k}^h \ni u^h \neq 0} \sup_{V_{p,k}^h \ni w^h \neq 0} \frac{|b(w^h, u^h)|}{\|w\|_V \|u\|_V} \geq \alpha_0, \tag{20}$$

where $\alpha_0 > 0$ is a constant. The proof of discrete inf-sup condition for other formulations is more involved and we omit it here.

For all the discrete variational formulations, we verify the following optimal error convergence rates numerically in the Sect. 4:

$$|u - u^h|_{1,\Omega} \leq C h^p, \tag{21}$$

where $C$ is a constant independent of $h$. For the approximate fluxes, we define the following errors in $L^2$ norm

$$\|q - q^h\|_{0,\Omega} = \|(\kappa \nabla u - \beta u) - q^h\|_{0,\Omega}. \tag{22}$$

The optimal convergence rate is

$$\|\boldsymbol{q} - \boldsymbol{q}^h\|_{0,\Omega} \le Ch^{p+1}, \tag{23}$$

where $C$ is a constant independent of $h$.

## 4  Numerical Experiments

The main result of these numerical tests is that the various formulations we discuss above are equivalent in the sense of resulting the same (optimal) error convergence rates. We focus on 2D and consider the problem (1) with $\kappa = 1, \gamma = 1, \boldsymbol{\beta} = (1,1)^T$ and a manufactured solution

$$u(x,y) = \sin(\pi x)\sin(\pi y)(2 - x + 3y). \tag{24}$$

$f$ is the corresponding forcing satisfying (1). The true flux $\boldsymbol{q}$ is calculated from (2). We apply all six variational formulations, namely, two primal and four mixed formulations, to solve this problem.

Both the primal and mixed trivial formulations do not involve any integration by parts in their variational formulations and their test functions are in $L^2$. The difference is that the primal formulation results from (1) while the mixed formulation results from (2). We use $C^{-1}$ basis functions for the test spaces for these formulations. Consequently, we obtain a matrix $G$ in (17) which is a block-diagonal matrix. The independence of each elemental block from the rest allows these methods to be computable efficiently (cheap elemental inversion) and thus relevant for practical purposes. All other formulations involve integration by parts to pass derivatives to the test functions, which in return results in a matrix $G$ in (17) which is not block-diagonal. This makes these formulations interesting from the theoretical point of view, but untractable in most general meshes, even though splitting schemes make these methods viable on tensor-product meshes [29]. Thus, once we show that all these formulations deliver equivalent results at the discrete level, we focus on these strong/trivial formulations as they are computationally advantageous. We chose not to compare against well established DPG technologies to brake test spaces, as the goal is simply to show the equivalence of the different variational forms rather than derive alternative computable methods. Lastly, to compare the primal trivial formulation with the first-order system least-square (FOSLS) method [4], the difference is that the primal trivial formulation does not lead to first order system and it introduces a Gramm matrix to solve for the residual errors simultaneously.

Figure 1 shows the $H^1$ semi-norm errors when using all six formulations for $p = 2, 3, 4, 5$. For the mixed formulations, we approximate the flux $\boldsymbol{q}^h$ using basis functions of the same polynomial order as for the solution $u^h$. The parameters of the bilinear form $g(\cdot, \cdot)$ are set to be the default values. The mesh configurations are $5 \times 5, 10 \times 10, 20 \times 20, 40 \times 40$. Herein, we plot the errors in natural logarithmic scale. As predicted, in all scenarios, the $H^1$ semi-norm errors converge in optimal rates, that is, order $p$ for all the variational formulations. This confirms the

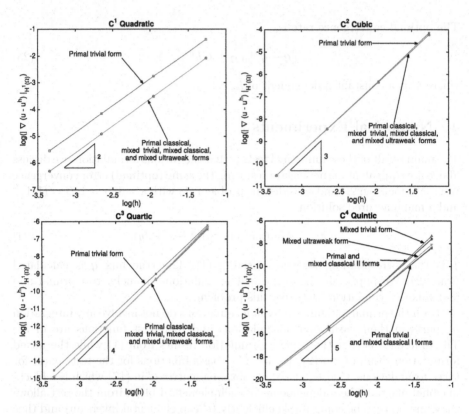

**Fig. 1.** $H^1$ semi-norm errors when using six formulations with isogeometric elements.

theoretical result (21). Therefore, all formulations are equivalent in the quality of the approximation they deliver. Interestingly, all primal trivial forms for even orders have optimal convergence rates, but the constants seem to be worse for this discretization for even orders than all the other weak forms we compared against. Nevertheless, for odd order polynomials both the rate of convergence and constant are comparable to those observed for all the other variational forms.

The mixed formulations introduce auxiliary variables for the fluxes. Thus, for the same mesh configuration, the resulting matrix systems of the mixed formulations have dimensions which are three times larger than the primal ones. However, the mixed formulations deliver more accurate fluxes approximations.

Figure 2 shows the flux errors in the $L^2$ norm when using the mixed forms for $p = 2, 3, 4, 5$. The fluxes are of optimal convergence rates $p + 1$, which confirms the error estimate (23). For $p = 2, 4$, we observe super-convergence rates approximately $p + 2$. Once again, these results numerically verify that these four formulations are equivalent. Therefore, for the following numerical results, we only show the results for the strong primal and mixed formulations. Finally, we not that the formulations with different choices of the auxiliary bilinear forms

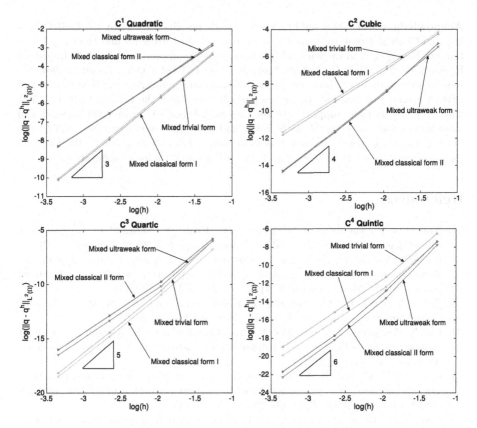

**Fig. 2.** $L^2$ errors of $q^h$ when using mixed formulations with isogeometric elements.

$g(\cdot,\cdot)$ in (14) and (16) also lead to optimal error convergence rates. Other inner products are also possible. The convergence rates in the $H^1$ semi-norm of all these scenarios are optimal.

## 5   Concluding Remarks

We introduce a residual minimization based mixed formulation for solving partial differential equations. A key feature of this method is the framework which uses highly-continuous B-splines for the trial spaces and basis functions with minimal regularity and lower order polynomials for the test spaces. The method shares with the discontinuous Petrov-Galerkin methodology the idea of stabilizing the formulation considering an adequate norm for the test space and unifies the interpretation of several methods such as the classical finite element method, isogeometric analysis, and isogeometric collocation methods. Under the standard assumption, the proposed variational formulations are stable and result in optimal approximation properties.

**Acknowledgement.** This publication was made possible in part by the CSIRO Professorial Chair in Computational Geoscience at Curtin University and the Deep Earth Imaging Enterprise Future Science Platforms of the Commonwealth Scientific Industrial Research Organisation, CSIRO, of Australia. The J. Tinsley Oden Faculty Fellowship Research Program at the Institute for Computational Engineering and Sciences (ICES) of the University of Texas at Austin has partially supported the visits of VMC to ICES. This work was partially supported by the European Union's Horizon 2020 Research and Innovation Program of the Marie Skłodowska-Curie grant agreement No. 777778, the Mega-grant of the Russian Federation Government (N 14.Y26.31.0013), the Institute for Geoscience Research (TIGeR), and the Curtin Institute for Computation.

# References

1. Auricchio, F., Da Veiga, L.B., Hughes, T., Reali, A., Sangalli, G.: Isogeometric collocation methods. Math. Models Methods Appl. Sci. **20**, 2075–2107 (2010)
2. Bazilevs, Y., Beirao da Veiga, L., Cottrell, J.A., Hughes, T.J.R., Sangalli, G.: Isogeometric analysis: approximation, stability and error estimates for h-refined meshes. Math. Models Methods Appl. Sci. **16**, 1031–1090 (2006)
3. Buffa, A., De Falco, C., Sangalli, G.: Isogeometric analysis: new stable elements for the stokes equation. Int. J. Numer. Methods Fluids (2010)
4. Cai, Z., Lazarov, R., Manteuffel, T.A., McCormick, S.F.: First-order system least squares for second-order partial differential equations: Part I. SIAM J. Numer. Anal. **31**, 1785–1799 (1994)
5. Calo, V.M., Collier, N.O., Niemi, A.H.: Analysis of the discontinuous petrov-galerkin method with optimal test functions for the reissner-mindlin plate bending model. Comput. Math. Appl. **66**, 2570–2586 (2014)
6. Calo, V.M., Deng, Q., Puzyrev, V.: Dispersion optimized quadratures for isogeometric analysis, arXiv preprint arXiv:1702.04540 (2017)
7. Calo, V.M., Romkes, A., Valseth, E.: Automatic variationally stable analysis for FE computations: an introduction, arXiv preprint arXiv:1808.01888 (2018)
8. Cohen, A., Dahmen, W., Welper, G.: Adaptivity and variational stabilization for convection-diffusion equations. ESAIM Math. Model. Numer. Anal. **46**, 1247–1273 (2012)
9. Collier, N., Dalcin, L., Calo, V.M.: On the computational efficiency of isogeometric methods for smooth elliptic problems using direct solvers. Int. J. Numer. Methods Eng. **100**, 620–632 (2014)
10. Collier, N., Dalcin, L., Pardo, D., Calo, V.M.: The cost of continuity: performance of iterative solvers on isogeometric finite elements. SIAM J. Sci. Comput. **35**, A767–A784 (2013)
11. Collier, N., Pardo, D., Dalcin, L., Paszynski, M., Calo, V.M.: The cost of continuity: a study of the performance of isogeometric finite elements using direct solvers. Comput. Methods Appl. Mech. Eng. **213**, 353–361 (2012)
12. Cottrell, J.A., Hughes, T.J.R., Bazilevs, Y.: Isogeometric Analysis: Toward Integration of CAD and FEA. Wiley, New York (2009)
13. Cottrell, J.A., Reali, A., Bazilevs, Y., Hughes, T.J.R.: Isogeometric analysis of structural vibrations. Comput. Methods Appl. Mech. Eng. **195**, 5257–5296 (2006)
14. Dahmen, W., Huang, C., Schwab, C., Welper, G.: Adaptive Petrov-Galerkin methods for first order transport equations. SIAM J. Numer. Anal. **50**, 2420–2445 (2012)
15. De Boor, C.: A Practical Guide to Splines, vol. 27. Springer, New York (1978)

16. Demkowicz, L.: Various variational formulations and closed range theorem. ICES Report **15** (2015)
17. Demkowicz, L., Gopalakrishnan, J.: A class of discontinuous Petrov-Galerkin methods. Part I: the transport equation. Comput. Methods Appl. Mech. Eng. **199**, 1558–1572 (2010)
18. Demkowicz, L., Gopalakrishnan, J.: A class of discontinuous Petrov-Galerkin methods. II. Optimal test functions. Numer. Methods Partial Differ. Eqn. **27**, 70–105 (2011)
19. Demkowicz, L., Gopalakrishnan, J.: Discontinuous Petrov-Galerkin (DPG) method, Encyclopedia of Computational Mechanics Second Edition, pp. 1–15 (2017)
20. Demkowicz, L., Gopalakrishnan, J., Niemi, A.H.: A class of discontinuous Petrov-Galerkin methods. Part III: adaptivity. Appl. Numer. Math. **62**, 396–427 (2012)
21. Deng, Q., Bartoň, M., Puzyrev, V., Calo, V.: Dispersion-minimizing quadrature rules for C1 quadratic isogeometric analysis. Comput. Methods Appl. Mech. Eng. **328**, 554–564 (2018)
22. Deng, Q., Calo, V.: Dispersion-minimized mass for isogeometric analysis. Comput. Methods Appl. Mech. Eng. **341**, 71–92 (2018)
23. Evans, J.A., Hughes, T.J.: Isogeometric divergence-conforming B-splines for the Darcy-Stokes-Brinkman equations. Mathe. Models Methods Appl. Sci. **23**, 671–741 (2013)
24. Garcia, D., Pardo, D., Dalcin, L., Calo, V.M.: Refined isogeometric analysis for a preconditioned conjugate gradient solver. Comput. Methods Appl. Mech. Eng. **335**, 490–509 (2018)
25. Garcia, D., Pardo, D., Dalcin, L., Paszyński, M., Collier, N., Calo, V.M.: The value of continuity: refined isogeometric analysis and fast direct solvers. Comput. Methods Appl. Mech. Eng. **316**, 586–605 (2017)
26. Hughes, T.J.R., Cottrell, J.A., Bazilevs, Y.: Isogeometric analysis: CAD, finite elements, NURBS, exact geometry and mesh refinement. Comput. Methods Appl. Mech. Eng. **194**, 4135–4195 (2005)
27. Hughes, T.J.R., Evans, J.A., Reali, A.: Finite element and NURBS approximations of eigenvalue, boundary-value, and initial-value problems. Comput. Methods Appl. Mech. Eng. **272**, 290–320 (2014)
28. Hundsdorfer, W., Verwer, J.: Numerical Solution of Time-Dependent Advection-Diffusion-Reaction Equations. Computational Mathematics, vol. 33. Springer, Heidelberg (2003)
29. Łoś, M., Deng, Q., Muga, I., Calo, V., Paszyński, M.: Isogeometric residual minimization method (iGRM) with direction splitting for implicit problems (2018, in preparation)
30. Niemi, A., Collier, N., Calo, V.M.: Discontinuous Petrov-Galerkin method based on the optimal test space norm for one-dimensional transport problems. Procedia Comput. Sci. (2011)
31. Niemi, A.H., Collier, N.O., Calo, V.M.: Automatically stable discontinuous Petrov-Galerkin methods for stationary transport problems: quasi-optimal test space norm. Comput. Math. Appl. **66**, 2096–2113 (2013)
32. Niemi, A.H., Collier, N.O., Calo, V.M.: Discontinuous Petrov-Galerkin method based on the optimal test space norm for steady transport problems in one space dimension. J. Comput. Sci. **4**, 157–163 (2013)
33. Pardo, D., Paszynski, M., Collier, N., Alvarez, J., Dalcin, L., Calo, V.M.: A survey on direct solvers for Galerkin methods. SeMA J. **57**, 107–134 (2012)

34. Piegl, L., Tiller, W.: The NURBS Book. Springer, Heidelberg (1997). https://doi.org/10.1007/978-3-642-97385-7
35. Puzyrev, V., Deng, Q., Calo, V.M.: Dispersion-optimized quadrature rules for isogeometric analysis: modified inner products, their dispersion properties, and optimally blended schemes. Comput. Methods Appl. Mech. Eng. **320**, 421–443 (2017)
36. Raviart, P.A., Thomas, J.M.: A mixed finite element method for 2-nd order elliptic problems. In: Galligani, I., Magenes, E. (eds.) Mathematical Aspects of Finite Element Methods. LNM, vol. 606, pp. 292–315. Springer, Heidelberg (1977). https://doi.org/10.1007/BFb0064470
37. Schillinger, D., Evans, J.A., Reali, A., Scott, M.A., Hughes, T.J.: Isogeometric collocation: cost comparison with Galerkin methods and extension to adaptive hierarchical NURBS discretizations. Comput. Methods Appl. Mech. Eng. **267**, 170–232 (2013)
38. Zitelli, J., Muga, I., Demkowicz, L., Gopalakrishnan, J., Pardo, D., Calo, V.M.: A class of discontinuous Petrov-Galerkin methods. Part IV: the optimal test norm and time-harmonic wave propagation in 1D. J. Comput. Phys. **230**, 2406–2432 (2011)

# CTCmodeler: An Agent-Based Framework to Simulate Pathogen Transmission Along an Inter-individual Contact Network in a Hospital

Audrey Duval[1]([⊠]) [iD], David Smith[1], Didier Guillemot[1],
Lulla Opatowski[1], and Laura Temime[2,3]

[1] unité B2PHI, Equipe PheMI, Université de Versailles Saint Quentin,
Institut Pasteur, Inserm, 25 rue du docteur roux, 75015 Paris, France
audrey.duval@ens.uvsq.fr
[2] Laboratoire MESuRS, Conservatoire national des Arts et Métiers,
292 rue Saint Martin, 75003 Paris, France
[3] unité PACRI, Institut Pasteur, Cnam,
25 rue du docteur roux, 75015 Paris, France

**Abstract.** Over the last decade, computational modeling has proved a useful tool to simulate the transmission dynamics of nosocomial pathogens and can be used to predict optimal control measures in healthcare settings. Nosocomial infections are a major public health issue especially since the increase of antimicrobial resistance. Here, we present CTCmodeler, a framework that incorporates an agent-based model to simulate pathogen transmission through inter-individual contact in a hospital setting. CTCmodeler uses real admission, swab and contact data to deduce its own parameters, simulate inter-individual pathogen transmission across hospital wards and produce weekly incidence estimates. Most previous hospital models have not accounted for individual heterogeneity of contact patterns. By contrast, CTCmodeler explicitly captures temporal heterogeneous individual contact dynamics by modelling close-proximity interactions over time. Here, we illustrate the use of CTCmodeler to simulate methicillin-resistant Staphylococcus aureus dissemination in a French long-term care hospital, using longitudinal data on sensor-recorded contacts and weekly swabs from the i-Bird study.

**Keywords:** Agent based model · Hospital · Multiresistant bacteria

## 1 Introduction

Healthcare-associated infections are an important public health threat in the hospital setting, in particular as a result of increasing selection for bacterial pathogens that are multi-resistant to antibiotics [20, 21]. Effective control strategies are needed to limit the global expansion of multi-resistant pathogens, yet better understanding of the transmission dynamics of nosocomial pathogens is necessary to design and implement optimal strategies.

© Springer Nature Switzerland AG 2019
J. M. F. Rodrigues et al. (Eds.): ICCS 2019, LNCS 11537, pp. 477–487, 2019.
https://doi.org/10.1007/978-3-030-22741-8_34

Mathematical models are useful tools to understand and predict the dynamics of pathogen spread in hospitals [1, 8, 10]. Such models have two main objectives: to describe the epidemiology of pathogen transmission and to assess the impact of control strategies such as hand hygiene, on endemic situations at the hospital scale [4, 12, 16, 17, 19]. The majority of published epidemiological mathematical models are so-called "compartmental" models, which group individuals into compartments according to their epidemic status (e.g. infectious, susceptible or recovered). However, because hospital settings typically host small populations with highly heterogeneous characteristics and behaviors, agent-based models (ABMs) may be more realistic, allowing for stochastic simulation of dynamics at the individual level [2, 3, 6, 9, 11, 13–15, 18].

In this paper, we present a new stochastic ABM set within a framework called CTCmodeler (Clinical Transmission and Contact modeler), which simulates the dissemination of pathogens at the hospital scale. The originality of our model is twofold. First, it explicitly accounts for networks of inter-individual within-hospital contacts. Second, it includes a module that computes model parameters from observed data. Here, we describe the ABM and illustrate the output of CTCmodeler using the example of methicillin-resistant *Staphylococcus aureus* (MRSA) spread in a French long-term stay hospital. For this illustration, the model is informed by detailed data on contacts and MRSA colonization collected over 6 months during the i-Bird study.

## 2    Methods

### 2.1    General Framework Description

CTCmodeler consists of three modules: learning, modeling and contact simulation. The learning module estimates parameters from observed data, including pathogen transmission rates, durations of colonization, probabilities of colonization at admission and daily swab frequencies. The modeling module is the ABM, detailed further below. Finally, the contact module simulates contact networks for future predictions. To simulate realistic contacts, this last module computes hourly probabilities of contact between different individual categories from observed contact data provided by the user. The whole framework was programmed in C++ with the repast HPC library 2.2.0 and is illustrated in a class diagram in Fig. 1.

### 2.2    Overview of the ABM Modeling Module

The modeling module of CTCmodeler has two options. The first option runs the ABM using user-provided data to reproduce and visualize real output. The second option simulates the transmission of a pathogen along the contact network using three sources of input described in further detail below: (i) input parameters that may be estimated from the learning module or directly provided as a csv file by the user, (ii) a contact file that may either be user-provided (based on observed data) or generated by the contact simulation module, and (iii) a patient admission file that can either by user-provided or obtained from the learning module based on real admission data. In the following, we describe the ABM following the ODD protocol of Grimm et al. [7]. R software (version 3.4.2) was used for graphics.

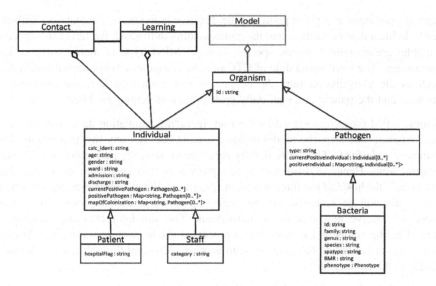

**Fig. 1.** Simplified class diagram for CTCmodeler. Patient and Staff objects inherit from Individual, which inherits all its common variables from Organism. Bacteria inherits from Pathogen, which also inherits from Organism.

**Purpose.** The CTCmodeler ABM module simulates nosocomial pathogen transmission between humans, using admission, contact and swab data from any hospital. The main goals are to trace the possible routes of transmission at the individual scale and to describe interaction dynamics between individuals. In the future, potential control measures can be implemented in the ABM.

**Entities, State Variables and Scales.** Figure 1 summarizes all agent variables in a simplified class diagram. The ABM has a main abstract class, the Organism agent, which is split into the Individual and Pathogen classes. Individual class has two children, Patient and Staff classes, which inherit some common variables from Individual, while Pathogen class is composed of a Bacteria class only. Each class has its own set of variables. The unique id variable characterizes each object during the simulation. Common variables for individuals are hospital anonymous number, age, gender, status (patient or staff), admission date, discharge date, allocated ward, a map of colonization, current positive pathogens and positive pathogens. Current positive pathogens lists all pathogens carried by the individual at a given point in time. Positive pathogens is a container with dates of swabs undergone by the individual and a list of pathogens carried at each swab date. The map of colonization contains decolonization dates for each pathogen carried by the individual. Patient class also includes a hospital flag variable, which describes the reason for hospitalization, while the staff class also includes a category variable, which describes that individual's occupation (nurse, hospital porter, physician, etc.).

Common variables for Pathogen class are the type (bacteria), positive individuals and current positive individuals. Current positive individuals lists all individuals who carry the pathogen at a given point in time. Positive individuals is a container with the last individual date of carriage and the corresponding individual. Bacteria class-specific variables are id, family, genus, species, spatype, MRB (type of multiresistance) and phenotypes. The learning module of CTCmodeler computes environmental parameters such as the transmission rate, the colonization at admission rate, the colonization duration and the patient and staff daily probabilities of getting swabbed.

**Scales.** ABM timescales are set by the user, including simulation dates and time step size. At each time step, the model builds a contact between any two individuals. The computer code for CTCmodeler is fully parallelized, with separate processes for each hospital ward. Hence, a distinct computer process is associated with each ward. For example, if the hospital has three wards, the program will have three processes in which individuals will be scattered. At the beginning of the model, corresponding ward-processes receive newly admitted individuals. The simulator also adds pathogens carried by the newly added individual's process if she/he acquires new ones. At each time step, an individual has contacts with another individual through a network space based on the contact file.

**Process Overview and Scheduling.** Simulations are run on three time levels: weeks, days and time steps.

*Weeks.* Each week, the simulation chooses which individuals to swab. Selected individuals only have a single swab day. This is based on the learning module of CTCmodeler, which computes the mean and standard deviation from the swab distribution of each swab day from provided data. On each simulation day, the model chooses a set of individuals thanks to a normal distribution deduced from mean and standard deviation of the corresponding day of the week.

*Days.* The model performs four actions on each day: add, remove, swab and decolonize individuals. Individuals are added or removed based on daily admissions and discharges listed in the admission file. Readmissions are possible during the simulation for both patients and staff. When individuals leave the simulation, they go into a transitory container to keep track of their information (e.g. colonization status). The model simulates swabs for each individual listed among a set determined weekly. Finally, colonization is cleared ("decolonization") for each pathogen carried by a given individual if the current simulation date matches the patient's decolonization date (stored in the map of colonization for all pathogens stored inside current positive pathogens).

*Time Step.* At each time step, the model simulates contacts and possible pathogen transmission events between individuals. Users choose the time step that corresponds to the contact frequency recorded in the contact file. The model builds an edge if two individuals are in contact. Colonized individuals can transmit all the pathogens they carry (i.e. all pathogens listed in current positive pathogen) with the transmission rate either (i) previously determined by the learning module of the framework or (ii) user-defined. The transmission rate is different for each couple (patient-to-patient, staff-to-patient, patient-to-staff and staff-to-staff). When a contact results in a new colonization, the duration of this colonization is randomly drawn from a Gamma distribution.

The shape and scale parameters of this distribution are deduced from the mean and standard deviation calculated from real data or user-defined in the learning module of the framework.

## 2.3 Design Concept

**Interaction.** Two types of interactions occur during the simulation.

*Inter-individual Contacts.* The ABM uses the contact file to link individuals during the simulation. At each time step, the model reads the file and adds an edge to the network.

*Inter-individual Transmission.* Colonized individuals can transmit their pathogens to a susceptible individual through the contact network. The transmission rate may either be user-defined or calculated from the learning module of CTCmodeler. In the latter case, it is computed from the observed data as,

$$T_{c1-c2} = \sum_{s=1}^{S} \frac{A_s}{\sum_{n=1}^{N} C_{c1-c2,s-n}} \tag{1}$$

Where $T_{c1-c2}$ represents the transmission rate between an individual from category 1 (e.g. patient) and another individual from category 2. $S$ is the total number of weeks, $A_s$ is the number of acquisitions observed during week $s$ and $C_{c1-c2,s-n}$ is the cumulative duration of contacts between colonized individuals from category 1 and non-colonized individuals from category 2 during weeks $s - n$. $N$ is the number of previous weeks that are taken into account.

For each interaction, the transmission rate is multiplied by the duration of the contact. If desired, the model can incorporate transmission rate saturation for longer contacts. This option (named duration threshold) is done by using $1 + \log\left(\sum_{n=1}^{N} C_{c1-c2,s-n}\right)$ as the denominator in $T_{c1-c2}$.

**Stochasticity.** Colonization duration, transmission rate, swab day of week probabilities and colonization at admission are randomly drawn from probability distributions during the simulation. Moreover, admission and discharge rates, as well as contacts, are randomly simulated when the user does not choose to read them directly from observed data.

**Observations.** The model predicts weekly incidence of pathogen acquisition, calculated as the number of new acquisitions at week $s$ divided by the number of negatively swabbed individuals during week $s - 1$. To account for possible lack of swab sensitivity, the user may also choose to define a new acquisition as a positive swab following two negative swabs. In this case, the number of new acquisitions at week $s$ is divided by the number of negatively swabbed individuals during both weeks $s - 1$ and $s - 2$.

## 2.4 Details

**Initialization.** Before the first time step, the initial number of colonized individuals is determined. When a swab file is available, the model observes the number of colonized swabbed individuals at time 0 and estimates the number of colonized non-swabbed individuals based on proportions of colonized swabbed patients and staff. When no swab data is available, the number of colonized individuals is computed from a user-defined number. Initially colonized individuals of each category are chosen at random.

**Input.** The three input files needed to run the ABM (admission, contact and parameter files) can either be user-provided or computed by the framework from the learning and contact modules. The admission file lists dates of hospital arrival and departure for all individuals included in a simulation. When not user-provided, admission files are randomly simulated based on daily admission and discharge rates for each category of individuals. The contact file lists all unsymmetrized contacts that occur between patients and staff over time during simulations. When not user-provided, contact files are randomly simulated using contact probabilities between individual categories. These contact probabilities may change depending on the hour of the day and on the type of day (weekday vs. weekend).The parameter file gathers all parameter values that the ABM needs. When not user-provided, these parameter values are computed from observed contact, admission and swab data.

## 3   Results

### 3.1   Data

Data from the i-Bird study of nosocomial pathogen transmission were used in this illustrative example of CTCmodeler. This study took place at Berck-sur-Mer in a French long-term care facility (LTCF) from the beginning of May to the end of October 2009, with the first two months as a pilot phase.

During July to October, 95% of patients and hospital staff carried an RFID sensor. Sensors recorded all close proximity (<1.5 m) interactions occurring in the 5 wards of the hospital every 30 s over this 4-month period. The 5 wards were specialized in neurology rehabilitation, geriatric rehabilitation, nutrition and post-operation rehabilitation. The staff population was heterogeneous and included nurses, auxiliary nurses, hospital porters, physicians, administration and animation staff, etc. This diversity entailed idiosyncratic contact dynamics, with many patient-to-patient, patient-to-physician and patient-to-hospital porters' contacts, as described in an earlier paper [5].

Moreover, over the entire study period, dedicated nurses swabbed patients and hospital staff weekly for MRSA colonization.

For this illustration, we used the i-Bird data to simulate MRSA spread in the hospital from July 6, 2009 to the September 28, 2009. During this period, sensors recorded 1,147,005 contacts and dedicated nurses swabbed 505 individuals (264 of whom were found to carry MRSA at some point during the study).

### 3.2   Input and Parameters

CTCmodeler was run using the i-Bird user-contact data file as the contact file to create the contact network. Individual admissions and discharges reproduced the observed data stored in the i-Bird user-admission file. The learning module calculated transmission rates, means and standard deviations of the colonization duration distribution and daily swab probabilities, based on i-Bird user-admission and i-Bird user-swab files. Finally, for these simple simulations, admission and contact file were user-provided and parameters file was deduced from the learning module. Table 1 summarizes all rates, probabilities and parameters deduced by the learning part (included in the parameters file) and used by the modeling part in this illustrative example. For this example, we assumed that an acquisition episode occurred when the individual had two negative swabs followed by a positive one. We also chose to focus on first acquisitions only (without possible recolonization). Second, the ABM used the computed patient-to-patient, patient-to-staff, staff-to-patient and staff-to-staff transmission rates to simulate pathogen transmission. We used a threshold to saturate patient-to-staff and staff-to-staff transmission rates with time. The i-Bird data is composed of 5 wards and 1 artificial ward (with all mobile workers) and hence 6 processes were used for simulations. Thanks to the Repast HPC library, processes were run in parallel and computationally efficient (approximate runtime of 10 min).

**Table 1.** Model parameters

| Variable | Value |
|---|---|
| *Transmission rates* | |
| Patient-to-patient | 0.000009 |
| Patient-to-staff | 0.133615 |
| Staff-to-patient | 0.000509 |
| Staff-to-staff | 0.136793 |
| *Colonized at admission rates* | |
| Patient | 0.21 |
| Staff | 0.33 |
| *Colonization duration (days)* | |
| Patient mean (sd) | 32 (771) |
| Staff mean (sd) | 27 (561) |
| *Patient daily swab (%)* | |
| Monday mean (sd) | 60.7 (15.1) |
| Tuesday mean (sd) | 34.9 (10.7) |
| Wednesday mean (sd) | 14.5 (10.4) |
| Thursday mean (sd) | 4.1 (2.8) |
| *Staff daily swab (%)* | |
| Monday mean (sd) | 24.2 (10.1) |
| Tuesday mean (sd) | 12.0 (4.1) |
| Wednesday mean (sd) | 7.4 (4.3) |
| Thursday mean (sd) | 10.2 (2.1) |

### 3.3 Output

Simulations started with 387 individuals, 151 patients and 236 hospital staff. Among them, 84 swabbed individuals (40 patients and 44 staff) were colonized with MRSA on the first day. Figure 2 depicts the predicted incidence of MRSA colonization over time within the entire hospital, compared with observed data collected during the i-Bird study. We computed 95% prediction bands, i.e. bands that cover 95% of model output obtained over 500 stochastic model simulation. These 95% prediction bands include all the observed data points (red points), thus validating the model predictions.

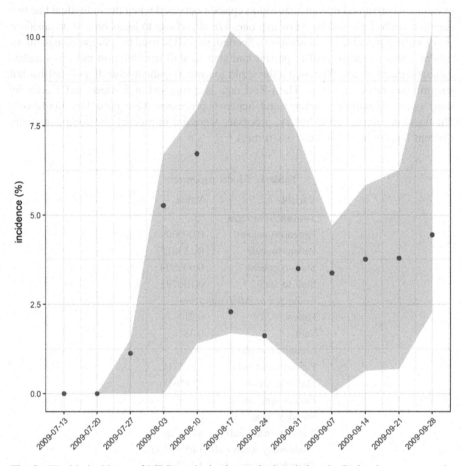

**Fig. 2.** Weekly incidence of MRSA colonization at the hospital scale. Red points represent the observed incidence in the i-Bird data. Grey bands correspond to the 95% prediction bands obtained with 500 ABM simulations. Incidence is in percentage of individuals found negative over the two previous weeks. Consequently, the first two data points are at 0.

## 4   Discussion

In this paper, we introduce CTCmodeler, a framework that simulates individual interactions and pathogen transmission in a hospital setting using an agent-based modeling approach. The main originality of this framework is to explicitly account for the inter-individual contact network within the hospital. This allows for tracing of transmission routes and makes the model realistic in terms of interaction mixing, as compared to most previously published models of nosocomial pathogen spread. Multiple options make the model flexible, such as the option for users to enter their own parameter values to input real data, allowing CTCmodeler to calculate relevant hospital parameters and run corresponding simulations.

The MRSA example we present illustrates how the model can be used to simulate the dissemination of a resistant bacteria through inter-individual contacts in a 5-ward long-term care facility. The framework calculated all needed parameters from real data and then simulated acquisition incidence over a 4-month period. The results obtained show that model predictions are consistent with the observed data. Indeed, the first two points are at zero because of the definition of acquisition chosen. The 500-simulations 95% prediction bands include all incidence points showing that the model can predict MRSA acquisition.

The current version of CTCmodeler has several limitations.

The first limitation is the assumption that pathogens spread exclusively through inter-individual contacts. This allows simulating the contact-mediated dissemination of pathogens such as MRSA or influenza, but would not work for pathogens spread through the environment for instance. Future extensions of the ABM should include the possibility of environmental contamination.

In addition, the model may be used to simulate multiple pathogens simultaneously, but without allowing for interactions between these pathogens (e.g. exchange of a resistance plasmid between two bacteria).

Lastly, although its ability to acquire parameters directly from observed data is among our model's strengths, it also makes it dependent on the precision of these data. When the ABM is used with parameters directly computed from real data files, any incorrect data entry will automatically affect simulations. Moreover, observed "contact" data often reflects close interactions, rather than actual skin-to-skin contacts, as was the case in the i-Bird study. In that case, these "contacts" may not necessarily facilitate pathogen transmission.

In conclusion, CTCmodeler has the potential to be a useful tool to predict pathogen transmission dynamics in a realistic, dynamic hospital setting. Future versions of the framework will include a wide array of control measures, such as hand hygiene, staff cohorting or antibiotic exposure and stewardship, allowing CTCmodeler to be used for public health decision making.

**Acknowledgments.** i-Bird Study Group: Anne Sophie Alvarez (AP-HP, Paris, France), Audrey Baraffe (AP-HP, Paris, France), Mariano Beiró (Universidad de Buenos Aires, Buenos Aires, Argentina), Inga Bertucci (AP-HP, Paris, France), Pierre-Yves Boëlle (Univ. Pierre et Marie Curie, Paris, France), Camille Cyncynatus (AbAg, Chilly-Mazarin, France), Florence Dannet (AP-HP, Paris, France), Marie Laure Delaby (AP-HP, Paris, France), Pierre Denys (AP-HP, Paris, France), Matthieu Domenech de Cellès (Univ. Pierre et Marie Curie, Paris, France), Eric Fleury (ENS Lyon, Lyon, France), Antoine Fraboulet (Insa, Lyon, France), Jean-Louis Gaillard (AbAg, Chilly-Mazarin, France), Boris Labrador (AP-HP, Paris, France), Jennifer Lasley (Inserm, Paris, France), Christine Lawrence (AP-HP, Paris, France), Judith Legrand (Univ. Paris Sud, Orsay, France), Odile Le Minor (Institut Pasteur, Paris, France), Caroline Ligier (Institut Pasteur, Paris, France), Lucie Martinet (Inria, Lyon, France), Karine Mignon (AbAg,

Chilly-Mazarin, France), Catherine Sacleux (AP-HP, Paris, France), Jérôme Salomon (Cnam, Paris, France), Thomas Obadia (Univ. Pierre et Marie Curie, Paris, France), Marie Perard (AP-HP, Paris, France), Laure Petit (Institut Pasteur, Paris, France) Laeticia Remy (AP-HP, Paris, France), Anne Thiebaut (Inserm, Paris, France), Damien Thomas (AbAg, Chilly-Mazarin, France), Philippe Tronchet (AP-HP, Paris, France), Isabelle Villain (AP-HP, Paris, France).

# References

1. Assab, R., et al.: Mathematical models of infection transmission in healthcare settings: recent advances from the use of network structured data. Curr. Opin. Infect. Dis. **30**(4), 410–418 (2017)
2. Barnes, S.L., et al.: The impact of reducing antibiotics on the transmission of multidrug-resistant organisms. Infect. Control Hosp. Epidemiol. **38**(6), 663–669 (2017)
3. Caudill, L., Lawson, B.: A unified inter-host and in-host model of antibiotic resistance and infection spread in a hospital ward. J. Theor. Biol. **421**, 112–126 (2017)
4. de Freitas DalBen, M., et al.: A model-based strategy to control the spread of carbapenem-resistant enterobacteriaceae: simulate and implement. Infect. Control Hosp. Epidemiol. **37**(11), 1315–1322 (2016)
5. Duval, A., et al.: Measuring dynamic social contacts in a rehabilitation hospital: effect of wards, patient and staff characteristics. Sci. Rep. **8**(1), 1686 (2018)
6. Ferrer, J., et al.: Nosolink: an agent-based approach to link patient flows and staff organization with the circulation of nosocomial pathogens in an intensive care unit. Proc. Comput. Sci. **18**, 1485–1494 (2013)
7. Grimm, V., et al.: A standard protocol for describing individual-based and agent-based models. Ecol. Model. **198**(1), 115–126 (2006)
8. Grundmann, H., Hellriegel, B.: Mathematical modelling: a tool for hospital infection control. Lancet Infect. Dis. **6**(1), 39–45 (2006)
9. Hotchkiss, J.R., et al.: An agent-based and spatially explicit model of pathogen dissemination in the intensive care unit. Crit. Care Med. **33**(1), 168–176 (2005). Discussion 253–254
10. van Kleef, E., et al.: Modelling the transmission of healthcare associated infections: a systematic review. BMC Infect. Dis. **13**, 294 (2013)
11. Laskowski, M., et al.: Agent-based modeling of the spread of influenza-like illness in an emergency department: a simulation study. IEEE Trans. Inf. Technol. Biomed. **15**(6), 877–889 (2011)
12. McBryde, E.S., et al.: A stochastic mathematical model of methicillin resistant Staphylococcus aureus transmission in an intensive care unit: predicting the impact of interventions. J. Theor. Biol. **245**(3), 470–481 (2007)
13. Meng, Y., et al.: An application of agent-based simulation to the management of hospital-acquired infection. J. Simul. **4**(1), 60–67 (2010)
14. Milazzo, L., et al.: Modelling of healthcare-associated infections: a study on the dynamics of pathogen transmission by using an individual-based approach. Comput. Methods Program. Biomed. **104**(2), 260–265 (2011)
15. Ong, B.S., Chen, M., Lee, V., Tay, J.C.: An individual-based model of influenza in nosocomial environments. In: Bubak, M., van Albada, G.D., Dongarra, J., Sloot, P.M.A. (eds.) ICCS 2008. LNCS, vol. 5101, pp. 590–599. Springer, Heidelberg (2008). https://doi.org/10.1007/978-3-540-69384-0_64

16. Pittet, D., et al.: Evidence-based model for hand transmission during patient care and the role of improved practices. Lancet Infect. Dis. **6**(10), 641–652 (2006)
17. Temime, L., et al.: Peripatetic health-care workers as potential superspreaders. Proc. Natl. Acad. Sci. U.S.A. **106**(43), 18420–18425 (2009)
18. Triola, M.M., Holzman, R.S.: Agent-based simulation of nosocomial transmission in the medical intensive care unit. In: Proceedings of the 16th IEEE Symposium Computer-Based Medical Systems, pp. 284–288 (2003)
19. Ueno, T., Masuda, N.: Controlling nosocomial infection based on structure of hospital social networks. J. Theor. Biol. **254**(3), 655–666 (2008)
20. Antimicrobial resistance: global report on surveillance 2014. WHO (2014)
21. Surveillance of antimicrobial resistance in Europe 2017. ECDC (2018)

# Socio-cognitive ACO
# in Multi-criteria Optimization

Aleksander Byrski[(✉)] [iD], Wojciech Turek, Wojciech Radwański,
and Marek Kisiel-Dorohinicki[iD]

AGH University of Science and Technology,
Al. Mickiewicza 30, 30-059 Krakow, Poland
{olekb,wojciech.turek,doroh}@agh.edu.pl, wojciech.radwanski@gmail.com

**Abstract.** In this paper a socio-cognitive ACO-type algorithm is proposed for multi-criteria TSP problem optimization. This algorithm is rooted in psychological inspirations and follows other socio-cognitive swarm intelligence methods proposed up to now. This paper presents the idea and shows the applicability of the proposed algorithm based on selected benchmark functions from the scope of well-known TSPLIB library.

**Keywords:** Multi-criteria optimization · Ant-colony algorithm · Socio-cognitive computing

## 1 Introduction

Providing sub-optimal solution to the variety of NP-hard optimization problems is one of the most popular research problems, which is still receiving enormous attention. The development of major metaheuristic methods, like simulated annealing, evolutionary algorithms or taboo search, with an uncountable set of variants and applications to particular problems, required tremendous effort of researchers worldwide. It is to note, that multi-criteria problems are especially interesting and difficult, because not only one solution is sought, but a number of it (called Pareto set). Other features should be also considered like even and dense coverage of the Pareto front by the solutions.

According to Wolpert and Macready "No free lunch theorems for optimization" [29] each metaheuristic algorithm must be tailored for a given problem. However not all of them were properly theoretically analyzed, as e.g. Simple Genetic Algorithm (works by Vose [28]). Therefore not only tailoring is needed, but the NFL theorem encourages the researchers to seek new metaheuristics, which can be better suited for particular problems. One should however remember that new algorithms should not be proposed for their own sake (cf. Sorensen [26]), and introducing new inspirations must be carefully justified.

One of the popular, bio-inspired method, dedicated for solving discrete optimization problems, is Ant Colony Optimization, ACO [15]. It is dedicated for

J. M. F. Rodrigues et al. (Eds.): ICCS 2019, LNCS 11537, pp. 488–501, 2019.
https://doi.org/10.1007/978-3-030-22741-8_35

solving single-criteria discrete problems, and it has proven efficiency in one of the most recognized problem of this class, the travelling salesperson problem, TSP. The natural consequence of its growing popularity was the appearance of variants dedicated for multi-criteria optimization.

Another promising inspiration from nature, which has already proven usefulness in several problems, is the socio-cognitive ACO [11]. It introduces several types of ants, which present different behaviour – use different decision function. The types of ants and their behaviours reflect the types of characters observed in humans and classified by cognitive psychologists. Previously observed features of the socio-cognitive ACO makes it a good candidate for solving complex problems of multi-criteria optimization.

In this paper a novel, socio-cognitive ACO algorithm for Multi-criteria Optimization is proposed. A set of experiments using multi-criteria TSP benchmark problems shows that the proposed method outperforms previously proposed methods in terms of Pareto set cardinality and solutions quality.

## 2    Swarm Intelligence in Multi-criteria Optimization

In this section, starting from giving basic information on Ant Colony Optimization, selected aspects of ACO in solving multi-criteria are presented and background in socio-cognitive metaheuristics is given, before describing the proposed algorithm in detail.

### 2.1    Ant Colony Optimization

Ant System, introduced in 1991 by Marco Dorigo in application to graph problems, is a progenitor of all the Ant Colony Optimization (ACO) techniques [15]. The classic ACO algorithm is an iterative process during which a certain number of agents (ants) gradually create solutions [17,18]. The main goal of the ant is to traverse the graph is search of a path with the lowest cost (usually meaning the shortest distance, but can also be the lowest fuel consumption, etc.).

In each step any particular ant selects a subsequent component of the solution (that is, a graph edge) with certain probability. This decision may be impacted by the levels of *pheromones*, which may have been deposited into the environment (onto the edges of the graph) by other ants. This interaction is guided by stigmergic relations (communication among individuals by means of an environment, instead of via direct contact) according to the rules proposed in [16]. The computation is finished once a feasible solution is found thanks to the cooperative efforts of all the ants.

Let us assume that one wants to solve a combinatiorial optimization problem, such as the Traveling Salesman Problem. In the basic ACO algorithm - the Ant System (AS, [15]) - the probability of moving from component $i$ to component $j$ for ant $k$ is defined as follows:

$$p_{ij}^k = \begin{cases} \frac{[\tau_{ij}]^\alpha \cdot [\eta_{ij}]^\beta}{\sum_{k \in allowed_k} [\tau_{ik}]^\alpha \cdot [\eta_{ik}]^\beta} & \text{if } j \in allowed_k \\ 0 & \text{otherwise} \end{cases} \tag{1}$$

where $\tau_{ij}$ is the intensity of the pheromone trail on edge $(i,j)$ and $\eta_{ij}(t)$ is the visibility of edge $(i,j)$, which in case of TSP can be defined as an inverted distance between the cities. $\alpha$ and $\beta$ are parameters that control the relative importance of the trail versus visibility. Finally, $allowed_k$ is a set of possible transitions for ant $k$. The tour ends when a feasible solution is found. After each iteration the ants update the pheromone trails on their paths based on its constructed solution. Furthermore, the pheromone also slowly evaporates to prevent premature convergence (cf. [16]).

In the classic AS version the update is performed at the end of a single iteration following the formula:

$$\tau'_{ij} = \rho\tau_{ij} + \sum_{k=1}^{m} \Delta\tau_{ij}^k \qquad (2)$$

where $\rho \in [0,1)$ is a pheromone persistence coefficient and $m$ is the number of the ants. The pheromone update value for each ant is defined as follows:

$$\Delta\tau_{ij}^k = \begin{cases} \frac{Q}{L_k} & \text{if } k\text{-th ant uses edge } (i,j) \\ 0 & \text{otherwise} \end{cases} \qquad (3)$$

where $L_k$ is the tour length of the $k$-th ant and $Q$ is a constant, often with value 1.

Since the first publication describing Ant System [17] researchers have put a lot of effort into improving the optimization performance. One of the most efficient modifications is Max-Min Ant System (MMAS, [27]). The algorithm introduced three major adjustments into the AS algorithm:

- only one ant leaves the pheromone trails after each iteration - either the best from the iteration or the best found so far by the algorithm,
- the pheromone trials values are limited by $\tau_{min}$ and $\tau_{max}$,
- the pheromone matrix is initialized with the $\tau_{max}$ value.

The effectiveness of the MMAS algorithm was further improved by the *pheromone trail smoothing* mechanism. When the algorithm is very close to convergence, the pheromone matrix gets smoothed by increasing the values proportionally to their difference to $\tau_{max}$. This mechanism can also be effectively applied to other elitist ant systems [8,16].

## 2.2   ACO in Multi-criteria Optimization

ACO algorithms are designed to solve single-criterion problems in their basic versions. However, over the last years there have been more and more attempts to adapt them to solve more complex problems. Designing such algorithms requires solving several difficulties. First of all, it should be specified how the deposition of the pheromone will be managed, in particular when several ant colonies are used at the same time. It should be determined how the ants will use this information in their decisions. Then you have to decide which ants and how they will modify this value.

There are three classic methods for solving multi-criteria problems using formic algorithms: using the weight function, the distance function or the min-max formula. These approaches are based on reducing the multi-criteria to a single-criterion problem by generating an artificial objective function combining individual criteria from the input problem.

The use of multi-criteria mapping to a single-criterion problem has several significant disadvantages and undesirable consequences. Therefore, several more advanced multi-criteria formulas have been developed that do not require combining different properties into a single one. In [22] quality tests for optimisation of various multicriteria implementations of ACO, we can find information on the basic method of categorizing these algorithms (Table 1).

**Table 1.** A taxonomy of multiple objective ACO algorithms proposed in [22]

|  | Single objective function | Multiple objective functions |
|---|---|---|
| Single pheromone matrix | MOACOM | MACS |
| Multiple pheromone matrices | P-ACO | BicriterionAnt |

The Multiple Objective ACO metaheuristic (MOACOM) [23] is an extension of the Ant System algorithm [15], which represents multiple criteria in a single distance matrix. Different criteria are aggregated into a single distance value using dedicated rules. The method introduces an order of the criteria – only the most important criteria is used for updating the pheromone matrix. The BicriterionAnt algorithm introduced two pheromone matrices, updated independently. Multiple pheromone matrices allow representing different criteria in ant's decisions. Te approach introduced in P-ACO algorithm [14] integrates values from several pheromone matrices, using different weights generated randomly for each ant.

The Multiple ant colony system (MACS) [1], a multiobjective version of the Ant System, uses three equally important objective functions. All three objectives share the same pheromone trails, which can be updated iteratively. The authors introduce the concept of different visibilities of each objective, which differentiate the behaviour of ants. The MACS algorithm will be used as a basis for the socio-cognitive solution presented in this paper.

### 2.3 Socio-cognitive Metaheuristics

In cognitive psychology, the character traits of egocentrism (taking one's own perspective) and altercentrism (taking another person's perspective into consideration) have long been recognized to play a key role in interpersonal relationships (see, for instance, [19,25]). Moreover, brain-imaging studies have shown that altercentricity and the strategy of perspective taking develop in parallel with brain maturation and psychosocial development during adolescence [5,13].

Perhaps mirroring this psychological development, in recent years, artificial intelligence researchers have started to incorporate altercentricity into robots and autonomous systems [24]. We also continue with utilizing the notions of ego- and altercentrism, adapting them appropriately to use in our computing system.

Typically, perspective taking is seen as a one-dimensional ability: the degree to which an agent can take another one's perspective. But recent research has explored a two-dimensional approach [6], where one distinguishes between the ability of an agent to handle conflict between its own and the other agent's perspectives, and the relative priority that an agent gives to his own perspective relative to the other's perspective. During social interactions, humans do not always share the same views. Being able to consider the other person's point of view therefore requires putting aside one's own perspective. This is particularly hard if one holds a strong view. Individuals endowed with good cognitive skills to manage conflicting information are therefore usually better perspective-takers [20]. In addition, however, humans also differ in terms of how much they are interested in or are willing to pay attention to others compared to themselves. Sometimes individuals focus only on their own perspective (egocentrism) while on other occasions individuals focus more on other people's perspective (altercentrism) [19,24].

The less a person focuses on her own perspective, the more that person will be motivated to engage in perspective taking [6]. Experimental research has suggested that these two dimensions (conflict handling and perspective priority) might be independent; and factors such as guilt or shame affect each of these dimensions individually [7]. This two-dimensional approach to perspective taking inspired us to define four types of individuals:

- Egocentric individuals, focusing on their own perspective and becoming creative thanks to finding their own new solutions to a given task. These individuals do not pay attention to the other ones and do not get inspired by the actions of other ones (or these inspirations do not become a main factor of their work).
- Altercentric individuals, focusing on the perspective of others and thus following the mass of others. Such individuals become less creative but they still can end up supporting good solutions by simply following them.
- Good-at-conflict-handling individuals, getting inspired in a complex way by the actions of other individuals, considering different perspectives and choosing the one considered as the best for them.
- Bad-at-conflict-handling individuals, acting purely randomly, following sometimes one perspective, sometimes another without any inner logic.

In this work we follow only the first three (although involving the fourth one is also feasible and will be tackled in the future work).

Based on those inspirations, novel socio-cognitive metaheuristics have been proposed, enhancing well-known swarm metaheuristics, like Ant Colony Optimization and Particle Swarm Optimization. The enhancement of these consisted in introducing different "species" of swarm individuals, inspired by egocentricity and altercentricity and making them perceive the others.

In the case of socio-cognitive ACO [11], different species of ants leaved different types of pheromones and they were able to perceive them and compute the probabilities of choosing the next edge, following the socio-cognitive inspirations. The egocentric ants focused on their own knowledge (perceived only the distance connected with the current edges to be chosen), the altercentric ants focused only on pheromone markings. Good-at-conflict-handling ants were inspired in a more complex way (they computed the attractiveness based on a weighted sum of information from distance and pheromone expressed by other species of ants). The structure of the ant population was static in the beginning (the percentage of different species). However later, going beyond these inspirations, the attractiveness was parameterized, and those parameters were explored along with considering automatic adaptation of the population structure [10].

In the case of socio-cognitive PSO [4], different species of particles were introduced into the basic PSO algorithm, making them perceive others, by modifying their cognitive abilities: the particles could perceive the best global solution gathered in the whole swarm, in their neighborhood and their own, historical best solution. In the case of socio-cognitive PSO, an effort has been made also to introduce auto-adaptation mechanism modifying the structure of population [3].

Both socio-cognitive ACO and PSO turned out to be better than classic reference algorithms when considering selected multi-dimensional benchmark functions and TSPLIB instances [9].

## 3   Multi-criteria Socio-cognitive ACO

In this section the reference algorithm, Multiple-Ant Colony System [2] is presented, being a base for extension towards introducing socio-cognitive mechanisms (i.e. different species of ants, similarly to [11].

### 3.1   Multiple-Ant Colony System

Multiple-Ant Colony System (MACS) was proposed by Barán and Schaerer in 2003 [2]. According to García-Martínez [22] this algorithm produces a very good coverage of the Pareto front. Because MACS uses one pheromone table, it is a very good candidate for extension using socio-cognitive inspirations.

The algorithm is a developed version of the MACS-VRPTW algorithm. The main difference is that it uses a single pheromone table $\tau$ and two heuristic functions which are used to make decisions about choosing the next edge. In the classic implementation of MACS-VRPTW, the algorithm uses two colonies, each with an independent array of pheromones. One minimizes the number of vehicles and the second distance travelled. Information between colonies is exchanged by maintaining the best solution for both colonies. The authors showed that the proposed methodology improves the quality of results for the problem under investigation [21].

Fist the value $\tau_0$, being a current starting pheromone value is computed as follows:

$$\tau_0 = \frac{1}{\hat{f}^1 \cdot \hat{f}^2} \tag{4}$$

where $\hat{f}^1, \hat{f}^2$ are values of the criteria function computed for the current solution.

The next vertex is chosen by the ant $h$ belonging to the set of $m$ ants according to the following equation:

$$j = \begin{cases} argmax_{j \in \Omega}(\tau_{ij} \cdot [\eta_{ij}^0]^{\beta\lambda_h} \cdot [\eta_{ij}^1]^{\beta(1-\lambda_h)}) & \text{if } q \leq q_0 \\ \hat{i} \end{cases} \tag{5}$$

where $\Omega$ is the set of the vertices adjacent to the current vertext the ant is located on; $\eta_{ij}^0, \eta_{ij}^1$ are visibility or distance perceived from the point of view of two criterion functions (0 and 1); $\lambda_h = \frac{h-1}{m-1}$, this value is used for enforcing the ant to explore new areas of the Pareto frontier; $q$ is a certain probability, while $q_0 = 0.98$ for this algorithm. In the above equation, $\hat{i}$ is a randomly-chosen vertex with the following probability:

$$p_i = \begin{cases} \dfrac{[\sum_{k=1}^{K} p_k \cdot \tau_{ij}^k]^\alpha \cdot [\eta_{ij}]^\beta}{\sum_{u \in \Omega} [\sum_{k=1}^{K} p_k \cdot \tau_{iu}]^\alpha \cdot [\eta_{iu}]^\beta} & \text{if } j \in allowed_k \\ 0 & \text{otherwise} \end{cases} \tag{6}$$

The pheromone at a certain edge is updated according to the following evaporation rule:

$$\tau_{ij} = (1 - \varrho) \cdot \tau_{ij} + \varrho\tau_0 \tag{7}$$

where $\varrho$ is a parameter of the algorithm.

The Pareto set is used to modify the value of $\tau_0$ during the life of the colony. The ant after finding a full, non-dominated solution adds them to the common set. Then after finding all solutions by ants in a given iteration for the first and second cost functions, their average values are calculated, which we substitute for the formula from the previous point. In this way, we get the value of $\tau_0'$. Then, if $\tau_0' > \tau_0$, all the pheromone values on the edges are converted into the value $\tau_0'$. Otherwise, the total pheromone table value update operation is performed by performing the following formula for each edge and each solution from the Pareto set $S$ (actually for each $s \in S$):

$$\tau_{ij}' = (1 - \varrho)\tau_{ij} + \frac{\varrho}{f^0(s) \cdot f^1(s)} \tag{8}$$

where $s$ is a solution belonging to the Pareto set $S$. Thus the criteria function are used to constantly modify the pheromone table.

## 3.2   Multi-criteria Socio-cognitive Ant Colony System

The proposed multicriteria algorithm merges the sociological inspirations together with the concepts introduced in MACS algorithm. The preliminary tests showed, that three types of ants should be used: the *Good-at-conflict-handling*, the *Egocentric* and *Altercentric*. The algorithm uses three independent pheromone matrices, one for each type of ant. The matrices are updated after each iteration, and the information contained in them is used depending on the ant's type. The following rules are used by the different types of ants:

- *Good-at-conflict-handling* ant considers pheromone in all three matrices. Each ant undertakes the decision about choosing the next vertex according to Eqs. 5 and 6. However the pheromone is now located in three tables, and the actual decision depends on the ant type (or species). Thus, assume that the $\tau_{ij}$ used in the above-cited equations is substituted by the $\hat{\tau}_{ij}$ computed as follows:

$$\hat{\tau}_{ij} = \kappa \cdot \tau_{ij}(GC) + \psi \cdot \tau_{ij}(EC) + \tau_{ij}(AC) \qquad (9)$$

where $\kappa = 1, \psi = 5, \omega = 7$ (these values were discovered experimentally) and may be further adapted; $\tau_{ij}(GC)$ is the value of pheromone for the vertex $(i, j)$ deposited in the pheromone table of good-at-conflict-handling ants $GC$, and respectively for $EC$ and $AC$.
- *Altercentric* ant considers only the pheromone values in all three matrices. Thus the next vertex is chosen according to the following equation:

$$j = \begin{cases} argmax_{j \in \Omega}(\hat{\tau}_{ij}^{\alpha}) \cdot & \text{if } q \leq q_0 \\ \hat{i} \end{cases} \qquad (10)$$

where the vertex $\hat{i}$ is chosen with the probability given by Eq. 6, with $\tau_{ij}$ substituted by $\hat{\tau}_{ij}$:

$$\hat{\tau}_{ij} = \tau_{ij}(GC) + \tau_{ij}(EC) + \tau_{ij}(AC) \qquad (11)$$

- *Egocentric* ant is only driven by the distance to the next vertex, ignoring the pheromone matrices, thus the next vertex is chosen as follows:

$$j = \begin{cases} argmax_{j \in \Omega}([\eta_{ij}^0]^{\beta \lambda_h} \cdot [\eta_{ij}^1]^{\beta(1-\lambda_h)}) & \text{if } q \leq q_0 \\ \hat{i} \end{cases} \qquad (12)$$

and $\hat{i}$ is chosen randmoly with the following probability:

$$p_{\hat{i}} = \begin{cases} \dfrac{[\sum_{k=1}^{K} p_k]^{\alpha} \cdot [\eta_{ij}]^{\beta}}{\sum_{u \in \Omega} [\sum_{k=1}^{K} p_k \cdot]^{\alpha} \cdot [\eta_{iu}]^{\beta}} & \text{if } j \in allowed_k \\ 0 & \text{otherwise} \end{cases} \qquad (13)$$

The update of the three pheromone matrices is done after the whole iteration is finished. Only the solutions from the identified Pareto set are used. Each solution $s \in S$ (belonging to the Pareto set) updates the matrix of the ant type, which found the solution, according to the following formula (for all $s \in S$):

$$\tau'_{i,j}(s) = \tau_{i,j}(s) + \frac{1}{0.5f^0(s) + 0.5f^1(s)} \tag{14}$$

this modification is realized for all the ant types (thus all the pheromone tables) separately.

Introducing of properly adapted ant species, namely egocentric, altercentric and good-at-conflict-handling follows the socio-cognitive computing paradigm proposed in [11].

## 4   Experimental Results

In order to verify the effectiveness of the proposed, socio-cognitive approach to the problem of multi-criteria optimization, the two-criteria travelling salesperson problem has been selected. The experiments were conducted using popular benchmark problems from the TSPLIB library[1]. Five variants of the problem have been selected: Kro100AB and Kro100ED contain graphs composed of 100, Kro200AB with 200 cities, and two problems, euclidA100 and euclidA300, generated with the program of the DIMACS competition[2], with 100 and 300 cities respectively. All tests have been executed on a single computer with Intel i7, 2.5 GHz CPU and 12 GB of RAM. Each of the experimental configurations was run 30 times and the results were gathered, consituting one Pareto front (either shown in the relevant figure or processed and described in a table in this section).

Two metrics were used to evaluate the quality of the solutions: the HyperVolume and the Cardinality of the computed Pareto front. The HyperVolume [12] value is a well-recognised metric for multi-criteria optimisation, which integrates the quality and the diversity of the Pareto front. It is computed as a volume of

**Table 2.** Results of search for optimal population structure using the problem Kroo100AB.

| Species % | | | |
|---|---|---|---|
| AC | EC | GC | HyperVolume |
| 20 | 20 | 60 | $1.998 \times 10^{10}$ |
| **20** | **40** | **40** | $2.005 \times 10^{10}$ |
| 20 | 60 | 20 | $1.994 \times 10^{10}$ |
| 40 | 20 | 40 | $1.99 \times 10^{10}$ |
| 40 | 40 | 20 | $1.99 \times 10^{10}$ |
| 60 | 20 | 20 | $1.82 \times 10^{10}$ |

---

[1] http://elib.zib.de/pub/mp-testdata/tsp/tsplib/tsplib.html.
[2] https://eden.dei.uc.pt/~paquete/tsp/.

a $n$-dimensional shape surrounded by the solutions of the Pareto front, where $n$ is the number of criteria.

The implementation of the proposed method (MSCACO) has been compared with the MACS algorithm. During the tests a constant composition of the ants types in the population of the MSCACO algorithm has been used: 40% of *Good-at-conflict-handling* ants, 40% of *Egocentric* and 20% of *Altercentric*. These values have been concluded from a series of preliminary tests (cf. Table 2).

We have also conducted a series of experiments in order to find an appropriate number of ants and number of iterations to get reasonable HV indicator values, finding that 100 ants and 50 iterations can be used as a good starting point for further research (see Table 3).

The most important results from the conducted experiments are collected in Table 4. In all cases a population of 100 ants were used. The algorithms executed 50 iterations.

The results clearly show the superiority of the proposed, socio-cognitive ACO algorithm over the MACS method. In all test cases the cardinality of the created Pareto Set is far greater. Also the values of the HyperVolume metrics are significantly better in all test cases, which is a valuable result.

More detailed results of the selected test cases are presented in Figs. 1 and 2, which present results from the Kro100AB and EuclidAB100 test cases. In both cases it is clearly visible, that the Pareto Front created by the MSCACO algorithm is more consistent and the particular results are better for vast majority of values.

**Table 3.** Results of search for optimal number of ants and iterations.

| Ant | Iterations | MACS HV | MSCACS HV |
|---|---|---|---|
| 10 | 10 | $1.898 \times 10^{10}$ | $1.952 \times 10^{10}$ |
| 20 | 10 | $1.952 \times 10^{10}$ | $1.979 \times 10^{10}$ |
| 50 | 10 | $1.964 \times 10^{10}$ | $1.993 \times 10^{10}$ |
| 10 | 20 | $1.925 \times 10^{10}$ | $1.97 \times 10^{10}$ |
| 10 | 30 | $1.933 \times 10^{10}$ | $1.966 \times 10^{10}$ |
| 10 | 50 | $1.95 \times 10^{10}$ | $1.977 \times 10^{10}$ |
| 50 | 20 | $1.972 \times 10^{10}$ | $2.01 \times 10^{10}$ |
| 20 | 50 | $1.967 \times 10^{10}$ | $1.999 \times 10^{10}$ |
| 30 | 50 | $1.967 \times 10^{10}$ | $2.012 \times 10^{10}$ |
| 50 | 30 | $1.957 \times 10^{10}$ | $2.011 \times 10^{10}$ |
| 50 | 50 | $1.977 \times 10^{10}$ | $2.017 \times 10^{10}$ |
| **100** | **50** | $1.971 \times 10^{10}$ | $2.026 \times 10^{10}$ |

**Table 4.** The comparison of the MSCACO and MACS optimisation results.

| | MACS | | MSCACS | |
|---|---|---|---|---|
| | Cardinality | HyperVolume | Cardinality | HyperVolume |
| Kro100AB | 75 | $1.971 \times 10^{10}$ | 140 | $2.026 \times 10^{10}$ |
| Kro100ED | 75 | $1.939 \times 10^{10}$ | 148 | $2.005 \times 10^{10}$ |
| Kro200AB | 102 | $7.399 \times 10^{10}$ | 186 | $7.504 \times 10^{10}$ |
| euclidAB100 | 91 | $1.819 \times 10^{10}$ | 142 | $1.882 \times 10^{10}$ |
| euclidAB300 | 142 | $1.925 \times 10^{10}$ | 218 | $1.970 \times 10^{10}$ |

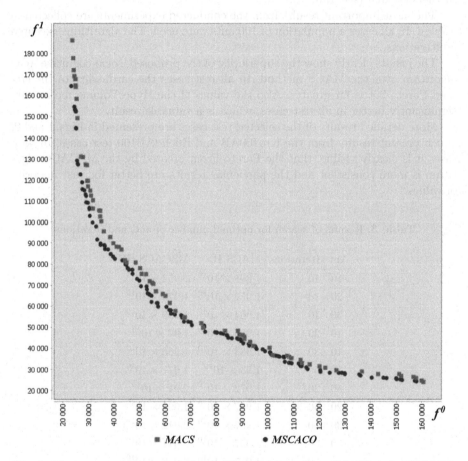

**Fig. 1.** The computed Pareto sets for the Kro100AB problem (MACS: red, MSCACO: blue) (Color figure online)

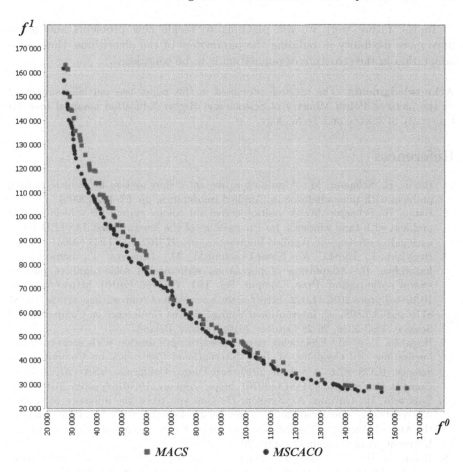

**Fig. 2.** The computed Pareto sets for the EuclidAB100 problem (MACS: red, MSCACO: blue) (Color figure online)

## 5    Conclusions and Future Work

In this paper a new metaheuristic algorithm for solving multi-criteria optimization, belonging to the class of socio-cognitive algorithms was presented. Following our previous work on incorporation of socio-cognitive inspirations, the population of ants has been divided into three species with different cognitive abilities. This new algorithm (MSCACO) has been built on the basis of the existing MACS. Introduction of socio-cognitive inspirations lead to increasing the efficiency of the algorithm applied to selected instances of multi-criteria TSP problems connected with TSPLIB library. It was shown, that socio-cognitive versions produced more solutions on the Pareto front, and finally the Hyper Volume metrics was also better.

In the future work we are planning to tackle new problems and introduce more flexibility in building the parameters of the algorithms, thus auto-adaptation in the structure of population is to be considered.

**Acknowledgments.** The research presented in this paper was partially supported by the funds of Polish Ministry of Science and Higher Education assigned to AGH University of Science and Technology.

# References

1. Barán, B., Schaerer, M.: A multiobjective ant colony system for vehicle routing problem with time windows. In: Applied Informatics, pp. 97–102 (2003)
2. Barán, B., Schaerer, M.: A multiobjective ant colony system for vehicle routing problem with time windows. In: Proceedings of the Twenty First IASTED International Conference on Applied Informatics, pp. 97–102. IASTED (2003)
3. Bugajski, I., Byrski, A., Kisiel-Dorohinicki, M., Lenaerts, T., Samson, D., Indurkhya, B.: Adaptation of population structure in socio-cognitive particle swarm optimization. Proc. Comput. Sci. **101**, 177–186 (2016). https://doi.org/10.1016/j.procs.2016.11.022. http://www.sciencedirect.com/science/article/pii/S1877050916326898. 5th International Young Scientist Conference on Computational Science, YSC 2016, 26-28 October 2016, Krakow, Poland
4. Bugajski, I., et al.: Enhancing particle swarm optimization with socio-cognitive inspirations. In: Connolly, M. (ed.) International Conference on Computational Science, ICCS 2016, 6–8 June 2016, San Diego, California, USA (2016). Proc. Comput. Sci. **80**, pp. 804–813 (2016). https://doi.org/10.1016/j.procs.2016.05.370
5. Bukowski, H., Curtain, A., Samson, D.: Can you resist the influence of others? Altercentrism, egocentrism and interpersonal personality traits. In: Proceedings of the Annual Meeting of the Belgian Association for Psychological Sciences (BAPS). Universite catholique de Louvain (2013)
6. Bukowski, H.: What influences perspective taking? A dynamic and multidimensional approach. Ph.D. thesis, Université catholique de Louvain (2014)
7. Bukowski, H., Samson, D.: Can emotions affect level 1 visual perspective taking? Cogn. Neurosci. (in press). https://doi.org/10.1080/17588928.2015.1043879
8. Bullnheimer, B., Hartl, R.F., Strauss, C.: A new rank based version of the ant system. A computational study (1997)
9. Byrski, A.: Socio-cognitive Metaheuristics Computing. AGH University of Science and Technology Press, Krakow (2018)
10. Byrski, A., et al.: Emergence of population structure in socio-cognitively inspired ant colony optimization. Comput. Sci. (AGH) **19**(1) (2018). https://doi.org/10.7494/csci.2018.19.1.2594
11. Byrski, A., et al.: Socio-cognitively inspired ant colony optimization. J. Comput. Sci. **21**, 397–406 (2017). https://doi.org/10.1016/j.jocs.2016.10.010
12. Cao, Y., Smucker, B.J., Robinson, T.J.: On using the hypervolume indicator to compare pareto fronts: applications to multi-criteria optimal experimental design. J. Stat. Plan. Infer. **160**, 60–74 (2015)
13. Choudhury, S., Blakemore, S.J., Charman, T.: Social cognitive development during adolescence. Soc. Cogn. Affect. Neurosci. **1**(3), 165–174 (2006)
14. Doerner, K., Gutjahr, W.J., Hartl, R.F., Strauss, C., Stummer, C.: Pareto ant colony optimization: a metaheuristic approach to multiobjective portfolio selection. Ann. Oper. Res. **131**(1–4), 79–99 (2004)

15. Dorigo, M., Di Caro, G.: Ant colony optimization: a new meta-heuristic. In: Proceedings of the 1999 Congress on Evolutionary Computation, CEC 1999, vol. 2, pp. 1470–1477. IEEE (1999)
16. Dorigo, M., Maniezzo, V., Colorni, A.: Ant system: optimization by a colony of cooperating agents. IEEE Trans. Syst. Man Cybern. Part B (Cybern.) **26**(1), 29–41 (1996)
17. Dorigo, M., Di Caro, G.: The ant colony optimization meta-heuristic. In: Corne, D., Dorigo, M., Glover, F. (eds.) New Ideas in Optimization, pp. 11–32. McGraw-Hill, New York (1999)
18. Dorigo, M., Caro, G.D., Gambardella, L.M.: Ant algorithms for discrete optimization. Technical report, IRIDIA/98-10, Université Libre de Bruxelles, Belgium (1999)
19. Feldman, H., Rand, M.E.: Egocentrism-altercentrism in the husband-wife relationship. J. Marriage Family **27**(3), 386–391 (1965)
20. Fizke, E., Barthel, D., Peters, T., Rakoczy, H.: Executive function plays a role in coordinating different perspectives, particularly when one's own perspective is involved. Cognition **130**(3), 315–334 (2014). https://doi.org/10.1016/j.cognition.2013.11.017
21. Gambardella, L.M., Taillard, E., Agazzi, G.: A multiple ant colony system for vehicle routing problems with time window. In: New Ideas in Optimization, pp. 63–76. McGraw-Hill (1999)
22. García-Martínez, C., Cordón, O., Herrera, F.: A taxonomy and an empirical analysis of multiple objective ant colony optimization algorithms for the bi-criteria tsp. Eur. J. Oper. Res. **180**(1), 116–148 (2007)
23. Gravel, M., Price, W.L., Gagné, C.: Scheduling continuous casting of aluminum using a multiple objective ant colony optimization metaheuristic. Eur. J. Oper. Res. **143**(1), 218–229 (2002)
24. Johnson, M., Demiris, Y.: Perceptual perspective taking and action recognition. Int. J. Adv. Robot. Syst. **2**(4), 301–308 (2005)
25. Nadel, J.: Some reasons to link imitation and imitation recognition to theory of mind. In: Doric, J., Proust, J. (eds.) Simulation and Knowledge of Action, pp. 119–135. John Benjamins, New York (2002)
26. Sörensen, K.: Metaheuristics the metaphor exposed. Int. Trans. Oper. Res. **22**(1), 3–18 (2015). https://doi.org/10.1111/itor.12001
27. Stützle, T., Hoos, H.H.: Max-min ant system. Future Gener. Comput. Syst. **16**(8), 889–914 (2000)
28. Vose, M.D.: The Simple Genetic Algorithm: Foundations and Theory. MIT Press, Cambridge (1998)
29. Wolpert, D.H., Macready, W.G.: No free lunch theorems for optimization. IEEE Trans. Evol. Comput. **1**(1), 67–82 (1997). https://doi.org/10.1109/4235.585893

# Reconfiguration of the Multi-channel Communication System with Hierarchical Structure and Distributed Passive Switching

Piotr Hajder$^{(\boxtimes)}$ ⓘ and Łukasz Rauch ⓘ

AGH University of Science and Technology, Krakow, Poland
{Phajder, lrauch}@agh.edu.pl

**Abstract.** One of the key problems in parallel processing systems is the architecture of internodal connections, thus affecting the computational efficiency of the whole. In this work authors describe proposition of a new multi-channel hierarchical computational environment with distributed passive switching. According to authors, improvement of communication efficiency should be based on grouping of system components. In the first type of clustering processing nodes are combined into independent groups that communicate using a dedicated channel group. The second type of clustering splits channels available in the system. In particular, they are divided into smaller independent fragments that can be combined into clusters that support selected users. In this work, a model of computational environment and basic reconfiguration protocol were described. The necessary components and management of reconfiguration, passive switching and hierarchization were discussed, highlighting related problems to be solved. Main reconfiguration restrictions were specified, using and combining together two structural complexity measures: Complexity B Index and Efficiency Index.

**Keywords:** Reconfiguration · Distributed systems · Multi-channel architecture

## 1 Introduction

In modeling modern scientific and technical problems there are often tasks with high computational complexity, such as simulations of quantum mechanics, multiscale modelling, weather forecasting, dynamics of chemical processes, 3D plasma modeling (e.g. in a nuclear fusion reactor) and so others. Their high time complexity makes it impossible to use sequential processing to solve them [1, 2]. In this case, the solution of such tasks must be based on the parallel (centralized or distributed) processing. In both cases, an important part of task solving cost is internodal communication, especially in the case of distributed computing. This means that methods of improving the efficiency of simulation task solving should be primarily seen in its parallelization, while improving communication architecture.

One of the basic methods of improving communication efficiency is to adapt the topology of the connection network to the traffic pattern, characteristic for task being solved. However, it should be remembered that these patterns will be radically different

© Springer Nature Switzerland AG 2019
J. M. F. Rodrigues et al. (Eds.): ICCS 2019, LNCS 11537, pp. 502–516, 2019.
https://doi.org/10.1007/978-3-030-22741-8_36

for various tasks. Taking into account the advances in information technology, it can be noticed that implementation of computational procedures should take place on the basis of MPP (*Massively Parallel Processing*) or CoW (*Cluster of Workstations*) processing models. In both cases, the efficiency of computational process requires a modern subsystem of internodal communication, which due to the variety of tasks to be solved should be tunable, i.e. it should be possible to reconfigure connections in preferably real time.

The increase in the number of nodes in large-scale computing systems reduces the quality of their use expressed by:

- Unacceptable increase in communication costs necessary to solve a computational task;
- Lack of uniformity of load on processing nodes and communication channels;
- Inefficiency of system resources reconfiguration in production environment.

The solution to above problem required solving following tasks:

1. Unification of terminology used in the area of multi-channel connection system with multi-channeling at the logical (virtual) level;
2. Designing of components for a parallel computational system, ensuring flexible reconfiguration of internodal connections;
3. Preparation and verification of analysis methods and synthesis of dedicated reconfigured computational architectures and their examination.

From the available literature [3–6] many interesting architectures of parallel computing systems are known. In most cases, they use factory multiprocessors computing modules, combined into larger, more efficient structures [7]. Scaling such a system is based on various virtualization and switching techniques. Similar and in many cases better results can be obtained by replacing centralized actives switching with distributed passive switching. The solutions proposed in the work are based on the following basic assumptions:

The basis of communication will be a multi-channel passive optical environment, similar in its idea (but not identical) to GPON (*Gigabyte Passive Optical Network*) technology [8]. The implementation of proposed solutions may also be based on multi-channel communication network using FDM (*Frequency Division Multiplexing*) or similar method, however, the obtained characteristics will be worse.

## 2 The Role of Multi-channeling and Hierarchization in Communication Efficiency Improvements

In the 1980s, K. Hwang formulated the hypothesis that the most influencing limitations of parallel and distributed system functioning result from internodal communication [9]. A similar thesis appears in other publications [10–12]. The solution of some of defined problems is the construction of computational architecture based on dynamically reconfigured connections, wherein each node should be supported by a set of logically independent channels. The basic ways to improve the efficiency of modern distributed are defined (Fig. 1). The bold font indicates the ones that can effectively take advantage of the new communication environment.

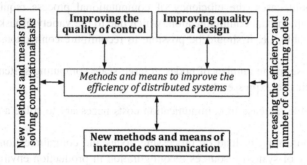

**Fig. 1.** Basic methods and means improving efficiency of distributed system

The concept of a reconfigurable computing environment is shown in Fig. 2. Separated computing environments called clusters can be autonomous or connected to each other, as in the case of clusters 1 and 3. Connections between clusters are carried out by using the same means as internal connections and can be made as two-point, group or broadcast connections. Individual nodes (PN – processing, SN – storage) can be utilized in one cluster or shared between them. This enables information exchange between clusters through inter-process communication or shared memory areas (operational or mass). Besides the main functionality, storage nodes can provide an execution environment for I/O operations, not directly related to computations.

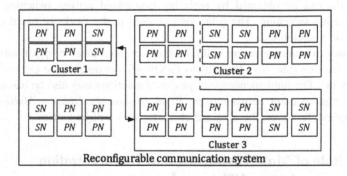

**Fig. 2.** Reconfigurable computing environment

According to Figs. 1 and 2, the improvement of communication efficiency should be based on grouping of system components (computing nodes or channels). In the first group, computing nodes, on the basis of the communication criterion, are combined into independent groups that communicate using a dedicated channel group. In addition to the traffic pattern generated by nodes forming a cluster, the division criteria may also include: classes of tasks solved within a given group, requirements regarding communication delays, error rates and others. The second type of clustering is based on grouping the channels available in the system. In particular, they are divided into smaller independent fragments that can be combined into clusters that support selected users. For example, a separate group may be created by users using services insensitive to communication delays, another generating low traffic, yet another characterized by high explosiveness of generated traffic. The third method is a modification of previous ones and relies on the hierarchization of logical connection channels, where channels are combined into groups that support different sets of processing nodes, which can also be grouped. Due to the diversity of proposed methods, the architecture of connections can be adapted to the traffic patterns and proper communication requirements. Because in optical systems the change in the wavelength used by transceiver element takes milliseconds, adjusting the connection architecture to the current requirement of users can be dynamic and be implemented in real time. The classification of efficiency improvement methods in multi-channel architecture is shown in Fig. 3. Each of the clustering methods is good and its selection depends on optimization criterion.

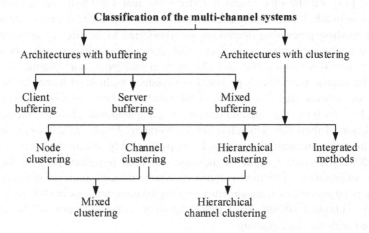

**Fig. 3.** Classification of efficiency improvement of multi-channel systems with clustering

In research on distributed computing systems, the concept of single- and multi-channeling are the most frequently occurring terms. Unfortunately, these terms are in many cases vague. Due to the fact that multi-channeling is the most important basic assumption, the own definition for both types of connections presented in this work were proposed.

**Definition 1**
*Let m and n be the physical nodes of the communication system, and p and q its logical nodes built on the basis of physical nodes. If in any physical channel $[\![m, n]\!]$ of the communication network only one logical channel $[\![p, q]\!]$ is used, then the connections of the communication system are called **single-channel**.*

**Definition 2**

*Let $m$ and $n$ be the physical nodes of a computer system, $p_i$, $(i = 1, \ldots, s_m^l)$, $q_j$ $(j = 1, \ldots, s_n^l)$ – sets of logical nodes built on the basis of physical nodes $m$ and $n$ respectively, where $s_m^l, s_n^l$ are logical degree of m-th and n-th physical vertex. If in any physical channel $[\![m, n]\!]$ more than one logical channel $[\![p_i, q_j]\!]$ is used, then the connections in the system are called* **multi-channel***.*

Definition 1 shows that if in a system based on a set of physical channels, logical channels dependent of the physical environment are built, then such a system remains terminologically single-channel. According to second definition, in multi-channel system the physical communication environment is divided into an ordered set of independent logical channels. Logical degree $s_n^l$ indicates the number of logical channels connected to $n$-th node.

The second basic solution used for the proposed communication environment is hierarchism, in particular hierarchical clustering. On the one hand, methods of the theory of hierarchical systems can be used to solve tasks of component selection and grouping [13], on the other hand it reflects the real conditions of functioning the backbone network. However, the effectiveness of existing methods may be irrelevant in a situation where processing operations are transferred to the edge of network.

In the available literature many successful applications of hierarchy can be found as means of improving communication efficiency in multiprocessor systems [14, 15]. In one of the papers, the author's proposal to introduce multi-level hierarchy of parallel system connections caused reduction in the number of point-to-point communication errors [16]. In turn, the hierarchization of computational clusters, including the manipulation of their size depending on the hierarchy level, and the separation (hierarchization) of communication channel groups favorably affects the level of latency, bandwidth and processing time of messages for each processor [17]. In the further work, a continuation of hierarchy approach can be noticed, with regard to reliability from the point of view of reconstructing interrupted computations in HPC architectures [18]. An additional advantage of this solution was the reduction of the overhead associated with message passing.

On the other hand, the hierarchy is also used in distributed systems, in particular the combination of Cloud, Fog and Edge Computing. However, entering in details, fog is a hierarchical structure [19], while the Edge is just heading in the direction where edge clusters are grouped against various criteria, including resource computational capabilities [20, 21]. It can therefore be expected that the hierarchy of system architecture components, i.e. nodes and communication channels, may bring beneficial effects due to operating parameters and structural measures of proposed architecture.

## 3　Reconfigurable Communication Environment

The model architecture presented in this work consists of two level hierarchy. The first level of computational system is a communication core with ring topology and consists of core nodes. They perform only communication operations and do not process information related to computational process. Following elements in each core node can be distinguished: communication processor, network interface (NI) and transceiver device. Each core node can be equipped with several transceiver devices supporting independent physical channels. In turn for the second level, computational system hierarchy consists of processing nodes (PN), equipped with network interface and transceiver device. It is assumed that one physical channel will reach each of processing nodes.

The basic requirement for proposed computing environment is reconfiguration, guaranteeing the flexibility of connection network. Therefore, it is required to propose devices that enable reconfiguration to the full extent, both at the core level and edge devices, starting from direct connections, ending with broadcast communication. The set of all devices can be divided into two groups: passive switching devices with the possibility of being reconfigured (reprogrammed) and transmitting devices.

The first group of elements consists of passive switching devices. Each component has to establish static connections between logical channels during the reconfiguration process. Through this operation, switching during exploitation is carried out passively [22]. To guarantee all three modes of communication (uni-, multi- and broadcast), four classes of switching devices are singled out: coupling (CD), splitting (SD) and interconnected (ID). The device performing filtration can be integrated with other components, unless it can disturb or limit reconfiguration process. These devices are shown in Fig. 4.

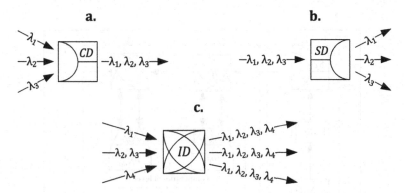

**Fig. 4.** Types of communication core devices. a. Coupling Device (CD); b. Splitting Device (SD); c. Interconnected Device (ID)

Presented elements belong to the network core, which mediates communication between processing devices. Regardless of their role in the system, each of them must be equipped with one or more communication processing units, a set of transceiver devices belonging to the second class of components in proposed architecture, and a network interface.

The next group are transceiver devices. All system processing components, including core elements, are equipped with a set of devices. Their main role is to send and receive packets. To enable communication in processing node, a dedicated network interface must be supplied, which has to appropriately convert the optical signal coming from many channels to an acceptable form. It is worth mentioning that in the general case the model is not limited only to the optical signal. It was only assumed that the transmission medium has to guarantee multi-channeling.

Among the transceiver element, there are two types of division:

- With respect to the transmission type: transmitting, receiving and transmitting-receiving (transceiver device);
- With respect to the nature of switching: fixed wavelength devices ($=$), adapted to a specific channel, and varying wavelength devices ($\sim$), capable of modulation.

Each of the transceiver devices currently supports exactly one logical channel, and therefore the number of logical channels of a physical device is equal to the number of transceiver devices available for a given node. It should be noted that from the point of view of connection flexibility, varying wavelength devices is more justified solution, but considering the construction costs and its reliability, fixed wave devices seems to be a more cost-effective solution.

With components defined, the relationships between them can be defined. The processing node having an appropriate communication module (transceiver devices and network interface NI) can be connected to the system. The general architecture of the communication system is presented in Fig. 5. The network core has ring topology, but it is not imposed. It is used for illustrating the relationship between components of the system.

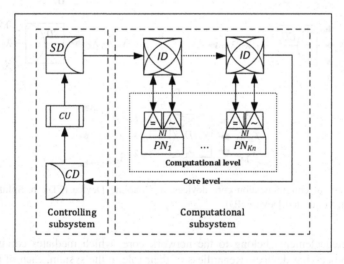

**Fig. 5.** Architecture of the computational system in a passive hierarchical communication environment

Two subsystems are singled out in the connection system: computational and control. In the computational subsystem $K_n$ computational nodes are connected to passive switching core elements, the number of which is directly dependent on the number of logical channels required. Due to the flexibility of internodal connections, each device can be both core (ID) and computational node (PN). In addition, the properties of the proposed architecture make it ideal for edge computing systems, including sensor networks and IoT [23, 24].

The proposed internodal connection architecture is not critical from the point of view of medium access protocols. These can be both autonomous and centrally controlled, deterministic or probabilistic. The protocols used can organize a channel with unicast, multicast or broadcast traffic. In previous work aimed at building a prototype of device, a central control unit (CU) connected by a common dedicated logical channel to each node of the system was used. From the CU level, decisions about the architecture of connections are made. Such a solution is particularly effective if the system dynamics is moderate, which is assured for tasks being solved now. In the future, it is planned to create a set of protocols ensuring autonomous implementation of the topology reconfiguration procedure. It is possible that processing nodes will be connected with a dedicated control channel, however the control unit will no longer be used in the system.

## 4 Basic Reconfiguration Algorithm

Reconfiguration in the architecture is distributed – both core and endpoint devices are the subject of the process. Core devices do not perform computational tasks, but only control and monitor the reconfiguration process and determine the optimal logical paths between devices.

For reconfiguration purposes, an additional channel may be allocated, in which information about changes that must be performed in each individual transceiver devices will be sent. Situations that may cause reconfiguration are, among other:

- Computational group requires an additional component (the component may be a single device, a group of devices with predefined topology or supercomputer);
- Detection of damage to the communication channel or device;
- Inefficiency of current connection architecture.

The occurrence of any of reasons for reconfiguration causes core devices to search for the optimal solution and as a result a set of logical paths is obtained, which must be added, changed or removed. Among the optimization criteria, one can distinguish in particular: communication delays, reliability of the communication between devices and minimal acceptable bandwidth.

Having a new set of logical paths prepared, a message is sent to each of devices, containing dedicated tuning rules for transceiver elements. The message reaches all devices so that they can assess whether the tuning process interferes with their communication paths.

When each of the physical nodes participating in reconfiguration reports its readiness, then in any logical path which includes any element from the set of reconfigured physical nodes, communication should be stopped. The standard solution used in classical network architectures in such case is buffering: during reconfiguration, each device can buffer traffic on the active parts of the path (Fig. 3). With low saturation of the physical channel, creating new connections may not require stopping communication in any of available paths. It can therefore be concluded that an important factor influencing the optimization of connection paths is the degree of introduced changes in relation to the underlying topology, which affects the selection of optimization methods.

After reconfiguration, each device reports its readiness to work. After obtaining mutual consent, the core sends information about the transition to the work state, fully restoring the communication to all nodes again. The core itself returns its original state of monitoring the connection architecture.

The basic reconfiguration algorithm in the proposed architecture has been presented in Fig. 6 in the form of a flowchart.

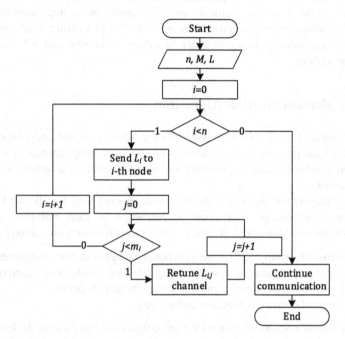

**Fig. 6.** Reconfiguration flowchart. Denotations: $n$ – number of physical nodes being retuned; $m_i$ – number of changes in $i$-th node, $m_i \in M$, where $M = (m_0, \ldots, m_{n-1})$; $L_i$ – list of changes in $i$-th node, $L_i \in L$, where $L = (L_0, \ldots, L_{n-1})$

Optimization process generates following data: list of n physical nodes to be reconfigured and $L_i$ lists of reconfiguration messages, describing changes in i-th node's communication channels. Each reconfiguration message $L_{i,j} = \{p, q, \lambda\}$ contains information about physical transmitting node p (entry point), physical receiving node q (exit point) and logical communication channel $\lambda$ between p and q nodes (e.g. a single wavelength or part of the band in optical fiber). There are three possible cases, which can take place during reconfiguration of a single node (retunePath in Fig. 6):

- Addition of new communication channel if $\lambda \neq \emptyset$ and path $[\![p,q]\!]$ between nodes does not exist;
- Modification of existing channel if $\lambda \neq \emptyset$ and path $[\![p,q]\!]$ between nodes exists;
- Deletion of existing connection between nodes if $\lambda = \emptyset$.

The message $L_{i,j}$ can be sent to any type of device. If any of the physical nodes $(p,q)$ is identical to the node being reconfigured, it means that the node is a computational device (endpoint). Otherwise, it is a core device.

## 5  Restrictions on Reconfiguration

The reconfiguration, however, has some technological and cost limitations. The maximum number of channels coexisting in the communication medium is technologically limited and at the moment it is about 2000 in the case of optical fiber [25]. The increasing number of channels also affects the complexity of architecture, which makes real-time reconfiguration difficult: more time is required to detect a triggering factor and to compute a new logical topology. In addition, the cost of constructing multi-channel architecture increases non-linearly with the increasing number of channels served.

The second important aspect is signal attenuation, occurring with an excessive number of devices connected to one channel or cluster. There may be a downtime in communication between two or more devices if a channel is overloaded, referring to low power signal source. The solution to this problem may require providing a signal regenerator, thus adversely affecting the costs of construction. The transceiver devices have a direct impact on the attenuation. For a low cost architecture, low energy transmitters, e.g. semiconductor lasers, should be used, which in turn limits the maximum range. The main conclusion from considerations is identification the number of channels as a bottleneck in the synthesis process of reconfigurable multi-channel architecture, which strongly influences complexity and construction costs of the system.

One of the possible approaches for complexity estimation are structural complexity measures. They can be applied during synthesis process of a given architecture. In this work, two hierarchical multi-channel architectures were proposed and analyzed for structural complexity using two measures: Complexity B Index (CBI) and Efficiency Index (EI) [26]. For indices calculation, the QuACN package in R was used [27]. Connection architectures were represented as graphs and generated as adjacency lists. Exemplary multi-channel topologies were presented in Figs. 7 and 8. Symmetrical architectures were analyzed for simplicity. Following denotations were applied: $B$ - number of channels, $\alpha$ – number of nodes per cluster, $n$ - number of clusters.

In the next step, structural complexity measures were calculated. Results are presented in Fig. 9. Each architecture consists of 1024 nodes, with changing number of nodes per cluster (and hence the number of clusters).

In the first topology complete connections within each clusters and between neighboring clusters were proposed. Appropriate neighborhood can have influence on load balance in the system. The second one is similar to the dragonfly topology, which

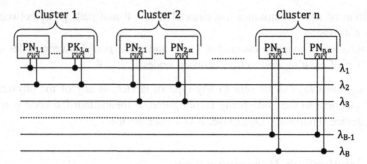

**Fig. 7.** Exemplary architecture topology with fully connected nodes between neighboring clusters, denoted as *Architecture A*

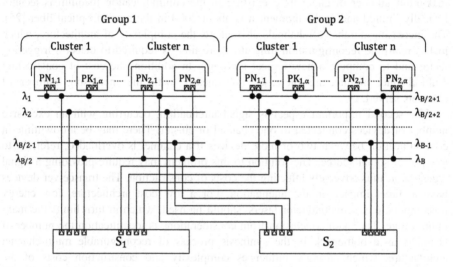

**Fig. 8.** Exemplary architecture topology with clustering of nodes and channels with two groups and two server devices, denoted as *Architecture B*

is used in MPP systems [28]. The devices are divided into two disjoint groups, thus implementing channel hierarchy. Intergroup communication is possible through server devices. On the other hand, hierarchization of nodes was carried out in the same way as in first topology – by node clustering.

CBI takes into account vertex degree and farness to calculate the measure. Complexity of the system is proportional to the CBI value. Branches and cycles increases CBI value, for clique is equal to number of nodes. Values close to zero are obtained for linear or monocyclic topologies. In the case of EI, its increasing value raises general communication efficiency, i.e. reliability, resilience, bandwidth, resulting from the occurrence of multiple paths connecting two nodes, but the complexity of the system also increases. Values is equal to one for clique and zero for isolated nodes. Both measures depend directly on the number of edges in the model graph. Thus, the

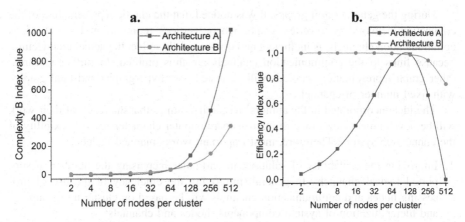

**Fig. 9.** Structural complexity measures calculated for proposed topologies: **a.** Complexity B Index B; **b.** Efficiency Index

increase in complexity of the communication system increases the number of $B$ logical channels, necessary for establishing the connections.

On the basis of the CBI value it can be concluded that for a small $\alpha$ architecture A is characterized by lower complexity. However, above $\alpha = 64$ architecture B has lower CBI value, reducing the number of channels $B$.

The main and noticeable difference in EI values is the complexity for small $\alpha$. Architecture B from the very beginning is characterized by high efficiency, expressed in length (cost) of the path between two nodes. Bearing in mind the CBI results, it can be determined that the system is characterized by complexity with an average number of edges. In turn, the low EI value for architecture A with small $\alpha$ results from the path length between nodes of cluster 1 and $n$.

Analyzing the dependence of structural measures on the number of clusters it can be noticed that the hierarchy decreases complexity of the connection system. It can be concluded that during the reconfiguration, clusters with a reduced number of nodes should be considered. Very complex structures are more difficult to reconfigure and require more time for detection of the triggering factor.

## 6   Summary

On the basis of conducted research, it can be concluded that the use of multi-channel environments in the implementation of hierarchy in computing system components can bring measurable benefits in the form of delay reduction in selected network segments by adapting the architecture to the current operating conditions and traffic pattern. Introduction of hierarchy has so far been successfully used, both in the case of parallel and distributed systems. Therefore a beneficial effect on the system's reliability and overall performance can be expected. However, before the implementation of such environment occurs, a number of tasks should be solved, e.g. reconfiguration conditions or formalization of the path selection process.

During the generation of graphs, it was noticed that the classic representation of the connection network as a vertex graph, where vertices represent nodes and edges communication channels, is inefficient and does not fully reflect the actual architecture, because links to the communication channels are thus omitted. In further research, other graph representation methods will be tested, e.g. hypergraphs and total graphs, with fixed number of channels.

The idea incorporated in the authors' research assumes that the result of their work will be development of a low cost micro-supercomputer characterized by flexibility of the connection system. Therefore, most important works planned include:

1. Improving the scalability of solution, in particular increasing the number of supported logical channels. In general, improving the efficiency of the use of processing nodes and communication channels will be based on clustering (grouping) and hierarchization of system components: nodes and channels;
2. Due to NP-hard complexity of reconfiguration task, it is necessary to develop optimization algorithms, enabling core devices optimal communication paths finding (regarding the appropriate optimization criteria), in particular considering number of changes in logical topology;
3. Computational nodes forming micro-supercomputer, based on Raspberry or Arduino technologies, will be equipped with tunable optical transceivers, settings of which will decide on the architecture of connections – its change will always involve the necessity of retuning the set of transceiver devices;
4. Preparation and implementation of algorithms for load balancing of computational nodes and channels;
5. Preparation of the basic scalable processing module used to build system of any size;
6. Expanding the scope of applications for proposed architecture.

The proposed environment can be particularly interesting in optimization of communication environment in tools dedicated for containerization and cluster management, such as OpenShift, Mesos, Docker or Kubernetes.

**Acknowledgement.** The work was realized as a part of fundamental research financed by the Ministry of Science and Higher Education, grant no. 16.16.110.663.

# References

1. Rauch, L., Bachniak, D.: Dynamic load balancing for CAFE multiscale modelling methods for heterogeneous hardware infrastructure. Procedia Comput. Sci. **108C**, 1813–1822 (2017)
2. Rauch, L., Bzowski, K., Pietrzyk, M., Krol, D., Slota, R., Kitowski, J.: Application of HPC based VirtRoll system to design laminar cooling of strips according to material properties in the coil. In: Tenth ACC Cyfronet AGH HPC Users' Conference, pp. 25–26 (2017)
3. Pajankar, A.: Raspberry Pi Supercomputing and Scientific Programming. Apress (2017)
4. Sterling, T., Anderson, M., Brodowicz, M.: High Performance Computing: Modern Systems and Practices. Morgan Kaufmann (2018)

5. InfiniBand Trade Association. https://www.infinibandta.org/. Accessed 5 Jan 2019
6. Gritzalis, D., Theocharidou, M., Stergiopoulos, G.: Critical Infrastructure Security and Resilience. Springer, Cham (2019). https://doi.org/10.1007/978-3-030-00024-0
7. Trobec, R., Vasiljević, R., Tomasević, M., Milutinović, R., Valero, M.: Interconnection networks in petascale computing systems: a survey. ACM Comput. Surv. **49**(3) (2016)
8. Radivojević, M., Matavulji, P.: The Emerging WDM EPON. Springer, Cham (2017). https://doi.org/10.1007/978-3-319-54224-9
9. Hwang, K., Ghosh, J.: Hypernet: a communication-efficient architecture for constructing massively parallel computers. IEEE Trans. Comput. **C-36**(12), 1450–1466 (1987)
10. Chang, B.-J., Hwang, R.-H.: Performance analysis for hierarchical multirate loss networks. IEEE J. Mag., 187–199 (2004)
11. Mesarovic, M.D.: Multilevel systems and concepts in process control. IEEE J. Mag. **58**(1), 111–125 (1970)
12. Gubko, M.V.: Mathematical models of optimization of hierarchical structures. LENAND (2006)
13. Korte, B., Vygen, J.: Combinatorial Optimization. Theory and Algorithms, 3rd edn. Springer, Heidelberg (2006). https://doi.org/10.1007/978-3-540-71844-4
14. Hajder, M., Bolanowski, M.: Connectivity analysis in the computational systems with distributed communications in the multichanel environment. Pol. J. Environ. Stud. **17**(2A), 14–18 (2008)
15. Rico-Gallego, J.A., Diaz-Martin, J.C., Manumachu, R.R., Lastovetsky, A.L.: A survey of communication performance models for high-performance computing. ACM Comput. Surv. **51**(6) (2019)
16. Yuan, L., Zhang, Y., Tang, Y., Rao, L., Sun, X.: LogGPH: a parallel computational model with hierarchical communication awareness. In: 13th IEEE International Conference on Computational Science and Engineering (2010)
17. Cappello, F., Fraigniaud, P., Mans, B., Rosenberg, A.L.: An algorithmic model for heterogeneous hyper-clusters: rationale and experience. Int. J. Found. Comput. Sci. **16**(2), 195–215 (2005)
18. Bautista-Gomez, L., Ropars, T., Maruyama, N., Cappello, F., Matsuoka, S.: Hierarchical clustering strategies for fault tolerance in large scale HPC systems. In: Proceedings of the 2012 IEEE International Conference on Cluster Computing (2012)
19. Bonomi, F., Milito, R., Zhu, J., Addepali, S.: Fog computing and its role in the internet of things. In: Proceedings of the First Edition of the MCC Workshop on Mobile Cloud Computing, New York (2012)
20. Ferrer, A.J., Marques, J.M., Jorba, J.: Towards the decentralized cloud: survey on approaches and challenges for mobile, ad hoc, and edge computing. ACM Comput. Surv. **51**(6) (2019)
21. Kafhali, S.E., Salah, K.: Efficient and dynamic scaling of fog nodes for IoT devices. J. Supercomput. **73**, 5261–5284 (2017)
22. Li, X., Shao, Z., Zhu, M., Yang, J.: Fundamentals of Optical Computing Technology. Springer, Singapore (2018). https://doi.org/10.1007/978-981-10-3849-5
23. Shi, W., Cao, J., Zhang, Q., Li, Y., Xu, L.: Edge computing: vision challenges. IEEE Internet Things J. **3**(5), 637–646 (2016)
24. Michail, O., Spirakis, P.G.: Elements of the theory of dynamic networks. Commun. ACM **61**(2), 72–81 (2018)
25. Agrawal, G.P.: Optical communication: its history and recent progress. In: Al-Amri, M.D., El-Gomati, M.M., Zubairy, M.S. (eds.) Optics in Our Time, pp. 177–199. Springer, Cham (2016). https://doi.org/10.1007/978-3-319-31903-2_8

26. Kamisiński, A., Chołda, P., Jajszczyk, A.: Assessing the structural complexity of computer and communication networks. ACM Comput. Surv. **47**(4) (2015)

27. QuACN: Quantitative Analysis of Complex Networks. https://rdrr.io/cran/QuACN/. Accessed 1 April 2019

28. Mubarak, M., Carothers, C.D., Ross, R.B., Carns, P.: Enabling parallel simulation of large-scale HPC network system. IEEE Trans. Parallel Distrib. Syst. **28**(1), 87–100 (2017)

# Multi-agent Environment for Decision-Support in Production Systems Using Machine Learning Methods

Jarosław Koźlak[1(✉)], Bartlomiej Sniezynski[1], Dorota Wilk-Kołodziejczyk[1,2], Albert Leśniak[1], and Krzysztof Jaśkowiec[2]

[1] AGH University of Science and Technology,
Al. Mickiewicza 30, 30-059 Kraków, Poland
{kozlak,sniezyn}@agh.edu.pl, albert@vp.pl
[2] Foundry Research Institute in Krakow, Al. Zakopiańska 73, Kraków, Poland
{dorota.wilk,krzysztof.jaskowiec}@iod.krakow.pl

**Abstract.** This paper presents a model and implementation of a multi-agent system to support decisions to optimize a production process in companies. Our goal is to choose the most desirable parameters of the technological process using computer simulation, which will help to avoid or reduce the number of much more expensive trial production processes, using physical production lines. These identified values of production process parameters will be applied later in a real mass production. Decision-making strategies are selected using different machine learning techniques that assist in obtaining products with the required parameters, taking into account sets of historical data. The focus was primarily on the analysis of the quality of prediction of the obtained product parameters for the different algorithms used and different sizes of historical data sets, and therefore different details of information, and secondly on the examination of the times necessary for building decision models for individual algorithms and data–sets. The following algorithms were used: Multilayer Perceptron, Bagging, RandomForest, M5P and Voting. The experiments presented were carried out using data obtained for foundry processes. The JADE platform and the Weka environment were used to implement the multi–agent system.

**Keywords:** Multi–agent systems · Production planning · Machine learning · Casting production

## 1 Introduction

Effective decision making in the management of the production process in an enterprise requires a proper selection of activities, some of which may present a significant complexity. Due to the high competitiveness within the industry markets, it is important to reduce necessary costs when offering the final product but to guarantee its required quality. During the production process, numerous

© Springer Nature Switzerland AG 2019
J. M. F. Rodrigues et al. (Eds.): ICCS 2019, LNCS 11537, pp. 517–529, 2019.
https://doi.org/10.1007/978-3-030-22741-8_37

decisions are made at various levels of detail, which in addition can be carried out either according to clearly justified procedures or using practical knowledge gained by previous experience and decision making taking into account the results of decisions which were made previously.

Decision support systems are developed since 1970s. Their goal is to help decision makers with use of models [13], data [12] and knowledge, based on AI approach [3]. Classically, it is implemented as a part of an enterprise-wide application like Enterprise Resource Planning (ERP). Along time, new technologies that can be used for decision support appear, such as genetic algorithms, machine learning, data mining and big data. It is now possible and advantageous to apply approaches such as Industry 4.0 [2,19] or Internet of Things [10] in the production process. Centralized approach becomes inefficient in integrating heterogeneous data sources, many data processing techniques and decision making methods. An approach especially designed for such a case is agent technology [20]. It allows to cooperate with various elements, and to enrich the developed systems with decentralized decision-making methods provided by multi-agent systems approach, and decision-making techniques based on knowledge learned by machine learning algorithms.

In this work, we deal with specific production assumptions in which not all of the campaigns are fully automated. This often happens due to the earlier purchase of components of production lines with limited functional ranges, whose full automatic integration in the future would require very high costs. Therefore, it is beneficial to use a multi–agent approach and data mining methods for a selection of basic decisions in certain selected fragments of the company production process, assuming that certain periods of the process cannot be precisely described and then analyze this information in order to draw specific conclusions that after can be aggregated to develop appropriate policy strategies.

Our scientific contribution is to propose and verify a multi–agent model which supports decision making during the production process and selection of different parameters crucial to the process. We focus on attempts to optimize activities related to the organization and control of the production process in the foundry. A simulator of such a system was prepared, which can predict the results of decisions made during the casting preparation process. Its initial version and preliminary results have been presented in [8]. The system has been expanded and much more experimental research has been carried out based on real data collected during the observation of the actual production process.

The paper is organized as follows. Section 2 contains a description of related research. In section 3 the model of our system is presented. In section 4 we describe main agents, their responsibilities and how they might be represented using the presented model. In section four we present our data-sets and results of our analyses. Section 5 concludes.

## 2    Related Research

In this section we tackle following research fields: application of multi–agent approaches to production systems, machine learning in production optimization and decision support systems in production systems, and in foundries particularly.

## 2.1 Multi-agent Approach in Production Systems

The multi-agent approach in production control processes is related to the assurance of decentralization of the decision-making process and the possibilities of using algorithms of collective decision making. At the same time, methodologies specific to the multi-agent approach are being developed that can be used to describe such systems. In [10] there is an analyses of various types of cyber–physical systems, composed of closely connected physical and software components. A classification of such systems was proposed while focusing on the advantages offered by a multi-agent approach in such systems: the autonomy, the use of artificial intelligence techniques or the search for operation strategies using agent simulations. The paper [19] is focused on the use of such important mechanisms used in agent systems which are the decision-making processes that are being developed and negotiations.

The paper [7] is focused on ensuring the fulfillment of requirements related to real-time systems and the use of low-cost technologies. In [23] an expert system based on agents using fuzzy calculation techniques is presented.

## 2.2 Machine Learning in Production Optimization

One of the important areas in production is machine scheduling problem. In [6] rule learning framework with creation of optimal solution is introduced. Support vector machines (SVM) algorithm is applied in [5]. The goal is to solve resource constraint scheduling problem. The main problem is labelling cost. It needs to be done by experts which is time-consuming. Therefore approach more often being used is based on reinforcement learning. In [14], reinforcement learning approach is proposed for scheduling in online flow-shop problem. Similar problem is solved in [15] using supervised learning approach.

## 2.3 Decision–Support Systems in Production Systems in Foundries

Taking into account the complexity of the decision-making process of production control, a number of computer decision support systems have been created to optimize these activities. In particular, various description models for such systems have been developed.

In [26] management issues of foundry enterprises and especially production management, are discussed. The authors propose several models such as: Single-piece management model (based on casting life-cycle), Process management model (based on task-driven technology), Duration monitoring model (based on surplus period), and Business intelligence data analysis model (based on data mining).

In [11] is presented a mathematical model based on the actual production process of a foundry flow shop. An improved genetic algorithm is proposed to solve the problem.

In [9], another meta-heuristic optimization method, a particle swarm optimization, is applied to find Pareto optimal solutions that minimize fuel consumption.

The paper [4] contains a model for ranking the suppliers for the industry using the multi-criteria decision making tool, where two hierarchies are distinguished: main criteria (price, quality, delivery and service are ranked based on the experts' opinions) and sub criteria (identified and ranked with respect to their associated main criteria).

In [16], one can find a model for production planning for dynamic cellular manufacturing system. The robust optimization approach is developed with deterministic non-linear mathematical model, which comprises cell formation, inter-cell layout design, production planning, operator assignment, machine reliability and alternative process routings, with the aim to minimize several parameters (machine breakdown cost, inter-intra cell part trip costs, machines relocation cost, inventory holding and backorder costs, operator's training and hiring costs).

Different particular production problems are solved using decision support systems. The paper [17] describes an automatic assist to estimate a production capacity in a casting company, In [25] there is a proposition of heat treatment batch plan, which is very important to ensure the quality of casting products. A charge plan for heat treatment is not a simple combination of different castings, castings with great difference in heat treatment process cannot be put into the same furnace, otherwise, the quality of castings can be adversely affected. On the other hand, furnace capacity and the delivery deadline of castings must be considered to maximize the use of resources and guarantee delivery on time. The charge plan for heat treatment is considered as a complex combinatorial optimization problem.

In [24], the key technology of intelligent factory is reviewed, and a new "Data + Prediction + Decision Support" mode of operation analysis and decision system based on data driven is applied in a die casting intelligent factory. Three layers of a cyber-physical system are designed. In order to form effective decision support, pre-processes the manufacturing data and uses the data mining technology to predict the key performance.

## 3    System Model

Let us define a multi-agent system for decision support with machine learning and data mining as the following pair:

$$System = (E, A), \tag{1}$$

where $E$ is environment and $A$ is a set of agents.

$$E = (P, C, I), \tag{2}$$

where $P$ is set of resources representing process parameters that are observed by agents, $C$ is a set controlled resources representing decisions about the process that may be taken by the agents, $I$ is a set resources representing intermediate results calculated by agents. We define $R$ as the set of all resources: $R = P \cup C \cup I$.

Every resource $r \in R$ has corresponding domain $D_r$ that contains a special empty value $\varepsilon$. We assume that in the time stamp $t$ the environment has specific values assigned to the resources. It is represented by a value function $f_t : R \ni r \to v_r \in D_r$. This function may be extended to assign values to set of resources. To represent a set of possible values of resources $X \subset R$ we write $D(X)$.

Agent definition is a simplified version of a learning agent proposed in [18]. The agent is the following tuple:

$$A_i = (X_i, Y_i, S_i, s_i, s_i^0, Act_i, \pi_i, L_i, E_i, K_i, Col_i). \tag{3}$$

$X_i$ represents observations of agent $i$, $X_i \subset R$, $Y_i$ represents variables calculated by the agent $i$, $Y_i \subset C \cup I$, $S_i$ is a set of states, $s_i \in S_i$ is a current state and $s_i^0 \in S_i$ is the initialization state, $Act_i$ is a set of actions, $\pi_i$ is a strategy, $\pi_i : D(X_i) \times S_i, \to Act_i$, $L_i$ is a possibly empty learning algorithm, that may modify $\pi_i$, $E_i$ represents experience collected by the agent that can be used by $L_i$ to learn the knowledge $K_i$. This knowledge can be used by $\pi_i$ to choose appropriate action or during action execution. $Col_i = \{(r, A_j) : r \in I, A_j \in A\}$ represents a set of agents that $A_i$ can ask to calculate given resource.

The type of the learning algorithm $L_i$ depends on the application. For example, reinforcement learning directly creates the strategy. The agent may also apply supervised learning algorithm to learn a model that is used to choose the best action (see [18]). In complex environments the agent could apply several learning algorithms to create more than one model; however, we did not need it in our experiments yet.

Relation between the agents is defined by a common use of resources. We say that agent $A_i$ depends on $A_j$ if $Y_j \cap X_i \neq \emptyset$. It means that $A_i$ observes resources calculated by $A_j$.

The behaviour of the agents is defined in Algorithm 1. After the initialization (lines 1–3) there is a main loop in which the agent observes the environment, processes communicates sent by other agents (see below) and updates its state (taking into account these communicates). Next (lines 8–11), it decides if results of the previous action or observed situation should be stored in the experience. If yes, appropriate example is added to the $E_i$. Next, an action $a_i$ is chosen according to the current strategy $\pi_i$ and it is executed (lines 12–13) using current knowledge $K_i$. Finally, the learning algorithm is executed (lines 14–16). In case of reinforcement learning, it may happen every turn. It results in update of the $K_i$ representing the $Q$ table used by the $\pi_i$ to chose the action. In case of supervised or unsupervised learning, a new model (or models), which is applied in $\pi_i$ is relearned or the old one is incrementally updated. This is more time consuming; therefore, it may be performed only occasionally (e.g. every 100 steps).

```
1  t = 0;
2  E_i = K_i = ∅;
3  s_i = s_i^0;
4  while agent A_i lives do
5  │  collect values of observations f_t(X_i);
6  │  process communicates received;
7  │  update state s_i;
8  │  if previous action have given results that should be learned then
9  │  │    generate an example e;
10 │  │    store the example e in the experience E_i;
11 │  end
12 │  chose the action using current strategy a_i = π_i(f_t(X_i), s_i);
13 │  execute action a_i using K_i;
14 │  if it is time for learning then
15 │  │    update the knowledge K_i using L_i(E_i);
16 │  end
17 end
```

**Algorithm 1.** Control algorithm of agent $A_i$

Two types of actions should be distinguished: ones that calculate resource values represented by the set $Act^c$ and ones that are related to a social behaviour $Act^s$. If $a_i \in Act^c$ then its execution calculates values of resources that are subset of $Y_i$ from values of $X_i$. Knowledge $K_i$ may be used during this calculation.

There should be at least three social actions: $ASKFOR(r,j), YES, NO \subset Act^s$. The action $ASKFOR(r,j)$ represents a request sent to the agent $A_j$, on which $A_i$ depends, to provide value of resource $r$. We assume that $(r, A_j) \in Col_i$. It is sent if the agent needs this value. The receiver may respond with answer YES and calculate this value or with answer NO.

## 4  System Realization

The implemented system uses multi–agent platform JADE [1] and machine learning and data mining environment Weka [22]. The developed system uses following external entities:

- Process – production process controlled by agents. Sensors of the Process send and receive data to/from Process Management Agents and controls the variables process parameters.
- User – obtains data regarding the processes and proposed recommendations as well as sends queries to the systems.
- Data base – stores historical data with parameters describing previous production processes.

The following agents are implemented in the system:

- *Production Management Agent* (PMA) - responsible for the management of the Process Management Agents and management of the data gathered from them, Its life cycle equals the life cycle of the whole system.

- *Process Controlling Agent* (PCA) – responsible for communication with the given production process, gathers information from sensor of the production process and sends them to the Production Management Agent, it is also responsible for setting the parameters of the production process. Its life cycle is limited to the duration of the given production process. There may be several PCA associated with various production stages.
- *Product Parameter Prediction Agent* (PPPA) – agent associated with Process Controlling Agent, it is possible to have multiple instances of such agent for various parameters and using different learning algorithms or different datasets of historical data. It functions during the whole life cycle of the system.
- *User Interface Agent* (UIA) – responsible for the interactions with user and subsequent communications with other agents. Its life cycle equals the life cycle of the whole system.

The first three agents are essential for the system. The last provide auxiliary functions. In our implementation process parameters, $P$ represents composition of the cast and casting parameters that are observed by agents, $C$ represents casting parameters that can be controlled, $I$ represent parameters of the product created in the casting process. $X_{PCA}$ correspond to the production process stage. Strategy $\pi_{PCA}$ is constant ($L_{PCA}$ is empty) and it executes actions $Act_{PCA}$ calculating high level parameters $Y_{PCA} \subset I$ of this stage. $X_{PPPA}$ is a subset of $R$ that have influence on a predicted cast parameters $Y_i$. This type of agent stores examples consisting of the production process description (independent variables) and achieved cast parameters (dependent variables) in $E_{PPPA}$. A learning algorithm $L_{PPPA}$ is used to create a knowledge $K_{PPPA}$ predicting cast parameters. We use regression algorithms for this purpose. $Col_{PPPA}$ consists of PCAs, because PPPA may ask PCA to calculate high level parameters. PMA is responsible for supervising the whole production process. $X_{PMA}$ is a subset of $R$ that are important. $Col_{PMA}$ consists of PCAs and PPPAs, because PMA make ask these agents for current values to recommend changes in in the process. Recommendations are presented to the user through (UIA) and if they are accepted, appropriate $Y_{PMA}$ values are set.

## 5   Experiments

### 5.1   Description of Scenarios

Casting production is connected with the completion of a number of simulations and test casts in advance, which aim at gathering data allowing for the subsequent launch of production. The tests were carried out on samples whose chemical composition was as presented in Table 1.

**Table 1.** Chemical composition of cast iron melts

| Melt | C | Si | Mn | P | S | Mg | Cu | Ni |
|------|------|------|------|-------|-------|-------|------|------|
| W1 | 3.65 | 2.59 | 0.18 | 0.052 | 0.014 | 0.060 | - | - |
| W2 | 3.41 | 2.62 | 0.30 | 0.046 | 0.016 | 0.056 | 0.48 | - |
| W3 | 3.39 | 2.62 | 0.29 | 0.042 | 0.010 | 0.036 | 0.51 | 0.72 |

The tests were carried out according to two different variants of heat treatment and data gatherer during real world experiments described in [21]. Heat treatment variants influence the final cast parameters. Samples prepared in this way have been tested in the field of breaking energy parameters for these casts. The research aims to determine the parameters of the production process. The tests are carried out during the preparation of the production of a cast element from a new material. Looking at the whole process, this can be placed in the scope of performing a trial series. The results of this work are supported by the designer and technologist at the stage of preparing the element for production (planning process parameters, the shape of the mold and casting), as well as at the stage of performing the test series and process simulation. In the absence of data in this area, there is a need for many technological trials. The developed tool allows you to limit their number.

The evaluation carried out for the purposes of the current article focuses on the following elements:

(i) determination of the correlation, we investigate correlations between parameters such as metal chemical composition and heat treatment of the casting. These processes have an effect on obtaining the parameters of the finished product, in our case of impact strength. The impact test was carried out on a machine called the Charpy hammer, type PSW 300, with a maximum impact energy of 300 J. During the examination of the predicted forces necessary to distort the considered alloys and the actual acquired information on the distorting forces of the considered sample we calculate also the prediction error of these quantities using different types of classifiers and different sizes of the considered set of samples.

(ii) analysis of the time costs of building the model for the larger artificially generated data sets describing samples, taking into account different classifiers and different sizes of the analyzed sets.

We decided to use for the prediction algorithms with a high level of complexity, of which a majority contain basic algorithms, which should increase the chances of obtaining high-quality results. We used the Weka [22] environment and following algorithms: *Bagging, RandomForest, M5P, MultilayerPerceptron* and *Vote*. The *Bagging* algorithm provides averaging of values returned by underlying REPTree algorithms using learning decision trees with reduced-pruning, which improves the stability and accuracy of results. The *RandomForest* algorithm calculates the results as an average obtained from several trees and it is considered that to significantly reduces the risk of over-fitting and limits the influence of a tree that does not achieve good results. The *M5P* algorithm generates 5 model trees. *MultilayerPerceptron* uses a supervised learning method called back propagation of error to set appropriate values of weights associated with neurons building the model. The *Vote* algorithm is a complex algorithm using Bagging, RandomForest, M5P and MultilayerPerceptron algorithms, from which the average value is calculated.

The considered data-sets are as follows:

- *Full* – 432 samples, for each considered set of parameter values it contains 3 records with different calculated results of sample breaking energy,
- *Sample1* – 144 samples, it contains the average value of sample breaking energy for each considered set of parameter values,
- *Sample2* – the smallest considered sample, it contains 14 randomly selected samples that were extracted from the reference pool containing 144 samples with mean values of breaking energy.

### 5.2   Results for Different Classifiers and Different Selection of Sample Data

Figure 1 shows that for different algorithms and for different sizes of data-sets it was possible to obtain a similar, high correlation level. It can be seen that usually for averaged values (*Sample1*) good results were obtained compared to other sets, especially in the case of MultilayerPerceptron. A very limited, randomly selected *Sample2* set also showed a fairly high correlation of predicted values.

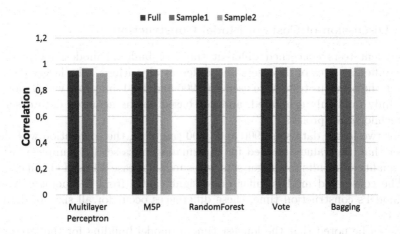

**Fig. 1.** Quality of prediction for selected classifiers and considered data-sets

In Fig. 2 the smallest relative absolute error for the Full set was obtained for the Vote algorithm, for the *Sample1* set - for the Vote algorithm as well, and for the *Sample2* set - for the RandomForest algorithm. It can be seen that for the majority of cases the smallest error is obtained for the data-sets with the most detailed data (and the largest set), only in the case of MultilayerPerceptron the best quality of prediction was obtained for *Sample1*, and for Random Forest for *Sample2*, which is definitely the smallest set. The smallest set (*Sample2*) is characterized by the highest error of the relative absolute error prediction, only in the case of Random Forest the analysis of data-set with the averaged values (*Sample1*) give the highest error.

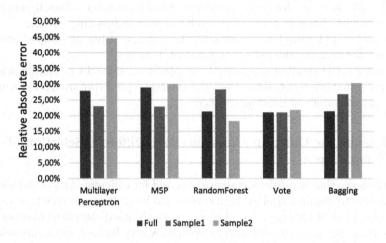

**Fig. 2.** Relative errors for selected classifiers and considered data-sets

## 5.3   Discussion of Costs of Model Construction

Subsequent tests concerned different times of building models for prediction using different sizes of data sets and different algorithms. The larger data sets than in the previous tests were used (90, 900 and 3600 samples), which, however, were fully randomly generated, and not based on the acquired data describing the technological process.

For two larger data-sets (900 and 3600 samples), the correlation was clearly higher than the value obtained for the smallest data-set (90 samples) (Fig. 3). The quality of results for larger sets was similar in case all used classifiers.

The considered models differ quite significantly from the point of view of the model's construction time, these differences occur for all sizes of data sets (Fig. 4).

It can be noted that the longest time of model building for the largest sets (3600 and 900 samples) has the Vote algorithm. This is in line with expectations because this algorithm not only uses several base algorithms, but they are of different kinds, which increases complications and resource consumption. For the smallest data-sets (90 samples) the RandomForest algorithm is the slowest. For a large set (3600) the RandomForest algorithm is the fastest, which may be

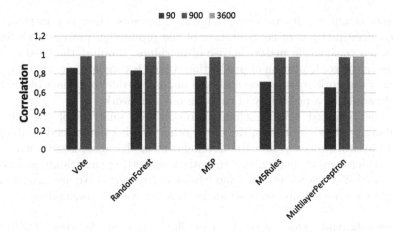

**Fig. 3.** Quality of prediction obtained for larger generated data-sets

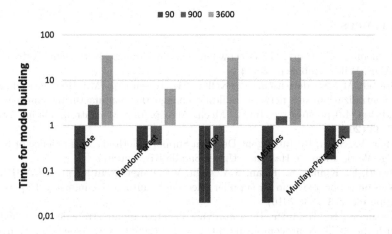

**Fig. 4.** Time for model building for different classifiers and data-set sizes [s]

due to the similarity of the underlying algorithms, for the average data-set (900 samples) the M5P wins, and for the smallest data-set (90 samples) M5P and M5Rules are the fastest.

## 6    Conclusions

In the paper a multi-agent system model for the optimization of the production process for the foundry companies and prediction of the production parameters using data-sets with different characteristics are presented.

The work carried out is aimed at obtaining a reduction in costs by performing a large number of virtual experiments and limiting the number of physical experiments, which is associated with the saving of materials and devices.

Experimental results are very promising. Therefore, these new methods may give good estimations of configuration of the production process with limited costs when considering materials, use of machines, avoiding the need to make changes in the configuration of production, which is cumbersome from the point of view of the work organization and even impossible due to the lack of free time, when it would be a possibility of carrying out such work.

Further planned works include more detailed analyses of the data-sets describing production process, the use of other existing data-sets describing the actual physical processes and the development of the agent system with more advanced interaction mechanisms (negotiations, setting constraints and norms, which particular agents building models should respect). We are also planning to apply machine learning algorithms in these interaction mechanisms.

**Acknowledgment.** This work is partially supported by the TECHMAT-STRATEG1/348072/2/NCBR/2017 Project.

# References

1. Bellifemine, F., Caire, G., Greenwood, D.: Developing Multi-agent Systems with JADE. Wiley, Hoboken (2007)
2. Brettel, M., Friederichsen, N., Keller, M., Rosenberg, M.: How virtualization, decentralization and network building change the manufacturing landscape: an industry 4.0 perspective. Int. J. Mech. Aerosp. Ind. Mechatron. Manuf. Eng. **8**, 37–44 (2014)
3. Dhar, V., Stein, R.: Intelligent Decision Support Methods: The Science of Knowledge Work. Prentice-Hall Inc., Upper Saddle River (1997)
4. Dweiri, F., Kumar, S., Khan, S.A., Jain, V.: Designing an integrated AHP based decision support system for supplier selection in automotive industry. Expert Syst. Appl. **62**, 273–283 (2016)
5. Gersmann, K., Hammer, B.: Improving iterative repair strategies for scheduling with the SVM. Neurocomputing **63**, 271–292 (2005). New Aspects in Neurocomputing: 11th European Symposium on Artificial Neural Networks
6. Ingimundardottir, H., Runarsson, T.P.: Supervised learning linear priority dispatch rules for job-shop scheduling. In: Coello, C.A.C. (ed.) LION 2011. LNCS, vol. 6683, pp. 263–277. Springer, Heidelberg (2011). https://doi.org/10.1007/978-3-642-25566-3_20
7. Kantamneni, A., Brown, L.E., Parker, G., Weaver, W.W.: Survey of multi-agent systems for microgrid control. Eng. Appl. Artif. Intell. **45**, 192–203 (2015)
8. Koźlak, J., Śnieżyński, B., Wilk-Kołodziejczyk, D., Kluska-Nawarecka, S., Jaśkowiec, K., Żabińska, M.: Agent-based decision-information system supporting effective resource management of companies. In: Nguyen, N.T., Pimenidis, E., Khan, Z., Trawiński, B. (eds.) ICCCI 2018. LNCS (LNAI), vol. 11055, pp. 309–318. Springer, Cham (2018). https://doi.org/10.1007/978-3-319-98443-8_28
9. Lee, H., Aydin, N., Choi, Y., Lekhavat, S., Irani, Z.: A decision support system for vessel speed decision in maritime logistics using weather archive big data. Comput. Oper. Res. **98**, 330–342 (2018)
10. Leitão, P., Karnouskos, S., Ribeiro, L., Lee, J., Strasser, T., Colombo, A.W.: Smart agents in industrial cyber-physical systems. Proc. IEEE **104**(5), 1086–1101 (2016)

11. Li, X., Guo, S., Liu, Y., Du, B., Wang, L.: A production planning model for make-to-order foundry flow shop with capacity constraint. Math. Probl. Eng. **2017**, 1–15 (2017)
12. Power, D.J.: Understanding data-driven decision support systems. Inf. Syst. Manag. **25**(2), 149–154 (2008)
13. Power, D.J., Sharda, R.: Model-driven decision support systems: concepts and research directions. Decis. Support Syst. **43**(3), 1044–1061 (2007)
14. Qu, S., Chu, T., Wang, J., Leckie, J., Jian, W.: A centralized reinforcement learning approach for proactive scheduling in manufacturing. In: ETFA, pp. 1–8. IEEE (2015)
15. Sadel, B., Sniezynski, B.: Online supervised learning approach for machine scheduling. Schedae Inf. **25**, 165–176 (2017)
16. Sakhaii, M., Tavakkoli-Moghaddam, R., Bagheri, M., Vatani, B.: A robust optimization approach for an integrated dynamic cellular manufacturing system and production planning with unreliable machines. Appl. Math. Model. **40**, 169–191 (2016)
17. Sika, R., et al.: Trends and Advances in InformationSystems and Technologies, vol. 747. Springer, Heidelberg (2018). https://doi.org/10.1007/978-3-319-77700-9
18. Sniezynski, B.: A strategy learning model for autonomous agents based on classification. Int. J. Appl. Math. Comput. Sci. **25**(3), 471–482 (2015)
19. Wang, S., Wan, J., Zhang, D., Li, D., Zhang, C.: Towards smart factory for industry 4.0: a self-organized multi-agent system with big data based feedback and coordination. Comput. Netw. **101**, 158–168 (2016). Industrial Technologies and Applications for the Internet of Things
20. Weiss, G. (ed.): Multiagent Systems: A Modern Approach to Distributed Artificial Intelligence, 2nd edn. MIT Press, Cambridge (2013)
21. Wilk-Kołodziejczyk, D., Regulski, K., Giętka, T., Gumienny, G., Kluska-Nawarecka, S., Jaśkowiec, K.: The selection of heat treatment parameters to obtain austempered ductile iron with the required impact strength. J. Mater. Eng. Perform. **27**, 5865–5878 (2018)
22. Witten, I.H., Frank, E., Hell, M.A.: Data Mining: Practical Machine Learning Tools and Techniques, 3rd edn. Elsevier, Amsterdam (2011)
23. Zarandi, M.H.F., Tarimoradi, M., Shirazi, M.A., Turksan, I.B.: Fuzzy intelligent agent-based expert system to keep information systems aligned with the strategy plans: a novel approach toward SISP. In: 2015 Annual Conference of the North American Fuzzy Information Processing Society (NAFIPS) Held Jointly with 2015 5th World Conference on Soft Computing (WConSC), pp. 1–5, August 2015
24. Zhao, Y., Qian, F., Gao, Y.: Data driven die casting smart factory solution. In: Wang, S., Price, M., Lim, M.K., Jin, Y., Luo, Y., Chen, R. (eds.) ICSEE/IMIOT-2018. CCIS, vol. 923, pp. 13–21. Springer, Singapore (2018). https://doi.org/10.1007/978-981-13-2396-6_2
25. Zhou, J., Ye, H., Ji, X., Deng, W.: An improved backtracking search algorithm for casting heat treatment charge plan problem. J. Intell. Manuf. **20**(3), 1335–1350 (2019)
26. Zhou, J., et al.: Research and application of enterprise resource planning system for foundry enterprises. Appl. Energy **10**, 7–17 (2013)

# Track of Applications of Matrix Methods in Artificial Intelligence and Machine Learning

# Biclustering via Mixtures of Regression Models

Raja Velu[1]([✉]), Zhaoque Zhou[1], and Chyng Wen Tee[2]

[1] Whitman School of Management, Syracuse University,
Syracuse, NY 13244, USA
{rpvelu,zzhou37}@syr.edu
[2] Lee Kong Chian School of Business, Singapore Management University,
Singapore 178899, Singapore
cwtee@smu.edu.sg

**Abstract.** Biclustering of observations and the variables is of interest in many scientific disciplines; In a single set of data matrix it is handled through the singular value decomposition. Here we deal with two sets of variables: Response and predictor sets. We model the joint relationship via regression models and then apply SVD on the coefficient matrix. The sparseness condition is introduced via Group Lasso; the approach discussed here is quite general and is illustrated with an example from Finance.

**Keywords:** Multivariate regression · Singular value decomposition · Dimension reduction · Mixture models

## 1 Introduction

In the area of data mining for high-dimensional data with '$m$' units and '$n$' variables, where '$m$' and '$n$' are very large, there is a great deal of interest to reduce the dimensions on both sides. For the reduction of large number of variables to a smaller number, the techniques based on Principal Component Analysis (PCA) are normally used. The interpretation of these components is usually improved by the sparseness constraints such as LASSO or RIDGE. On the other hand, the reduction in the large number of units is handled by clustering techniques or through finite mixtures models. In the former, the focus is on correlation between the variables, and in the latter the focus is on the distance between units with appropriate standardization of the variables. In this paper, we want to study the relationship between two sets of variables $Y_t$ and $X_t$, which are of dimensions '$m$' and '$n$' collected over '$T$' time periods via multivariate regression model and we want to reduce the dimension of the $(m \times n)$ coefficient matrix on both units and variables. Biclustering methods generally focus on a single data matrix; here we focus on the estimated coefficient matrix that relates $Y_t$ to $X_t$, that represents both the data matrix $Y$ and $X$. But this has to be properly weighted in for the estimation errors.

© Springer Nature Switzerland AG 2019
J. M. F. Rodrigues et al. (Eds.): ICCS 2019, LNCS 11537, pp. 533–549, 2019.
https://doi.org/10.1007/978-3-030-22741-8_38

In Sect. 2, we introduce the model, review the methodology of clustering of regression models. The following section will cover the estimation methods and some approximate solutions for biclustering. In Sect. 4, we define the problem of biclustering of the regression models and evaluate the procedures using both simulated and real data. The final section will outline the future work.

## 2    Clustering of Regression Models

In microarray experiments, the objectives are twofold: to detect differentially expressed genes and to group genes with similar expression. It is believed that genes with different expression can provide disease-specific markers and co-expressed genes can contribute to understanding the regulatory network aspect of the gene expression (see Qin and Self [5]). In finance, there is a great deal of interest in assessing the commonality in returns among stocks and how this commonality can be related to commonality in order flow (see Hasbrouck and Seppi [3]). This research has led to fundamental questions related to diversification—a central tenet of Markowitz's portfolio theory. The basic regression model can be written as:

$$\underset{T \times 1}{Y_i} = \underset{T \times n}{X_i} \cdot \underset{n \times 1}{c_i} + e_i, \ \ i = 1, \ldots, m, \tag{1}$$

where the times series of responses (returns) on unit $i$ are given in $Y_i$ and the corresponding (order flow) variables related to $i$th response is given in $X_i$.

If $e_i$'s are assumed to be correlated over '$i$', then the above model set-up is called seemingly unrelated regression equations (SURE). The focus on the clustering methods is on in-grouping '$c_i$' into gene-based groups, or in the finance context into trading-pattern based or into industry groups.

**Finite Mixture Regression Models:** We will assume that there are '$K$' groups and each set of models in (1) can come from one of these groups. The random indicator is given by $S_i$ is distributed with unknown probability distribution $\eta = (\eta_1, \ldots, \eta_K)$. The unknown parameters are given in $\theta = (\eta_1, \ldots, \eta_K, c_1, \ldots, c_K, \Sigma_1, \ldots, \Sigma_K)$. The marginal distribution of $Y_i$ given $X_i$ and $\theta$ can be written as:

$$
\begin{aligned}
f(y_i \mid X_i, \theta) &= \sum_{k=1}^{K} f(y_i \mid X_i, S_i, \theta) f(S_i = k \mid \theta) \\
&= \sum_{k=1}^{K} \eta_k f(y_i, X_i c_k, \Sigma_k),
\end{aligned}
\tag{2}
$$

as stated in Frühwirth-Schalter [2, p. 243]. It is important to note that the regression coefficients are identifiable if and only if the number of clusters is less than the number of distinct $(n-1)$ dimensional hyperplanes generated by the covariates. If the covariates show not much variability, there may be identifiability problems which indicates that the groups may not be that distinctive.

Because the cluster membership is not known a priori, treating them as missing information and then using the EM algorithm is what is typically followed. The initial cluster centers are formed from the least squares estimate of $c$'s and grouping them via empirical clustering procedures such as $K$-means clustering. The number of clusters, the choice of '$K$' is made through some criterion of reproducibility over two or more clustering samples. The following BIC criterion is suggested for the selection of '$K$':

$$\text{BIC}_K = 2\text{loglikelihood} - n \ln T. \tag{3}$$

Qin and Self [5] also suggest a measure for the stability of cluster centers:

$$\text{BMV}_K = \max_{k=1...,K}(\text{volume}(\hat{\Sigma}_k)), \tag{4}$$

where BMV is the bootstrapped maximum value and the volume is measured by largest eigenvalue. The joint use of BIC and BMV will lead to a selection of '$K$' value that has large BIC and a small BMV.

The clustering of regression models is fit by applying the EM algorithm as follows: stack the data as $Y = (Y_1', \ldots, Y_m')'$ and $X = (X_1', \ldots, X_m')'$, $S = (S_1, \ldots, S_m)'$ and $\theta = (c_1', \ldots, c_K', \eta_1, \ldots, \eta_K)$ be the stacked parameter vector. The E-step finds the expectation of unknown variables, such as cluster membership and the M-step maximizes the likelihood to obtain parameter estimates. These result in the following steps:

$$
\begin{aligned}
\hat{s}_{ik} &= \frac{\eta_k \Pr(y_i \mid s_{ik} = 1; \theta)}{\sum_{k=1}^{K} \eta_k \Pr(y_i \mid s_{ik} = 1; \theta)} \\
\hat{\eta}_k &= \sum_{i=1}^{m} \hat{s}_{ik} \\
\hat{c}_k &= \left(\sum_{i=1}^{m} \hat{s}_{ik} X_i' X_i\right)^{-1} \sum_{i=1}^{m} \hat{s}_{ik} X_i' Y_i \\
\hat{\sigma}_k^2 &= \sum_{i=1}^{m} \hat{s}_{ik}(Y_i - X_i\hat{c}_k)'(Y_i - X_i\hat{c}_k) / \sum_{i=1}^{m} T\hat{s}_{ik}.
\end{aligned}
\tag{5}
$$

The convergence of these estimates depend on the choice of initial values.

From the structure of the estimates of '$c$' coefficients in (5), it is useful to note that the cluster '$c_k$' coefficients are weighted average of the models entertained with weights being assigned by the chance of their belongings to the cluster membership. Note

$$\hat{c}_k = \left(\sum_{i=1}^{m} w_i\right)^{-1}\left(\sum_{i=1}^{m} w_i \tilde{c}_i\right) \tag{6}$$

where $w_i = \hat{s}_{ik} X_i' X_i$ and $\tilde{c}_i = (X_i' X_i)^{-1} X_i' Y_i$, the least squares estimate. If the design matrix $X_i$'s are same, then the above simplifies to,

$$\hat{c}_k = \sum_{i=1}^{m} \hat{s}_{ik} \tilde{c}_i \tag{7}$$

which implies the likelihood equations based on the data can be replaced by the likelihood equations using the least squares estimates and their distributions.

## 3   Biclustering of Observations and Variables

In the last section the focus was on clustering subjects but clustering of variables is also very useful in practice. The similarity in variables is captured through PCA. The concept of biclustering was discussed earlier by Rao [6] who advocated computing the singular value decomposition (SVD) of the $m \times n$ data matrix $Y$ where '$m$' is the number of subjects and '$n$' is the number of variables and also using the SVD of $Y'$. The biclustering procedure simultaneously clusters both the subjects and the variables thus providing a way to be selective of both subjects and the variables. This is quite useful if any investments in follow-up studies are expensive.

The sparse singular value decomposition (SSVD) is proposed as an exploratory tool for biclustering in Lee, Shen, Huang and Marron [4]. The requirement is that for a '$m \times n$' data matrix, $Y$, use the following penalized sum-of-squares criterion.

$$\|Y - suv'\|_F^2 + \lambda_u P_1(su) + \lambda_v P_2(sv) \tag{8}$$

where the first term is the squared Frobenius norm, $P_1$ and $P_2$ are sparse-inducing penalty terms. Then penalty terms considered are LASSO-based and the solutions are based on soft-thresholding rule. Because of our intent in biclus-tering of regression models, we want to discuss the regression approach to solve (8), as given in Lee et al. [4]. For fixed '$u$', minimization of (8) with respect to $(s, v)$ is equivalent to minimizing (8) with respect to $\tilde{v} = sv$ if

$$\|Y - u\tilde{v}'\|_F^2 + \lambda_v P_2(\tilde{v}) = \|vec(Y') - (I_n \otimes u)\tilde{v}\|_F^2 + \lambda_v P_2(\tilde{v}) \tag{9}$$

where $\otimes$ denotes the Kronecker product. Note for a given '$u$', $I_n \otimes u$ acts as the design matrix and $\tilde{v}$ is the regression coefficient. Note $P_2(\tilde{v}) = \sum_{j=1}^{n} |\tilde{v}_j|$ is the LASSO-penalty.

To solve for '$v$', fix $\tilde{u} = su$ and

$$\|Y - \tilde{u}v'\|_F^2 + \lambda_u P_1(\tilde{u}) = \|vec(Y) - (I_m \otimes v)\tilde{u}\|_F^2 + \lambda_u P_1(\tilde{u}) \tag{10}$$

with the LASSO penalty, $P_1(\tilde{u}) = \sum_{i=1}^{m} |\tilde{u}_i|$. Interestingly this regression app-roach is inherent in the solution of the classical Eckart-Young theorem, that is applied to arrive at the components of singular value decomposition.

## 4   Parsimonious Regression Models

The main interest in biclustering is to discover some desirable unit-variable asso-ciation. In the analysis of microarray data, the goal is to identify biologically relevant genes that are significantly expressed for certain cancer types. Initially the biclustering is used as an unsupervised learning tool; the measurements, the expression levels are for thousands of genes, '$m$' over a small number of subjects, '$n$'. The information on cancer types is used a posterior to interpret and evaluate the performance of SSVD (see Lee et al. [4]). Visually the low-rank approximations can reveal 'checkerboard' structure resulting from gene and subjects grouping. This feature can be appreciated in many different settings and scientific areas as well.

It is well understood that supervised learning tools fare better as they use the additional information which would also support validating the results. Thus in the context of microarray data, the use of information on the presence or absence of cancer will lead to better clustering; if so, where and how do we look for 'checkerboard' structure? In the regression context the covariances between the response and the predictor variables play an important role and thus the form of regression coefficient matrix matters. We suggest two approaches. Stack up the '$c$' coefficients in a matrix and use the SVD on it; we have the $m \times n$ matrix,

$$C \equiv (c_1, \ldots, c_m)' = U\Lambda V' = A \cdot B \tag{11}$$

with rows of $U$ refer to '$m$' subjects and the rows of $V'$ refer to the combinations of the predictors, the '$X$' variables in the model. When the design matrices $X_i$ are identical, the solution to (11) can be related to reduced-rank regression but when they are different, the calculation of $A(m \times r)$ and $B(r \times n)$ is not straightforward. (See (Chapter 7) in Reinsel and Velu [7]). We provide some essential details here.

With the model as stated in (1) and with the condition on the stacked coefficient matrix, C as stated in (11), it follows that we want to estimate C under the condition,

$$Rank(C) = r \leq min(m, n) \tag{12}$$

which implies that A and B are full-rank matrices. Notice that the model (1) can be rewritten as

$$Y_i = X_i B' a_i + e_i \tag{13}$$

where each $a_i$ is of dimension $r \times 1$. Also notice that $BX_i'$ provides the most useful linear combinations of the predictors and the distinction among the regression equations in (1) are reflected in the coefficients $a_i$. The dimension-reductions through reduced rank seemed reasonable because of the 'similarity' of the predictor variable sets ($x_i$) among the $m$ different regression equations in (1). To move forward, we need to impose some normalization conditions (as in normalization required in SVD on U and V):

$$A'\Gamma A = I_r, B\hat{\Sigma}_{xx}B' = \Lambda_r^2 \tag{14}$$

where $\Gamma = \hat{\Sigma}_{ee}^{-1}$ and $\hat{\Sigma}_{xx} = \frac{1}{mT}\sum_{i=1}^m X_i'X_i$. To set up the estimation criterion and the resulting estimates, we closely follow the details given in Reinsel and Velu [7].

Observe that the model (1) can be expressed in the vector form as

$$\mathbf{y} = \bar{X}\mathbf{c} + \mathbf{e} \tag{15}$$

where $\mathbf{y} = (Y_1', \ldots, Y_m')'$, $\bar{X} = Diag(X_1, \ldots, X_m)$ and $\mathbf{c} = vec(C')$ with $\mathbf{e} = (e_1', \ldots, e_m')'$, $Cov(\mathbf{e}) = \Sigma_{ee} \otimes I_T$. The generalized least squares (GLS) estimator is given as

$$\hat{\mathbf{c}} = [\bar{X}'(\Sigma_{ee}^{-1} \otimes I_T)\bar{X}]^{-1}\bar{X}'(\Sigma_{ee}^{-1} \otimes I_T)\mathbf{y} \tag{16}$$

The covariance matrix of $\hat{c}$ that is useful to make inferences on $c$ is given as

$$Cov(\hat{\mathbf{c}}) = [\bar{X}'(\Sigma_{ee}^{-1} \otimes I_T)\bar{X}]^{-1} \tag{17}$$

The error covariance matrix is estimated by stocking the residuals $\hat{e}_i = Y_i - X_i\hat{c}_i$ as $\hat{e} = (\hat{e}_1, \ldots, \hat{e}_m)'$ and $\hat{\Sigma}_{ee} = \frac{1}{T}\hat{e}\hat{e}'$.

To obtain the estimates of $A$ and $B$, we need to resort to iterative procedures as there is no one step solution as in SVD. Denote $\alpha = vec(A')$ and $\beta = vec(B')$ and $\theta = (\alpha', \beta')'$. The criterion to be minimized is

$$S_T(\theta) = \frac{1}{2T} \cdot \mathbf{e}'(\Sigma_{ee}^{-1} \otimes I_T)\mathbf{e} \tag{18}$$

subject to the normalizing constraints in (14). Note that $\mathbf{c} = vec(C') = (A \otimes I_n)vec(B') = (I_m \otimes B')vec(A')$ and so $\mathbf{e} = \mathbf{y} - \bar{X}(A \otimes I_n)\beta = \mathbf{y} - \bar{X}(I_m \otimes B')\alpha$. The first order equations that result from minimizing (18) leads to the following iterative solutions—given $\alpha$ solve for $\beta$ and vice versa.

$$\begin{aligned}
\hat{\alpha} &= [\bar{X}(B)'(\Sigma_{ee} \otimes I_T)\bar{X}(B)]^{-1}\bar{X}(B)'(\Sigma_{ee}^{-1} \otimes I_T)\mathbf{y} \\
\hat{\beta} &= [\bar{X}(A)'(\Sigma_{ee} \otimes I_T)\bar{X}(A)]^{-1}\bar{X}(A)'(\Sigma_{ee}^{-1} \otimes I_T)\mathbf{y}
\end{aligned} \tag{19}$$

In the above, $\bar{X}(B) = \bar{X}(I_m \otimes B')$ and $\bar{X}(A) = \bar{X}(A \otimes I_n)$.

If the design matrices $X_i$ are the same, the solution is rather straight-forward. We will comment on this model later.

Some observations are in order: Note because $c_i = B'\alpha_i$, which from the estimation point of view implies that,

$$X_i'Y_i = (X_i'X_i)c_i = (X_i'X_i)B'\alpha_i \tag{20}$$

will lead to some simplifications in the cluster '$\theta$' estimates if we follow the mixtures model approach taken in Sect. 2. Before formulating the problem as finite mixtures with constraints on the regression coefficient matrix, we want to discuss briefly an approach to introduce sparseness in the estimated A and B matrices.

In the first approach we discuss here in similar to Lee et al. [4] where the spareness structure on A and B matrices are introduced and the simplified structure would be used for identifying both the clusters of units and the clusters of the variables. While sparseness studies in the context of reduced-rank regression is of recent origin, many tend to use norms other than Frobenius. In order to keep the focus and for continuity, we will consider Frobenius norm and in that setting, we discuss the decomposition of $C$-matrix with spareness constraints. Chen and Huang [1] discuss the simultaneous dimension reduction and variable selection. The penalited regression version imposes group-LASSO type penalty that assumes that each '$c_k$' as a group. Note that the range of '$c_k$' over various '$k$' is not linear space and has certain manifold structure. The methodology provided in this paper is quite straightforward and is easy to implement.

The optimization methods related to reduced-rank regression exploit the bilinear nature of the rank decomposition in (11). Note the coefficient matrix, '$C$' is bilinear in the component matries, '$A$' and '$B$' because given either one, the '$C$' matrix is linear function of the other. This leads to the simplified iterative solutions given in (19). Before we formulate the penalized version of the problem, observe from (15)

$$\mathbf{y} = \bar{X}vec(C') + \mathbf{e} = \bar{X}(A \otimes I)B + \mathbf{e} \tag{21}$$

Using (21), the criterion $S_T(\theta)$ in (18) can be restated in terms of approximating full-rank estimate of '$C$' matrix by a matrix of reduced rank.

With (16), (17) and (21), the penalized version of the reduced-rank regression model can be stated as,

$$\underset{A,B}{Min} S_T(\theta) + \lambda_A P_1(A) + \lambda_B P_2(B) \tag{22}$$

similar to what is given in (8) for the SSVD of the data matrix $Y$ for one set of variables. As argued in Chen and Huang [1], we will reduce the problem in (22) to minimizing over '$B$' for a given '$A$', thus resulting in some simplification. First note that minimizing the criterion $S_T(\theta)$ in (18) under the full rank for '$C$' matrix is equivalent to

$$\underset{\theta}{Min} S_T^*(\theta) = \frac{1}{2T}(\hat{\mathbf{c}} - \mathbf{c})'[\bar{X}'(\Sigma_{ee}^{-1} \otimes I_T)\bar{X}](\hat{\mathbf{c}} - \mathbf{c}) \tag{23}$$

and thus for a given '$A$', it is the same as,

$$\underset{B}{Min} S_T^*(\theta) + \sum_{i=1}^{n} \lambda_i \left\| \beta^i \right\| \tag{24}$$

where '$\beta^i$' denotes the $i^{th}$ row of vector of '$B''$' matrix and $\lambda_i$'s are penalty factors that are positive. The condition is known as group-LASSO which implies that the $i^{th}$ predictor can be taken out of the regression framework. The formulation in (24) is more direct and relates how even in the extended set-up, the problem simplifies to calculating SVD of appropriately weighed full-rank regression coefficient matrix. Observe that minimizing criterion in (24) can be further simplified to, because $\hat{\mathbf{c}} - \mathbf{c} = \hat{\mathbf{c}} - (A \otimes I)\beta$, for a given $A$, (24) reduces to,

$$\underset{\beta}{Min}[(A'\Sigma_{ee}^{-1} \otimes I)\hat{\mathbf{c}} - \beta]'[\bar{X}'(\Sigma_{ee}^{-1} \otimes I_T)\bar{X}]$$
$$[(A'\Sigma_{ee}^{-1} \otimes I)\hat{\mathbf{c}} - \beta] + \sum_{i=1}^{n} \lambda_i \left\| \beta^i \right\| \tag{25}$$

Here $\beta^i$ is the transpose of $B^i$ and denotes the $i$-th subpart of the '$\beta$' vector.

Solving (25) requires iterative procedures. Before we describe and apply these methods, we want to observe that certain simplifications that occur when the design matrix, $X_i = X$, in a commonly used multivariate regression model. With $\bar{X} = I_T \otimes X$, the optimization in (25) reduces to

$$\underset{\beta}{Min}[(A'\Sigma_{ee}^{-1} \otimes I)\hat{c} - \beta]'[\Sigma_{ee}^{-1} \otimes XX']$$

$$[(A'\Sigma_{ee}^{-1} \otimes I)\hat{c} - \beta] + \sum_{i=1}^{n} \lambda_i \left\| \beta^i \right\|, \tag{26}$$

the criterion given in Chen and Huang [1]. The solution is easier to obtain because of the simplified structure of

$$\hat{c} = (\Sigma_{ee}^{-1} \otimes XX')^{-1}(\Sigma_{ee}^{-1} \otimes XY') = I_m \otimes (X'X)^{-1}X'Y \tag{27}$$

Our model although appears to be more complex, but it is solvable by numerical routines.

The subgradient solution we suggest here follows Yuan and Lin [8]. The subgradient equations are:

$$[(A'\Sigma_{ee}^{-1} \otimes I)\hat{c} - \beta]_l + \lambda_l s_l = 0 \tag{28}$$

where $[\cdot]$ denotes the $l^{th}$ subvector of $\beta$, where $l = 1, 2, \cdots, n$.

Here $s_l = \beta^l / \left\| \beta^l \right\|$ if $\beta^l \neq 0$ and $s_l$ is a vector with $\left\| s_l \right\|_2 < 1$ otherwise. When $\left\| \beta^l \right\| = 0$, to obtain the subgradient equations, inpute '$\beta$' without the '$l^{th}$' variable and go through the same process of computations and the soft-threshold estimator is given as:

$$\hat{\beta}^l = \left(1 - \frac{\lambda_l}{\|s_l\|}\right)_+ s_l \tag{29}$$

In our practice, the subgradient algorithm converges in several iterations (less than five) and is not so computationally costly for big datasets.

There may be other ways to obtain this and the research is underway to explore these methods.

## 5   Parsimonious Finite Mixtures Biclustering

The parsimonious modeling proposed in (22) where both '$A$' and '$B$' matrices are shrunk through LASSO type penalty may be appropriate to reduce the number of coefficients. But it is not clear how this can be used for grouping. The finite mixture model has the natural appeal as it can be used systematically to decide the number of clusters as well as the probabilities of each unit belonging to various clusters. Recall that in the decomposition the regression coefficient matrix, '$C$', the part '$A$' represents the units' side and the part '$B$' represents variables' side. While the parsimonious or sparse representation may be appropriate, it is not clear how the spareness in 'A' can easily lead to clustering of observations. So we begin with model (13), where the 'r' dimensional '$a_i$' corresponds to unit '$i$'. Assume that there are '$K$' groups and for a given '$B$', thus given $x_i^* = x_i B'$, we postulate that, $a_i$'s come from one of these groups. Thus from (2),

$$
\begin{aligned}
f(y_i \mid X_i^*, \theta) &= \sum_{k=1}^{K} f(y_i \mid X_i^*, S_i, \theta) f(S_i = k \mid \theta) \\
&= \sum_{k=1}^{K} \eta_k^* f(y_i, X_i^* a_k, \Sigma_k),
\end{aligned}
\tag{30}
$$
$$
where \quad \theta = (\eta_1, \ldots, \eta_K, a_1, \ldots, a_K, \Sigma_1, \ldots, \Sigma_K).
$$

This implies that although the row rank of '$A$' is taken to be '$m$', based on the distance between the rows, it can be reduced to '$K$' independent rows.

The EM algorithm (5) can be applied to estimate $\theta$. This is conditional on '$B$' given; thus replace '$X_i$' in (5) by $X_i^*$. Now given '$A$', estimate '$B$' via the second equation in (19). This process can be iterated back and forth between (5) and the estimation of '$\beta$'.

**Sparseness:** To introduce further sparseness in '$B$', we can follow the logic given following equation (21). For a given '$A$', use equation (25) and the subgradient method ((28) and (29)) to solve for '$\beta$'. With this the joint modeling of clustering and the dimension reduction are both achieved.

The appropriate model selection as one can see is going to depend upon a number of parameters: rank of the '$C$' matrix, '$r$' and the row rank of the '$A$' matrix, $K$ and the associated other parameters. We will use the grid search via:

$$
\begin{aligned}
\mathrm{BIC}(r, K) &= 2\text{loglikelihood} - n \ln T \\
\mathrm{BMV}(r, K) &= \max_{r, K}(\text{volume}(\hat{\Sigma}_{r,K}))
\end{aligned}
\tag{31}
$$

This approach is novel and combines both the clustering via finite mixtures and the dimension reduction via SVD. Its performance and its properties are to be studied in depth.

# 6 Numerical Illustration

The illustration given here is not meant to draw any serious implications toward economic theory but our goal is eventually relate the methodology to draw some useful inferences on the trading behavior on major stocks. Since the early 2000s, both academics and practitioners have paid more attention to the magnitudes of cross-sectional interactions among stocks. However, the study of commonality in short-horizon returns, order flows, and liquidity is still of interest in the microstructure analysis of equity markets. Hasbrouck and Seppi [3] note that both short-horizon returns and order flows are characterized by common factors. With this, two research questions have emerged.

Liquidity commonality can easily arise even when trading activity runs in different directions for different stocks, because sizable buy or sell motivated trading can strain liquidity. But commonality in returns can arise because of less firm-specific and more market-wide factor. For instance public information flows and correlation in order imbalances across market, that may affect all stocks. Furthermore, commonality in order flows may be influenced by the differential liquidity of individual stocks as well as by other factors such as asymmetric information, idiosyncratic risks, transaction costs and other forms of market imperfections.

If commonality in stocks' order flows account for the covariance structure of short-term returns, how should we characterize relationships involving commonality in both returns and order flows? Microstructure research focuses on how individual asset price adjusts to new idiosyncratic information. If the market is efficient, new information would be disseminated and interpreted immediately by all market participants, thus prices would quickly adjust to a new equilibrium value determined by the content of the information. But in practice, the price adjustment does not seem to be processed at the same speed for all stocks. Therefore, the price discovery and order flow dynamics have more complex relationship when we consider multiple assets at the same time.

We chose the Dow stocks as our sample because first, the rapid pace of trading provides frequently updated prices and allows us to construct some high-frequency trading measures. Of the 30 firms in the index, we excluded Kraft Foods, for which data are not available for some duration. Second, these 29 stocks, considered as the large capitalization stocks, are normally categorized in the same style and traded mainly by institutional traders, that is, we can expect more correlated trading on these stocks such as index arbitrage, dynamic hedging strategies and naive momentum trading.

Our sample covers 252 trading days in 2015. We establish a standard time frame for the data series using 15-min intervals covering 9:30–9:45, 9:45–10:00, ..., 15:45–16:00 for a total of 26 intervals per trading day. The 15-min time resolution represents a compromise between, on the one hand, needing to look at correlations in contemporaneous order flows across stocks at very short intervals and, on the other hand, requiring enough time for feedback effects from prices into subsequent order submissions. Such data smoothing is essential in handling high frequency data.

We define the stock returns as the difference between the log end-of-interval quote midpoint and the log begin-of-interval quote midpoint. We also consider the following eight order flow measures: (1) Total number of trades; (2) Total share volume; (3) Total dollar volume using log price as stock price (4) The square root of the dollar volume defined as the sum of the square root of each trade's dollar volume using regular stock price; (5) Signed trades defined as the difference between buy trades and sell trades; (6) Signed share volume defined as the difference between buying share volume and selling share volume; (7) Signed dollar volume defined as the difference between buying dollar volume and selling dollar volume using log price as stock price; (8) The square root of the signed dollar volume defined as the difference between the sum of the square root of each buy trade's dollar volume and the same measure for each sell trade using regular stock price. These measures are in the 15-min intervals of trading and as the stocks considered are high capitalization stocks, the time intervals always have active observations. These measures generally reflect two dimensions of order flows (See Fig. 1). Table 1 presents two corresponding eigenvectors and shows that the first eigenvector is dominated by "aggregate measures" and the second one is by "signed measures". Thus the first component represents a 'sum' measure and the second component, a 'contrast' measure. The contrast is between two sides, buy and sell.

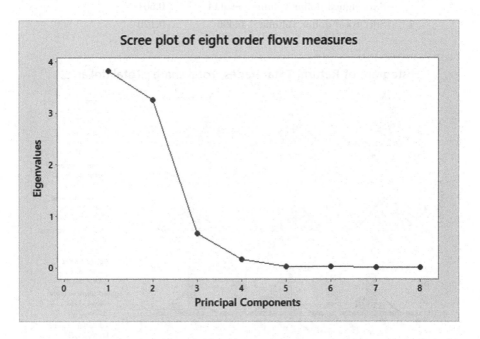

**Fig. 1.** Scree plot of eight standardized order flows measures

Figure 2 presents the histogram of average returns and averages of eight order flow measures for 29 stocks in the sample. Because these variables have significantly different order of magnitude, we standardize our variables to have unit variance and to remove the time-of-day effects. For a representative variable "$z$", let $z_{i,d,k}$ denote the observation from firm $i$ on the $k$-th 15-miniute subperiod of day $d$. Then the standardized variable becomes $z^*_{i,d,k} = (z_{i,d,k} - \mu_{i,k})/\sigma_{i,k}$, where $\mu_{i,k}$ and $\sigma_{i,k}$ are the mean and standard deviation for firm $i$ and subperiod $k$, estimated across days.

**Table 1.** Eigenvectors corresponding to first two largest eigenvalues of eight standardized order flows measures

|  | Eigenvector 1 | Eigenvector 2 |
|---|---|---|
| Signed trades | −0.099 | 0.470 |
| Num of trades | 0.482 | 0.117 |
| Signed share volume | −0.127 | 0.487 |
| Total share volume | 0.487 | 0.113 |
| Signed dollar volume | −0.126 | 0.487 |
| Total dollar volume | 0.486 | 0.115 |
| Sqrt signed dollar volume | −0.114 | 0.501 |
| Sqrt total dollar volume | 0.490 | 0.120 |

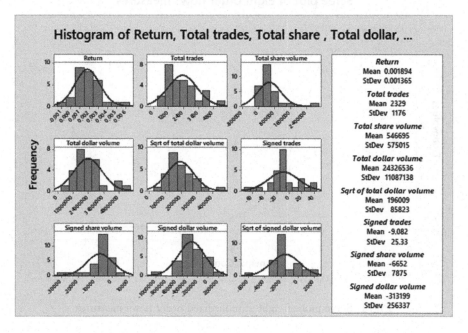

**Fig. 2.** The histograms of returns mean and 8 order flow measures mean for 29 stocks

We run the following regression for each stock $i$ and set the coefficient equals to 0 if it is insignificant at 0.1 level. The whole $29 \times 8$ coefficient matrix is shown in Table 5.

$$r_{i,t} = \sum_{k=1}^{8} c_k x_{k,i,t} + e_i \qquad (32)$$

where $r_{i,t}$ is the return for stock $i$ at time $t$, $x_{k,i,t}$ is the $k$-th order flow measure for stock $i$ at time $t$. The coefficient matrix clearly indicates that not all order flow variables are significant. This is an ad-hoc calculation that does not account for any commonality in the stocks. The methodology developed in this paper provides a more comprehensive framework.

**Table 2.** Two centroids generated by the $K$-means clustering ($K = 2$)

| Variables | Centorid 1 (1) | Centorid 2 (2) | Difference (2)–(1) |
|---|---|---|---|
| Signed trades | 0.203 | 0.041 | −0.162 |
| Num of trades | −0.109 | −0.109 | 0.000 |
| Signed share volume | −0.155 | 0.811 | 0.966** |
| Total share volume | 0.020 | 0.063 | 0.043 |
| Signed dollar volume | 0.159 | −0.840 | −0.999** |
| Total dollar volume | −0.029 | −0.018 | 0.011 |
| Sqrt signed dollar volume | 0.150 | 0.454 | 0.305* |
| Sqrt total dollar volume | 0.097 | 0.090 | −0.007 |

*, ** denote significant at 5% and 1% level, respectively.

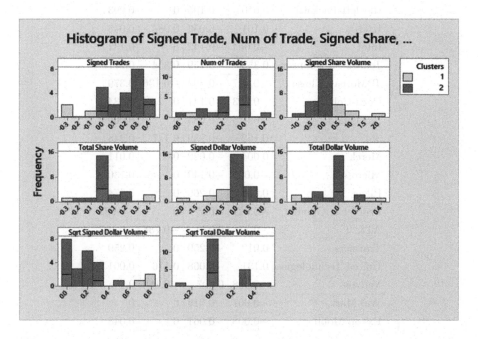

**Fig. 3.** The histograms of coefficients with 2 clusters

First, financial theory confirms that there are two types traders in the market: noise traders and informed trader, therefore we simply use $K$-means clustering ($K = 2$) to group the stocks based on the coefficient matrix. Two centroids are shown in Table 2. We observe that the signed share volume and signed dollar

**Table 3.** Rank two approximation of coefficient matrix using SVD and SSVD algorithm (U matrix)

| Company name | SVD | | SSVD | |
|---|---|---|---|---|
| | $U^{(1)}$ | $U^{(2)}$ | $U^{(1)}$ | $U^{(2)}$ |
| Alcoa | −0.139 | −0.134 | −0.122 | 0.114 |
| American Express | 0.217 | −0.211 | 0.208 | 0.221 |
| Boeing | 0.011 | −0.241 | 0 | 0.246 |
| Bank of America | −0.212 | −0.123 | −0.199 | 0.099 |
| Caterpillar | 0.103 | −0.050 | 0.087 | 0.041 |
| Cisco | −0.275 | −0.206 | −0.277 | 0.186 |
| Chevron | −0.010 | −0.048 | 0 | 0.035 |
| Du Pont | −0.006 | −0.082 | 0 | 0.070 |
| Disney | 0.001 | −0.402 | 0 | 0.418 |
| General Electric | −0.253 | −0.363 | −0.255 | 0.364 |
| Home Depot | 0.005 | −0.100 | 0 | 0.093 |
| Hewlett-Packard | 0.010 | −0.105 | 0 | 0.093 |
| IBM | 0.101 | −0.054 | 0.084 | 0.045 |
| Intel | −0.004 | −0.142 | 0 | 0.131 |
| Johnson & Johnson | 0.352 | −0.259 | 0.354 | 0.277 |
| JPMorgan Chase | −0.147 | −0.383 | −0.136 | 0.379 |
| Coca-Cola | −0.003 | −0.141 | 0 | 0.134 |
| McDonald | −0.015 | −0.083 | 0 | 0.069 |
| 3M | 0.162 | −0.239 | 0.147 | 0.248 |
| Merck | 0.004 | −0.022 | 0 | 0.011 |
| Microsoft | −0.022 | −0.143 | 0 | 0.130 |
| Pifzer | 0.006 | −0.269 | 0 | 0.274 |
| Procter & Gamble | −0.021 | 0.028 | 0 | −0.022 |
| AT& T | −0.716 | 0.070 | −0.738 | −0.107 |
| Travelers | 0.012 | −0.259 | 0 | 0.259 |
| United Technologies | 0.131 | −0.068 | 0.116 | 0.061 |
| Verizon | −0.158 | 0.007 | −0.150 | 0 |
| Wal-Mart | −0.007 | −0.016 | 0 | 0 |
| Exxon Mobil | −0.008 | −0.061 | 0 | 0.048 |

volume have significant difference in these two centroids. It implies that these two variables represent important features for classifying the stocks with different trading behaviors. Figure 3 supports our observation because only the histograms of signed share volume and signed dollar volume have no overlap.

Furthermore, we compute the singular value decomposition (SVD) and the sparse singular value decomposition (SSVD) on the regression coefficients matrix as in (8), (9) and (10). Table 3 shows the results for the $U$ matrix using SVD and SSVD algorithm. When we use $K$-means clustering($K=2$) to cluster these stocks based on $U$ matrix we get from SVD, we obtain the same clustering results as the one using the whole coefficient matrix. It implies that rank-two approximation matrix may have captured most of the information in the structure and the coefficient matrix. From the elements of $U^{(1)}$ from SSVD, we can refer that there may be three groups.

**Table 4.** Rank two approximation of coefficient matrix using SVD and SSVD algorithm (V matrix)

| Variables | SVD | | SSVD | |
|---|---|---|---|---|
| | $V^{(1)}$ | $V^{(2)}$ | $V^{(1)}$ | $V^{(2)}$ |
| Signed trades | 0.063 | −0.391 | 0.039 | 0.383 |
| Num of trades | −0.031 | 0.516 | −0.010 | −0.534 |
| Signed share volume | −0.678 | 0.015 | −0.683 | 0 |
| Total share volume | −0.026 | −0.165 | −0.006 | 0.145 |
| Signed dollar volume | 0.699 | −0.005 | 0.705 | 0 |
| Total dollar volume | −0.043 | 0.275 | −0.026 | −0.267 |
| Sqrt signed dollar volume | −0.203 | −0.448 | −0.182 | 0.426 |
| Sqrt total dollar volume | 0.054 | −0.526 | 0.033 | 0.543 |

Table 4 presents the '$V$' matrix using SVD and SSVD algorithm, we notice that "signed share volume" and "signed dollar volume" dominate all the other variables in both SVD and SSVD cases. Furthermore, it can be observed that more weights are assigned to these two variables when we use the SSVD algorithm.

**Table 5.** Estimated coefficients of the regressions (32) for all 29 stocks

| Company name | Signed trades | Num of trades | Signed share volume | Total share volume | Signed dollar volume | Total share volume | Sqrt signed dollar volume | Sqrt total dollar volume |
|---|---|---|---|---|---|---|---|---|
| Alcoa | −0.10* | 0 | 0.32* | 0 | −0.37* | 0 | 0.70* | 0 |
| American Express | 0.23* | −0.23* | −0.69* | 0 | 0.70* | 0 | 0.11* | 0.28* |
| Boeing | 0.30* | −0.35* | 0 | 0 | 0 | 0 | 0 | 0.34* |
| Bank of America | −0.29* | 0 | 0.59* | 0 | −0.53* | 0 | 0.82* | 0 |
| Caterpillar | 0.24* | 0 | −0.30* | 0 | 0.35* | 0 | 0 | 0 |
| Cisco | 0.41* | −0.18* | 0.92* | 0.40* | −0.93* | −0.40* | 0 | 0 |
| Chevron | 0 | 0 | 0 | 0 | 0 | 0 | 0.21* | 0 |
| Du Pont | 0.19* | 0 | 0 | 0 | 0 | 0 | 0.20* | 0 |
| Disney | 0.07 | −0.59* | 0 | 0 | 0 | −0.22 | 0.28* | 0.55* |
| General Electric | 0.21* | −0.47* | 0.80* | 0 | −0.86* | 0 | 0.28* | 0.55* |
| Home Depot | 0.27* | −0.18 | 0 | 0 | 0 | 0 | 0 | 0 |
| Hewlett-Packard | 0.36* | 0 | −0.07* | 0.07 | 0 | 0 | 0.13* | 0 |
| IBM | 0.26* | 0 | −0.28 | 0 | 0.35* | 0 | 0 | 0 |
| Intel | 0.30* | 0 | 0 | 0.16 | 0 | −0.18* | 0.20* | 0 |
| Johnson & Johnson | 0 | −0.32* | −1.18* | 0 | 1.17* | 0 | 0.33* | 0.34* |
| JPMorgan Chase | 0 | −0.31* | 0.41* | 0.35* | −0.44* | −0.45* | 0.54* | 0.35* |
| Coca-Cola | 0.28* | −0.17 | 0 | 0 | 0 | 0 | 0.19* | 0 |
| McDonald | 0 | 0 | 0 | 0.18 | 0 | 0 | 0.30* | 0 |
| 3M | 0.37* | −0.23* | −0.46* | 0 | 0.52* | −0.20 | 0 | 0.28* |
| Merck | 0.34* | 0 | 0 | −0.18 | 0 | 0.21 | 0 | 0 |
| Microsoft | 0 | 0 | 0 | 0.12 | 0 | −0.14 | 0.50* | 0 |
| Pifzer | 0.29* | −0.35* | 0 | 0 | 0 | 0 | 0.11 | 0.36* |
| Procter & Gamble | 0 | 0 | 0 | 0 | 0 | 0.20* | 0.34* | −0.29* |
| AT& T | −0.31* | 0.20* | 2.13* | 0 | −2.21* | 0.40* | 0.84* | −0.27* |
| Travelers | 0.41* | −0.16 | 0 | 0.20 | 0 | −0.32* | 0 | 0.28* |
| United Technologies | 0.32* | 0 | −0.41* | 0 | 0.41* | 0 | 0 | 0 |
| Verizon | 0.36* | 0 | 0.51 | −0.32* | −0.54 | 0.33* | 0 | 0 |
| Wal-Mart | 0.17* | 0.20 | 0 | −0.12 | 0 | 0 | 0.20* | 0 |
| Exxon Mobil | 0.09 | 0 | 0 | 0 | 0 | 0 | 0.20* | 0 |

* coefficients presented are significant at 5% level.

# References

1. Chen, L., Huang, J.Z.: Sparse reduced-rank regression for simultaneous dimension reduction and variable selection. J. Am. Stat. **107**(500), 1533–1545 (2012)
2. Frühwirth-Schnatter, S.: Finite Mixture and Markov Switching Models. Springer, New York (2006). https://doi.org/10.1007/978-0-387-35768-3
3. Hasbrouck, J., Seppi, D.J.: Common factors in prices, order flows, and liquidity. J. Financ. Econ. **59**(3), 383–411 (2001)

4. Lee, M., Shen, H., Huang, J.Z., Marron, J.S.: Biclustering via sparse singular value decomposition. Biometrics **66**(4), 1087–1095 (2010)
5. Qin, L.-X., Self, S.G.: The clustering of regression models method with applications in gene expression data. Biometrics **62**(2), 526–533 (2006)
6. Rao, C.R.: The use and interpretation of principal component analysis in applied research. JSankhyā: Indian J. Stat. Ser. A **26**, 329–358 (1964)
7. Reinsel, G.C., Velu, R.: Multivariate Reduced-Rank Regression: Theory and Applications. Springer, New York (1998). https://doi.org/10.1007/978-1-4757-2853-8
8. Yuan, M., Lin, Y.: Model selection and estimation in regression with grouped variables. J. R. Stat. Soc.: Ser. B (Stat. Methodol.) **68**(1), 49–67 (2006)

# An Evaluation Metric for Content Providing Models, Recommendation Systems, and Online Campaigns

Kourosh Modarresi[✉] and Jamie Diner[✉]

Adobe Inc., San Jose, USA
kouroshm@alumni.stanford.edu, jamdin@gmail.com

**Abstract.** Creating an optimal digital experience for users require providing users desirable content and also delivering these contents in optimal time as user's experience and interaction taking place. There are multiple metrics and variables that may determine the success of a "user digital experience". These metrics may include accuracy, computational cost and other variables. Many of these variables may be contradictory to one another (as explained later in this submission) and their importance may depend on the specific application the digital experience optimization may be pursuing. To deal with this intertwined, possibly contradicting and confusing set of metrics, this work introduces a generalized index entailing all possible metrics and variables – that may be significant in defining a successful "digital experience design model". Besides its generalizability, as it may include any metric the marketers or scientists consider to be important, this new index allows the marketers or the scientists to give different weights to the corresponding metrics as the significance of a specific metric may depends on the specific application. This index is very flexible and could be adjusted as the objective of" user digital experience optimization" may change.

Here, we use "recommendation" as equivalent to "content providing" throughout the submission. One well known usage of "recommender systems" is in providing contents such as products, ads, goods, network connections, services, and so on. Recommender systems have other wide and broad applications and – in general – many problems and applications in AI and machine learning could be converted easily to an equivalent "recommender system" one. This feature increases the significance of recommender systems as an important application of AI and machine learning.

The introduction of internet has brought a new dimension on the ways businesses sell their products and interact with their customers. Ubiquity of the web and consequently web applications are soaring and as a result much of the commerce and customer experience are taking place on line. Many companies offer their products exclusively or predominantly online. At the same time, many present and potential customers spend much time on line and thus businesses try to use efficient models to interact with online users and engage them in various desired initiatives. This interaction with online users is crucial for businesses that hope to see some desired outcome such as purchase, conversions of any types, simple page views, spending longer time on the business pages and so on.

J. M. F. Rodrigues et al. (Eds.): ICCS 2019, LNCS 11537, pp. 550–563, 2019.
https://doi.org/10.1007/978-3-030-22741-8_39

Recommendation system is one of the main tools to achieve these outcomes. The basic idea of recommender systems is to analyze what is the probability of a desires action by a specific user. Then, by knowing this probability, one can make decision of what initiatives to be taken to maximize the desirable outcomes of the online user's actions. The types of initiatives could include, promotional initiatives (sending coupons, cash, …) or communication with the customer using all available media venues such as mail, email, online ad, etc. the main goal of recommendation or targeting model is to increase some outcomes such as "conversion rate", "length of stay on sites", "number of views" and so on. There are many other direct or indirect metrics influenced by recommender systems. Examples of these could include an increase of the sale of other products which were not the direct goal of the recommendations, an increase the chance of customer coming back at the site, increase in brand awareness and the chance of retargeting the same user at a later time.

**Keywords:** Recommendation systems · Machine learning · Artificial intelligence

# 1 The Formulation of a New Metric for Recommendation Systems

## 1.1 An Overview of This Work

At first, we demonstrate the problem we want to address, and we do it by using many models, data sets and multiple metrics. Then, we propose our unified and generalized metric to address the problems we observe in using different multiple and separate metrics.

Thus, we use several models and multiple data sets to evaluate our approach. First, we use all these data sets to evaluate performances of the different models using different performance metrics which are "the state of the art". Then, we are observing the difficulties of any evaluation using these performance metrics. That is because dealing with different performance metrics, which often make contradictory conclusion, it'd be hard to decide which model has the best performance (so to use the model for the targeting campaign in mind). Therefore, we create our performance index which produces a single, unifying performance metric evaluation a targeting model.

## 1.2 The Data

Machine learning is the science of data driven modeling approach and as such, there is no one single model could work for all types of data [11–25]. As explained above, we use multiple of models, since there are models that may work the best only on some specific data sets. Also, to use as many diverse data sets, for this work, we use 15 publicly available data sets. The features of the datasets are described in Table 1.

To create a higher degree of variation in the choice of data, we created 100 bootstrap samples from each of the datasets by selecting a random number of rows, a random number of columns, and a random number of missing values. The number of

**Table 1.** The features of some of the different data sets used in this work

| Name | numRows | numCols | numNumericalCol | numCategoricalCol |
|---|---|---|---|---|
| abalone | 4176 | 9 | 8 | 1 |
| air_quality | 9357 | 13 | 13 | 0 |
| batting | 947 | 16 | 16 | 0 |
| bike | 731 | 14 | 9 | 5 |
| boston | 506 | 14 | 13 | 1 |

missing values for the rrecsys package was selected at random between 1% and 97%, while the number of missing values for mice, softImpute, missMDA, missForest, and impute was selected between 1% and 75%, as higher number of missing values caused runtime errors. The number of missing values for the Amelia package was selected between 1% and 20%, with only 20 bootstrap samples per dataset. The reason for this was because running with more sparse data or a larger number of bootstraps caused segmentation errors that crashed R. Before any analysis, we can see that for very sparse matrices, the all packages except rrecsys are not a good option.

## 1.3    The Models for Recommendation Systems

Several models in seven packages, rrecsys, mice, Amelia, softImpute, missMDA, missForest, and impute, all in R, have been used to compare the results. These models are [44, 46, 53, 55, 59–61]:

**Rrecsys:** The package rrecsys had the following models implemented:

1. **itemAverage**: impute the average rating of the item in the missing values
2. **userAverage**: impute the average rating of the user in the missing values
3. **globalAverage**: impute the overall average rating in the missing values
4. **IBKNN**: imputes the weighted average of the k-nearest neighbors of the item in the missing values

**Mice:** In mice, we varied the imputation method available for **numerical** variables with the following:

1. **pmm**: imputes the predictive mean matching in the missing values. One of the biggest advantages is that it maintains the distribution of the data, and one of the biggest disadvantages is that it does not handle well very sparse matrices.
2. **norm**: fits a bayesian linear regression model and imputes the data by predicting missing values.
3. **norm.nob**: fits a linear regression model that ignores the model error.
4. **mean**: imputes the mean of the column in the missing samples.
5. **sample**: finds a random item that is not missing in the column where there is a missing value and imputes it.

**Amelia:** The Amelia package works by performing multiple imputation on bootstrap samples based on the EMB algorithm (Expectation-Maximization with Bayesian Hierarchical classification). It allows for multiprocessing, making it fast and robust. It makes the following assumptions:

1. All the variables have Multivariate Normal Distribution.
2. Missing data is random (MAR).

**softImpute** [44, 46]: The package softImpute implements matrix completion using nuclear norm regularization. It offers two algorithms:

1. **SVD**: iteratively computes the soft-thresholded Singular Value Decomposition (SVD) of a filled in matrix
2. **ALS**: uses alternating ridge regression to fill in the missing entries.

**missMDA:** The package missMDA imputes the missing values of a mixed dataset (with continuous and categorical variables) using the principal component method "factorial analysis for mixed data" (FAMD). It implements 2 methods:

1. **EM**: Expectation Maximization algorithm where missing values are imputed with initial values such as the mean of the variable for the continuous variables and the proportion of the category for each category using the non-missing entries. Then, it performs the FAMD algorithm on the completed dataset until convergence.
2. **Regularized**: Similar to the EM but adds a regularization parameter to avoid overfitting.

**missForest:** The package missForest creates one Random Forest model for each of the columns of the dataset, where the training set is given by the non-missing rows. Then, it uses that model to impute the missing values. It iterates until convergence. Even though Random Forest may deal with missing values in theory, this implementation first imputes the mean of the column into the missing values.

**Impute:** The package impute only has one method that fills in missing values using the K-nearest neighbors. It only works on *numerical* datasets.

## 1.4    The Results

In this section, we use the performance (execution time and accuracy) of the models (Sect. 1.3) with respect to different variables such as size and sparsity of the data sets (Sect. 1.2).

**Execution Time:** We want to compare the computational complexity of the packages because it is of interest to find if a model will perform well when scaling.

We can see that two packages stand out from the rest, missMDA and missForest, followed by softImpute. This is an expected result, as both packages create one model for each column of the dataset and then iterates into convergence. Because we need to create $k \times m$ (*number of iterations* $\times$ *number of columns*), the execution time is greater for these models.

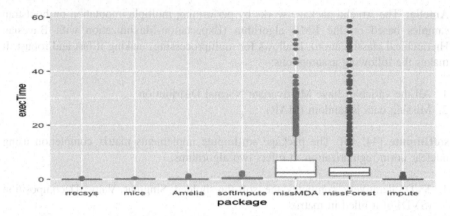

**Fig. 1.** We can see the comparison of the execution time (in seconds) of the packages.

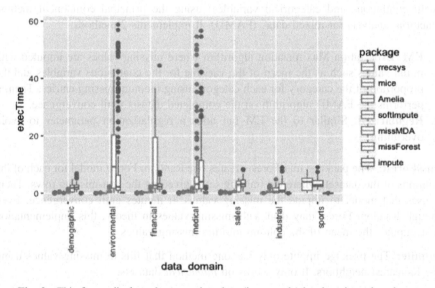

**Fig. 2.** This figure displays the execution time (in seconds) by domain and package.

We can see that missMDA and missForest stand out from the rest in all of the packages regardless of the domain. As expected, the domains with the largest datasets (environmental, financial, and medical) took the longest time to run. The high variance in some of the runs is because the execution time increased exponentially in some models, so small datasets got executed relatively fast while large datasets took an exponentially larger amount of time to execute.

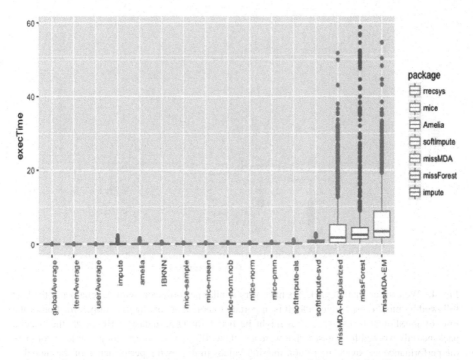

**Fig. 3.** The comparison of the execution time (in seconds) by model and package.

We can see that the **EM** model from missMDA was the one that took the longest to execute, followed by **missForest** and the **Regularized** method from missMDA and the **SVD** method from softImpute. This is in line with the results obtained in the previous points.

**Accuracy Performance:** We measure the error using the Normalized Root Mean Squared Error (NRMSE), which is similar to the RMSE but it divides the error by the variance of the dataset. The reason we used this metric is to be able to compare the performance of difference datasets regardless of the range or variance it has.

$$NRMSE = \sqrt{mean\left(\left(x_{true} - x_{pred}\right)^2\right)\Big/var(x_{true})}$$

**Model Performance with Respect Sparsity:** Below we can see how the model's performance changed by how sparse the matrix was in every domain.

Here, the lines represent a fitted linear model and the shaded region the 95% confidence interval. We can see that the difference in performance between the models is not statistically significant in the demographic, medical, and environmental domain for most levels of sparsity. We can see that the worst performing model in the environmental and financial domain is IBKNN for all levels of sparsity, while the missMDA-EM is the worst performing model for the sales, industrial and sports

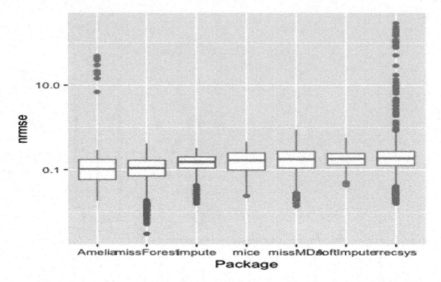

**Fig. 4.** We can see that all packages have very similar performance, with Amelia being the best followed by missForest and impute. It is important to note that Amelia took only a fraction of the time of missForest to execute, so it might be better for large datasets. However, the Amelia package only worked for datasets that were less than 20% sparse, so it is not possible to compare the performance as having more non-missing values improves the performance of the model.

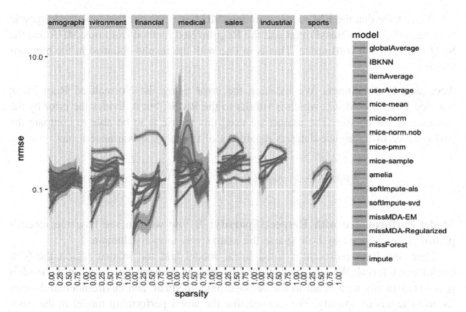

**Fig. 5.** As we can see, the performance of any model is a function of the sparsity of data.

domains with a statistically significant difference. The best performing model in most of the domains and sparsity levels is the missForest, which is an expected result as Random Forest work extremely well in terms of accuracy but are not scalable for large datasets. We can also see an upward trend in most of the models that shows that the model's performance decline if the matrix is sparser. This is an expected result, as having a sparser matrix implies a smaller training sample size and may cause the model to have high bias.

**Model Performance with Respect to the Size of the Data:** Below we can see how the model's performance (measured by the median NRMSE) is a function of the size of the data matrix.

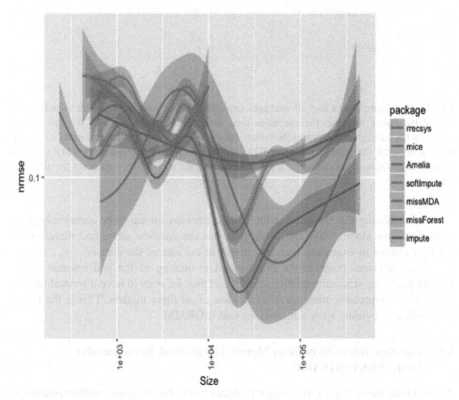

**Fig. 6.** Also, as it can be seen, the accuracy of any model depends on the size of the data as this figure demonstrates that point.

**Execution Time with Respect to the Size of the Data:** Below we can see how the model's performance (measured by the execution time in seconds) changed by size the matrix (rows × columns).

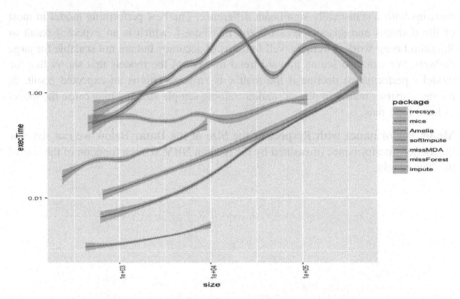

**Fig. 7.** This figure shows that all packages experience an upward trend, where increasing the size of the matrix increases the execution time which is an expected result. We can see that rrecsys is the package that executes the fastest, followed by impute, Amelia, and mice. At the top, we can see that the softImpute, missMDA, and missForest take considerably more time to execute in any size of the dataset. All these differences are statistically significant for all matrix sizes.

It is interesting that the trend in the Amelia package appear to be completely linear with a positive slope, while the softImpute has the smallest slope and therefore the smallest change in execution time by change in the size of the dataset.

All these issues point to the difficulty of evaluating of the performance of any content providing (recommendation) model and thus we need to have a generalized and physically interpretable measure in evaluation of all these models. This is the reason this work is providing such a metric, we call it GRMM.

## 1.5    The New Recommendation Metric, Generalized Recommender Model Metric (GRMM)

As we observed in Sect. 1.4, a major problem in the building and implementation of recommender systems (or content providing model) is the problem of evaluation of such systems. The reasons for the issue of evaluation is due to the presence of many various – and sometime conflicting – performance criteria. As an example, when considering accuracy as one of the performance criteria, we could have a recommender system that may be highly accurate for some quantitative data sets while displaying less desirable accuracy for non-quantitative (qualitative) data sets and so this makes it difficult to come up with a clear description for the accuracy of the recommender system model and/or implementation. The same is true for another important performance measure for the recommender systems, i.e., the execution time. Given all different (and often conflicting) performance measures, it is very difficult to have a correct evaluation of any recommender system.

This work addresses the problem by creating a unified performance measure for Recommender Systems, "Generalized Recommender Model Metric, GRMM". This metric is a normalized measure between 0–1 with 1 to be the optimal value or the highest performance measure index. This is an extremely useful tool for an AI and machine learning scientist (model design), also for marketer and engineer (in charge of implementation of the recommender systems as different system implementation have different performances also) to have a single number indicating the overall performance of any recommender system, including and encompassing all possible performance measures and sub-measures. Thus, this new recommender system index (GRMM) helps all involved in the process of designing, implementing and using recommender systems to have a single number (from 0–1) to evaluate the performance of their systems and to choose the best system (of the highest GRMM metric).

This index metric, GRMM is a flexible and most generalized index allowing everyone dealing with content creation and content providing (scientists, marketers, consultants, engineers and so on) to evaluate the model of providing content using any specific metric or a set of metrics and by giving correct weights (reflecting the business importance).

In general, for any data set X, we could have many performance measures such as accuracy (for qualitative data, accuracy for quantitative data, and so on execution time (as a function of sparsity or dimensions and so on).

In the most general case, for any recommender system X, we could have as many performance measures that may be required, let's say "n" different measures $m_i$ for i = 1:n i.e., $M(X) = [m_1(X), m_2(X), \ldots, m_n(X)]$.

It is important to mention that each performance measure of $m_i(X)$ is a sigmoid function with a continuous value of zero to one. As an example, if $m_1(X)$ is the accuracy measure for the recommender system X, then,

$$m_1(X) = \frac{1}{1 + e^{-\sum_{j=1}^{s} b_j a_j(X)}}$$

Where each function $a_j(X)$ is the specific accuracy (sub-measure of accuracy) performance of the corresponding recommender system model with respect to the factor j. There could be as many as "s" different sub-measure for accuracy or factors in here such as accuracy of the recommender system for continuous or quantitative variables, or the accuracy of the same recommender system for qualitative variables, accuracy with respect to size, sparsity and so on. The coefficients $b_j$ are weights given (by scientists, marketers or engineers) to each sub-measure and are normalized,

$$\sum_{j=1}^{s} b_j = 1$$

Each coefficient $b_j$ could be chosen depending on the specific application and/or specific weights chosen for each factor (accuracy factor for this measure, $m_1(X)$).

Thus, for any given recommender system X, our Generalized Recommender Model Metric, GRMM(X) is,

$$GRMM(X) = \frac{n}{\frac{1}{m_1(X)} + \frac{1}{m_2(X)} + \ldots + \frac{1}{m_n(X)}}$$

Though, a marketer, AI scientist or an engineer may like to give different weights to different performance measures, $m_i(X)$. This is done by considering the corresponding weights, $w_i$ for each performance measure of $m_i(X)$. We have these weights to be normalized so that,

$$\sum_{i=1}^{n} w_i = 1$$

Hence, the most general form of our index, GRMM, becomes,

$$GRMM(X) = \frac{1}{w_1 * \frac{1}{m_1(X)} + w_2 * \frac{1}{m_2(X)} + \ldots + w_n * \frac{1}{m_n(X)}}$$

This is the most general form of our index. This completely covers all possible performance measures (n number of them, with n to be any natural number larger or equal to 1) for any given recommender system. Also, the index includes the weights $w_i$ to allow the marketers, scientists or engineers to choose the corresponding significance or weights for each performance measure. In this most general case, we also consider having different weights for any sub- performance measure. As explained above, for the example of accuracy measure, the weights $b_j$ gives also the same flexibility for the marketers, scientists and engineers to give different importance or weights for different accuracy measures of any recommender system.

Surely, this is the most generalized version of GRMM index.

There is always value for simplicity and brevity in creating any model. That is the reason we tried to find ways to make the index a bit simpler yet covering the main points and performance measure and sub-measures. Thus, in our analysis, we have found that the pseudo-optimal value for n is 2 with the two measures to be accuracy and execution time (time complexity). We also found it's also very good pseudo-optimal choice of setting all weights, *to be equal. That is,*

$$w_i = \frac{1}{n}$$

That is also the same procedure, that for all suboptimal measure the practical value of all weights is to be equal. For the example above, for accuracy measure and its sub-measures,

$$b_j = \frac{1}{s}$$

Obviously, we have the option of using both the most general case of our GRMM index or the "simplified version". The latter version has some extra explanations and descriptions in the original submission.

GRMM is a new and unique metric addressing a major shortcoming in this domain of content providing and content offering. It is physically interpretable and understandable and easy to use for all individuals including marketers, product managers, engineers and scientists.

# References

1. Bjorck, A.: Numerical Methods for Least Squares Problems. SIAM, Philadelphia (1996)
2. Boyd, S., Vandenberghe, L.: Convex Optimization. Cambridge University Press, Cambridge (2004)
3. Breese, J.S., Heckerman, D., Kadie, C.: Empirical analysis of predictive algorithms for collaborative filtering. In: Proceedings of Fourteenth Conference on Uncertainty in Artificial Intelligence. Morgan Kaufmann (1998)
4. Cai, J.-F., Candès, E.J., Shen, Z.: A singular value thresholding algorithm for matrix completion. SIAM J. Optim. **20**(4), 1956–1982 (2008)
5. Candès, E.J., Recht, B.: Exact matrix completion via convex optimization. Found. Comput. Math. **9**, 717–772 (2008)
6. Candès, E.J.: Compressive sampling. In: Proceedings of the International Congress of Mathematicians, Madrid, Spain (2006)
7. Chen, P.-Y., Wu, S.-Y., Yoon, J.: The impact of online recommendations and consumer feedback on sales. In: Proceedings of the 25th International Conference on Information Systems, pp. 711–724 (2004)
8. Cho, Y.H., Kim, J.K., Kim, S.H.: A personalized recommender system based on web usage mining and decision tree Induction. Expert Syst. Appl. **23**, 329–342 (2002)
9. Claypool, M., Gokhale, A., Miranda, T., Murnikov, P., Netes, D., Sartin M.: Combining content-based and collaborative filters in an online newspaper. In: Proceedings of the ACM SIGIR 1999 Workshop on Recommender Systems (1999)
10. Çoba, L., Zanker, M.: rrecsys: an R-package for prototyping recommendation algorithms. In: RecSys 2016 Poster Proceedings (2016)
11. Abalone Data Set. https://archive.ics.uci.edu/ml/datasets/abalone
12. Air Quality Data Set. https://archive.ics.uci.edu/ml/datasets/Air+Quality
13. Batting Data Set. http://www.tgfantasybaseball.com/baseball/stats.cfm
14. Bike Sharing Dataset Data Set. https://archive.ics.uci.edu/ml/datasets/bike+sharing+dataset
15. Boston Data Set. https://archive.ics.uci.edu/ml/datasets/housing
16. Physicochemical Properties of Protein Tertiary Structure Data Set. https://archive.ics.uci.edu/ml/datasets/Physicochemical+Properties+of+Protein+Tertiary+Structure
17. Data: Census: click on the "Compare Large Cities and Towns for Population, Housing, Area, and Density" link on Census 2000. https://factfinder.census.gov/faces/nav/jsf/pages/community_facts.xhtml
18. Concrete Compressive Strength Data Set. https://archive.ics.uci.edu/ml/datasets/Concrete+Compressive+Strength
19. ISTANBUL STOCK EXCHANGE Data Set. https://archive.ics.uci.edu/ml/dtasets/ISTANBUL+STOCK+EXCHANGE#
20. Parkinsons Data Set. https://archive.ics.uci.edu/ml/datasets/parkinsons

21. S&P 500 Index Options. http://www.cboe.com/products/stock-index-options-spx-rut-msci-ftse/s-p-500-index-options/s-p-500-index/spx-historical-data
22. seeds Data Set. http://archive.ics.uci.edu/ml/datasets/seeds
23. Waveform Database Generator (Version 1) Data Set. https://archive.ics.uci.edu/ml/datasets/Waveform+Database+Generator+(Version+2)
24. Breast Cancer Wisconsin (Prognostic) Data Set. https://archive.ics.uci.edu/ml/datasets/Breast+Cancer+Wisconsin+%28Prognostic%29
25. Yacht Hydrodynamics Data Set. http://archive.ics.uci.edu/ml/datasets/yacht+hydrodynamics
26. d'Aspremont, A., El Ghaoui, L., Jordan, M.I., Lanckriet, G.R.G.: A direct formulation for sparse PCA using semidefinite programming. SIAM Rev. **49**(3), 434–448 (2007)
27. Davies, A.R., Hassan, M.F.: Optimality in the regularization of ill-posed inverse problems. In: Sabatier, P.C. (ed.) Inverse Problems: An Interdisciplinary Study. Academic Press, London (1987)
28. DeMoor, B., Golub, G.H.: The restricted singular value decomposition: properties and applications. SIAM J. Matrix Anal. Appl. **12**(3), 401–425 (1991)
29. Donoho, D.L., Tanner, J.: Sparse nonnegative solutions of underdetermined linear equations by linear programming. Proc. Natl. Acad. Sci. **102**(27), 9446–9451 (2005)
30. Efron, B., Hastie, T., Johnstone, I., Tibshirani, R.: Least angle regression. Ann. Stat. **32**, 407–499 (2004)
31. Elden, L.: Algorithms for the regularization of ill-conditioned least squares problems. BIT **17**, 134–145 (1977)
32. Elden, L.: A note on the computation of the generalized cross-validation function for ill-conditioned least squares problems. BIT **24**, 467–472 (1984)
33. Engl, H.W., Hanke, M., Neubauer, A.: Regularization methods for the stable solution of inverse problems. Surv. Math. Ind. **3**, 71–143 (1993)
34. Engl, H.W., Hanke, M., Neubauer, A.: Regularization of Inverse Problems. Kluwer, Dordrecht (1996)
35. Engl, H.W., Groetsch, C.W. (eds.): Inverse and Ill-Posed Problems. Academic Press, London (1987)
36. Gander, W.: On the linear least squares problem with a quadratic Constraint. Technical report STAN-CS-78-697, Stanford University (1978)
37. Golub, G.H., Van Loan, C.F.: Matrix computations. In: Computer Assisted Mechanics and Engineering Sciences, 4th edn. Johns Hopkins University Press, US (2013)
38. Golub, G.H., Van Loan, C.F.: An analysis of the total least squares problem. SIAM J. Numer. Anal. **17**, 883–893 (1980)
39. Golub, G.H., Kahan, W.: Calculating the singular values and pseudo-inverse of a matrix. SIAM J. Numer. Anal. Ser. B **2**, 205–224 (1965)
40. Golub, G.H., Heath, M., Wahba, G.: Generalized cross-validation as a method for choosing a good ridge parameter. Technometrics **21**, 215–223 (1979)
41. Guo, S., Wang, M., Leskovec, J.: The role of social networks in online shopping: information passing, price of trust, and consumer choice. In: ACM Conference on Electronic Commerce (EC) (2011)
42. Häubl, G., Trifts, V.: Consumer decision making in online shopping environments: the effects of interactive decision aids. Market. Sci. **19**, 4–21 (2000)
43. Hastie, T., Tibshirani, R., Friedman, J.: The Elements of Statistical Learning; Data mining, Inference and Prediction. Springer, New York (2001). https://doi.org/10.1007/978-0-387-84858-7
44. Hastie, T., Mazumder, R.: Matrix completion via iterative soft-thresholded SVD (2015)
45. Hastie, T., Tibshirani, R., Narasimhan, B., Chu, G.: Package 'impute'. CRAN (2017)
46. Honaker, J., King, G., Blackwell, M.: Amelia II: a program for missing data (2012)

47. Hua, T.A., Gunst, R.F.: Generalized ridge regression: a note on negative ridge parameters. Commun. Statist. Theory Methods **12**, 37–45 (1983)
48. Iyengar, V.S., Zhang, T.: Empirical study of recommender systems using linear classifiers. In: Cheung, D., Williams, G.J., Li, Q. (eds.) PAKDD 2001. LNCS (LNAI), vol. 2035, pp. 16–27. Springer, Heidelberg (2001). https://doi.org/10.1007/3-540-45357-1_5
49. Jeffers, J.: Two case studies in the application of principal component. Appl. Stat. **16**, 225–236 (1967)
50. Jolliffe, I.: Principal Component Analysis. Springer, New York (1986). https://doi.org/10.1007/b98835
51. Jolliffe, I.T.: Rotation of principal components: choice of normalization Constraints. J. Appl. Stat. **22**, 29–35 (1995)
52. Jolliffe, I.T., Trendafilov, N.T., Uddin, M.: A modified principal component technique Based on the LASSO. J. Comp. Graph. Stat. **12**, 531–547 (2003)
53. Josse, J., Husson, F.: missMDA: a package for handling missing values in multivariate data analysis. J. Stat. Softw. **70**(1), 1–31 (2016)
54. Linden, G., Smith, B., York, J.: Amazon.com recommendations: item-to-item collaborative filtering. Internet Comput. **7**(1), 76–80 (2003)
55. Mazumder, R., Hastie, T., Tibshirani, R.: Spectral regularization algorithms for learning large incomplete matrices. JMLR **2010**(11), 2287–2322 (2010)
56. McCabe, G.: Principal variables. Technometrics **26**, 137–144 (1984)
57. Modarresi, K.: A local regularization method using multiple regularization levels, Stanford, April 2007
58. Modarresi, K., Golub, G.H.: An efficient algorithm for the determination of multiple regularization parameters. In: Proceedings of Inverse Problems Design and Optimization Symposium (IPDO), 16–18 April 2007, Miami Beach, Florida, pp. 395–402 (2007)
59. Modarresi, K., Diner, J.: An efficient deep learning model for recommender systems. In: Shi, Y., et al. (eds.) ICCS 2018. LNCS, vol. 10861, pp. 221–233. Springer, Cham (2018). https://doi.org/10.1007/978-3-319-93701-4_17
60. Modarresi, K.: Recommendation system based on complete personalization. Procedia Comput. Sci. **80**, 2190–2204 (2016)
61. Modarresi, K.: Computation of recommender system using localized regularization. Procedia Comput. Sci. **51**, 2407–2416 (2015)
62. Modarresi, K.: Algorithmic approach for learning a comprehensive view of online users. Procedia Comput. Sci. **80**, 2181–2189 (2016)
63. Sedhain, S., Menon, A.K., Sanner, S., Xie, L.: Autorec: autoencoders meet collaborative filtering (2015)
64. Stekhoven, D.: Using the missForest Package. CRAN (2012)
65. Strub, F., Mary, J., Gaudel, R.: Hybrid collaborative filtering with autoencoders (2016)
66. Van Buuren, S., Groothuis-Oudshoorn, K.: MICE: multivariate imputation by chained equations in R. J. Stat. Softw. **45**(3), 1–67 (2011)

# Tolerance Near Sets and tNM Application in City Images

Deivid de Almeida Padilha da Silva[1], Daniel Caio de Lima[2],
and José Hiroki Saito[1,2(✉)]

[1] UNIFACCAMP – University Center of Campo Limpo Paulista,
São Paulo, Brazil
`deivid.aps@gmail.com`, `saito@cc.faccamp.br`
[2] UFSCar - Federal University of São Carlos, São Carlos, Brazil
`daniel.lima@dc.ufsca.br`

**Abstract.** The Tolerance Near Set theory - is a formal basis for the observation, comparison and classification of objects, and tolerance Nearness Measure (tNM) is a normalized value, that indicates how much two images are similar. This paper aims to present an application of the algorithm that performs the comparison of images based on the value of tNM, so that the similarities between the images are verified with respect to their characteristics, such as Gray Levels and texture attributes extracted using Gray Level Co-occurrence Matrix (GLCM). Images of the center of some selected cities around the world, are compared using tNM, and classified.

**Keywords:** tNM · Near Sets · tolerance Near Sets · Gray level · Statistical attributes

## 1 Introduction

The image processing is complex, and in some cases the human eyes can not identify image details. The attribute extraction task to compare two images, is inherent to the Near Sets theory [1]. Generally, each image has its attributes that can be used for classification, and the computational algorithms are indispensable to extract those attributes, for classification. The Near Sets (NS), and the tolerance Near Sets (TNS), are theories that provide the formal basis for observation, comparisons, and classification of objects, using n-dimensional attribute vectors [2]. Using these theories, the tolerance Nearness Measure (tNM) can be obtained considering two images.

The TNS have been applied in many areas, and shows to be very promising, in image analysis, comparing gray level values of pixels, or texture attributes. This paper refers to the tNM implementation to obtain the similarity index between two images. The tNM implementation has an advantage, such as the possibility of using parallel processing, during the tolerance classification of image objects or subimages [3]. To obtain the texture attributes from images, Gray-Level Co-occurrence Matrix (GLCM) is used [4].

© Springer Nature Switzerland AG 2019
J. M. F. Rodrigues et al. (Eds.): ICCS 2019, LNCS 11537, pp. 564–579, 2019.
https://doi.org/10.1007/978-3-030-22741-8_40

The rest of this paper is organized as follows. At Sect. 2, the mathematical description of NS, TNS, and tNM, is presented; followed by Sect. 3, of tNM implementation. The Sect. 4 is referred to the applications and results; and the Sect. 5, conclusions and future works.

## 2  NS, TNS, tNM

Perceptive objects are objects that can be detected by humans. Perceptive systems are referred to the perceptive objects associated with a set of probe functions that describes these objects, and is formally defined as follows.

**Definition 1:** A perceptive system $(O, F)$ consists of a nonempty set $O$ of perceptive objects, and a nonempty set $F$, of real valued functions $\Phi$, such that: $\Phi \in F \,|\, \Phi : O \rightarrow R$.

### 2.1  Object Description

If $(O, F)$ is a perceptive system, and $B \subseteq F$ is a set of probe functions, a description of a perceptual object is obtained by a vector, such as of Eq. (1):

$$\Phi_B(x) = (\Phi_1(x), \Phi_2(x), \ldots, \Phi_i(x), \ldots, \Phi_l(x)), \tag{1}$$

where: $l$ is the dimension of the vector $\Phi_B$, and each $\Phi_i(x)$ is a probe function.

Then considering a perceptive system $(O, F)$, and $B$ a subset of $F(B \subseteq F)$, $O$ is a set of objects with its characteristics described by vector $\Phi_B$.

An important definition related to NS the indiscernibility relation, which results in the classification of objects in equivalence classes [2], so that the properties of reflectivity, symmetry, and transitivity are satisfied by all the objects in a class. The equivalence relation, in a given set A, is satisfied if, $\forall a \in A$, aRa, (reflectivity); $\forall a, b \in A$, if $aRb$ then $bRa$ (symmetry); $\forall a, b, c \in A$, if aRb and  bRc then aRc (transitivity). It follows the definition of the indiscernibility relation.

### 2.2  Perceptual Indiscernibility Relation

**Definition 2:** Let $(O, F)$ be a perceptual system. For each $B \subseteq F$, the perceptual indiscernibility relation $\sim_B$ is defined as Eq. (2):

$$\sim_B = \{(x, y)O \times O : \forall \Phi i \in B . \Phi i(x) = \Phi i(y)\}, \tag{2}$$

meaning that, two perceptual objects $x$ and $y$, are indiscernibly related if they have the same value for all probe functions of $B$.

This perceptual indiscernibility relation is a modification of the relation described by Pawlak [5], in his rough set theory, provided that in NS, it is always considered a pair of sets that are close each other.

## 2.3   Weak Perceptual Indiscernibility Relation

**Definition 3:** Let $(O, F)$ be a perceptual system, and $X, Y \subseteq O$. The set $X$ is weakly near from the set $Y$ if there are $x \in X$, $y \in Y$, and $\Phi i \in F$, such that $x \simeq_B y$, as defined in Eq. (3) of the relation of weak perceptual indiscernibility relation, $\simeq_B$:

$$\simeq_B = \{(x, y)\, O \times O : \exists\, \Phi i \in F . \Phi i(x) = \Phi i(y)\}, \tag{3}$$

The previous definitions are related to the NS theory, and as its improvement, applying tolerance to measure the relation between perceptual objects, tolerance NS was proposed, as follows.

## 2.4   Tolerance Near Sets

TNS is characterized by the tolerance relation between the perceptual objects, so that it can be defined as follows.

**Definition 4:** Let $(O, F)$ be a perceptual system. For $B \subseteq F$, the perceptual tolerance relation $\cong_{B,\epsilon}$ is defined by Eq. (4), where the $L^2$ norm is denoted by "$\| . \|$".

$$\cong_{B,\epsilon} = \{(x, y)\, O \times O : \| \Phi(x) - \Phi(y) \|_2 \leq \epsilon\}, \tag{4}$$

The great difference between NS and TNS is that the objects in TNS classes are subjected to reflectivity, and symmetry, but not to transitivity, properties.

It is stated that the tolerance concept is inherent to the idea of proximity between objects [6], such that it is possible to identify image segments that are similar, each other, with a tolerable difference between them. In TNS, these images are considered in the same classes. If two image segments are similar, with tolerance, the TNS classification can result in two different classes, when two image segments are similar to a third image segment, but not similar, from each other, and in consequence, the transitivity property can not be satisfied to all objets.

Given the perceptual tolerance relation definition, it is possible to observe that the transitivity property can not be present to all perceptual objects in TNS. Another characteristic of TNS is the use of a tolerance value ε, that is the threshold value of the distance between perceptual objects, such that if the distance is below or equal this value, they are considered in the same class.

According to Poli et al. [6], the basic structure of TNS, in the case of images used as perceptual objects, is consisted of a nonempty set of images, and a finite set of probe functions. Each object description consists of several measurements obtained by image processing techniques. TNS provides a quantitative approach, by the use of these measurements, as probe functions, and the threshold value ε, to determine the similarity of objects, without the claim for the objects to be exactly the same [3].

## 2.5   Tolerance Nearness Measure

The tNM was introduced by Henry and Peters [3], from the necessity to determine the degree of similarity between objects, during the application of NS, in content based image retrieval.

**Definition 5:** Considering $(O, F)$, a perceptual system, with two disjunct sets $X$ and $Y$, such that, $Z = X \cup Y$, the similarity measure $tNM_{\cong B, \varepsilon}(X, Y)$, between $X$ and $Y$, can be resumed as Eq. (5):

$$tNM_{\cong B, \varepsilon}(X, Y) = \left(\sum_{C \in H \cong B \varepsilon(Z)} |C|\right)^{-1} \times \sum_{C \in H \cong B, \varepsilon(Z)} |C| \frac{min(|C \cap X|, |C \cap Y|)}{max(|C \cap X|, |C \cap Y|)} \quad (5)$$

with $C$ denoting a TNS class, and $H$, the set of all classes in $Z$.

Equation (5) has as the first term of the product, the inverse of the addition of the modulus of all classes in $Z$. The second term, is the addition to all classes in $H$, of the ratio between minimum and maximum intersection, of the class $C$ with $X$ and $Y$, multiplied by the modulus of $C$.

## 3 Methodology of tNM Implementation

In this section, the TNS classification algorithm and tNM implementation are described.

### 3.1 TNS Classification

The TNS classification algorithm in a perceptive system $(O, F)$, with $B \subseteq F$, a set of probe functions, and a set of $n$ objects , is described as the following Algorithm 1.

**Algorithm-1: TNS Classification**
**Input**: X, $\varepsilon$ // X is the set of objects, and $\varepsilon$, tolerance;
**Output**: H // H is the set of TNS classes
    Step 1 –  for each object $x_i$ in X, calculate its probe function value;
    Step 2 – for each object $x_i$ of X, obtain  its set of neighbours, $N_i$, whose euclidean distance is below or equal $\varepsilon$, measured between its vectors of probe functions;
    Step 3 - i = 1 ; start for the first index value;
    Step 4 – **while** (i < = n)
        - for each object $x_i$ criate a pre-class $C_i$, including the object $x_i$; and
        denote a local set of neighbours $N_i' = N_i$ ;
    **repeat**
      **while** $N_i'$ (non empty)
        chose a neighbour $x_j \in N_i'$, with the minimum distance from $x_i$;
        include this neighbour in  the pre-class $C_i$;
        make $N_i' = N_i' - \{x_j\}$;
        make $N_i' = N_i' - \{x_k\}$ for all $x_k$ whose distance is above $\varepsilon$ from $x_j$;
        include the new class obtained in H;
        if  some neighbour $x_k \in N_i$ was not included in $C_i$; create a new
        neighbourhood $N_i'$with these objects; and a new pre-class $C_i$, including
        the  object $x_i$;
    **until** all objects of the neighbourhood $N_i$ are include in some class;
    i = i + 1;
    Step 5 – all the classes in H must be verified, and the redundat classes must be eliminated, as well as, subsets of classes.
    **end.**

### 3.2   tNM Algorithm

Basically, the algorithm to compute tNM, starts with the input of two images X, and Y, and two approaches can be selected. The first one, denoted GL, uses the gray level of pixels as probe functions. In this case, the tolerance represents the quantity of different gray levels considered in the same class. If only one gray level is considered in the same class, the tolerance is zero. The objects in this approach are the pixels of the images. The second approach, denoted SA, divides the image in subimages, which become objects of the perceptual system. Then, statistical attributes of each subimage are obtained using Gray Level Co-occurrence Matrix, GLCM, which describes the occurrence of patterns of pixel pairs in the image [4].

After computation of GLCM, statistical attributes such as correlation, energy, contrast and homogeneity, can be obtained. These attributes are considered as probe functions, and they can be used to compute the Euclidean distance of the pair of subimages. Two subimages with distance below or equal the tolerance $\varepsilon$, are included in the same class, applying Algorithm 1.

**Algorithm-2: Procedure** tNM $(X,Y,\varepsilon, n, K)$   // subimage approach
**Input**: X, Y, $\varepsilon$, n   // input images, tolerance, and number of subimages
**Output**: tNM   // tNM value between two input images
  **begin**                  // initialization, when the two images are read
**1**   K is initialized with zero
**2**   subimage creation, dividing the input images in n subimages each
**3**   obtaining of atributes, as probe functions for each subimage
**4**   TNS classification of the subimages
  **Repeat**       // tNM computation
**5**       for each class $C_i$ is calculated its minimum and maximum intersection ratio
      with images X and Y
**6**       obtain the product of the ratio of (5)  with the class module
**7**       addition of the result of (6) in K
  **until** não haja mais classes
**8**   obtain the final value of tNM dividing K by the module of $X \cup Y$
  **end**
  **return (tNM)**
  **end_procedure**

Algorithm 2 corresponds to the tNM computation, dividing each input image in $n$ subimages. At the step (1), the variable K is initialized with zero. Then, at (2) both input images are divided into $n$ subimages each. In the case of GL approach, a subimage is a pixel. In (3) the attributes for each subimage are computed, for the probe function vector. Then, in (4), the subimages are classified using the TNS classification Algorithm 1. After these four steps, the computation of tNM is started. For each class $C_i$ obtained previously, the ratio of the minimum intersection between input images and all objects in a class, i.e., min $(X \cap C_i, Y \cap C_i)$, and the maximum intersection between the same sets, i.e., max $(X \cap C_i, Y \cap C_i)$, are computed, at step (5). At step (6) the previously obtained ratio is multiplied by the modulus of the class, $|C_i|$, and at

(7) the obtained product is accumulated in $K$. These three steps are repeated until all classes are computed. Then, at step (8), the final value of $K$ is divided by the modulus $|X \cap Y|$, resulting in tNM value.

# 4 Application and Results

The relevance of the city comparison is confirmed by recent articles, such as Domingues et al. [7], that described the previous studies about city structures using complex networks, contributing for the understanding and improvements in transit systems, growth, and planning of the cities. In this paper it is described the application of tNM in downtown images, considering aspects related to the satellite image textures, and gray level, indicating how much each city is similar to the other cities, in aspects, such as structures, paving, and vegetation.

The tNM System was developed in Python, and in this section, it is first described experiments for parameter determination in city image pair tNM computation. After that, an experiment of classification of 26 cities around the world is described.

## 4.1 Determination of Parameters

To determine the values of parameters such as tolerance, and subimage size, in both approaches, an experiment of tNM applied to two city images, with $600 \times 600$ pixels, of Mexico City and Frankfurt, were realized. The images were obtained from Google Maps, Fig. 1.

The first tNM measure in GL approach, used the tolerance of 1% or 25 gray levels in a class, the number of generated classes was 10, execution time, 10 s, and tNM of 0.747. The second tNM measure in GL approach, used the tolerance of 50% or 124 gray levels in a class. The number of generated classes was 2, execution time, 5 s, and tNM of 0.816.

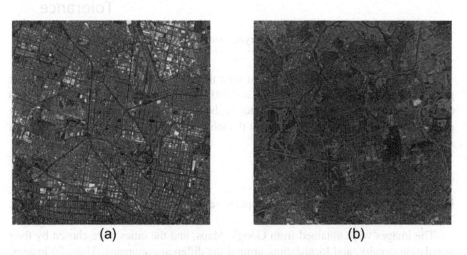

**Fig. 1.** Pair of city images. (a) Mexico City. (b) Frankfurt.

The first tNM measure in SA approach, used the tolerance of 0.9. The dimension of a subimage was $10 \times 10$, then 7,200 subimages were generated, classified in two classes, and execution time of 10 min, resulting in tNM of 0.999. In this case the value of tNM does not indicate the high similarity between two images, but the low resolution of tNM, with this high value of tolerance.

In the second tNM measure, SA approach, it was used the tolerance of 0.5. The dimension of a subimage was $10 \times 10$, then 7,200 subimages were generated, classified in four classes, and execution time of 10 min, resulting in tNM of 0.991. In this case the value of tNM was close to the previous experiment, indicating the same high degree of generalization of the compared images.

In the next experiment with SA approach, it was used the tolerance of 0.1. The dimension of subimage was $10 \times 10$, and 7,200 subimages were also generated, classified in 28 classes. The execution time remained the same, resulting in tNM of 0.661. This value seems realistic considering the two images.

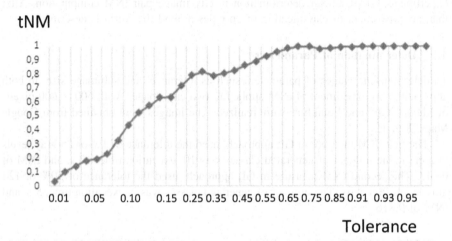

**Fig. 2.** Graphic of tNM, varying with tolerance, in SA approach.

Figure 2 illustrates how tNM varied with the tolerance in SA approach, using the pair of images of Fig. 1. If the tolerance is 0.01, tNM is near zero. When tolerance is 0.10, tNM is near 0.5, and when tolerance is above 0.65, tNM value is near 1, showing generalization. This figure indicates that the tolerance value suitable to the experiments in SA approach can be defined as 0.1.

### 4.2    Comparing City Images

In the following experiment, it was compared several images of cities around the world, Table 1.

The images were obtained from Google Maps, and the cities were chosen by their population density, and localization, around the different continents. Then, 26 images,

**Table 1.** Cities around the continents

| American | Brazil | Cuiaba |
| | | Campo Grande |
| | Mexico | Mexico City |
| | Canada | Winnipeg |
| | | Regina |
| | | Edmonton |
| European | Germany | Frankfurt |
| | Portugal | Lisbon |
| | France | Lyon |
| | Italy | Naples |
| | | Palermo |
| African | Liberia | Monrovia |
| | Congo | Pointe Noire |
| | Ivory Coast | Abobo |
| | Egipt | Porto Said |
| | Mozambique | Matola |

| Asian | Japan | Sakai |
| | | Niigata |
| | India | Nav Mumbai |
| | Kazakhstan | Astana |
| | Nepal | Kathmandu |

| Australian | Australia | Newcastle |
| | | Adelaide |
| | | Canberra |
| | | Gold Coast |
| | New Zealand | Wellington |

6 from the America continent; and 5 from each other continents, Europe, Asia, Africa, and Oceanian. In this experiment, the images were fixed to $256 \times 256$ pixels, and a tolerance of 10% was used for GL approach, and 0.1 for SA.

### 4.3   Highest and Lowest Values of tNM Obtained Comparing City Images

In Table 2, it is shown the top thirty highest tNM values obtained when comparing the considered cities around the world, using GL approach. The highest value of tNM, 0.950, was obtained between Regina and Edmonton, both from Canada, in American Continent. The images of these two cities are showed in Fig. 3(a) and (b), respectively. In Table 3, it is shown the thirty highest tNM values, obtained, when it was used the SA approach, and in this case the highest tNM value, 0.936, was obtained comparing Regina and Pointe Noire, from American and African Continents, respectively. The image of Pointe Noire city is shown in Fig. 3(c). The tNM value between Regina and Edmonton in SA approach, was of 0.647, not so high, showing the difference between GL an SA approach; and the tNM value between Regina and Pointe Noire in GL approach was of 0.831.

**Table 2.** Highest tNM values obtained in GL approach.

| City 1 | City 2 | tNM | | City 1 | City 2 | tNM |
|--------|--------|-----|---|--------|--------|-----|
| Regina | Edmonton | 0.950 | | Adelaide | Canberra | 0.919 |
| Mexico City | Pointe-Noire | 0.944 | | Lisbon | Astana | 0.917 |
| Matola | Nav Mumbai | 0.943 | | Manitoba | Astana | 0.913 |
| Edmonton | Lisboa | 0.942 | | Campo Grande | Canberra | 0.912 |
| Palermo | Adelaide | 0.938 | | Naples | Adelaide | 0.911 |
| Regina | Naples | 0.935 | | Porto Said | Newcastle | 0.910 |
| Frankfurt | Monrovia | 0.931 | | Kathmandu | Adelaide | 0.909 |
| Mexico City | Canberra | 0.928 | | Kathmandu | Canberra | 0.909 |
| Naples | Palermo | 0.927 | | Astana | GoldCoast | 0.909 |
| Edmonton | Naples | 0.924 | | Regina | Lisbon | 0.909 |
| Pointe-Noire | Canberra | 0.924 | | Lisbon | Naples | 0.905 |
| Edmonton | Astana | 0.922 | | Cuiaba | Lisbon | 0.903 |
| Lisbon | Gold Coast | 0.921 | | Pointe-Noire | Adelaide | 0.902 |
| Campo Grande | Pointe-Noire | 0.921 | | Abobo | Porto Said | 0.900 |
| Campo Grande | Mexico City | 0.921 | | Palermo | Pointe-Noire | 0.900 |

**Table 3.** Highest tNM values obtained in SA approach.

| City 1 | City 2 | tNM | | City 1 | City2 | tNM |
|--------|--------|-----|---|--------|-------|-----|
| Regina | Pointe-Noire | 0.936 | | Palermo | Astana | 0.883 |
| Regina | Gold Coast | 0.936 | | Lyon | Porto Said | 0.880 |
| Lisbon | Naples | 0.936 | | Winnipeg | Wellington | 0.878 |
| Pointe-Noire | Gold Coast | 0.927 | | Edmonton | Lisbon | 0.877 |
| Naples | Lyon | 0.922 | | Campo Grande | Astana | 0.877 |
| Edmonton | Naples | 0.918 | | Abobo | Kathmandu | 0.874 |
| Campo Grande | Palermo | 0.914 | | Cuiabá | Campo Grande | 0.873 |
| Monrovia | Newcastle | 0.911 | | Winnipeg | Palermo | 0.871 |
| Winnipeg | Gold Coast | 0.911 | | Frankfurt | Niigata | 0.869 |
| Winnipeg | Regina | 0.902 | | Mexico City | Porto Said | 0.867 |
| Sakai | Newcastle | 0.902 | | Naples | Sakai | 0.866 |
| Winnipeg | Pointe-Noire | 0.896 | | Gold Coast | Wellington | 0.863 |
| Frankfurt | Adelaide | 0.893 | | Palermo | Pointe-Noire | 0.862 |
| Cuiabá | Winnipeg | 0.890 | | Porto Said | Adelaide | 0.861 |
| Edmonton | Lyon | 0.886 | | Campo Grande | Winnipeg | 0.859 |

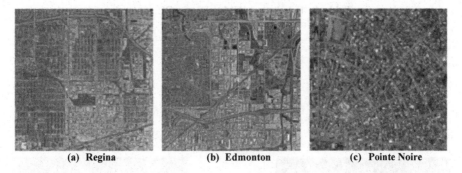

**(a) Regina**          **(b) Edmonton**          **(c) Pointe Noire**

**Fig. 3.** City images: (a) Regina, (b) Edmonton, (c) Pointe Noire, with highest tNM values for GL approach (Regina x Edmonton); and for AS approach (Regina x Pointe Noire).

In Fig. 4 it is showed the tNM obtained when Regina is compared with all other cities considered in this experiment, using both approaches, where the highest tNM in GL and AE are highlighted. It is also observed that the tNM values in both approach are not close in most cities, but the behavior of these values are quite similar, showing the difference between GL and statistical approaches.

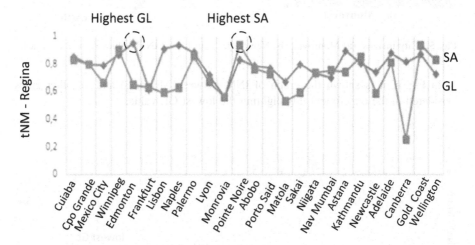

**Fig. 4.** tNM values for GL and AS, obtained when Regina is computed with all the other cities considered, showing the highest value in both approaches.

In Table 4, it is shown the five lowest tNM values obtained, in GL approach, and the lowest value, 0.349, was obtained comparing Monrovia and Newcastle. In SA approach, the value of 0.911, was obtained between these two cities, showing that in SA approach, both cities are very similar, because the gray level is not considered, as can be seen in the images shown in Figs. 5(a) and (b), respectively.

**Table 4.** Five lowest tNM values in GL approach.

| City 1 | City 2 | tNM |
|--------|--------|------|
| Frankfurt | Porto Said | 0.402 |
| Monrovia | Abobo | 0.396 |
| Frankfurt | Newcastle | 0.388 |
| Monrovia | Porto Said | 0.362 |
| Monrovia | Newcastle | **0.349** |

(a) **Monrovia**                    (b) **Newcastle**

**Fig. 5.** City images: (a) Monrovia, (b) Newcastle, with lowest tNM values for GL approach.

In Fig. 6, it is shown the graph of tNM between Monrovia and all other cities considered in the experiment, highlighting the lowest GL value.

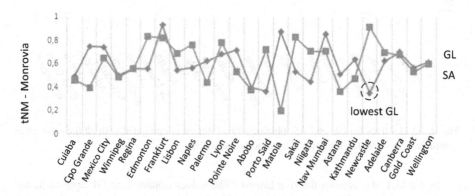

**Fig. 6.** tNM values for GL and SA, obtained when Monrovia is computed with all the other cities considered, showing the lowest GL value.

In Table 5, it is shown the five lowest tNM values obtained, in SA approach, and the lowest value, 0.062, was obtained comparing Matola and Canberra, from African and Australian Continents, respectively. It is noted that the tNM between these two cities in GL approach was of 0.802, not so low value such as in AS approach. The images of these two cities are shown in Figs. 7(a) and (b), respectively.

**Table 5.** Five lowest tNM values obtained in SA approach.

| City 1 | City 2 | tNM |
| --- | --- | --- |
| Cuiaba | Camberra | 0.154 |
| Astana | Camberra | 0.129 |
| Catmandu | Camberra | 0.126 |
| Abobo | Camberra | 0.125 |
| Matola | Camberra | **0.062** |

**(a) Matola**

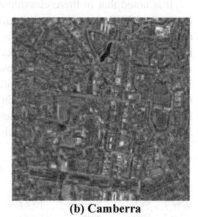

**(b) Camberra**

**Fig. 7.** City images: (a) Matola, (b) Canberra, with lowest tNM values for SA approach.

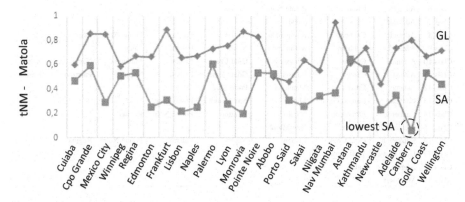

**Fig. 8.** tNM values for GL and SA, obtained when Matola is computed with all the other cities considered, showing the lowest SA value.

In Fig. 8, it is shown the graph of tNM between Matola and all other cities considered in the experiment, highlighting the lowest GL value. It can be noted that in this case almost all GL values was above SA values, showing that the gray levels of the images were similar to Matola image, although the statistical attributes were different.

## 4.4    City Image Classification

In Tables 6 and 7, it is showed the TNS classification of the city images, using tNM results. If tNM is a measure of similarity, and in Algorithm 1 it is used the distance from the objects compared with a tolerance $\varepsilon$, it was defined a tNM distance, denoted $d_{tNM}$, obtained as Eq. (6):

$$d_{tNM} = (1-tNM) \tag{6}$$

Using $d_{tNM}$, with tNM values obtained for GL approach, it was generated the classes shown in Table 6; and for SA approach, in Table 7.

It is noted that in these classifications, one city can be present in different classes, because of TNS classes are not equivalent classes. One class that Regina, Edmonton, and Pointe Noire cities are present is the Class 14, in Table 6. These cities showed the highest tNM in GL and SA approaches, as showed in Fig. 4. The Monrovia city is alone in Class 19, since it has the lowest GL tNM, as showed in Fig. 6. In SA, Regina is present in several classes with Pointe Noire, the highest value of tNM, such as: Class 6, Class 10, and Class 12, but Edmonton, is not present in these classes, although Regina and Edmonton had the highest GL value of tNM. Canberra that had the lowest SA value of tNM, is alone in SA class 8.

**Table 6.**  Classes in GL approach.

| Classes | Cities | | | | | | | |
|---|---|---|---|---|---|---|---|---|
| Class 1 | Abobo | Newcastle | Porto Said | | | | | |
| Class 2 | Abobo | Astana | Cuiaba | Edmonton | Lisbon | | | |
| Class 3 | Adelaide | Canberra | Campo Grande | Kathmandu | Mexico City | Lyon | | |
| Class 4 | Adelaide | Canberra | Campo Grande | Kathmandu | Mexico City | Naples | Palermo | |
| Class 5 | Adelaide | Canberra | Campo Grande | Kathmandu | Naples | Palermo | Pointe Noire | |
| Class 6 | Adelaide | Lisbon | Pointe Noire | Regina | | | | |
| Class 7 | Adelaide | Lyon | Sakai | | | | | |
| Class 8 | Adelaide | Canberra | Kathmandu | Naples | Palermo | Pointe Noire | Regina | |
| Class 9 | Adelaide | Canberra | Kathmandu | Naples | Palermo | Sakai | | |
| Class 10 | Astana | Cuiaba | Edmonton | Gold Coast | Lisbon | Manitoba | Naples | Regina |
| Class 11 | Astana | Cuiaba | Edmonton | Gold Coast | Lisbon | Naples | Palermo | Regina |
| Class 12 | Astana | Cuiaba | Manitoba | Porto Said | | | | |
| Class 13 | Canberra | Campo Grande | Matola | Nav Mumbay | Point Noire | | | |
| Class 14 | Canberra | Edmonton | Naples | Palermo | Pointe Noire | Regina | | |
| Class 15 | Mexico City | Matola | Nav Mumbay | | | | | |
| Class 16 | Edmonton | Gold Coast | Lisbon | Naples | Palermo | Pointe Noire | Regina | |
| Class 17 | Gold Coast | Wellington | | | | | | |
| Class 18 | Niigata | Sakai | | | | | | |
| Class 19 | Monrovia | | | | | | | |

**Table 7.** Classes in SA approach.

| Classes | Cities | | | | | | |
|---------|--------|--|--|--|--|--|--|
| Class 1 | Abobo | Kathmandu | | | | | |
| Class 2 | Adelaide | Mexico City | Frankfurt | Porto Said | | | |
| Class 3 | Adelaide | Mexico City | Lyon | Porto Said | | | |
| Class 4 | Adelaide | Frankfurt | Lyon | Porto Said | | | |
| Class 5 | Adelaide | Frankfurt | Niigata | | | | |
| Class 6 | Adelaide | Pointe Noire | Regina | | | | |
| Class 7 | Astana | Campo Grande | Palermo | | | | |
| Class 8 | Canberra | | | | | | |
| Class 9 | Campo Grande | Cuiaba | Manitoba | Pointe Noire | | | |
| Class 10 | Kathmandu | Gold Coast | Palermo | Pointe Noire | Regina | | |
| Class 11 | Mexico City | Edmonton | Lisbon | Lyon | Porto Said | | |
| Class 12 | Cuiaba | Gold Coast | Manitoba | Pointe Noire | Palermo | Regina | Wellington |
| Class 13 | Edmonton | Frankfurt | Monrovia | Newcastle | | | |
| Class 14 | Edmonton | Lisbon | Lyon | Naples | Porto Said | | |
| Class 15 | Frankfurt | Monrovia | Newcastle | Sakai | | | |
| Class 16 | Frankfurt | Nav Mumbay | | | | | |
| Class 17 | Matola | | | | | | |
| Class 18 | Naples | Sakai | | | | | |
| Class 19 | Niigata | Sakai | | | | | |

### 4.5  Average tNM Between Continents

In Table 8, it is illustrated the average values and standard deviation of tNM calculated between cities of the same continent, in the GL approach. It can be noted that the average tNM value between different continents was close to 0.700, as showed at the last row, average$^2$, where the corresponding value is the average of the column values, excluding the average tNM value in the same continent, showed at the diagonal. The average tNM value in the same continent was above the average value between different continents, only in American Continent, showing that in the other continents the kind of cities was diversified.

Table 9, corresponds to the average and standard deviation of tNM between cities of the same continent, in the SA approach. It can be noted that the average tNM value between different continents was close to 0.600, as showed at the last row, average$^2$. In this approach, the average tNM value in the same continent was above the average value between different continents, in majority of the continents, with exception of the African Continent, in which the average tNM value was 0.630.

**Table 8.** Average tNM between Continents in GL approach.

| Continents | American | European | African | Asian | Australian |
|---|---|---|---|---|---|
| American | $0.808 \pm 0.082$ | $0.788 \pm 0.188$ | $0.719 \pm 0.128$ | $0.763 \pm 0.089$ | $0.777 \pm 0.090$ |
| European | $0.788 \pm 0.188$ | $0.706 \pm 0.169$ | $0.704 \pm 0.141$ | $0.754 \pm 0.108$ | $0.748 \pm 0.138$ |
| African | $0.719 \pm 0.128$ | $0.704 \pm 0.141$ | $0.509 \pm 0.177$ | $0.682 \pm 0.135$ | $0.696 \pm 0.141$ |
| Asian | $0.763 \pm 0.089$ | $0.754 \pm 0.108$ | $0.682 \pm 0.135$ | $0.652 \pm 0.129$ | $0.726 \pm 0.113$ |
| Australian | $0.777 \pm 0.090$ | $0.748 \pm 0.138$ | $0.696 \pm 0.141$ | $0.726 \pm 0.113$ | $0.594 \pm 0.215$ |
| Average[2] | $0.761 \pm 0.123$ | $0.748 \pm 0.143$ | $0.700 \pm 0.136$ | $0.731 \pm 0.111$ | $0.736 \pm 0.120$ |

**Table 9.** Average tNM between Continents in SA approach.

| Continents | American | European | African | Asian | Australian |
|---|---|---|---|---|---|
| American | $0.729 \pm 0.134$ | $0.693 \pm 0.142$ | $0.622 \pm 0.179$ | $0.648 \pm 0.125$ | $0.642 \pm 0.222$ |
| European | $0.693 \pm 0.142$ | $0.752 \pm 0.120$ | $0.600 \pm 0.219$ | $0.624 \pm 0.161$ | $0.644 \pm 0.167$ |
| African | $0.622 \pm 0.179$ | $0.600 \pm 0.219$ | $0.630 \pm 0.197$ | $0.612 \pm 0.161$ | $0.569 \pm 0.237$ |
| Asian | $0.648 \pm 0.125$ | $0.624 \pm 0.161$ | $0.612 \pm 0.161$ | $0.728 \pm 0.096$ | $0.613 \pm 0.219$ |
| Australian | $0.642 \pm 0.222$ | $0.644 \pm 0.167$ | $0.569 \pm 0.237$ | $0.613 \pm 0.219$ | $0.748 \pm 0.090$ |
| Average[2] | $0.651 \pm 0.167$ | $0.640 \pm 0.172$ | $0.600 \pm 0.199$ | $0.624 \pm 0.166$ | $0.617 \pm 0.211$ |

## 5    Conclusions

In this work, it was developed two approaches for tNM, in images. The GL approach considers an object, or subimage, a pixel with its corresponding gray level; and SA approach considers an object, a subimage with its statistical attributes. The experiments showed that reasonable tolerance value is 10% for GL approach, and 0.1 for SA. With these values of tolerance, 26 downtown images of cities around the world, distributed in five continents, was compared using tNM based distance, $d_{tNM}$, to classification. The results showed that the two approaches present in some situations, very different values of tNM, depending on the gray level of the image in GL approach; and statistical attributes in SA approach. This can also be explained by the use of tolerance values in GL approach, and the size of subimage in SA approach. As future works, experiments should be suggested using more than one image from the same cities, to verify how the tNM varies in the same city images, varying the tolerance, and subimage size.

**Acknowledgement.** Deivid de Almeida Padilha da Silva acknowledges CAPES, Coordenação de Aperfeiçoamento de Pessoal de Nível Superior, the Brazilian Federal Government Agency under the Ministry of Education, for granting his Master Degree scholarship.

## References

1. Pawlak, Z., Peters, J.: Systemy Wspomagania Decyzji I, vol. 57 (2007)
2. Peters, J.F.: Tolerance near sets and image correspondence. Int. J. Bio-Inspired Comput. **1**(4), 239–245 (2009)

3. Henry, C.J.: Near sets: theory and applications. University of Manitoba, Canada, Ph.D. thesis (2010)
4. Haralick, R., Shanmugam, K., Dinstein, I.: Textural features for image classification. IEEE Trans. Syst. Man Cybern. SMC-3 **6**, 610–621 (1973)
5. Pawlak, Z.: Classification of objects by means of attributes. Polish Academy of Sciences [PAS]. Institute of Computer Science (1981)
6. Poli, G., et al.: Solar flare detection system based on tolerance near sets in a GPU–CUDA framework. Knowl. Based Syst. **70**, 345–360 (2014)
7. Domingues, G.S., Silva, F.N., Comin, C.H., Costa, L.F.: Topological characterization of world cities. J. Stat. Mech: Theory Exp. **083212**, 1–18 (2018)

# Meta-Graph Based Attention-Aware Recommendation over Heterogeneous Information Networks

Feifei Dai[1,2], Xiaoyan Gu[1(✉)], Bo Li[1], Jinchao Zhang[1], Mingda Qian[1], and Weiping Wang[1]

[1] Institute of Information Engineering, Chinese Academy of Sciences, Beijing, China
{daifeifei,guxiaoyan,libo,zhangjinchao,qianmingda,wangweiping}@iie.ac.cn
[2] School of Cyber Security, University of Chinese Academy of Sciences, Beijing, China

**Abstract.** Heterogeneous information network (HIN), which involves diverse types of data, has been widely used in recommender systems. However, most existing HINs based recommendation methods equally treat different latent features and simply model various feature interactions in the same way so that the rich semantic information cannot be fully utilized. To comprehensively exploit the heterogeneous information for recommendation, in this paper, we propose a Meta-Graph based Attention-aware Recommendation (MGAR) over HINs. First of all, MGAR utilizes rich meta-graph based latent features to guide the heterogeneous information fusion recommendation. Specifically, in order to discriminate the importance of latent features generated by different meta-graphs, we propose an attention-based feature enhancement model. The model enables useful features and useless features contribute differently to the prediction, thus improves the performance of the recommendation. Furthermore, to holistically exploit the different interrelation of features, we propose a hierarchical feature interaction method which consists three layers of second-order interaction to mine the underlying correlations between users and items. Extensive experiments show that MGAR outperforms the state-of-the-art recommendation methods in terms of RMSE on Yelp and Amazon Electronics.

**Keywords:** Recommendation · Meta-graph ·
Heterogeneous information networks · Attention model ·
Feature interaction

## 1 Introduction

In the era of information explosion, recommender systems, which help users discover items by interests, play a pivotal role in various online services [1,9,10,16]. Traditional recommendation methods, based on matrix factorization, mainly focus on characterizing the user-item interaction while neglecting abundant

© Springer Nature Switzerland AG 2019
J. M. F. Rodrigues et al. (Eds.): ICCS 2019, LNCS 11537, pp. 580–594, 2019.
https://doi.org/10.1007/978-3-030-22741-8_41

auxiliary data. However, as auxiliary data is likely to contain useful information, neglecting it may degrade the performance of recommendation. To solve this drawback, heterogeneous information network (HIN), consisting of multiple types of entities and relations, has been adopted in recommender systems to flexibly model the heterogeneity of rich auxiliary data. As an emerging topic, HINs based recommendation methods have attracted intensive research interests.

Existing HINs based recommendation approaches have evolved mainly through two dimensions: meta-path based recommendation models [5,13] and meta-graph based recommendation methods [18]. The former predefine a set of meta-paths for generating the semantic relatedness constrained by entity types. Then they treat the semantic relatedness as latent features for recommendation. This category of models, however, can only express simple relationship between two different types of entities that hardly capturing complicated semantics. Meta-graph based recommendation methods, in contrast, leverage meta-graph (or meta-structure) [6,17] to generate the complex relevance of two entities, and thus can capture more complex semantics that meta-path cannot. Although these methods have achieved performance improvement to some extent, they still face two major challenges. First, most existing methods simply treat all latent features indiscriminately for recommendation. However, as not all latent features are equally useful, the useless features may even introduce noises and degrade the performance. Furthermore, as the way of assembling features has significant influence on the accuracy of recommendation, factorization machine (FM) [4,10,11], which leverages the second-order feature interaction, is widely adopted in the recommender systems. Nevertheless, most existing FM based recommendations are hindered by modelling all feature interactions roughly in the same way, neglecting the fact that not all interactions are equally predictive.

To address the above challenges, we propose a Meta-Graph based Attention-aware Recommendation (MGAR) method. Figure 2 illustrates the overall view of our proposed method which utilizes rich meta-graph based latent features to make heterogeneous information fusion recommendation. Firstly, we propose a feature extracting module to generate different features. Furthermore, we propose an attention-based feature enhancement module to learn the importance of the latent features generated by different meta-graphs. The useless features are learned to set a lower weight as they may introduce noises and degrade the performance, while the useful features are enhanced. Finally, a hierarchical feature interaction method is introduced for recommendation, which considers three layers of second-order feature interaction. The method can mine the underlying correlations between users and items so that the features could be efficiently assembled and thoroughly used to predict the rating for recommendation. Our contributions can be summarized as follows:

- We propose a novel attention-based feature enhancement module to learn the importance of different latent features for enabling feature interactions contribute differently to the prediction.

- We design a hierarchical feature interaction method to make full use of the generated features so that the performance of recommendation could be improved.
- Extensive experiments conducted on two widely used datasets demonstrate that MGAR outperforms several state-of-the-art recommendation methods.

## 2  Preliminary

In this section, we introduce the definitions of heterogeneous information network [14] and meta-graph [18] which are frequently used in this paper.

**Definition 1. Heterogeneous Information Network.** *An information network is a directed graph $\mathcal{G} = (\mathcal{V}, \mathcal{E})$ with a node type mapping $\varphi : \mathcal{V} \to \mathcal{A}$ and an edge type mapping $\phi : \mathcal{E} \to \mathcal{R}$. Particularly, when the number of node types $|\mathcal{A}| > 1$ or the number of edge types $|\mathcal{R}| > 1$, the network is called a **heterogeneous information network (HIN)**.*

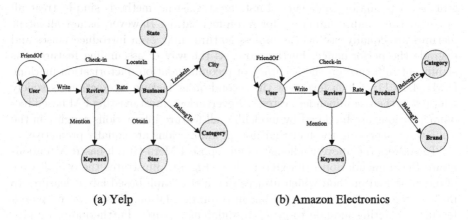

(a) Yelp                (b) Amazon Electronics

**Fig. 1.** Examples of heterogeneous information network schema.

*Example 1.* HIN contains diverse node types and edge types. As shown in Fig. 1, the dataset of Yelp consists of 8 types of nodes (User, Review, Business, City, State, Star, Category, Keyword) and 8 types of edges (FriendOf, Write, Rate, Check-in, Mention, LocateIn, Obtain, BelongTo). Similarly, the dataset of Amazon Electronics consists of 6 types of nodes and 6 types of edges as well.

**Definition 2. Meta-graph.** *A meta-graph M is a directed acyclic graph (DAG) with a single source node $n_s$ and a single target node $n_t$, defined on an HIN $\mathcal{G} = (\mathcal{V}, \mathcal{E})$ with schema $\mathcal{T}_\mathcal{G} = (\mathcal{A}, \mathcal{R})$, where $\mathcal{V}$ is the node set, $\mathcal{E}$ is the edge set, $\mathcal{A}$ is the node type set, and $\mathcal{R}$ is the edge type set. Then we define a meta-graph $M = (\mathcal{V}_M, \mathcal{E}_M, \mathcal{A}_M, \mathcal{R}_M, n_s, n_t)$, where $\mathcal{V}_M \subseteq \mathcal{V}$, $\mathcal{E}_M \subseteq \mathcal{E}$, constrained by $\mathcal{A}_M \subseteq \mathcal{A}$ and $\mathcal{R}_M \subseteq \mathcal{R}$, respectively.*

*Example 2.* The example of meta-graph is shown in Fig. 3. It is a schema both in Yelp and Amazon Electronics datasets. We can see that the meta-graph is a DAG with U (User) as the source node and I (Business in Yelp and Product in Amazon Electronics) as the target node.

## 3    The Proposed Model

Given HIN containing $m$ users and $n$ items, the goal of the proposed method is to predict the user-item rating matrix, denoted as $\hat{R} \in \mathbb{R}^{m \times n}$. As shown in Fig. 2, our framework consists of three main phases. (1) Feature extractor: in order to generate different features, we propose a feature extractor which not only learns the user and item embeddings but also generates the latent features based on different meta-graphs. (2) Attention-based feature enhancement: to learn the importance of the generated latent features, we design a two-layer attention network to set a lower weight on the useless features, while the useful features are enhanced. (3) Hierarchical feature interaction for rating: to efficiently assemble and thoroughly use the features for rating, it considers three layers of second-order feature interaction which could mine the underlying correlations between users and items.

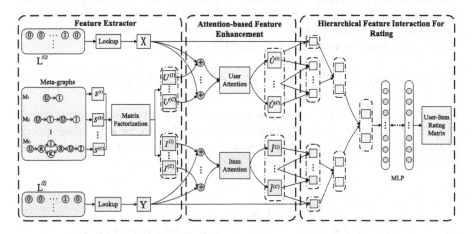

**Fig. 2.** The overall architecture of the proposed model.

### 3.1    Feature Extractor

**User and Item Embedding.** To learn the representation of users and items in HIN, we leverage the lookup-based embedding method following [5]. Firstly, we represent the users and items by one-hot encodings, denoted as $L^{(U)} \in \mathbb{R}^{m \times m}$ and $L^{(I)} \in \mathbb{R}^{n \times n}$ individually. As the one-hot encodings may result in corse of dimensionality and enormous sparsity, we then adopt the lookup operation in projecting them into dense embeddings. We use $X = \{x\}^m$ and $Y = \{y\}^n$ denote the embeddings of users and items, where $x, y \in \mathbb{R}^d$ denotes the embedding of user $u$ and item $i$ with $d$ denotes the embedding dimensionality ($d \ll m, d \ll n$).

**Meta-graph Based Latent Features Generating.** In order to thoroughly explore the semantic information between users and items, we learn their latent features based on different meta-graphs. Following [18], we first compute the user-item similarity matrices and then generate the latent features of users and items. More specifically, we compute the similarity matrix of the meta-graph by adjacency matrix. Take Fig. 3 as an example, the user-item similarity matrix is calculated by:

$$S = W_{UR} \cdot ((W_{RI} \cdot W_{RI}^T) \odot (W_{RK} \cdot W_{RK}^T)) \cdot W_{UR}^T \cdot W_{UI}, \quad (1)$$

where $\odot$ is the Hadamard product and $W_*$ is the adjacency matrix between two different node types. Following similarity steps, by designing C different meta-graphs, we can get C different user-item similarity matrices, denoted by $\{S^{(i)}\}^C$. In addition, we use matrix factorization (MF) to generate the latent features of users and items from the user-item similarity matrix $S^{(i)}$. First, we let $U^{(i)}$ and $I^{(i)}$ denote the $i$-th user-specific matrix and item-specific matrix, respectively. Then, MF factorizes $S^{(i)}$ into two low-rank user-specific and item-specific matrices ($U^{(i)} \in \mathbb{R}^{m \times d}$ and $I^{(i)} \in \mathbb{R}^{n \times d}$, $i \in \{1, 2, ..., C\}$). MF model maps both user and item to a joint latent factor space of dimensionality $d$, such that user-item interactions are modeled as inner products in that space and the interaction can be captured by $S^{(i)} = U^{(i)}I^{(i)T}$. As the sparsity of $S^{(i)}$, we only model the observed interaction while avoiding overfitting through a regularized model. Therefore, we can learn $U^{(i)}$ and $I^{(i)}$ by:

$$\min_{U^{(i)}, I^{(i)}} \frac{1}{2} ||S^{(i)} - U^{(i)}I^{(i)T}||_d^2 + \frac{\lambda_{U^{(i)}}}{2} ||U^{(i)}||_d^2 + \frac{\lambda_{I^{(i)}}}{2} ||I^{(i)}||_d^2, \quad (2)$$

where $\lambda_{U^{(i)}}$ and $\lambda_{I^{(i)}}$ are hyper-parameters that used to avoid overfitting. As obtaining C different user-item similarity matrices before, we got C user-specific matrices (latent feature matrices of users) and C item-specific matrices (latent feature matrices of items), denoted as $\{U^{(i)}\}^C$, $\{I^{(i)}\}^C$.

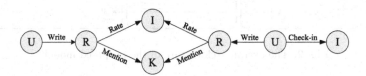

**Fig. 3.** An example of meta-graph. (U: user, R: review, I: item, K: keyword)

## 3.2   Attention-Based Feature Enhancement

As different latent features of user and item obtained from different meta-graphs contribute differently to recommendation, lacks of differentiating the importance of latent features may result in suboptimal prediction. To solve this problem, we propose a novel attention-based feature enhancement module to discriminate the importance of different latent features. The flowchart of the module is shown in Fig. 4. Specifically, let $E^{(i)} \in \mathbb{R}^{m \times d}$ and $F^{(i)} \in \mathbb{R}^{n \times d}$ denote the attention weight matrices of user and item latent features generated by the $i$-th meta-graph, respectively. The corresponding weighted user latent features $\hat{U}^{(i)}$ and weighted item latent features $\hat{I}^{(i)}$ can be given as follows:

$$\hat{U}^{(i)} = E^{(i)} \odot U^{(i)}, \tag{3}$$

$$\hat{I}^{(i)} = F^{(i)} \odot I^{(i)}, \tag{4}$$

where $i \in \{1, 2, ..., C\}$ and the attention weight matrix $E^{(i)}$ is learned by a two-layer attention network:

$$E_1^{(i)} = ReLU(W_1^{(i)}(X \oplus U^{(i)}) + b_1^{(i)}), \tag{5}$$

$$E_2^{(i)} = ReLU(W_2^{(i)} E_1^{(i)} + b_2^{(i)}), \tag{6}$$

where $W_*^{(i)}$ and $b_*^{(i)}$, $* \in \{1, 2\}$ denote the weight matrix and the bias vector for the $*$-th fully connected layer, $\oplus$ denotes the feature concatenation operation. Since the range of the attention weights is uncertain, we restrict the distance to (0, 1) by normalizing the above attention weight matrix with the softmax function:

$$E^{(i)} = \frac{\exp(E_2^{(i)})}{\sum\limits_{c=1}^{C} \exp(E_2^{(c)})}. \tag{7}$$

We learn the attention weight matrix $F^{(i)}$ with the same way of $E^{(i)}$, so that we do not repeat here.

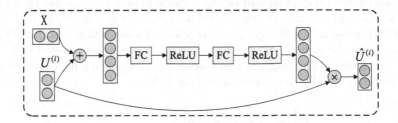

**Fig. 4.** Demonstration of the attention-based feature enhancement module.

### 3.3 Hierarchical Feature Interaction for Rating

To efficiently assemble and thoroughly use the features for rating, we design a hierarchical feature interaction method with factorization machine to model complicated interactions between user $u$, item $i$ and their weighted latent features. The predicted rating of user $u$ on item $i$ is defined as:

$$\hat{R}_{u,i} = w_0 + \sum_{j=1}^{d} w_j x_j + \sum_{j=1}^{d} \hat{w}_j \hat{U}_{u,j} + \sum_{j=1}^{d} w'_j y_j + \sum_{j=1}^{d} \hat{w}'_j \hat{I}_{i,j} + H_{u,i}, \tag{8}$$

where $w_0$ is the global bias. $w_j$, $\hat{w}_j$, $w'_j$ and $\hat{w}'_j$ represent the weight for $x_j$, $\hat{U}_{u,j}$, $y_j$ and $\hat{I}_{i,j}$, respectively. In which, $x_j$ denotes the $j$-th feature of user $u$, $\hat{U}_{u,j}$ denotes the $j$-th weighted latent feature of user $u$, $y_j$ denotes the $j$-th feature of item $i$ and $\hat{I}_{i,j}$ denotes the $j$-th weighted latent feature of item $i$. Furthermore, $\hat{U}_{u,j} \in \hat{U}$ and $\hat{U} = \sum_{i=1}^{C} \hat{U}^{(i)}$, $\hat{I}_{i,j} \in \hat{I}$ and $\hat{I} = \sum_{j=1}^{C} \hat{I}^{(j)}$. $H_{u,i}$ is the result of interactions between user $u$, item $i$ and their weighted latent features. $H_{u,i} \in H$ and $H$ denotes the result of interactions between users, items and their weighted latent features which is the core component for modelling hierarchical feature interactions. In order to obtain $H$, we first consider to model the pairwise correlation between users and their weighted latent features, and all nested correlations among weighted latent features. Therefore, the user representations $\hat{X}$ can be defined as follows:

$$\hat{X} = \sum_{i=1}^{C} X \odot \hat{U}^{(i)} + \sum_{i=1}^{C} \sum_{j=i+1}^{C} \hat{U}^{(i)} \odot \hat{U}^{(j)}. \tag{9}$$

By employing the same operation on items and their weighted latent features, we can similarly define the item representations $\hat{Y}$ as follows:

$$\hat{Y} = \sum_{i=1}^{C} Y \odot \hat{I}^{(i)} + \sum_{i=1}^{C} \sum_{j=i+1}^{C} \hat{I}^{(i)} \odot \hat{I}^{(j)}. \tag{10}$$

Then, we employ a stack of fully connected layers to capture the correlation among users and items. We first merge user representations $\hat{X}$ and item representations $\hat{Y}$ with the Hadamard product, which models the two-way interaction between users and items. We then place a multi-layer perception (MLP) above the Hadamard product. By specifying non-linearity activation function (ReLU) which is more biologically plausible and proven to be non-saturated [3,12], we allow the model to learn higher-order feature interactions in a non-linearity way. The definition of fully connected layers is as follows:

$$\begin{cases} Z_1 = ReLU(W_1(\hat{X} \odot \hat{Y}) + b_1), \\ Z_2 = ReLU(W_2 Z_1 + b_2), \\ \quad\quad \ldots\ldots \\ Z_t = ReLU(W_t Z_{t-1} + b_t), \end{cases} \tag{11}$$

where $W_t$ and $b_t$ denote the weight matrix and bias vector of the $t$-th layer, respectively. At last, the final result of interactions between users, items and their weighted latent features is defined as:

$$H = W_Z Z_t, \tag{12}$$

where $W_Z$ denote the neuron weights.

## 3.4    Optimization

To estimate the parameters of MGAR, we specify an objective function $L$. The overall objective to be optimized is defined as:

$$L = \sum_{(u,i)\in\tau} \left(R_{u,i} - \hat{R}_{u,i}\right)^2 + \lambda||W||^2, \tag{13}$$

where $\tau$ denotes the set of training instances, the target $R_{u,i}$ is a real value in our datasets, $\lambda$ controls the strength of regularization and W is the set of weights in our implementation. Moreover, we use Adam as the optimizer so that the learning rate can be self adapted during the training phase.

**Overfitting Prevention.** Besides the regularization, we also use dropout technique to prevent overfitting as neural network models are easy to overfit on training data. We employ dropout in attention networks and randomly drop $\rho$ percent of latent factors, where $\rho$ is termed as the dropout ratio.

**Batch Normalization.** One difficulty of training multi-layered neural networks is caused by the fact of covariance shift [7]. If the parameters of the previous layer change, the distribution of the later layer's inputs needs to change accordingly. To address this problem, we employ batch normalization (BN), which normalizes the input to a zero-mean unit-variance Gaussian distribution for each training mini-batch. It can lead to faster convergence and better performance in MGAR.

## 4    Experiments

In order to evaluate the effectiveness of the proposed method, MGAR, we conduct extensive experiments comparing with several state-of-the-art recommendation methods on two widely used datasets. The experiments are implemented with TensorFlow and run in a Linux server with NVIDIA Tesla M40 GPU.

### 4.1    Datasets and Evaluation Metric

**Datasets.** Experiments are conducted on two datasets, Yelp and Amazon Electronics, which are provided in [18] and have rich heterogeneous information. The detailed description of our datasets is shown in Table 1. For Yelp[1] which has 191,506 user-item interaction records, it contains 8 entity types and 9 relations. There are many domains in Amazon[2] and we use the electronics domain for our experiments. For Amazon Electronics which has 183,807 user-item interaction records, it contains 6 entity types and 6 relations. Moreover, ratings in our two datasets are from 1 to 5.

---

[1] https://www.yelp.com/datasetchallenge.
[2] http://jmcauley.ucsd.edu/data/amazon/.

**Evaluation Metric.** To evaluate the performance of the recommendation, we adopt Root Mean Square Error (RMSE) as evaluation metric where a lower RMSE score indicates a better performance. It is defined as follows:

$$RMSE = \sqrt{\frac{\sum_{(u,i)\in R_{test}} (R_{u,i} - \hat{R}_{u,i})^2}{|R_{test}|}}, \tag{14}$$

where $R_{test}$ is the set of all user-item pairs $(u, i)$ in the test set, and $R_{u,i}$ is the observed rating in the test set.

**Table 1.** The detailed description of Yelp and Amazon Electronics datasets.

| Datasets | Relations (A-B) | Number of A | Number of B | Number of (A-B) |
|---|---|---|---|---|
| Yelp | User-Business | 36,105 | 22,496 | 191,506 |
| | User-Review | 36,105 | 191,506 | 191,506 |
| | Review-Business | 191,506 | 22,496 | 191,506 |
| | Review-Keyword | 191,506 | 10 | 955,041 |
| | User-User | 17,065 | 17,065 | 140,344 |
| | Business-Category | 22,496 | 869 | 67,940 |
| | Business-City | 22,496 | 215 | 22,496 |
| | Business-State | 22,496 | 18 | 22,496 |
| | Business-Star | 22,496 | 9 | 22,496 |
| Amazon Electronics | User-Product | 59,297 | 20,216 | 183,807 |
| | User-Review | 59,297 | 183,807 | 183,807 |
| | Review-Product | 183,807 | 20,216 | 183,807 |
| | Review-Keyword | 183,807 | 10 | 796,392 |
| | Product-Category | 20,216 | 682 | 87,587 |
| | Product-Brand | 9,533 | 2,015 | 9,533 |

## 4.2 Experimental Settings

To evaluate the effectiveness of our method, we compare it with several traditional recommendation methods and deep learning models. They are introduced as follows:

- **MF** [8]: Matrix factorization is a traditional model for recommendation which only uses the user-item rating matrix.
- **LibFM** [11]: LibFM has shown its strong performance for personalized recommendation and rating prediction task which is released by Rendle[3]. In this paper, we only use the user-item rating matrix for comparison.
- **FMG** [18]: FMG is designed for rating prediction which gets latent features of users and items by factorizing the similarity matrices of different metagraphs. It simply concatenates all the latent features and embeddings as the input features.

---

[3] http://www.libfm.org/.

- **AFM** [2]: This is an attention-driven factor model which integrates item features driven by user's attention.
- **NFM** [4]: NFM combines neural network with factorization machine for rating prediction. It converts each log (i.e., user ID, item ID and all context variables) to a feature vector by using one-hot encoding.

To demonstrate the capability of our model, we randomly split the datasets into training set (80%) and testing set (20%). For fairly comparing the performance of our model, we use the meta-graphs designed in [18] that the number of meta-graph types in Yelp and Amazon Electronics is 9 and 6, respectively. We randomly initialize the parameters with a Gaussian distribution, where the mean and the standard deviation is 0 and 0.01, respectively. We trained the proposed model for 6000 batches in total and the batch size was set to 512. The learning rate was searched in [0.001, 0.003, 0.005, 0.1].

### 4.3  Comparison with Existing Models

The RMSE results on Yelp and Amazon Electronics datasets are shown in Table 2 which presents the best effect achieved by MGAR and comparative methods. From Table 2, we have the following observations:

- MF and LibFM have worse results than FMG. The reason is that FMG employs HINs based meta-graphs to characterize the rich semantic information while MF and LibFM only use the user-item rating matrix which have insufficient information.
- The performance of AFM is better than FMG. FMG simply concatenates all the corresponding features as the input of rating prediction without differentiating the importance of the latent features generated by different meta-graphs. Compared with FMG, AFM learns what feature of an item a user cares most about which demonstrates the design of attention-based feature enhancement module in our work is significant.
- NFM obtains a better performance than other baselines. It seamlessly combines the linearity of factorization machine in modelling second-order feature interactions and the non-linearity of neural network in modelling higher-order feature interactions which improves the result effectively.
- MGAR observably outperforms all baselines in all cases and well demonstrates its effectiveness. The reason is that MGAR proposes not only a novel attention-based feature enhancement module to discriminate the importance of different latent features but also a hierarchical feature interaction method with factorization machine to efficiently assemble and thoroughly use the features for rating prediction.

**Table 2.** The RMSE results on Yelp and Amazon Electronics of MGAR and all baseline methods. The best results are in boldface. Percentages in the brackets are the reduction of RMSE comparing MGAR with the corresponding approaches in the table header.

| Methods | Yelp | Amazon Electronics |
|---|---|---|
| MF [8] | 2.5141 (+53.7%) | 2.9656 (+61.8%) |
| LibFM [11] | 1.7637 (+34.0%) | 1.3462 (+15.9%) |
| FMG [18] | 1.2456 (+6.6%) | 1.1864 (+4.6%) |
| AFM [2] | 1.2193 (+4.6%) | 1.1681 (+3.1%) |
| NFM [4] | 1.1988 (+2.9%) | 1.1627 (+2.6%) |
| **MGAR** | **1.1637** | **1.1322** |

## 4.4  Further Analysis on MGAR

**Ablation Study.** To verify the impact of different modules on the performance of MGAR, we design three diverse baselines for experiments: (a) MGAR-att is built by removing the attention-based feature enhancement module of MGAR; (b) MGAR-inter is built by removing the hierarchical feature interaction module; (c) MGAR-fm is built by removing the factorization machine. The results are shown in Table 3. It can be observed that MGAR is most affected by the attention-based feature enhancement module. It proves that latent features of different meta-graphs make different contributions to prediction. Furthermore, the hierarchical feature interaction module also plays a key role as it could holistically exploit the different interrelations of features. In addition, the factorization machine improves the performance of MGAR, to some extent, as the way of assembling features has significant influence on the accuracy of recommendation.

**Impact of MLP with Different Layers.** We also verify the impact of MLP with different fully connected layers. MLP-0, MLP-1, MLP-2, MLP-3 and MLP-4 denote MLP with 0–4 layers, respectively. In which, MLP-0 denotes we only merge user representations and item representations with the Hadamard product which has no fully connected layer. The results are shown in Table 4. We can observe that the performance of MLP-0 is worse than MLP-1 as MLP-0 only learns the linearity interaction of second-order features and neglects

**Table 3.** The RMSE results on Yelp and Amazon Electronics for ablation analysis.

| Method | Yelp | Amazon Electronics |
|---|---|---|
| MGAR-att | 1.4974 | 1.2142 |
| MGAR-inter | 1.3041 | 1.1816 |
| MGAR-fm | 1.2532 | 1.1934 |
| **MGAR** | **1.1637** | **1.1322** |

non-linearity interaction of higher-order features which is important in capturing useful information for rating prediction. Furthermore, we can find that more layers do not improve the performance and the best one is only one layer. The reason is that the hierarchical feature interaction module has encoded informative second-order feature interactions, and based on which, a simple non-linearity function is sufficient to capture higher-order feature interactions.

**Table 4.** Impact of MLP with different layers.

| Method | Yelp | Amazon Electronics |
|--------|--------|--------------------|
| MLP-0 | 1.1678 | 1.1336 |
| **MLP-1** | **1.1637** | **1.1322** |
| MLP-2 | 1.1655 | 1.1331 |
| MLP-3 | 1.1673 | 1.1337 |
| MLP-4 | 1.1709 | 1.1345 |

**Effect of Model Parameters.** We conducted experiments to further analyse the influence of parameters $\rho$ and $\lambda$ on performance. The dropout ratio $\rho$ is used to prevent MGAR from overfitting. The trade-off parameter $\lambda$ controls

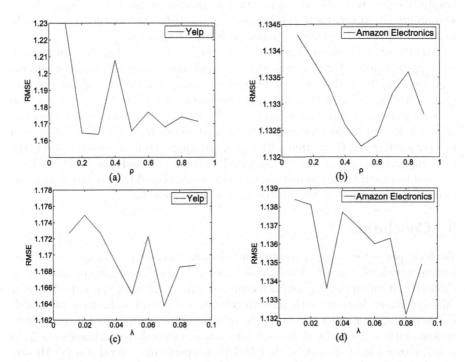

**Fig. 5.** Parameter sensitivity analysis of $\rho$ and $\lambda$ on performance.

the regularization strength as Eq. (13) shows. The RMSE results on Yelp and Amazon Electronics are shown in Fig. 5. We can observe that MGAR is sensitive to $\rho$ and $\lambda$. A too small or too large value may lead to a bad learning of the model. The best result is obtained with $\rho = 0.3$ and $\lambda = 0.07$ on Yelp, $\rho = 0.5$ and $\lambda = 0.08$ on Amazon Electronics.

## 5    Related Work

Over the last few years, various recommendation methods have been proposed, which can be roughly divided into traditional methods [1,10] and HINs based methods [13,16,18]. In traditional methods, they only use historical interactions for recommendation. For example, MF [8] is a standard collaborative filtering method which only uses the user-item rating matrix with $L_2$ regularization. LibFM [11] only models the latent feature interactions of users and items for personalized recommendation and rating prediction.

Since traditional methods can only use specific information and their performances are not well, many works attempt to leverage additional information for recommendation, such as heterogeneous information [6,13]. As a newly emerging direction, heterogeneous information network can naturally model complex entities and rich relations in recommender system. HeteRec [16] measures similarity between users and items by meta-paths and the weighted ensemble model is learned from the latent features of users and items. FMG [18] employs meta-graphs to characterize the rich semantic information so that it has strong expressive ability. However, their performance is unsatisfactory as they cannot fully utilize the captured features by a simple feature interaction.

To fully use the captured features, some works [3,12,15] start to explore deep learning techniques for recommendation which have achieved immense success and been widely used on speech recognition, computer vision and natural language processing. NFM [4] combines neural network with factorization machine to learn linearity interaction of second-order features and non-linearity interaction of higher-order features. AFM [2] is used to integrate item features driven by user's attention. Compared with them, our approach uses attention networks to distinguish the importance of different latent features and adopts a novel hierarchical feature interaction method to model complicated feature interactions so that the performance of MGAR has greatly improvement.

## 6    Conclusion

We have presented a novel recommendation method over heterogeneous information networks, named Meta-Graph based Attention-aware Recommendation (MGAR). It enhances the useful features by discriminating the importance of different latent features with the attention-based feature enhancement model. Furthermore, we predict the rating by modelling the complicated feature interactions with a hierarchical feature interaction method. Experiments on Yelp and Amazon Electronics show that MGAR outperforms several state-of-the-art recommendation algorithms.

**Acknowledgement.** This work was supported by the Strategic Priority Research Program of the Chinese Academy of Sciences (XDC02050200) and Beijing Municipal Science and Technology Project (Z181100002718004).

# References

1. Anyosa, S.C., Vinagre, J., Jorge, A.M.: Incremental matrix co-factorization for recommender systems with implicit feedback. In: Proceedings of the 27th International Conference on World Wide Web, pp. 1413–1418 (2018)
2. Chen, J., Zhuang, F., Hong, X., Ao, X., Xie, X., He, Q.: Attention-driven factor model for explainable personalized recommendation. In: The 41st International ACM SIGIR Conference on Research & Development in Information Retrieval, pp. 909–912 (2018)
3. Covington, P., Adams, J., Sargin, E.: Deep neural networks for YouTube recommendations. In: Proceedings of the 10th ACM Conference on Recommender Systems, pp. 191–198 (2016)
4. He, X., Chua, T.S.: Neural factorization machines for sparse predictive analytics. In: Proceedings of the 40th International ACM SIGIR conference on Research and Development in Information Retrieval, pp. 355–364 (2017)
5. Hu, B., Shi, C., Zhao, W.X., Yu, P.S.: Leveraging meta-path based context for top-n recommendation with a neural co-attention model. In: Proceedings of the 24th ACM SIGKDD International Conference on Knowledge Discovery and Data Mining, pp. 1531–1540 (2018)
6. Huang, Z., Zheng, Y., Cheng, R., Sun, Y., Mamoulis, N., Li, X.: Meta structure: computing relevance in large heterogeneous information networks. In: Proceedings of the 22nd ACM SIGKDD International Conference on Knowledge Discovery and Data Mining, pp. 1595–1604 (2016)
7. Ioffe, S., Szegedy, C.: Batch normalization: accelerating deep network training by reducing internal covariate shift. In: International Conference on Machine Learning, pp. 1–9 (2015)
8. Koren, Y., Bell, R., Volinsky, C.: Matrix factorization techniques for recommender systems. Computer **8**, 30–37 (2009)
9. Lian, J., Zhou, X., Zhang, F., Chen, Z., Xie, X., Sun, G.: xDeepFM: combining explicit and implicit feature interactions for recommender systems. arXiv preprint arXiv:1803.05170 (2018)
10. Liu, H., He, X., Feng, F., Nie, L., Zhang, H.: Discrete factorization machines for fast feature-based recommendation. arXiv preprint arXiv:1805.02232 (2018)
11. Rendle, S.: Factorization machines with libFM. ACM Trans. Intell. Syst. Technol. (TIST) **3**(3), 57 (2012)
12. Shan, Y., Hoens, T.R., Jiao, J., Wang, H., Yu, D., Mao, J.: Deep crossing: web-scale modeling without manually crafted combinatorial features. In: Proceedings of the 22nd ACM SIGKDD International Conference on Knowledge Discovery and Data Mining, pp. 255–262 (2016)
13. Shi, C., Zhang, Z., Luo, P., Yu, P.S., Yue, Y., Wu, B.: Semantic path based personalized recommendation on weighted heterogeneous information networks. In: Proceedings of the 24th ACM International on Conference on Information and Knowledge Management, pp. 453–462 (2015)
14. Sun, Y., Han, J., Yan, X., Yu, P.S., Wu, T.: PathSim: meta path-based top-k similarity search in heterogeneous information networks. Proc. VLDB Endow. **4**(11), 992–1003 (2011)

15. Xu, J., He, X., Li, H.: Deep learning for matching in search and recommendation. In: The 41st International ACM SIGIR Conference on Research & Development in Information Retrieval, pp. 1365–1368 (2018)
16. Yu, X., et al.: Personalized entity recommendation: a heterogeneous information network approach. In: Proceedings of the 7th ACM international conference on Web Search and Data Mining, pp. 283–292 (2014)
17. Zhang, D., Yin, J., Zhu, X., Zhang, C.: MetaGraph2Vec: complex semantic path augmented heterogeneous network embedding. In: Pacific-Asia Conference on Knowledge Discovery and Data Mining, pp. 196–208 (2018)
18. Zhao, H., Yao, Q., Li, J., Song, Y., Lee, D.L.: Meta-graph based recommendation fusion over heterogeneous information networks. In: Proceedings of the 23rd ACM SIGKDD International Conference on Knowledge Discovery and Data Mining, pp. 635–644 (2017)

# Determining Adaptive Loss Functions and Algorithms for Predictive Models

Michael C. Burkhart[✉][iD] and Kourosh Modarresi[iD]

Adobe Inc., 345 Park Avenue, San José, CA 95110, USA
{mburkhar,modarres}@adobe.com

**Abstract.** We consider the problem of training models to predict sequential processes. We use two econometric datasets to demonstrate how different losses and learning algorithms alter the predictive power for a variety of state-of-the-art models. We investigate how the choice of loss function impacts model training and find that no single algorithm or loss function results in optimal predictive performance. For small datasets, neural models prove especially sensitive to training parameters, including choice of loss function and pre-processing steps. We find that a recursively-applied artificial neural network trained under $L_1$ loss performs best under many different metrics on a national retail sales dataset, whereas a differenced autoregressive model trained under $L_1$ loss performs best under a variety of metrics on an e-commerce dataset. We note that different training metrics and processing steps result in appreciably different performance across all model classes and argue for an adaptive approach to model fitting.

## 1 Introduction

We develop time series estimators for current datasets and use them to iteratively forecast a few steps into the future. We consider the effects of using different training loss functions and different evaluation metrics. The training methodology, including the choice of loss function, seems to impact model performance more than typically reported, potentially due to the relatively smaller amounts of available training data.

## 2 Datasets

We used time series datasets from the FRED Economic Data portal provided by the Federal Reserve Bank of St. Louis.

### 2.1 Advance Retail Sales

The Advance Retail Sales: Retail and Food Services Total (RSAFSNA) dataset is a monthly accounting of retail and food services sales totals in the US, provided by the U.S. Bureau of the Census [21]. The data begins in January 1992 and includes

© Springer Nature Switzerland AG 2019
J. M. F. Rodrigues et al. (Eds.): ICCS 2019, LNCS 11537, pp. 595–607, 2019.
https://doi.org/10.1007/978-3-030-22741-8_42

a final estimated value for October 2018. This monthly data is strongly periodic. We split the time series into two contiguous sets: a training set containing 310 observations from Jan. 1992 to Oct. 2017 and a testing set of 12 observations from Nov. 2017 to Oct. 2018. Models learned to use the previous $\ell = 11$ observations to predict the next datapoint. All training and validation was conducted on the training set. See Fig. 1 (left) for a plot of the data and Fig. 2 for its autocorrelation and partial autocorrelation plots.

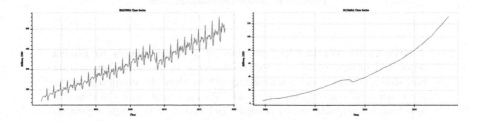

**Fig. 1.** (Left) Plot of the Advance Retail Sales dataset. Note the strong seasonality for this monthly data. The recession of 2008 is clearly visible in the data. (Right) Plot of the E-Commerce Retail Sales dataset. This coarser, quarterly data displays less seasonality than the Retail Sales numbers. While the 2008 recession remains visible, its relative effect seems less pronounced. Even as total consumption dropped, the share of online consumption increased.

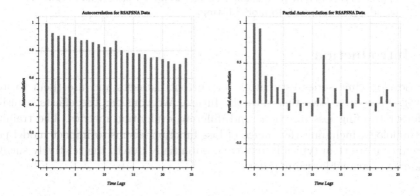

**Fig. 2.** Diagnostic plots for the Advance Retail Sales dataset. (Left) Autocorrelation function; (Right) Partial Autocorrelation function. The two bumps in the ACF correspond to lags at times 12 and 24, exactly one and two years prior, respectively. Similarly, the bump in the PACF corresponds to the lag at time 12, exactly one year prior.

## 2.2   E-Commerce Retail Sales

The E-Commerce Retail Sales (ECOMSA) dataset is a quarterly accounting of goods and services sales totals where orders were placed or prices were negotiated online. The U.S. Bureau of the Census provides data from Q4 1999 to Q3

2018 [22]. We split the time series into two contiguous sets: a training set containing 72 observations from Q4 1999 to Q3 2017 and a testing set of 4 observations from Q4 2017 to Q3 2018. Models used the previous 7 observations to predict the next datapoint. All training and validation was conducted on the training set. See Fig. 1(right) for a plot of the data and Fig. 3 for its autocorrelation and partial autocorrelation plots.

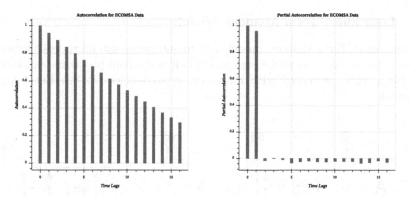

**Fig. 3.** Diagnostic plots for the E-Commerce Retail Sales dataset. (Left) Autocorrelation function; (Right) Partial Autocorrelation function. Periodicity may be less apparent for this data due to the general strong trend of growth.

## 3    Methodology

We describe some of the modeling approaches used on this data. Consider a real-valued time series $X_1, X_2, \ldots$. Our problem is to use the historical data $X_{t-\ell}, X_{t-\ell+1}, \ldots, X_{t-1}$ to predict the next realization of the series $X_t$. The length $\ell$ of the historical data used by our model is often referred to as number of lags in the model. Once we have learned a model for $X_t$ given $X_{t-\ell:t-1}$, we can use it iteratively to predict multiple steps into the future by using our predictions in place of known historical data. For example, if we predict $\hat{X}_t$ given $X_{t-\ell:t-1}$, we can then predict $\hat{X}_{t+1}$ given $X_{t-\ell+1:t-1}, \hat{X}_t$. Note however, that the uncertainty associated to this second prediction $\hat{X}_{t+1}$ will be higher than that for $\hat{X}_t$, because we are now using uncertain data to make predictions.

### 3.1    Autoregressive Model

An autoregressive model with $\ell$ lags models

$$X_t = \alpha_1 X_{t-1} + \cdots + \alpha_\ell X_{t-\ell} + \epsilon_t$$

where $\epsilon_t \sim \mathcal{N}(\mu, \sigma^2)$ for parameters $\alpha_1, \ldots, \alpha_\ell, \mu, \sigma$. We often denote such a model $AR(\ell)$. The hyperparameter $\ell$ is often selected by considering the partial autocorrelation function, that measures the correlation between $X_t$ and $X_{t-i}$ after accounting for the linear dependence of $X_t$ on $X_{t-1}, \ldots, X_{t-i+1}$.

## 3.2   Artificial Neural Network Model

We can use a fully-connected multiple layer artificial neural network $f_\theta : \mathbb{R}^\ell \to \mathbb{R}$ to model

$$X_t = f_\theta(X_{t-1}, \ldots, X_{t-\ell})$$

We learn the parameters $\theta$ by passing batches of data to any standard optimizer.

## 3.3   Long Short-Term Memory Model

A Long Short-Term Memory model [7,9] maintains a stateful representation of history to overcome the vanishing and exploding gradient problems encountered by RNN's when presented with long-term dependencies [2]. See Fig. 4 for an illustration.

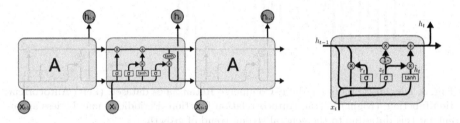

**Fig. 4.** Schematics for recurrent neural network variants LSTM (left, shown in sequence) and GRU (right, cell only). Images courtesy of Chris Olah; re-used with permission.

## 3.4   Gated Recurrent Unit Model

A variant of the LSTM, the Gated Recurrent Unit (GRU) model simplifies LSTM architecture [3]. Chung et al. suggested that GRU's can outperform LSTM's on smaller datasets [4]. See Fig. 4 for a side-by-side comparison.

## 3.5   Temporal Convolutional Network Model

The Temporal Convolutional Network (TCN) model presents an alternative neural network-based approach to time series modeling [1,12]. This convolutional architecture creates connections that are causal (dependent only on previous time steps) and dilated (periodic in time). See Fig. 5 for a schematic. Sometimes such models employ residual blocks [8] that stack networks and include a copy of the first layer in the final layer.

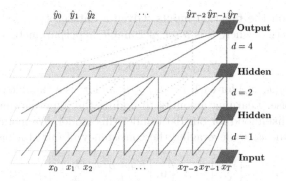

**Fig. 5.** Schematic for TCN model. Causal convolutions are those that only go backward in time. Dilated convolutions use only every $k$-th entry. Stacking dilated convolutions increases the receptive field multiplicatively. Diagram re-used from [1].

### 3.6    Gaussian Mixture Model

A Gaussian Mixture Model fits the model

$$p(x) = \sum_{i=1}^{k} \varphi_i \cdot \eta(x; \mu_i, \Sigma_i)$$

where $\eta(\cdot; \mu_i, \Sigma_i)$ denotes the p.d.f. of a normal distribution with mean $\mu_i$ and covariance $\Sigma_i$. This model consists of a mixture of $k$ Gaussian distributions. As both the latent parameters and mixture memberships must be inferred from the dataset, the Expectation Maximization (EM) algorithm is commonly used to fit such a model [5]. Expectation Maximization alternates between computing soft assignments (expectations) of datapoints to mixtures and finding the latent parameters that maximize model likelihood under these assignments. Its estimates converge to the Maximum Likelihood (MLE) parameters [23].

Note that GMM is a generative model. We distinguish generative probabilistic models that learn the joint distribution $p(x, y)$ from discriminative probabilistic models that learn the conditional distribution $p(y|x)$ [14]. For example, we consider most regressive models (including the autoregressive and Gaussian process models described here) to be discriminative. To apply this model for our purposes here, we learned the joint distribution of $(X_{t-\ell:t-1}, X_t)$ as a Gaussian mixture and then for a given test sequence $x_{t-\ell:t-1}$, we predicted

$$\hat{x}_t = \arg\max_{x_t} \{p_\theta(x_{t-\ell:t-1}, x_t)\}$$

under our trained model $p_\theta$.

### 3.7    Gaussian Process Regression Model

A Gaussian Process (GP) is an infinite collection of random variables for which every finite subset has a multivariate Gaussian distribution [17]. For the purposes of a regression, the GP specifies a covariance structure on function outputs that depends on the function inputs. Given a dataset $\{(x_i, y_i)\}_{i=1}^n$, and a kernel function $k_\theta(\cdot, \cdot)$ where $\theta$ is a set of tunable hyperparameters, the GP model specifies

$$Y_{1:n}|x_{1:n} \sim \mathcal{N}(\mathbf{0}, K + \sigma^2 I_n)$$

where $K_{ij} = k_\theta(x_i, x_j)$ for each $1 \le i, j \le n$, and $\sigma^2$ is a tunable hyperparameter for process noise. Learning amounts to finding the hyperparameters $\theta, \sigma^2$ that maximize the likelihood for our data under this model. We made predictions at a new test point $x_*$ by leveraging the consistency of our stochastic model. We consider

$$\begin{bmatrix} Y_{1:n} \\ Y_* \end{bmatrix} \sim \mathcal{N}\left(\mathbf{0}, \begin{bmatrix} K_\theta + \sigma^2 I_n & (\mathbf{k}_*)^\top \\ \mathbf{k}_* & k_{**} \end{bmatrix}\right)$$

where $(\mathbf{k}_*)_i = k_\theta(x_i, x_*)$ and $k_{**} = k_\theta(x_*, x_*)$. This implies that

$$Y_*|x_{1:n}, Y_{1:n}, x_* \sim \mathcal{N}\left((\mathbf{k}_*)^\top K^{-1} Y_{1:n}, k_{**} - (\mathbf{k}_*)^\top K^{-1} \mathbf{k}_*\right)$$

We used the mean of this conditional distribution for predictive purposes, with a squared exponential kernel.

## 4   Results

We tuned the artificial neural networks by hand. The following choices were made for each network:

- Model architecture: For recurrent neural networks, the size of the hidden layer and whether to include fully connected layers above or below the recurrent layer can significantly alter model performance.
- Regularization: Common approaches include dropout [18], added Gaussian noise [15], batch normalization [10], and Tikhonov regularization [19,20].
- Optimizer: There are a variety of modern optimization strategies that adaptively vary learning rates and add momentum. For this project, we used the Adam optimizer [11].
- Optimization parameters: Learning rate and training batch size were hand-selected.
- Synergistic effects: Choices in one category impact choices in another. For example, learning rate and training batch size are interdependent: adjustments to learning rate should often accompany adjustments to batch size, and vice versa. While the literature presents many individual solutions for deep learning, less is known about how they interact. For example, there has been considerable discussion on where to place dropout in a recurrent architecture [6,16,24]. In TCN's, we saw a clear connection between architecture size and learning rate as well.

For all neural models, we preprocessed data by subtracting the training mean and dividing by the training standard deviation for each datapoint. We trained and predicted on these scaled data, and inverted the transformation before calculating prediction error.

For the models denoted "$\Delta$," we learned and predicted on the time series of differences, and then took a cumulative sum of predicted differences to tabulate our final predictions. Differencing can remove time-based effects in the data (non-stationarities).

## 4.1   Advance Retail Sales

The AR predictions appear to be overly conservative. Training under the $L_1$ objective loss results in uniformly worse performance than $L_2$ (for the non-differenced model) (Fig. 6).

The GRU and LSTM architectures generally perform less well than the ANN architecture (Fig. 7).

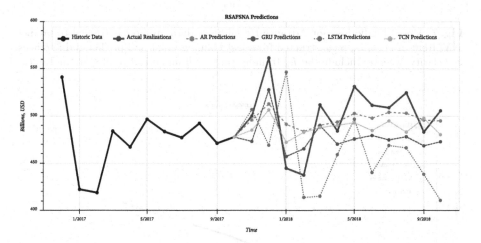

**Fig. 6.** Model predictions for Advance Retail Sales data on models trained to predict $X_t$ from $X_{t-\ell:t-1}$ For clarity, we only include the $L_2$ training runs. Full numerical results are in Table 1.

Due to the way we tested the models, a single missed prediction feeds back into the model and compounds the error for subsequent results. For example, the LSTM model badly misses its second prediction and then appears to be off by a time step for the rest of the test. These models have not been given the month of the year as input, only the 11 previous values for retail sales. We see how a single poor prediction that is fed back into the model as input can disrupt the model's sense of seasonality.

Differencing proved especially effective for the recurrent neural network models.

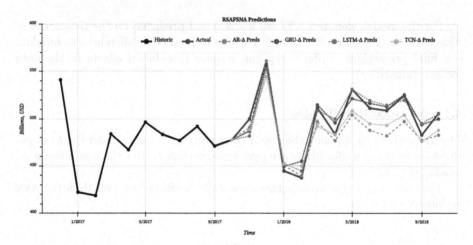

**Fig. 7.** Model predictions for Advance Retail Sales data on models trained to predict $\Delta X_t$ from $\Delta X_{t-\ell:t-1}$, reconstructed from last training point and predicted differences. For clarity, we only include the $L_2$ training runs. Full numerical results are in Table 1.

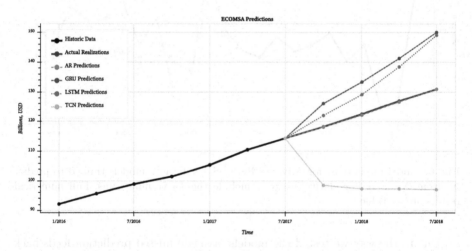

**Fig. 8.** Model predictions for E-Commerce Retail Sales data on models trained to predict $X_t$ from $X_{t-\ell:t-1}$; For clarity, we only include the $L_2$ training runs. Full numerical results are in Table 2.

## 4.2   E-Commerce Retail Sales

For $L_2$, we used least squares optimization, for which the optimal parameter values have explicit solutions. For $L_1$ and $L_4$, we performed gradient descent with momentum to find parameters that minimized training error on the full training set. We initialized at the $L_2$ optimum and checked for convergence.

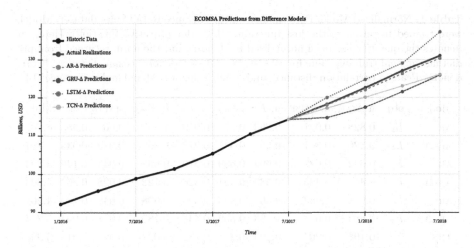

**Fig. 9.** Model predictions for E-Commerce Retail Sales data on models trained to predict $\Delta X_t$ from $\Delta X_{t-\ell:t-1}$, reconstructed from last training point and predicted differences. For clarity, we only include the $L_2$ training runs. Full numerical results are in Table 2.

Finally, it is worth noting that the performance of the TCN (non-differenced) on this data set was extremely poor (for the non-differenced model), under all objective functions. For any metric we consider, it would be much better to use the null predictions than the TCN predictions. It is possible that the TCN architecture requires a different regularization or training strategy in this case of a very small dataset. One might imagine that the TCN is reverting to the mean of the training data. (We note in Fig. 1 the strong upward trend of this data.) In the future, it may be advisable to include a separate linear trend for datasets like this, and fit the neural networks to residuals. Differencing appears to ameliorate this issue (Figs. 8 and 9).

For the AR model, differencing effectively increases the lag by one time step and removes any linear dependence on time. We see fairly similar results for the differenced and non-differenced versions of the AR and ANN models. However, for the recurrent models (LSTM & GRU) and the TCN, taking differences can prove to be a very effective way to boost model performance.

# A    Implementation Details

For this project, we used the Miniconda v.4.5.11 package manager with Python v.3.6.7. All neural networks were constructed in Keras v.2.2.4 with a Tensorflow v1.12.0 backend running Intel MKL optimizations.

**Table 1.** Normalized Model Performance on the Advance Retail Sales dataset. Models were trained to optimize the loss function "obj." (for objective function). Here, "n" denotes the quantity has been normalized by dividing out the norm returned from predicting the final training value for all test data, "Cos." denotes cosine distance, "Corr." denotes Pearson correlation distance, and "Mah." denotes Mahalanobis distance [13].

| Model | obj. | n.-MAE | n.-RMSE | n.-$L_3$ | n.-$L_4$ | n.-$L_\infty$ | n.-Cos. | MAPE | Corr. | Mah. |
|---|---|---|---|---|---|---|---|---|---|---|
| AR | $L_1$ | 0.828 | 0.921 | 0.963 | 0.97 | 0.919 | 1.138 | 0.06 | 0.955 | 6.467 |
| AR-$\Delta$ | $L_1$ | 0.326 | 0.351 | 0.35 | 0.34 | 0.278 | 0.069 | 0.022 | 0.033 | 4.897 |
| AR | $L_2$ | 0.642 | 0.681 | 0.694 | 0.689 | 0.579 | 0.668 | 0.045 | 0.146 | 3.894 |
| AR-$\Delta$ | $L_2$ | 0.41 | 0.435 | 0.433 | 0.421 | 0.326 | 0.122 | 0.028 | 0.062 | 4.803 |
| AR | $L_4$ | 0.638 | 0.663 | 0.664 | 0.656 | 0.593 | 0.608 | 0.045 | 0.184 | 4.26 |
| AR-$\Delta$ | $L_4$ | 0.604 | 0.6 | 0.583 | 0.562 | 0.447 | 0.187 | 0.041 | 0.094 | 5.244 |
| ANN | $L_1$ | 0.198 | 0.204 | 0.204 | 0.2 | 0.18 | 0.039 | 0.014 | 0.017 | 3.998 |
| ANN-$\Delta$ | $L_1$ | 0.281 | 0.296 | 0.293 | 0.284 | 0.22 | 0.06 | 0.019 | 0.03 | 5.209 |
| ANN | $L_2$ | 0.261 | 0.272 | 0.274 | 0.271 | 0.242 | 0.065 | 0.018 | 0.03 | 3.594 |
| ANN-$\Delta$ | $L_2$ | 0.288 | 0.299 | 0.294 | 0.285 | 0.223 | 0.065 | 0.02 | 0.033 | 4.824 |
| ANN | $L_4$ | 0.636 | 0.643 | 0.638 | 0.626 | 0.583 | 0.2 | 0.043 | 0.1 | 4.971 |
| ANN-$\Delta$ | $L_4$ | 0.316 | 0.33 | 0.328 | 0.319 | 0.256 | 0.142 | 0.022 | 0.073 | 4.88 |
| GRU | $L_1$ | 1.056 | 1.055 | 1.061 | 1.069 | 1.085 | 1.076 | 0.072 | 1.473 | 5.682 |
| GRU-$\Delta$ | $L_1$ | 0.384 | 0.367 | 0.351 | 0.336 | 0.277 | 0.076 | 0.026 | 0.034 | 4.668 |
| GRU | $L_2$ | 0.843 | 0.793 | 0.761 | 0.735 | 0.663 | 0.443 | 0.058 | 0.212 | 6.044 |
| GRU-$\Delta$ | $L_2$ | 0.216 | 0.221 | 0.222 | 0.22 | 0.209 | 0.068 | 0.015 | 0.028 | 5.076 |
| GRU | $L_4$ | 0.703 | 0.694 | 0.674 | 0.654 | 0.604 | 0.643 | 0.05 | 0.396 | 6.059 |
| GRU-$\Delta$ | $L_4$ | 0.489 | 0.476 | 0.459 | 0.442 | 0.37 | 0.063 | 0.034 | 0.03 | 4.408 |
| LSTM | $L_1$ | 1.066 | 1.027 | 0.991 | 0.956 | 0.809 | 0.336 | 0.074 | 0.137 | 6.0 |
| LSTM-$\Delta$ | $L_1$ | 0.281 | 0.276 | 0.268 | 0.259 | 0.225 | 0.061 | 0.019 | 0.025 | 4.291 |
| LSTM | $L_2$ | 1.652 | 1.635 | 1.593 | 1.538 | 1.214 | 2.761 | 0.115 | 1.04 | 6.898 |
| LSTM-$\Delta$ | $L_2$ | 0.222 | 0.234 | 0.236 | 0.235 | 0.215 | 0.078 | 0.016 | 0.036 | 3.948 |
| LSTM | $L_4$ | 0.441 | 0.526 | 0.583 | 0.614 | 0.642 | 0.332 | 0.032 | 0.183 | 4.828 |
| LSTM-$\Delta$ | $L_4$ | 0.218 | 0.258 | 0.285 | 0.298 | 0.309 | 0.085 | 0.015 | 0.043 | 5.735 |
| TCN | $L_1$ | 0.915 | 0.917 | 0.911 | 0.898 | 0.793 | 1.182 | 0.064 | 1.108 | 6.752 |
| TCN-$\Delta$ | $L_1$ | 0.321 | 0.3 | 0.284 | 0.27 | 0.226 | 0.084 | 0.022 | 0.009 | 3.651 |
| TCN | $L_2$ | 0.798 | 0.775 | 0.755 | 0.734 | 0.656 | 0.739 | 0.055 | 0.376 | 5.108 |
| TCN-$\Delta$ | $L_2$ | 0.401 | 0.38 | 0.358 | 0.339 | 0.268 | 0.1 | 0.028 | 0.02 | 3.786 |
| TCN | $L_4$ | 1.004 | 0.991 | 0.977 | 0.962 | 0.916 | 0.902 | 0.069 | 0.57 | 6.123 |
| TCN-$\Delta$ | $L_4$ | 0.411 | 0.396 | 0.377 | 0.359 | 0.29 | 0.142 | 0.028 | 0.017 | 4.126 |
| GMM | MLE | 0.219 | 0.22 | 0.218 | 0.214 | 0.203 | 0.063 | 0.015 | 0.028 | 4.516 |
| GMM-$\Delta$ | MLE | 0.225 | 0.233 | 0.233 | 0.228 | 0.196 | 0.042 | 0.016 | 0.02 | 4.829 |
| GPR | MLE | 0.729 | 0.674 | 0.627 | 0.588 | 0.435 | 0.097 | 0.051 | 0.034 | 6.062 |
| GPR-$\Delta$ | MLE | 0.209 | 0.237 | 0.262 | 0.279 | 0.299 | 0.082 | 0.015 | 0.025 | 4.203 |

**Table 2.** Normalized Model Performance on the E-Commerce Retail Sales dataset. Models were trained to optimize the loss function "obj." (for objective function). Here, "n" denotes the quantity has been normalized by dividing out the norm returned from predicting the final training value for all test data, "Cos." denotes cosine distance, "Corr." denotes Pearson correlation distance, and "Mah." denotes Mahalanobis distance.

| Model | obj. | n.-MAE | n.-RMSE | n.-$L_3$ | n.-$L_4$ | n.-$L_\infty$ | n.-Cos. | MAPE | Corr. | Mah. |
|---|---|---|---|---|---|---|---|---|---|---|
| AR | $L_1$ | 0.058 | 0.056 | 0.054 | 0.053 | 0.049 | 0.002 | 0.005 | 0.001 | 0.676 |
| AR-$\Delta$ | $L_1$ | 0.015 | 0.014 | 0.013 | 0.012 | 0.011 | 0.001 | 0.001 | 0.0 | 0.088 |
| AR | $L_2$ | 0.017 | 0.02 | 0.02 | 0.02 | 0.019 | 0.001 | 0.001 | 0.0 | 0.432 |
| AR-$\Delta$ | $L_2$ | 0.051 | 0.05 | 0.049 | 0.048 | 0.043 | 0.002 | 0.004 | 0.0 | 0.472 |
| AR | $L_4$ | 0.02 | 0.027 | 0.031 | 0.033 | 0.036 | 0.003 | 0.002 | 0.0 | 0.383 |
| AR-$\Delta$ | $L_4$ | 0.069 | 0.071 | 0.073 | 0.074 | 0.077 | 0.006 | 0.006 | 0.0 | 0.593 |
| ANN | $L_1$ | 0.283 | 0.281 | 0.285 | 0.29 | 0.305 | 0.062 | 0.023 | 0.003 | 0.847 |
| ANN-$\Delta$ | $L_1$ | 0.082 | 0.083 | 0.083 | 0.083 | 0.079 | 0.007 | 0.007 | 0.0 | 0.36 |
| ANN | $L_2$ | 0.206 | 0.193 | 0.185 | 0.18 | 0.171 | 0.011 | 0.017 | 0.005 | 0.663 |
| ANN-$\Delta$ | $L_2$ | 0.215 | 0.221 | 0.224 | 0.225 | 0.226 | 0.055 | 0.017 | 0.0 | 1.038 |
| ANN | $L_4$ | 0.614 | 0.645 | 0.64 | 0.632 | 0.593 | 0.784 | 0.051 | 0.152 | 5.266 |
| ANN-$\Delta$ | $L_4$ | 0.377 | 0.396 | 0.404 | 0.407 | 0.412 | 0.21 | 0.031 | 0.0 | 1.58 |
| GRU | $L_1$ | 0.724 | 0.681 | 0.659 | 0.647 | 0.635 | 0.123 | 0.059 | 0.001 | 0.692 |
| GRU-$\Delta$ | $L_1$ | 0.778 | 0.839 | 0.879 | 0.905 | 0.957 | 0.909 | 0.063 | 0.007 | 2.434 |
| GRU | $L_2$ | 1.263 | 1.204 | 1.176 | 1.162 | 1.153 | 0.491 | 0.104 | 0.002 | 1.361 |
| GRU-$\Delta$ | $L_2$ | 0.462 | 0.425 | 0.402 | 0.386 | 0.33 | 0.019 | 0.038 | 0.007 | 1.976 |
| GRU | $L_4$ | 1.705 | 1.629 | 1.592 | 1.576 | 1.567 | 0.872 | 0.14 | 0.003 | 1.791 |
| GRU-$\Delta$ | $L_4$ | 0.117 | 0.12 | 0.12 | 0.12 | 0.111 | 0.015 | 0.01 | 0.001 | 0.507 |
| LSTM | $L_1$ | 1.25 | 1.208 | 1.19 | 1.183 | 1.182 | 0.642 | 0.102 | 0.001 | 1.678 |
| LSTM-$\Delta$ | $L_1$ | 0.248 | 0.242 | 0.237 | 0.233 | 0.223 | 0.05 | 0.021 | 0.012 | 2.538 |
| LSTM | $L_2$ | 0.97 | 1.002 | 1.028 | 1.047 | 1.093 | 0.959 | 0.079 | 0.006 | 2.06 |
| LSTM-$\Delta$ | $L_2$ | 0.293 | 0.312 | 0.332 | 0.347 | 0.375 | 0.129 | 0.024 | 0.016 | 1.904 |
| LSTM | $L_4$ | 2.677 | 2.836 | 3.01 | 3.142 | 3.387 | 7.211 | 0.218 | 0.11 | 5.676 |
| LSTM-$\Delta$ | $L_4$ | 0.732 | 0.922 | 1.029 | 1.088 | 1.178 | 1.903 | 0.059 | 0.032 | 2.898 |
| TCN | $L_1$ | 2.978 | 2.749 | 2.61 | 2.523 | 2.299 | 1.554 | 0.246 | 1.999 | 3.625 |
| TCN-$\Delta$ | $L_1$ | 0.284 | 0.288 | 0.29 | 0.29 | 0.287 | 0.084 | 0.023 | 0.0 | 1.132 |
| TCN | $L_2$ | 2.625 | 2.428 | 2.308 | 2.234 | 2.041 | 1.284 | 0.216 | 1.874 | 3.07 |
| TCN-$\Delta$ | $L_2$ | 0.299 | 0.301 | 0.302 | 0.301 | 0.297 | 0.088 | 0.024 | 0.0 | 1.219 |
| TCN | $L_4$ | 3.148 | 2.904 | 2.756 | 2.662 | 2.406 | 1.723 | 0.26 | 1.968 | 3.627 |
| TCN-$\Delta$ | $L_4$ | 0.332 | 0.335 | 0.337 | 0.337 | 0.335 | 0.112 | 0.027 | 0.0 | 1.287 |
| GMM | MLE | 0.02 | 0.021 | 0.021 | 0.022 | 0.021 | 0.001 | 0.002 | 0.001 | 0.425 |
| GMM-$\Delta$ | MLE | 0.026 | 0.029 | 0.031 | 0.033 | 0.035 | 0.005 | 0.002 | 0.001 | 0.177 |
| GPR | MLE | 0.19 | 0.201 | 0.209 | 0.214 | 0.225 | 0.052 | 0.015 | 0.001 | 0.499 |
| GPR-$\Delta$ | MLE | 0.345 | 0.336 | 0.33 | 0.327 | 0.322 | 0.072 | 0.028 | 0.0 | 1.709 |

# References

1. Bai, S., Kolter, J.Z., Koltun, V.: An empirical evaluation of generic convolutional and recurrent networks for sequence modeling. ArXiv e-prints (2018). arXiv:1803.01271
2. Bengio, Y., Simard, P., Frasconi, P.: Learning long-term dependencies with gradient descent is difficult. IEEE Trans. Neural Netw. **5**(2), 157–166 (1994)
3. Cho, K., et al.: Learning phrase representations using RNN encoder-decoder for statistical machine translation. In: Conference on Empirical Methods in Natural Language Processing, pp. 1724–1734 (2014)
4. Chung, J., Gülçehre, Ç., Cho, K., Bengio, Y.: Empirical evaluation of gated recurrent neural networks on sequence modeling. In: NIPS Workshop on Deep Learning (2014)
5. Dempster, A.P., Laird, N.M., Rubin, D.B.: Maximum likelihood from incomplete data via the em algorithm. J. R. Stat. Soc. Ser. B (Methodol.) **39**(1), 1–38 (1977)
6. Gal, Y., Ghahramani, Z.: A theoretically grounded application of dropout in recurrent neural networks. In: Advances in Neural Information Processing Systems, pp. 1019–1027 (2016)
7. Gers, F.A., Schmidhuber, J., Cummins, F.: Learning to forget: continual prediction with lstm. Neural Comput. **12**(10), 2451–2471 (2000)
8. He, K., Zhang, X., Ren, S., Sun, J.: Deep residual learning for image recognition. In: IEEE Conference on Computer Vision and Pattern Recognition, pp. 770–778 (2016)
9. Hochreiter, S., Schmidhuber, J.: Long short-term memory. Neural Comput. **9**(8), 1735–1780 (1997)
10. Ioffe, S., Szegedy, C.: Batch normalization: accelerating deep network training by reducing internal covariate shift. In: International Conference on Machine Learning, vol. 37, pp. 448–456 (2015)
11. Kingma, D.P., Ba, J.: Adam: a method for stochastic optimization. In: International Conference on Learning Representations (2015)
12. Lea, Colin, Vidal, René, Reiter, Austin, Hager, Gregory D.: Temporal convolutional networks: a unified approach to action segmentation. In: Hua, Gang, Jégou, Hervé (eds.) ECCV 2016. LNCS, vol. 9915, pp. 47–54. Springer, Cham (2016). https://doi.org/10.1007/978-3-319-49409-8_7
13. Mahalanobis, P.C.: On the generalized distance in statistics. Proc. Nat. Inst. Sci. India **2**(1), 49–55 (1936)
14. Ng, A.Y., Jordan, M.I.: On discriminative vs. generative classifiers: a comparison of logistic regression and naive Bayes. In: Advances in Neural Information Processing Systems, pp. 841–848 (2002)
15. Noh, H., You, T., Mun, J., Han, B.: Regularizing deep neural networks by noise: its interpretation and optimization. In: Advances in Neural Information Processing Systems, pp. 5109–5118 (2017)
16. Pham, V., Bluche, T., Kermorvant, C., Louradour, J.: Dropout improves recurrent neural networks for handwriting recognition. In: International Conference on Frontiers in Handwriting Recognition, pp. 285–290 (2014)
17. Rasmussen, C.E., Williams, C.K.I.: Gaussian Processes for Machine Learning. MIT Press, Cambridge (2006)
18. Srivastava, N., Hinton, G., Krizhevsky, A., Sutskever, I., Salakhutdinov, R.: Dropout: a simple way to prevent neural networks from overfitting. J. Mach. Learn. Res. **15**, 1929–1958 (2014)

19. Tikhonov, A.N.: On the stability of inverse problems. Doklady Akademii Nauk SSSR **39**(5), 195–198 (1943)
20. Tikhonov, A.N.: Solution of incorrectly formulated problems and the regularization method. Doklady Akademii Nauk SSSR **151**(3), 501–504 (1963)
21. U.S. Bureau of the Census: Advance retail sales: Retail and food services, total [RSAFSNA] dataset. FRED, Federal Reserve Bank of St. Louis (2018)
22. U.S. Bureau of the Census: E-commerce retail sales [ECOMSA] dataset. FRED, Federal Reserve Bank of St. Louis (2018)
23. Wu, C.F.J.: On the convergence properties of the EM algorithm. Annal. Stat. **11**(1), 95–103 (1983)
24. Zaremba, W., Sutskever, I., Vinyals, O.: Recurrent neural network regularization. ArXiv e-prints http://arxiv.org/abs/1409.2329 (2014)

# Adaptive Objective Functions and Distance Metrics for Recommendation Systems

Michael C. Burkhart$^{(\boxtimes)}$ and Kourosh Modarresi

Adobe Inc., 345 Park Ave, San José, CA 95110, USA
{mburkhar,modarres}@adobe.com

**Abstract.** We describe, develop, and implement different models for the standard matrix completion problem from the field of recommendation systems. We benchmark these models against the publicly available Netflix Prize challenge dataset, consisting of users' ratings of movies on a 1–5 scale. While the original competition concentrated only on RMSE, we experiment with different objective functions for model training, ensemble construction, and model/ensemble testing.

Our best-performing estimators were (1) a linear ensemble of base models trained using linear regression (see ensemble $e_1$, RMSE: 0.912) and (2) a neural network that aggregated predictions from individual models (see ensemble $e_4$, RMSE: 0.912). Many of the constituent models in our ensembles had yet to be developed at the time the Netflix competition concluded in 2009. To our knowledge, not much research has been done to establish best practices for combining these models into ensembles. We consider this problem, with a particular emphasis on the role that the choice of objective function plays in ensemble construction.

## 1 Background

In 2006, Netflix released a dataset containing 100,480,507 movie ratings (on a 1–5 scale) from $m = 480,189$ users on $n = 17,770$ movies [10]. The set was divided into 99,072,112 training points and 1,408,395 probe points for contestants to train and validate models. Netflix's in-house algorithm Cinematch scored an RMSE (root mean square error) of 0.9514 on the probe set. The prize of one million dollars went to the first contestants to improve RMSE on a hidden test set by 10%. A combined team of "Bellkor in BigChaos" and "Pragmatic Theory" accomplished this in 2009 with an RMSE of 0.8558 on the probe data [9,61,79]. As one of the largest real-life datasets available, the Netflix Prize data remains a benchmark for innovations in recommender systems today.

### 1.1 Results

Tables 1 and 2 catalogue the final results. Full details and methods follow in the subsequent sections.

© Springer Nature Switzerland AG 2019
J. M. F. Rodrigues et al. (Eds.): ICCS 2019, LNCS 11537, pp. 608–621, 2019.
https://doi.org/10.1007/978-3-030-22741-8_43

**Table 1.** Normalized model performance under different metrics.

| Model | Name | n. MAE | n. RMSE | n. $L_3$ | n. $L_4$ | n. Cos. Dist. |
|---|---|---|---|---|---|---|
| $m_0$ | Baseline | 0.878 | 0.901 | 0.913 | 0.923 | 0.807 |
| $m_1$ | irlba 5 | 0.860 | 0.888 | 0.903 | 0.916 | 0.784 |
| $m_2$ | irlba 7 | 0.857 | 0.886 | 0.902 | 0.915 | 0.781 |
| $m_3$ | irlba 13 | 0.853 | 0.883 | 0.899 | 0.913 | 0.775 |
| $m_4$ | softImpute 5 | 0.787 | 0.841 | 0.876 | 0.904 | 0.702 |
| $m_5$ | softImpute 7 | 0.780 | 0.837 | 0.874 | 0.903 | 0.695 |
| $m_6$ | softImpute 13 | 0.773 | 0.836 | 0.876 | 0.909 | 0.692 |
| $m_7$ | softImpute 100 | 0.824 | 0.903 | 0.952 | 0.989 | 0.806 |
| $m_8$ | Movie k-NN | 0.855 | 0.901 | 0.924 | 0.942 | 0.808 |
| $m_9$ | User k-means | 0.845 | 0.893 | 0.922 | 0.947 | 0.794 |
| $m_{10}$ | Avg of $m_8, m_9$ | 0.832 | 0.869 | 0.890 | 0.906 | 0.751 |
| $m_{11}$ | Cross k-NN | 0.816 | 0.862 | 0.891 | 0.915 | 0.738 |
| $m_{12}$ | Time-aware cross k-NN | 0.793 | 0.862 | 0.907 | 0.943 | 0.735 |
| $m_{13}$ | Neural v5 | 0.784 | 0.854 | 0.897 | 0.932 | 0.723 |
| $m_{14}$ | Neural v11 | 0.780 | 0.830 | **0.861** | **0.886** | 0.684 |
| $m_{15}$ | Neural v13 | 0.771 | 0.836 | 0.879 | 0.915 | 0.695 |
| $m_{16}$ | Time-aware neural | **0.764** | **0.829** | 0.87 | 0.904 | **0.681** |
| $m_{17}$ | Neural one-hot | 0.791 | 0.842 | 0.874 | 0.901 | 0.702 |
| $m_{18}$ | Time-binned SVD | 0.827 | 0.873 | 0.900 | 0.922 | 0.757 |
| $m_{19}$ | WALS | 0.778 | 0.837 | 0.876 | 0.909 | 0.695 |
| $m_{20}$ | xgboost | 0.79 | 0.845 | 0.879 | 0.908 | 0.708 |
| $m_{21}$ | Gaussian factorization | 0.769 | 0.839 | 0.885 | 0.922 | 0.696 |
| $m_{22}$ | Sparse tensor factorization | 0.87 | 0.901 | 0.917 | 0.931 | 0.808 |
| $m_{23}$ | Poisson factorization | 1.013 | 1.002 | 0.971 | 0.942 | 0.741 |
| $m_{24}$ | Factorization machine | 0.862 | 0.892 | 0.908 | 0.921 | 0.792 |
| $m_{25}$ | VAE | 0.881 | 0.905 | 0.917 | 0.927 | 0.814 |

**Table 2.** Normalized ensemble performance under different metrics.

| Ensemble | Name | n. MAE | n. RMSE | n. $L_3$ | n. $L_4$ | n. Cos. Dist. |
|---|---|---|---|---|---|---|
| $e_0$ | Subset avg | 0.762 | 0.814 | 0.849 | 0.878 | 0.658 |
| $e_1$ | Linear regression | 0.750 | **0.808** | **0.847** | 0.879 | 0.649 |
| $e_2$ | Random forest | 0.757 | 0.811 | **0.847** | **0.877** | 0.653 |
| $e_3$ | xgboost | 0.754 | 0.809 | 0.848 | 0.880 | 0.649 |
| $e_4$ | One-hot NN | **0.749** | **0.808** | 0.848 | 0.881 | **0.648** |
| $e_5$ | Neural network | 0.752 | 0.811 | 0.851 | 0.885 | 0.653 |

## 2    Problem Description

Let $M$ be an $m \times n$ matrix where entry $M_{ij} \in \{1, \ldots, 5\}$ contains user $i$'s rating of movie $j$. The matrix completion problem aims to recover the matrix $M$ from a subset $\Omega \subset [m] \times [n]$ of its entries. We let $\mathcal{P}_\Omega(M)$ denote the projection of

$M$ onto this subset, which amounts to zeroing out unobserved elements of $M$. We further divide the probe data randomly into a test set of 1,000,000 points and a validation set of the remaining 408,395 points. Our aim is to construct ensembles of predictors. We build individual predictors using the training set and build ensemble estimators using the validation set. We then report RMSE performance on the test set.

As the de facto standard error for regression tasks, RMSE penalizes larger errors more than MAE (mean absolute error). However, the extent to which RMSE accurately represents misclassification loss remains debatable. For example, correctly distinguishing between 3- and 5-star ratings may prove much more important than distinguishing between 1- and 3-star ratings [30]. If the ultimate goal is to produce a user's top-N movies, [17] argue precision- and recall-based metrics much better characterize success.

## 3    SVD

SVD methods play an important role in matrix completion. We discuss this approach first because many of the subsequent methods benefit greatly from leveraging a learned SVD decomposition, either by using SVD parameters for initialization or to estimate user-user and movie-movie similarities.

Under the common additional assumption that $M$ is of low rank $r$, where $r \ll \min\{m, n\}$, we can consider the SVD of $M$ given by $M = U\Sigma V^{\top}$, where $U$ is an $m \times r$ matrix with orthonormal columns, $\Sigma$ is an $r \times r$ diagonal matrix of positive entries, and $V$ is an $r \times n$ matrix with orthonormal columns. Such a matrix has $\mathcal{O}(mr)$ degrees of freedom (assuming, as in our case, that $m > n$). Considering uniform sampling with replacement as a coupon collector's problem, we then require at least $\mathcal{O}(mr \log m)$ entries from $M$ in order for a successful reconstruction [15]. We hypothesize that users rate movies based upon $r$ underlying features, having importance relative to the singular values of $M$. The rows of $U$ correspond to feature weightings for each user and the rows of $V$ correspond to feature weightings for each item.

### 3.1    Low Rank Recovery

[14,15] described sufficient conditions on the incoherence of the rows of $U$ and $V$ such that $M$ could be recovered with high probability through nuclear norm-minimization. Recall that the nuclear norm is given $\|M\|_* = \sum_{k=1}^{r} \sigma_k(M)$ where $\sigma_k(M)$ denotes the $k$th singular value of $M$. The precise optimization problem was to minimize $\|\hat{M}\|_*$ subject to $\mathcal{P}_\Omega(M) = \mathcal{P}_\Omega(\hat{M})$, where $\hat{M}$ denotes the estimate for $M$. Such a choice of objective function makes the problem convex (indeed, all norms are convex), as opposed to minimizing $\text{rank}(\hat{M})$. In a similar vein, [52] developed "soft-thresholded" SVD to find $\hat{M}$ minimizing $\frac{1}{2}\|\mathcal{P}_\Omega(\hat{M} - M)\|_F^2 + \lambda\|\hat{M}\|_*$, where $\|\cdot\|_F$ denotes the Frobenius matrix norm.

# 4  Factorization Models

We implemented numerous approaches to matrix factorization. In this approach, we estimate $\hat{M} = UV^\top$ where $U$ is an $n \times k$ matrix of user factors and $V$ is an $m \times k$ matrix of item factors. Unconstrained matrix factorization [75,76] simply finds $\hat{U}, \hat{V} = argmin_{U,V} \|\mathcal{P}_\Omega(UV^\top - M)\|^2$, where the matrix norm is the Frobenius norm, and predicts $\hat{M} = \hat{U}\hat{V}^\top$. The straightforward mathematical description leaves a handful of implementation choices. We can initialize $U, V$ either randomly or from the SVD decomposition by taking, for example, $U\sqrt{\Sigma}, \sqrt{\Sigma}V$ from a learned SVD model. We can perform optimization with the whole dataset in memory or perform batch-based optimization by iterating over subsets of the training data.

We also implemented Tikhonov-regularized [77,78] matrix factorization [66, 73] to solve $\hat{U}, \hat{V} = argmin_{U,V} \|\mathcal{P}_\Omega(UV^\top - M)\|^2 + \lambda(\|U\|^2 + \|V\|^2)$ and non-negative matrix factorization [46,47], with the constraint that all entries of $U, V$ be non-negative. In our experience, model performance seemed tightly coupled with initialization.

## 4.1  Accounting for Implicit Preferences

Hu et al. developed a weighted matrix factorization method that accounts for the implicit preference a user gives to a movie through the act of watching and rating it [38]. Called weighted alternating least squares (WALS), this method seeks $\hat{U}, \hat{V} = argmin_{U,V} \|\mathcal{P}_\Omega(\sqrt{W} \odot (UV^\top - M))\|^2 + \lambda(\|U\|^2 + \|V\|^2)$ where $W$ denotes the number of movies a user has rated and $\odot$ denotes element-wise multiplication. We used the contributed TensorFlow model and initialized with SVD output.

## 4.2  Neural Network Matrix Factorizaton

This estimator constituted a feedforward fully-connected neural network mapping learned representation vectors for users and movies through the network to predict the corresponding rating. Such an approach is commonly referred to as neural network matrix factorization [21,32]. Learned model parameters consist of $m$ user vectors in $\mathbb{R}^r$, $n$ movie vectors in $\mathbb{R}^r$, and all parameters for the neural network. We initialized user vectors with the $U$ matrix and the movie vectors with the $V$ matrix from the soft-thresholded SVD model. Neural network parameters received Glorot uniform initialization [28]. For each batch, we performed three training steps: neural network parameters were updated, user representations were updated, and movie representations were updated. We applied Tikhonov $L_2$-regularization to $U$ and $V$. For all parameter updates, we used the Adam optimizer [41] that maintains different learning rates for each parameter like AdaGrad [20] and allows these rates to sometimes increase like Adadelta [84] but adapts them based on the first two moments from recent gradient updates. We used a leaky rectified linear unit (ReLU) activation [50,57], and applied dropout after the first hidden layer to prevent overfitting [74]. Neither Nesterov momentum-aided Adam [19] nor Batch Normalization [39] appeared to improve our results. Many different versions of neural networks were developed with minor variations in architecture, initialization, and training.

### 4.3 Probabilistic Matrix Factorization

Factorization can also be performed in a probabilistic setting, by specifying a generative graphical model and then finding the maximum a posteriori (MAP) parameters [70] or by performing Gibbs sampling in a Bayesian setting [69]. In Gaussian matrix factorization, we model $M_{ij} \sim^{\text{i.i.d.}} \mathcal{N}(U_i V_j^\top + b_{ij}, \sigma^2)$ where, as before, $U$ is an $n \times k$ matrix of user factors and $V$ is an $m \times k$ matrix of item factors. The $b_{ij}$ denote the average of the mean rating from user $i$ and the mean rating of movie $j$, and account for user- and movie- effects. We learn $U$ and $V$ to maximize the log likelihood of the observed data under this model. In Poisson matrix factorization [29], $M_{ij} \sim^{\text{i.i.d.}} \text{Poisson}(U_i V_j^\top + b_{ij})$. We predict $\hat{M}_{ij} = \mathbb{E}[\hat{X}_{ij} | \hat{X}_{ij} \in \{1, \ldots, 5\}]$ where $\hat{X}_{ij} \sim \text{Poisson}(\hat{U}_i \hat{V}_j^\top + b_{ij})$ and $\hat{U}, \hat{V}$ denote our learned model parameters.

### 4.4 Factorization Machine

A factorization machine [64,65] of second degree learns a regression model $\hat{y}(x) = w_0 + \sum_{i=1}^{\ell} w_i x_i + \sum_{1 \leq i < j \leq \ell} \langle v_i, v_j \rangle x_i x_j$ for parameters $w_k \in \mathbb{R}$ and $v_k \in \mathbb{R}^\ell$, $k = 1 \ldots, \ell$. Such models were designed for sparsity, and in our case, we let $x \in \mathbb{R}^{m+n}$ denote a one-hot vector representation for the user concatenated with a one-hot vector representation for the movie.

## 5   Neighborhood Models

Neighborhood-based techniques prove to be useful ingredients in a Netflix ensemble [42,80]. A $k$-NN model estimates a function's value at a test point by averaging the values of the $k$ nearest training points. In our case, we can predict the rating for a (user, movie)-pair by averaging the user's ratings for the $k$ nearest movies or averaging the $k$ nearest users' ratings of the movie. To do so, we must first set $k$ and a metric for comparing two movies (or users). Given the sparseness of the Netflix dataset, we must also account for the cases where we have no training data on any of the $k$ nearest neighbors of a given test point. In this case, we typically fall back to the baseline global rating. A lower value of $k$ tends to increase the accuracy of the ratings we can calculate, but also increases the number of test points for which we have insufficient training data.

In addition to comparing the performance yielded by choice of metric, it may also be illustrative to consider how the choice of metric impacts training data availability. Suppose, given a (user, movie)-pair, the user tends to have rated many more of the 20 nearest movies under the cosine metric than of the 20 nearest under the Euclidean metric. Under the hypothesis that the act of expressing a rating indicates preference (the ratings matrix is not revealed uniformly at random), this fact might also provide us with information, independent of how closely the ratings aligned to the target.

In [6], Bell, Koren, and Volinksy remark that neighborhood-based kernel regression approaches may fail to account for relationships in the similarity space. They describe Lord of the Rings as an example, where a neighborhood in movie-movie similarity space might include all three movies from the trilogy, causing the underlying effect from the trilogy to be counted three times. To account for this, they optimize weights with shrinkage instead of relying on a predefined similarity metric [5,7]. They also allow a neighborhood based method to defer judgement, when provided insufficient or low-quality neighborhood information [8]. There are also factorized versions of learned user similarity [80].

## 5.1    k-NN on SVD Latent Space

Consider the $r$-dimensional rows of $U$ from the soft-thresholded SVD decomposition of $\mathcal{P}_\Omega(M)$. These vectors give a dense, low-dimensional (we let $r = 5$) representation for each user. We use a k-d tree [11] to find the $k = 15$ nearest neighbors for each user according to the Euclidean metric. For a given (user, movie)-pair in the probe set, we determine if any of the user's neighbors rated the queried movie, and if so, calculate a weighted average over these ratings, where the weights are proportional to the exponentiated negative distance between the user and her neighbors (*cf.* Nadaraya–Watson kernel-regression [56,82]).

A smaller value for $k$ restricts to only the most similar neighbors, and so decreases the bias of this estimate. However, it also increases the chance that very few (or none) of the neighbors will have expressed a rating for the given movie. In the case that fewer than three of the $k = 15$ nearest neighbors to a user expressed a preference for a queried movie, then this method does not return a rating, and the average ensemble is taken over the remaining estimators. This allows the estimator to abstain from rating when it is not sufficiently confident, and elegantly fall back to estimators that will be more reliable for a given (user, movie)-pair. In a similar vein, we can cluster users according to k-means and use the above approach with cluster members in place of neighbors. Clustering can yield improved efficiency through memoization [54], as ratings for a given cluster need only be computed once, and can then be applied to subsequent queries. We also created a similar estimator that instead operates on the latent representation for movies.

## 5.2    Crossing User Neighborhoods with Movie Neighborhoods

The original k-NN approach would find neighbors for either users or movies and then aggregate ratings along a vector in the dual dimension (movies or users, respectively). It is possible instead to use k-NN to find neighbors for both rows and columns, and then aggregate along the sub-matrix consisting of the cross product between neighboring users and neighboring movies. In other words, to predict on a rating for user $i$ on movie $j$, we would find indices $\mathcal{N}_{u_i} \subset [m]$

corresponding to the neighbors of user $i$, and indices $\mathcal{N}_{v_j} \subset [n]$ corresponding to the neighbors of movie $j$, and compute a weighted average over the available rankings in $\mathcal{N}_{u_i} \times \mathcal{N}_{v_j}$, where the weights account for distances in user-space, movie-space, and the difference in time between the ratings. This allows us to leverage ratings of similar movies provided by similar users.

# 6   Gradient Boosted Trees

Ensemble methods combine multiple weak learners (estimators) into a single strong estimator [18,22,31,71]. Breiman's bootstrap aggregating ("bagging") approach trains multiple learners (in parallel) on bootstrapped samples. In contrast, boosting algorithms like Adaboost [25] and gradient boosting [26,27,51] iteratively add weak learners to improve an ensemble, concentrating effort on currently misclassified examples. (Note that concentrating on currently misclassified examples is not required of a boosting algorithm: see, for example, Boost by Majority [23] and Brown Boost [24]). In gradient boosting, we supply a loss function $L(\cdot, \cdot)$ and a method to train new weak learners $h_i$; in our case, we use regression trees [13]. For our application, we learned ratings for a (user, movie) pair as a function of their representations in thresholded SVD feature space. We used XGBoost [16] to build the estimator.

# 7   Variational Autoencoding

Variational autoencoding learns parameters for an autoencoder using a common Bayesian technique known as variational inference. An autoencoder models the identity function with a neural network [33,68]. Its architecture includes a hidden layer of relatively small dimensionality that serves as an information bottleneck. Upon training, the output from this layer yields a lower-dimensional representation of the original data. If we restrict to linear maps and impose $L_2$ loss on our reconstruction, autoencoding solves for the principal components from PCA [37,59]. In this way, we can consider autoencoding to be a nonlinear extension of PCA.

In Bayesian statistics, variational inference approximates intractable integrals (expectations) through optimization, by substituting the integrand (probability distribution) for the closest member of a parametrized family of distributions [40]. When optimization is batch-based, this process is known as stochastic variational inference [36,67].

A variational autoencoder learns a probabilistic autoencoding model, as two conditional distributions described by neural networks. The encoding distribution $q_\theta(z|x)$ describes how to sample the latent low-dimensional representation from an observation $x$ and the decoding distribution $p_\varphi(x|z)$ describes how to sample a reconstructed $x$ from the latent representation. Optimization aims to maximize $\mathcal{L}(\theta, \varphi) = \mathbb{E}_{q_\theta(z|x)}[\log p_\varphi(x|z)] - D_{\mathrm{KL}}\big(q_\theta(z|x)||p(z)\big)$ For computational expediency, the expectations above are often approximated via single-sample Monte Carlo integration [53]. In particular, for the $i$th data-point $X_i$, we sample $Z_i \sim q_\theta(\cdot|X = x_i)$and form the unbiased approximations $\mathbb{E}_{q_\theta(z|x)}[\log p_\varphi(x,z)] \approx \log p_\varphi(X_i, Z_i)$ and $\mathbb{E}_{q_\theta(z|x)}[\log q_\theta(z|x)] \approx \log q_\theta(Z_i|X_i)$.

Multiple authors have implemented VAE's for collaborative filtering. [48] learned item representations from known content data. [49] concentrated on implicit ratings data; they consider observations in the form of a single user's (sparse) vector counts for item consumption, and argued that their two adjustments, using a multinomial likelihood and adjusting the VAE objective, were key to their performance.

We take $X_i \in \mathbb{R}^{17770}$ to be user $i$'s ratings for each movie (more precisely, the residual ratings after subtracting off half of user i and movie j's mean ratings). We model $q_\theta(z|x) = \eta_{10}(z; f_1(x), \exp(f_2(x)I_{10}))$ where $\eta_{10}$ denotes a 10-dimensional Gaussian, $I_{10}$ is the identity matrix, and $f_1, f_2$ are leaky relu-activated neural networks. Here, $\theta$ corresponds to the parameters for the neural networks $f_1, f_2$. We model $p_\phi(z_i|x) = \eta_{m_i}(z; g(x), I_{m_i})$ where $m_i$ denotes the number of movies user $i$ rated, $g$ is a leaky relu-activated neural networks with a single hidden layer, and $\phi$ denotes the parameters for $g$.

# 8  Incorporating Rating Time

The Netflix training and probe sets include a time stamp for each rating event. We consider ways to leverage the effect of time in our model.

## 8.1  Time-Aware Neural Factorization

Building on the success of Neural Network Matrix Factorization, this model added two time components as inputs to the neural network: (1) the time of rating, normalized to lie in $[0, 1]$ and (2) the approximate number of years between the movie's release and the time of rating. As updates to U and V are sparse (any given row only updates a handful of times for each run through the data set), we used a Nesterov Momentum optimizer [58,62] to train them, while continuing to apply the Adam optimizer for the neural network parameters (all of which are updated at each training step). Newer optimizers such as Adam tweak the learning rate for each parameter depending on a window of previous gradients for each parameter. This approach may not be best when updates to a given parameter occur only sporadically [63], so here we use Nesterov.

## 8.2  Neural One-Hot Factorization with a Time Component

In this approach, we designed a neural network that takes user- and movie-representations, and time features "movie release year" and "time of rating" and "time of user's first rating", and outputs a probability distribution on $\{1, \ldots, 5\}$. Training minimizes the cross-entropy between the (point-mass, or slightly modified point-mass) distribution on the underlying label and the model's predicted distribution. In addition to providing estimates for (user, movie, time)-ratings, this model allows us to predict the variance or uncertainty of our estimate.

## 8.3   Time-Binned SVD

We can partition the training and probe data into approximately equally sized bins based on the time stamps associated to them (so ratings that occur around the same time will be placed in the same or a neighboring bin) and learn a separate SVD model for each time window. Each of these models can then be used to predict a rating for a given (user, movie, time) probe pair, and a weighted average formed over all such predictions, with a higher weight given to the bin into which the query was placed.

## 8.4   Tensor Factorization

After partitioning our data into time bins (in the same way as for time-binned SVD), we can view our training data as a ratings tensor, where the users by movies matrix now extends along a third, temporal dimension. This allows us to perform time-aware factorization into three tensors, one for each dimension. Hitchcock pioneered a generalization of SVD to tensors, known as the minimal canonical polyadic (CP) decomposition [34], yielding the model $M = \sum_{i=1}^{r} \lambda_i a_i^1 \otimes a_i^2 \otimes a_i^3$. We initialize with higher-order SVD [35,45,81] and use alternating least squares to fit the model.

# 9   Ensembling

Famously, the winning solution to the Netflix Prize challenge consisted of a blend of 107 different models [9,61,79]. Our best-performing models, too, aggregated predictions from other models. After building individual models using only data from the training set, we built ensembles of models using data only from the validation set. We distinguish ensembles with the letter $e$ from our base models lettered $m$.

## 9.1   Average over a Selected Subset

Considering all $\binom{19}{10}$ size 10 subsets of $\{m_1, \ldots, m_{19}\}$, we found the subset whose simple average produced the smallest RMSE on the validation set. The average of these models was then computed for the test set. We performed the brute-force search with the Numba package that allows for just-in-time compilation (to LLVM) and parallelized for-loops.

## 9.2   Linear Regression

We performed stepwise variable selection using the Bayesian Information Criterion [72] (see also Akike's Information Criterion [2]) for a multiple linear regression model. We also used linear regression to select the most informative subset of 10 predictors [55].

## 9.3   Random Forests for Regression

We used Breiman's random forest regression algorithm [12] on the 10 top predictors, as determined by linear regression in the above section. We also tried a boosting approach for ensembling, the XGBoost algorithm [16]. The importance matrix calculated from boosting gives the top 10 models, in order, as: $m_{14}$, $m_{16}$, $m_{15}$, $m_{19}$, $m_7$, $m_{13}$, $m_{17}$, $m_{12}$, $m_{18}$, $m_{10}$.

## 9.4   Neural Network Regression

We trained neural networks on the validation set using predictions from individual models as inputs and true values on the validation set as outputs. We adopted two main architectures: (1) a direct continuous-valued function that was trained to minimize RMSE, and (2) a distributional function that was trained to minimize cross-entropy between its pdf-outputs and one-hot representations of the true values. These models were then applied to the test set to measure performance.

## 9.5   Impact of Objective Function on Ensemble Building

To illustrate the role that the choice of objective function plays in ensemble building, we took the boosted ensemble model and optimized it on the validation data under a range of different objective functions. See Table 3 for results. As extreme gradient boosting uses second derivatives of the objective function, we do not include $L_1$ or Huber loss.

**Table 3.** RMSE performance after optimizing $e_5$ under different objective functions.

| Objective | n. MAE | n. RMSE | n. $L_3$ | n. $L_4$ | n. $L_5$ | n. Cos. Dist. |
|-----------|--------|---------|----------|----------|----------|---------------|
| $L_2$ | **0.752** | **0.811** | 0.851 | 0.885 | 0.912 | **0.653** |
| $L_3$ | 0.780 | 0.817 | **0.842** | 0.863 | 0.881 | 0.655 |
| $L_4$ | 0.810 | 0.836 | 0.846 | **0.855** | 0.864 | 0.664 |
| $L_5$ | 0.844 | 0.859 | 0.858 | 0.857 | **0.857** | 0.677 |
| Cosh | 0.766 | 0.813 | 0.843 | 0.868 | 0.889 | **0.653** |

# 10   Conclusions

The vast majority of benchmarks against the Netflix dataset report only RMSE performance, in line with the prize's original objective. For estimators that aim to select a user's top $n$ movies, people sometimes consider precision-recall metrics. We summarize our results under numerous metrics in Table 1 for base models and Table 2 for ensembles. We normalize the $L_p$ metrics by diving by the loss obtained by predicting the mean training value for all test points. For example, predicting the training mean (3.67 stars) for all movies in the test set yields an RMSE of 1.127, so all reported normalized RMSE's correspond to standard RMSE divided by 1.127.

Ensembles proved essential to the winning solution for the Netflix Grand Prize. In 2009 when the competition concluded, matrix factorization and neighborhood methods provided the fundamental components from which the ensembles were built. Since 2009, researchers introduced many new machine learning approaches for recommender systems including neural network matrix factorization, factorization machines, extreme gradient boosting, and variational autoencoding. These methods have been tested individually, but little work has been done to consider how these new approaches can be combined to form more effective ensemble estimators. This paper provides a first step in that direction.

# A    Implementation Details

In Python v3.6.0, we used Scikit-learn for k-NN and k-means [60] and Tensorflow for neural network training [1]. We performed some just-in-time for-loop optimization with Numba [43]. We used Cython [4] to precompile some Python functions for additional performance. The ttfm package provided our factorization machine implementation. We compiled Tensorflow from source with MKL (Intel) support and additional instructions.

In R v3.5.1, we used 'irlba' for truncated SVD [3] (*cf.* [44]) and 'softImpute' for soft-thresholded SVD [52]. While increasing the rank $r$ tended to result in improved performance for both algorithms, returns were modest. We used 'MASS' for linear regression and 'leaps' for subset variable selection. We trained random forests with 'ranger' [83] and performed boosting with 'xgboost.' We used the R package 'Reticulate' and the Python package 'rp2' to transfer data between the two languages.

# References

1. Abadi, M., et al.: TensorFlow: large-scale machine learning on heterogeneous systems. In: USENIX, pp. 265–283 (2016)
2. Akaike, H.: A new look at the statistical model identification. IEEE Trans. Autom. Contr. **19**(6), 716–723 (1974)
3. Baglama, J., Reichel, L.: Augmented implicitly restarted Lanczos bidiagonalization methods. SIAM J. Sci. Comput. **27**(1), 19–42 (2005)
4. Behnel, S., Bradshaw, R., Citro, C., Dalcín, L., Seljebotn, D.S., Smith, K.: Cython: the best of both worlds. Comput. Sci. Eng. **13**(2), 31–39 (2011)
5. Bell, R., Koren, Y.: Scalable collaborative filtering with jointly derived neighborhood interpolation weights. In: IEEE International Conference on Data Mining, pp. 43–52 (2007)
6. Bell, R., Koren, Y., Volinsky, C.: Modeling relationships at multiple scales to improve accuracy of large recommender systems. In: SIGKDD, pp. 95–104 (2007)
7. Bell, R.M., Koren, Y.: Improved neighborhood-based collaborative filtering. In: SIGKDD (2007)

8. Bell, R.M., Koren, Y.: Lessons from the Netflix prize challenge. SIGKDD Explor. **9**(2), 75–79 (2007)
9. Bell, R.M., Koren, Y., Volinsky, C.: The BellKor solution to the Netflix prize (2009)
10. Bennett, J., Lanning, S.: The Netflix prize. In: KDD Cup and Workshop (2007)
11. Bentley, J.L.: Multidimensional binary search trees used for associative searching. Commun. ACM **18**(9), 509–517 (1975)
12. Breiman, L.: Bagging predictors. Mach. Learn. **24**(2), 123–140 (1996)
13. Breiman, L., Friedman, J., Stone, C.J., Olshen, R.: Classification and Regression Trees. Chapman & Hall/CRC in Boca Raton, FL (1984)
14. Candès, E.J., Recht, B.: Exact matrix completion via convex optimization. Found. Comput. Math. **9**(6), 717 (2009)
15. Candès, E.J., Tao, T.: The power of convex relaxation: near-optimal matrix completion. IEEE Trans. Inf. Theory **56**(5), 2053–2080 (2010)
16. Chen, T., Guestrin, C.: XGBoost: a scalable tree boosting system. In: SIGKDD, pp. 785–794 (2016)
17. Cremonesi, P., Koren, Y., Turrin, R.: Performance of recommender algorithms on top-n recommendation tasks. In: RecSys, pp. 39–46 (2010)
18. Dasarathy, B.V., Sheela, B.V.: A composite classifier system design: concepts and methodology. Proc. IEEE **67**(5), 708–713 (1979)
19. Dozat, T.: Incorporating Nesterov momentum into Adam. In: International Conference on Learning Representations (2016)
20. Duchi, J., Hazan, E., Singer, Y.: Adaptive subgradient methods for online learning and stochastic optimization. J. Mach. Learn. Res. **12**, 2121–2159 (2011)
21. Dziugaite, G.K., Roy, D.M.: Neural network matrix factorization (2015). arXiv:1511.06443
22. Efron, B.: Bootstrap methods: another look at the jackknife. Ann. Stat. **7**(1), 1–26 (1979)
23. Freund, Y.: Boosting a weak learning algorithm by majority. Inf. Comput. **121**(2), 256–285 (1995)
24. Freund, Y.: An adaptive version of the boost by majority algorithm. Mach. Learn. **43**(3), 293–318 (2001)
25. Freund, Y., Schapire, R.E.: A decision-theoretic generalization of on-line learning and an application to boosting. J. Comput. Syst. Sci. **55**(1), 119–139 (1997)
26. Friedman, J.H.: Greedy function approximation: a gradient boosting machine. Ann. Stat. **29**(5), 1189–1232 (2001)
27. Friedman, J.H.: Stochastic gradient boosting. Comput. Stat. Data Anal. **38**(4), 367–378 (2002)
28. Glorot, X., Bengio, Y.: Understanding the difficulty of training deep feedforward neural networks. In: International Conference on Artificial Intelligence and Statistics, vol. 9, pp. 249–256 (2010)
29. Gopalan, P., Hofman, J.M., Blei, D.M.: Scalable recommendation with hierarchical poisson factorization. In: Uncertainty in Artificial Intelligence, pp. 326–335 (2015)
30. Gunawardana, A., Shani, G.: A survey of accuracy evaluation metrics of recommendation tasks. J. Mach. Learn. Res. **10**, 2935–2962 (2009)
31. Hansen, L.K., Salamon, P.: Neural network ensembles. IEEE Trans. Pattern Anal. Mach. Intell. **12**(10), 993–1001 (1990)
32. He, X., Liao, L., Zhang, H., Nie, L., Hu, X., Chua, T.S.: Neural collaborative filtering. In: International World Wide Web Conference, pp. 173–182 (2017)
33. Hinton, G.E., Salakhutdinov, R.R.: Reducing the dimensionality of data with neural networks. Science **313**(5786), 504–507 (2006)

34. Hitchcock, F.L.: The expression of a tensor or a polyadic as a sum of products. J. Math. Phys. **6**, 164–189 (1927)
35. Hitchcock, F.L.: Multiple invariants and generalized rank of a p-way matrix or tensor. J. Math. Phys. **7**, 39–79 (1928)
36. Hoffman, M., Blei, D.M., Wang, C., Paisley, J.: Stochastic variational inference. J. Mach. Learn. Res. **14**, 1303–1347 (2013)
37. Hotelling, H.: Relations between two sets of variates. Biometrika **28**, 321–377 (1936)
38. Hu, Y., Koren, Y., Volinsky, C.: Collaborative filtering for implicit feedback datasets. In: IEEE International Conference on Data Mining, pp. 263–272 (2008)
39. Ioffe, S., Szegedy, C.: Batch normalization: accelerating deep network training by reducing internal covariate shift. In: International Conference on Machine Learning, pp. 448–456 (2015)
40. Jordan, M.I., Ghahramani, Z., Jaakkola, T.S., Saul, L.K.: An introduction to variational methods for graphical models. Mach. Learn. **37**(2), 183–233 (1999)
41. Kingma, D.P., Ba, J.: Adam: a method for stochastic optimization. In: International Conference on Learning Representations (2015)
42. Koren, Y.: Factorization meets the neighborhood: a multifaceted collaborative filtering model. In: SIGKDD, pp. 426–434 (2008)
43. Lam, S.K., Pitrou, A., Seibert, S.: Numba. In: Proceedings of the Workshop LLVM Compiler (2015)
44. Lanczos, C.: An iteration method for the solution of the eigenvalue problem of linear differential and integral operators. J. Res. NIST **45**(4), 255–282 (1950)
45. Lathauwer, L.D., Moor, B.D., Vandewalle, J.: A multilinear singular value decomposition. SIAM J. Matrix Anal. Appl. **21**(4), 1253–1278 (2000)
46. Lee, D.D., Seung, H.S.: Learning the parts of objects by non-negative matrix factorization. Nature **401**, 788–791 (1999)
47. Lee, D.D., Seung, H.S.: Algorithms for non-negative matrix factorization. In: Advances in Neural Information Processing Systems, pp. 556–562 (2000)
48. Li, X., She, J.: Collaborative variational autoencoder for recommender systems. In: ACM SIGKDD, pp. 305–314 (2017)
49. Liang, D., Krishnan, R.G., Hoffman, M.D., Jebara, T.: Variational autoencoders for collaborative filtering. In: International World Wide Web Conference, pp. 689–698 (2018)
50. Maas, A.L., Hannun, A.Y., Ng, A.Y.: Rectifier nonlinearities improve neuralnetwork acoustic models. In: International Conference on Machine Learning, vol. 30 (2013)
51. Mason, L., Baxter, J., Bartlett, P.L., Frean, M.R.: Boosting algorithms as gradient descent. In: Advances in Neural Information Processing Systems, pp. 512–518 (2000)
52. Mazumder, R., Hastie, T., Tibshirani, R.: Spectral regularization algorithms for learning large incomplete matrices. J. Mach. Learn. Res. **11**, 2287–2322 (2010)
53. Metropolis, N., Ulam, S.M.: The Monte Carlo method. J. Am. Stat. Assoc. **44**(247), 335–341 (1949)
54. Michie, D.: Memo functions and machine learning. Nature **218**, 19–22 (1968)
55. Miller, A.J.: Selection of subsets of regression variables. J. Roy. Stat. Soc. Ser. A **147**(3), 389–425 (1984)
56. Nadaraya, E.A.: On estimating regression. Teor. Veroyatnost. i Primenen. **9**(1), 157–159 (1964)
57. Nair, V., Hinton, G.: Rectified linear units improve restricted Boltzmann machines. In: International Conference on Machine Learning (2010)

58. Nesterov, Y.: A method of solving a convex programming problem with convergence rate o(1/sqr(k)). Soviet Math. Dokl. **27**, 372–376 (1983)
59. Pearson, K.: On lines and planes of closest fit to systems of points in space. Philos. Mag. **2**(11), 559–572 (1901)
60. Pedregosa, F., et al.: Scikit-learn: machine learning in Python. J. Mach. Learn. Res. **12**, 2825–2830 (2011)
61. Piotte, M., Chabbert, M.: The pragmatic theory solution to the Netflix grand prize (2009)
62. Polyak, B.T.: Some methods of speeding up the convergence of iteration methods. USSR Comput. Math. Math. Phys. **4**(5), 1–17 (1964)
63. Reddi, S.J., Kale, S., Kumar, S.: On the convergence of Adam and Beyond. In: International Conference on Learning Representations (2018)
64. Rendle, S.: Factorization machines. In: IEEE International Conference on Data Mining, pp. 995–1000 (2010)
65. Rendle, S.: Factorization machines with LibFM. ACM Trans. Intell. Syst. Technol. **3**(3), 57 (2012)
66. Rennie, J.D.M., Srebro, N.: Fast maximum margin matrix factorization for collaborative prediction. In: International Conference on Machine Learning (2005)
67. Robbins, H., Monro, S.: A stochastic approximation method. Ann. Math. Stat. **22**(3), 400–407 (1951)
68. Rumelhart, D.E., Hinton, G.E., Williams, R.J.: Learning internal representations by error propagation, vol. 1, pp. 318–362 (1986)
69. Salakhutdinov, R., Mnih, A.: Bayesian probabilistic matrix factorization using Markov chain Monte Carlo. In: International Conference on Machine Learning, pp. 880–887 (2008)
70. Salakhutdinov, R.R., Mnih, A.: Probabilistic matrix factorization. In: Advances in Neural Information Processing Systems, pp. 1257–1264 (2008)
71. Schapire, R.E.: The strength of weak learnability. Mach. Learn. **5**(2), 197–227 (1990)
72. Schwarz, G.: Estimating the dimension of a model. Ann. Stat. **6**(2), 461–464 (1978)
73. Srebro, N., Rennie, J., Jaakkola, T.S.: Maximum-margin matrix factorization. In: Advances in Neural Information Processing Systems, pp. 1329–1336 (2005)
74. Srivastava, N., Hinton, G., Krizhevsky, A., Sutskever, I., Salakhutdinov, R.: Dropout: a simple way to prevent neural networks from overfitting. J. Mach. Learn. Res. **15**, 1929–1958 (2014)
75. Tenenbaum, J.B., Freeman, W.T.: Separating style and content. In: Advances in Neural Information Processing Systems, pp. 662–668 (1997)
76. Tenenbaum, J.B., Freeman, W.T.: Separating style and content with bilinear models. Neural Comput. **12**(6), 1247–1283 (2000)
77. Tikhonov, A.N.: On the stability of inverse problems. Proc. USSR Acad. Sci. **39**(5), 195–198 (1943)
78. Tikhonov, A.N.: Solution of incorrectly formulated problems and the regularization method. Proc. USSR Acad. Sci. **151**(3), 501–504 (1963)
79. Töscher, A., Jahrer, M.: The BigChaos solution to the Netflix grand prize (2009)
80. Töscher, A., Jahrer, M., Legenstein, R.: Improved neighborhood-based algorithms for large-scale recommender systems. In: KDD Cup (2008)
81. Tucker, L.R.: Some mathematical notes on three-mode factor analysis. Psychometrika **31**(3), 279–311 (1966)
82. Watson, G.S.: Smooth regression analysis. Sankhyā Ser. A **26**(4), 359–372 (1964)
83. Wright, M., Ziegler, A.: Ranger. J. Stat. Softw. **77**(1), 1–17 (2017)
84. Zeiler, M.D.: ADADELTA: an adaptive learning rate method (2012). arXiv:1212.5701

# An Early Warning Method for Agricultural Products Price Spike Based on Artificial Neural Networks Prediction

Jesús Silva[1]([✉]), Mercedes Gaitán Angulo[2], Jenny Romero Borré[3],
Liliana Patricia Lozano Ayarza[3], Omar Bonerge Pineda Lezama[4],
Zuleima del Carmen Martínez Galán[2], and Jorge Navarro Beltran[5]

[1] Universidad Peruana de Ciencias Aplicadas, Lima, Peru
jesussilvaucp@gmail.com
[2] Corporación Universitaria Empresarial de Salamanca (CUES),
Barranquilla, Colombia
{M_gaitan689,viceacademica}@cuesw.edu.co
[3] Universidad de la Costa, St. 58 #66, Barranquilla, Atlántico, Colombia
{jromero58,llozano2}@cuc.edu.co
[4] Universidad Tecnológica Centroamericana (UNITEC),
San Pedro Sula, Honduras
omarpineda@unitec.edu
[5] Corporación Universitaria Latinoamericana, Barranquilla, Colombia
jorgeelbacan05@gmail.com

**Abstract.** In general, the agricultural producing sector is affected by the diversity in supply, mostly from small companies, in addition to the rigidity of the demand, the territorial dispersion, the seasonality or the generation of employment related to the rural environment. These characteristics differentiate the agricultural sector from other economic sectors. On the other hand, the volatility of prices payed by producers, the high cost of raw materials, and the instability of both domestic and international markets are factors which have eroded the competitiveness and profitability of the agricultural sector. Because of the advance in technology, applications have been developed based on Artificial Neural Networks (ANN) which have helped the development of sales forecast on consumer products, improving the accuracy of traditional forecasting systems. This research uses the RNA to develop an early warning system for facing the increase in agricultural products, considering macro and micro economic variables and factors related to the seasons of the year.

**Keywords:** Predictive model · Multilayer perceptron ·
Multiple Input Multiple Output · Forecast · Support vector machines ·
Cyclic variation

## 1 Introduction

In Colombia, there are legally established institutions for the promotion and defense of consumer rights. Article 75 of the Consumer Statute, established the National Consumer Protection Network, consisting of multiple entities such as the Consumer's

J. M. F. Rodrigues et al. (Eds.): ICCS 2019, LNCS 11537, pp. 622–632, 2019.
https://doi.org/10.1007/978-3-030-22741-8_44

Protection Councils, the Superintendence of Industry and Commerce, the Superinten-dence of Home Public Services, the Financial Superintendence, the National Health Superintendence, the Ports and Transportation Superintendence, the National Institute for Surveillance of Medicines and Foods, the National Television Commission, the Civil Aeronautics, the Mayoralties, and the consumer leagues [1].

For its part, inflation in Colombia has not been particularly innocuous to the agricultural sector. It has been assigned the dual character of victim and propagator of inflation [2]. Victim in several ways [3]: by credit constraints that arise as a control measure of inflation; in the elevation of both wages and the cost of modern inputs, which restricts their use and therefore affects productivity. In turn, the joint effect of these factors, tends to reduce the profitability of agriculture but, on the other hand, as an activity characterized by a composition of most the fixed assets such as land, inflation tends to significantly enhance the heritage of all producers.

There are traditional models that are probabilistic, deterministic, and hybrids, such as [4, 5]: Simple Moving Average, Weighted Moving Average, Exponential Smoothing, Regression Analysis, Box-Jenkins method (ARIMA), trend projections, etc., which have been used to generate forecasts, providing certain advantages and disadvantages compared to the others. However, these models are yet unable to offer good results in an environment of high uncertainty and constant changes. To this end, new paradigms based on numeric modeling of nonlinear systems are necessary, such as the Artificial Neural Networks (ANN), and the Support Vector Regression (SVR).

The present study proposes to Multi-Layer Perceptron with Multiple-Input Multiple-Output (MLP and MIMO) as a model for agricultural products price pre-diction together with the variation coefficient of the price reference from Colombian State as the warning level criterion, as well as macro and micro economic variables. Due to the agricultural nature of the product, the seasonal factor is integrated.

## 2    Theoretical Review

### 2.1    Artificial Neural Networks

Artificial Neural Networks (ANNs) can learn from data and can be used to construct reasonable input-output mapping, with no prior assumptions are made on the statistical model of the input data [6]. ANNs have nonlinear modeling capability with a data-driven approach so that the model is adaptively formed based on the features presented from the data [7].

An introduction to ANNs model specifications and implementation and their approximation properties has been provided from an econometric perspective [8]. Numerous studies have shown that ANNs can solve a variety of challenging compu-tational problems, such as pattern classification, clustering or categorization, function approximation, prediction or forecasting, optimization (travelling salesman problem), retrieval by content, and control [9].

Some studies of ANN application related to financial early warning models have been conducted by [10] as well as [11] who used ANN as a classifier with a categorical output. Other authors used ANNs as financial forecasting models with continuous value.

Some of them are [12] as well as [13], who implemented ANNs with a single-step prediction output. A previous study on ANNs forecasting model was also proposed by [14] for a multi-step prediction with a direct strategy, so the number of models is equal to the number of the prediction horizon. In the context of basic commodities price, the need for prediction is not limited to one-step forward but could be extended to include multi-step ahead predictions. Three strategies to tackle the multi-step forecasting problem can be considered, namely recursive, direct, and multiple output strategies [15]. The Multiple Input Multiple Output (MIMO) techniques train a single prediction model f that produces vector outputs of future prediction values.

The present study proposes to Multi-Layer Perceptron with Multiple Input and Multiple Output (MLP-MIMO) as to agricultural products price prediction model coupled with the coefficient of variation from the Colombian state price reference to the criteria of warning level.

## 2.2   Consideration of Microeconomic and Macroeconomic Indicators

The correct evaluation and interpretation of macroeconomic indicators is essential for any country, because they help to make decisions on fiscal or monetary policy and are signals that gives the market for the economic agents to take precautions. Likewise, the economic indicators are a way to predict and anticipate the phenomena, that is why it will be implemented in the study of the main economic indicators of Colombia. According to [16], the following elements are considered the main economic indicators:

- Gross Domestic Product
- The consumer price index
- Producer Price Index
- Employment indicators
- Retail Sales
- Consumer Confidence

## 2.3   Garson's Algorithm to Determine the Level of Importance

Garson's algorithm was developed to determine the degree or level of importance of an entry indicator in an ANN. In many cases related to the measurement of the variables, the weights in the hidden layer and their interactions in the output network are considered. A measure proposed by Garson [17] consists of dividing the weights of the hidden layer into components associated with each input node and then assigning each of them a percentage of the total weights.

Several studies show the effectiveness of the Garson algorithm to evaluate the importance of an entry in the RNA [18, 19, 20]. The certainty of the algorithm of Garson was experimentally determined, concluding that the measure is applied successfully under a wide variety of conditions. As a result of this analysis, the Garson's algorithm, on a scale from 0 to 1, determines a unique value for each explanatory variable that describes the relationship with the response variable in the model.

# 3 Material and Methods

## 3.1 Data

In this study, the raw data were obtained from the National Administrative Department of Statistics of Colombia (DANE – Departamento Administrativo Nacional de Estadística), which provided a sales database of 115,022 farmers from the main agricultural regions of Colombia in the time period from 2015 to 2017 [21]. The macroeconomic variables considered in this study range from food inflation, GDP, employment rate, minimum wage to commercial balance and capital flow of the nation [22]. Internal factors such as demand, and substitute and complementary products were also analyzed. Seasonal factors were incorporated to adjust the forecasts obtained [23].

## 3.2 Methods

The used method is summarized as follows. For the selection of products, an order is emitted according to the f1-score harmonic average [24], which is defined as the reciprocal of the arithmetic mean of the reciprocals. This value is used to determine the average of variations with respect to time. It is mostly used to find the average values of efficiency and error. This study allows to obtain a factor considering that both sales and quantity are important. In total, 10 products were selected, Eq. (1) [25].

$$f1 = 2 * \frac{Q * V}{Q + V} \tag{1}$$

Where:

F1 = harmonic mean f1-score
Q = quantity sold in units
V = sales in dollars.

With the sample of products to be considered, the interactivity with other products is analyzed to select the substitutes and complementary products. A quantitative alternative for selecting substitute and complementary products is the cross-elasticity of demand, which analyzes two products and, according to their sign, establishes whether it is substitute (+) or complementary (−). Then, the resulting maximum or minimum value represents greater interaction between both products, that is, the maximum positive value is chosen for substitute, and the maximum negative value for complementary. Cross-elasticity is defined as the proportional variation in the demanded quantity for a good or service caused by a proportional variation in the price of other goods [24].

The early warning model consists of three main components, namely preprocessing, predictive model, and post-processing, as depicted in Fig. 1 [26].

Preprocessing: Before all raw data about commodity prices are presented to the predictive model, the preprocessing operations are applied on the data. The price surveys were conducted by local government at working days, so the commodity price data represent a daily basis with missing values in weekends and holidays. The data are

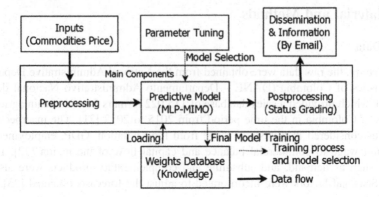

**Fig. 1.** Model of the early warning system [26].

therefore added to weekly data with the mean function to reduce the volume of the data for computational efficiency [26].

Predictive models: The predictive models are built from the trained final MLP-MIMO models obtained from the parameter tuning. Every single commodity in every city has its own model parameter structure. The weights after training are stored in a weight database so the model can be reloaded at any time [7].

Post-processing: The output of the predictive model is a normalized price prediction for eight weeks ahead. The post-processing is responsible for denormalizing the predicted price and determining early warning status based on the maximum predicted price according to Section 'Price Spike and Early Warning Status Leveling'. An alert will be sent to the stakeholders when the price is above a given threshold (on status 'watch' or 'monitoring') by an email service [10].

Finally, the Garson's algorithm for determining the level of importance was developed to determine the degree or level of importance of an input indicator in ANN. In many cases related to the measurement of the variables, the weights in the hidden layer and their interactions in the output network are considered. A measure proposed by [27] consists of dividing the weights of the hidden layer into components linked to each input node and then assigning each of them a percentage of the total weights.

## 4    Results and Discussion

### 4.1    Macroeconomic Indicators Considered

The indicators described in Tables 1 and 2 are the result of the availability analysis and correlation. In the first column, CODE indicates the codes that were used to facilitate the manipulation of data and related tables. In column two, INDICATOR, there are both microeconomic and macroeconomic indicators. Its value was analyzed with the units represented in parentheses. Because of the little space, the most important indicators out of the 25 macroeconomic resulting indicators are shown.

**Table 1.** Internal indicators used [23]

| Code | Indicator | Source |
|------|-----------|--------|
| Internal/Microeconomic indicators | | |
| I1 | Year | Company |
| I2 | Month | Company |
| I3 | Price (USD) | Company |
| Quant | Demand (Units) | Company |
| I4 | Price of the substituted product (USD) | Company |
| I5 | Price of the complementary product (USD) | Company |

**Table 2.** External indicators used [22].

| | External/Macroeconomic indicators | |
|------|-----------------------------------|--------|
| I6 | Currency (US Dollar Index) | www.tradingeconomics.com |
| I7 | GDP annual growth rate (%) | World Bank |
| I8 | GDP growth rate (%) | Central Bank of Colombia |
| I9 | Employment rate (%) | Dane |
| I10 | Labor participation rate (%) | Central Bank of Colombia |
| I11 | Minimum wage (USD/MONTH) | Central Bank of Colombia |
| I12 | Unemployment rate (%) | Central Bank of Colombia |
| I13 | Minimum wage in manufacture (USD/MONTH) | Central Bank of Colombia |
| I14 | Consumer Price Index (CPI) (Index points) | Dane |

## 4.2 Main Products Considered

To select the products on which the forecasts were made, the f1-score criterion is used. The ordering by this factor considered both the quantity and the value of sales for the selection of the most important products. Table 3 presents the values of the f1-score factor for each selected product.

**Table 3.** Prioritization and selection of products

| Code | Product | Quantity | F1-score |
|------|---------|----------|----------|
| P1 | Potato | 88318 | 131939.957 |
| P2 | Coffee | 67785 | 66264.5132 |
| P3 | Cocoa | 64673 | 62848.756 |
| P4 | Tomato | 66604 | 51443.7233 |
| P5 | Onion | 55489 | 45963.3935 |
| P6 | Lettuce | 50960 | 41689.3872 |
| P7 | Paprika | 46320 | 32567.0352 |
| P8 | Coriander | 37773 | 30722.393 |
| P9 | Parsley | 35126 | 30206.8012 |
| P10 | Lemon | 17232 | 29903.9777 |

## 4.3    Model of the Early Warning System

According to [16], the increase in the price is considered normal when it is below a certain threshold. The threshold is derived from the government reference price established by the Ministry of Commerce and the variation coefficient noted (CVtarget). There are four degrees of warning status: normal, advisory, monitoring, and warning, whose criteria are presented in Table 4.

**Table 4.** Levels of warning status and their criteria.

| Level | Status | Interval price increase related to the reference price |
|---|---|---|
| I | Normal | $\leq 1.5$ CVtarget |
| II | Advisory | (1.5 CVtarget, 2 CVtarget] |
| III | Monitoring | (2 CVtarget, 3.3 CVtarget] |
| IV | Warning | >3.3 CVtarget |

According to [16], these levels should be adjusted according to the seasons of rain and drought in Colombia by a factor of 0,952 and 1,025 respectively.

The reference price of each product and the threshold for each warning status used in this study are presented in Table 5, according to [16] together with DANE [21].

**Table 5.** Reference price, interval price increase, and the levels of early warning status.

| Commodity | Reference price[a] Pesos per lb | Interval percentage of price increase relative to reference price | | | |
|---|---|---|---|---|---|
| | | Normal | Advisory | Watch | Warning |
| Potato | 1500 | $\leq 5\%$ | (5%, 10%] | (10%,1 5%] | >15% |
| Coffee | 3000 | $\leq 25\%$ | (25%, 50%] | (50%, 75%] | >75% |
| Cocoa | 3500 | $\leq 10\%$ | (10%, 20%] | (20%, 30%] | >30% |
| Tomato | 1200 | $\leq 5\%$ | (5%, 10%] | (5%, 10%] | >15% |
| Onion | 1000 | $\leq 5\%$ | (5%, 10%] | (10%, 15%] | >15% |
| Lettuce | 2350 | $\leq 25\%$ | (25%, 50%] | (50%, 75%] | >75% |
| Paprika | 1500 | $\leq 10\%$ | (10%, 20%] | (20%, 30%] | >30% |
| Coriander | 800 | $\leq 5\%$ | (5%, 10%] | (5%, 10%] | >15% |
| Parsley | 700 | $\leq 10\%$ | (10%, 20%] | (20%, 30%] | >30% |
| Lemon | 950 | $\leq 25\%$ | (25%, 50%] | (50%, 75%] | >75% |

[a]DANE, 2019

All models are trained using backpropagation algorithm with learning rate = 0.9 according to Eq. (2). The log-sigmoid function is used at the hidden layer(s) and the linear/purelin function is used at the output layer [27].

$$W(t+1) = W(t) - \eta \frac{\partial E_{av}}{\partial W} \qquad (2)$$

Where $W$ is the weights vector, $t$ is the iteration number, $\eta$ is the learning rate, is the partial derivative.

The combination of two activation functions gives a minimum average of mean squared error (MSE) on the validation set and a minimum training epoch compared to any other combinations. In some cases of the training in the cross-validation, the optimum learning rate value that gives a minimum of epoch training without increasing the MSE is 0.6, 0.7, or 0.8, but, in most cases, learning rate = 0.9 gives the minimum epoch, especially in the last fold of the cross validation.

The last fold uses 5/6 data training (hundreds of data), which takes the longest training time among all folds of the cross-validation. Therefore, we determine to use 0.9 as the learning rate for all the training process. All training processes are stopped when the MSE training threshold is reached.

Figure 2 shows a graph that displays the performance of the prediction of the final model in the validation set. The next example is the case of the price prediction for the coffee product in Colombia at time t + 1 (next week) and t + 8 (eight weeks ahead).

a.                                          b.

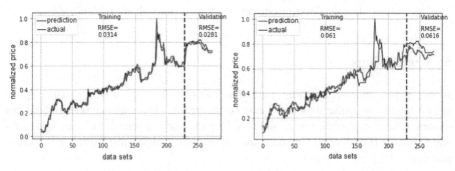

**Fig. 2.** The final model's prediction performance. (a) time t + 1 (next week). (b) t + 8 (eight weeks ahead).

### 4.4   Importance of Macroeconomic Indicators

Table 6 presents the level of importance that was obtained in the RNA with the Garson's algorithm. The coefficient shown in the last column presents the degree of importance that generates each indicator, it should be emphasized that a different coefficient is derived for each product in the same indicator, so an average Garson's coefficient is shown among the 10 products.

According to the Garzon's coefficient, the most relevant internal variables are the price and year, while in the macroeconomic policies are the foreign investment and the range of corruption, due to that both destabilize the price of the dollar.

**Table 6.** Summary of most important indicators.

| CODE | Indicator description | Garson coef. |
|------|----------------------|--------------|
| Internal - Microeconomic | | |
| I5 | Price of the complementary product | 0.013257374 |
| I2 | Month | 0.016043408 |
| I4 | Price of the substitute product | 0.016370018 |
| I3 | Price | 0.021072393 |
| I1 | Year | 0.021388844 |
| External - Macroeconomic | | |
| I20 | Change of the producer price index (%) | 0.023568662 |
| I39 | Changes in stock (USD THO) | 0.023690179 |
| I37 | Public spending (USD THO) | 0.023876179 |
| I49 | Vat - value added tax (%) | 0.024390901 |
| I38 | Business Confidence Index points | 0.024481111 |
| I42 | Ease of doing business in Ecuador | 0.025104695 |
| I18 | Rate of inflation, MoM (%) | 0.025132292 |
| I26 | Trade balance (USD THO) | 0.025274287 |
| I41 | Range in corruption | 0.025759514 |
| I31 | Foreign direct investment, net flows (USD million) | 0.026626361 |

## 5 Conclusions

In this paper, an early warning method is proposed for agricultural products price spikes based on artificial neural network prediction. The method is favorable for modeling basic commodity price and may help the State to predict the price spikes for some specific commodities on some specific regions to determine further anticipatory policies.

Agricultural products are analyzed obtaining the following 10 relevant products: Potato, coffee, cocoa, tomato, onion, lettuce, paprika, coriander, parsley and lemon. The early warning system can be adjusted according to seasonality (drought or rain) by multiplying the ranges by the respective coefficients of seasonality. Micro and macro variables were evaluated according to their importance in the affectation of the prices according to the Garzon's coefficient.

## References

1. Fonseca, Z., et al.: Encuesta Nacional de la Situación Nutricional en Colombia 2010. Da Vinci, Bogotá (2011)
2. Instituto Colombiano de Bienestar Familiar (ICBF): Ministerio de Salud y Protección Social, Instituto Nacional de Salud (INS) Departamento Administrativo para la Prosperidad Social, Universidad Nacional de Colombia. The National Survey of the Nutritional Situation of Colombia (ENSIN) (2015)

3. Food and Agriculture Organization of the United Nations (FAO): Pan American Health Organization (PAHO), World Food Programme (WFP), United nations International Children's Emergency Fund (UNICEF). Panorama of Food and Nutritional Security in Latin America and the Caribbean. Inequality and Food Systems, Santiago (2018)

4. Frank, R.J., Davey, N., Hunt, S.P.: Time series prediction and neural networks. J. Intell. Rob. Syst. **31**(3), 91–103 (2001)

5. Haykin, S.: Neural Networks and Learning Machines. Prentice Hall International, New Jersey (2009)

6. Jain, A.K., Mao, J., Mohiuddin, K.M.: Artificial neural networks: a tutorial. IEEE Comput. **29**(3), 1–32 (1996)

7. Kulkarni, S., Haidar, I.: Forecasting model for crude oil price using artificial neural networks and commodity future prices. Int. J. Comput. Sci. Inf. Secur. **2**(1), 81–89 (2008)

8. McNelis, P.D.: Neural Networks in Finance: Gaining Predictive Edge in the Market, vol. 59, pp. 1–22. Elsevier Academic Press, Massachusetts (2005)

9. Mombeini, H., Yazdani-Chamzini, A.: Modelling gold price via artificial neural network. J. Econ. Bus. Manag. **3**(7), 699–703 (2015)

10. Sevim, C., Oztekin, A., Bali, O., Gumus, S., Guresen, E.: Developing an early warning system to predict currency crises. Eur. J. Oper. Res. **237**(1), 1095–1104 (2014)

11. Zhang, G.P.: Time series forecasting using a hybrid ARIMA and neural network model. Neurocomputing **50**(1), 159–175 (2003)

12. Horton, N.J., Kleinman, K.: Using R For Data Management, Statistical Analysis, and Graphics. CRC Press, Clermont (2010)

13. Chang, P.C., Wang, Y.W.: Fuzzy Delphi and backpropagation model for sales forecasting in PCB industry. Expert Syst. Appl. **30**(4), 715–726 (2006)

14. Lander, J.P.: R for Everyone: Advanced Analytics and Graphics. Addison-Wesley Professional, Boston (2014)

15. Chopra, S., Meindl, P.: Supply Chain Management: Strategy, Planning and Operation. Prentice Hall, NJ (2001)

16. Izquierdo, N.V., Lezama, O.B.P., Dorta, R.G., Viloria, A., Deras, I., Hernandez-Fernandez, L.: Fuzzy logic applied to the performance evaluation. Honduran coffee sector case. In: Tan, Y., Shi, Y., Tang, Q. (eds.) Advances in Swarm Intelligence. ICSI 2018. LNCS, vol. 10942, p. 164–173. Springer, Cham (2018). https://doi.org/10.1007/978-3-319-93818-9_16

17. Babu, C.N., Reddy, B.E.: A moving-average filter based hybrid ARIMA–ANN model for forecasting time series data. Appl. Soft Comput. **23**(1), 27–38 (2014)

18. Cai, Q., Zhang, D., Wu, B., Leung, S.C.: A novel stock forecasting model based on fuzzy time series and genetic algorithm. Procedia Comput. Sci. **18**(1), 1155–1162 (2013)

19. Egrioglu, E., Aladag, C.H., Yolcu, U.: Fuzzy time series forecasting with a novel hybrid approach combining fuzzy c-means and neural networks. Expert Syst. Appl. **40**(1), 854–857 (2013)

20. Kourentzes, N., Barrow, D.K., Crone, S.F.: Neural network ensemble operators for time series forecasting. Expert Syst. Appl. **41**(1), 4235–4244 (2014)

21. Departamento Administrativo Nacional de Estadística-DANE.: Manual Técnico del Censo General. DANE, Bogotá (2018)

22. Fajardo-Toro, C.H., Mula, J., Poler, R.: Adaptive and hybrid forecasting models—a review. In: Ortiz, Á., Andrés Romano, C., Poler, R., García-Sabater, J.-P. (eds.) Engineering Digital Transformation. LNMIE, pp. 315–322. Springer, Cham (2019). https://doi.org/10.1007/978-3-319-96005-0_38

23. Deliana, Y., Rum, I.A.: Understanding consumer loyalty using neural network. Pol. J. Manag. Stud. **16**(2), 51–61 (2017)

24. Scherer, M.: Waste flows management by their prediction in a production company. J. Appl. Math. Comput. Mech. **16**(2), 135–144 (2017)
25. Chang, O., Constante, P., Gordon, A., Singana, M.: A novel deep neural network that uses space-time features for tracking and recognizing a moving object. J. Artif. Intell. Soft Comput. Res. **7**(2), 125–136 (2017)
26. Sekmen, F., Kurkcu, M.: An early warning system for turkey: the forecasting of economic crisis by using the artificial neural networks. Asian Econ. Financ. Rev. **4**(1), 529–543 (2014)
27. Ke, Y., Hagiwara, M.: An English neural network that learns texts, finds hidden knowledge, and answers questions. J. Artif. Intell. Soft Comput. Res. **7**(4), 229–242 (2017)

# Predicting Heart Attack Through Explainable Artificial Intelligence

Mehrdad Aghamohammadi[1](✉), Manvi Madan[2], Jung Ki Hong[3], and Ian Watson[2]

[1] Department of Mechanical Engineering, The University of Auckland,
Auckland, New Zealand
magh798@aucklanduni.ac.nz
[2] Department of Computer Science, The University of Auckland,
Auckland, New Zealand
Mmad610@aucklanduni.ac.nz, ian@cs.auckland.ac.nz
[3] Department of Applied Mathematics, The University of Auckland,
Auckland, New Zealand
Jhon360@aucklanduni.ac.nz

**Abstract.** A novel classification technique based on combined Genetic Algorithm (GA) and Adaptive Neural Fuzzy Inference System (ANFIS) for diagnosis of heart Attack is reported. Exploiting the combined advantages of neural networks, fuzzy logic and GA, the performance of the proposed system is investigated by evaluation functions such as sensitivity, specificity, precision, accuracy and Root Mean Squared Error (RMSE). Also, the efficiency of the algorithm is evaluated by employing 9-fold cross validation. To address the explainability of the predictions, explainable graphs are provided. The results show that the performance of the proposed algorithm is quite satisfactory. Furthermore, the importance of various symptoms in diagnosis of heart attack is investigated through defining and employing an importance evaluation function. It is shown that some symptoms have key roles in effective prediction of heart Attack.

**Keywords:** Explainable artificial intelligence · ANFIS · GA · Heart attack

## 1 Introduction

The advancements in healthcare infrastructure of a nation play a key role in determining the health-related quality of life of its residents. The emerging trends in artificial intelligence has boosted the utility of computational resources to improve the healthcare infrastructure globally [1–4]. One such condition that has gained a lot of attention from research scientists is heart attack [5–8], owning to the fact that it affects 33% of deaths annually. However, the decision making process in this domain still relies upon the expertise or guesswork of the cardiovascular expert. This makes it highly ambiguous to assure the quality of the decisions [9]. Consequently, there is a need for a system to aid the process of cardiovascular risk assessment.

The increase in the number of artificial intelligence (AI) enthusiasts have helped in the process by generating considerable funding for AI-based research in medical

© Springer Nature Switzerland AG 2019
J. M. F. Rodrigues et al. (Eds.): ICCS 2019, LNCS 11537, pp. 633–645, 2019.
https://doi.org/10.1007/978-3-030-22741-8_45

decision making. Nevertheless, the most significant concern in the process has remained the explainability of the results generated by AI-based systems. Particularly, in the case of medical diagnosis, it is extremely important for the decisions to have a plausible justification.

There are numerous artificial intelligence techniques that can support healthcare decision support systems such as genetic fuzzy systems, neural networks, and genetic programming. Three main aspects of every AI systems need to be taken into account: reliability which pertains to the accuracy of the results, explainability which accounts for providing relevant justifications of the results and understandability that focuses on the usability.

This paper concerns the effective diagnosis of heart attack based on a novel classification technique. Exploiting the advantages of Adaptive Neuro Fuzzy Inference System (ANFIS) and Genetic Algorithm (GA) in learning and optimization, an effective classification technique is proposed. The proposed system takes the medical data of a patient to predict how likely the patient might be at a risk of having heart attack. The algorithm is trained using a neural network to determine fuzzy parameters by processing the dataset fed to it. The membership functions (MFs) generated by the Fuzzy Inference System (FIS) are to be optimised by Genetic Algorithm (GA).

The system has an innovative interface which explains the results generated by the algorithm. The possibility of having heart attack is determined in five levels: no risk, slight risk, average risk, high risk and very high risk. Additionally, the explainable interface of the system reflects the relevance of various attributes such as heart rate, chest pain type and other factors that contribute to the predicted results. This makes the decision-making process intuitive as well as reliable. The explainable interface of the system intends to provide insight into how reasonable the predictions are. This gives the clinicians the ability to trace back a prognosis of heart attack to the associated symptoms, and helps in making the process of medical diagnosis trustworthy, safe and transparent. The system's reliability is evaluated by evaluation functions such as sensitivity, specificity, precision, accuracy, Root Mean Squared Error (RMSE) and k-fold cross validation.

A brief narrative of the hypothesis considered to be evaluated is provided in Sect. 2. Section 3, illuminates the existing literature in the domain. In Sect. 4, the methodology adopted to investigate the hypothesis is discussed. Section 5, focuses on the results achieved. Finally, conclusions are presented in Sect. 6

## 2 Hypothesis Description

The most of the studies in the realm of AI-based systems have focused on reliability, and explainability is addressed infrequently [10]. Diagnosis of heart attack corresponds to a number of related symptoms which can explain the predictions made by the system. This supports the principle of explainability of the results generated by AI algorithms. In this paper, a potential methodology to predict heart attack in a patient is discussed based on the values of various symptoms such as age, sex, type of chest pain and resting heart rate. An effective justification of the predictions is attained through providing evaluation graphs explaining the predictions made by the algorithm. In fact, the paper represents an effort to mimic the diagnosis process followed by clinicians through providing explainable graphs and giving an assessment based on the previous experience or existing precedence.

## 3  State of the Art

There is a plethora of work currently done in the domain of artificial intelligence for the medical diagnosis. An application to predict heart attack was presented in [8]. The authors used the Gini index to measure the impurity of training dataset and incorporated it into ANFIS to predict heart attack. Guessing the first approximation of the solutions, their proposed algorithm was employed to tune the parameters and normalize the weights. However, a further investigation is needed to overcome the problem of missing features in large data sets. A convolutional neural network for multi-label classification was presented in [11]. Defining importance values for the input data, the authors tried to present explainable results for each clinical code in the ICD-9 taxonomy and verified their results through evaluating by a physician. Their results showed high precisions along with high Macro-F1 scores comparing to previous works. Employing ANFIS, the prediction of renal failure time frame of Chronic Kidney Disease (CKD) was studied in [12]. A threshold value of glomerular filtration rate (GFR) was defined as the criterion to determine the renal failure. The proposed system could effectively capture the vagueness of the process to predict well-grounded GFR values. In addition, it could predict GFR values over increasing forecasting period with accuracy. Even so, not taking urine protein into account dampers the precision of the system.

Various studies have tried to address the explainability of AI-based algorithms. A rule-based fuzzy inference system was embedded into deep learning networks to make it explainable in [13]. The fuzzy rules were transformed into symbolic functions to initialize the neural network. The system could automatically identify and analyse the input pattern and adjust the behaviour of the predictors so that it could be adaptable. The challenges were overcoming the problem of lack of labelled data and also increasing the reliability of the output data and reducing the required time. In another study, the possible methodologies to approach explainable AI systems in medical domain were discussed in [9]. The authors addressed two types of explainability including post-hoc and ante-hoc explainabilities. The post-hoc explainability relates to explanation after the event in question such as the local interpretable model-agnostic explanations. On the other hand, the ante-hoc explainability corresponds to explanation before the event in question such as fuzzy inference systems. The authors suggested that integrating knowledge-based and neural techniques could provide a text-based explainable model.

The potential of ANFIS has been explored in various research applications. One such application was proposed in [14], where the authors discussed the efficiency of ANFIS and rough sets to predict tuberculosis (TB). Having four output variables related to the chances of having TB and 30 input variables related to the conditions of the patients, the authors reported close to 97% accuracy of the results obtained by ANFIS model compared to 92% accuracy of the rough sets results. A new ANFIS model using Levenberg-Marquardt algorithm for the diagnosis of heart disease was presented in [5]. The authors reported that the proposed model was more reliable and accurate comparing to grid partitioning classification and the case when ANFIS was trained with least squares estimation and backpropagation method. However, the proposed algorithm required calculating the Jacobian matrix for each iteration which could be a challenge for large data sets. Another brilliant employment of ANFIS for diagnosis of heart disease was presented in [7]. Using a 5 layer ANFIS and a hybrid algorithm of forward and backward pass, the authors reported the accuracy of 92.3%. Additionally in an attempt to integrate ANFIS with optimisation algorithms, GA was employed to train ANFIS for diagnosis of heart disease in [6]. The authors used 252 out of 297 cases of the dataset for training and the remained for testing, and reported the accuracy of about 98%.

## 4  Methodology

Adaptive Neuro Fuzzy Inference System (ANFIS) is a suitable machine learning technique for the classification problems. Combining the features of fuzzy system and neural network, the system can classify and predict unknown inputs. ANFIS exploits the advantages of the human expert knowledge based on the fuzzy linguistic rules, and the ability to learn and adapt based on the neural network. The fuzzy logic creates the flexibility of defining linguistic rules which provides the ability to use and deal with natural languages and human expert knowledge. Fuzzy inference system (FIS) constitutes of three main components including selection of fuzzy rules in order to form a rule-based fuzzy system, membership functions which create a database, and output which is related to the rules and membership functions, and creates an explainable conclusion [15].

The proposed ANFIS-GA algorithm takes the medical data of a patient to predict the risk of having heart attack. The dataset used in this paper is the Cleveland dataset which includes medical reports of 297 patients [16]. It represents 14 features including age, sex, chest pain type, resting blood pressure, cholesterol, fasting blood sugar, resting electrocardiographic results, maximum heart rate, exercise induced angina, depression induced by exercise relative to rest, the slope of peak exercise, number of major vessels, the heart status, and the diagnosis of heart disease as shown in Table 1. 161 out of 297 patients provided in the dataset have no heart disease and 136 patients have some level of heart disease. The diagnosis of heart attack is categorized into five different possibilities including at no risk (level 0), slight risk (level 1), average risk (level 2), high risk (level 3) and very high risk (level 4) of having heart attack. 257 instances of the dataset are used for training and 40 instances are used for testing.

The block diagram of the proposed algorithm is presented in Fig. 1. First, using Takagi Sugeno type fuzzy inference system, grid partitioning method is used to create initial membership functions based on the training set. Using Gaussian membership functions, if-then rules are generated. After creating an initial FIS, the algorithm is trained using ANFIS to determine fuzzy parameters by processing the dataset fed to it. ANFIS used in the algorithm generates a single-output Sugeno FIS and integrates the least-squares and the backpropagation gradient decent methods for training FIS membership function parameters. The system is trained for 1000 epochs. The training process stops whenever the designated epoch number is reached. In next step and in a novel technique, the membership function parameters are optimized by GA. As a result of combining these three algorithms (the fuzzy logic, the neural network and the genetic algorithm), a new classification algorithm for predicting heart attack is obtained. GA is set to minimize the cost function (CF) which is defined as root mean squared error (RMSE) between the desired (expected) and the predicted heart attack level.

$$CF = RMSE = \sqrt{\frac{1}{N} \sum_{i=1}^{N} (DO_i - PO_i)^2}, \tag{1}$$

where $DO$ is the desired output and $PO$ is the predicted output achieved by the algorithm and $N$ is the number of features provided in the dataset. GA parameters are: crossover probability 0.9, mutation probability 0.2, initial population 100 and number of iterations 200. The predictions made by the algorithm can be categorized into four classes. YN which is the case where a patient at a risk of having heart attack is predicted to be at no risk, YY for the case where the algorithm successfully predicts a patient being at a risk of having heart attack, NN which corresponds to the case where the algorithm truly predicts no risk of having heart attack for a patient at level 0, and the class NY, which is related to the case where the algorithm predicts being at a risk of having heart attack for a patient who is healthy.

To evaluate the proposed evolutionary ANFIS-GA system, four evaluation functions including sensitivity, specificity, precision and accuracy are defined to be used along with RMSE

$$Sensitivity = \frac{YY}{YY + YN}, \tag{2}$$

$$Specificity = \frac{NN}{NN + NY}, \tag{3}$$

$$Precision = \frac{YY}{YY + NY}, \tag{4}$$

$$Accuracy = \frac{YY + NN}{YY + YN + NN + NY}. \tag{5}$$

To investigate the reliability of the proposed system, evaluation functions and RMSE are calculated and compared for both training set and testing set. To address the explainability issue, the algorithm is designed in such a way that it provides graphs related to the symptoms of the predicted patients. The results are generated for the patients who are successfully predicted to be at no risk, at high risk (level 3 and 4) and at a risk of having heart attack. Additionally, to study the robustness of the proposed system and to provide an explainable criterion for the obtained results, the performance of the system is investigated for the cases where one of the features is discarded from the data set. Features are removed one by one and the performance of the system is evaluated and compared with the case when all the features are considered.

To detect the features with highest importance in diagnosis of heart attack, an Importance Evaluation Function (IEF) is defined based on the evaluation functions

$$IEF = 1 - \frac{x}{400}, \tag{6}$$

where $x$ is the summation of sensitivity, specificity, precision and accuracy percentages of testing set. IEF varies within zero and one, where value one shows the highest importance. If a feature with high importance value is removed, the evaluation scores of the testing set will decrease significantly. Top features with highest IEF are to be recognized and the performance of the proposed system would be evaluated based on these features for training and testing data sets. As a result, a simpler model consisting of the recognized features would be proposed and the performance of the system would be evaluated. The results would help to validate the explainability provided for the predictions.

**Table 1.** Features of the Cleveland dataset.

| Feature number | Name | Definition |
|---|---|---|
| 1 | age | Age in years |
| 2 | sex | Sex (0 = Female; 1 = Male) |
| 3 | cp | Chest pain type (1: Typical angina; 2: Atypical angina; 3: Non-angina pain; 4: Asymptomatic) |
| 4 | trestbps | Resting blood pressure (in mm Hg on admission to the hospital) |
| 5 | chol | Serum cholesterol in mg/dl |
| 6 | fbs | Fasting blood sugar greater than 120 mg/dl (0 = false; 1 = true) |
| 7 | restecg | Resting electrocardiographic results: 0: normal, 1: having ST-T wave abnormality (T wave inversions and/or ST elevation or depression greater than 0.05 mV), 2: showing probable or definite left ventricular hypertrophy by Estes' criteria |
| 8 | thalach | Maximum heart rate achieved |
| 9 | exang | Exercise induced angina (0 = no; 1 = yes) |
| 10 | oldpeak | ST depression induced by exercise relative to rest |
| 11 | slope | The slope of the peak exercise ST segment (1: upsloping; 2: flat; 3: downsloping) |
| 12 | ca | Number of major vessels (0–3) colored by flourosopy |
| 13 | thal | The heart status (3 = normal; 6 = fixed defect; 7 = reversable defect) |
| 14 | num | Diagnosis of heart disease (0,1,2,3,4 where 0 = healthy; 4 = high risk) |

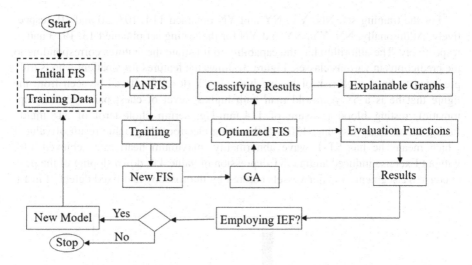

**Fig. 1.** The proposed ANFIS-GA flow chart.

## 5    Results and Discussion

To investigate the performance of the proposed ANFIS-GA algorithm in predicting heart attack, the algorithm was tested on the Cleveland dataset. As illustrated in Fig. 2, the cost function (RMSE) of the training set reduced from 0.82754 (output of ANFIS) to 0.75372 (output of GA) after 200 iterations. Consequently, the evaluation functions for the training set obtained as the sensitivity 91.1504%, the specificity 79.1667%, the precision 77.4436% and the accuracy 84.4358%. Additionally, the sensitivity, specificity, precision and accuracy for the testing set achieved 79.1667%, 81.25%, 86.3636% and 80%, respectively.

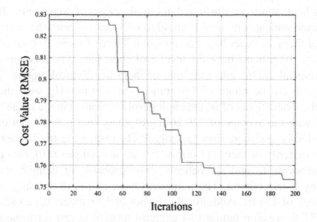

**Fig. 2.** Cost value (RMSE) as a function of iterations obtained by GA.

For the training set, NN, YY, NY and YN obtained 114, 103, 30 and 10, respectively. Additionally, NN, YY, NY and YN for the testing set obtained 13, 19, 3 and 5, respectively. The algorithm has the capability to illustrate the features corresponding to the predictions in various classes. Figure 3, shows the features for a patient who is truly predicted to be at high risk of having heart attack (level 3). It can be seen from the figure that he is a 58 years old man with highest level of chest pain type (asymptomatic), resting blood pressure of 114 mm-Hg, serum cholesterol of 318 mg/dl, fasting blood sugar less than 120 mg/dl, resting electrocardiographic result of value 1 which means he has ST-T wave abnormality, maximum heart rate achieved 140, without Exercise induced angina, ST depression of value 4.4, down sloping of the peak exercise ST segment, 3 major vessels colored by fluoroscopy and fixed defect of heart.

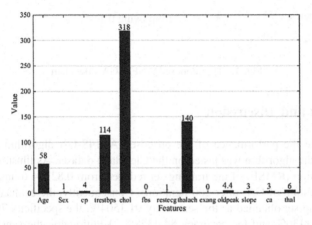

**Fig. 3.** The features corresponding to a patient who is successfully predicted by the proposed algorithm to be at level three risk of having heart attack.

The corresponding features of the patients who are successfully predicted to be at a risk of having heart attack are illustrated in Fig. 4 (19 out of 40 patients). It can be seen that the predicted patients are aged from 40 to 68 years old with the average of 55.68. Nearly 74% of them are older than 55 years old. Three of these 19 patients are women and 16 are men. 16 patients have the level 4 of chest pain type (Asymptomatic), 2 of them have Atypical angina and only one of them has Typical angina. Their resting blood pressure is in the range of 110 mm-Hg to 170 mm-Hg with the average of 136.95 mm-Hg. The amount of serum cholesterol for them is in the range of 131 mg/dl to 335 mg/dl with the average of 226.26 mg/dl. Only four patients out of 19 patients have fasting blood sugar more than 120 mg/dl. The resting electrocardiographic results show that the majority of them (11 out of 19) are normal (level 0), two patients have ST-T wave abnormality and six patients show probable or definite left ventricular hypertrophy by Estes' criteria. The maximum heart rate is in the range of 90 to 181 with the average of 135.89. Additionally, 11 out of 19 patients have exercise induced angina. The ST depression induced by exercise relative to rest for these patients is in the range of 0 to 4.4 with the average of 2.005. For the majority of them (13 out of 19) the slope of the peak exercise ST segment is flat, four of them have upsloping and two have downsloping. Only one of the patients has three major colored by fluoroscopy, five of them have two, seven of them have one, and six of them have no major vessels. Also, 12 out of 19 patients have reversible defect related to their heart status, six of them have fixed defect, and only one of them is normal. These graphs provide explainability for the predictions and are useful for clinicians to interpret the results.

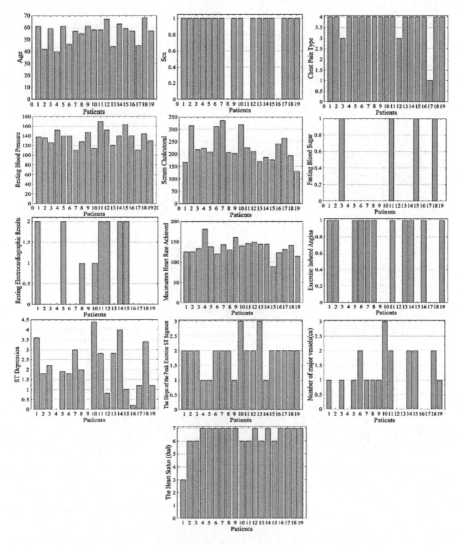

**Fig. 4.** The features corresponding to the patients who are successfully predicted by the proposed algorithm to be at a risk of having heart attack.

Furthermore, the features corresponding to the patients who are successfully predicted to be at no risk of having heart attack are provided in Fig. 5. As can be seen in this figure, the patients (13 patients) have less than 120 mg/dl fasting blood sugar. The result of resting electrocardiographic for 11 patients is normal and two patients have probable

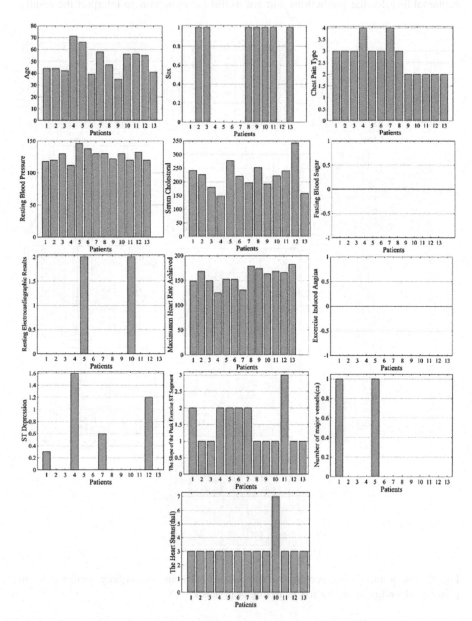

**Fig. 5.** The features corresponding to the patients who are successfully predicted by the proposed algorithm to be at no risk of having heart attack.

or definite left ventricular hypertrophy by Estes' criteria. None of the patients have exercise induced angina. For nine of the patients ST depression is zero, 11 patients have no major vessels colored by fluoroscopy and two patients have one vessel. 12 patients have normal heart status and one patient has reversible defect of heart.

The features were discarded and the performance of the model was evaluate (Table 2). The top six features with highest values of IEF were: ST depression induced by exercise relative to rest with IEF 0.2556, number of major vessels colored by fluoroscopy with IEF 0.245, maximum heart rate achieved with IEF 0.2329, and exercise induced angina, resting electrocardiographic results, age with IEF values of 0.2143. Consequently, the algorithm was tested for the smaller model of dataset containing these features and the performance of the proposed algorithm was evaluated. The sensitivity, specificity, precision and accuracy for the new testing set achieved by the proposed algorithm were 79.1667%, 87.5%, 90.4762% and 82.5%, respectively. Additionally, RMSE obtained 0.85158. 19 out of 40 patients are successfully predicted to be at a risk of having heart attack. Six of the patients have one major vessel colored by fluoroscopy, seven of them are recognized of having two major vessels, one of them have three major vessels and the remained five patients have no major vessels. Only two patients have no ST depression and the remained patients have ST depression in the range of 0.2 to 4.4. Additionally, 14 patients are successfully predicted to be at no risk of having heart attack. All of these patients have no exercise induced angina. One patient has one major vessel and the remained patients have no major vessels. Nine patients out of 14 recognized patients are without ST depression and three patients have probable or definite left ventricular hypertrophy by Estes' criteria.

**Table 2.** Evaluation results achieved by the proposed ANFIS-GA algorithm for the training set (Tr) and the testing set (Ts).

| Removed feature number | Evaluation functions | | | | | | | | | |
|---|---|---|---|---|---|---|---|---|---|---|
| | Sensitivity (%) | | Specificity (%) | | Precision (%) | | Accuracy (%) | | RMSE | |
| | Tr | Ts | Tr | Ts | Tr | Ts | Tr | Ts | Tr | Ts |
| – | 91.15 | 79.17 | 79.17 | 81.25 | 77.44 | 86.36 | 84.44 | 80 | 0.75 | 0.91 |
| 1 | 92.92 | 79.17 | 76.39 | 75 | 75.54 | 82.61 | 83.66 | 77.5 | 0.77 | 0.87 |
| 2 | 91.15 | 83.33 | 74.31 | 75 | 73.57 | 83.33 | 81.71 | 80 | 0.77 | 0.85 |
| 3 | 91.15 | 87.5 | 71.53 | 81.25 | 71.53 | 87.5 | 80.16 | 85 | 0.80 | 0.79 |
| 4 | 92.92 | 87.5 | 76.39 | 81.25 | 75.54 | 87.5 | 83.66 | 85 | 0.77 | 0.82 |
| 5 | 92.92 | 83.33 | 71.53 | 81.25 | 71.92 | 86.96 | 80.93 | 82.5 | 0.78 | 0.84 |
| 6 | 92.92 | 87.5 | 71.53 | 81.25 | 71.92 | 87.5 | 80.93 | 85 | 0.76 | 0.77 |
| 7 | 92.92 | 79.17 | 74.31 | 75 | 73.94 | 82.61 | 82.49 | 77.5 | 0.76 | 0.88 |
| 8 | 92.04 | 75 | 72.92 | 75 | 72.73 | 81.82 | 81.32 | 75 | 0.76 | 0.89 |
| 9 | 92.92 | 79.17 | 75 | 75 | 74.47 | 82.61 | 82.88 | 77.5 | 0.76 | 0.87 |
| 10 | 92.92 | 83.33 | 68.75 | 62.5 | 70 | 76.92 | 79.38 | 75 | 0.77 | 0.88 |
| 11 | 92.04 | 79.17 | 82.64 | 81.25 | 80.62 | 86.36 | 86.77 | 80 | 0.75 | 0.91 |
| 12 | 93.81 | 79.17 | 75 | 68.75 | 74.65 | 79.17 | 83.27 | 75 | 0.79 | 0.92 |
| 13 | 92.04 | 79.17 | 69.44 | 81.25 | 70.27 | 86.36 | 79.38 | 80 | 0.79 | 0.84 |

Furthermore, to validate the efficiency of the proposed ANFIS-GA algorithm, 9-fold cross validation was used. The dataset including 297 instances was partitioned into nine equal groups of 33 instances. In each step, a unique group was taken as the testing dataset while the remaining groups were taken as the training dataset. Therefore, the evaluation functions for the training and testing datasets were calculated. Consequently, the evaluation results are provided as the average of evaluation functions scores as presented in Table 3. As can be seen, the performance of the proposed algorithm as a result of employing 9-fold cross validation is quite satisfactory.

**Table 3.** The average of evaluation functions for training sets (Trs) and testing sets (Tss) obtained from employing 9-fold cross validation.

| Sensitivity (%) | | Specificity (%) | | Precision (%) | | Accuracy (%) | |
|---|---|---|---|---|---|---|---|
| Trs | Tss | Trs | Tss | Trs | Tss | Trs | Tss |
| 91.51 | 88.38 | 78.31 | 76.89 | 81.67 | 78.40 | 81.12 | 78.3 |

## 6  Conclusion

A novel Adaptive Neural Fuzzy Inference System (ANFIS)-Genetic Algorithm (GA) algorithm for predicting heart attack was proposed. Using the explainablity of fuzzy rules, training capability of neural network and the optimization power of GA, the algorithm was employed on the Cleveland dataset. The trained Fuzzy Inference System (FIS) obtained from ANFIS was fed into GA to optimize the membership function parameters. As a result, the Root mean Squared Error (RMSE) between the expected results and the predicted ones obtained from the algorithm reduced from 0.82754 to 0.75372. The performance of the proposed algorithm was evaluated by evaluation functions such as sensitivity, specificity, precision, accuracy and RMSE. Also, 9-fold cross validation was employed to evaluate the efficiency of the algorithm. The results showed that the performance of the proposed algorithm in predicting heart attack was quite satisfactory.

To provide explainable results to be useful for clinicians and patients, the algorithm was designed in such a way that it provides matrices corresponding to the predicted patients in various classes of having heart attack. Additionally, an Importance Evaluation Function (IEF) was proposed to investigate the importance of various features in predicting heart attack. Features were discarded one by one and the performance of the model was evaluated based on the evaluation functions. Calculating IEF for all features, the top six features with highest importance were determined. As a result, a new model based on the top important features was generated. Consequently, the performance of the algorithm on the new model was evaluated. The sensitivity, specificity, precision, accuracy and RMSE of the new testing set were obtained 79.1667%, 87.5%, 90.4762%, 82.5% and 0.85158, respectively. Detecting the features with highest importance and showing the satisfactory performance of the algorithm on the new model may be considered as a good criterion for providing explainable predictions based on the importance of the features.

The performance of the proposed algorithm needs to be evaluated through being employed on other datasets as the future work. Additionally, based on the proposed IEF, it was observed that some features have key roles in predicting heart attack. A cooperation with medical experts would be beneficial to validate the predictions.

# References

1. Shortliffe, E., Cimino, J.: Biomedical Informatics. Springer, New York (2006). https://doi.org/10.1007/0-387-36278-9
2. Allahverdi, N., Tunali, A., Işik, H., Kahramanli, H.: A Takagi-Sugeno type neuro-fuzzy network for determining child anemia. Expert Syst. Appl. **38**, 7415–7418 (2011)
3. Chattopadhyay, S.: A neuro-fuzzy approach for the diagnosis of depression. Appl. Comput. Inf. **13**, 10–18 (2017)
4. Bright, T., et al.: Effect of clinical decision-support systems. Ann. Intern. Med. **157**, 29 (2012)
5. Sagir, A.M., Sathasivam, S.: A novel adaptive neuro fuzzy inference system based classification model for heart disease prediction. Pertanika J. Sci. Technol. **25**(1) (2017)
6. Uyar, K., İlhan, A.: Diagnosis of heart disease using genetic algorithm based trained recurrent fuzzy neural networks. Proc. Comput. Sci. **120**, 588–593 (2017)
7. Ziasabounchi, N., Askerzade, I.: ANFIS based classification model for heart disease prediction. Int. J. Electr. Comput. Sci. IJECS-IJENS **14**(02), 7–12 (2014)
8. Richard, D., Mala, D.: GI-ANFIS approach for envisage heart attack disease using data mining techniques. Int. J. Innov. Res. Adv. Eng. (IJIRAE) **5** (2018)
9. Holzinger, A., Biemann, C., Pattichis, C., Kell, D.: What do we need to build explainable AI systems for the medical domain? https://arxiv.org/abs/1712.09923
10. DARPA Agency, Program Information: Explainable Artificial Intelligence. https://www.darpa.mil/program/explainable-artificial-intelligence
11. Mullenbach, J., Wiegreffe, S., Duke, J., Sun, J., Eisenstein, J.: Explainable prediction of medical codes from clinical text. https://arxiv.org/abs/1802.05695
12. Norouzi, J., Yadollahpour, A., Mirbagheri, S., Mazdeh, M., Hosseini, S.: Predicting renal failure progression in chronic kidney disease using integrated intelligent fuzzy expert system. Comput. Math. Methods Med. **2016**, 1–9 (2016)
13. Bonanno, D., Nock, K., Smith, L., Elmore, P., Petry, F.: An approach to explainable deep learning using fuzzy inference. In: Next-Generation Analyst V (2017)
14. Uçar, T., Karahoca, A., Karahoca, D.: Tuberculosis disease diagnosis by using adaptive neuro fuzzy inference system and rough sets. Neural Comput. Appl. **23**, 471–483 (2012)
15. Zadeh, L.: Fuzzy sets. Inf. Control **8**, 338–353 (1965)
16. Index of /ml/machine-learning-databases/heart-disease. http://archive.ics.uci.edu/ml/machine-learning-databases/heart-disease/

# Track of Architecture, Languages, Compilation and Hardware support for Emerging and Heterogeneous Systems

# Dynamic and Distributed Security Management for NoC Based MPSoCs

Siavoosh Payandeh Azad[1]([✉]) [ID], Gert Jervan[1] [ID], and Johanna Sepulveda[2] [ID]

[1] Tallinn University of Technology, Tallinn, Estonia
{siavoosh.azad,gert.jervan}@taltech.ee
[2] Institute for Security in Information Technology,
Technical University of Munich, Munich, Germany
johanna.sepulveda@tum.de

**Abstract.** Multi-Processors System-on-Chip (MPSoCs) have emerged as the enabler technology for new computational paradigms such as Internet-of-Things (IoT) and Machine Learning. Network-on-Chip (NoC) communication paradigm has been adopted in several commercial MPSoCs as an effective solution for mitigating the communication bottleneck. The widespread deployment of such MPSoCs and their utilization in critical and sensitive applications, turns security a key requirement. However, the integration of security into MPSoCs is challenging. The growing complexity and high hyper-connectivity to external networks expose MPSoC to Malware infection and code injection attacks. Isolation of tasks to manage the ever-changing and strict mixed-criticality MPSoC operation is mandatory. Hardware-based firewalls are an effective protection technique to mitigate attacks to MPSoCs. However, the fast reconfiguration of these firewalls impose a huge performance degradation, prohibitive for critical applications. To this end, this paper proposes a lightweight broadcasting mechanism for firewall reconfiguration in NoC-based MPSoC. Our solution supports efficient and secure creation of dynamic security zones in the MPSoC through the communication management while avoiding deadlocks. Results show that our approach decreases the security reconfiguration process by a factor of 7.5 on average when compared to the state of the art approaches, while imposing negligible area overhead.

**Keywords:** MPSoCs · Network-on-Chip · Security

## 1 Introduction

Multi-Processors Systems-on-Chip (MPSoCs) integrate a variety of different Intellectual Property (IP) hardware cores (e.g., processor, memory, interfaces) that communicates through a Network-on-Chip (NoC). The NoC integrates routers and links to exchange the IP cores data encapsulated as packets. MPSoCs enable the development of cutting edge technologies. The flexibility and computational power has turned MPSoCs into a perfect solution for many critical

© Springer Nature Switzerland AG 2019
J. M. F. Rodrigues et al. (Eds.): ICCS 2019, LNCS 11537, pp. 649–662, 2019.
https://doi.org/10.1007/978-3-030-22741-8_46

applications. They are widely adopted in several critical applications (e.g., automotive, avionics) and deployed in different commercial products such as TileMx100 from Tilera [1], MPPA from Kalray [2] or SCC from Intel [3]. However, the mixed-criticality nature of the MPSoCs has turned security into an important requirement. Integrating security at MPSoCs is challenging.

High VLSI integration levels, cache hierarchies and complex MPSoC architecture bring on a wide surface of attacks. For instance, MPSoCs applications are stored in external non-volatile memories which are prone to malicious modifications. Moreover, MPSoCs are able to support several applications, usually characterized by different levels of criticality. When executed simultaneously, these applications are forced to share several MPSoC resources, such as processors, memories and communication structure. Code injection attacks exploit these two aforementioned MPSoC characteristics to retrieve secret information (i.e., cryptographic keys, passwords), gain control over the critical applications or deny the completion of critical tasks. In order to mitigate code injection attacks, physical and temporal isolation of the applications is required.

Previous works have proposed isolation through software security mechanisms. However, it has been shown the effectiveness of the micro-architectural attacks even under restrictive software-controlled isolation scenarios [18,20]. Thus, in order to guarantee the security of a system, hardware security must be also considered [25]. Hardware firewalls integrated in the network interface of the MPSoCs, between the IP core and the NoC routers, have shown to be an effective isolation mechanism. These firewalls store the security policy (set of security rules) of the system and verify packet-wise the communication rights. When the packet content matches the security policy, the transaction takes place. Otherwise, the transaction is discarded and an alarm is triggered to activate a recovery mechanism. In order to be compliant with the MPSoC protection requirements, firewalls should be dynamically reconfigured based on the criticality and security requirements of the tasks being executed on the system. Moreover, the reconfiguration latency should meet the performance requirements of the system. Previous works built security zones using a single manager and uni-casting communication mechanism. However, these techniques do not scale with larger NoC sizes and impose large performance penalty due the reconfiguration overhead. In order to overcome these drawbacks, in this work we propose: (i) a new method for partitioning the network for reconfiguration; (ii) a light-weight broadcasting solution for secure firewall reconfiguration; and (iii) a distributed solution for MPSoCs security managers.

The rest of this paper is organized as follows: Sect. 2 describes the related works. Section 3 provides details of the proposed multi-zone reconfiguration broadcasting mechanism. Section 4 evaluates the proposed mechanism and finally Sect. 5 concludes the paper.

## 2   Related Work

Security integration in NoC-based MPSoCs through hardware firewalls has been target of wide research [5,7–11,13,16,26]. According to the reconfigurability and granularity characteristics of the firewalls, they can be classified as static/dynamic and single/multi-level firewalls. Static and single-level firewalls were proposed in [8,10,11]. These firewalls are appropriate for MPSoCs that support fixed and static applications characterized by a single level of criticality and will fail to protect dynamic and mixed-critically MPSoCs (firewall policies should be updated and refreshed at run-time). In order to protect these MPSoCs enhanced firewall management is required. The work of [17,19] proposed the integration of static and multi-level firewalls. Despite the good results for mixed-critical applications, the dynamic nature of some of these applications is not yet supported. Protection of MPSoCs with dynamic behaviour was proposed in the works of [24,26]. These works presented reconfigurable firewalls able to built and reshape in run-time security zones that isolate a set of IP cores with the same criticality. The dynamic reconfigurable firewall-based MPSoC protection presents a wide design space exploration. The design of the dynamic firewalls structure has a huge impact on the system performance. These parameters include: (I) location and the number of security managers (SM) that control the firewalls; (II) firewalls size; (III) granularity of the security control; (IV) number of security levels; (V) number of security zones; and (vi) management of the reconfiguration traffic. The works of [12,15,22,23,27] explore the number of security zones and number of security levels. Despite the good results, the above-mentioned works have not considered the effect of firewalls size, granularity of the security control, distribution of security managers and management of the reconfiguration traffic. This exploration was partially addressed by the authors of [6]. However, the effect of impact of the location and number of Security Managers and the management of the reconfiguration traffic are still open questions. The goal of this work is to answer the above mentioned open issues. We propose and evaluate the impacts of a dynamic, lightweight, multi-layer, multi-manager, multi-zone security infrastructure in NoC-based MPSoC using packet broadcasting.

## 3  Multi-zone Security Broadcasting

NoC-based MPSoCs are able to support several applications characterized by different levels of criticality. These application can be divided into smaller pieces of code, called tasks, which are split and mapped (dynamically) into the different MPSoC computation and storage resources. This fact, speeds up the execution of the application, but forces the peer IP interaction through the NoC. Collision among critical tasks may leak information regarding the critical and secret data. Moreover, the dynamic and mix-criticality nature of the different tasks require frequent updating capabilities of the security mechanisms. Previous attacks, such as Meltdown and Spectre [18,20], have shown that efficient isolation among critical applications is mandatory. Hard and inefficient updating security checks may be exploited to gain control over the system resources and to leak the secret information. They also show that hardware firewall technology may be efficient in such a threats, once no operation is executed before the permissions are verified.

Hardware-based security provides an efficient mean for access control enforcement. Firewalls, customized network protocols and customized routers are used to build security zones. These zones encapsulate the sensitive traffic into trusted areas. They are constituted by a set of trusted IPs and routers, in which only sensitive and trusted traffic is exchanged. Firewall information should be able to be updated, once the permissions of the applications being executed in the MPSoC may change. There are different alternatives to update the firewall information. One alternative is the uni-cast updating, however it incurs into a high system performance and power degradation.

In this work we propose a flexible and efficient security zone creation through the fast configuration of hardware-based firewall structure embedded into the Network Interface (NI) and security routers. The efficient update capability of the firewalls and routing mechanism is achieved through two techniques: (I) utilization of local security managers at each zone; and (II) hybrid communication, which uses a lightweight broadcasting approach for the security reconfiguration traffic and uni-casting communication for the remaining traffic. As a consequence efficient and elastic security zones are created into the MPSoC.

### 3.1  Multi-zone Security Broadcasting Mechanism

NoC-based broadcasting enables the transmission of information to all the IP cores that belong to the network. Classical NoC-based communication is carried out through uni-cast data exchange between a sender-receiver pair. Information is wrapped as packets. Figure 1 shows the packet format used in this work, which is compatible with the Open Core Protocol (OCP). It embodies information that specifies the criticality of the transaction. Note that any other format can be used. Our secure MPSoC architecture is composed of security routers, security managers and network interfaces equipped with firewalls. Firewalls match the packet content with the security rules stored in LUTs at the NIs. Figure 2 shows the firewall fields. A matching packet is allowed to communicated, otherwise, it is discarded. Firewalls are classified as *initiator firewall*, which checks injected packets to the NoC, and *target firewall* which check ejected packets from the NoC. The values stored in this tables are configured by a security manager via reconfiguration packets. Upon receiving a reconfiguration packet, the NI searches for the valid entries in the firewall table which match the node-ID field. If the entry exists in the table, it is updated, otherwise a replacement table policy will take place. The replacement policy will use empty entries in the table if they exist, otherwise, will evict an existing entry. The replacement policy will have an effect on the performance of the system if the number of the entries in the firewall tables is too small. The network is partitioned into several security zones, each being organized by a local security manager. A single system-wide security manager, mapped on a secure node manages secure system booting and provides the policies for the local managers. The boundaries of such zones are defined and updated by the system manager depending on the application characteristics. The reconfiguration events happen once a new task is mapped on a node which has a different privilege than the previously defined value.

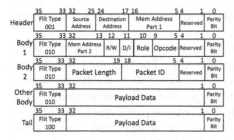

**Fig. 1.** Packet format

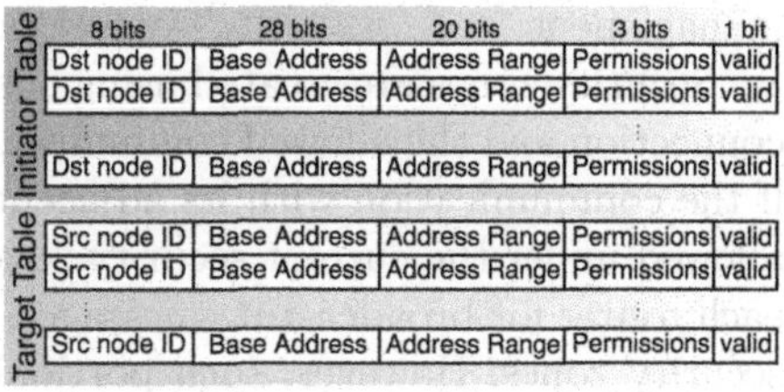

**Fig. 2.** Organization of firewall tables

Reconfiguration information is broadcasted from the local security manager to the nodes of the security zone.

The security router supports uni-casting and broadcasting mechanisms. Security routers are built upon a 5 input port wormhole switching and 3 pipeline stages as shown in Fig. 3. It is composed of four elements: (i) input buffers, which store the input data arriving from an input port implemented as circular buffer FIFO; (ii) routing unit, which selects the router output port in which incoming packets will be redirected, using Logic Based Distributed Routing (LBDR) mechanism; (iii) switch allocator (SA), that grants the utilization of the crossbar switch to one of the input buffers; and (iv) crossbar switch (Xbar), which links input to output ports of the router. Note that the security router can be built upon any other architecture. By managing the routing, security zones can be built. These zones apply only to the security management and has no effect on the network's packets (there will be no physical isolation for the normal packets). In order to have this property, we would need separate zone boundary and routing algorithm definition for security packets.

Previous works have implemented NoC broadcasting mechanisms [21]. They use a set of 8 bits $R_x y$ (routing bits) to define the underlying routing and two sets of 4 bits $C_x$ (connectivity bits) to express the connectivity of a router to its neighbors and regions. This information is used to compute the packet routing. In this work, an extra set of routing ($R_{xy\_bc}$) and connectivity bits ($C_{x\_bc}$) are used

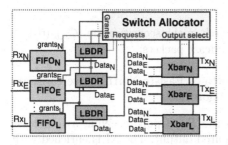

**Fig. 3.** Router architecture

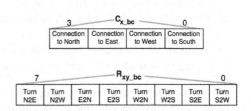

**Fig. 4.** Organization of $C_{x\_bc}$ and $R_{xy\_bc}$ bits (Color figure online)

to implement the broadcasting and create security zones through the routing management.

Security zones are created using connectivity bits. They set the existence of a connection and the allowed communication direction. The $C_{x\_bc}$ bits are set to '1' if the communication with its adjacent neighbor router in a specific direction is allowed under the current security zone planning. Using the connectivity bits of each router for broadcasting messages, the NoC can be partitioned into different security zones. However, such portioning should be applied with care to avoid overlapping partitions. The connectivity bits for the broadcast messages can be dynamically reconfigured using the following approaches:

- To change the connectivity bits by the manager via reconfiguration packets.
- To modify the connectivity bits through OSR-lite [28].

However, the details of the reconfiguration mechanism for the dynamic update of the security zones are out of scope of this paper.

The routing unit uses the incoming input port signal (North, East, West, South, Local) information to select the output port. A set of signals describe the input direction of the routing logic. For example, signal $N' = $ '1' means that the routing logic is located in the North input (hence shows the incoming direction of the broadcast packets).

The $R_{xy\_bc}$ bits represent the allowed turns for broadcasting packets (see Fig. 4). We define turn $X2Y$, as a turn from input $X$ to $Y$ of the router. Using this notation, the requests for broadcasting ($R_{N\_bc}, R_{E\_bc}, R_{W\_bc}, R_{S\_bc}, R_{L\_bc}$ for North, East, West, South and Local respectively) would be generated as in Eq. 1.

$$R_{N\_bc} = [S' + L' + [E'.R_{xy}(2) + W'.R_{xy\_bc}(4)]].C_{x\_bc}(0) \qquad (1)$$
$$R_{E\_bc} = [W' + L' + [N'.R_{xy\_bc}(0) + S'.R_{xy\_bc}(6)]].C_{x\_bc}(1)$$
$$R_{W\_bc} = [E' + L' + [N'.R_{xy\_bc}(1) + S'.R_{xy\_bc}(7)]].C_{x\_bc}(2)$$
$$R_{S\_bc} = [N' + L' + [E'.R_{xy\_bc}(3) + W'.R_{xy\_bc}(5)]].C_{x\_bc}(3)$$
$$R_{L\_bc} = \neg L'$$

Signals $N', E', W', S'$ and $L'$ describe the direction of the incoming packet and are hardwired in each routing unit. For example, the request to north $R_{N\_bc}$ in Eq. 1 is set to one if:

1. The router has connectivity towards north in the security zone ($C_{x\_bc}(0)$)
2. The routing logic is processing the broadcast packets:
   - from south or local packets; or
   - from West or East and West-to-North or East-to-North (these turns are allowed under the current routing algorithm).

These request signals are propagated to the $SA$. Upon receiving a broadcast message, $SA$ awaits the release of the required output. The NI is enriched with a mechanism that suppresses the injection of broadcast messages from non-privileged nodes.

## 3.2    Discussion of Deadlock-Free

In order to avoid deadlocks, two cases should be considered: (1) a single broad-cast message cannot contribute to the creation of deadlocks; and (2) multiple broadcast messages can not create deadlocks. To avoid deadlocks, the following assumptions must be met:

1. The broadcast routing algorithm must not add additional turns to the current networks routing.
2. Only security managers are allowed to broadcast.
3. Each network zone has one and only one manager.
4. The broadcast flits can only be forwarded if all destination FIFOs are empty (the router should block them until they become empty)

The first assumption guarantees that the allowed set of turns is kept. The work in [21] proves the deadlock freeness with the concurrent use of adaptive routing for broadcast and uni-cast messages. However, this approach requires the use of Virtual-Cut-Through (VCT) switching. This results in a poor buffer management. In order to avoid deadlocks and reduce the impact of VCT, in this work, we employ Dimension Ordered Routing (DOR) for the uni-cast and broadcasting traffics, supported with wormhole switching.

*Example:* Figure on the right shows a deadlock scenario for wormhole switching without use of DOR. Packets A and B are broadcast messages and can only advance together. However, Packet C (a unicast packet) is waiting on A and blocking B. Since B is stuck, so is A, hence deadlock occurs.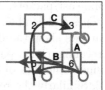

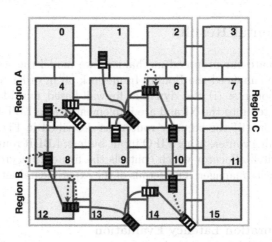

**Fig. 5.** Example of complex deadlock scenario formation using several regions (Color figure online)

**Table 1.** Experimental setup details

| Network Size | 4X4 | NI FIFO depth | 32 flits |
|---|---|---|---|
| Routing algorithm | XY | Network traffic packet size | 8 flits |
| Network FIFO depth | 4 flits | Network Traffic PIR | 0.03 |

Second and third assumptions ensure that at each security zone only the security manager has the right to broadcast which avoids race conditions between multiple input ports of the same router.

The forth assumption ensures that the destination buffers for the broadcasting messages are balanced. The imbalance in the buffers would cause a situation where some broadcast flits will advance further while others can get stuck behind other packets which can cause a deadlock situation. This problem is caused by the fact that the normal uni-cast packets are not bounded to the security zones and can be transported among different zones. Figure 5 illustrates an example of such deadlocks due to imbalance of the buffers. The managers of Region A and B (node 5 and node 13) broadcast the reconfiguration packets (marked in Blue). The reconfiguration packets are broadcast packets and can be forwarded at the same time. The other packets (marked in Red and Green) are normal uni-cast packets. Note that the uni-cast packets are not bound by the security zones and can be forwarded according to the base routing algorithm. The dotted arrows show the dependency between the packets. Note the dependency of all reconfiguration packets in each region; The local port of the node 4 is occupied since blue flits in node 4 can not advance due to the flits being stuck in node 6. Similarly the reconfiguration packets in node 14 can not advance due to the flits getting stuck in node 12. Use of VCT, removes this problem entirely and ensures that there is no chance of starvation on the network.

## 4    Experimental Results

In order to evaluate our proposed mechanism, we model the secure MPSoC in VHDL-RTL. We enhanced the Bonfire framework [4] in order to support three additional components: (i) the dynamic firewall-based multi-level protection. They are integrated into the NI and are implemented as a LUT-based structure and a comparator; (ii) the security router, as described in Fig. 3, a five input wormhole switching router, using FIFO input buffers, LBDR routing mechanism [14], a single switch allocator which controls the five output crossbar switches; and (iii) the security manager. The details of the experimental setup are provided in Table 1.

### 4.1    Reconfiguration Latency Evaluation

In order to evaluate the reconfiguration latency of the proposed broadcasting scheme we employed two different scenarios: (I) single security zone and a single

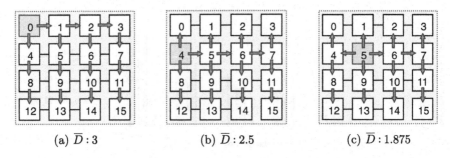

(a) $\overline{D}:3$          (b) $\overline{D}:2.5$          (c) $\overline{D}:1.875$

**Fig. 6.** Different mapping scenarios of the security manager (marked in red) using broadcasting in a single zone (Color figure online)

manager, as in Fig. 6; (II) three security zones and three security managers. Results are compared to the work of [6]. The first scenario organizes the system in a single security broadcasting zone and a single manager (marked in red). The experiments explore the reconfiguration impact of three different mappings of the security manager node. We use the notation of average distance presented in [6] for distinguishing different manager locations as in Eq. 2.

$$\overline{D} = \frac{\sum_{i=0}^{N-1} shortest\_path(ID_M, ID_i)}{N} \qquad (2)$$

Where $N$ is the number of nodes in the network, $ID_i$ is the node identifier for the network tile $i$ and $M$ is the $id$ of the tile where the security manager task is mapped.

Figure 7 depicts the network reconfiguration latency results for different system manager locations and different firewall table size, using broadcast mechanism in a network with a single zone under different traffic patterns (no-traffic, Uniform Random, Hot-Spot and Bit-reversal). In the "No Traffic" scenario, The network nodes do not send packets to the network and only reconfiguration packets are injected. This has been used as a baseline to compare other traffic pattern's behaviour. The results show that different manager placements can affect the dynamic reconfiguration latency. A distance of 3 increases the latency by 11% compared to distance of 1.875 and 2.76% compared to distance of 2.5 on average. In comparison with the work presented in [6], the current approach decreases the reconfiguration latency for different firewall sizes by a factor of 7.5 on average (776 cycles on average compared to 5807 cycles on average in [6]).

Figure 8 describes the effect of the size of the reconfiguration packets in the total network reconfiguration latency for different manager locations and under different load conditions. Similar to [6], the results confirm that larger reconfiguration packets will reduce the configuration latency.

Figure 9 describes the organization of a $4 \times 4$ mesh-NoC using in the second scenario where three zones and three managers are implemented. Red arrows denote the broadcast path in each zone. Figure 10 shows the configuration of the connectivity bits in each router under this second scenario (see Fig. 9). The links

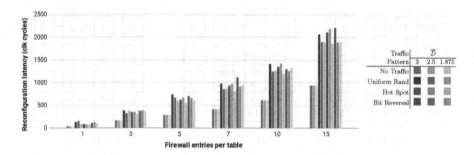

**Fig. 7.** Network reconfiguration latency for different security manager location and different firewall size under different load conditions (reconfiguration packet carries a single firewall entry).

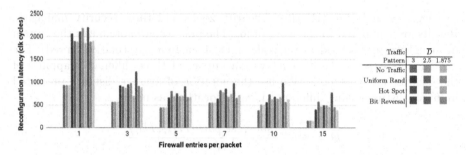

**Fig. 8.** Network reconfiguration latency for different security manager location and different reconfiguration packet size under different load conditions (firewall has 15 entries in all cases).

marked in red are disconnected in order to form the different security zones for broadcast mechanism. Partial reconfiguration (for selective dynamic reconfiguration) of a firewall or smaller firewall size (in case of area constrained embedded MPSoCs) has an impact on the reconfiguration latency. Figure 11 depicts the network reconfiguration latency for different firewall table sizes, using broadcast mechanism in the second scenario (described in Fig. 9) under different load conditions. In this experiment each reconfiguration packet only updates one firewall location. Another factor that has an impact on the reconfiguration latency is the number of firewall entries that each reconfiguration packet can update. Figure 12 depicts the network reconfiguration latency for different reconfiguration packet sizes, using broadcast mechanism in scenario II, described in Fig. 9 under different load conditions. The decreasing reconfiguration latency trend is clear with the growing number of firewall entries in per reconfiguration packet. The experimental results show that the proposed approach reconfigure the entire network firewalls on average in approximately 1602 clock cycles based on the reconfiguration policy and the traffic pattern.

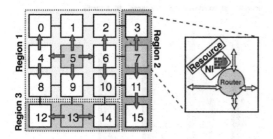

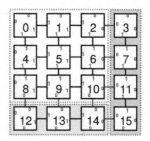

**Fig. 9.** Scenario II: Organization of different zones with distributed security managers (marked in red). (Color figure online)

**Fig. 10.** Connectivity bits ($C_{x\_bc}$) configuration in Scenario II.

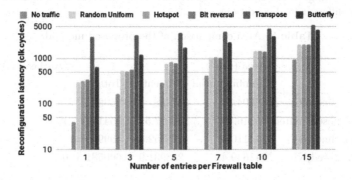

**Fig. 11.** Reconfiguration latency (scenario II) using different firewall table sizes under different load conditions.

### 4.2 Overhead Evaluation

In order to evaluate the impact of our approach, the baseline router (packet switching) and NI where compared with the security router and firewall-enhanced NI (with 16 entries per NI). These models have been synthesized using AMS $180nm$ CMOS technology standard cell library in Synopsys Design Compiler. Table 2 describes the imposed area overhead on the system using the proposed method. The area overhead on the simple router at hand is less than 11% and for the network interface is completely negligible. However, the baseline router size is very small. Considering a $4\,mm^2$ chip with 16-core NoC-based MPSoC, the overhead of our approach on the system area would be roughly 0.2%, which is negligible. Table 3 shows the critical path delay overhead which shows that the proposed method does not impact the performance of the original system.

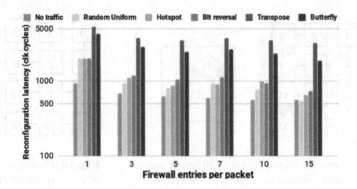

**Fig. 12.** Reconfiguration latency (scenario II) using different reconfiguration packet size under different load conditions.

**Table 2.** Area comparison of the proposed method

|  | Area ($\mu$m$^2$) | | | Overhead (%) |
|---|---|---|---|---|
|  | Combinatorial | Sequential | Total | |
| Baseline router | 38604.7 | 47388 | 85992.8 | – |
| Proposed router | 45316.45 | 50097.5 | 95414.0 | 10.9 |
| Baseline NI | 311110.3 | 288852.8 | 599963.2 | – |
| Proposed NI | 311649.4 | 288931.9 | 600581.3 | 0.1 |

**Table 3.** Critical-path comparison of the proposed method

|  | Critical-path delay ($ns$) | | Overhead (%) |
|---|---|---|---|
|  | Baseline | Proposed | |
| Router | 4.85 | 4.81 | – |
| NI | 4.79 | 4.82 | 0.6 |

## 5    Conclusions

Dynamic isolation of mixed-critical application is mandatory in current MPSoCs. Security zones can be used to isolate applications. Hardware-based firewalls have been proposed as a effective mean for application isolation and security zone creation. However, the efficient reconfiguration of firewall is mandatory. In this paper we have proposed a hybrid communication mechanism that support unicast and broadcast communication to achieve a good performance and efficient firewall update. Our security MPSoC infrastructure is based on a lightweight broadcasting and partitioning mechanism for firewall reconfiguration. Experimental results show that our approach provides a speed up by a factor of 7.5 compared to security updating using uni-casting approaches while imposing negligible area overhead on the system.

**Acknowledgements.** The authors would like to thank Prof. Dr. Ing. Thomas Hollstein for our discussion regarding deadlock-freeness challenges in broadcasting in NoC based systems.

# References

1. Accelerating the Data Plane With the TILE-Mx Manycore Processor. http://www.tilera.com/files/drim_EZchip_LinleyDataCenterConference_Feb2015_7671.pdf. Accessed 13 Mar 2017
2. KALRAY MPPA: A New Era of processing. https://de.slideshare.net/infokalray/kalray-sc13-external3. Accessed 13 Mar 2017
3. Using Intel's Single-Chip Cloud Computer (SCC). https://communities.intel.com/docs/DOC-19269. Accessed 13 Mar 2017
4. Project Bonfire (2016). https://github.com/Project-Bonfire
5. Achballah, A.B., et al.: FW_IP: a flexible and lightweight hardware firewall for NoC-based systems. In: 2018 International Conference on Advanced Systems and Electric Technologies (IC_ASET) (2018). https://doi.org/10.1109/ASET.2018.8379868
6. Azad, S.P., Niazmand, B., Jervan, G., Sepúlveda, J.: Enabling secure MPSoC dynamic operation through protected communication. In: 25th IEEE International Conference on Electronics Circuits and Systems (ICECS), pp. 1–4 (2018)
7. Diguet, J.P., et al.: Noc-centric security of reconfigurable SoC. In: First International Symposium on Networks-on-Chip (NOCS 2007) (2007)
8. Evain, S., et al.: From NoC security analysis to design solutions. In: IEEE Workshop on Signal Processing Systems Design and Implementation, pp. 166–171 (2005). https://doi.org/10.1109/SIPS.2005.1579858
9. Fernandes, R., et al.: A non-intrusive and reconfigurable access control to secure NoCs. In: 2015 IEEE International Conference on Electronics, Circuits, and Systems (ICECS), pp. 316–319 (2015). https://doi.org/10.1109/ICECS.2015.7440312
10. Fiorin, L., et al.: Security aspects in networks-on-chips: overview and proposals for secure implementations. In: 10th Euromicro Conference on Digital System Design Architectures, Methods and Tools (DSD 2007), pp. 539–542 (2007). https://doi.org/10.1109/DSD.2007.4341520
11. Fiorin, L., et al.: Implementation of a reconfigurable data protection module for NoC-based MPSoCs. In: 2008 IEEE International Symposium on Parallel and Distributed Processing, pp. 1–8 (2008). https://doi.org/10.1109/IPDPS.2008.4536514
12. Fiorin, L., et al.: Secure memory accesses on networks-on-chip. IEEE Trans. Comput. **57**(9), 1216–1229 (2008). https://doi.org/10.1109/TC.2008.69
13. Fiorin, L., et al.: A security monitoring service for NoCs. In: Proceedings of the 6th IEEE/ACM/IFIP International Conference on Hardware/Software Codesign and System Synthesis, CODES+ISSS 2008, pp. 197–202. ACM, New York (2008). https://doi.org/10.1145/1450135.1450180
14. Flich, J., Duato, J.: Logic-based distributed routing for NoCs. IEEE Comput. Architect. Lett. **7**(1), 13–16 (2008). https://doi.org/10.1109/L-CA.2007.16
15. Grammatikakis, M.D., et al.: High-level security services based on a hardware NoC firewall module. In: 2015 12th International Workshop on Intelligent Solutions in Embedded Systems (WISES), pp. 73–78 (2015)
16. Grammatikakis, M.D., et al.: Security in MPSoCs: a NoC firewall and an evaluation framework. IEEE Trans. Comput. Aided Des. Integr. Circuits Syst. **34**(8), 1344–1357 (2015). https://doi.org/10.1109/TCAD.2015.2448684

17. Hu, Y., et al.: Automatic ILP-based firewall insertion for secure application-specific networks-on-chip. In: 2015 Ninth International Workshop on Interconnection Network Architectures: On-Chip, Multi-Chip, pp. 9–12 (2015). https://doi.org/10.1109/INA-OCMC.2015.9

18. Kocher, P., et al.: Spectre attacks: exploiting speculative execution. In: 40th IEEE Symposium on Security and Privacy (S&P 2019) (2019)

19. Kornaros, G., Tomoutzoglou, O., Coppola, M.: Hardware-assisted security in electronic control units: secure automotive communications by utilizing one-time-programmable network on chip and firewalls. IEEE Micro **38**(5), 63–74 (2018). https://doi.org/10.1109/MM.2018.053631143

20. Lipp, M., et al.: Meltdown: reading kernel memory from user space. In: 27th USENIX Security Symposium (USENIX Security 18) (2018)

21. Rodrigo, S., Flich, J., Duato, J., Hummel, M.: Efficient unicast and multicast support for CMPs. In: Proceedings of the 41st Annual IEEE/ACM International Symposium on Microarchitecture, MICRO 41, pp. 364–375. IEEE Computer Society, Washington, DC (2008). https://doi.org/10.1109/MICRO.2008.4771805

22. Sepúlveda, J., Fernandes, R., Marcon, C., Florez, D., Sigl, G.: A security-aware routing implementation for dynamic data protection in zone-based MPSoC. In: 2017 30th Symposium on Integrated Circuits and Systems Design (SBCCI), pp. 59–64, August 2017

23. Sepúlveda, J., Gogniat, G., Pedraza, C., Pires, R., Chau, W.J., Strum, M.: Hierarchical NoC-based security for MP-SoC dynamic protection. In: 2012 IEEE 3rd Latin American Symposium on Circuits and Systems (LASCAS), pp. 1–4 (2012). https://doi.org/10.1109/LASCAS.2012.6180312

24. Sepúlveda, J., et al.: Efficient and flexible NoC-based group communication for secure MPSoCs. In: 2015 International Conference on ReConFigurable Computing and FPGAs (ReConFig), pp. 1–6 (2015). https://doi.org/10.1109/ReConFig.2015.7393301

25. Sepúlveda, J., Diguet, J.P., Reinbrecht, C.: Security of emerging architectures. In: IEEE/ACM International Conference On Computer Aided Design (ICCAD) (2018). https://hal.archives-ouvertes.fr/hal-01911907

26. Sepúlveda, J., et al.: Dynamic NoC-based architecture for MPSoC security implementation. In: Proceedings of the 24th Symposium on Integrated Circuits and Systems Design, SBCCI 2011, pp. 197–202. ACM, New York (2011). https://doi.org/10.1145/2020876.2020921

27. Sepúlveda, J., et al.: QoSS hierarchical NoC-based architecture for MPSoC dynamic protection. Int. J. Reconfig. Comput. **2012**, 3:3 (2012). https://doi.org/10.1155/2012/578363

28. Strano, A., Bertozzi, D., Trivio, F., Snchez, J.L., Alfaro, F.J., Flich, J.: OSR-Lite: fast and deadlock-free NoC reconfiguration framework. In: 2012 International Conference on Embedded Computer Systems (SAMOS), pp. 86–95, July 2012. https://doi.org/10.1109/SAMOS.2012.6404161

# Scalable Fast Multipole Method
# for Electromagnetic Simulations

Nathalie Möller[1(✉)], Eric Petit[2], Quentin Carayol[1], Quang Dinh[1],
and William Jalby[3]

[1] Dassault Aviation, Saint-Cloud, France
{nathalie.moller,quentin.carayol,quang.dinh}@dassault-aviation.com
[2] Intel, Meudo, France
eric.petit@intel.com
[3] LI-PaRAD, University of Versailles, Versailles, France
william.jalby@uvsq.fr

**Abstract.** To address recent many-core architecture design, HPC applications are exploring hybrid parallel programming, mixing MPI and OpenMP. Among them, very few large scale applications in production today are exploiting asynchronous parallel tasks and asynchronous multithreaded communications to take full advantage of the available concurrency, in particular from dynamic load balancing, network, and memory operations overlapping. In this paper, we present our first results of ML-FMM algorithm implementation using GASPI asynchronous one-sided communications to improve code scalability and performance. On 32 nodes, we show an 83.5% reduction on communication costs over the optimized MPI+OpenMP version.

**Keywords:** CEM · MLFMM · MPI · PGAS · Tasks

## 1 Introduction

The stability of the architecture paradigm makes more predictable the required effort to port large industrial high performance codes from a generation of supercomputers to another. However, recent advances in hardware design result in an increasing number of nodes and an increasing number of cores per node: one can reasonably foresee thousands of nodes mustering thousands of cores, with a subsequent decrease of memory per core. This shift from latency oriented design to throughput oriented design requires application developers to reconsider their parallel programming usage outside the current landscape of production usage. Due to the large and complex code base in HPC, this tedious code modernization process requires careful investigation. Our objective is to expose efficiently fundamental properties such as concurrency and locality.

In this case-study, this is achieved thanks to asynchronous communication and overlapping with computation. We demonstrate our methodology on a Dassault Aviation production code for Computational Electromagnetism (CEM)

© Springer Nature Switzerland AG 2019
J. M. F. Rodrigues et al. (Eds.): ICCS 2019, LNCS 11537, pp. 663–676, 2019.
https://doi.org/10.1007/978-3-030-22741-8_47

implementing the Multi-Level Fast Multipole Method (ML-FMM) algorithm presented in Sect. 2. This complex algorithm is already using a hybrid MPI and OpenMP implementation which provides state of the art performance compared to similar simulations [6], as discussed in related work Sect. 2.2. To put priorities in the modernization process, we evaluate the potential of our optimization with simple measurements that can be reproduced on other applications. This evaluation is presented in Sect. 3. With this algorithm, the three problems to address are load-balancing, communication scalability, and overlapping. A key common aspect of these issues is the lack of asynchronism in all levels of parallelism. In this paper, we will focus on exploring the impact of off-line load-balancing strategies in Sect. 4 and introducing asynchronism in communications in Sect. 5.

The result Sect. 6 shows, on 32 nodes, an 83.5% improvement in communication time over the optimized MPI+OpenMP version. Further optimizations to explore are discussed in future work. To demonstrate our load balancing and communications improvement for ML-FMM, we are releasing *FMM-lib* library [1] under LGPL-3 license.

## 2    SPECTRE and MLFMM

SPECTRE is a Dassault Aviation simulation code for electromagnetic, and acoustic applications. It is intensively used for RCS (Radar Cross-Section) computations, antenna design and external acoustic computations. These problems can be described using the Maxwell equations. With a Galerkin discretization, the equations result in a linear system with a dense matrix: the Method of Moments (MoM). Direct resolution leads to an $O(N^3)$ complexity, where N denotes the number of unknowns. A common approach presented in Sect. 2.1 to solve larger systems is to use the Multi-Level Fast Multipole Method (MLFMM) to reduce the complexity to $O(NlogN)$ [10]. In SPECTRE, the MLFMM is implemented with hybrid MPI+OpenMP parallelism. For the time being, all MPI communications are blocking.

### 2.1    The FMM Algorithm

The FMM (Fast Multipole Method) has been introduced in 1987 by L. Greengard and V. Rohklin [1] and is part of the 20th century top ten algorithm [5]. The algorithm relies on the properties of the Green kernel. Let us consider a set of n points $x_p$ and a function $f$ with known values at each of these points. The Fast Multipole Method (FMM) is an algorithm which allows, for all $p \leq n$, fast computation of the sums:

$$\sigma(p) = \sum_{q \leq n \ q \neq p} G(x_p - x_q) f(x_q),$$

where G(.) is the Green kernel. In two dimensions, a naive computation of these sums would require a large number of operations in $O(n^2)$, whereas FMM yields a result in $O(n)$ operations.

The FMM relies on an accurate approximation of the Green kernel, hierarchical space decomposition and a criterion of well separation. The rationale is to model the distant point to point interactions by hierarchically grouping the points into a single equivalent point. Therefore the FMM relies on the accuracy of this far-field low-rank approximation. To make the approximation valid, the group of points has to be far enough from the target point: this is the *well-separated* property.

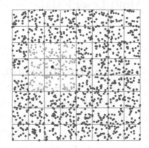

**Fig. 1.** Near field and far field characterization in 2D. (Color figure online)

Figure 1 shows a quadtree covering a two-dimensional space and defining the near field and the far field of the particles in the green square. The red particles are *well-separated*: they form the far field and interactions are computed using the FMM. The grey particles outside the green square are too close: they form the near field and interactions are computed using the MoM.

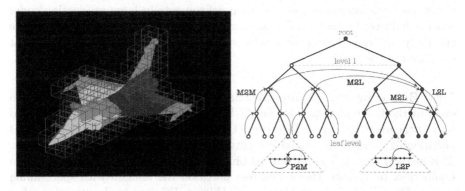

**Fig. 2.** Left: Hierarchically 3D partitioning, Right: 2D quadtree with FMM operators

As illustrated in Fig. 2, the 3D computational domain is hierarchically partitioned using octrees, and the FMM algorithm is applied using a two-step tree traversal. The upward pass aggregates children's contributions into larger parent nodes, starting from the leaf level. The downward pass collects the contributions from same level source cells, which involves communications, and translates the values from parents down to children. The operators are commonly called P2M (Particle to Multipole) and M2M (Multipole to Multipole) for the first phase and M2L (Multipole to Local), L2L (Local to Local) and L2P (Local to Particle) for the second one.

Compared to n-body problems, electromagnetic simulation introduces a major difference: at each level, the work to compute a multipole is doubling, which leads the complexity to $O(NlogN)$ instead of $O(N)$. Moreover, FMM is referred to as MLFMM, for Multi-Level FMM.

## 2.2  Related Work

**Parallel Programming and Core Optimization.** At node level, [11], Chandramowlishwaran et al. have done extensive work to optimize the computations with manual SIMD, data layout modifying to be vectorization friendly, replacement of matrix storage by on-the-fly computations, use of FFTs and OpenMP parallelization. These optimizations have been implemented in the KIFMM code. In [23], Yokota et al. also achieve parallelisation through OpenMP and vectorization with inline assembly. In [18] and [16], they discuss data driven asynchronous versions using both tasks, at node level with Quark, and on distributed systems with TBB and sender-initiated communications using Charm++. Milthorpe et al. use the X10 programming language [19] which supports two levels of parallelism: a partitioned global address space for communications and a dynamic tasks scheduler with work stealing. In [8] and [9], Agullo et al. automatically generate a directed acyclic graph and schedule the kernels with StarPU within a node that may expose heterogeneous architectures. Despite being impressive demonstrators opening a lot of opportunities, to our best knowledge, none of these codes has been integrated with all their refinements in large production codes to solve industrial use-cases. Furthermore, they require adaptation to match the specific properties of electromagnetic simulation with doubling complexity at each level of the tree.

**Load Balancing and Communication Pattern.** In the literature [16], two main methods are identified for fast N-Body partitioning, and can be classified into Orthogonal Recursive Bisection (ORB) or Hashed Octrees (HOT). ORB partitioning consists in recursively cutting the domain into two sub-domains. This method creates a well-balanced binary tree, but is limited to power of two numbers of processors. Hashed octrees partition the domain with space-filling curves. The most known are Morton and Hilbert. Efficient implementation relies on hashing functions and are, therefore, limited to 10 levels depth with 32 bits or 21 levels with 64 bits.

In [17], Lakshuk et al. propose to load balance the computational work with a weighted Morton curve. Weights, based on the interaction lists, are computed and assigned to each leaf. In a similar approach, Milthorpe et al. [19] propose a formula to evaluate at runtime the computational work of the two main kernels:

P2P and M2L. Nevertheless current global approaches on space filling curve do not consider the communication costs.

A composite approach is proposed by Yokota et al. [16]. They use a modified ORB (Orthogonal Recursive Bisection) method, combined with a local Morton key so that the bisections align with the tree structure. For the local Morton key, particles are weighted considering computation and communication using ad-hoc formula. This is complemented by another work of the same author in [24] and [15] about FMM communication complexity. However, their model cannot be generalized to our use-cases.

In [12], Cruz et al. elaborate a graph-based approach to load balance the work between the nodes while minimizing the total amount of communications. They use vertices' weights proportional to the computational work and edges proportional to communications' volume. They compute the partition using PARMETIS [3]. First result demonstrate interesting strong scaling results on small clusters.

## 3   Profiling

The FMM's behavior has been examined in terms of execution time, scalability, communications, load-balance, and data locality. To this end, two test cases have been used, as well as different profiling tools like ScoreP [4] and Maqao [2]. The test cases are a generic metallic UAV (Unmanned Aerial Vehicle) [7], with 95 524 nodes, 193 356 elements and 290 034 dofs, and a Dassault Aviation Falcon F7X airplane with 765 187 nodes, 1 530 330 elements and 2 295 495 dofs. All experiments, for profiling or results, are run on a Dassault Aviation cluster, composed by 32 nodes of two Intel Sandy Bridge E5-2670 (8 cores@2.60 GHz) interconnected with InfiniBand FDR. Binaries are compiled with Intel 2015 and run with bullxmpi-1.2.9.1.

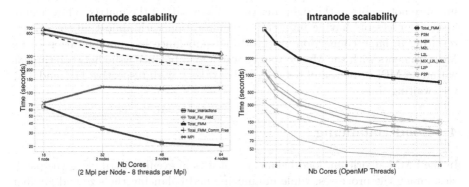

**Fig. 3.** Internode and intranode strong scaling analysis on UAV

*Scalability.* In the FMM algorithm, a common difficulty is the scalability of the distributed memory, due to the large number of communications required. Therefore, we are interested in separating communication and computational costs. In order to evaluate the potential gain by overlapping the communications with computations without other changes in the algorithm, we create and measure an artificial communication-free *Total FMM Comm Free* version of the code. Of course this code does not converge numerically anymore, but the performance of a significant number of iteration can be compared with the same number of original iterations. In the current implementation, the communications consist in exchanging the vector of unknowns and far-field contributions before and after the FMM tree traversal. Unknowns are exchanged via MPI broadcast and allreduce communications, while far fields are sent via MPI two-sided point to point communications and one allreduce at the top of the tree. The left part of Fig. 3 shows the strong scaling analysis. We use 4 nodes, and according to the best performing execution mode of SPECTRE/MLFMM, we launch one MPI process per socket with 8 OpenMP threads. Communications are not scaling: with 8 processes, the gap (log scale) between the *Total FMM* and *Total FMM comm-free* represents 35%. Further experiments with the larger F7X test case, run on 32 nodes, with one MPI process per node, show that the time spent in the communications grows considerably and reaches 59%. The right part of Fig. 3 focuses on intranode scalability. It highlights the lack of shared memory parallelism: over eight threads, parallel efficiency falls under 56%. In its current status, this last measurement prevents efficient usage of current manycore architectures and is a risk, for the future increase in the number of cores, which must be addressed.

*Data Locality.* Figure 4 shows the communication matrix of the original application, which reflects the quantity of data exchanged. A gradient from blue to red towards the diagonal would represent a good locality pattern. Despite communications being more important around the diagonal, vertical and horizontal lines are noticeable in the right and the bottom part of the matrix. They denote a set of processors having intensive communication with almost all the others, resulting in large connection overhead and imbalance in communication costs. Measurements with the Maqao tool expose a load balance issue between threads within an MPI worker, and between MPI workers, with respective 17% and 37% idle/waiting time.

## 4    Load Balancing

In the FMM algorithm, load-balancing consists in evenly distributing the particles among the processes, while minimizing and balancing the size and number of communications of the far-fields. There is no general and optimal solution to this problem. Each application of the FMM algorithm and more specifically each use case, may benefit from different strategies. In the case of CEM for structures, the particles are Gauss points that are not moving, and therefore one can pay

**Fig. 4.** Communication matrix, UAV test case (Color figure online)

for off-line elaborated approaches. In our library, we propose two load balancing strategies using Morton space-filling curves and Histograms. Both methods compute separators matching the underlying octree grid. Having the separator frontier aligned on the regular octree is of prime importance for the cutoff distance test computation in the remaining of the FMM algorithm. The Morton version relies on a depth-first traversal, and isn't limited by any depth, and the Histogram version is an ameliorated ORB with multi-sections. Our implementations are generic and freely available under LGPLv3.0 [1].

### 4.1  SPECTRE's Load Balancing

In the initial version of SPECTRE, the load balancing relies on distributing an array of leaves among the processors. The array is sorted in a breadth-first way. Thus, this method results in being equivalent to a Morton ordering.

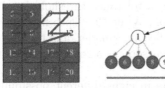
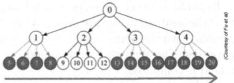

**Fig. 5.** Morton tree traversal

As shown in Fig. 5, drawing a Morton curve on the quadtree grid represented on the left side corresponds to picking the cells in ascending order, i.e. a depth-first tree traversal. While considering only leaves, breadth-first traversal results in the same ordering.

### 4.2  Morton

The study of the existing load balancing scheme implemented in SPECTRE has left little room for improvement using the Morton load balancing strategy. Nevertheless, Morton ordering benefits of good locality in the lower levels, but on the highest level, the first cut can cause spatial discontinuities and therefore generate communications. In order to observe the results obtained with the different load balancing strategies, an interactive visualization tool, based on OpenGL, has been developed to produce the views. In Fig. 6, each process displays its own scene: the vertical plane cuts the UAV in several pieces, generating neighboring and communications.

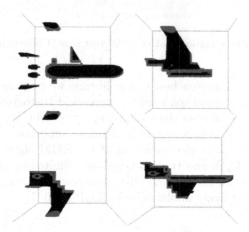

**Fig. 6.** UAV cut into discontiguous pieces with Morton ordering

Our implementation of Morton ordering uses a depth-first traversal on the underlying regular octree taking into account the number of elements present in each node. To avoid over-decomposition, a tolerance threshold has been introduced to tune the load balancing precision. While the depth-first traversal algorithm progresses down the octree, nodes become smaller, and the load balancing precision improves. The algorithm stops as soon as a separator meeting the precision threshold is computed.

In 3D, rotating the axis order results in eight different possibilities, which we have implemented and tested. Figure 7 shows the communication matrices obtained for each case. One can see that the most interesting communication matrices result from Z-axis first orderings. However, from 1 to 64 processes with our previous experimental setup, Morton ZYX ordering, obtains a limited 5% to 10% performance improvement compared with the original ordering.

### 4.3  Histograms (Multisection Kd-Tree)

Multi-sections allow to overpass the ORB limit of power of two numbers of processors: the targeted number of domains is decomposed into prime numbers and the multi-sections are computed with global histograms. This is a costly and complex algorithm. This method, also referred as *implicit kd-tree*, has already been explored in other domains such as ray tracing [13]:

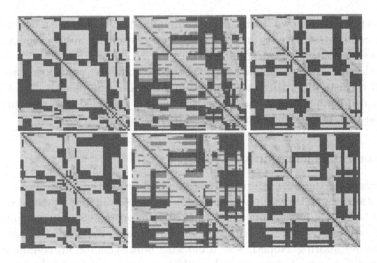

**Fig. 7.** Axis order influence on communication matrices. First column: X-axis first: XYZ and XZY - Second column: Y-axis first: YXZ and YZX, - Third column: Z-axis first: ZXY and ZYX.

We developed two scalable distributed parallel versions: the first one is a complete computation of the multi-sections leading to an exact distribution, and the second one is an approximation where the separator is rounded to match the octree grid. The second method is less accurate, but requires less computation and communication. Our implementation is fully distributed. It can handle very large inputs, which do not fit in a single node, by managing the particle exchanges, to produce the final sorted set.

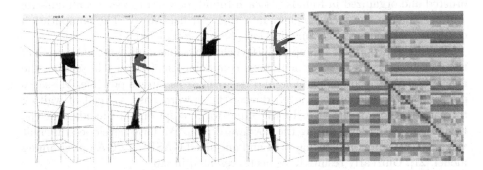

**Fig. 8.** Left: OpenGL visualization of histogram load balancing, UAV, 8 domains - Right: Communication matrix on UAV with 64 domains

Left part of Fig. 8 shows the UAV distributed among eight processes with red points representing the neighboring cells. One can see that they represent a large part of the points and come from the vertical cut along the Z-axis. The right part of Fig. 8 shows the resulting communication matrix with 64 processes: the communication locality has been worsened, resulting in an execution time 1.7 times longer than the original version.

In electromagnetic simulations, another commonly used load balancing method consists in cutting the airplane in slices along its length [22]. The histogram algorithm allows to easily test this method by realizing all the cuts along only one axis, but the experiments did not show any improvement.

### 4.4   Load Balancing Future Work

Load balancing the work is critical between and inside the nodes. For the work distribution among the computation nodes, we tested different classical methods without achieving any significant performance improvement. Further investigation is required. A good distribution should balance the work but also minimize the neighboring. Nonetheless, a blocking bulk-synchronous communication model induces many barriers, which could tear down any load balancing effort. Therefore, introducing asynchronism, overlapping, and fine grain task parallelism inside the nodes may influence our conclusion and must be fixed before exploring new load-balancing strategies.

## 5   Communications

The FMM's communications consist of two-sided point to point exchanges of far-field terms. In the current version, all communications are executed at the top of the octree when the upward pass is completed. The highest level of the tree is exchanged via a blocking MPI_Allreduce call. The other level communications are executed with blocking MPI_Send, MPI_Recv or MPI_Sendrecv. They are ordered and organized in rounds. At each round, pair of processes communicate and accumulate the received data into their far-field array. The computation continues only once the communication phase is completed.

We aim at proposing a completely asynchronous and multithreaded version of these exchanges. Efficient asynchronous communications consist in sending the available data as soon as possible and receiving it as late as possible. Messages are sent at the end of each level, during the upward pass, instead of waiting to reach the top of the tree. In the same way, receptions are handled at the beginning of each level during the downward pass. We have developed different versions based on non-blocking MPI and one-sided GASPI, a PGAS programing model [21]. Our early results have been presented in [20].

PGAS (Partitioned Global Address Space) is a programming model based on a global memory address space partitioned among the distributed processes. Each process owns a local part and directly accesses to the remote parts both in read and write modes. Communications are one-sided: the remote process does not participate. Data movement, memory duplications and synchronizations are reduced. Moreover, PGAS languages are well suited to clusters exploiting RDMA

(Remote Direct Memory Access) and letting the network controllers handle the communications. We use GPI, an implementation of the GASPI API [21].

The FMM-lib library entirely handles the GASPI communications outside the Fortran original code. At the initialization step, the GASPI segments are created. All necessary information to handle the communications needs to be precomputed. Indeed, when a FMM level is locally terminated, the corresponding process sends the available information by writing remotely into the recipient's memory at a pre-computed offset without requiring action from the recipient. This write is followed by a notification containing a notifyID and a notifyValue which are used to indicate which data has been received. When the recipient needs information to pursue its computation, it checks for notifications and waits only if necessary.

A similar pattern can be built using the MPI 2-sided non-blocking communications via MPI_Isend and MPI_Irecv. These calls return immediately, and let the communication complete while the computation pursues. However, the lack of communication progression is a well-known problem. Some methods, like manual progression, help to force the completion. It consists in calling MPI_Test on the communication corresponding MPI_Request. The MPI standard ensures that the MPI_Test calls trigger the communications [14].

## 6    Communication Results

We use the generic metallic UAV, place one MPI/GASPI process per node, and increase the number of nodes from 1 to 32. Each node is fully used with 16 OpenMP threads.

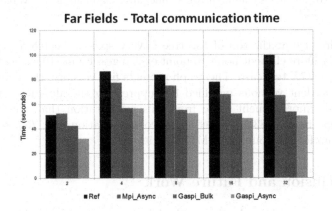

**Fig. 9.** Far fields communication time

Figure 9 shows the execution time of the different MPI and GASPI versions. We compare four different versions: *Ref, MPI Async, Gaspi Bulk*, and *Gaspi Async*. The first three ones handle all the exchanges at the top of the tree. The idea is to measure the improvement, without any algorithmic modification.

*Ref* uses blocking MPI calls, *MPI Async* uses non-blocking MPI calls and manual progression, and *Gaspi Bulk* uses one-sided GASPI writes. The *Gaspi Async* version sends data as soon a level is complete, receives as late as possible, and relies on hardware progression. One can see that, without introducing any overlapping, on 32 nodes, the *Gaspi Bulk* version already reaches 46% speedup over *Ref*. MPI Async version achieves 36% speedup, but still remains slower than the synchronous *Gaspi Bulk* version. Finally, introducing overlapping enables the *Gaspi Async* version to gain three more percentage points over the *Ref* version, with a total of 49% speedup on communications.

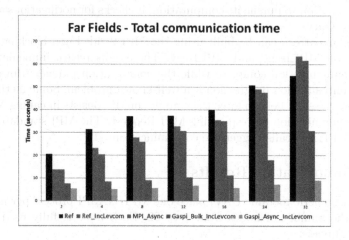

**Fig. 10.** Far fields communication time, after eliminating the Allreduce.

The allreduce at the top of the tree is very sparse. Therefore, we tried to replace it by more efficient point to point exchanges. Figure 10 shows the results on the larger F7X test case. The graph presents five versions: the reference, and all four precedent versions modified by suppressing the allreduce. All GASPI versions take more benefit more from this modification than the MPI ones. *Gaspi Async* reaches 83.5% speedup on communication over *Ref*, resulting in 29% speedup for the complete FMM algorithm.

# 7   Conclusion and Future Work

In this paper, we are investigating methodologies to introduce load balancing and asynchronous communications, in a large industrial Fortran MLFMM application. Applying the different load balancing strategies has demonstrated that it is possible to improve the communication pattern, but it would require more refined options in the future work to further improve the solution. Furthermore, since our load balancing may be harmed by the bulk synchronous communication scheme, based on blocking MPI communications, we prioritized the implementation of a fully asynchronous communication model. We are exploring the use of non-blocking MPI and multithreaded asynchronous one-sided GASPI, and have already obtained a significant 83.5% speedup.

At the present time, we are working on optimizing the intranode scalability introducing fine grain parallelism with the use of tasks. The next step is to make the algorithm fully asynchronous to expose the maximum parallelism: all the barriers between levels will be broken into fine grain task dependencies management.

**Acknowledgements.** The optimized SPECTRE application described in this article is the sole property of Dassault Aviation.

# References

1. FMM-lib. https://github.com/EXAPARS/FMM-lib
2. Maqao. https://www.maqao.org
3. Parmetis. http://glaros.dtc.umn.edu/gkhome/metis/parmetis/overview
4. Score-p. https://www.vi-hps.org/projects/score-p/
5. Top 10 algorithm. https://archive.siam.org/pdf/news/637.pdf
6. Workshop-EM-ISAE.    https://websites.isae-supaero.fr/workshop-em-isae-2018/workshop-em-isae-2018
7. Workshop-EM-ISAE.    https://websites.isae-supaero.fr/workshop-em-isae-2016/accueil
8. Agullo, E., Bramas, B., Coulaud, O., Darve, E., Messner, M., Takahashi, T.: Task-based FMM for multicore architectures. Technical Report RR-8277, INRIA, March 2013
9. Agullo, E., Bramas, B., Coulaud, O., Darve, E., Messner, M., Takahashi, T.: Task-based FMM for heterogeneous architectures. Research Report RR-8513, Inria, April 2014
10. Carayol, Q.: Development and analysis of a multilevel multipole method for electromagnetics. Ph.D. thesis, Paris 6 (2002)
11. Chandramowlishwaran, A., Madduri, K., Vuduc, R.: Diagnosis, tuning, and redesign for multicore performance: A case study of the fast multipole method. In: Proceedings of the 2010 ACM/IEEE International Conference for High Performance Computing, Networking, Storage and Analysis, SC 2010, Washington, DC, USA, pp. 1–12. IEEE Computer Society (2010)
12. Cruz, F.A., Knepley, M.G., Barba, L.A.: PetFMM-a dynamically load-balancing parallel fast multipole library. CoRR, abs/0905.2637 (2009)
13. Groß, M., Lojewski, C., Bertram, M., Hagen, H.: Fast implicit kd-trees: accelerated isosurface ray tracing and maximum intensity projection for large scalar fields. In: Proceedings of the Ninth IASTED International Conference on Computer Graphics and Imaging, CGIM 2007, Anaheim, CA, USA, pp. 67–74. ACTA Press (2007)
14. Hoefler, T., Lumsdaine, A.: Message progression in parallel computing - to thread or not to thread? IEEE Computer Society, October 2008
15. Ibeid, H., Yokota, R., Keyes, D.: A performance model for the communication in fast multipole methods on HPC platforms. CoRR, abs/1405.6362 (2014)
16. Jabbar, M.A., Yokota, R., Keyes, D.: Asynchronous execution of the fast multipole method using charm++. CoRR, abs/1405.7487 (2014)

17. Lashuk, I., et al.: A massively parallel adaptive fast-multipole method on heterogeneous architectures. In: Proceedings of the Conference on High Performance Computing Networking, Storage and Analysis, SC 2009, New York, NY, USA, pages 58:1–58:12. ACM (2009)
18. Ltaief, H., Yokota, R.: Data-driven execution of fast multipole methods. CoRR, abs/1203.0889 (2012)
19. Milthorpe, J., Rendell, A.P., Huber, T.: PGAS-FMM: implementing a distributed fast multipole method using the x10 programming language. CCPE **26**(3), 712–727 (2014)
20. Moller, N., Petit, E., Carayol, Q., Dinh, Q., Jalby, W.: Asynchronous one-sided communications for scalable fast multipole method in electromagnetic simulations. In: 2017 Short Paper presented at COLOC workshop, Euro-Par 2017, Santiago de Compostela, 29 August 2017
21. Simmendinger, C., Jägersküpper, J., Machado, R., Lojewski, C.: A PGAS-based implementation for the unstructured CFD solver TAU. In: PGAS11, USA (2011)
22. Sylvand, G.: La méthode multipôle rapide en électromagnétisme. Performances, parallélisation, applications. Theses, Ecole des Ponts ParisTech, June 2002
23. Yokota, R., Barba, L.A.: A tuned and scalable fast multipole method as a preeminent algorithm for exascale systems. CoRR, abs/1106.2176 (2011)
24. Yokota, R., Turkiyyah, G., Keyes, D.: Communication complexity of the fast multipole method and its algebraic variants. CoRR, abs/1406.1974 (2014)

# Author Index